Applied Drilling Engineering

Adam T. Bourgoyne Jr.
Professor of Petroleum Engineering
Louisiana State U.

Keith K. Millheim
Manager—Critical Drilling Facility
Amoco Production Co.

Martin E. Chenevert
Senior Lecturer of Petroleum Engineering
U. of Texas

F.S. Young, Jr.
President
Woodway Energy Co.

SPE Textbook Series, Volume 2

Henry L. Doherty Memorial Fund of AIME
Society of Petroleum Engineers
Richardson, TX USA

Dedication

This book is dedicated to the many students who were forced to study from the trial drafts of this work.

Disclaimer

This book was prepared by members of the Society of Petroleum Engineers and their well-qualified colleagues from material published in the recognized technical literature and from their own individual experience and expertise. While the material presented is believed to be based on sound technical knowledge, neither the Society of Petroleum Engineers nor any of the authors or editors herein provide a warranty either expressed or implied in its application. Correspondingly, the discussion of materials, methods, or techniques that may be covered by letters patents implies no freedom to use such materials, methods, or techniques without permission through appropriate licensing. Nothing described within this book should be construed to lessen the need to apply sound engineering judgment nor to carefully apply accepted engineering practices in the design, implementation, or application of the techniques described herein.

ISBN 978-1-55563-001-0

14 15 16 17 18 19 / 21 20 19 18 17 16

Society of Petroleum Engineers
222 Palisades Creek Drive
Richardson, TX 75080-2040 USA

http://store.spe.org
books@spe.org
1.972.952.9393

SPE Textbook Series

The Textbook Series of the Society of Petroleum Engineers was established in 1972 by action of SPE Board of Directors. The Series is intended to ensure availability of high-quality textbooks for the use in undergraduate courses in areas clearly identified as being within the petroleum engineering field. The work is directed by the Society's Books Committee, one of more than 40 Society-wide standing committees. Members of the Books Committee provide technical evaluation of the book. Below is a listing of those who have been most closely involved in the final preparation of this book.

Book Editors

Jack F. Evers, U. of Wyoming
David S. Pye, Union Geothermal Div.

Books Committee (1984)

Acknowledgments

The authors would like to acknowledge the help of individuals and companies in the oil- and gas-producing industry that are too numerous to mention. Without the unselfish help of so many, this book would not have been possible. In particular, the American Petroleum Inst., the Intl. Assn. of Drilling Contractors, and the Petroleum Extension Service of the U. of Texas were of tremendous assistance in providing background material for several chapters.

Special thanks are due numerous individuals who have served on the SPE Textbook Committee during the past decade for their help and understanding. In particular, a large contribution was made by Jack Evers of the U. of Wyoming, who served for several years on the Textbook Committee as senior reviewer and coordinator for this work. Finally, the authors would like to recognize the contribution of Dan Adamson, SPE Executive Director, who constantly prodded the authors to "finish the book."

Adam T. Bourgoyne Jr.

When I accepted the challenge of writing part of this textbook, I had no idea of how much of my free time would be consumed. There were many evenings, weekends, and even holidays and vacations when I was busy writing, correcting, or editing. I thank Valerie, my wife, for the understanding and patience in letting me complete this monumental task.

I would like to extend my gratitude to Allen Sinor for his dedicated effort in helping me with our part of the textbook. If it were not for Allen, I doubt I could have completed it. I would also like to thank John Horeth II, Warren Winters, Mark Dunbar, and Tommy Warren for their assistance with the problems and examples; Amoco Production Co. for permission to write part of this textbook; and the research staff in Tulsa that helped with the typing and drafting.

Keith K. Millheim

It is impossible for me to list the many people to whom I am indebted for their assistance in the preparation of my part of this book. The many meetings, discussions, and work sessions I had with my drilling industry associates span a period of 8 years and are too numerous to recall. For their assistance I am thankful.

I would also particularly like to thank the U. of Texas and SPE for their encouragement and support.

Martin E. Chenevert

The Society of Petroleum Engineers Textbook Series is made possible in part by grants from the Shell Companies Foundation and the SPE Foundation.

Preface

This book was written for use as a college textbook in a petroleum engineering curriculum. The material was organized to present engineering science fundamentals first, followed by example engineering applications involving these fundamentals. The level of engineering science gradually advances as one proceeds through the book.

Chap. 1 is primarily descriptive material and intended as an introduction to drilling engineering. It is suitable for use as a text in a freshman- or sophomore-level introductory petroleum engineering course. Chaps. 2 and 3 are designed for use in a drilling-fluids and cements laboratory course and are aimed at the sophomore or junior level. Chaps. 4 through 7 are suitable for a senior-level drilling engineering course. Chap. 8 provides additional material that could be covered in a more advanced course at the senior level or in a masters-degree program.

Because the text was designed for use in more than one course, each chapter is largely independent of previous chapters, enabling an instructor to select topics for use in a single course. Also, the important concepts are developed from fundamental scientific principles and are illustrated with numerous examples. These principles and examples should allow anyone with a general background in engineering or the physical sciences to gain a basic understanding of a wide range of drilling engineering problems and solutions.

Contents

Chapter 1
Rotary Drilling Process

The objectives of this chapter are (1) to familiarize the student with the basic rotary drilling equipment and operational procedures and (2) to introduce the student to drilling cost evaluation.

1.1 Drilling Team

The large investments required to drill for oil and gas are made primarily by oil companies. Small oil companies invest mostly in the shallow, less-expensive wells drilled on land in the United States. Investments in expensive offshore and non-U.S. wells can be afforded only by large oil companies. Drilling costs have become so great in many areas that several major oil companies often will form groups to share the financial risk.

Many specialized talents are required to drill a well safely and economically. As in most complex industries, many different service companies, contractors, and consultants, each with its own organization, have evolved to provide necessary services and skills. Specialized groups within the major oil companies also have evolved. A staff of drilling engineers is generally identifiable as one of these groups.

A well is classified as a *wildcat* well if its purpose is to discover a new petroleum reservoir. In contrast, the purpose of a *development* well is to exploit a known reservoir. Usually the geological group recommends wildcat well locations, while the reservoir engineering group recommends development well locations. The drilling engineering group makes the preliminary well designs and cost estimates for the proposed well. The legal group secures the necessary drilling and production rights and establishes clear title and right-of-way for access. Surveyors establish and stake the well location.

Usually the drilling is done by a drilling contractor. Once the decision to drill the well is made by management, the drilling engineering group prepares a more detailed well design and writes the bid specifications. The equipment and procedures that the operator will require, together with a well description, must be included in the bid specifications and drilling contract. In areas where previous experience has shown drilling to be routine, the bid basis may be the cost per foot of hole drilled. In areas where costs cannot be estimated with

reasonable certainty, the bid basis is usually a contract price per day. In some cases, the bid is based on cost per foot down to a certain depth or formation and cost per day beyond that point. When the well is being financed by more than one company, the well plan and drilling contract must be approved by drilling engineers representing the various companies involved.

Before the drilling contractor can begin, the surface location must be prepared to accommodate the specific rig. Water wells may have to be drilled to supply the requirements for the drilling operation. The surface preparation must be suited to local terrain and supply problems; thus, it varies widely from area to area. In the marshland of south Louisiana, drilling usually is performed using an inland barge. The only drillsite preparation required is the dredging of a *slip* to permit moving the barge to location. In contrast, drillsite preparation in the Canadian Arctic Islands requires construction of a manmade *ice platform* and extensive supply and storage facilities. Fig. 1.1 shows an inland barge on location in the marsh area of south Louisiana and Fig. 1.2 shows a drillsite in the Canadian Arctic Islands.

After drilling begins, the manpower required to drill the well and solve any drilling problems that occur are provided by (1) the drilling contractor, (2) the well operator, (3) various drilling service companies, and (4) special consultants. Final authority rests either with the drilling contractor when the rig is drilling on a cost-per-foot basis or with the well operator when the rig is drilling on a cost-per-day basis.

Fig. 1.3 shows a typical drilling organization often used by the drilling contractor and well operator when a well is drilled on a cost-per-day basis. The drilling engineer recommends the drilling procedures that will allow the well to be drilled as safely and economically as possible. In many cases, the original well plan must be modified as drilling progresses because of unforeseen circumstances. These modifications also are the responsibility of the drilling engineer. The company representative, using the well plan, makes the on-site decisions concerning drilling operations and other services needed. The rig operation and rig personnel supervision are the responsibility of the tool pusher.

Fig. 1.1 – Texaco drilling barge *Gibbens* on location in Lafitte field, Louisiana.

Fig. 1.2 – Man-made ice platform in deep water area of the Canadian Arctic Islands.

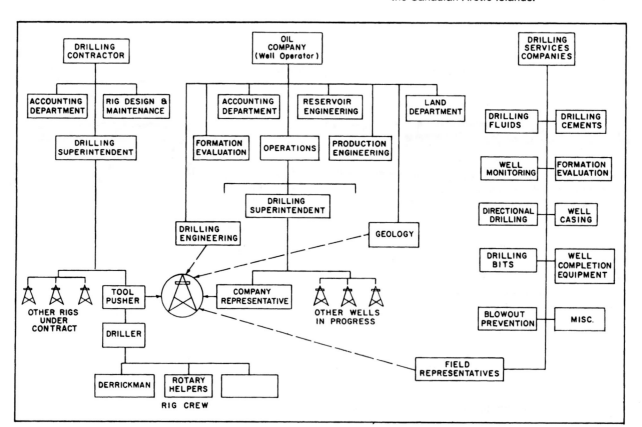

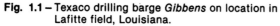

Fig. 1.3 – Typical drilling rig organizations.

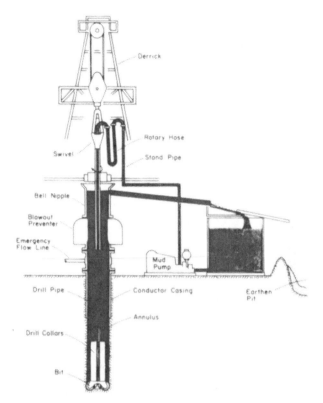

Fig. 1.4 – The rotary drilling process.

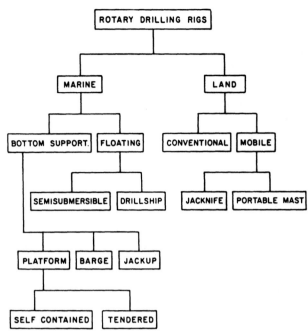

Fig. 1.5 – Classification of rotary drilling rigs.

1.2 Drilling Rigs

Rotary drilling rigs are used for almost all drilling done today. A sketch illustrating the rotary drilling process is shown in Fig. 1.4. The hole is drilled by rotating a bit to which a downward force is applied. Generally, the bit is turned by rotating the entire drillstring, using a rotary table at the surface, and the downward force is applied to the bit by using sections of heavy thick-walled pipe, called drill collars, in the drillstring above the bit. The cuttings are lifted to the surface by circulating a fluid down the drillstring, through the bit, and up the annular space between the hole and the drillstring. The cuttings are separated from the drilling fluid at the surface.

As shown in Fig. 1.5, rotary drilling rigs can be classified broadly as land rigs or marine rigs. The main design features of land rigs are portability and maximum operating depth. The derrick of the conventional land rig must be built on location. In many cases the derrick is left over the hole after the well is completed. In the early days of drilling, many of these standard derricks were built quite close together as a field was developed. However, because of the high cost of construction, most modern land rigs are built so that the derrick can be moved easily and reused. The various rig components are skid-mounted so that the rig can be moved in units and connected easily. The jackknife, or cantilever, derrick (Fig. 1.6) is assembled on the ground with pins and then raised as a unit using the rig-hoisting equipment. The portable mast (Fig. 1.7), which is suitable for moderate-depth wells, usually is mounted on wheeled trucks or trailers that incorporate the hoisting machinery, engines, and

derrick as a single unit. The telescoped portable mast is raised to the vertical position and then extended to full height by hydraulic pistons on the unit.

The main design features of marine rigs are portability and maximum water depth of operation. Submersible drilling barges generally are used for inland water drilling where wave action is not severe and water depths are less than about 20 ft. The entire rig is assembled on the barge, and the unit is towed to the location and sunk by flooding the barge. Once drilling is completed, the water is pumped from the barge, allowing it to be moved to the next location. After the well is completed, a platform must be built to protect the wellhead and to support the surface production equipment. In some cases, the operating water depth has been extended to about 40 ft by resting the barge on a shell mat built on the seafloor.

Offshore exploratory drilling usually is done using self-contained rigs that can be moved easily. When water depth is less than about 350 ft, bottom-supported rigs can be used. The most common type of bottom-supported mobile rig is the jackup (Fig. 1.8). The jackup rig is towed to location with the legs elevated. On location, the legs are lowered to the bottom and the platform is "jacked up" above the wave action by means of hydraulic jacks.

Semisubmersible rigs that can be flooded similar to an inland barge can drill resting on bottom as well as in a floating position. However, modern semisubmersible rigs (Fig. 1.9) are usually more expensive than jackup rigs and, thus, are used mostly in water depths too great for resting on bottom. At present, most semisubmersible rigs are anchored over the hole. A few semisubmersible rigs employ large engines to position the rig over the hole dynamically. This can extend greatly the maximum operating water depth. Some of these rigs can be used in water depths as great as 6,000 ft. The shape of a semisubmersible rig tends to dampen wave motion greatly regardless of wave direction. This allows its use in areas such as the North Sea where wave action is severe.

A second type of floating vessel used in offshore drilling is the drillship (Fig. 1.10). Drillships are usually much less costly than semisubmersibles unless they are designed to be positioned dynamically. A few drillships being planned will be able to operate in water depths up to 13,000 ft. Some are designed with

the rig equipment and anchoring system mounted on a central turret. The ship is rotated about the central turret using thrusters so that the ship always faces incoming waves. This helps to dampen wave motion. However, the use of drillships usually is limited to areas where wave action is not severe.

Offshore development drilling usually is done from fixed platforms. After the exploratory drilling program indicates the presence of sufficient petroleum reserves to justify construction costs, one or more platforms from which many directional wells

Fig. 1.6 – Jackknife rig on location in Port Hudson field, Louisiana.

Fig. 1.8 – Jackup rig *Mr. Mel* on location in the Eugene Island area, offshore Louisiana.

Fig. 1.7 – Portable mast being transported.

Fig. 1.9 – A semisubmersible drilling rig on location.

can be drilled are built and placed on location. The platforms are placed so that wellbores fanning out in all directions from the platform can develop the reservoir fully. The various rig components usually are integrated into a few large modules that a derrick barge quickly can place on the platform.

Large platforms allow the use of a self-contained rig – i.e., all rig components are located on the platform (Fig. 1.11). A platform/tender combination can be used for small platforms. The rig tender, which is a floating vessel anchored next to the platform, contains the living quarters and many of the rig components (Fig. 1.12). The rig-up time and operating cost will be less for a platform/tender operation. However, some operating time may be lost during severe weather.

Platform cost rises very rapidly with water depth. When water depths are too great for the economical use of development platforms, the development wells can be drilled from floating vessels, and the wellhead equipment installed on the ocean floor. Underwater completion technology is still relatively new and experimental.

Although drilling rigs differ greatly in outward appearance and method of deployment, all rotary rigs have the same basic drilling equipment. The main component parts of a rotary rig are the (1) power system, (2) hoisting system, (3) fluid-circulating system, (4) rotary system, (5) well control system, and (6) well monitoring system.

1.3 Rig Power System

Most rig power is consumed by the hoisting and fluid circulating systems. The other rig systems have much smaller power requirements. Fortunately, the hoisting and circulating systems generally are not used simultaneously, so the same engines can perform both functions. Total power requirements for most rigs are from 1,000 to 3,000 hp. The early drilling rigs were powered primarily by steam. However, because of high fuel consumption and lack of portability of the large boiler plants required, steam-powered rigs have become impractical. Modern rigs are powered by internal-combustion diesel engines and generally subclassified as (1) the diesel-electric type or (2) the direct-drive type, depending on the method used to transmit power to the various rig systems.

Diesel-electric rigs are those in which the main rig engines are used to generate electricity. Electric power is transmitted easily to the various rig systems, where the required work is accomplished through use of electric motors. Direct-current motors can be wired to give a wide range of speed-torque characteristics that are extremely well-suited for the hoisting and circulating operations. The rig components can be packaged as portable units that can be connected with plug-in electric cable connectors. There is considerable flexibility of equipment placement, allowing better space utilization and weight distribution. In addition, electric power allows the use of a relatively simple and flexible control system. The driller can apply power smoothly to various rig

Fig. 1.10 – An offshore drillship.

Courtesy of Texaco Inc.

Fig. 1.11 – A self-contained platform rig on location in the Eugene Island area, offshore Louisiana.

Fig. 1.12 – A tendered platform rig.[12]

components, thus minimizing shock and vibration problems.

Direct-drive rigs accomplish power transmission from the internal combustion engines using gears, chains, belts, and clutches rather than generators and motors. The initial cost of a direct-drive power system generally is considerably less than that of a comparable diesel-electric system. The development of hydraulic drives has improved greatly the performance of this type of power system. Hydraulic drives reduce the shock and vibrational problems of the direct-drive system. Torque converters, which are hydraulic drives designed so that output torque increases rapidly with output load, are now used to extend the speed-torque characteristic of the internal combustion engine over greater ranges that are better suited to drilling applications. The use of torque converters also allows the selection of engines based on running conditions rather than starting conditions.

Power-system performance characteristics generally are stated in terms of output horsepower, torque, and fuel consumption for various engine speeds. As illustrated in Fig. 1.13, the shaft power developed by an engine is obtained from the product of the angular velocity of the shaft, ω, and the output torque T:

$$P = \omega T. \quad\dotfill (1.1)$$

The overall power efficiency determines the rate of fuel consumption w_f at a given engine speed. The heating values H of various fuels for internal combustion engines are shown in Table 1.1. The heat energy input to the engine, Q_i, can be expressed by

$$Q_i = w_f H. \quad\dotfill (1.2)$$

Since the overall power system efficiency, E_t, is defined as the energy output per energy input, then

$$E_t = \frac{P}{Q_i}. \quad\dotfill (1.3)$$

TABLE 1.1 – HEATING VALUE OF VARIOUS FUELS

Fuel Type	Density (lbm/gal)	Heating Value (Btu/lbm)
diesel	7.2	19,000
gasoline	6.6	20,000
butane	4.7	21,000
methane	–	24,000

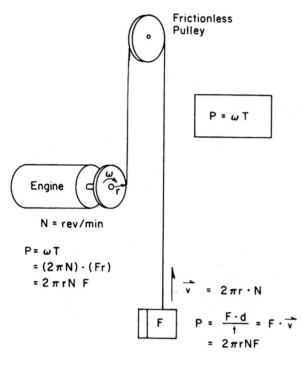

Fig. 1.13 – Engine power output.

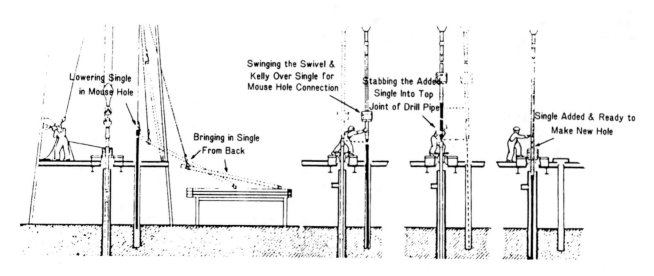

Fig. 1.14 – Making a connection.[12]

Example 1.1. A diesel engine gives an output torque of 1,740 ft-lbf at an engine speed of 1,200 rpm. If the fuel consumption rate was 31.5 gal/hr, what is the output power and overall efficiency of the engine?

Solution. The angular velocity, ω, is given by

$$\omega = 2\pi(1,200) = 7,539.8 \text{ rad/min.}$$

The power output can be computed using Eq. 1.1:

$$P = \omega T$$

$$= \frac{7,539.8 \,(1740) \text{ ft-lbf/min}}{33,000 \text{ ft-lbf/min/hp}} = 397.5 \text{ hp.}$$

Since the fuel type is diesel, the density ρ is 7.2 lbm/gal and the heating value H is 19,000 Btu/lbm (Table 1.1). Thus, the fuel consumption rate w_f is

$$w_f = 31.5 \text{ gal/hr } (7.2 \text{ lbm/gal})\left(\frac{1 \text{ hour}}{60 \text{ minutes}}\right)$$

$$= 3.78 \text{ lbm/min.}$$

The total heat energy consumed by the engine is given by Eq. 1.2:

$$Q_i = w_f H$$

$$= \frac{3.78 \text{ lbm/min}(19,000 \text{ Btu/lbm})(779 \text{ ft-lbf/Btu})}{33,000 \text{ ft-lbf/min/hp}}$$

$$= 1,695.4 \text{ hp.}$$

Thus, the overall efficiency of the engine at 1,200 rpm given by Eq. 1.3 is

$$E_t = \frac{P}{Q_i} = \frac{397.5}{1695.4} = 0.234 \text{ or } 23.4\%.$$

1.4 Hoisting System

The function of the hoisting system is to provide a means of lowering or raising drillstrings, casing strings, and other subsurface equipment into or out of the hole. The principal components of the hoisting system are (1) the derrick and substructure, (2) the block and tackle, and (3) the drawworks. Two routine drilling operations performed with the hoisting system are called (1) *making a connection* and (2) *making a trip*. Making a connection refers to the periodic process of adding a new joint of drillpipe as the hole deepens. This process is described in Fig. 1.14. Making a trip refers to the process of removing the drillstring from the hole to change a portion of the downhole assembly and then lowering the drillstring back to the hole bottom. A trip is made usually to change a dull bit. The steps involved in coming out of the hole are shown in Fig. 1.15

1.4.1 Derrick or Portable Mast. The function of the derrick is to provide the vertical height required to raise sections of pipe from or lower them into the hole. The greater the height, the longer the section of

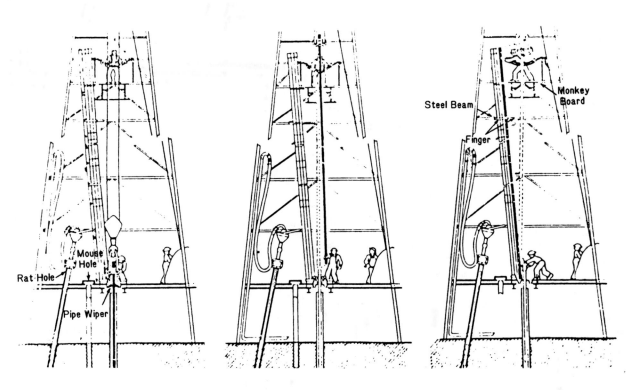

Fig. 1.15 – Pulling out of the hole.[12]

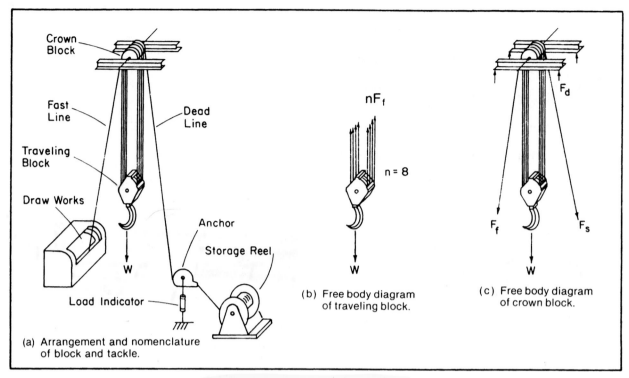

Fig. 1.16 – Schematic of block and tackle.

pipe that can be handled and, thus, the faster a long string of pipe can be inserted in or removed from the hole. The most commonly used drillpipe is between 27 and 30 ft long. Derricks that can handle sections called *stands*, which are composed of two, three, or four joints of drillpipe, are said to be capable of pulling *doubles*, *thribbles*, or *fourbles*, respectively.

In addition to their height, derricks are rated according to their ability to withstand compressive loads and wind loads. Allowable wind loads usually are specified both with the drillstring in the hole and with the drillstring standing in sections in the derrick. When the drillstring is standing in the derrick resting against the pipe-racking platform, an overturning moment is applied to the derrick at that point. Wind ratings must be computed assuming wind loading is in the same direction as this overturning moment. Anchored guy wires attached to each leg of the derrick are used to increase the wind rating of small portable masts. The American Petroleum Institute (API) has published standards dealing with derrick specifications and ratings. [1-3]

To provide working space below the derrick floor for pressure control valves called *blowout preventers*, the derrick usually is elevated above the ground level by placement on a substructure. The substructure must support not only the derrick with its load but also the weight of other large pieces of equipment. API *Bull. D10*[4] recommends rating substructure load-supporting capacity according to (1) the maximum pipe weight that can be set back in the derrick, (2) the maximum pipe weight that can be suspended in the rotary table (irrespective of setback load), and (3) the corner loading capacity (maximum supportable load at each corner). Also, in API

Standard 4A,[1] three substructure types have been adopted. In addition, many non-API designs are available. The choice of design usually is governed by blowout preventer height and local soil conditions.

1.4.2 Block and Tackle. The block and tackle is comprised of (1) the crown block, (2) the traveling block, and (3) the drilling line. The arrangement and nomenclature of the block and tackle used on rotary rigs are shown in Fig. 1.16a. The principal function of the block and tackle is to provide a *mechanical advantage*, which permits easier handling of large loads. The mechanical advantage M of a block and tackle is simply the load supported by the traveling block, W, divided by the load imposed on the drawworks, F_f:

$$M = \frac{W}{F_f}. \qquad \qquad \qquad \qquad \qquad \ldots \ldots (1.4)$$

The load imposed on the drawworks is the tension in the fast line.

The *ideal mechanical advantage*, which assumes no friction in the block and tackle, can be determined from a force analysis of the traveling block. Consider the free body diagram of the traveling block as shown in Fig. 1.16b. If there is no friction in the pulleys, the tension in the drilling line is constant throughout. Thus, a force balance in the vertical direction yields

$$nF_f = W,$$

where n is the number of lines strung through the

TABLE 1.2 – AVERAGE EFFICIENCY FACTORS FOR BLOCK-AND-TACKLE SYSTEM

Number of Lines (n)	Efficiency (E)
6	0.874
8	0.841
10	0.810
12	0.770
14	0.740

traveling block. Solving this relationship for the tension in the fast line and substituting the resulting expression in Eq. 1.4 yields

$$M_i = \frac{W}{W/n} = n,$$

which indicates that the ideal mechanical advantage is equal to the number of lines strung between the crown block and traveling block. Eight lines are shown between the crown block and traveling block in Fig. 1.16. The use of 6, 8, 10, or 12 lines is common, depending on the loading condition.

The input power P_i of the block and tackle is equal to the drawworks load F_f times the velocity of the fast line, v_f:

$$P_i = F_f v_f. \quad \dots \dots \dots \dots \dots \dots \dots (1.5)$$

The output power, or *hook power*, P_h is equal to the traveling block load W times the velocity of the traveling block, v_b:

$$P_h = W v_b. \quad \dots \dots \dots \dots \dots \dots \dots (1.6)$$

For a frictionless block and tackle, $W = nF_f$. Also, since the movement of the fast line by a unit distance tends to shorten each of the lines strung between the crown block and traveling block by only $1/n$ times the unit distance, then $v_b = v_f/n$. Thus, a frictionless system implies that the ratio of output power to input power is unity:

$$E = \frac{P_h}{P_i} = \frac{(nF_f)(v_f/n)}{F_f v_f} = 1.$$

Of course, in an actual system, there is always a power loss due to friction. Approximate values of block and tackle efficiency for roller-bearing sheaves are shown in Table 1.2.

Knowledge of the block and tackle efficiency permits calculation of the actual tension in the fast line for a given load. Since the power efficiency is given by

$$E = \frac{P_h}{P_i} = \frac{W v_b}{F_f v_f} = \frac{W v_f/n}{F_f v_f} = \frac{W}{F_f n},$$

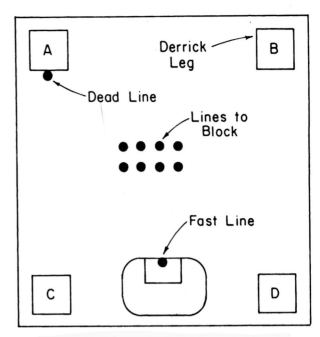

Fig. 1.17 – Projection of drilling lines on rig floor.

HW #1

then the tension in the fast line is

$$F_f = \frac{W}{En}. \quad \dots \dots \dots \dots \dots \dots \dots (1.7)$$

Eq. 1.7 can be used to select drilling line size. However, a safety factor should be used to allow for line wear and shock loading conditions.

The line arrangement used on the block and tackle causes the load imposed on the derrick to be greater than the hook load. As shown in Fig. 1.16c, the load F_d applied to the derrick is the sum of the hook load W, the tension in the dead line, F_s, and the tension in the fast line, F_f:

$$F_d = W + F_f + F_s. \quad \dots \dots \dots \dots \dots (1.8a)$$

If the load, W, is being hoisted by pulling on the fast line, the friction in the sheaves is resisting the motion of the fast line and the tension in the drilling line increases from W/n at the first sheave (deadline) to W/En at the last sheave (fast line). Substituting these values for F_f and F_s in Eq. 1.8a gives

$$F_d = W + \frac{W}{En} + \frac{W}{n} = \left(\frac{1 + E + En}{En}\right)W. \quad \dots (1.8b)$$

The total derrick load is not distributed equally over all four derrick legs. Since the drawworks is located on one side of the derrick floor, the tension in the fast line is distributed over only two of the four derrick legs. Also, the dead line affects only the leg to which it is attached. The drilling lines usually are arranged as in the plan view of the rig floor shown in Fig. 1.17. For this arrangement, derrick Legs C and D would share the load imposed by the tension in the fast line and Leg A would assume the full load imposed by the tension in the dead line. The load

TABLE 1.3 – EXAMPLE CALCULATION OF DERRICK LEG LOAD

Load Source	Total Load	Load on Each Derrick Leg			
		Leg A	Leg B	Leg C	Leg D
hook load	W	$W/4$	$W/4$	$W/4$	$W/4$
fast line	W/En			$W/2En$	$W/2En$
dead line	W/n	W/n			
		$W(n+4)/(4n)$	$W/4$	$W(En+2)/4En$	$W(En+2)/4En$

TABLE 1.4 – NOMINAL BREAKING STRENGTH OF 6 × 19* CLASSIFICATION WIRE ROPE, BRIGHT (UNCOATED) OR DRAWN-GALVANIZED WIRE, INDEPENDENT WIRE-ROPE CORE (IWRC)[7]

Nominal Diameter (in.)	Approximate Mass (lbm/ft)	Nominal Strength	
		Improved Plow Steel (lbf)	Extra Improved Plow Steel (lbf)
1/2	0.46	23,000	26,600
9/16	0.59	29,000	33,600
5/8	0.72	35,800	41,200
3/4	1.04	51,200	58,800
7/8	1.42	69,200	79,600
1	1.85	89,800	103,400
1 1/8	2.34	113,000	130,000
1 1/4	2.89	138,800	159,800
1 3/8	3.50	167,000	192,000
1 1/2	4.16	197,800	228,000
1 5/8	4.88	230,000	264,000
1 3/4	5.67	266,000	306,000
1 7/8	6.50	304,000	348,000
2	7.39	344,000	396,000

*Six strands having 19 wires per strand.

distribution for each leg has been calculated in Table 1.3.

Note that for $E \geq 0.5$, the load on Leg A is greater than the load on the other three legs. Since if any leg fails, the entire derrick also fails, it is convenient to define a *maximum equivalent derrick load*, F_{de}, which is equal to four times the maximum leg load. For the usual drilling line arrangement shown in Fig. 1.17,

$$F_{de} = \left(\frac{n+4}{n}\right)W. \quad \ldots\ldots\ldots\ldots\ldots\ldots (1.9)$$

A parameter sometimes used to evaluate various drilling line arrangements is the *derrick efficiency factor*, defined as the ratio of the actual derrick load to the maximum equivalent load. For a maximum equivalent load given by Eq. 1.9, the derrick efficiency factor is

$$E_d = \frac{F_d}{F_{de}} = \frac{\left(\dfrac{1+E+En}{En}\right)W}{\left(\dfrac{n+4}{n}\right)W} = \frac{E(n+1)+1}{E(n+4)}.$$

For the block and tackle efficiency values given in Table 1.2, the derrick efficiency increases with the number of lines strung between the crown block and traveling block.

The drilling line is subject to rather severe service during normal tripping operations. Failure of the drilling line can result in (1) injury to the drilling personnel, (2) damage to the rig, and (3) loss of the drillstring in the hole. Thus, it is important to keep drilling line tension well below the nominal breaking strength and to keep the drilling line in good condition. The nominal breaking strength (new) for one type of wire rope commonly used for drilling line is shown in Table 1.4 for various rope diameters. The correct method for measuring wire rope diameter is illustrated in Fig. 1.18.

Drilling line does not tend to wear uniformly over its length. The most severe wear occurs at the *pickup points* in the sheaves and at the *lap points* on the drum of the drawworks. The pickup points are the points in the drilling line that are on the top of the crown block sheaves or the bottom of the traveling block sheaves when the weight of the drillstring is lifted from its supports in the rotary table during tripping operations. The rapid acceleration of the heavy drillstring causes the

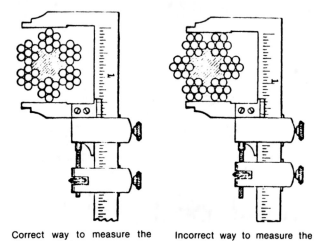

Correct way to measure the diameter of wire rope.

Incorrect way to measure the diameter of wire rope.

Fig. 1.18 – Measurement of wire rope diameter.[7]

most severe stress at these points. The lap points are the points in the drilling line where a new layer or lap of wire begins on the drum of the drawworks.

Drilling line is maintained in good condition by following a scheduled *slip-and-cut* program. Slipping the drilling line involves loosening the dead line anchor and placing a few feet of new line in service from the storage reel. Cutting the drilling line involves removing the line from the drum of the drawworks and cutting off a section of line from the end. Slipping the line changes the pickup points, and cutting the line changes the lap points. The line is sometimes slipped several times before it is cut. Care must be taken not to slip the line a multiple of the distance between pickup points. Otherwise, points of maximum wear are just shifted from one sheave to the next. Likewise, care must be taken when cutting the line not to cut a section equal in length to a multiple of the distance between lap points.

API[18] has adopted a slip-and-cut program for drilling lines. The parameter adopted to evaluate the amount of line service is the *ton-mile*. A drilling line is said to have rendered one ton-mile of service when the traveling block has moved 1 U.S. ton a distance of 1 mile. Note that for simplicity this parameter is independent of the number of lines strung. Ton-mile records must be maintained in order to employ a satisfactory slip-and-cut program. Devices that automatically accumulate the ton-miles of service are available. The number of ton-miles between cutoffs will vary with drilling conditions and drilling line diameter and must be determined through field experience. In hard rock drilling, vibrational problems may cause more rapid line wear than when the rock types are relatively soft. Typical ton-miles between cutoff usually range from about 500 for 1-in.-diameter drilling line to about 2,000 for 1.375-in.-diameter drilling line.

Example 1.2. A rig must hoist a load of 300,000 lbf. The drawworks can provide an input power to the block and tackle system as high as 500 hp. Eight lines are strung between the crown block and traveling block. Calculate (1) the static tension in the fast line when upward motion is impending, (2) the maximum hook horsepower available, (3) the maximum hoisting speed, (4) the actual derrick load, (5) the maximum equivalent derrick load, and (6) the derrick efficiency factor. Assume that the rig floor is arranged as shown in Fig. 1.17.

Solution.

1. The power efficiency for $n = 8$ is given as 0.841 in Table 1.2. The tension in the fast line is given by Eq. 1.7.

$$F_f = \frac{W}{En} = \frac{300,000}{0.841\,(8)} = 44,590 \text{ lbf.}$$

2. The maximum hook horsepower available is

$$P_h = E \cdot p_i = 0.841\,(500) = 420.5 \text{ hp.}$$

3. The maximum hoisting speed is given by

$$v_b = \frac{P_h}{W} = \frac{420.5 \text{ hp} \left(\dfrac{33,000 \text{ ft-lbf/min}}{\text{hp}} \right)}{300,000 \text{ lbf}}$$

$$= 46.3 \text{ ft/min.}$$

To pull a 90-ft stand would require

$$t = \frac{90 \text{ ft}}{46.3 \text{ ft/min}} = 1.9 \text{ min.}$$

4. The actual derrick load is given by Eq. 1.8b:

$$F_d = \left(\frac{1 + E + En}{En} \right) W$$

$$= \left(\frac{1 + 0.841 + 0.841(8)}{0.841(8)} \right)(300,000)$$

$$= 382,090 \text{ lbf.}$$

5. The maximum equivalent load is given by Eq. 1.9:

$$F_{de} = \left(\frac{n+4}{n} \right) W = \frac{8+4}{8}(300,000) = 450,000 \text{ lbf.}$$

6. The derrick efficiency factor is

$$E_d = \frac{F_d}{F_{de}} = \frac{382,090}{450,000} = 0.849 \text{ or } 84.9\%.$$

1.4.3 Drawworks. The drawworks (Fig. 1.19) provide the hoisting and braking power required to raise or lower the heavy strings of pipe. The principal parts of the drawworks are (1) the drum, (2) the brakes, (3) the transmission, and (4) the catheads. The drum transmits the torque required for hoisting or braking. It also stores the drilling line required to move the traveling block the length of the derrick.

The brakes must have the capacity to stop and sustain the great weights imposed when lowering a string of pipe into the hole. Auxiliary brakes are used to help dissipate the large amount of heat generated during braking. Two types of auxiliary brakes commonly used are (1) the hydrodynamic type and (2) the electromagnetic type. For the hydrodynamic type, braking is provided by water being impelled in a direction opposite to the rotation of the drum. In the electromagnetic type, electrical braking is provided by two opposing magnetic fields. The magnitude of the magnetic fields is dependent on the speed of rotation and the amount of external excitation current supplied. In both types, the heat developed must be dissipated by a liquid cooling system.

The drawworks transmission provides a means for easily changing the direction and speed of the traveling block. Power also must be transmitted to *catheads* attached to both ends of the drawworks.

Fig. 1.19 – Example drawworks used in rotary drilling.

Fig. 1.20 – Friction-type cathead.[12]

Fig. 1.21 – Tongs powered by chain to cathead.

Friction catheads shown in Fig. 1.20 turn continuously and can be used to assist in lifting or moving equipment on the rig floor. The number of turns of rope on the drum and the tension provided by the operator controls the force of the pull. A second type of cathead generally located between the drawworks housing and the friction cathead can be used to provide the torque needed to screw or unscrew sections of pipe. Fig. 1.21 shows a joint of drillpipe being tightened with tongs powered by a chain from the cathead. Hydraulically or air-powered spinning and torquing devices also are available as alternatives to the conventional tongs. One type of power tong is shown in Fig. 1.22.

1.5 Circulating System

A major function of the fluid-circulating system is to remove the rock cuttings from the hole as drilling progresses. A schematic diagram illustrating a typical rig circulating system is shown in Fig. 1.23. The drilling fluid is most commonly a suspension of clay and other materials in water and is called *drilling mud*. The drilling mud travels (1) from the steel tanks to the mud pump, (2) from the pump through the high-pressure surface connections to the drillstring, (3) through the drillstring to the bit, (4) through the nozzles of the bit and up the annular space between the drillstring and hole to the surface, and (5) through the contaminant-removal equipment back to the suction tank.

The principal components of the rig circulating system include (1) mud pumps, (2) mud pits, (3) mud-mixing equipment, and (4) contaminant-removal equipment. With the exception of several experimental types, mud pumps always have used reciprocating positive-displacement pistons. Both two-cylinder (duplex) and three-cylinder (triplex) pumps are common. The duplex pumps generally are *double-acting* pumps that pump on both forward and backward piston strokes. The triplex pumps generally are *single-acting* pumps that pump only on forward piston strokes. Triplex pumps are lighter and more compact than duplex pumps, their output pressure pulsations are not as great, and they are cheaper to operate. For these reasons, the majority of new pumps being placed into operation are of the triplex design.

The advantages of the reciprocating positive-displacement pump are the (1) ability to move high-solids-content fluids laden with abrasives, (2) ability to pump large particles, (3) ease of operation and maintenance, (4) reliability, and (5) ability to operate over a wide range of pressures and flow rates by changing the diameters of the *pump liners* (compression cylinders) and pistons. Example duplex and triplex mud pumps are shown in Fig. 1.24.

The overall efficiency of a mud-circulating pump is the product of the mechanical efficiency and the volumetric efficiency. Mechanical efficiency usually is assumed to be 90% and is related to the efficiency of the prime mover itself and the linkage to the pump drive shaft. Volumetric efficiency of a pump whose suction is adequately charged can be as high as 100%. Most manufacturers' tables rate pumps using a mechanical efficiency, E_m, of 90% and a volumetric efficiency, E_v, of 100%.

Generally, two circulating pumps are installed on the rig. For the large hole sizes used on the shallow portion of most wells, both pumps can be operated in parallel to deliver the large flow rates required. On the deeper portions of the well, only one pump is needed, and the second pump serves as a standby for use when pump maintenance is required.

A schematic diagram showing the valve arrangement and operation of a double-acting pump is shown in Fig. 1.25. The theoretical displacement from a double-acting pump is a function of the piston rod diameter d_r, the liner diameter d_l, and the stroke length L_s. On the forward stroke of each piston, the volume displaced is given by

$$\frac{\pi}{4} d_l^2 L_s.$$

Similarly, on the backward stroke of each piston, the volume displaced is given by

$$\frac{\pi}{4} \left(d_l^2 - d_r^2 \right) L_s.$$

Thus, the total volume displaced per complete pump cycle by a pump having two cylinders is given by

$$F_p = \frac{\pi}{4} (2) L_s (2 d_l^2 - d_r^2) E_v, \quad \dots \dots \dots (1.10)$$
$$\text{(duplex)}$$

where E_v is the volumetric efficiency of the pump. The pump displacement per cycle, F_p, is commonly called the *pump factor*.

For the single-acting (triplex) pump, the volume displaced by each piston during one complete pump cycle is given by

$$\frac{\pi}{4} d_l^2 L_s.$$

Thus, the pump factor for a single-acting pump having three cylinders becomes

$$F_p = \frac{3\pi}{4} L_s E_v d_l^2. \quad \dots \dots \dots \dots \dots (1.11)$$
$$\text{(triplex)}$$

The flow rate q of the pump is obtained by

Fig. 1.22 – Drillpipe tongs.

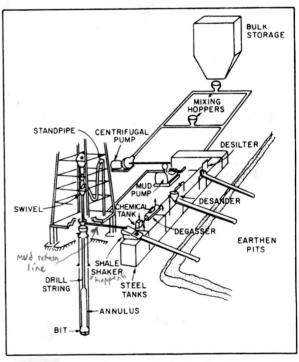

Fig. 1.23 – Schematic of example rig circulating system for liquid drilling fluids.

(a) Duplex design.

(b) Triplex design.

Fig. 1.24 – Example mud circulating pumps.

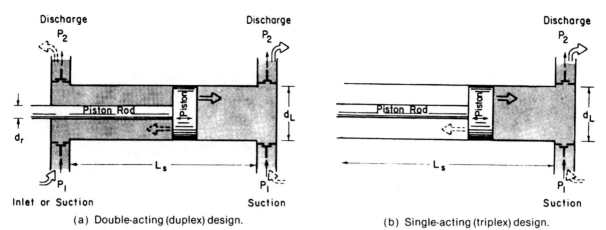

Fig. 1.25 – Schematic of valve operation of single- and double-acting pumps.[10]

multiplying the pump factor by N, the number of cycles per unit time. In common field usage, the terms *cycle* and *stroke* often are used interchangeably to refer to one complete pump revolution.

Pumps are rated for (1) hydraulic power, (2) maximum pressure, and (3) maximum flow rate. If the inlet pressure of the pump is essentially atmospheric pressure, the increase in fluid pressure moving through the pump is approximately equal to the discharge pressure. The hydraulic power output of the pump is equal to the discharge pressure times the flow rate. In field units of hp, psi, and gal/min, the hydraulic power developed by the pump is given by

$$P_H = \frac{\Delta p q}{1714}. \qquad \dots \dots \dots \dots \dots \dots (1.12)$$

For a given hydraulic power level, the maximum discharge pressure and flow rate can be varied by changing the stroke rate and liner size. A smaller liner will allow the operator to obtain a higher pressure, but at a lower rate. Due to equipment maintenance problems, pressures above about 3,500 psig seldom are used.

The flow conduits connecting the mud pumps to the drillstring include (1) a *surge chamber*, (2) a 4- or 6-in. heavy-walled pipe connecting the pump to a pump manifold located on the rig floor, (3) a *standpipe* and *rotary hose*, (4) a *swivel*, and (5) a *kelly*. The surge chamber (see Fig. 1.26) contains a gas in the upper portion, which is separated from the drilling fluid by a flexible diaphragm. The surge chamber greatly dampens the pressure surges developed by the positive-displacement pump. The discharge line also contains a pressure relief valve to prevent line rupture in the event the pump is started against a closed valve. The standpipe and rotary hose provide a flexible connection that permits vertical movement of the drillstring. The swivel contains roller bearings to support the rotating load of the drillstring and a rotating pressure seal that allows fluid circulation through the swivel. The kelly, which is a pipe rectangular or hexagonal in cross section, allows the drillstring to be rotated. It normally has a

3-in.-diameter passage for fluid circulation to the drillstring.

Example 1.3. Compute the pump factor in units of barrels per stroke for a duplex pump having 6.5-in. liners, 2.5-in. rods, 18-in. strokes, and a volumetric efficiency of 90%.

Solution. The pump factor for a duplex pump can be determined using Eq. 1.10:

$$F_p = \frac{\pi}{2} L_s E_v \left(2 d_l^2 - d_r^2 \right)$$

$$= \frac{\pi}{2} (18)(0.9) \left[2(6.5)^2 - (2.5)^2 \right]$$

$$= 1991.2 \text{ in.}^3/\text{stroke.}$$

Recall that there are 231 in.^{3} in a U.S. gallon and 42 U.S. gallons in a U.S. barrel. Thus, converting to the desired field units yields

$$1991.2 \text{ in.}^3/\text{stroke} \times \text{gal}/231 \text{ in.}^3 \times \text{bbl}/42 \text{ gal}$$

$$= 0.2052 \text{ bbl/stroke.}$$

Mud pits are required for holding an excess volume of drilling mud at the surface. This surface volume allows time for settling of the finer rock cuttings and for the release of entrained gas bubbles not mechanically separated. Also, in the event some drilling fluid is lost to underground formations, this fluid loss is replaced by mud from the surface pits. The settling and suction pits sometimes are dug in the earth with a bulldozer but more commonly are made of steel. A large earthen reserve pit is provided for contaminated or discarded drilling fluid and for the rock cuttings. This pit also is used to contain any formation fluids produced during drilling and well-testing operations.

Dry mud additives often are stored in sacks, which are added manually to the suction pit using a mud-mixing hopper. However, on many modern rigs bulk storage is used and mud mixing is largely automated. Liquid mud additives can be added to the suction pit from a chemical tank. Mud jets or motor-driven agitators often are mounted on the pits for auxiliary mixing.

The contaminant-removing equipment includes mechanical devices for removing solids and gases from the mud. The coarse rock cuttings and cavings are removed by the *shale shaker*. The shale shaker is composed of one or more vibrating screens over which the mud passes as it returns from the hole. A shale shaker in operation is shown in Fig. 1.27. Additional separation of solids and gases from the mud occurs in the settling pit. When the amount of finely ground solids in the mud becomes too great, they can be removed by *hydrocyclones* and *decanting centrifuges*. A hydrocyclone (Fig. 1.28) is a cone-shaped housing that imparts a whirling fluid motion much like a tornado. The heavier solids in the mud are thrown to the housing of the hydrocyclone and fall through the apex at the bottom. Most of the liquid and lighter particles exit through the vortex finder at the top. The decanting centrifuge (Fig. 1.29) consists of a rotating cone-shaped drum which has a screw conveyor attached to its interior. Rotation of the cone creates a centrifugal force that throws the heavier particles to the outer housing. The screw conveyor moves the separated particles to the discharge.

When the amount of entrained formation gas leaving the settling pit becomes too great, it can be separated using a degasser. A vacuum chamber degasser is shown in Fig. 1.30. A vacuum pump mounted on top of the chamber removes the gas from the chamber. The mud flows across inclined flat surfaces in the chamber in thin layers, which allows the gas bubbles that have been enlarged by the reduced pressure to be separated from the mud more easily. Mud is drawn through the chamber at a reduced pressure of about 5 psia by a mud jet located in the discharge line.

A gaseous drilling fluid can be used when the formations encountered by the bit have a high strength and an extremely low permeability. The use of gas as a drilling fluid when drilling most sedimentary rocks results in a much higher penetration rate than is obtained using drilling mud. An order-of-magnitude difference in penetration rates may be obtained with gas as compared with drilling mud. However, when formations are encountered that are capable of producing a significant volume of water, the rock cuttings tend to stick together and no longer can be easily blown from the hole. This problem sometimes can be solved by injecting a mixture of surfactant and water into the gas to make a foam-type drilling fluid. Drilling rates with foam are generally less than with air but greater than with water or mud. As the rate of water production increases, the cost of maintaining the foam also increases and eventually offsets the drilling rate improvement.

A second procedure that often is used when a water-producing zone is encountered is to seal off the

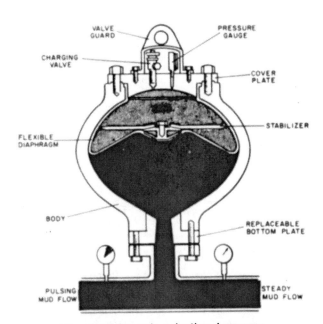

Fig. 1.26 – Example pulsation dampener.

Fig. 1.27 – Shale shaker in operation.

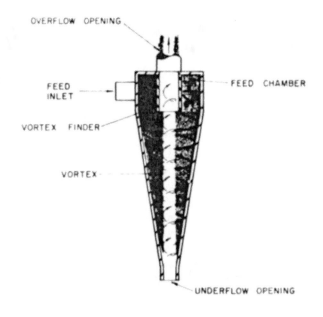

Fig. 1.28 – Schematic of hydrocyclone.

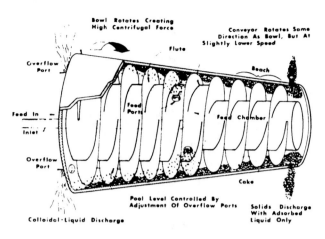

Fig. 1.29 – Schematic of a decanting centrifuge.

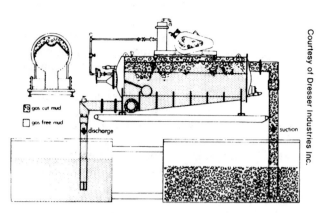

Fig. 1.30 – Schematic of a vacuum chamber degasser.

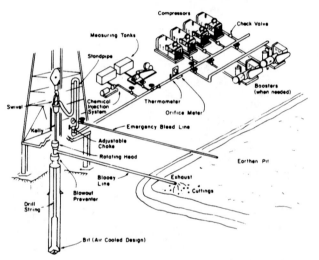

Fig. 1.31 – Schematic of circulating system for air drilling.

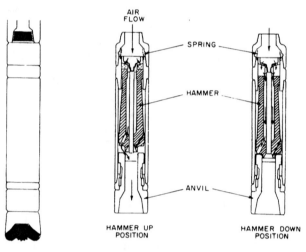

Fig. 1.32 – Percussion tool used in air drilling.

permeable zone. The water-producing zones can be plugged by use of (1) low-viscosity plastics or (2) silicon tetrafluoride gas. A catalyst injected with the plastic causes the plastic to begin to solidify when it contacts the hot formation. Silicon tetrafluoride gas reacts with the formation water and precipitates silica in the pore spaces of the rock. Best results are obtained when the water-producing formation is isolated for fluid injection by use of packers. Also, sufficient injection pressure must be used to exceed the formation pressure. Since this technique requires expending a considerable amount of rig time, the cost of isolating numerous water zones tends to offset the drilling rate improvement.

Both air and natural gas have been used as drilling fluids. An air compressor or natural gas pressure regulator allows the gas to be injected into the standpipe at the desired pressure. An example rig circulating system used for air drilling is shown in Fig. 1.31. The injection pressure usually is chosen so that the minimum annular velocity is about 3,000 ft/min. Also shown are small pumps used to inject water and surfactant into the discharge line. A

rotating head installed below the rig floor seals against the kelly and prevents the gas from spraying through the rig floor. The gas returning from the annulus then is vented through a *blooey* line to the reserve pit, at least 200 ft from the rig. If natural gas is used, it usually is burned continuously at the end of the blooey line. Even if air is used, *care must be taken to prevent an explosion.* Small amounts of formation hydrocarbons mixed with compressed air can be quite dangerous.

The subsurface equipment used for drilling with air is normally the same as the equipment used for drilling with mud. However, in a few areas where the compressive rock strength is extremely high, a percussion tool may be used in the drillstring above the bit. A cutaway view of an example percussion device is shown in Fig. 1.32. Gas flow through the tool causes a hammer to strike repeatedly on an anvil above the bit. The tool is similar in operation to the percussion hammer used by construction crews to break concrete. Under a normal operating pressure of 350 psia, the percussion tool causes the bit to hammer the formation about 1,800 blows/min in

Courtesy of ARMCO Natl. Supply Co.

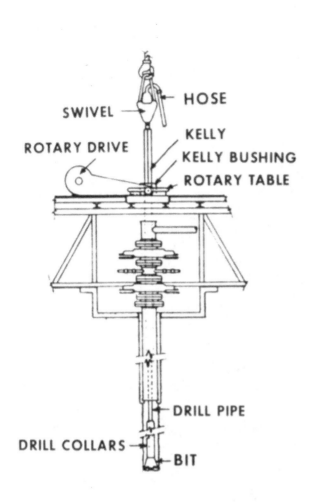

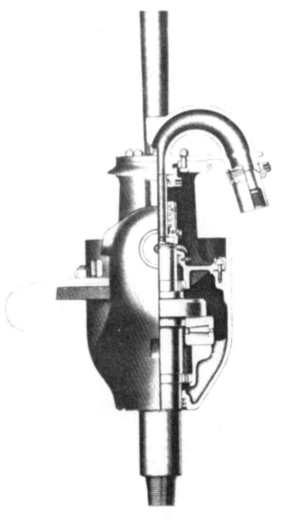

Fig. 1.33 – Schematic of rotary system.[12]

Fig. 1.34 – Cutaway view of example swivel.

addition to the normal rotary action. Penetration rates in extremely hard formations have been improved significantly by use of this tool.

1.6 The Rotary System

The rotary system includes all of the equipment used to achieve bit rotation. A schematic diagram illustrating the arrangement and nomenclature of the rotary system is shown in Fig. 1.33. The main parts of the rotary system are the (1) swivel, (2) kelly, (3) rotary drive, (4) rotary table, (5) drillpipe, and (6) drill collars.

The swivel (Fig. 1.34) supports the weight of the drillstring and permits rotation. The *bail* of the swivel is attached to the hook of the traveling block, and the *gooseneck* of the swivel provides a downward-pointing connection for the rotary hose. Swivels are rated according to their load capacities.

The kelly is the first section of pipe below the swivel. The outside cross section of the kelly is square or hexagonal to permit it to be gripped easily for turning. Torque is transmitted to the kelly through *kelly bushings*, which fit inside the master bushing of the *rotary table*. The kelly must be kept as straight as possible. Rotation of a crooked kelly causes a

whipping motion that results in unnecessary wear on the crown block, drilling line, swivel, and threaded connections throughout a large part of the drillstring.

A view of a kelly and kelly bushings in operation is shown in Fig. 1.35. The kelly thread is right-handed on the lower end and left-handed on the upper end to permit normal right-hand rotation of the drillstring. A *kelly saver sub* is used between the kelly and the first joint of drillpipe. This relatively inexpensive short section of pipe prevents wear on the kelly threads and provides a place for mounting a rubber protector to keep the kelly centralized.

An example rotary table is shown in Fig. 1.36. The opening in the rotary table that accepts the kelly bushings must be large enough for passage of the largest bit to be run in the hole. The lower portion of the opening is contoured to accept slips that grip the drillstring and prevent it from falling into the hole while a new joint of pipe is being added to the drillstring. A lock on the rotary prevents the table from turning when pipe is unscrewed without the use of backup tongs.

Power for driving the rotary table usually is provided by an independent rotary drive. However, in some cases, power is taken from the drawworks. A hydraulic transmission between the rotary table and

Fig. 1.35 – View of kelly and kelly bushings.

Fig. 1.36 – Example rotary table.

Fig. 1.37 – Example power sub.

TABLE 1.5 – DIMENSIONS AND STRENGTH OF API SEAMLESS INTERNAL UPSET DRILLPIPE

Size of Outer Diameter (in.)	Weight per Foot With Coupling (lbf)	Internal Diameter (in.)	Internal Diameter At Full Upset (in.)	Collapse Pressure*				Internal Yield Pressure*				Tensile Strength*			
				D (psi)	E (psi)	G** (psi)	S-135** (psi)	D (psi)	E (psi)	G** (psi)	S-135 (psi)	D 1,000 (lbf)	E 1,000 (lbf)	G** 1,000 (lbf)	S-135** 1,000 (lbf)
2⅜	4.85	1.995	1.437	6,850**	11,040	13,250	16,560	7,110**	10,500	14,700	18,900	70	98	137	176
2⅜	6.65	1.815	1.125	11,440	15,600	18,720	23,400	11,350	15,470	21,660	27,850	101	138	194	249
2⅞	6.85	2.441	1.875	–	10,470	12,560	15,700	–	9,910	13,870	17,830	–	136	190	245
2⅞	10.40	2.151	1.187	12,110	16,510	19,810	24,760	12,120	16,530	23,140	29,750	157	214	300	386
3½	9.50	2.992	2.250	–	10,040	12,110	15,140	–	9,520	13,340	17,140	–	194	272	350
3½	13.30	2.764	1.875	10,350	14,110	16,940	21,170	10,120	13,800	19,320	24,840	199	272	380	489
3½	15.50	2.602	1.750	12,300	16,770	20,130	25,160	12,350	16,840	23,570	30,310	237	323	452	581
4	11.85	3.476	2.937	–	8,410	10,310	12,820	–	8,600	12,040	15,470	–	231	323	415
4	14.00	3.340	2.375	8,330	11,350	14,630	17,030	7,940	10,830	15,160	19,500	209	285	400	514
4½	13.75	3.958	3.156	–	7,200	8,920	10,910	–	7,900	11,070	14,230	–	270	378	486
4½	16.60	3.826	2.812	7,620	10,390	12,470	15,590	7,210	9,830	13,760	17,690	242	331	463	595
4½	20.00	3.640	2.812	9,510	12,960	15,560	19,450	9,200	12,540	17,560	22,580	302	412	577	742
5	16.25	4.408	3.750	–	6,970	8,640	10,550	–	7,770	10,880	13,980	–	328	459	591
5	19.50	4.276	3.687	7,390	10,000	12,090	15,110	6,970	9,500	13,300	17,100	290	396	554	712
5½	21.90	4.778	3.812	6,610	8,440	10,350	12,870	6,320	8,610	12,060	15,500	321	437	612	787
5½	24.70	4.670	3.500	7,670	10,460	12,560	15,700	7,260	9,900	13,860	17,820	365	497	696	895
5 9/16	19.00**	4.975	4.125	4,580	5,640	–	–	5,090	6,950	–	–	267	365	–	–
5 9/16	22.20**	4.859	3.812	5,480	6,740	–	–	6,090	8,300	–	–	317	432	–	–
5 9/16	25.25**	4.733	3.500	6,730	8,290	–	–	7,180	9,790	–	–	369	503	–	–
6⅝	22.20**	6.065	5.187	3,260	4,020	–	–	4,160	5,530	–	–	307	418	–	–
6⅝	25.20	5.965	5.000	4,010	4,810	6,160	6,430	4,790	6,540	9,150	11,770	359	489	685	881
6⅝	31.90**	5.761	4.625	5,020	6,170	–	–	6,275	8,540	–	–	463	631	–	–

*Collapse, internal yield, and tensile strengths are minimum values with no safety factor. D, E, G, S-135 are standard steel grades used in drillpipe.[5]
**Not API standard; shown for information only.

the rotary drive often is used. This greatly reduces shock loadings and prevents excessive torque if the drillstring becomes stuck. Excessive torque often will result in a *twist-off* – i.e., a torsional failure due to a break in the subsurface drillstring.

Power swivels or power subs installed just below a conventional swivel can be used to replace the kelly, kelly bushings, and rotary table. Drillstring rotation is achieved through a hydraulic motor incorporated in the power swivel or power sub. These devices are available for a wide range of rotary speed and torque combinations. One type of power sub is shown in Fig. 1.37.

The major portion of the drillstring is composed of drillpipe. The drillpipe in common use is hot-rolled, pierced, seamless tubing. API has developed specifications for drillpipe. Drillpipe is specified by its outer diameter, weight per foot, steel grade, and range length. The dimensions and strength of API drillpipe of grades D, E, G, and S-135 are shown in Table 1.5. Drillpipe is furnished in the following API length ranges.

Range	Length (ft)
1	18 to 22
2	27 to 30
3	38 to 45

Range 2 drillpipe is used most commonly. Since each joint of pipe has a unique length, the length of each joint must be measured carefully and recorded to allow a determination of total well depth during drilling operations.

The drillpipe joints are fastened together in the drillstring by means of *tool joints* (Fig. 1.38). The female portion of the tool joint is called the *box* and the male portion is called the *pin*. The portion of the drillpipe to which the tool joint is attached has thicker walls than the rest of the drillpipe to provide for a stronger joint. This thicker portion of the pipe is called the *upset*. If the extra thickness is achieved by decreasing the internal diameter, the pipe is said to have an internal upset. A rounded-type thread is used now on drill pipe. The U.S. Standard V thread was used in early drillpipe designs, but thread failure was frequent because of the stress concentrations in the thread root. A tungsten carbide hard facing sometimes is manufactured on the outer surface of the tool joint box to reduce the abrasive wear of the tool joint by the borehole wall when the drillstring is rotated.

The lower section of the rotary drillstring is composed of *drill collars*. The drill collars are thick-walled heavy steel tubulars used to apply weight to the bit. The buckling tendency of the relatively thin-walled drillpipe is too great to use it for this purpose. The smaller clearance between the borehole and the drill collars helps to keep the hole straight. *Stabilizer subs* (Fig. 1.39) often are used in the drill collar string to assist in keeping the drill collars centralized.

In many drilling operations, a knowledge of the volume contained in or displaced by the drillstring is required. The term *capacity* often is used to refer to the cross-sectional area of the pipe or annulus expressed in units of contained volume per unit length. In terms of the pipe diameter, d, the capacity of pipe, A_p, is given by

$$A_p = \frac{\pi}{4} d^2. \qquad (1.13)$$

Similarly, the capacity of an annulus, A_a, in terms of the inner and outer diameter, is

$$A_a = \frac{\pi}{4} (d_2^2 - d_1^2). \qquad (1.14)$$

The term *displacement* often is used to refer to the cross-sectional area of steel in the pipe expressed in units of volume per unit length. The displacement, A_s, of a section of pipe is given by

$$A_s = \frac{\pi}{4}(d_1^2 - d^2). \qquad \ldots \ldots \ldots \ldots \ldots (1.15)$$

Displacements calculated using Eq. 1.15 do not consider the additional fluid displaced by the thicker steel sections at the tool joints or couplings. When a more exact displacement calculation is needed, tables provided by the tool joint or coupling manufacturer can be used. Table 1.6 gives average displacement values for Range 2 drillpipe, including tool joint displacements.

Example 1.4. A drillstring is composed of 7,000 ft of 5-in., 19.5-lbm/ft drillpipe and 500 ft of 8-in. OD by 2.75-in. ID drill collars when drilling a 9.875-in. borehole. Assuming that the borehole remains in gauge, compute the number of pump cycles required to circulate mud from the surface to the bit and from

SIZES AND DIMENSIONS FOR GRADE E "XTRAHOLE" DESIGN						
Dimension Symbol	3½" Drillpipe		4½" Drillpipe		5" Drillpipe	
	Inches	mm	Inches	mm	Inches	mm
A	3⅝	92	4¹¹⁄₁₆	119	5⅛	130
B	2⅛	54	3¼	83	3¾	95
C	4¾	121	6¼	159	6⅜	162
D	.438	11	.672	17	.531	13
E	4¹⁷⁄₃₂	115	5²³⁄₃₂	145	5⁵⁹⁄₆₄	150
LP	6½	165	7	178	7	178
LB	9½	241	10	254	10	254

Courtesy of Hughes Tool Co.

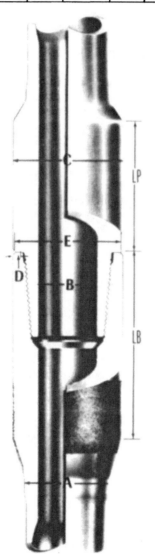

Fig. 1.38 – Cutaway view and dimensions for example tool joint.

Courtesy of HOMCO

Fig. 1.39 – Example stabilizer.

TABLE 1.6 – AVERAGE DISPLACEMENTS FOR RANGE 2 DRILL PIPE

Size of Outer Diameter (in.)	Nominal Weight (lbm/ft)	Tool-Joint Type	Actual Weight in Air (lbm/ft)	Displacement (ft/bbl)	Displacement (bbl/ft)	Displacement bbl/90 ft Stand
2⅜	6.65	internal flush	6.90	398.4	0.00251	0.23
2⅞	10.40	internal flush	10.90	251.9	0.00397	0.36
		slim hole	10.40	263.0	0.00379	0.34
3½	13.30	full hole	13.90	197.6	0.00506	0.46
		slim hole	13.40	204.9	0.00488	0.44
		internal flush	13.80	199.2	0.00502	0.45
	15.50	internal flush	16.02	171.5	0.00583	0.52
4	14.00	full hole	15.10	181.8	0.00550	0.50
		internal flush	15.10	176.1	0.00568	0.51
4½	16.60	full hole	17.80	154.3	0.00648	0.58
		xtrahole	18.00	152.7	0.00655	0.59
		slim hole	17.00	161.6	0.00619	0.56
		internal flush	17.70	155.3	0.00644	0.58
	20.00	xtrahole	21.40	128.5	0.00778	0.70
		full hole	21.30	129.0	0.00775	0.70
		slim hole	20.50	134.0	0.00746	0.67
		internal flush	21.20	129.5	0.00772	0.69
	22.82	xtrahole	24.10	114.0	0.00877	0.79
	32.94	xtrahole	36.28	75.7	0.01320	1.19
5	19.50	xtrahole	20.60	133.3	0.00750	0.68
	25.60	xtrahole	26.18	107.4	0.00932	0.84
	42.00	xtrahole	45.2±	60.8±	0.0165±	1.48±

the bottom of the hole to the surface if the pump factor is 0.1781 bbl/cycle.

Solution. For field units of feet and barrels, Eq. 1.13 becomes

$$A_p = \left(\frac{\pi}{4}d^2\right)\text{in.}^2\left(\frac{\text{gal}}{231\ \text{in.}^3}\right)\left(\frac{\text{bbl}}{42\ \text{gal}}\right)\left(\frac{12\ \text{in.}}{\text{ft}}\right)$$

$$= \left(\frac{d^2}{1,029.4}\right)\text{bbl/ft.}$$

Using Table 1.5, the inner diameter of 5-in., 19.5 lbm/ft drillpipe is 4.276 in.; thus, the capacity of the drillpipe is

$$\frac{4.276^2}{1,029.4} = 0.01776\ \text{bbl/ft}$$

and the capacity of the drill collars is

$$\frac{2.75^2}{1,029.4} = 0.00735\ \text{bbl/ft.}$$

The number of pump cycles required to circulate new mud to the bit is given by

$$\frac{[0.01776\,(7,000) + 0.00735\,(500)]\text{bbl}}{0.1781\ \text{bbl/cycle}} = 719\ \text{cycles.}$$

Similarly, the annular capacity outside the drillpipe is given by

$$\frac{9.875^2 - 5^2}{1,029.4} = 0.0704\ \text{bbl/ft}$$

and the annular capacity outside the drill collars is

$$\frac{9.875^2 - 8^2}{1,029.4} = 0.0326\ \text{bbl/ft.}$$

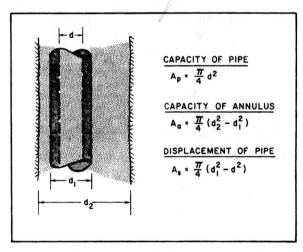

CAPACITY OF PIPE
$$A_p = \frac{\pi}{4}d^2$$

CAPACITY OF ANNULUS
$$A_a = \frac{\pi}{4}(d_2^2 - d_1^2)$$

DISPLACEMENT OF PIPE
$$A_s = \frac{\pi}{4}(d_1^2 - d^2)$$

Fig. 1.40 – Capacity and displacement nomenclature.

The pump cycles required to circulate mud from the bottom of the hole to the surface is given by

$$\frac{[0.0704\,(7,000) + 0.0326\,(500)]\text{bbl}}{0.1781\ \text{bbl/cycle}}$$

$$= 2,858\ \text{cycles.}$$

1.7 The Well Control System

The well control system prevents the uncontrolled flow of formation fluids from the wellbore. When the bit penetrates a permeable formation that has a fluid pressure in excess of the hydrostatic pressure exerted by the drilling fluid, formation fluids will

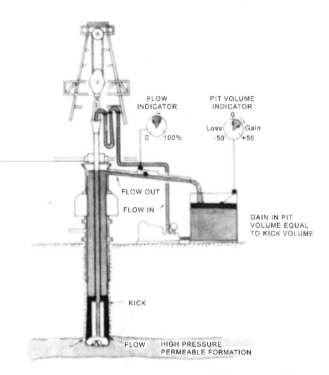

Fig. 1.41 – Kick detection during drilling operations.

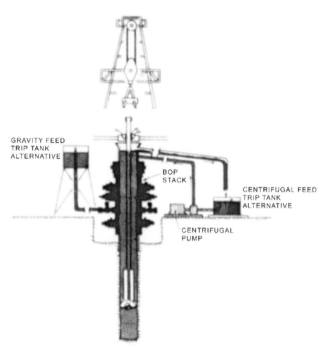

Fig. 1.42 – Two alternative trip-tank arrangements for kick detection during tripping operations.

begin displacing the drilling fluid from the well. The flow of formation fluids into the well in the presence of drilling fluid is called a *kick*. The well control system permits (1) detecting the kick, (2) closing the well at the surface, (3) circulating the well under pressure to remove the formation fluids and increase the mud density, (4) moving the drillstring under pressure, and (5) diverting flow away from rig personnel and equipment.

Failure of the well control system results in an uncontrolled flow of formation fluids and is called a *blowout*. This is perhaps the worst disaster that can occur during drilling operations. Blowouts can cause loss of life, drilling equipment, the well, much of the oil and gas reserves in the underground reservoir, and damage to the environment near the well. Thus, the well control system is one of the more important systems on the rig.

Kick detection during drilling operations usually is achieved by use of a pit-volume indicator or a flow indicator. The operation of these devices is illustrated in Fig. 1.41. Both devices can detect an increase in the flow of mud returning from the well over that which is being circulated by the pump.

Pit volume indicators usually employ floats in each pit that are connected by means of pneumatic or electrical transducers to a recording device on the rig floor. The recording device indicates the volume of all active pits. High- and low-level alarms can be preset to turn on lights and horns when the pit volume increases or decreases significantly. An increase in surface mud volume indicates that formation fluids may be entering the well. A decrease indicates that drilling fluid is being lost to an underground formation.

Mud flow indicators are used to help detect a kick

more quickly. The more commonly used devices are somewhat similar in operation to the pit level indicators. A paddle-type fluid level sensor is used in the flowline. In addition, a pump stroke counter is used to sense the flow rate into the well. A panel on the rig floor displays the flow rate into and out of the well. If the rates are appreciably different, a gain or loss warning will be given.

While making a trip, circulation is stopped and a significant volume of pipe is removed from the hole. Thus, to keep the hole full, mud must be pumped into the hole to replace the volume of pipe removed. Kick detection during tripping operations is accomplished through use of a hole fill-up indicator. The purpose of the hole fill-up indicator is to measure accurately the mud volume required to fill the hole. If the volume required to fill the hole is less than the volume of pipe removed, a kick may be in progress.

Small trip tanks provide the best means of monitoring hole fill-up volume. Trip tanks usually hold 10 to 15 bbl and have 1-bbl gauge markers. Two alternative trip-tank arrangements are illustrated in Fig. 1.42. With either arrangement, the hole is maintained full as pipe is withdrawn from the well. Periodically, the trip tank is refilled using the mud pump. The top of a gravity-feed type trip tank must be slightly lower than the bell nipple to prevent mud from being lost to the flowline. The required fill-up volume is determined by periodically checking the fluid level in the trip tank. When a trip tank is not installed on the rig, hole fill-up volume should be determined by counting pump strokes each time the hole is filled. The level in one of the active pits should not be used since the active pits are normally too large to provide sufficient accuracy.

Fig. 1.43 – Example ram-type blowout preventer.

The flow of fluid from the well caused by a kick is stopped by use of special pack-off devices called *blowout preventers* (BOP's). Multiple BOP's used in a series are referred to collectively as a *BOP stack*. The BOP must be capable of terminating flow from the well under all drilling conditions. When the drillstring is in the hole, movement of the pipe without releasing well pressure should be allowed to occur. In addition, the BOP stack should allow fluid circulation through the well annulus under pressure. These objectives usually are accomplished by using several ram preventers and one annular preventer.

An example of a ram preventer is shown in Fig. 1.43. Ram preventers have two packing elements on opposite sides that close by moving toward each other. *Pipe rams* have semicircular openings which match the diameter of pipe sizes for which they are designed. Thus the pipe ram must match the size of pipe currently in use. If more than one size of drillpipe is in the hole, additional ram preventers must be used in the BOP stack. Rams designed to close when no pipe is in the hole are called *blind rams*. Blind rams will flatten drillpipe if inadvertently closed with the drillstring in the hole but will not stop the flow from the well. *Shear rams* are blind rams designed to shear the drillstring when closed. This will cause the drillstring to drop in the hole and will stop flow from the well. Shear rams are closed on pipe only when all pipe rams and annular preventers have failed. Ram preventers are available for working pressures of 2,000, 5,000, 10,000, and 15,000 psig.

Annular preventers, sometimes called bag-type preventers, stop flow from the well using a ring of synthetic rubber that contracts in the fluid passage. The rubber packing conforms to the shape of the pipe in the hole. Most annular preventers also will

close an open hole if necessary. A cross section of one type of annular preventer is shown in Fig. 1.44. Annular preventers are available for working pressures of 2,000, 5,000, and 10,000 psig.

Both the ram and annular BOP's are closed hydraulically. In addition, the ram preventers have a screw-type locking device that can be used to close the preventer if the hydraulic system fails. The annular preventers are designed so that once the rubber element contacts the drillstring, the well pressure helps hold the preventer closed.

Modern hydraulic systems used for closing BOP's are high-pressure fluid *accumulators* similar to those developed for aircraft fluid control systems. An example vertical accumulator is shown in Fig. 1.45. The accumulator is capable of supplying sufficient high-pressure fluid to close all of the units in the BOP stack at least once and still have a reserve. Accumulators with fluid capacities of 40, 80, or 120 gal and maximum operating pressures of 1,500 or 3,000 psig are common. The accumulator is maintained by a small pump at all times, so the operator has the ability to close the well immediately, independent of normal rig power. For safety, stand-by accumulator pumps are maintained that use a secondary power source. The accumulator fluid usually is a noncorrosive hydraulic oil with a low freezing point. The hydraulic oil also should have good lubricating characteristics and must be compatible with synthetic rubber parts of the well-control system.

The accumulator is equipped with a pressure-regulating system. The ability to vary the closing pressure on the preventers is important when it is necessary to *strip* pipe (lower pipe with the preventer closed) into the hole. If a kick is taken during a trip, it is best to strip back to bottom to allow efficient circulation of the formation fluids from the well. The

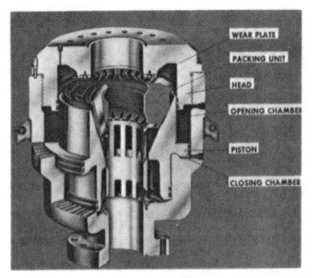

Courtesy of Hydril Co.

Fig. 1.44 – Example annular-type blowout preventer.

Courtesy of Koomey Inc.

Fig. 1.45 – Example accumulator system.

Courtesy of Koomey Inc.

Fig. 1.46 – Example remote control panel for operating blowout preventers.

application of too much closing pressure to the preventer during stripping operations causes rapid wear of the sealing element. The usual procedure is to reduce the hydraulic closing pressure during stripping operations until there is a slight leakage of well fluid.

Stripping is accomplished most easily using the annular preventer. However, when the surface well pressure is too great, stripping must be done using two pipe ram preventers placed far enough apart for external upset tool joints to fit between them. The upper and lower rams must be closed and opened alternately as the tool joints are lowered through.

Space between ram preventers used for stripping operations is provided by a *drilling spool*. Drilling spools also are used to permit attachment of high-pressure flowlines to a given point in the stack. These high-pressure flowlines make it possible to pump into the annulus or release fluid from the annulus with the BOP closed. A conduit used to pump into the annulus is called a *kill line*. Conduits used to release fluid from the annulus may include a *choke line*, a *diverter line*, or simply a flowline. All drilling spools must have a large enough bore to permit the next string of casing to be put in place without removing the BOP stack.

The BOP stack is attached to the casing using a *casing head*. The casing head, sometimes called the *braden head*, is welded to the first string of casing

cemented in the well. It must provide a pressure seal for subsequent casing strings placed in the well. Also, outlets are provided on the casing head to release any pressure that might accumulate between casing strings.

The control panel for operating the BOP stack usually is placed on the derrick floor for easy access by the driller. The controls should be marked clearly and identifiably with the BOP stack arrangement used. One kind of panel used for this purpose is shown in Fig. 1.46.

The arrangement of the BOP stack varies considerably. The arrangement used depends on the magnitude of formation pressures in the area and on the type of well control procedures used by the operator. API presents several recommended arrangements of BOP stacks. Fig. 1.47 shows typical arrangements for 10,000- and 15,000-psi working pressure service. Note that the arrangement nomenclature uses the letter "A" to denote an annular preventer, the letter "R" to denote a ram preventer, and the letter "S" to denote a drilling spool. The arrangement is defined starting at the casing head and proceeding up to the bell nipple. Thus, Arrangement RSRRA denotes the use of a BOP stack with a ram preventer attached to the casing head, a drilling spool above the ram preventer, two ram preventers in series above the drilling spool,

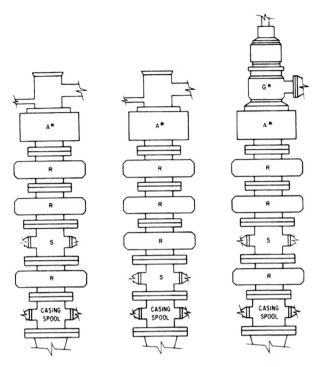

Fig. 1.47 – Typical surface stack blowout preventer arrangements for 10,000- and 15,000-psi working pressure service.

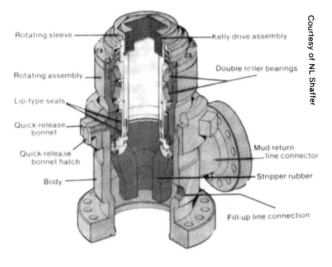

Fig. 1.48 – Cutaway view of rotating blowout preventer.

and an annular preventer above the ram preventers.

In some cases, it may be desirable to conduct drilling operations with a slight surface pressure on the annulus. A *rotating head*, which seals around the kelly at the top of the BOP stack, must be used when this is done. A rotating-type BOP is shown in Fig. 1.48. Rotating heads most commonly are employed when air or gas is used as a drilling fluid. They also can be used when formation fluids are entering the well very slowly from low-permeability formations. However, *this practice is dangerous unless the formation being drilled has a very low permeability.* This must be established from experience gained in drilling in the local area. For example, this practice is known to be safe in the Ellenberger formation in some areas of west Texas.

When the drillstring is in the hole, the BOP stack can be used to stop only the flow from the annulus. Several additional valves can be used to prevent flow from inside the drillstring. These valves include *kelly cocks* and *internal blowout preventers*. Shown in Fig. 1.49 is an example kelly cock. Generally, an *upper kelly cock* having left-hand threads is placed above the kelly and a *lower kelly cock* having right-hand threads is placed below the kelly. The lower kelly cock also is called a *drillstem valve*. Two kelly cocks are required because the lower position might not be accessible in an emergency if the drillstring is stuck in the hole with the kelly down.

An internal BOP is a valve that can be placed in the drillstring if the well begins flowing during tripping operations. Ball valves similar to the valve shown in Fig. 1.49 also can be used as an internal BOP. In addition, dart-type (check-valve) internal BOP's (Fig. 1.50) are available. This type of internal BOP should be placed in the drillstring before drillpipe is

stripped back in the hole because it will permit mud to be pumped down the drillstring after reaching the bottom of the well. Internal BOP's are installed when needed by screwing into the top of an open drillstring with the valve or dart in the open position. Once the BOP is installed, the valve can be closed or the dart released.

A high-pressure circulating system used for well control operations is shown in Fig. 1.51. The kick normally is circulated from the well through an *adjustable choke*. The adjustable choke is controlled from a remote panel on the rig floor. An example choke and a control panel are shown in Figs. 1.52 and 1.53. Sufficient pressure must be held against the well by the choke so that the bottomhole pressure in the well is maintained slightly above the formation pressure. Otherwise, formation fluids would continue to enter the well.

Mechanical stresses on the emergency high-pressure flow system can be quite severe when handling a kick. The rapid pressure release of large volumes of fluid through the surface piping frequently is accompanied by extreme vibrational stresses. Thus, care should be taken to use the strongest available pipe and to anchor all lines securely against reaction thrust. Also, some flexibility in the piping to and from the wellhead is required. The weight of all valves and fittings should be supported on structural members so that bending stresses are not created in the piping. Because of fluid abrasion, the number of bends should be minimized. The bends required should be sweep-turn bends rather than sharp "L" turns, or have an abrasion-resistant target at the point of fluid impingement in the bend.

API[8] presents several recommended choke

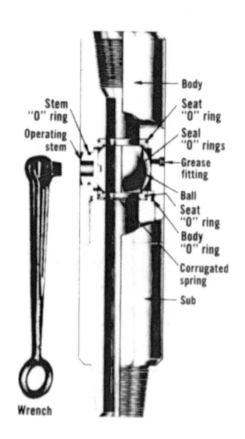

Courtesy of Texas Iron Works

Courtesy of NL Shaffer

Fig. 1.49 – Example kelly cock.

Fig. 1.50 – Example dart-type internal blowout preventer.

PARTS LIST

Item	Part
1	Main Sub
2	Sealing Sub
3	Setting Tool Assy.
4	Spider with Guide
5	Spring
6	Dart
7	Dart Rubber
8	Hold Down Bar
9	Base Stand
10	Releasing Pin
11	Setting Tool Handle

Suggested Extra Parts:
1 each: Spider
    w/Guide
    Spring
    Dart Rubber

Shaffer Inside
Blowout Preventer
in Holder

manifold arrangements for 2,000, 3,000, 5,000, 10,000, and 15,000 psig working pressure systems. In addition to these recommendations, well operators have developed many other optional designs. The arrangement selected must be based on the magnitude of the formation pressures in the area and the well control procedures used by the operator. Shown in Fig. 1.51 is one of the alternative API arrangements. In this arrangement, a hydraulically controlled valve separates the BOP stack from the choke manifold. This valve normally is closed during drilling operations to prevent drilling mud solids from settling in the choke system. The controls that operate this valve are placed on the BOP control panel so that the BOP can be operated easily. Two adjustable chokes would allow kick circulation to continue in the event one of the adjustable chokes fails.

A mud gas separator permits any produced formation gases to be vented. Also, valves are arranged so that the well fluids can be diverted easily to the reserve pit to prevent excessive pressure from fracturing shallow formations below a short casing string.

The kill line permits drilling fluid to be pumped down the annulus from the surface. This procedure is used only under special circumstances and is not part of a normal well control operation. The kill line most frequently is needed when subsurface pressure during a kick causes an exposed formation to fracture and to begin rapidly taking drilling fluid from the upper portion of the hole.

1.8 Well-Monitoring System

Safety and efficiency considerations require constant monitoring of the well to detect drilling problems quickly. An example of a driller's control station is shown in Fig. 1.54. Devices record or display parameters such as (1) depth, (2) penetration rate, (3) hook load, (4) rotary speed, (5) rotary torque, (6) pump rate, (7) pump pressure, (8) mud density, (9) mud temperature, (10) mud salinity, (11) gas content of mud, (12) hazardous gas content of air, (13) pit level, and (14) mud flow rate.

In addition to assisting the driller in detecting drilling problems, good historical records of various aspects of the drilling operation also can aid geological, engineering, and supervisory personnel. In some cases, a centralized well-monitoring system housed in a trailer is used (Fig. 1.55). This unit provides detailed information about the formation being drilled and fluids being circulated to the surface in the mud as well as centralizing the record keeping of drilling parameters. The *mud logger* carefully inspects rock cuttings taken from the shale shaker at regular intervals and maintains a log describing their appearance. Additional cuttings are labeled according to their depth and are saved for further study by the paleontologist. The identification of the microfossils present in the cuttings assists the geologist in correlating the formations being drilled. Gas samples removed from the mud are analyzed by the mud logger using a gas chromatograph. The presence of a hydrocarbon reservoir often can be detected by this type of analysis.

Recently, there have been significant advances in subsurface well-monitoring and data-telemetry systems. These systems are especially useful in monitoring hole direction in nonvertical wells. One of the most promising techniques for data telemetry from subsurface instrumentation in the drillstring to the surface involves the use of a mud pulser that sends information to the surface by means of coded pressure pulses in the drilling fluid contained in the drillstring. One system, illustrated in Fig. 1.56, uses a bypass valve to the annulus to create the needed pressure signal.

1.9 Special Marine Equipment

Special equipment and procedures are required when drilling from a floating vessel. The special equipment is required to (1) hold the vessel on location over the borehole and (2) compensate for the vertical, lateral, and tilting movements caused by wave action against the vessel. Vessel motion problems are more severe for a drillship than for a semisubmersible. However, drillships usually are less expensive and can be moved rapidly from one location to the next.

A special derrick design must be used for drillships because of the tilting motion caused by wave action. The derrick of a drillship often is designed to withstand as much as a 20° tilt with a full load of drillpipe standing in the derrick. Also, special pipe-handling equipment is necessary to permit tripping operations to be made safely during rough weather. This equipment permits drillpipe to be laid down quickly on a pipe rack in doubles or thribbles rather than supported in the derrick. A block guide track also is used to prevent the traveling block from swinging in rough weather.

Most floating vessels are held on location by anchors. When the ocean bottom is too hard for conventional anchors, anchor piles are driven or cemented in boreholes in the ocean floor. The vessel is moored facing the direction from which the most severe weather is anticipated. A drillship has been designed that can be moored from a central turret containing the drilling rig. The ship is rotated about the turret using thrusters mounted in the bow and stern so that it always faces incoming waves. Most mooring systems are designed to restrict horizontal vessel movement to about 10% of the water depth for the most severe weather conditions; however, horizontal movement can be restricted to about 3% of the water depth for the weather conditions experienced 95% of the time. As many as 10 anchors are used in a mooring system. Several common anchor patterns are shown in Fig. 1.57.

A few vessels have large thrust units capable of holding the drilling vessel on location without anchors. This placement technique is called *dynamic positioning*. The large fuel consumption required for

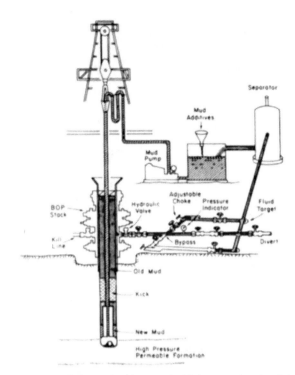

Fig. 1.51 – Schematic of example high-pressure circulating system for well control operations.

Fig. 1.52 – Example choke manifold showing 15,000-psi hand-adjustable choke and 15,000-psi remote adjustable choke.

Courtesy of Louisiana State U. Blowout Prevention Center

Fig. 1.53—Example control panels for remote adjustable choke.

Courtesy of Louisiana State U. Blowout Prevention Center

Fig. 1.54 – Example driller's control unit.

Fig. 1.55 – Example well monitoring unit.

dynamic positioning is economically feasible only when (1) frequent location changes are required or (2) the lengths of the anchor lines required are excessive. Also, the range of weather conditions that can be sustained is more limited for dynamic positioning. Dynamic positioning generally is not used in water depths of less than 3,000 ft.

The position of the vessel with reference to the borehole must be monitored at all times. Excessive wear on the subsea equipment will result if the vessel is not aligned continuously over the hole. Two types of alignment indicators in common use are (1) the

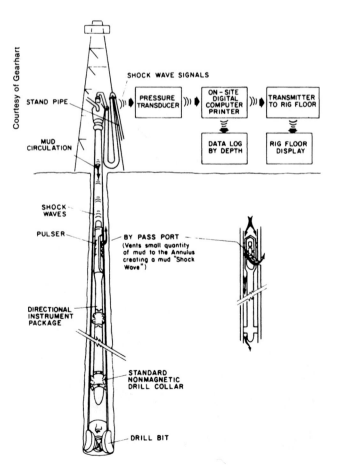

Fig. 1.56 – Example subsurface well monitoring system.

mechanical type and (2) the acoustic type. The mechanical type system uses a dual-axis inclinometer attached to a cable running from the wellhead to the ship. It is assumed that sufficient tension is maintained in the line to keep it straight. In addition, an inclinometer may be attached to the flow conduit that conducts the drilling fluid from the ocean floor to the drilling vessel. The acoustic-type position indicator uses beacon transmitters on the ocean floor and hydrophones on the ship. Doppler sonar may be used also. This system is more accurate than the taut-line system in deep water and does not depend on a mechanical link with the vessel.

Part of the equipment used to compensate for the horizontal and vertical movement of the vessel during normal drilling operations is shown in Fig. 1.58. A *marine riser* conducts the drilling fluid from the ocean floor to the drilling vessel. A *flex joint* at the bottom of the marine riser allows lateral movement of the vessel. The vertical movement of the vessel is allowed by a *slip joint* placed at the top of the marine riser. The riser is secured to the vessel by a pneumatic tensioning system. The tension requirements can be reduced by adding buoyant sections to the riser system.

The vertical movement of the drillstring can be absorbed by a *bumper sub* between the drillpipe and drill collars. However, many problems result from this arrangement, since vertical vessel movement causes the entire length of drillpipe to reciprocate relative to the casing and hole. Also, it is not possible to vary bit weight when bumper subs are used. Surface motion-compensating equipment called *heave compensators* have been developed in order to eliminate this problem. A constant hook load is maintained through use of a pneumatic tensioning device on the traveling block as shown in Fig. 1.59.

The BOP stack for a floating drilling operation is placed on the ocean floor below the marine riser. This ensures that the well can be closed even in severe weather, such as a hurricane, when it may become necessary to disconnect the marine riser. Also, it

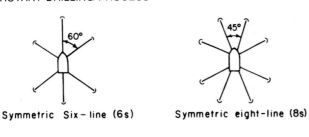

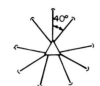

Symmetric Six-line (6s)   Symmetric eight-line (8s)

Symmetric nine-line (9s)   Symmetric ten-line (10s)

Symmetric twelve-line (12s)   45°-90° eight-line (8a)

30°-60° eight-line (8b)   45°-90° ten-line (10a)

Fig. 1.57 – Example spread mooring patterns.[17]

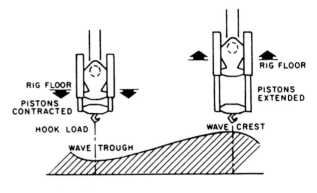

Fig. 1.59 – Operation of a heave compensator.[19]

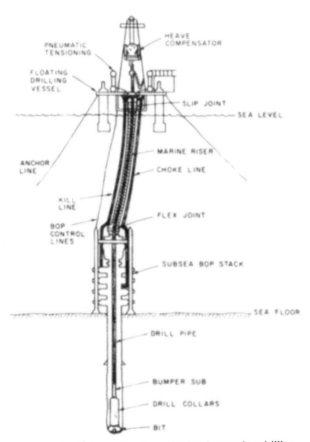

Fig. 1.58 – Schematic of equipment for marine drilling operations.

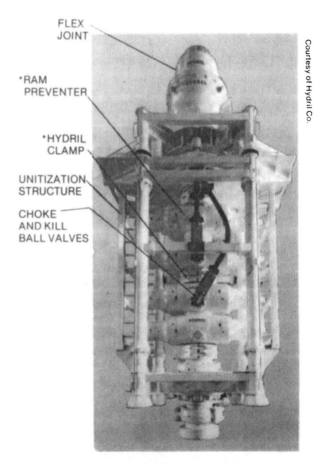

Fig. 1.60 – Example subsea blowout preventer stack.

would be extremely difficult to design a marine riser and slip joint assembly capable of withstanding high annular pressures. Identical hydraulically operated connectors often are used above and below the BOP stack. This makes it possible to add on an additional BOP stack above the existing one in an emergency.

An example subsea BOP stack is shown in Fig. 1.60. The kill line and choke line to the BOP stack are attached to the marine riser. Shown in Fig. 1.61 are cutaway views of example upper and lower marine riser equipment with the choke and kill lines integrally attached. The hydraulic lines required to operate the BOP stack, side valves, and connectors are attached to a cable guide. They are stored and

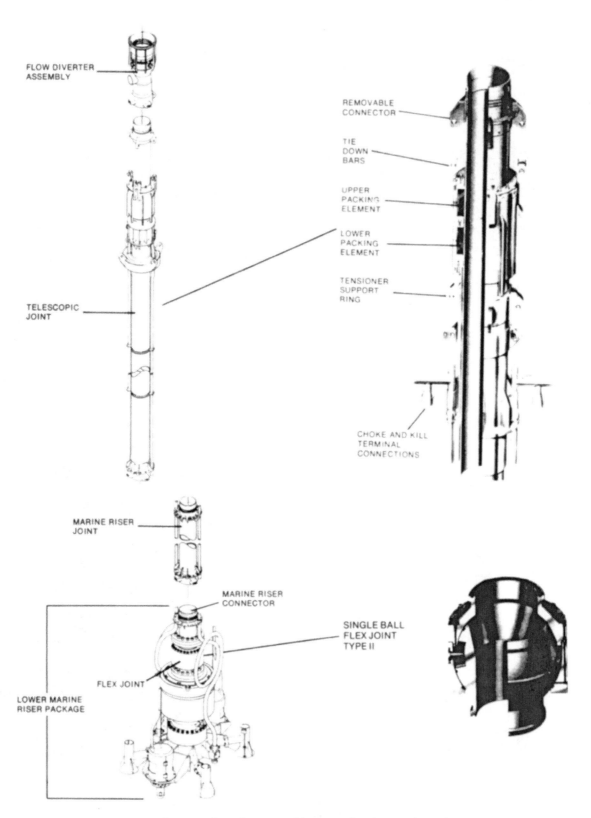

FLOW DIVERTER ASSEMBLY

TELESCOPIC JOINT

MARINE RISER JOINT

MARINE RISER CONNECTOR

FLEX JOINT

LOWER MARINE RISER PACKAGE

REMOVABLE CONNECTOR

TIE DOWN BARS

UPPER PACKING ELEMENT

LOWER PACKING ELEMENT

TENSIONER SUPPORT RING

CHOKE AND KILL TERMINAL CONNECTIONS

SINGLE BALL FLEX JOINT TYPE II

Fig. 1.61 – Cutaway view of upper and lower marine riser equipment.

handled on the drill vessel by air-driven hose reels. A direct hydraulic system can be used for water depths less than about 300 ft. The direct system is similar to the system used on land rigs and has individual power oil lines to each control. An *indirect* system must be used for deep water. The indirect system has one source of power oil to the subsea BOP stack. Accumulator bottles are mounted on the subsea stack to store an adequate volume of pressurized hydraulic oil at the seafloor. Flow of the pressurized power oil is distributed to the various functions by pilot valves on the ocean floor. Smaller hydraulic lines, which allow much faster response time, are used to actuate the pilot valves. Electric and acoustic actuators also are available. A cross section of a control hose bundle for an indirect system is shown in Fig. 1.62. The large hose in the center is the power-oil hose.

Various schemes have been developed for installing subsea equipment. The diagram shown in Fig. 1.63 illustrates one approach. A guide base assembly is the initial piece of equipment lowered to the ocean floor. Four cables surrounding the central hole in the guide base extend back to the ship where a constant tension is maintained in the cables. Equipment then can be lowered into position over the hole using a guide assembly that rides on the guide lines. Two extra guide lines attached to one side of the guide base allow a television camera to be lowered to the ocean floor when desired. The first sections of hole are drilled without a BOP stack on the ocean floor. When a marine riser is used, a rotating head at the surface allows formation fluids to be diverted away from the rig in an emergency. The conductor casing is lowered into the hole with the subsea wellhead attached to the top. The

wellhead assembly latches into the guide base structure. The casing is cemented in place with returns back to the ocean floor. The wellhead assembly is designed so that all future casing and tubing strings are landed in the wellhead. The BOP stack is lowered and latched into the top of the wellhead. The marine riser then can be deployed and latched into the BOP.

Pneumatic tensioning devices have had wide application in floating drilling operations. They largely have replaced the use of counterweights for cable tensioning. Fig. 1.64 illustrates the operating principal of a pneumatic tensioning device. The desired tension is obtained by regulating the air pressure exerted on a piston. Hydraulic fluid on the opposite side of the piston serves to dampen the

Hose Bundle   Hose Bundle Reel

Fig. 1.62 – Cross section of control hose bundle for an indirect system.[17]

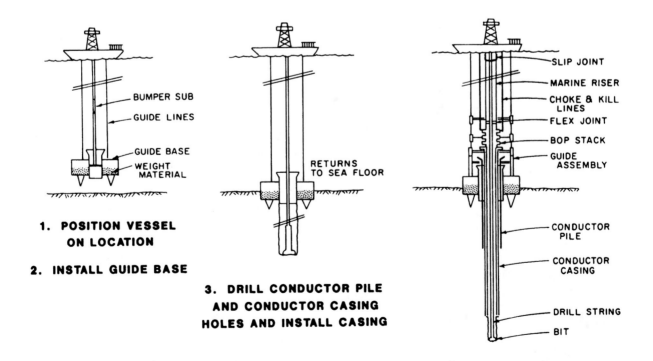

BUMPER SUB
GUIDE LINES

GUIDE BASE
WEIGHT MATERIAL

1. POSITION VESSEL ON LOCATION

2. INSTALL GUIDE BASE

RETURNS TO SEA FLOOR

3. DRILL CONDUCTOR PILE AND CONDUCTOR CASING HOLES AND INSTALL CASING

SLIP JOINT
MARINE RISER
CHOKE & KILL LINES
FLEX JOINT
BOP STACK
GUIDE ASSEMBLY

CONDUCTOR PILE

CONDUCTOR CASING

DRILL STRING
BIT

4. INSTALL BOP STACK AND MARINE RISER

Fig. 1.63 – Example subsea equipment installation procedure.

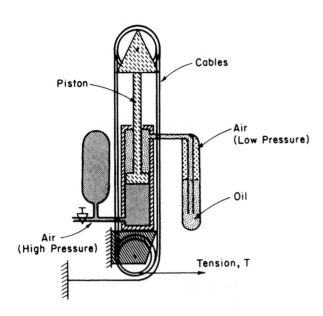

Fig. 1.64 – Schematic of a pneumatic tensioning device.

action of the piston and lubricate the packing. A block-and-tackle system allows the use of a shorter piston stroke. Pneumatic tensioning devices often are used on the marine riser, the various guide lines to the subsea wellhead, and on surface motion compensators for the drillstring.

1.10 Drilling Cost Analysis
The main function of the drilling engineer is to recommend drilling procedures that will result in the successful completion of the well as safely and inexpensively as possible. The drilling engineer must make recommendations concerning routine rig operations such as drilling fluid treatment, pump operation, bit selection, and any problems encountered in the drilling operation. In many cases, the use of a drilling cost equation can be useful in making these recommendations. The usual procedure is to break the drilling costs into (1) variable drilling costs and (2) fixed operating expenses that are independent of alternatives being evaluated.

1.10.1 Example Drilling Cost Formula.
The most common application of a drilling cost formula is in evaluating the efficiency of a bit run. A large fraction of the time required to complete a well is spent either drilling or making a trip to replace the bit. The total time required to drill a given depth, ΔD, can be expressed as the sum of the total rotating time during the bit run, t_b, the nonrotating time during the bit run, t_c, and trip time, t_t. The drilling cost formula is

$$C_f = \frac{C_b + C_r(t_b + t_c + t_t)}{\Delta D}, \quad \dots \dots (1.16)$$

where C_f is drilled cost per unit depth, C_b is the cost of bit, and C_r is the fixed operating cost of the rig per unit time independent of the alternatives being evaluated.

Since this drilling cost function ignores risk fac-

tors, the results of the cost analysis sometimes must be tempered with engineering judgment. Reducing the cost of a bit run will not necessarily result in lower well costs if the risk of encountering drilling problems such as stuck pipe, hole deviation, hole washout, etc., is increased greatly.

Example 1.5. A recommended bit program is being prepared for a new well using bit performance records from nearby wells. Drilling performance records for three bits are shown for a thick limestone formation at 9,000 ft. Determine which bit gives the lowest drilling cost if the operating cost of the rig is $400/hr, the trip time is 7 hours, and connection time is 1 minute per connection. Assume that each of the bits was operated at near the minimum cost per foot attainable for that bit.

Bit	Bit Cost ($)	Rotating Time (hours)	Connection Time (hours)	Mean Penetration Rate (ft/hr)
A	800	14.8	0.1	13.8
B	4,900	57.7	0.4	12.6
C	4,500	95.8	0.5	10.2

Solution. The cost per foot drilled for each bit type can be computed using Eq. 1.16. For Bit A, the cost per foot is

$$C_f = \frac{800 + 400(14.8 + 0.1 + 7)}{13.8(14.8)} = \$46.81/\text{ft}.$$

Similarly, for Bit B,

$$C_f = \frac{4,900 + 400(57.7 + 0.4 + 7)}{12.6(57.7)} = \$42.56/\text{ft}.$$

Finally, for Bit C,

$$C_f = \frac{4,500 + 400(95.8 + 0.5 + 7)}{10.2(95.8)} = \$46.89/\text{ft}.$$

The lowest drilling cost was obtained using Bit B.

1.10.2 Drilling Cost Predictions.
The drilling engineer frequently is called upon to predict the cost of a well at a given location. These predictions are required so that sound economic decisions can be made. In some cases, such as the evaluation of a given tract of land available for lease, only an approximate cost estimate is required. In other cases, such as in a proposal for drilling a new well, a more detailed cost estimate may be required.

Drilling cost depends primarily on well location and well depth. The location of the well will govern the cost of preparing the wellsite, moving the rig to the location, and the daily operating cost of the drilling operation. For example, an operator may find from experience that operating a rig on a given lease offshore Louisiana requires expenditures that will average about $30,000/day. Included in this daily operating cost are such things as rig rentals, crew boat rentals, work boat rentals, helicopter

**TABLE 1.7 – AVERAGE 1978 COSTS OF DRILLING AND EQUIPPING WELLS
IN THE SOUTH LOUISIANA AREA**

Depth Interval (ft)	Dry Holes			Completed Wells		
	Number of Wells, n_i	Mean Depth, D_i (ft)	Cost, C_i ($)	Number of Wells, n_i	Mean Depth, D_i (ft)	Cost, C_i ($)
0 to 1,249	1	1,213	64,289	0	–	–
1,250 to 2,499	1	1,542	65,921	9	1,832	201,416
2,499 to 3,749	8	3,015	126,294	20	3,138	212,374
3,750 to 4,999	11	4,348	199,397	20	4,347	257,341
5,000 to 7,499	43	6,268	276,087	47	6,097	419,097
7,500 to 9,999	147	8,954	426,336	117	9,070	614,510
10,000 to 12,499	228	11,255	664,817	165	11,280	950,971
12,500 to 14,999	125	13,414	1,269,210	110	13,659	1,614,422
15,000 to 17,499	54	16,133	2,091,662	49	16,036	2,359,144
17,500 to 19,999	21	18,521	3,052,213	17	18,411	3,832,504
20,000 and more	7	21,207	5,571,320	11	20,810	5,961,053

rentals, well monitoring services, crew housing, routine maintenance of drilling equipment, drilling fluid treatment, rig supervision, etc. The depth of the well will govern the lithology that must be penetrated and, thus, the time required to complete the well.

An excellent source of historical drilling-cost data presented by area and well depth is the annual joint association survey on drilling costs published by API. Shown in Table 1.7 are data for the south Louisiana area taken from the 1978 joint association survey. Approximate drilling cost estimates can be based on historical data of this type.

Drilling costs tend to increase exponentially with depth. Thus, when curve-fitting drilling cost data, it is often convenient to assume a relationship between cost, C, and depth, D, given by

$$C = ae^{bD}, \qquad\qquad\qquad (1.17)$$

where the constants a and b depend primarily on the well location. Shown in Fig. 1.65a is a least-square curve fit of the south Louisiana completed well data given in Table 1.7 for a depth range of 7,500 ft to about 21,000 ft. For these data, a has a value of about 1×10^5 dollars and b has a value of 2×10^{-4} ft^{-1}. Shown in Fig. 1.65b is a more conventional cartesian representation of this same correlation.

When a more accurate drilling cost prediction is needed, a cost analysis based on a detailed well plan must be made. The cost of tangible well equipment (such as casing) and the cost of preparing the surface location usually can be predicted accurately. The cost per day of the drilling operations can be estimated from considerations of rig rental costs, other equipment rentals, transportation costs, rig supervision costs, and others. The time required to drill and complete the well is estimated on the basis of rig-up time, drilling time, trip time, casing placement time, formation evaluation and borehole survey time, completion time and trouble time. Trouble time includes time spent on hole problems such as stuck pipe, well control operations, formation fracture, etc. Major time expenditures always are required for drilling and tripping operations.

An estimate of drilling time can be based on historical penetration rate data from the area of interest. The penetration rate in a given formation varies inversely with both compressive strength and shear strength of the rock. Also, rock strength tends to increase with depth of burial because of the higher confining pressure caused by the weight of the overburden. When major unconformities are not present in the subsurface lithology, the penetration rate usually decreases exponentially with depth. Under these conditions, the penetration rate can be related to depth, D, by

$$\frac{dD}{dt} = K\,e^{-2.303a_2 D}, \qquad\qquad (1.18)$$

where K and a_2 are constants. The drilling time, t_d, required to drill to a given depth can be obtained by separating variables and integrating. Separating variables gives

$$K\int_0^{t_d} dt = \int_0^D e^{2.303a_2 D}\,dD.$$

Integrating and solving for t_d yields

$$t_d = \frac{1}{2.303a_2 K}\left(e^{2.303a_2} - 1\right). \qquad (1.19)$$

As experience is gained in an area, more accurate predictions of drilling time can be obtained by plotting depth vs. drilling time from past drilling operations. Plots of this type also are used in evaluating new drilling procedures designed to reduce drilling time to a given depth.

Example 1.6. The bit records for a well drilled in the South China Sea are shown in Table 1.8. Make plots of depth vs. penetration rate and depth vs. rotating time for this area using semilog paper. Also, evaluate the use of Eq. 1.19 for predicting drilling time in this area.

Solution. The plots obtained using the bit records are shown in Fig. 1.66. The constants K and a_2 can be determined using the plot of depth vs. penetration rate on semilog paper. The value of $2.303a_2$ is 2.303 divided by the change in depth per log cycle:

$$2.303a_2 = \frac{2.303}{6,770} = 0.00034.$$

The constant 2.303 is a convenient scaling factor since

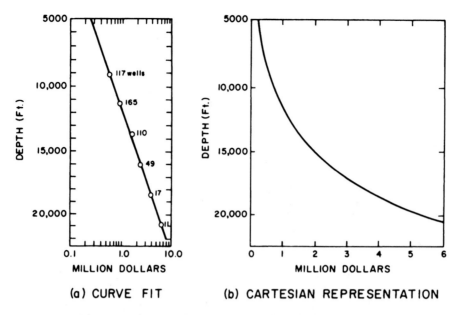

(a) CURVE FIT   (b) CARTESIAN REPRESENTATION

Fig. 1.65 – Least-square curve fit of 1978 completed well costs for wells below 7,500 ft in the south Louisiana area.

semilog paper is based on common logarithms. The value of K is equal to the value of penetration rate at the surface. From depth vs. penetration rate plot, $K=280$. Substitution of these values of a_2 and K in Eq. 1.19 gives

$$t_d = 10.504 \, (e^{0.00034D} - 1).$$

The line represented by this equation also has been plotted on Fig. 1.66. Note that the line gives good agreement with the bit record data over the entire depth range.

A second major component of the time required to drill a well is the trip time. The time required for tripping operations depends primarily on the depth of the well, the rig being used, and the drilling practices followed. The time required to change a bit and resume drilling operations can be approximated using the relation

$$t_t = 2\left(\frac{t_s}{\bar{l}_s}\right)D, \quad \ldots\ldots\ldots\ldots\ldots\ldots (1.20)$$

where t_t is the trip time required to change bits and resume drilling operations, t_s is the average time required to handle one stand of the drillstring, and $\bar{l}_s$ is the average length of one stand of the drillstring. The time required to handle the drill collars is greater than for the rest of the drillstring, but this difference usually does not warrant the use of an additional term in Eq. 1.20. Historical data for the rig of interest are needed to determine t_s.

The previous analysis shows that the time required per trip increases linearly with depth. In addition, the footage drilled by a single bit tends to decrease with depth, causing the number of trips required to drill a given depth increment also to increase with depth. The footage drilled between trips can be estimated if the approximate bit life is known. Integrating Eq. 1.18 between D_i, the depth of the last trip, and D, the depth of the next trip, gives the following equation:

$$D = \frac{1}{2.303a_2}\ln(2.303a_2 K t_b + e^{2.303a_2 D_i}). \quad . \, (1.21)$$

The total bit rotating time, t_b, generally will vary with depth as the bit size and bit type are changed. Eqs. 1.20 and 1.21 can be used to estimate the total trip time required to drill to a given depth using estimated values of t_s, t_b, a_2 and K. As experience is gained in an area using a particular rig, more accurate predictions of trip time can be obtained by plotting depth vs. trip time data from past drilling operations.

Example 1.7. Construct an approximate depth vs. trip time plot for the South China Sea area if the rig can handle a 90-ft stand in an average time of 2.7 minutes. Assume an average bit life of 10.5 hours for the entire depth interval. Use the values of a_2 and K obtained in Example 1.6. Also, the casing program calls for casing set at 500, 2,000, and 7,500 ft. The planned well depth is 9,150 ft.

Solution. The time required per round trip at a given depth is given by Eq. 1.20:

$$t_t = 2\left(\frac{2.7/60}{90}\right)D = 0.001\,D.$$

The approximate depth of each trip can be obtained from the casing program and Eq. 1.21. The use of Eq. 1.21 gives

$$D = \frac{1}{0.00034}\ln\,[0.00034\,(280)\,(10.5) + e^{0.00034D_i}]$$

$$= 2,941\,\ln\,(0.9996 + e^{0.00034D_i})$$

The first bit will drill to the first casing depth. Thus, the first trip will occur at 500 ft. Subsequent trips are predicted as shown in Table 1.9. Col. 2 is obtained by selecting the smaller of the two depths shown in Cols. 5 and 6. Col. 3 is obtained using Eq. 1.20, and Col. 4 is the cumulative obtained by summing Col. 3. Col. 5 is obtained using Eq. 1.21, and Col. 6 is obtained from the planned casing program. The results of Table 1.9 have been plotted in Fig. 1.67.

TABLE 1.8 – BIT RECORDS FROM SOUTH CHINA SEA AREA

Bit No.	Depth Out (ft)	Mean Depth (ft)	Bit Time (hours)	Total Drilling Time (hours)	Average Penetration Rate (ft/hr)	Hole Size (in.)
1	473	237	1.0	1.0	473	15.00
2	1,483	978	5.0	6.0	202	15.00
3	3,570	2,527	18.5	24.5	113	12.25
4	4,080	3,825	8.0	32.5	64	12.25
5	4,583	4,332	7.0	39.5	72	12.25
6	5,094	4,839	7.0	46.5	73	12.25
7	5,552	5,323	14.0	60.5	32	12.25
8	5,893	5,723	11.5	72.0	30	12.25
9	6,103	5,998	9.0	81.0	23	12.25
10	6,321	6,212	11.5	92.5	19	12.25
11	6,507	6,414	9.0	101.5	21	12.25
12	6,773	6,640	9.0	110.5	30	12.25
13	7,025	6,899	9.5	120.0	27	12.25
14	7,269	7,147	8.0	128.0	31	12.25
15	7,506	7,388	16.0	144.0	15	8.5
16	7,667	7,587	12.0	156.0	13	8.5
17	7,948	7,808	14.0	170.0	20	8.5
18	8,179	8,064	8.0	178.0	29	8.5
19	8,404	8,292	10.5	188.5	21	8.5
20	8,628	8,516	11.0	199.5	20	8.5
21	8,755	8,692	7.0	206.5	18	8.5
22	8,960	8,858	10.0	216.5	21	8.5
23	9,145	9,053	11.0	227.5	17	8.5

Formations with high strength require the use of a greater number of bits to drill a given depth interval. In some cases, the number of trips required to drill a well is too great to treat each trip individually with convenience as was done in Example 1.7. The time required per round trip is relatively constant over a 1,000-ft interval. Thus the total trip time required per 1,000 ft is approximately equal to the time per round trip times the number of trips per 1,000 ft. The number of bits required per 1,000 ft, N_b', at a given depth can be approximated by dividing the drilling time per 1,000 ft, t_d', by the average bit life for that depth interval:

$$N_b' = \frac{t_d'}{t_b} .$$

The drilling time required to drill from D to $(D+1,000)$ can be obtained using Eq. 1.19.

$$t_d' = \frac{1}{2.303a_2K}[e^{2.303a_2(D+1,000)} - 1]$$

$$- \frac{1}{2.303a_2K}(e^{2.303a_2 D} - 1).$$

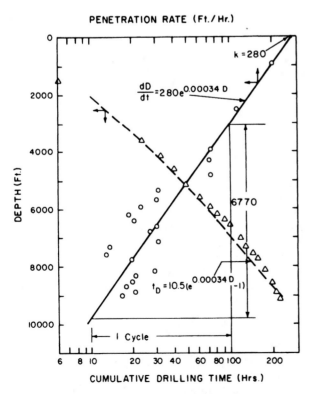

Fig. 1.66 – Example drilling-time plots for South China Sea area.

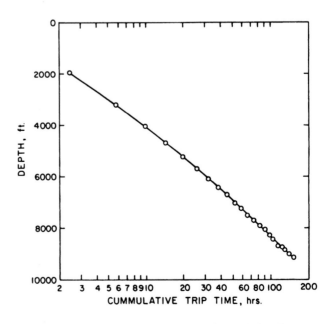

Fig. 1.67 – Example trip-time plot for South China Sea area.

**TABLE 1.9 – EXAMPLE TRIP-TIME COMPUTATION
FOR SOUTH CHINA SEA AREA**

(1) Trip No.	(2) Depth, D_i (ft)	(3) Trip Time (hours)	(4) Cumulative Trip Time (hours)	(5) Depth, D (ft)	(6) Next Casing Depth (ft)
1	500	0.5	0.5	2,299	2,000
2	2,000	2.0	2.5	3,205	7,500
3	3,205	3.2	5.7	4,057	7,500
4	4,057	4.1	9.8	4,717	7,500
5	4,717	4.7	14.5	5,256	7,500
6	5,256	5.3	19.8	5,711	7,500
7	5,711	5.7	25.5	6,105	7,500
8	6,105	6.1	31.6	6,452	7,500
9	6,452	6.5	38.1	6,762	7,500
10	6,762	6.8	44.9	7,043	7,500
11	7,043	7.0	51.9	7,299	7,500
12	7,299	7.3	59.2	7,534	7,500
13	7,500	7.5	66.7	7,721	9,150
14	7,721	7.7	74.4	7,926	9,150
15	7,926	7.9	82.3	8,118	9,150
16	8,118	8.1	90.4	8,298	9,150
17	8,298	8.3	98.7	8,467	9,150
18	8,467	8.5	107.2	8,627	9,150
19	8,627	8.6	115.8	8,779	9,150
20	8,779	8.8	124.6	8,923	9,150
21	8,923	8.9	133.5	9,061	9,150
22	9,061	9.1	142.6	9,192	9,150
23	9,150	9.2	151.8		

This equation simplifies to

$$t'_d = \frac{e^{2.303a_2 D}}{2.303a_2 K}(e^{2.303a_2} - 1). \qquad \ldots \ldots \ldots \ldots (1.22)$$

Multiplying the number of bits per 1,000 ft, N'_b, by the time per round trip yields trip time per 1,000 ft.

Example 1.8. Compute the trip time requirements for the South China Sea area between 8,000 and 9,000 ft. Use the conditions stated in Example 1.7.

Solution. The average trip time can be estimated using Eq. 1.20 for a mean depth of 8,500 ft.

$$\bar{t}_t = \frac{2(2.7/60)}{90}(8,500) = 8.5 \text{ hours.}$$

The drilling time required to drill from 8,000 to 9,000 ft is determined using Eq. 1.22.

$$\bar{t}_d = \frac{e^{0.00034(8,000)}}{(0.00034)280}(e^{0.34} - 1) = 64.6 \text{ hours.}$$

Thus, the number of bits required between 8,000 and 9,000 ft is

$$N'_b = \frac{t'_d}{t_b} = \frac{64.6}{10.5} = 6.15.$$

Multiplying the trip time per trip by the number of trips required yields

$$8.5(6.15) = 52.3 \text{ hours.}$$

This compares favorably with the trip time required between 7,926 and 8,923 ft computed in Example

1.7. From Table 1.9, the trip time per 1,000 ft is shown to be

$$133.5 - 82.3 = 51.2 \text{ hours.}$$

In addition to predicting the time requirements for drilling and tripping operations, time requirements for other planned drilling operations also must be estimated. These additional drilling operations usually can be broken into the general categories of (1) wellsite preparation, (2) rig movement and rigging up, (3) formation evaluation and borehole surveys, (4) casing placement, (5) well completion, and (6) drilling problems. The cost associated with wellsite preparation and moving the rig on location depends primarily on the terrain, the distance of the move, and the type of rig used. The cost of formation evaluation depends on the number and cost of the logs and tests scheduled, plus rig time required to condition the drilling fluid and run the logs and tests. The time required to run, cement, and test the casing depends primarily on the number of casing strings, casing depths, diameters, and weights per foot. These costs also must include the rig time required for running and cementing the casing strings, rigging up the surface equipment on each casing size, and perhaps changing the drillpipe or drill collar sizes to accommodate the new hole size. The cost of completing the well depends on the type of completion used, and this cost estimate is often made by the production engineer.

On many wells, a large fraction of the well cost may be because of unexpected drilling problems such as mud contamination, lost circulation, stuck drillstring, broken drillstring, ruptured casing, etc. These unforeseen costs cannot be predicted with any degree of accuracy and in some cases are not included in an original cost estimate. Requests for additional funds then must be submitted whenever a significant problem is encountered. However, long-range economic decisions concerning a drilling program in a given area should include average well costs due to drilling problems.

In areas where formation strength is low, time spent drilling and tripping may account for only about one-half to one-third of the total time needed to finish the well. Shown in Table 1.10 is a detailed time breakdown for an offshore Louisiana well drilled to 10,000 ft using a small platform rig tender. Only about 36% of the time required to drill and complete this well was spent drilling and tripping to change bits. About 7% of the time was spent "fishing" parts of the drillstring from the hole.

TABLE 1.10 – EXAMPLE RIG TIME ANALYSIS FOR TENDERED RIG

Drilling Operation	Total Required (hours)	Time Fraction
Drilling	351	0.17
Tripping	388	0.19
Rigging up	348	0.17
Formation evaluation and borehole surveys	103	0.05
Casing placement	199	0.10
Well completion	211	0.10
Drilling problems (total)	450	0.22
Mud conditioning. 143		
Well control operations. . . 12		
Fishing operations 152		
Severe weather 97		
Rig repairs 20		
Logistics. 26		
Total	2,050	1.00

Exercises

1.1 The following data were obtained on a diesel engine operating in a prony brake.

Engine Speed (rpm)	Torque (ft-lbf)	Fuel Consumption (gal/hr)
1,200	1,400	25.3
1,000	1,550	19.7
800	1,650	15.7
600	1,700	12.1

a. Compute the brake horsepower at each engine speed. *Answer: 319.9, 295.1, 251.3, and 194.2 hp.*

b. Compute the overall engine efficiency at each engine speed. *Answer: 0.235, 0.278, 0.297, and 0.298.*

c. Compute the fuel consumption in gallons per day for an average engine speed of 800 rpm and a 12-hour work day. *Answer: 188.4 gal/D.*

1.2 Compute the tension in the fast line when lifting a 500,000 lbf load for 6, 8, 10, and 12 lines strung between the crown block and traveling block. *Answer: 95,347; 74,316; 61,728; and 54,113 lbf.*

1.3 A rig must hoist a load of 200,000 lbf. The drawworks can provide a maximum input power of 800 hp. Ten lines are strung between the crown block and the traveling block and the dead line is anchored to a derrick leg on one side of the v-door (Fig. 1.17).

a. Calculate the static tension in the fast line when upward motion is impending. *Answer: 24,691 lbf.*

b. Calculate the maximum hook horsepower available. *Answer: 648 hp.*

c. Calculate the maximum hoisting speed. *Answer: 106.9 ft/min.*

d. Calculate the derrick load when upward motion is impending. *Answer: 244,691 lbf.*

e. Calculate the maximum equivalent derrick load. *Answer: 280,000 lbf.*

f. Calculate the derrick efficiency factor. *Answer: 0.874.*

1.4 Compute the minimum time required to reel a 10,000-ft cable weighing 1 lbf/ft to the surface using a 10-hp engine. *Answer: 151.5 min.*

1.5 A 1.25-in. drilling line has a nominal breaking strength of 138,800 lbf. A hook load of 500,000 lbf is anticipated on a casing job and a safety factor based on static loading conditions of 2.0 is required. Determine the minimum number of lines between the crown block and traveling block that can be used. *Answer: 10.*

1.6 A driller is pulling on a stuck drillstring. The derrick is capable of supporting a maximum equivalent derrick load of 500,000 lbf, the drilling line has a strength of 51,200 lbf, and the strength of the drillpipe in tension is 396,000 lbf. If eight lines are strung between the crown block and traveling block and safety factors of 2.0 are required for the derrick, drillpipe, and drilling line, how hard can the driller pull trying to free the stuck pipe? *Answer: 166,667 lbf.*

1.7 A rig accelerates a load of 200,000 lbf from zero to 60 ft/min in 5 seconds. Compute the load shown on the hook load indicator. *Answer: 201,242 lbf.*

1.8 A load of 400,000 lbf is lowered a distance of 90 ft using the auxiliary drawworks brakes. Compute the heat that must be dissipated by the brake cooling system. *Answer: 46,213 Btu.*

1.9 A drawworks drum has a diameter of 30 in., a width of 56.25 in., and contains 1.25-in. drilling line. Calculate the approximate length of line to the first lap point. *Answer: 368.2 ft.*

1.10 For the drawworks drum dimensions given in Exercise 1.9 and a fast line tension of 50,000 lbf, compare the drawworks torque when the drum is almost empty to the drawworks torque when the drum contains five laps. *Answer: 65,104 ft-lbf empty; 85,938 ft-lbf with five laps.*

1.11 Consider a triplex pump having 6-in. liners and 11-in. strokes operating at 120 cycles/min and a discharge pressure of 3,000 psig.

a. Compute the pump factor in units of gal/cycle at 100% volumetric efficiency. *Answer: 4.039 gal/cycle.*

b. Compute the flow rate in gal/min. *Answer: 484.7 gal/min.*

c. Compute the energy expended by each piston

during each cycle and the pump power developed. *Answer: 77,754 ft-lbf/cycle/cylinder; 848 hp.*

1.12 A double-acting duplex pump with 6.5-in. liners, 2.5-in. rods, and 18-in. strokes was operated at 3,000 psig and 20 cycles/min. for 10 minutes with the suction pit isolated from the return mud flow. The mud level in the suction pit, which is 7 ft wide and 20 ft long, was observed to fall 18 in. during this period. Compute the pump factor, volumetric pump efficiency, and hydraulic horsepower developed by the pump. *Answer: 7.854 gal/cycle; 0.82; 274.9 hp.*

1.13 A 1,000-hp pump can operate at a volumetric efficiency of 90%. For this pump, the maximum discharge pressure for various liner sizes is:

Liner Size (in.)	Maximum Discharge Pressure (psig)
7.50	1,917
7.25	2,068
7.00	2,229
6.75	2,418
6.50	2,635
6.00	3,153

Plot the pump pressure flow rate combinations possible at maximum hydraulic horsepower using cartesian coordinate paper. Repeat this using log-log paper.

1.14 A drillstring is composed of 9,000 ft of 5-in. 19.5-lbm/ft drillpipe and 1,000 ft of drill collars having a 3.0-in. ID. Compute these items:

a. Capacity of the drillpipe in barrels. *Answer: 159.8 bbl.*

b. Capacity of the drill collars in barrels. *Answer: 8.7 bbl.*

c. Number of pump cycles required to pump surface mud to the bit. The pump is a duplex double-acting pump with 6-in. liners, 2.5-in. rods, 16-in. strokes, and operates at a volumetric efficiency of 85%. *Answer: 1,164 cycles.*

d. Displacement of the drillpipe in bbl/ft. *Answer: 0.0065 bbl/ft (neglects tool joints).*

e. Displacement of the drill collars in bbl/ft. The OD of the collars is 8.0 in. *Answer: 0.0534 bbl/ft.*

f. Loss in fluid level in the well if 10 stands (thribbles) of drillpipe are pulled without filling the hole. The ID of the casing in the hole is 10.05 in. *Answer: 64 ft.* 59

g. Loss in fluid level in the well if one stand of drill collars is pulled without filling the hole. *Answer: 108 ft.* 48

h. Change in fluid level in the pit if the pit is 8 ft wide and 20 ft long, assuming that the hole is filled after pulling 10 stands of drillpipe. *Answer: 2.5 in.*

i. Change in fluid level in a 3- × 3-ft trip tank assuming that the hole is filled from the trip tank after pulling 10 stands of drillpipe. *Answer: 3.6 ft.*

1.15 The mud logger places a sample of calcium carbide in the drillstring when a connection is made. The calcium carbide reacts with the mud to form acetylene gas. The acetylene is detected by a gas detector at the shale shaker after pumping 4,500 strokes. The drillstring is composed of 9,500 ft of 5-in., 19.5-lbm/ft drillpipe and 500 ft of drill collars having an ID of 2.875 in. The pump is a double-

acting duplex pump with 6-in. liners, 2-in. rods, and 14-in. strokes and operates at a volumetric efficiency of 80%.

a. Estimate the number of pump cycles required to move the gas from the surface to the bit. *Answer: 1,400 cycles (neglects gas slip).*

b. Estimate the number of pump cycles required to move the gas from the bit to the shale shaker. *Answer: 3,100 cycles.*

c. If the penetration rate of the bit is 20 ft/hr and the pump speed is 60 cycles/min., how many feet are drilled by the bit before formation gas expelled from the rock destroyed by the bit travels from the bit to the surface? *Answer: 17.2 ft.*

1.16 Discuss the functions of this marine drilling equipment: marine riser, ball joint, pneumatic tensioning device, bumper sub, slip joint, and taut-line inclinometer.

1.17 A pneumatic riser tensioning device is arranged as shown in Fig. 1.64 and has a 10-ft piston stroke. How much vertical ship movement is allowed using this device? *Answer: 40 ft.*

1.18 A recommended bit program is being prepared for a new well, using bit performance records from nearby wells. Drilling records for three bits are shown below for a thick shale section encountered at 12,000 ft. Determine which bit gives the lowest drilling cost if the hourly operating cost of the rig is $1,000/hr and the trip time is 10 hours. The connection time is included in the rotating time shown below.

Bit	Bit Cost ($)	Interval Drilled (ft)	Rotating Time (hours)
A	700	106	9
B	4,000	415	62
C	8,000	912	153

Answer: Bit B ($183.13/ft).

1.19 The penetration rates using gas, foam, and mud in an area are 10 ft/hr, 5 ft/hr, and 1 ft/hr, respectively. If gas is used, each water zone encountered must be sealed off. The cost of the plugging treatment is $2,000, and 25 hours of rig time are required to complete the sealing operation. The normal operating cost for air drilling is $200/hr. The use of foaming agents requires an additional $60/hr. The normal operating cost when using mud is $160/hr. Regardless of the drilling fluid used, the average bit cost is $1,000. The average bit life is 25 hours and the average trip time is 6 hours. Determine which drilling fluid yields the lowest drilling cost if one water zone is encountered per 1,000 ft drilled and if five water zones are encountered per 1,000 ft drilled. *Answer: gas is best for both assumptions ($35.80/ft and $63.80/ft).*

1.20 Pipe being recovered from an interval of borehole has a value of $30/ft. On the average, 20 hours of rig time must be expended to recover 200 ft of pipe. The cost per foot to sidetrack the well and redrill the junked interval of borehole would be about $150/ft. Do the fishing operations appear profitable if the average operating cost is $500/hr and the cost of abandoning the junked hole would be approximately $5/ft? *Answer: fishing is the best alternative ($50/ft).*

1.21 Assume that C_i represents the average cost for n_i wells drilled to a mean depth, D_i, and that C_i varies approximately exponentially with depth such that an expression

$$C = ae^{bD}$$

can be used to curve fit N observed values of n_i, C_i, and D_i. If we define a residual error, r_i, as

$$r_i = n_i \ln \frac{C_i}{C},$$

it is possible to determine the constants a and b using the N observed values of n_i, D_i, and C_i such that the sum of the residuals squared has a minimum value. Derive expressions for a and b that result in a minimum value of

$$\sum_{i=1}^{N} r_i^2.$$

1.22 Apply the expressions for a and b derived in Exercise 1.21 to obtain a least-square curve fit of the south Louisiana completed well cost data given in Table 1.7 for well depths below 7,500 ft.

1.23 Complete the following using the cost vs. depth data given in Table 1.7 for dry holes drilled in south Louisiana in 1978.

a. A plot of cost vs. depth on cartesian paper.

b. A plot of cost vs. depth on semilog paper.

c. Determine a set of constants a and b of Eq. 1.17 that allow a curve fit of these data. *Answer: $65,513 and 0.000212.*

1.24 The following bit records were obtained on a well drilled in Maverick County, Texas.

Bit No.	Depth Out (ft)	Time (hours)	Bit Size (in.)
1	500	2.0	17.5
2	1,925	15.0	17.5
3	2,526	14.9	17.5
4	2,895	20.2	17.5
5	3,177	26.3	17.5
6	3,452	23.2	17.5
7	3,937	29.7	17.5
8	4,286	27.3	17.5
9	4,621	28.2	17.5
10	4,973	31.3	17.5
11	5,171	19.4	17.5
12	5,298	15.9	17.5
13	5,424	15.9	17.5
14	5,549	15.7	17.5
15	5,625	13.8	17.5
16	5,743	15.8	17.5
17	5,863	18.9	17.5
18	6,006	16.2	17.5
19	6,158	18.4	17.5
20	6,340	27.3	17.5
21	6,602	29.8	17.5
22	6,783	23.9	17.5
23	6,978	26.4	17.5
24	7,165	27.3	17.5
25	7,292	21.5	17.5
26	7,386	20.5	17.5
27	7,528	26.5	17.5
28	7,637	22.8	17.5
29	7,741	23.8	17.5
30	7,795	17.2	12.0
31	7,855	26.4	12.0
32	7,917	26.9	12.0
33	7,988	26.8	12.0
34	8,060	25.8	12.0
35	8,494	270.0	12.0
36	8,614	35.1	12.0
37	8,669	19.0	12.0
38	8,737	29.7	8.5
39	8,742	3.4	8.5
40	8,778	8.5	8.5
41	9,661	179.3	8.5
42	9,874	65.0	8.5
43	9,973	30.0	8.5
44	10,016	11.8	8.5
45	10,219	64.7	8.5
46	10,408	57.2	8.5
47	10,575	61.2	8.5
48	10,661	36.1	8.5

a. Plot a depth vs. rotating time curve for this well.

b. Evaluate the use of Eq. 1.19 in this area.

c. Assuming that the rig can pull thribbles at an average time per stand of 4 minutes, plot the trip time per trip vs. depth. *Answer: $t_t = 0.00148 D$.*

·d. Using the bit records, make a plot of total trip time vs. depth.

e. Compare the performance of Bits 34 and 35. Assume a daily operating cost of $24,000/D, a bit cost of $3,000 for Bit 34, and a bit cost of $12,000 for Bit 35. *Answer: $565/ft and $679/ft.*

1.25 Two rigs are available for drilling a well in southern California. One rig costs $800/hr but can only pull doubles. The other rigs costs $1,000/hr but can pull thribbles. In this area K is 200 ft/hr and $2.303a_2$ is 0.0004. The time required to pull one stand is about 4 minutes for both rigs. Considering only the cost of the tripping operations, which rig would be best for a well drilled to 7,000 ft? Assume an average bit life of 10 hours for all bits and casing setting depths of 500 and 2,000 ft. *Answer: Thribble rig ($151,200 vs. $181,400).*

References

1. Spec. for Steel Derricks, Std. 4A, API, Dallas (April 1967).
2. Spec. for Portable Masts, Std. 4D, API, Dallas (March 1967).
3. Spec. for Rotary Drilling Equip., Std. 7, API, Dallas (May 1979).
4. "Procedure for Selecting Rotary Drilling Equipment," *Bull. D10,* API, Dallas (Aug. 1973).
5. Spec. for Casing, Tubing and Drill Pipe, Spec. 5A and 5AX, API, Dallas (March 1976).
6. "Performance Properties of Casing, Tubing, and Drill Pipe," *Bull. 5C2,* API, Dallas (March 1975).
7. Spec. for Wire Rope, Spec. 9A, API, Dallas (Jan. 1976).
8. "Recommended Practices for Blowout Prevention Equipment Systems," RP53, API, Dallas (Feb. 1976).
9. *Drilling Operations Manual,* F.W. Cole and P.L. Moore, (eds.) Petroleum Publishing Co., Tulsa (1965).
10. *Petroleum Engineering – Drilling and Well Completions,* C. Gatlin (ed.), Prentice-Hall Inc., Englewood Cliffs, NJ (1960).
11. *Lessons in Rotary Drilling,* U. of Texas, Unit II, Lesson 3.
12. *A Primer of Oil Well Drilling,* third and fourth editions, U. of Texas.
13. *Drilling Practices Manual,* P.L. Moore (ed.), Petroleum Publishing Co., Tulsa (1974).
14. *Tool Pusher's Manual,* Intl. Assoc. of Oil Well Drill. Contractors, Dallas.
15. *Rotary Drilling Handbook,* sixth edition, J.E. Brantly (ed.) Palmer Pub., New York City.
16. *Oil Well Drilling Technology,* A. McCray and F. Cole (eds.) U. of Oklahoma Press, Norman.
17. *Design for Reliability in Deepwater Floating Drilling*

Operations, L.M. Harris (ed.), Petroleum Publishing Co., Tulsa (1972).

18. Recommended Practice on Application, Care, and Use of Wire Rope for Oil-Field Service, RP 9B, API, Dallas (Dec. 1972).
19. Woodall-Mason, N. and Tilbe, J.R.: "Value of Heave Compensators to Floating Drilling," *J. Pet. Tech.* (Aug. 1976) 938-946.
20. Spec. for Drilling and Well Servicing Structures, Std. 4E, API, Dallas (March 1974).

Nomenclature

a = constant used in curve-fitting drilling cost vs. depth

a_2 = constant

A = capacity

A_s = displacement of section of pipe

b = constant used in curve-fitting drilling cost vs. depth

C = cost

C_f = cost per interval drilled

C_i = mean cost of n_i wells drilled to a mean depth, D_i

C_r = fixed operating cost of rig per unit time

d = diameter

d_1 = inner diameter of annulus

d_2 = outer diameter of annulus

d_l = diameter of liner in pump

d_r = diameter of rod in pump

D = depth

D_i = initial drilled depth of bit run; also mean depth of n_i wells of mean cost, C_i

ΔD = depth interval drilled during bit run

e = base of natural logarithm

E = efficiency

F = force

F_f = force in fast line

F_{de} = maximum equivalent force on derrick

F_p = pump factor

F_s = force in static line

H = heating value of fuel

K = constant

ln = natural logarithm to base e

$\bar{l}_s$ = average length of one stand of drillstring

L = length of level arm on prony brake

L_s = stroke length on pump

M = mechanical advantage

M_i = ideal mechanical advantage of frictionless system

n = number of lines strung between crown block and traveling block

n_i = number of wells included in average cost computation

N = number of cycles per unit time

N_b' = number of bits per 1,000 ft

N_c = number of cylinders used in pump

Δp = pressure change

P = power

P_i = input power

q = flow rate

Q_i = power (heat) input from fuel consumption

r = radius

r_d = drum radius

r_i = residual error for observation i

t_b = rotating time on bit during bit run

t_c = nonrotating time on bit during bit run (such as connection time)

t_d = total drilling time to depth of interest

t_d' = drilling time per 1,000 ft

$\bar{t}_b$ = average bit life

$\bar{t}_s$ = average time required to handle one stand of drillstring during tripping operations

t_t = time of tripping operations required to change bit

T = torque

v = velocity

v_b = velocity of block

v_f = velocity of fast line

V = volume

w_f = mass rate of fuel consumption

W = load supported by block-and-tackle system

ρ = density

ω = angular velocity

Subscripts

a = annulus

b = bit; block

c = cylinders

d = derrick; drum; drilling

e = equivalent

f = fast; fuel

h = hook

H = hydraulic

i = inner; mean; indicated; ideal; input

m = mechanical

p = pipe; pump

r = rig

s = static; stand; stroke

t = overall

v = volumetric

SI Metric Conversion Factors

bbl $\times$ 1.589 873	E$-$01	= m^3
bbl/ft $\times$ 5.216 119	E$-$01	= m^3/m
Btu/lbm $\times$ 2.326*	E$+$03	= J/kg
ft $\times$ 3.048*	E$-$01	= m
ft/bbl $\times$ 1.917 134	E$+$00	= m/m^3
ft-lbf $\times$ 1.355 818	E$+$00	= J
gal $\times$ 3.785 412	E$-$03	= m^3
hp $\times$ 7.460 43	E$-$01	= kW
in. $\times$ 2.54*	E$+$00	= cm
lbf $\times$ 4.448 222	E$+$00	= N
lbm/ft $\times$ 1.488 164	E$+$00	= kg/m
lbm/gal $\times$ 1.198 264	E$+$02	= kg/m^3
lbm/min $\times$ 7.559 873	E$-$03	= kg/s
psi $\times$ 6.894 757	E$+$00	= kPa

*Conversion factor is exact.

Chapter 2
Drilling Fluids

The purposes of this chapter are to present (1) the primary functions of the drilling fluid, (2) the test procedures used to determine whether the drilling fluid has suitable properties for performing these functions, and (3) the common additives used to obtain the desirable properties under various well conditions. The mathematical modeling of the flow behavior of drilling fluids is not discussed in this chapter but is presented in detail in Chapter 4.

Drilling fluid is used in the rotary drilling process to (1) clean the rock fragments from beneath the bit and carry them to the surface, (2) exert sufficient hydrostatic pressure against subsurface formations to prevent formation fluids from flowing into the well, (3) keep the newly drilled borehole open until steel casing can be cemented in the hole, and (4) cool and lubricate the rotating drillstring and bit. In addition to serving these functions, the drilling fluid should not (1) have properties detrimental to the use of planned formation evaluation techniques, (2) cause any adverse effects upon the formation penetrated, or (3) cause any corrosion of the drilling equipment and subsurface tubulars.

The drilling engineer is concerned with the selection and maintenance of the best drilling fluid for the job. The drilling fluid is related either directly or indirectly to most drilling problems. If the drilling fluid does not perform adequately the functions listed above, it could become necessary to abandon the well. Also, the additives required to maintain the drilling fluid in good condition can be quite expensive. Drilling fluid cost often exceeds $1 million on a single deep well in some areas. A drilling fluid specialist called a *mud engineer* frequently is kept on duty at all times to maintain the drilling fluid in good condition at the lowest possible cost.

A broad classification of drilling fluids is shown in Fig. 2.1. The main factors governing the selection of drilling fluids are (1) the types of formations to be drilled, (2) the range of temperature, strength, permeability, and pore fluid pressure exhibited by the formations, (3) the formation evaluation procedure used, (4) the water quality available, and (5) ecological and environmental considerations. However, to a large extent, the drilling fluid composition that yields the lowest drilling cost in an area must be determined by trial and error. Waterbase muds are the most commonly used drilling fluids. Oil-base muds are generally more expensive and require more stringent pollution control procedures than water-base muds. Their use usually is limited to drilling extremely hot formations or formations that are affected adversely by water-base muds. The use of gases as drilling fluids is limited to areas where the formations are competent and impermeable. Gas/liquid mixtures can be used when only a few formations capable of producing water at significant rates are encountered.

Fig. 2.2 shows the composition of a typical 11-lbm/gal water-base mud. Water-base muds consist of a mixture of solids, liquids, and chemicals, with water being the continuous phase. Some of the solids react with the water phase and dissolved chemicals and, therefore, are referred to as *active solids*. Most of the active solids present are hydratable clays. The chemicals added to the mud restrict the activity of such solids, thereby allowing certain drilling fluid properties to be maintained between desired limits. The other solids in a mud do not react with the water and chemicals to a significant degree and are called *inactive solids*. The inactive solids vary in specific gravity, which therefore complicates analyses and control of the solids in the muds. Any oil added to water-base mud is emulsified into the water phase and is maintained as small, discontinuous droplets. This type of fluid mixture is called an *oil-in-water emulsion*.

Fig. 2.3 shows the composition of a typical 11-lbm/gal oil-base mud. Oil-base muds are similar in composition to water-base muds, except the continuous phase is oil instead of water and water droplets are emulsified into the oil phase. This type of fluid is called a *water-in-oil emulsion*. Another

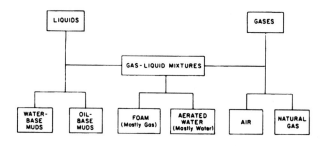

Fig. 2.1—Classification of drilling fluids.

major difference is that all solids are considered inactive because they do not react with the oil.

2.1 Diagnostic Tests

The American Petroleum Inst.[1] (API) has presented a recommended practice for testing liquid drilling fluids. These tests were devised to help the mud engineer determine whether the drilling fluid is performing its functions properly. By running these tests at regular intervals, it is often possible to identify and correct potential drilling problems early and prevent a serious loss of rig time. The test equipment needed to perform the diagnostic tests recommended by the API include (1) a *mud balance* for determining drilling fluid density, (2) a *Marsh funnel* for checking drilling fluid consistency, (3) a *rotational viscometer* for determining gel strength and apparent viscosity at various shear rates, (4) a *filter press* for determining mud filtration rate and mudcake characteristics, (5) a *high-pressure, high-temperature filter press* for determining mud filtration rate and mudcake characteristics at elevated temperature and pressure, (6) a *pH meter* for determining H^+ concentration, (7) a *sand screen* for determining sand content, (8) a *mud still* for determining solids, oil, and water contents, and (9) a *titration apparatus* for chemical analysis.

The student is referred to Ref. 1 for detailed instruction on the proper use of this test equipment. Shown in Fig. 2.4 is the standard API drilling mud report form usually used to present the results of the diagnostic tests. Information presented on this report is used by almost everyone involved with the drilling operations. Also, this information can be quite helpful when planning for future wells in the area.

2.1.1 The Mud Balance.
The nomenclature used to describe a mud balance is shown in Fig. 2.5.[2] The test consists essentially of filling the cup with a mud sample and determining the rider position required for balance. The balance is calibrated by adding lead shot to a calibration chamber at the end of the scale. Water usually is used for the calibration fluid. The density of fresh water is 8.33 lbm/gal. The drilling fluid should be degassed before being placed in the mud balance to ensure an accurate measurement.

2.1.2 The Marsh Funnel.
The time required for a mud sample to flow through a Marsh funnel (Fig. 2.6) is a rapid test of the consistency of a drilling fluid. The test consists essentially of filling the funnel with a mud sample and then measuring the time

required for 1 quart of the sample to flow from the initially full funnel into the mud cup. The funnel viscosity is reported in units of seconds per quart. Fresh water at 75°F has a funnel viscosity of 26 s/qt.

The flow rate from the marsh funnel changes significantly during the measurement of funnel viscosity because of the changing fluid level in the funnel. This causes the test results to become less meaningful for *non-Newtonian fluids*, which exhibit different apparent viscosities at different flow rates for a given tube size. Unfortunately, most drilling fluids exhibit a non-Newtonian behavior. Thus, while the funnel viscometer can detect an undesirable drilling fluid consistency, additional tests usually must be made before an appropriate mud treatment can be prescribed.

2.1.3 The Rotational Viscometer.
The rotational viscometer can provide a more meaningful measurement of the rheological characteristics of the mud than the marsh funnel. The mud is sheared at a constant rate between an inner bob and an outer rotating sleeve. Six standard speeds plus a variable speed setting are available with the rotational viscometer shown in Fig. 2.7. Only two standard speeds are possible on most models designed for field use. The dimensions of the bob and rotor are chosen so that the dial reading is equal to the apparent Newtonian viscosity in centipoise at a rotor speed of 300 rpm. At other rotor speeds, the apparent viscosity, μ_a, is given by

$$\mu_a = \frac{300\,\theta_N}{N}, \qquad \qquad (2.1)$$

where θ_N is the dial reading in degrees and N is the rotor speed in revolutions per minute.

The viscometer also can be used to determine rheological parameters that describe non-Newtonian fluid behavior. At present, the flow parameters of the *Bingham plastic* rheological model are reported on the standard API drilling mud report. Two parameters are required to characterize fluids that follow the Bingham plastic model. These parameters are called the *plastic viscosity* and *yield point* of the fluid. The plastic viscosity, μ_p, in centipoise normally is computed using

$$\mu_p = \theta_{600} - \theta_{300}, \qquad \qquad (2.2)$$

where θ_{600} is the dial reading with the viscometer operating at 600 rpm and θ_{300} is the dial reading with the viscometer operating at 300 rpm. The yield point, τ_y, in lbf/100 sq ft normally is computed using

$$\tau_y = \theta_{300} - \mu_p. \qquad \qquad (2.3)$$

A third non-Newtonian rheological parameter called the *gel strength*, in units of lbf/100 sq ft, is obtained by noting the maximum dial deflection when the rotational viscometer is turned at a low rotor speed (usually 3 rpm) after the mud has remained static for some period of time. If the mud is allowed to remain static in the viscometer for a period of 10 seconds, the maximum dial deflection obtained when the viscometer is turned on is reported as the *initial gel* on the API mud report form. If the mud is allowed to

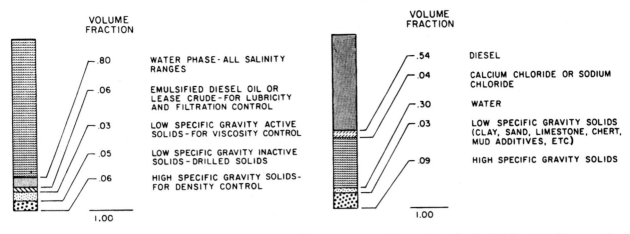

VOLUME FRACTION

- .80   WATER PHASE - ALL SALINITY RANGES
- .06   EMULSIFIED DIESEL OIL OR LEASE CRUDE - FOR LUBRICITY AND FILTRATION CONTROL
- .03   LOW SPECIFIC GRAVITY ACTIVE SOLIDS - FOR VISCOSITY CONTROL
- .05   LOW SPECIFIC GRAVITY INACTIVE SOLIDS - DRILLED SOLIDS
- .06   HIGH SPECIFIC GRAVITY SOLIDS - FOR DENSITY CONTROL

1.00

Fig. 2.2—Composition of typical 11-lbm/gal water-base mud.

VOLUME FRACTION

- .54   DIESEL
- .04   CALCIUM CHLORIDE OR SODIUM CHLORIDE
- .30   WATER
- .03   LOW SPECIFIC GRAVITY SOLIDS (CLAY, SAND, LIMESTONE, CHERT, MUD ADDITIVES, ETC)
- .09   HIGH SPECIFIC GRAVITY SOLIDS

1.00

Fig. 2.3—Composition of typical 11-lbm/gal oil-base mud.

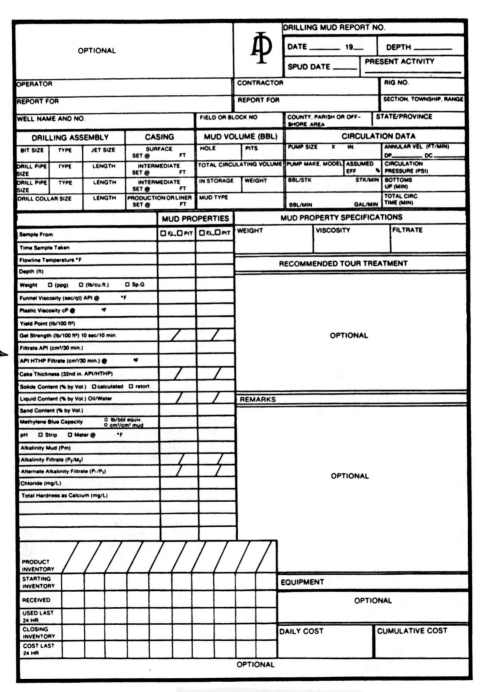

Fig. 2.4—Standard API drilling mud report form.

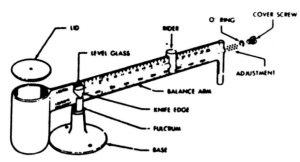

Fig. 2.5—Example mud balance.[2]

$$\mu_a = \frac{300\,\theta_N}{N} = \frac{300\,(28)}{300} = 28 \text{ cp} .$$

Similarly, use of Eq. 2.1 for the 600-rpm dial reading gives

$$\mu_a = \frac{300\,(46)}{600} = 23 \text{ cp} .$$

Note that the apparent viscosity does not remain constant but decreases as the rotor speed is increased. This type of non-Newtonian behavior is shown by essentially all drilling muds.

The plastic viscosity of the mud can be computed using Eq. 2.2:

$$\mu_p = \theta_{600} - \theta_{300} = 46 - 28 = 18 \text{ cp} .$$

The yield point can be computed using Eq. 2.3:

$$\tau_y = \theta_{300} - \mu_p = 28 - 18 = 10 \text{ lbf/100 sq ft} .$$

remain static for 10 minutes, the maximum dial deflection is reported as the *10-min gel*. These as well as other non-Newtonian parameters are discussed in detail in Chapter 4. However, it is sufficient that the beginning student view these parameters as diagnostic indicators that must be kept within certain ranges.

Example 2.1. A mud sample in a rotational viscometer equipped with a standard torsion spring gives a dial reading of 46 when operated at 600 rpm and a dial reading of 28 when operated at 300 rpm. Compute the apparent viscosity of the mud at each rotor speed. Also compute the plastic viscosity and yield point.

Solution. Use of Eq. 2.1 for the 300-rpm dial reading gives

2.1.4 pH Determination. The term pH is used to express the concentration of hydrogen ions in an aqueous solution. pH is defined by

$$pH = -\log[H^+], \quad \dots\dots\dots\dots\dots\dots \quad (2.4)$$

where $[H^+]$ is the hydrogen ion concentration in moles per liter. At room temperature, the *ion product constant of water*, K_w, has a value of 1.0×10^{-14} mol/L. Thus, for water

Fig. 2.6—Marsh funnel.

Fig. 2.7—Rotational viscometer.

TABLE 2.1—RELATIONS BETWEEN pH, [H⁺] AND [OH⁻] IN WATER SOLUTIONS

$[H^+]$	pH	$[OH]$	pOH	Reaction
1.0×10^0	0.00	1.0×10^{-14}	14.00	
1.0×10^{-1}	1.00	1.0×10^{-13}	13.00	
1.0×10^{-2}	2.00	1.0×10^{-12}	12.00	
1.0×10^{-3}	3.00	1.0×10^{-11}	11.00	Acidic
1.0×10^{-4}	4.00	1.0×10^{-10}	10.00	
1.0×10^{-5}	5.00	1.0×10^{-9}	9.00	
1.0×10^{-6}	6.00	1.0×10^{-8}	8.00	
1.0×10^{-7}	7.00	1.0×10^{-7}	7.00	Neutral
1.0×10^{-8}	8.00	1.0×10^{-6}	6.00	
1.0×10^{-9}	9.00	1.0×10^{-5}	5.00	
1.0×10^{-10}	10.00	1.0×10^{-4}	4.00	
1.0×10^{-11}	11.00	1.0×10^{-3}	3.00	Alkaline
1.0×10^{-12}	12.00	1.0×10^{-2}	2.00	
1.0×10^{-13}	13.00	1.0×10^{-1}	1.00	
1.0×10^{-14}	14.00	1.0×10^0	0.00	

Fig. 2.8—Two methods for measuring pH: pH paper (left) and pH meter (right).

$$H_2O \rightleftharpoons H^+ + OH^-$$

$$K_w = [H^+][OH^-] = 1.0 \times 10^{-14}.$$

For pure water, $[H^+] = [OH^-] = 1.0 \times 10^{-7}$, and the pH is equal to 7. Since in any aqueous solution the product $[H^+][OH^-]$ must remain constant, an increase in $[H^+]$ requires a corresponding decrease in $[OH^-]$. A solution in which $[H^+] > [OH^-]$ is said to be *acidic*, and a solution in which $[OH^-] > [H^+]$ is said to be *alkaline*. The relation between pH, $[H^+]$, and $[OH^-]$ is summarized in Table 2.1.

The pH of a fluid can be determined using either a special *pH paper* or a *pH meter* (Fig. 2.8). The pH paper is impregnated with dyes that exhibit different colors when exposed to solutions of varying pH. The pH is determined by placing a short strip of the paper on the surface of the sample. After the color of the test paper stabilizes, the color of the upper side of the paper, which has not contacted the mud, is compared with a standard color chart provided with the test paper. When saltwater muds are used, caution should be exercised when using pH paper. The solutions present may cause the paper to produce erroneous values.

The pH meter is an instrument that determines the pH of an aqueous solution by measuring the electropotential generated between a special glass electrode and a reference electrode. The electromotive force (EMF) generated across the specially formulated glass membrane has been found empirically to be almost linear with the pH of the solution. The pH meter must be calibrated using buffered solutions of known pH.

Example 2.2. Compute the amount of caustic required to raise the pH of water from 7 to 10.5. The molecular weight of caustic is 40.

Solution. The concentration of OH^- in solution at a given pH is given by

$$K_w = [H^+][OH^-] = 1.0 \times 10^{-14}$$

$$[OH^-] = \frac{1.0 \times 10^{-14}}{[H^+]} = \frac{10^{-14}}{10^{-pH}}$$

$$= 10^{(pH-14)}.$$

The change in OH^- concentration required to increase the pH from 7 to 10.5 is given by: $\Delta[OH^-] = [OH^-]_2 - [OH^-]_1$.

$$\Delta[OH^-] = 10^{(10.5-14)} - 10^{(7-14)}$$

$$= 3.161 \times 10^{-4} \text{ mol/L}.$$

Since caustic has a molecular weight of 40, the weight of caustic required per liter of solution is given by

$$40(3.161 \times 10^{-4}) = 0.0126 \text{ g/L}.$$

2.1.5 The API Filter Press — Static Filtration. The filter press (Fig. 2.9) is used to determine (1) the filtration rate through a standard filter paper and (2) the rate at which the mudcake thickness increases on the standard filter paper under standard test conditions. This test is indicative of the rate at which permeable formations are sealed by the deposition of a mudcake after being penetrated by the bit.

The flow of mud filtrate through a mudcake is described by Darcy's law. Thus, the rate of filtration is given by

$$\frac{dV_f}{dt} = \frac{k A \Delta p}{\mu h_{mc}}, \quad \dots \dots \dots \dots \dots \dots (2.5)$$

where

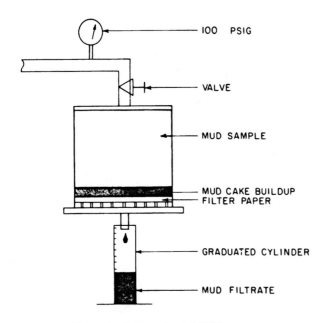

Fig. 2.9—Schematic of API filter press.

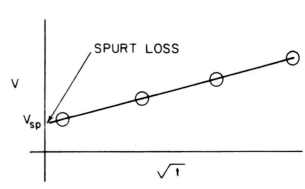

Fig. 2.10—Example filter press data.

dV_f/dt = the filtration rate, cm^3/s,

k = the permeability of the mudcake, darcies,

A = the area of the filter paper, cm^2,

Δp = the pressure drop across the mudcake, atm,

μ = the viscosity of the mud filtrate, cp, and

h_{mc} = the thickness of the filter (mud) cake, cm.

At any time, t, during the filtration process, the volume of solids in the mud that has been filtered is equal to the volume of solids deposited in the filter cake:

$$f_{sm}V_m = f_{sc}h_{mc}A \ ,$$

where f_{sm} is the volume fraction of solids in the mud and f_{sc} is the volume fraction of solids in the cake, or

$$f_{sm}(h_{mc}A + V_f) = f_{sc}h_{mc}A \ .$$

Therefore,

$$h_{mc} = \frac{f_{sm}V_f}{A(f_{sc}-f_{sm})} = \frac{V_f}{A\left(\dfrac{f_{sc}}{f_{sm}}-1\right)} \ \ldots . (2.6)$$

Inserting this expression for h_{mc} into Eq. 2.5 and integrating,

$$\int_o^{V_f} V_f dV_f = \int_o^t \frac{kA\Delta p}{\mu} A\left(\frac{f_{sc}}{f_{sm}}-1\right)dt,$$

$$\frac{V_f^2}{2} = \frac{k}{\mu} A^2 \left(\frac{f_{sc}}{f_{sm}}-1\right)\Delta p\, t \ ,$$

or

$$V_f = \sqrt{2k\,\Delta p\left(\frac{f_{sc}}{f_{sm}}-1\right)}\ A\frac{\sqrt{t}}{\sqrt{\mu}} \ \ldots \ldots (2.7)$$

The standard API filter press has an area of 45 cm^2 and is operated at a pressure of 6.8 atm (100 psig). The filtrate volume collected in a 30-min time period is reported as the standard water loss. Note that Eq. 2.7 indicates that the filtrate volume is proportional to the square root of the time period used. Thus, the filtrate collected after 7.5 min should be about half the filtrate collected after 30 min. It is common practice to report twice the 7.5-min filtrate volume as the API water loss when the 30-min filtrate volume exceeds the capacity of the filtrate receiver. However, as shown in Fig. 2.10, a spurt loss volume of filtrate, V_{sp}, often is observed before the porosity and permeability of the filter cake stabilizes and Eq. 2.7 becomes applicable. If a significant spurt loss is observed, the following equation should be used to extrapolate the 7.5-min water loss to the standard API water loss.

$$V_{30} = 2(V_{7.5} - V_{sp}) + V_{sp}. \ \ldots \ldots \ldots \ldots (2.8)$$

The best method for determining spurt loss is to plot V vs. $\sqrt{t}$ and extrapolate to zero time as shown in Fig. 2.10.

In addition to the standard API filter press, a smaller filter press capable of operating at elevated temperature and pressure also is commonly used. The filtration rate increases with temperature because the viscosity of the filtrate is reduced. Pressure usually has little effect on filtration rate because the permeability of the mudcake tends to decrease with pressure and the term $\sqrt{k\Delta p}$ in Eq. 2.7 remains essentially constant. However, an elevated pressure is required to prevent boiling when operating above 212°F. The area of the filter paper used in the high-temperature high-pressure (HTHP) filter press is one-half the area of the standard filter press. Thus, the volume of filtrate collected in 30 min must be doubled before reporting as API water loss. An example HTHP filter press is shown in Fig. 2.11.

Fig. 2.11—HTHP filter press.

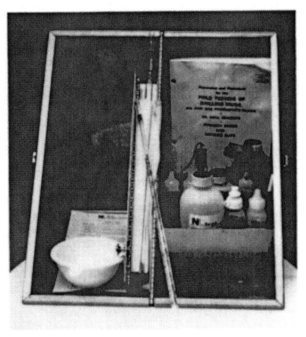

Fig. 2.12—Titration apparatus.

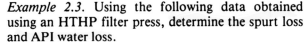

Example 2.3. Using the following data obtained using an HTHP filter press, determine the spurt loss and API water loss.

Time (min)	Filtrate Volume (cm³)
1.0	6.5
7.5	14.2

Solution. The spurt loss of the cell can be obtained by extrapolating to zero time using the two data points given:

$$6.5 - \frac{14.2 - 6.5}{\sqrt{7.5} - \sqrt{1}} \sqrt{1} = 2.07 \text{ cm}^3 .$$

However, since the standard API filter press has twice the cross-sectional area of the HTHP filter press, the corrected spurt loss is 4.14 cm³. The 30-min filtrate volume can be computed using Eq. 2.8:

$$V_{30} = 2(V_{7.5} - V_{sp}) + V_{sp}$$

$$= 2(14.2 - 2.07) + 2.07 = 26.33 \text{ cm}^3 .$$

Adjusting for the effect of filter press cross-sectional area, we obtain an API water loss of 52.66 cm³ at the elevated temperature and pressure of the test.

Both low-temperature and high-pressure API filter presses are operated under static conditions – that is, the mud is not flowing past the cake as filtration takes place. Other presses have been designed to model more accurately the filtration process wherein mud is flowed past the cake, as it does in the wellbore. Such presses that model *dynamic filtration* have shown that after a given period of time the mudcake thickness remains constant – that is, the cake is eroded as fast as it is being deposited. Thus, dynamic-filtration rates are higher than static filtration rates. With a constant thickness cake, integrating Eq. 2.5, we have

$$V_f = \frac{kA \Delta p t}{\mu \, h_{mc}} \quad \dots \dots \dots \dots \dots \dots \dots \dots (2.9)$$

A standard dynamic filtration test has not been developed to date. Field mud testing uses the static filtration test to characterize the filtration quality of the mud. Unfortunately, there are no reliable guidelines for correlating static and dynamic filtration rates. Our ability to predict quantitatively filtration rates in the wellbore during various drilling operations remains questionable.

2.1.6 Chemical Analysis. Standard chemical analyses have been developed for determining the concentration of various ions present in the mud. Tests for the concentration of OH^-, Cl^-, and Ca^{++} are

TABLE 2.2—INTERNATIONAL ATOMIC TABLE

Element	Symbol	Atomic Number	Atomic Weight	Valence	Element	Symbol	Atomic Number	Atomic Weight	Valence
ACTINIUM	Ac	89	227.0	—	MOLYBDENUM	Mo	42	95.95	3,4,6
ALUMINUM	Al	13	26.97	3	NEODYMIUM	Nd	60	144.27	3
ANTIMONY	Sb	51	121.76	3,5	NEON	Ne	10	20.183	0
ARGON	A	18	39.944	0	NICKEL	Ni	28	58.69	2,3
ARSENIC	As	33	74.91	3,5	NITROGEN	N	7	14.008	3,5
BARIUM	Ba	56	137.36	2	OSMIUM	Os	76	190.2	2,3,4,8
BERYLLIUM	Be	4	9.02	2	OXYGEN	O	8	16.000	2
BISMUTH	Bi	83	209.00	3,5	PALLADIUM	Pd	46	106.7	2,4
BORON	B	5	10.82	3	PHOSPHORUS	P	15	30.98	3,5
BROMINE	Br	35	79.916	1,3,5,7	PLATINUM	Pt	78	195.23	2,4
CADMIUM	Cd	48	112.41	2	POLONIUM	Po	84	210.0	—
CALCIUM	Ca	20	40.08	2	POTASSIUM	K	19	39.096	1
CARBON	C	6	12.01	2,4	PRASEODYMIUM	Pr	59	140.92	3
CERIUM	Ce	58	140.13	3,4	PROTOACTINIUM	Pa	91	231.0	—
CESIUM	Cs	55	132.91	1	RADIUM	Ra	88	226.05	2
CHLORINE	Cl	17	35.457	1,3,5,7	RADON	Rn	86	222.0	0
CHROMIUM	Cr	24	52.01	2,3,6	RHENIUM	Re	75	186.31	—
COBALT	Co	27	58.94	2,3	RHODIUM	Rh	45	102.91	3
COLUMBIUM	Cb	41	92.91	3,5	RUBIDIUM	Rb	37	85.48	1
COPPER	Cu	29	63.57	1,2	RUTHENIUM	Ru	44	101.7	3,4,6,8
DYSPROSIUM	Dy	66	162.46	3	SAMARIUM	Sm, Sa	62	150.43	3
ERBIUM	Er	68	167.2	3	SCANDIUM	Sc	21	45.10	3
EUROPIUM	Eu	63	152.0	2,3	SELENIUM	Se	34	78.96	2,4,6
FLUORINE	F	9	19.000	1	SILICON	Si	14	28.06	4
GADOLINIUM	Gd	64	156.9	3	SILVER	Ag	47	107.880	1
GALLIUM	Ga	31	69.72	2,3	SODIUM	Na	11	22.997	1
GERMANIUM	Ge	32	72.60	4	STRONTIUM	Sr	38	87.63	2
GOLD	Au	79	197.2	1,3	SULFUR	S	16	32.06	2,4,6
HAFNIUM	Hf	72	178.6	4	TANTALUM	Ta	73	180.88	5
HELIUM	He	2	4.003	0	TELLURIUM	Te	52	127.61	2,4,6
HOLMIUM	Hð	67	164.94	3	TERBIUM	Tb	65	159.2	3
HYDROGEN	H	1	1.0080	1	THALLIUM	Tl	81	204.39	1,3
INDIUM	In	49	104.76	3	THORIUM	Th	90	232.12	4
IODINE	I	53	126.92	1,3,5,7	THULIUM	Tm	69	169.4	3
IRIDIUM	Ir	77	193.1	3,4	TIN	Sn	50	118.70	2,4
IRON	Fe	26	55.85	2,3	TITANIUM	Ti	22	47.90	3,4
KRYPTON	Kr	36	83.7	0	TUNGSTEN	W	74	183.92	6
LANTHANUM	La	57	138.92	3	URANIUM	U	92	238.07	4,6
LEAD	Pb	82	207.21	2,4	VANADIUM	V	23	50.95	3,5
LITHIUM	Li	3	6.940	1	VIRGINIUM	Vi	87	224.0	1
LUTECIUM	Lu	71	174.99	3	XENON	Xe	54	131.3	0
MAGNESIUM	Mg	12	24.32	2	YTTERBIUM	Yb	70	173.04	3
MANGANESE	Mn	25	54.93	2,3,4,6,7	YTTRIUM	Yt	39	88.92	3
MASURIUM	Ma	43	—	—	ZINC	Zn	30	65.38	2
MERCURY	Hg	80	200.61	1,2	ZIRCONIUM	Zr	40	91.22	4

required to complete the API drilling mud report form. A titration apparatus used to perform these tests is shown in Fig. 2.12.

Titration involves the reaction of a known volume of sample with a standard solution of known volume and concentration. The concentration of the ion being tested then can be determined from a knowledge of the chemical reaction taking place. Several terms used to describe the concentration of a given substance in solution are (1) *molality* − the number of gram-moles of solute per kilogram of solvent, (2) *molarity* − the number of gram-moles of solute per liter of solution, (3) *normality* − the number of gram equivalents of the solute per liter of solution [one gram equivalent weight (gew) is the weight of the substance that would react with one gram-mole of hydrogen], (4) *parts per million* (ppm) − the number of grams of solute per million grams of solution, (5) *milligrams per liter* − the number of milligrams of solute per liter of solution, and (6) *percent by weight* − the number of grams of solute per 100 grams of solution.

It is unfortunate that so many terms are used to express concentration. It is even more unfortunate that some of the terms are used inconsistently by many people in the petroleum industry. For example, the term parts per million sometimes is used interchangeably with milligrams per liter, even at high concentrations.

Example 2.4. A $CaCl_2$ solution is prepared at 68°F by adding 11.11 g of $CaCl_2$ to 100 cm^3 of water. At this temperature, water has a density of 0.9982 g/cm^3 and the resulting solution has a density of 1.0835 g/cm^3. Express the concentration of the solution using (1) molality, (2) molarity, (3) normality, (4) parts per million, (5) milligrams per liter, and (6) percent by weight.

Solution. The atomic weight of Ca and Cl are shown to be 40.08 and 35.457, respectively, in Table 2.2. Thus, the molecular weight of $CaCl_2$ is

$$40.08 + 2(35.457) = 111.$$

1. For a water density of 0.9982 g/cm^3, the molality of the solution is

$$\frac{11.11\,g}{(0.9982\,g/cm^3)(100\,cm^3)} \times \frac{1\,g\,mol}{111\,g} \times \frac{1,000\,g}{kg}$$

$$= 1.003\,g\,mol/kg.$$

The volume of the solution can be computed from the mass of solute and solvent and the density of the solution. Since 11.11 g of $CaCl_2$ added to 100 cm^3 of water gave a solution density of 1.0835 g/cm^3, the solution volume is

$$\frac{(11.11 + 99.82)g}{1.0835\,g/cm^3} = 102.38\,cm^3 .$$

2. Thus, the molarity of the solution is

$$\frac{11.11\,g}{102.38\,cm^3} \times \frac{1\,g\,mol}{111\,g} \times \frac{1,000\,cm^3}{1\,L}$$

$$= 0.978\,g\,mol/L.$$

3. Since 0.5 mol of $CaCl_2$ would tend to react with 1 mol of hydrogen, the gram-equivalent weight of $CaCl_2$ is half the molecular weight. The normality of the solution is

$$\frac{11.11\,g}{102.38\,cm^3} \times \frac{1\,gew}{55.5\,g} \times \frac{1,000\,cm^3}{1\,L}$$

$$= 1.955\,gew/L.$$

4. The concentration of $CaCl_2$ in parts per million is given by

$$\frac{11.11\,g}{(11.11 + 99.82)g} \times \frac{10^6\,g}{million\,g}$$

$$= 100,153\,ppm.$$

5. Concentration of $CaCl_2$ in milligrams per liter is

$$\frac{11.11\,g}{102.38\,cm^3} \times \frac{1,000\,mg}{g} \times \frac{1,000\,cm^3}{1\,L}$$

$$= 108,517\,mg/L.$$

6. Finally, the concentration of $CaCl_2$ as a percent by weight is

$$\frac{11.11\,g}{(11.11 + 99.82)g} \times 100\% = 10.02\,wt\%.$$

2.1.7 Alkalinity.

Alkalinity refers to the ability of a solution or mixture to react with an acid. The *phenolphthalein alkalinity* refers to the amount of acid required to reduce the pH to 8.3, the phenolphthalein endpoint. The phenolphthalein alkalinity of the mud and mud filtrate is called the P_m and P_f, respectively. The P_f test includes the effect of only dissolved bases and salts while the P_m test includes the effect of both dissolved and suspended bases and salts. The *methyl orange alkalinity* refers to the amount of acid required to reduce the pH to 4.3, the methyl orange endpoint. The methyl orange alkalinity of the mud and mud filtrate is called the M_m and M_f, respectively. The API diagnostic tests include the determination of P_m, P_f, and M_f. All values are reported in cubic centimeters of 0.02 N (normality = 0.02) sulfuric acid per cubic centimeter of sample.

The P_f and M_f tests are designed to establish the concentration of hydroxyl, bicarbonate, and carbonate ions in the aqueous phase of the mud. At a pH of 8.3, the conversion of hydroxides to water and carbonates to bicarbonates is essentially complete. The bicarbonates originally present in solution do not enter the reactions. Thus, at a pH of 8.3,

$$OH^- + H^+ \rightarrow HOH,$$

and

$$CO_3^{2-} + H^+ \rightarrow HCO_3^- .$$

As the pH is further reduced to 4.3, the acid then reacts with the bicarbonate ions to form carbon dioxide and water:

$$HCO_3^- + H^+ \rightarrow CO_2 \uparrow + HOH .$$

Unfortunately, in many mud filtrates, other ions and organic acids are present that affect the M_f test.

The P_f and P_m test results indicate the reserve alkalinity of the suspended solids. As the $[OH^-]$ in solution is reduced, the lime and limestone suspended in the mud will go into solution and tend to stabilize the pH. This reserve alkalinity generally is expressed as an equivalent lime concentration. Converting the $Ca(OH)_2$ concentration from 0.02 N to field units of lbm/bbl yields

$$\frac{0.02\,gew}{1\,L} \times \frac{37.05\,g}{gew} \times \frac{0.35\,lbm/bbl}{g/L}$$

$$= 0.26\,lbm/bbl .$$

Thus, the free lime is given by 0.26 $(P_m - f_w \cdot P_f)$, where f_w is the volume fraction of water in the mud.

Example 2.5. A drilling mud is known to contain $Ca(OH)_2$. The alkalinity tests are conducted to determine the amount of undissolved lime in suspension in the mud. When 1 cm^3 of mud filtrate is titrated using 0.02 N H_2SO_4, 1.0 cm^3 of H_2SO_4 is required to reach the phenolphthalein endpoint and 1.1 cm^3 of H_2SO_4 is required to reach the methyl orange endpoint. When 1 cm^3 of mud is diluted with 50 cm^3 of water before titration so that any suspended lime can go into solution, 7.0 cm^3 of H_2SO_4 is required to reach the phenolphthalein endpoint. Compute the amount of free lime in suspension in the mud if the mud has a total solids content of 10%.

Solution. Since both the P_f and M_f have approximately the same value, an absence of any carbonates or bicarbonates is indicated. Thus, the alkalinity of the filtrate is mainly due to the presence of hydroxides. The free lime in lbm/bbl is given by

$$0.26(P_m - f_w P_f) = 0.26[7.0 - 0.9(1.0)]$$

$$= 1.59 \text{ lbm/bbl} .$$

2.1.8 Chloride Concentration. Salt can enter and contaminate the mud system when salt formations are drilled and when saline formation water enters the wellbore. The chloride concentration is determined by titration with silver nitrate solution. This causes the chloride to be removed from the solution as AgCl, a white precipitate:

$$Ag^+ + Cl^- \rightarrow AgCl \downarrow .$$

The endpoint of the titration is detected using a potassium chromate indicator. The excess Ag^+ present after all Cl^- has been removed from solution reacts with the chromate to form Ag_2CrO_4, an orange-red precipitate:

$$2\,Ag^+ + CrO_4 \rightarrow Ag_2CrO_4 \downarrow .$$

Since AgCl is less soluble than Ag_2CrO_4, the latter cannot form permanently in the mixture until the precipitation of AgCl has reduced the $[Cl^-]$ to a very small value. A 0.0282 N $AgNO_3$ concentration usually is used for the titration. Since the equivalent weight of Cl^- is 35.46, the concentration of Cl^- in the filtrate is given by

$$\text{mg/L } Cl^- = V_{tf}\left(0.0282 \frac{\text{gew}}{1 \text{ L}}\right)\left(35.46 \frac{\text{g}}{\text{gew}}\right)$$

$$\cdot \left(1,000 \frac{\text{mg}}{\text{g}}\right) = 1,000\, V_{tf} ,$$

where V_{tf} is the volume of $AgNO_3$ required to reach the endpoint per cubic centimeter of mud filtrate used in the titration. If the Cl^- ions are produced by NaCl, the NaCl content of the mud is determined by the relationship

$$\text{mg/L NaCl} = \frac{(23 + 35.46)\text{g NaCl}}{35.46 \text{ g } Cl^-} \cdot \text{mg/L } Cl^-$$

$$= 1.65 \cdot 1,000\, V_{tf}$$

$$= 1,650\, V_{tf}, \ \dots\dots\dots\dots (2.10)$$

since the atomic weight of sodium is 23 and the atomic weight of chlorine is 35.46.

Example 2.6. One cm^3 of mud filtrate is titrated using 0.0282 N $AgNO_3$. Nine cm^3 of $AgNO_3$ solution are required to reach the endpoint of the titration as indicated by the potassium chromate indicator. Compute the concentration of Cl^- present expressed in milligrams of Cl^- per liter. Also, assuming that only sodium chloride was present, compute the salinity of the filtrate in milligrams of NaCl per liter.

Solution. The Cl^- concentration is given by

$$1,000\, V_{tf} = 1,000 \times 9$$

$$= 9,000 \text{ mg } Cl^-/\text{L} ,$$

and the NaCl concentration is given by

$$1,650\, V_{tf} = 1,650\,(9)$$

$$= 14,850 \text{ mg NaCl/ L} .$$

2.1.9 Water Hardness. Water containing large amounts of Ca^{2+} and Mg^{2+} ions is known as *hard water.* These contaminants are often present in the water available for use in the drilling fluid. In addition, Ca^{2+} can enter the mud when anhydrite ($CaSO_4$) or gypsum ($CaSO_4 \cdot 2H_2O$) formations are drilled. Cement also contains calcium and can contaminate the mud. The total Ca^{2+} and Mg^{2+} concentration is determined by titration with a standard (0.02 N) Versenate (EDTA) solution. The standard Versenate solution contains Sodium Versenate, an organic compound capable of forming a chelate with Ca^{++} and Mg^{++}.

The chelate ring structure is quite stable and essentially removes the Ca^{++} and Mg^{++} from solution. Disodium ethylenediaminetetraacetic acid (EDTA) plus calcium yields the EDTA chelate ring:

Magnesium ions form a wine-red complex with the dye Eriochrome Black T. If a solution containing both Ca^{2+} and Mg^{2+} is titrated in the presence of this dye, the Versenate first forms a calcium complex. After the $[Ca^{2+}]$ has been reduced to a very low level, the Versenate then forms a complex with the magnesium ions. The depletion of the available Mg^{2+} ions from the dye Eriochrome Black T causes the color of the solution to change from wine-red to blue. The amount of Versenate used depends on the total concentration of Ca^{2+} and Mg^{2+}. A small amount of Mg^{2+} is included in the dye indicator solution to ensure the proper color action in the event no Mg^{2+} is present in the sample.

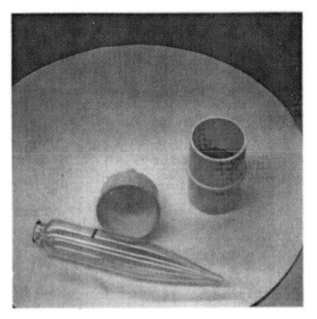

Fig. 2.13—Sand content apparatus.

Fig. 2.14—Mud still.

The hardness test sometimes is performed on the mud as well as the mud filtrate. The mud hardness indicates the amount of calcium suspended in the mud as well as the calcium in solution. This test usually is made on gypsum-treated muds to indicate the amount of excess $CaSO_4$ present in suspension. To perform the hardness test on mud, a small sample of mud is first diluted to 50 times its original volume with distilled water so that any undissolved calcium or magnesium compounds can go into solution. The mixture then is filtered through hardened filter paper to obtain a clear filtrate. The total hardness of this filtrate then is obtained using the same procedure used for the filtrate from the low-temperature low-pressure API filter press apparatus. Since the mud was diluted to 50 times the original volume, a 50-cm^3 sample would have to be titrated to determine the calcium and magnesium present in 1 cm^3 of mud. The usual procedure is to titrate a 10-mL sample and multiply the titration volume by five. The mud hardness often is reported as an equivalent calcium sulfate concentration. The equivalent weight of $CaSO_4$ is 68.07. Converting the $CaSO_4$ concentration from 0.02 N to field units of pounds mass per barrel yields

$$0.02 \text{ gew/L} \times 68.07 \text{ g/gew} \times \frac{0.35 \text{ lbm/bbl}}{1 \text{ g/L}}$$

$$= 0.477 \text{ lbm/bbl} .$$

Thus, the total $CaSO_4$ concentration in pounds mass per barrel is given by

$$0.477 \ V_{tm} \ ,$$

where V_{tm} is the titration volume in cubic centimeters of 0.02 N Versenate solution required per cubic centimeter of mud sample. The free $CaSO_4$ in pounds mass per barrel is given by

$$0.477 \ (V_{tm} - f_w \ V_{tf}) \ ,$$

where V_{tf} is the titration volume in cubic centimeters of 0.02 N Versenate solution required per cubic centimeter of mud filtrate.

Example 2.7. Compute the total calcium concentration of the mud expressed as pounds per barrel of $CaSO_4$ if 10 mL of 0.02 N Versenate solution was required to titrate a 1-cm^3 sample of mud that had been diluted and filtered as described above.

Solution. The total $CaSO_4$ concentration is given by

$$0.477 \ V_{tm} = 0.477(10) = 4.77 \text{ lbm/bbl} .$$

2.1.10 Sand Content. The sand content of the mud is measured using a 200-mesh sieve and a glass tube calibrated to read directly the percentage of sand by volume. Sand is abrasive to the fluid circulating system, and desanders usually are used when necessary to maintain the sand content at a low level. The standard apparatus used to determine the sand content of the mud is shown in Fig. 2.13.

2.1.11 The Mud Retort. The mud retort (Fig. 2.14) is used to determine the volume fraction of oil, water, and solids in a mud. A calibrated mud is placed in the retort cup; then the liquids are distilled into a graduate cylinder. The solids fraction of the mud, f_s, is determined by

$$f_s = 1 - f_w C_f - f_o , \quad\quad\quad\quad\quad\quad\quad (2.10)$$

where f_w is the volume fraction of distilled water collected in the graduated cylinder, f_o is the volume fraction of distilled oil, and C_f is the volume increase factor due to the loss of dissolved salt during retorting. The volume correction applied to the distilled water fraction, C_f, is obtained from Tables 2.3 and 2.4.

TABLE 2.3—DENSITIES OF NaCl SOLUTIONS AT 68°F

Specific Gravity	Percent NaCl by Weight of		Weight of Solution per		Pounds of NaCl Added to Water per			Volume* of Solution (bbl)
	Solution	Water	Gallon	Cubic Foot	Gallon	Cubic Foot	Barrel	
0.9982	0	0.00	8.33	62.32	—	—	—	1.000
1.0053	1	1.01	8.39	62.76	0.084	0.63	3.53	1.003
1.0125	2	2.04	8.45	63.21	0.170	1.27	7.14	1.006
1.0268	4	4.17	8.57	64.10	0.347	2.60	14.59	1.013
1.0413	6	6.38	8.69	65.01	0.531	3.98	22.32	1.020
1.0559	8	8.70	8.81	65.92	0.725	5.42	30.44	1.028
1.0707	10	11.11	8.93	66.84	0.925	6.92	38.87	1.036
1.0857	12	13.64	9.06	67.78	1.136	8.50	47.72	1.045
1.1009	14	16.28	9.19	68.73	1.356	10.15	56.96	1.054
1.1162	16	19.05	9.31	69.68	1.587	11.87	66.65	1.065
1.1319	18	21.95	9.45	70.66	1.828	13.68	76.79	1.075
1.1478	20	25.00	9.58	71.65	2.083	15.58	87.47	1.087
1.1640	22	28.21	9.71	72.67	2.350	17.58	98.70	1.100
1.1804	24	31.58	9.85	73.69	2.631	19.68	110.49	1.113
1.1972	26	35.13	9.99	74.74	2.926	21.89	122.91	1.127

*Final volume of solution after adding specified quantity of sodium chloride to 1 bbl of fresh water.

TABLE 2.4—DENSITIES OF CaCl$_2$ SOLUTIONS AT 68°F

Specific Gravity	Percent CaCl$_2$ by Weight of		Weight of Solution per		Pounds of CaCl$_2$ Added to Water per			Volume* of Solution (bbl)
	Solution	Water	Gallon	Cubic Foot	Gallon	Cubic Foot	Barrel	
0.9982	0	0.00	8.33	62.32	—	—	—	1.000
1.0148	2	2.04	8.47	63.35	0.170	1.27	7.14	1.004
1.0316	4	4.17	8.61	64.40	0.347	2.60	14.59	1.008
1.0486	6	6.38	8.75	65.46	0.531	3.98	22.32	1.013
1.0659	8	8.70	8.89	66.54	0.725	5.42	30.44	1.019
1.0835	10	11.11	9.04	67.64	0.925	6.92	38.87	1.024
1.1015	12	13.64	9.19	68.71	1.136	8.50	47.72	1.030
1.1198	14	16.28	9.34	69.91	1.356	10.15	56.96	1.037
1.1386	16	19.05	9.50	71.08	1.587	11.87	66.65	1.044
1.1578	18	21.95	9.66	72.28	1.828	13.68	76.79	1.052
1.1775	20	25.00	9.83	73.51	2.083	15.58	87.47	1.059
1.2284	25	33.33	10.25	76.69	2.776	20.77	116.61	1.084
1.2816	30	42.86	10.69	80.01	3.570	26.71	149.95	1.113
1.3373	35	53.85	11.16	83.48	4.486	33.56	188.40	1.148
1.3957	40	66.67	11.65	87.13	5.554	41.55	233.25	1.192

*Final volume of solution after adding specified quantity of calcium chloride to 1 bbl of fresh water.

TABLE 2.5—NaCl CONCENTRATIONS AS wt%, ppm, AND mg/l

Salt (wt%)	ppm	mg/L
0.5	5,000	5,020
1	10,000	10,050
2	20,000	20,250
3	30,000	30,700
4	40,000	41,100
5	50,000	52,000
6	60,000	62,500
7	70,000	73,000
8	80,000	84,500
9	90,000	95,000
10	100,000	107,100
11	110,000	118,500
12	120,000	130,300
13	130,000	142,000
14	140,000	154,100
15	150,000	166,500
16	160,000	178,600
17	170,000	191,000
18	180,000	203,700
19	190,000	216,500
20	200,000	229,600
21	210,000	243,000
22	220,000	256,100
23	230,000	270,000
24	240,000	283,300
25	250,000	297,200
26	260,000	311,300

TABLE 2.6—TYPICAL CATION EXCHANGE CAPACITIES OF SEVERAL SOLIDS

Solid	Milliequivalents of Methylene Blue per 100 g of Solids
attapulgite	15 to 25
chlorite	10 to 40
gumbo shale	20 to 40
illite	10 to 40
kaoline	3 to 15
montmorillonite	70 to 150
sandstone	0 to 5
shale	0 to 20

TABLE 2.7—DENSITY OF SEVERAL MUD ADDITIVES

Material	Specific Gravity	Density lbm/gal	Density lbm/bbl
attapulgite	2.89	24.1	1,011
water	1.00	8.33	350
diesel	0.86	7.2	300
bentonite clay	2.6	21.7	910
sand	2.63	21.9	920
average drilled solids	2.6	21.7	910
API barite	4.2	35.0	1,470
CaCl$_2$*	1.96	16.3	686
NaCl*	2.16	18.0	756

*Highly water soluble (do not assume ideal mixing).

Example 2.8. A 12-lbm/gal saltwater mud is retorted and found to contain 6% oil and 74% distilled water. If the chloride test shows the mud to have a chloride content of 79,000 mg Cl^-/L, what is the solids fraction of the mud? Assume the mud is a sodium chloride mud.

Solution. A mud with a Cl^- content of 79,000 mg Cl^-/L has a NaCl content of

$$mg\ NaCl/L = 1.65 \times Cl^-\ mg/L$$
$$= 1.65 \times 79,000$$
$$= 130,350\ mg\ NaCl/L\ .$$

From Table 2.5, a 130,350-mg NaCl/L solution has a salinity of 12%. From Table 2.3, a 12% solution of NaCl has a volume increase factor of 1.045. From Eq. 2.10,

$$f_s = 1 - f_w C_f - f_o = 1 - (0.74)(1.045)$$
$$- 0.06 = 0.167\ .$$

2.1.12 Cation Exchange Capacity of Clays. In addition to determining the volume fraction of the low specific gravity solids, it is often desirable to determine the amount of easily hydrated clay present in these solids. *Sodium montmorillonite* is a good hydratable clay often added to the mud to increase viscosity and cutting carrying capacity. The sodium ion is held loosely in the clay structure and is exchanged readily for other ions and certain organic compounds. The organic dye, methylene blue ($C_{16}H_{18}N_3SCl \cdot 3H_2O$), readily replaces the exchangeable cations in montmorillonite and certain other clays. The methylene blue test is used on drilling muds to determine the approximate montmorillonite content. The test is only qualitative because organic material and some other clays present in the mud also will adsorb methylene blue. The mud sample usually is treated with hydrogen peroxide to oxidize most of the organic material. The cation exchange capacity is reported in milliequivalent weights (meq) of methylene blue per 100 ml of mud. The methylene blue solution used for titration is usually 0.01 N, so that the cation exchange capacity is numerically equal to the cubic centimeters of methylene blue solution per cubic centimeter of sample required to reach an endpoint. If other adsorptive materials are not present in significant quantities, the montmorillonite content of the mud in pounds per barrel is five times the cation exchange capacity.

The methylene blue test can also be used to determine the adsorptive tendencies of various solids. When this is done, the results are reported per 100 g of solids rather than per 100 mL of mud. Table 2.6 lists typical results obtained.

Example 2.9. One cubic centimeter of mud is diluted using 50 cm^3 of distilled water and treated with sulfuric acid and hydrogen peroxide to oxidize any organic material present. The sample then is titrated using 0.01 N methylene blue solution. Compute the approximate montmorillonite content of the mud if 5.0 cm^3 of methylene blue is needed to reach an endpoint.

Solution. The montmorillonite content is approximately five times the cation exchange capacity. Since for a 0.01 N methylene blue solution, the cation exchange capacity is equal to the cubic centimeters of solution used per cubic centimeter of mud sample, the montmorillonite content is

$$5(5.0) = 25\ lbm/bbl\ .$$

2.2 Pilot Tests

The drilling fluid specialist uses the API diagnostic tests discussed in Section 2.1 to detect potential problems and identify their cause. Alternative mud treatments then must be evaluated using small samples. This ensures that the mixture used will provide the desired results at the lowest possible cost before treatment of the active mud system is started. The units of measure most commonly used when treating the active drilling fluid system are pounds for weight and barrels for volume. The units of measure most commonly used for pilot tests are grams for weight and cubic centimeters for volume. Converting from lbm/bbl to g/cm^3 gives

$$\frac{1.0\ lbm}{1\ bbl} \times \frac{454\ g}{lbm} \times \frac{1\ bbl}{42\ gal} \times \frac{1\ gal}{3785\ mL}$$
$$= \frac{1\ g}{350\ mL}\ .$$

Thus, adding 1 g of material to 350 mL of fluid is equivalent to adding 1 lbm of material to 1 bbl of fluid.

Pilot testing frequently involves evaluation of mixtures of given concentrations and densities. It generally is assumed in mixing calculations that the resulting mixture is ideal — i.e., the total volume is equal to the sum of component volumes:

$$V_t = V_1 + ... + V_n\ . \quad\quad\quad\quad (2.11)$$

Also, it is frequently necessary to compute the volume of solids added to a mixture from a knowledge of its mass and density. The volume V_i of a given mass, m_i, of an additive having a density, ρ_i, is given by

$$V_i = \frac{m_i}{\rho_i}\ . \quad\quad\quad\quad\quad\quad (2.12)$$

Typical densities of several materials often present in drilling fluid are shown in Table 2.7. The mixture density can be computed from a knowledge of the total mass and total volume added to the mixture. Thus, the mixture density is given by

$$\rho = \frac{m_1 + m_2 + ... + m_n}{V_1 + V_2 + ... + V_n}\ , \quad\quad\quad (2.13)$$

where the volume of the solid components is computed using Eq. 2.12.

Example 2.10. Compute the volume and density of a mud composed of 25 lbm of bentonite clay, 60 lbm of API barite, and 1 bbl of fresh water.

Solution. Using Table 2.7, the densities of clay and API barite are 910 lbm/bbl and 1470 lbm/bbl, respectively. The total volume is given by

$$V_t = V_1 + V_2 + V_3 = 1.0 + \frac{25}{910} + \frac{60}{1470}$$

$$= 1.0683 \text{ bbl} .$$

The mixture density is the total mass per unit volume. From Table 2.7, the density of water is 350 lbm/bbl. Thus, the mixture density is

$$\rho = \frac{350 + 25 + 60}{1.0683} = \frac{407 \text{ lbm}}{\text{bbl}} \text{ or } 9.7 \text{ lbm/gal} .$$

Concentrated solutions of NaCl or CaCl$_2$ sometimes are used in drilling fluids. The assumption of ideal mixing is usually valid only for mixtures and is not accurate for solutions. The resulting total volume and fluid density obtained by adding various weights of NaCl and CaCl$_2$ to fresh water are shown in Tables 2.3 and 2.4.

Example 2.11. Determine the volume and density of a brine composed of 110.5 lbm of NaCl and 1 bbl of fresh water at 68°F.

Solution. From Table 2.3, the total volume is 1.113 bbl and the solution density is 9.85 lbm/gal. Note that if ideal mixing is assured, the total volume calculated is given by

$$V_t = V_1 + V_2 = 1.0 + \frac{110.5}{756} = 1.1462 \text{ bbl} .$$

This volume corresponds to a density of

$$\rho = \frac{350 + 110.5}{1.1462} = 401.76 \text{ lbm/bbl}$$

$$\text{or } 9.57 \text{ lbm/gal} .$$

This value does not compare favorably with the true value shown in Table 2.3.

Pilot test results are dependent to some extent upon the mixing procedures used. The ability to obtain representative test results can depend on (1) the order in which the chemicals are added, (2) whether a material is added as a dry solid or in solution, (3) the amount of sample agitation used, (4) the sample aging period, and (5) the sample temperature. Care should be taken to add chemicals to the sample in the same order and manner used under field conditions. Chemicals normally added to the mud system using the chemical barrel should be added to the pilot sample in solution. Also, when possible, a representative temperature and aging period should be used.

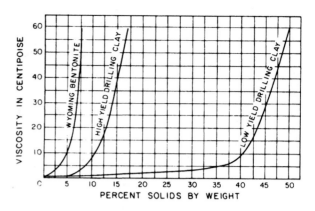

Fig. 2.15 – Effect of clay concentration on viscosity of fresh water.[3]

2.3 Water-Base Muds

Water is the basic component of most drilling fluids. Many wells are begun using the natural water available in the area. As drilled solids become entrained in the water, a natural mud is formed. Some clays hydrate readily in water and greatly increase the viscosity of the mud. This increase in viscosity enhances the ability of the drilling fluid to carry the rock cuttings to the surface, especially in the larger hole sizes where the annular velocity developed by the pump is relatively low. The clay particles also form a mudcake on the hole wall opposite permeable formations. This greatly reduces the amount of water loss to these zones and helps to prevent the hole wall from caving into the hole. Because of these beneficial effects, clays that will hydrate readily with the available water often are added at the surface if they are not present in the formations drilled.

The presence of hydrated clays in the water has undesirable as well as desirable effects on the rotary drilling process. A reduction in penetration rate and an increase in frictional pressure losses are observed when the clay content of the drilling fluid increases. When drilling relatively small holes in hard, competent formations, these undesirable effects may be more important than the beneficial effects. When this is the case, water alone can be used as the drilling fluid. Equipment capable of removing finely divided solids must be used continually to prevent the formation of a natural mud.

2.3.1 Clays Encountered in Drilling Fluids. A large number of clay minerals with widely different properties are present in nature. Not all clay minerals hydrate readily in water. In general, the high-swelling clays are desirable and are added to the mud for viscosity and filtration control. The low-swelling clays enter the mud as cuttings and cavings, and are referred to as contaminants or drilled solids.

2.3.1.1 Commercial Clays. Commercial clays used in drilling fluids are graded according to their ability to increase the viscosity of water. The *yield* of a clay is defined as the number of barrels of mud that can be produced using 1 ton of clay if the mud has an apparent viscosity of 15 cp when measured in a rotational viscometer at 600 rpm. The most common commercial clay mined for use in drilling fluids is called *Wyoming bentonite*. It has a yield of about 100 bbl/ton when used with pure water. A less expensive commercial clay called *high-yield clay* has a yield of about 45 bbl/ton. It is not uncommon for native clays to yield less than 10 bbl/ton. A comparison of mud viscosities obtained from various concentrations of Wyoming bentonite, high-yield clay, and an example native clay in pure water is shown in Fig. 2.15.[3] Note that regardless of clay type, once sufficient clay has been added to obtain a 15-cp mud, the mud viscosity increases rapidly with further increases in clay content.

Wyoming bentonite is composed primarily of sodium montmorillonite. The name montmorillonite originally was applied to a mineral found near Montmorillon, France. The term now is reserved usually for hydrous aluminum silicates approximately represented by the formula: $4 SiO_2 \cdot Al_2O_3 \cdot H_2O$ + water; but with some of the aluminum cations, Al^{3+}, being replaced by magnesium cations, Mg^{2+}. This replacement of Al^{3+} by Mg^{2+} causes the montmorillonite structure to have an excess of electrons. This negative charge is satisfied by loosely held cations from the associated water. The name sodium montmorillonite refers to a clay mineral in which the loosely held cation is the Na^+ ion.

API and the European Oil Companies Materials Assn. (OCMA) have set certain specifications for bentonites that are acceptable for use in drilling fluids. These specifications are listed in Table 2.8.

A model representation of the structure of sodium montmorillonite is shown in Fig. 2.16.[4] A central alumina octahedral sheet has silica tetrahedral sheets on either side. These sheetlike structures are stacked with water and the loosely held cations between them. Polar molecules such as water can enter between the unit layers and increase the interlayer spacing. This is the mechanism through which montmorillonite hydrates or swells. A photomicrograph of montmorillonite particles in water is shown in Fig. 2.17.[4] Note the platelike character of the particles.

In addition to the substitution of Mg^{2+} for Al^{3+} in the montmorillonite lattice, many other substitutions are possible. Thus, the name montmorillonite often is used as a group name including many specific mineral structures. However, in recent years, the name smectite has become widely accepted as the group name, and the term montmorillonite has been reserved for the predominantly aluminous member of the group shown in Fig. 2.16. This naming convention has been adopted in this text.

The salinity of the water greatly affects the ability of the commercial smectite clays to hydrate. A fibrous clay mineral called attapulgite can be used when the water salinity is too great for use of the smectite clays. The name attapulgite originally was applied to a clay mineral found near Attapulgus, GA. Attapulgite is approximately represented by the

TABLE 2.8—SPECIFICATIONS FOR BENTONITE

	Specified Values	
	API	OCMA
Minimum yield	91 bbl/ton	16 m^3 tonne
Maximum moisture	10 wt%	10 wt%
Wet-screen analysis		
(residue on No. 200 sieve)	2.5 wt%	2.5 wt%
Maximum API water loss		
22.5 lbm/bbl H$_2$O	15.0 mL	—
26.3 lbm/bbl H$_2$O	—	15 mL
Minimum yield point	3 × plastic	—
(22.5 lbm/bbl H$_2$O)	viscosity	
Minimum dial reading, 600 rpm		
(22.5 lbm/bbl H$_2$O)	30	—

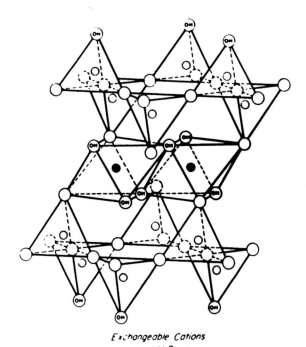

Exchangeable Cations
$n H_2O$

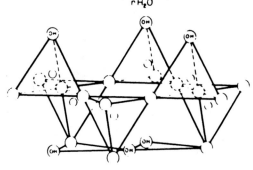

Fig. 2.16—Structure of sodium montmorillonite.[4]

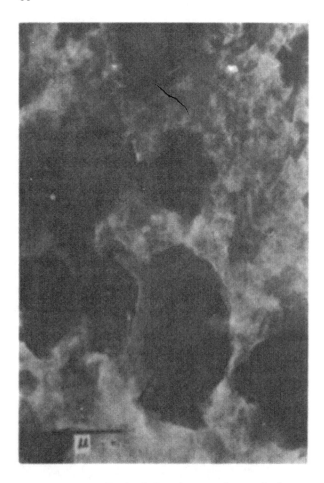

Fig. 2.17—Transmission electron micrograph of montmorillonite.[4]

formula: $(OH_2)_4 (OH)_2 Mg_5Si_8O_{20} \cdot 4H_2O$, but with some pairs of the magnesium cations, $2Mg^{2+}$, being replaced by a single trivalent cation. A photomicrograph of attapulgite in water is shown in Fig. 2.18. The ability of attapulgite to build viscosity is thought to be due to interaction between the attapulgite fibers rather than hydration of water molecules. A longer period of agitation is required to build viscosity with attapulgite than with smectite clays. However, with continued agitation, viscosity

decreases eventually are observed due to mechanical breakage of the long fibers. This can be offset through the periodic addition of a new attapulgite material to the system.

The clay mineral sepiolite, a magnesium silicate with a fibrous texture, has been proposed as a high-temperature substitute for attapulgite. The idealized formula can be written $Si_{12}Mg_8O_{32} \cdot nH_2O$. X-ray diffraction techniques and scanning electron microscope studies have established that the crystalline structure of this mineral is stable at temperatures up to 800°F. Slurries prepared from sepiolite exhibit favorable rheological properties over a wide range of temperatures.

2.3.1.2 Low-Swelling Clays. As formations are drilled, many different minerals enter the mud system and are dispersed throughout the mud by mechanical crushing and chemical hydration. Various types of low-swelling clays enter the mud, which contributes to the total cation exchange capacity of the mud. These clays are very similar to montmorillonite in that they have alumina octahedral sheets and silica tetrahedral sheets (Fig. 2.19). The major difference in such clays is the presence of different ions within the lattice of the sheets that were introduced during clay deposition.

2.3.2 Cation Exchange in Smectite Clays. The smectite clays have the ability to exchange readily the loosely held cations located between the sheetlike structures for other cations present in the aqueous solution. A well-known application of the ion exchange reaction is the softening of water. Ion exchange reactions in drilling fluids are important because the ability of the clay particles to hydrate depends greatly on the loosely held cations present. The ability of one cation to replace another depends on the nature of the cations and their relative concentrations. The common cations will replace each other when present in the same concentration in this order:

$$Al^{3+} > Ba^{2+} > Mg^{2+} > Ca^{2+} > H^+ > K^+ > Na^+$$

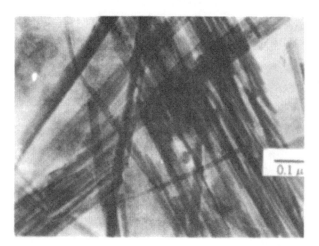

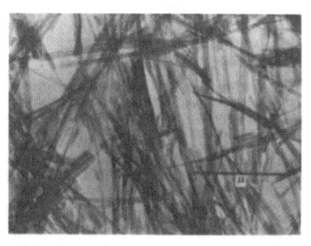

Fig. 2.18—Transmission electron micrographs of attapulgite (left) and sepiolite (right).[4]

However, this order can be changed by increasing the concentration of the weaker cation present. Many organic compounds also will adsorb between the sheetlike clay structures. As discussed in Section 2.12, the adsorption of methylene blue is the standard test for the cation exchange capacity of the mud.

2.3.3 Effect of Montmorillonite and Drilled Solids on Drilling Fluids Density. Solids in the drilling fluid cause an increase in density as well as viscosity. Since the specific gravity of all clays is near 2.6, the density of a clay/water mixture of a given viscosity depends on the yield of the clay used. A clay with a high yield must be used if a mud having a density near water is desired. If a mud having a higher density is desired, a clay with a lower yield can be used. In many cases, a natural buildup of low yield drilled solids in the mud as drilling progresses provides the desired fluid density. Also, API barite, a dense, inert mineral having a specific gravity near 4.2, can be added to any clay/water mixture to increase the density. However, the clay/water mixture must have a gel strength of about 3 lbf/100 sq ft to hold the barium sulfate in suspension.

Example 2.12 Using the data provided in Fig. 2.15, compute the density of a mud having an apparent viscosity of 20 cp (as measured in a rotational viscometer operated at 600 rpm) for each of the three clay types shown.

Solution. The percent solids by weight required to obtain a 20-cp mud for each of the clay types given in Fig. 2.15 are as follows.

Clay Type	Solids (wt%)
Wyoming bentonite	6.3
High-yield clay	12.6
Low-yield native clay	43.5

From Table 2.7, the density of clay is approximately 21.7 lbm/gal, and the density of fresh water is 8.33 lbm/gal. If we choose 100 lbm of mud as a basis for the density calculation, and let x represent the weight fraction of clay, then the density can be expressed as follows.

$$\rho = \frac{m_1 + m_2}{V_1 + V_2} = \frac{100}{\dfrac{100x}{21.7} + \dfrac{100(1-x)}{8.33}}$$

Using this equation and the values for weight fraction shown above, the following mud densities are computed.

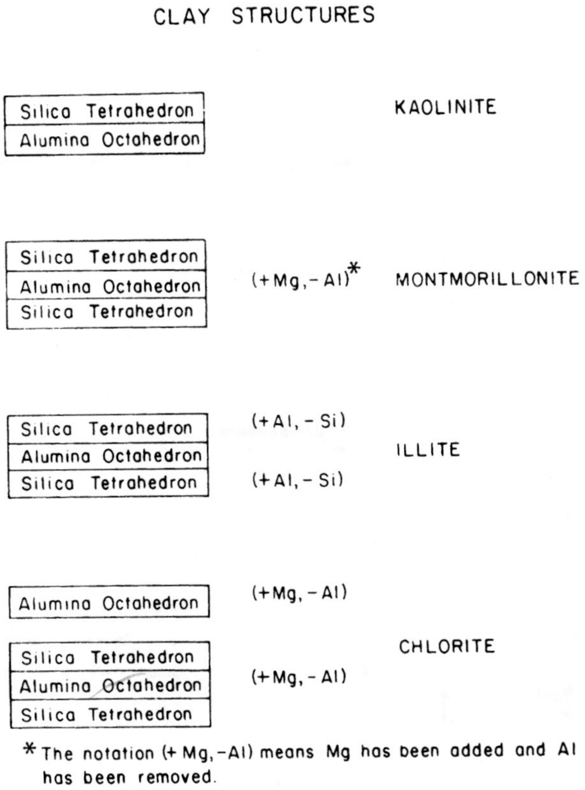

CLAY STRUCTURES

Fig. 2.19—Typical clays found in drilling muds.

Clay Type	Mud Density at 20 cp (lbm/gal)
Wyoming bentonite	8.67
High-yield clay	9.03
Low-yield native clay	11.38

2.3.4 Solids Control for Unweighted Muds. Several strings of steel casing may have to be cemented in the well as drilling progresses to complete the drilling operation successfully. Since each string of casing requires a subsequent reduction in hole size, the first bit size used on a well is often relatively large. The volume of rock fragments generated by the bit per hour of drilling is given in consistent units by

$$V_s = \frac{\pi(1-\phi)d^2}{4} \frac{dD}{dt}, \quad \ldots\ldots\ldots\ldots\ldots (2.14)$$

where

V_s = the solids volume of rock fragments entering the mud,
ϕ = the average formation porosity,
d = the bit diameter, and
$\dfrac{dD}{dt}$ = the penetration rate of the bit.

The first few thousand feet of hole drilled in the U.S. gulf coast area usually has a diameter of about 15 in. and is drilled in excess of 100 ft/hr. Thus, for an average formation porosity of 0.25, V_s would be given approximately by

$$V_s = \frac{\pi(1-0.25)(15)^2}{4(231 \text{ in.}^3/\text{gal})(42 \text{ gal/bbl})}$$

$$\cdot (100)(12 \text{ in./ft}) = 16.4 \text{ bbl/hr}.$$

From Table 2.7 the average density of drilled solids is approximately 910 lbm/bbl. At 16.4 bbl/hr, this results in

$$910 \frac{\text{lbm}}{\text{bbl}} \times 16.4 \frac{\text{bbl}}{\text{hr}} \times \frac{1 \text{ ton}}{2,000 \text{ lbm}} = 7.5 \text{ tons/hr}.$$

Thus, the volume of drilled solids that must be removed from the mud can be quite large.

The solids in a mud often are classified as either inert or active. The inert solids are those that do not hydrate or otherwise react with other components of the mud. The inert solids include such minerals as sand, silt, limestone, feldspar, and API barite. With the exception of API barite, which is used to increase the mud density, these inert solids usually are considered undesirable in a mud. They increase the frictional pressure drop in the fluid system but do not greatly increase the ability to carry the rock cuttings to the surface. The filter cake formed from these solids is thick and permeable rather than thin and relatively impermeable. This has a direct bearing on many drilling problems including stuck pipe, excessive pipe torque and drag, loss of circulation, and poor cement bonding to the formation.

There are four basic methods used to prevent the concentration of solids in the mud from increasing to an undesirable level. These are (1) screening, (2) forced settling, (3) chemical flocculation, and (4) dilution. The particle-size range for both the desirable and undesirable solids in the mud and the particle-size range that can be rejected by screening and forced settling are shown in Fig. 2.20. Screening always is applied first in processing the annular mud stream. Recent developments in screening equipment have made possible the use of extremely fine screens. This allows the removal of most of the solids before their size has been reduced to the size of the API barite particles. API specifications for commercial barium sulfate require that 97% of the particles pass through a 200-mesh screen. A 200-mesh screen has 200 openings per inch. Particles less than about 74 microns in diameter will pass through a typical 200-mesh screen. Screen sizes below 200 mesh cannot be used with weighted muds because of the cost of replacing the API barite discarded with the solids.

The natural settling rate of drilled solids is much too low for settling pits to be effective. Thus, devices such as hydrocyclones (Fig. 1.28) and centrifuges (Fig. 1.29) are used to increase the gravitational force acting on the particles (see Sec. 1.5). At present, both the hydrocyclones and high-speed centrifuges are being used as forced settling devices with unweighted muds. The cut point (Fig. 2.21) of a hydrocyclone is the particle size at which half the particles of that size are discarded. The rated cut points of several common hydrocyclones are shown in Table 2.9 and Fig. 2.22. Since the particle-size range of API barite is usually about 2 to 80 microns, hydrocyclones cannot be used with weighted muds unless they are

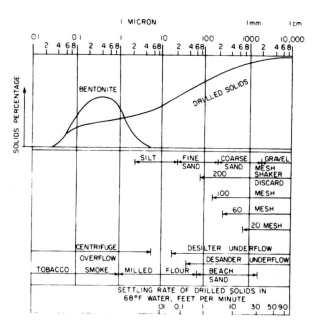

Fig. 2.20—Particle size range for common solids found in unweighted water-base muds (after Ref. 5).

used in series with a screen. Centrifuges that operate at high revolutions per minute and have a contoured bowl rather than a conical bowl have been developed for use on unweighted mud systems downstream of the small hydrocyclones. The contoured bowl increases the path length of the solids in the centrifuge and allows finer solids to be separated. The centrifuge overflow primarily contains solids less than 6 microns in diameter.

The removal of fine active clay particles can be facilitated by adding chemicals that cause the clay particles to *flocculate* or agglomerate into larger units. Once the agglomeration of the clay particles has been achieved, separation can be accomplished more easily by settling. Flocculation is discussed in more detail in Section 2.3.5.

The concentration of the solids not removed by screening or forced settling can be reduced by dilution. Because of the limited storage capacity of the active mud pits, dilution requires discarding some of the mud to the reserve pit. Dilution, thus, requires discarding a portion of the additives used in previous mud treatments. In addition, the new mud created by the addition of water must be brought to the desired density and chemical content. To keep the cost of dilution low, the mud volume should be kept small. Old mud should be discarded before dilution rather than after dilution. Also, the cost of a large one-step dilution is less than frequent small dilutions. The cost of dilution increases rapidly with mud density.

An example arrangement of the solids control equipment for an unweighted clay/water mud is shown in Fig. 2.23.[5] The various components are arranged in decreasing order of clay size removal to prevent clogging. Dilution water is introduced upstream of the hydrocyclones to increase their separation efficiency. Each device is arranged to prevent newly processed mud from cycling back to the input of the device. Chemical treatment normally is made downstream of all separation equipment.

TABLE 2.9—RATED CUT POINT OF HYDROCYCLONES

Hydrocyclone Size (in.)	Rated Cut Point (microns)
6	40
4	20
2	10

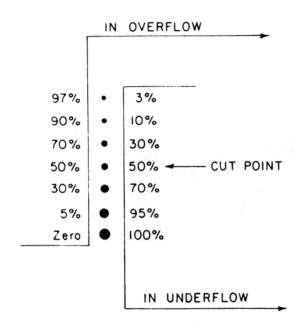

IN OVERFLOW

97%	•	3%
90%	•	10%
70%	•	30%
50%	•	50% ← CUT POINT
30%	●	70%
5%	●	95%
Zero	●	100%

IN UNDERFLOW

SEPARATION DIAGRAM
(Not to Scale)

Fig. 2.21—Solids distribution and removal using hydrocyclones.

TABLE 2.10—COMMON DEFLOCCULANTS USED TO LOWER YIELD POINT AND GEL STRENGTH

Deflocculant	pH of Deflocculant in a 10-wt% Solution	Optimal Mud pH	Approximate Maximum Effective Temperature (°F)
Phosphates		9.0	175
Sodium acid pyrophosphate	4.8		
Sodium hexa-metaphosphate	6.8		
Sodium tetraphosphate	7.5		
Tetra sodium pyrophosphate	10.0		
Tannins		11.5	250
Quebracho	3.8		
Alkaline tannate			
Hemlock tannin			
Desco			300
Lignins			400
Processed lignite	4.8		
Alkaline lignite	9.5		
Chrome lignite	10.0		
Lignosulfonates		10.0	350
Calcium lignosulfonate	7.2		
Chrome lignosulfonate	7.5		

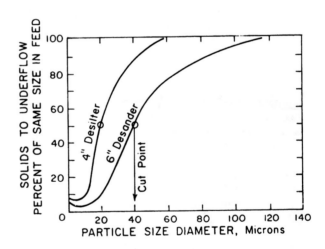

Fig. 2.22—Solids distribution and removal using hydrocyclones.

Example 2.13 A mud cup is placed under one cone of a hydrocyclone being used to process an unweighted mud. Thirty seconds were required to collect 1 qt of ejected slurry. The density of the slurry was determined using a mud balance to be 17.4 lbm/gal. Compute the mass of solids and volume of water being ejected by the cone per hour.

Solution. The density of the slurry ejected from the desilter can be expressed in terms of the volume fraction of low-specific-gravity solids by

$$\bar{\rho} = \frac{m_s + m_w}{V_s + V_w} = \frac{\rho_s V_s + \rho_w V_w}{V_s + V_w} = \rho_s f_s + \rho_w f_w.$$

Using the values given in Table 2.7, the average density of low-specific-gravity solids is 21.7 lbm/gal

and the density of water is 8.33 lbm/gal. Substituting these values in the above equation yields

$$17.4 = 21.7 f_s + 8.33(1 - f_s).$$

Solving for the volume fraction of solids gives

$$f_s = \frac{17.4 - 8.33}{21.7 - 8.33} = 0.6784.$$

Since the slurry is being ejected at a rate of 1 qt/30 s, the mass rate of solids is

$$0.6784 \frac{(1 \text{ qt})}{30 \text{ seconds}} \times \frac{\text{gal}}{4 \text{ qt}} \times \frac{21.7 \text{ lbm}}{\text{gal}}$$

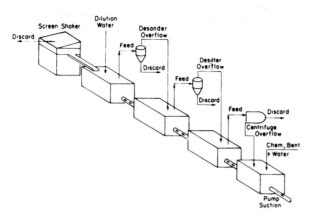

Fig. 2.23 – Schematic arrangement of solids-control equipment for unweighted mud systems.[5]

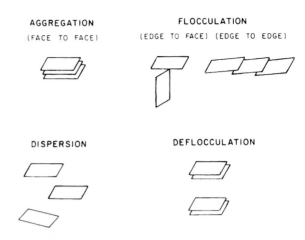

Fig. 2.24—Association of clay particles.[3]

$$\times \frac{3{,}600 \text{ seconds}}{\text{hr}} = 441.6 \frac{\text{lbm}}{\text{hr}} ,$$

and the volume rate of water ejected is

$$(1.0 - 0.6784) \frac{1 \text{ qt}}{30 \text{ seconds}} \times \frac{\text{gal}}{4 \text{ qt}}$$

$$\times 3{,}600 \text{ s/hr} = 9.65 \text{ gal/hr}.$$

Note that to prevent the gradual loss of water from the mud, 9.65 gal of water must be added each hour to make up for the water ejected by this single cone.

2.3.5 Chemical Additives.
Unweighted clay/water muds are controlled primarily by removing inert solids, diluting, and adding bentonite when required to keep the active solids at the proper concentration. However, after using all available methods of solids control, one or more of the mud properties may be at an undesirable value and require selective adjustment. Chemical additives commonly are used for (1) pH control, (2) viscosity control, and (3) filtrate control.

Caustic (NaOH) almost always is used to alter the mud pH. A high mud pH is desirable to suppress (1) corrosion rate, (2) hydrogen embrittlement, and (3) the solubility of Ca^{2+} and Mg^{2+}. In addition, the high pH is a favorable environment for many of the organic viscosity control additives. The pH of most muds is maintained between 9.5 and 10.5. An even higher pH may be used if H_2S is anticipated.

Flocculation refers to a thickening of the mud due to edge-to-edge and edge-to-face associations of clay platelets. These arrangements of platelets are shown in Fig. 2.24. Flocculation is caused by unbalanced electrical charges on the edge and surface of the clay platelets. When the mud is allowed to remain static or is sheared at a very low rate, the positive and negative electrical charges of different clay platelets begin to link up to form a "house of cards" structure.

The hydrated clay platelets normally have an excess of electrons and, thus, a net negative charge. Since like charges repel, this tends to keep the clay platelets dispersed. The local positive and negative charges on the edge of the clay platelets do not have a chance to link up. Anything that tends to overcome the repelling forces between clay platelets will increase the tendency of a mud to flocculate. The common causes of flocculation are (1) a high active solids concentration, (2) a high electrolyte concentration, and (3) a high temperature.

The concentration of flocculated particles in the mud is detected primarily by an abnormally high yield point and gel strength. At high shear rates, the "house of cards" structure is destroyed. The plastic viscosity, which describes fluid behavior at high shear rates, usually is not affected greatly by flocculation.

The normal range of plastic viscosity for an unweighted mud is from 5 to 12 cp measured at 120°F. The yield point is descriptive of the low shear rates present in the annulus and greatly affects the cuttings carrying capacity and annular frictional pressure drop. A yield point in the range of 30 to 30 lbf/100 sq ft often is considered acceptable for unweighted clay/water muds in large-diameter holes. This yield point range will enhance the ability of a mud to carry the cuttings to the surface without increasing the frictional pressure drop in the annulus enough to cause formation fracture. The gel strength is descriptive of the mud behavior when the pump is stopped. The gel strength of the mud prevents settling of the solids during tripping operations. However, an excessive gel requires a large pump pressure to be applied to start the fluid moving and could cause formation fracture. A *progressive* gel increases with time and is less desirable than a *fragile* gel (Fig. 2.25).

2.3.5.1 Deflocculants. Deflocculants (thinners) are materials that will reduce the tendency of a mud to flocculate. The deflocculants are thought to render ineffective the positive charges located on the edge of the clay platelets and, thus, destroy the ability of the platelets to link together. A large number of deflocculants are available. The common types are listed in Table 2.10. None of the deflocculants are totally effective against all causes of

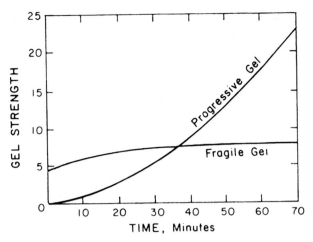

Fig. 2.25—Fragile and progressive gel strength.[3]

TABLE 2.11—pH OF 10% AQUEOUS SOLUTIONS (BY WEIGHT) OF CHEMICALS COMMONLY USED IN WATER BASE MUDS

Chemical	pH of Solution
Calcium hydroxide, $Ca(OH)_2$ (slaked lime)	12.0
Calcium sulfate, $CaSO_4 \cdot 2H_2O$ (gypsum)	6.0
Sodium carbonate, Na_2CO_3 (soda ash)	11.1
Sodium hydroxide, $NaOH$ (caustic soda)	12.9

flocculation. Since many of the deflocculants are acidic and only slightly soluble in the acid form, they must be used with caustic ($NaOH$) to increase the pH.

Any of the deflocculants can be used to lower yield point and gel strength when flocculation is caused by excessive solids. The phosphates are used for the unweighted clay/water muds that are not exposed to high salt concentrations and high temperature. Phosphates also are effective against flocculation caused by Ca^{2+} or Mg^{2+} ions. The phosphates remove the Ca^{2+} or Mg^{2+} by *sequestering* or tying up the ions much like the formation of a complex ion. Sodium acid pyrophosphate (SAPP) is excellent for treating flocculation caused by cement contamination. The Ca^{2+} ions are removed by sequestering, and the excessive mud pH is reduced at the same time.

The use of phosphates must be discontinued before a depth is reached at which the mud would be exposed to temperatures in excess of 175°F. At high temperatures, the phosphates revert to orthophosphates and become flocculants rather than deflocculants. The more temperature-stable tannins, lignites, or lignosulfonates are used in place of the phosphates at greater depths. However, in some cases, sodium acid pyrophosphates still are used for treating cement contamination.

Tannin extracts from the quebracho tree and hemlock tree are used as deflocculants. Quebracho was the most commonly used thinner during the 1950's. Tannins are effective against flocculation caused by moderate concentrations of Ca^{2+} or Cl^- but do not perform well at high electrolyte concentrations, especially at temperatures above 250°F. Tannins also are effective against cement contamination since they perform best at a high pH (above 11).

Lignites are obtained from a mineral mined in North Dakota called leonardite. The humic acid contained in the lignite functions as a deflocculant. Lignites remove calcium by precipitation. The more acidic processed lignite is effective against cement contamination since it also tends to reduce the pH. Concentrations of lignites above 10 lbm/bbl should

not be used. A rigid gel can form if calcium enters a mud having a high concentration of lignites. Alkaline lignite is a blend of one part caustic to six parts lignite and can be used when it is undesirable to add caustic separately. Chrome lignite is obtained by complexing lignite with chromates. Also, zinc sometimes is used in place of chrome. The heavy metal complexes are intended to add temperature stability. However, they are more expensive and are detrimental to filtration control.

Lignosulfonates are obtained from spent sulfite liquor generated in the manufacture of paper. They are basically sulfonated lignites complexed with chrome. Lignosulfonates are economical because they are obtained from a waste product. They are more soluble than lignites and are more effective than lignites at values of pH below 11. The pH of lignosulfonate-treated muds can be reduced to as low as 9.5 without greatly reducing the deflocculation action or causing lignosulfonate degradation. Lignosulfonates produce CO_2 as they degrade and can cause carbonate contamination over a long period of time. At temperatures above 350°F, H_2S also is liberated in small quantities.

2.3.5.2 pH of Mud Additives. When certain mud additives enter a mud they affect the pH of the system. Table 2.11 lists such additives.

2.3.6 Chemical Removal of Contaminants. The addition of chemical contaminants to the drilling fluid, either at the surface or through the wellbore, produces an imbalance in the chemical equilibrium of the fluid, which can cause serious rheological or drilling problems to develop. The common contaminants are calcium, magnesium, carbon dioxide, hydrogen sulfide, and oxygen.

2.3.6.1 Calcium. When calcium enters the mud, sodium montmorillonite will convert to calcium montmorillonite, which first produces flocculation and eventually aggregation of the montmorillonite. Thus, it is often desirable to remove the calcium by chemical treatment.

Calcium can enter the mud in many different forms: soft cement, gypsum, saltwater flows, hard-water additions, or lime.

In most cases calcium is removed from the system by adding soda ash (Na_2CO_3), which forms insoluble calcium carbonate:

$$Ca^{2+} + 2OH^- + Na_2CO_3 \rightarrow CaCO_3\downarrow$$
$$+ 2Na^+ + 2OH^-.$$

If cement or lime gets into the mud, the pH usually rises to unacceptable levels because hydroxyl ions as well as calcium have been added. In such cases either SAPP (sodium acid pyrophosphate – $Na_2H_2P_2O_7$) or sodium bicarbonate ($NaHCO_3$) usually is added.

When SAPP is added,

$$2Ca^{2+} + 4OH^- + Na_2H_2P_2O_7 \rightarrow Ca_2P_2O_7\downarrow$$
$$+ 2Na^+ + 2OH^- + 2H_2O.$$

In this reaction, calcium is removed and the four hydroxyl ions on the left side of the equation are reduced to two hydroxyl ions on the right side.

When sodium bicarbonate is added,

$$Ca^{2+} + 2OH^- + NaHCO_3 \rightarrow CaCO_3\downarrow + Na^+$$
$$+ OH^- + H_2O.$$

In this reaction, calcium is removed as a calcium carbonate precipitate.

2.3.6.2 Hard Water.
The term *hard water* refers to water containing calcium and magnesium ions. These ions produce unwanted calcium or magnesium montmorillonite in the mud and usually are removed by chemical treatment. Magnesium is removed first by the addition of sodium hydroxide (NaOH).

$$Mg^{2+} + 2NaOH \rightarrow Mg(OH)_2 + 2Na^+.$$

Additional caustic then is added to bring the pH to about 10.5. Calcium is removed, as listed above, by the addition of soda ash. Raising the pH to 10.5 decreases the solubility of $CaCO_3$ and, thus, increases the effectiveness of the soda ash treatment.

If the Ca^{2+} and Mg^{2+} content of the water is not known, the total hardness can be obtained using the hardness test described in Sec. 2.1.9. The Mg^{2+} ion then can be removed by treating with excess NaOH and the hardness test repeated to establish the Ca^{2+} concentration alone. The Mg^{2+} concentration then can be computed as the difference between the total hardness and the Ca^{2+} content. The Mg^{2+} usually is removed from the water first.

2.3.6.3 Carbon Dioxide.
Many formations drilled contain carbon dioxide, which when mixed with the mud can produce carbonate ions (CO_3^{2-}) and bicarbonate ions (HCO_3^-):

$$CO_2 + H_2O \rightleftharpoons H^+ + HCO_3^- \rightleftharpoons 2H^+ + CO_3^{2-}.$$

The presence of such ions produces a drilling fluid that has unacceptable filtration and gelation characteristics that cannot be corrected by normal chemical additive methods until the carbonate and bicarbonate ions are removed from the mud.

The carbonate and bicarbonate ions are removed from the mud by the addition of calcium hydroxide, which raises the pH, converts bicarbonate ions to carbonate ions, then precipitates the carbonate ions as insoluble calcium carbonate:

$$CO_3^{2-} + 2Na^+ + Ca(OH)_2 \rightarrow CaCO_3\downarrow + 2NA^+$$
$$+ 2OH^-.$$

Note that in the previous case calcium was removed by the addition of the carbonate ion and in this case the carbonate ion is removed by the addition of calcium. The presence of calcium is preferred over the presence of carbonate; therefore, most operators maintain a 50 to 75 ppm calcium level in water-base muds where the carbonate ion has been known to cause a problem.

2.3.6.4 Hydrogen Sulfide.
When hydrogen sulfide enters the mud it must be removed to prevent hydrogen embrittlement of the steel or harm to personnel working at the surface. The products produced when hydrogen sulfide enters the mud are

$$H_2S \rightleftharpoons H^+ + HS^-$$

and

$$HS^- \rightleftharpoons H^+ + S^{2-}.$$

It is thought that the hydrogen ion, H^+, penetrates into the high-strength steels and forms pockets of H_2 gas that expand and cause the pipe to become brittle and fail. One method of treatment is to keep the pH high and, thereby, form sodium bisulfite (NaHS) and sodium sulfide (Na_2S):

$$H_2S + NaOH \rightarrow NaHS + H_2O$$

and

$$NaHS + NaOH \rightarrow Na_2S + H_2O.$$

If the pH is allowed to drop, the sulfides revert to H_2S, which is foul smelling and deadly.

A common method to remove the sulfide ion is by the addition of zinc ions to the mud in the form of basic zinc carbonate:

$$S^{-2} + Zn^{+2} \rightarrow ZnS\downarrow.$$

2.3.6.5 Oxygen.
The presence of oxygen in the drilling fluid causes an acceleration of corrosion rates that can produce pitting of the pipe and loss of excessive amounts of metal. Detection of such adverse corrosion conditions usually is determined by use of corrosion rings placed in the tool joints of the drillpipe and by galvanic probes placed in the stand pipe on the rig floor.

Most oxygen enters the mud in the surface pits. Rate of entrainment can be reduced greatly by proper inspection and operation of mud-stirring and other surface equipment. In some cases it may be necessary to remove the oxygen from the mud as sodium sulfate. This is accomplished by the addition of sodium sulfite at the suction of the mud pumps:

$$O_2 + 2Na_2SO_3 \rightarrow 2Na_2SO_4\downarrow.$$

**TABLE 2.12—QUANTITY OF TREATING AGENT PER BARREL TREATED (lbm/bbl)
NEEDED FOR EACH PPM OF CONTAMINANT**

Contaminant	Contaminating Ion	To Remove Add	Amount of Treating Agent to Add to Remove 1 mg/L Contaminant (lbm/bbl)
Gypsum or anhydrite	calcium (Ca^{2+})	soda Ash if pH okay	0.000928 lbm/bbl/mg/L
		SAPP if pH too high	0.000971 lbm/bbl/mg/L
		sodium bicarbonate if pH too high	0.00147 lbm/bbl/mg/L
Cement	calcium (Ca^{2+})	SAPP or	0.000971 lbm/bbl/mg/L
	hydroxide (OH^-)	sodium bicarbonate	0.00147 lbm/bbl/mg/L
Lime	calcium (Ca^{2+})	sodium bicarbonate	0.00147 lbm/bbl/mg/L
		SAPP or	0.000535 lbm/bbl/mg/L
	hydroxide (OH^-)	sodium bicarbonate	0.000397 lbm/bbl/mg/L
Hard water	magnesium (Mg^{2+})	caustic soda to pH 10.5 then add soda ash	0.00116 lbm/bbl/mg/L
	calcium (Ca^{2+})	soda ash	0.000928 lbm/bbl/mg/L
Hydrogen sulfide	sulfide (S^{2-})	keep pH above 10 and add zinc basic carbonate	0.00123 lbm/bbl/mg/L
Carbon dioxide	carbonate ($CO_3{}^{2-}$)	gypsum if pH okay	0.00100 lbm/bbl/mg/L
		lime if pH too low	0.000432 lbm/bbl/mg/L
	bicarbonate ($HCO_3{}^-$)	lime	0.000424 lbm/bbl/mg/L

2.3.6.6 Concentration of Chemicals for Contamination Removal. After it has been determined that a certain chemical contaminant has entered the mud and the type of chemical to be added (i.e., treating agent) has been evaluated, the final step is to decide how much agent to add.

A basic chemical principle is that for complete chemical removal of contaminants it takes a concentration of treating agent, expressed in equivalent weights per unit volume, equal to the concentration of contaminant, also expressed in equivalent weights per unit volume.

If $[A]$ denotes the concentration of treating agent in mole/L and $[C]$ denotes the concentration of contaminant in mole/L,

$$[A]V_a = [C]V_c , \quad\dots\dots\dots\dots\dots\dots (2.15)$$

where V_a and V_c are the effective valences exhibited by Agent A and Contaminant C, respectively, in the chemical reaction involved. Using this concept, Table 2.12 was prepared. This table lists the amount of treating agent required to remove 1 mg/L from 1 bbl of mud for various contaminants present.

Example 2.14. A titration test has shown that a drilling mud contains 100 mg/L of calcium. The mud engineer plans to add enough soda ash (Na_2CO_3) to his 1,500-bbl system to reduce the calcium concentration to 50 mg/L. Determine the amount of soda ash he should add to each barrel of mud for each mg/L of calcium present in the mud. Also determine the total mass of soda ash needed in the desired treatment.

Solution. Table 2.2 shows calcium to have a molecular weight of 40 and a valence of two. Thus, for a desired change in contaminant concentration of 1 mg/L,

$$\Delta[Ca^{2+}] = \frac{1.0 \text{ mg/L}}{40 \text{ g/g mole} (1{,}000 \text{ mg/g})}$$

$$= 2.5 \times 10^{-5} \text{ g mole/L}$$

Since Ca^{2+} has a valence of two, the concentration expressed as a normality is

$$\Delta[Ca^{2+}]V_c = 2.5 \times 10^{-5}(2)$$

$$= 5.0 \times 10^{-5} \text{ gew/L}$$

The reaction of interest is

$$Ca^{2+} + CO_3{}^{2-} \rightarrow CaCO_3\downarrow$$

and the concentration of treating agent $CO_3{}^{2-}$ is given by Eq. 2.15:

$$\Delta[CO_3{}^{2-}]V_a = \Delta[Ca^{2+}]V_c$$

$$= 5.0 \times 10^{-5} \text{ gew/L}$$

Solving for $\Delta[CO_3{}^{2-}]$ with V_c equal to 2 yields

$$\Delta[CO_3{}^{2-}] = 2.5 \times 10^{-5} \text{ g mole/L}.$$

The molecular weights of Na, C, and O are 23, 12, and 16, respectively. Thus, the concentration of Na_2CO_3 needed is 2.5×10^{-5} g mole/L $\cdot [2(23) + 12 + 3(16)]$ g/g mole $= 2.65 \times 10^{-3}$ g/L.

Recall that one equivalent barrel for pilot testing is 350 mL. The Na_2CO_3 concentration expressed in gm/eq. bbl or lbm/bbl is given by

$$2.65 \times 10^{-3} \text{ g/L} \cdot \frac{0.35 \text{ L}}{\text{eq. bbl}}$$

$$= 0.000928 \text{ g/eq. bbl.}$$

Thus, a treatment of 0.000928 lbm of Na_2CO_3 per barrel of mud per mg/L change in Ca^{2+} concentration is indicated. Note that this value is already given in Table 2.12. To reduce the Ca^{2+} concentration of 1,500 bbl of mud from 100 mg/L to 50 mg/L requires the addition of

$$0.000928 \, (1,500)(100-50)$$

$$= 69.6 \text{ lbm of } Na_2CO_3.$$

2.3.7 Filtration Control.

Several types of materials are used to reduce filtration rate and improve mud cake characteristics. Since filtration problems usually are related to flocculation of the active clay particles, the deflocculants also aid filtration control. When clay cannot be used effectively, water-soluble polymers are substituted. The common water-soluble polymers used for filtration control are (1) starch, (2) sodium carboxymethylcellulose, and (3) sodium polyacrylate. Polymers reduce water loss by increasing the effective water viscosity.

Starch, unlike clay, is relatively unaffected by water salinity or hardness. It is primarily used in muds with high salt concentrations. Since thermal degradation begins at about 200°F, it cannot be used in muds exposed to high temperatures. Also, it is subject to bacterial action and must be used with a preservative except in saturated saltwater muds or muds with a pH above 11.5.

Sodium carboxymethylcellulose (CMC) can be used at temperatures up to approximately 300°F but is less effective at salt concentrations above 50,000 ppm. CMC tends to deflocculate clay at low concentrations and, thus, lowers gel strength and yield point in addition to water loss. CMC polymers are available in a number of grades. In order of increasing effectiveness, these grades are technical, regular, and high viscosity.

Sodium polyacrylate is even more temperature-stable than CMC but is extremely sensitive to calcium. At low concentrations it is a flocculant and must be added at concentrations above 0.5 lbm/bbl to act as a filtration control agent. Both calcium and clay solids must be kept at a minimum when using this type polymer.

All of the above filtration control additives have limitations at high temperatures, salinity, or hardness, and also increase mud viscosity. Lignites and modified lignites have been used extensively as thinners and filtration control agents. Lignite products are designed to take advantage of the excellent temperature stability of lignite molecules. Lignite is a low rank of coal between peat and subbituminous. The major active component is humic acid, which is not well-defined chemically.

Humic acid is a mixture of high-molecular-weight polymers containing aromatic and heterocyclic structures with many functional groups such as carboxylic acid groups. Humic acid also can chelate calcium or magnesium ions in mud. Ground, causticized, chromium-treated, sulfonated, or polymer-reacted lignites are used to reduce fluid loss without increasing viscosity.

Example 2.15. The unweighted freshwater mud in the pit appears too viscous, and several diagnostic tests are performed to determine the problem. The test results are as follows.

Density, lbm/gal	9.0
Plastic viscosity, cp	10
Yield point, lbf/sq ft	90
pH	11.5
API filtrate, cm^3	33
Cake thickness, in.	3/32
Chloride, ppm	200
Methylene blue capacity	5
Temperature, °F	120
Estimated bottomhole temperature, °F	145

Using the results of the test, prescribe any additional test that should be performed or a recommended mud treatment. Refer to Figs. 2.30 and 2.31 for normal acceptable ranges of plastic viscosity and yield point.

Solution. The test results given show a normal plastic viscosity but an abnormally high yield point and API filtrate for an unweighted mud. Thus, the mud is flocculated. The normal plastic viscosity indicates the concentration of inert solids in the mud is not excessive, and this is verified by the fact that the mud density is within the normal range for an unweighted mud. The common causes of flocculation are (1) a high active solids or clay concentration, (2) a high electrolyte concentration, and (3) a high temperature. A methylene blue capacity of 5 indicates the active clay content is roughly equivalent to 25 lbm/bbl sodium montmorillonite, which is not excessive. In addition, the temperature of the mud is not excessive. Thus, the cause of the flocculation is probably a high electrolyte concentration. Common sources of a high electrolyte concentration in the mud are (1) salt (Na^+ and Cl^-), (2) anhydrite or gypsum (Ca^{2+} and SO_4^{2-}), (3) cement (Ca^{2+} and OH^-), and (4) acid-forming gases (H_2S and CO_2). The low salinity indicates the absence of salt while the abnormally high pH indicates the absence of acid-forming gases and probable presence of cement or lime contamination.

In summary, the test results indicate that the mud is flocculated due to cement or lime contamination. One possible prescription would be to perform pilot tests using a deflocculant suitable for treating these contaminants. Since the bottomhole temperature is low, sodium acid pyrophosphate would be suitable. This deflocculant is known to be effective at concentrations as low as 0.5 lbm/bbl for treating cement contamination. The acidic nature of this deflocculant also tends to reduce the high pH of the contaminated mud. A filtra-

TABLE 2.13—COMMONLY USED LOST CIRCULATION ADDITIVES

Material	Type	Description	Concentration (lbm/bbl)	Largest Fracture Sealed (in.)
Nut shell	granular	50%—$^3/_{16}$-10 mesh 50%—10-100 mesh	20	0.25
Plastic	granular	50%—10-100 mesh	20	0.25
Limestone	granular	50%—10-100 mesh	40	0.12
Sulfur	granular	50%—10-100 mesh	120	0.12
Nut shell	granular	50%—10-16 mesh 50%—10-100 mesh	20	0.12
Expanded perlite	granular	50%—$^3/_{16}$-10 mesh 50%—10-100 mesh	60	0.10
Cellophane	lamellated	$^3/_4$-in. flakes	8	0.10
Sawdust	fibrous	$^1/_4$-in. particles	10	0.10
Prairie hay	fibrous	$^1/_2$-in. fibers	10	0.10
Bark	fibrous	$^3/_8$-in. fibers	10	0.07
Cotton seed hulls	granular	fine	10	0.06
Prairie hay	fibrous	$^3/_8$-in. particles	12	0.05
Cellophane	lamellated	$^1/_2$-in. flakes	8	0.05
Shredded wood	fibrous	$^1/_4$-in. fibers	8	0.04
Sawdust	fibrous	$^1/_{16}$-in. particles	20	0.02

tion control additive also may be required to reduce the API filtration loss. This can be determined using pilot tests.

2.3.8 Lost Circulation Control. *Lost circulation* material is added to a mud to control loss of mud into highly permeable sandstones, natural fractures, cavernous formations, and induced fractures. Before a mud filter cake can be deposited, lost circulation additives must bridge across the large openings and provide a base upon which the mud cake can be built.

A list of commonly used lost circulation additives is given in Table 2.13. Note that the largest fracture or opening that may be sealed is 0.25 in. Larger openings must be sealed using special techniques.

2.3.9 Clay Extenders. Good filtration characteristics often are not required when drilling hard, competent, low-permeability formations. When high annular velocities can be developed by the pump, clear water frequently is used as the drilling fluid. In larger hole sizes, high annular velocities are not achieved easily and the effective viscosity in the annulus must be increased to improve the carrying capacity of the drilling fluid. The use of common grades of commercial clay to increase viscosity can cause a large decrease in the drilling rate. To obtain a high viscosity at a much lower clay concentration, certain water-soluble vinyl polymers called *clay extenders* can be used. In addition to increasing the yield of sodium montmorillonite, clay extenders serve as flocculants for other clay solids. The flocculated solids are much easier to separate using solids control equipment.

The vinyl polymers are thought to increase viscosity by adsorbing on the clay particles and linking them together. The performance of commercially available polymers varies greatly as a result of differences in molecular weight and degree of hydrolysis. Also, since vinyl polymers act through adsorption, their effectiveness is reduced greatly in brackish or hard water. However, it is not un-common to double the yield of commercial clays such as Wyoming bentonite using clay extenders in fresh water. The API filtration rate is also approximately twice that which would be obtained using a conventional clay/water mud having the same apparent viscosity (measured at 600 rpm).

2.3.10 Clay Flocculants. Certain polymers, when added to a mud, increase the electrochemical attractive forces between particles and produce some degree of aggregation. This, in turn, reduces the viscosity of the mud. Polymers that act in this manner are the partially hydrolyzed polyacrylates of which Cyfloc™ is an example.

2.3.11 Clay Substitutes. A recently developed biopolymer produced by a particular strain of bacteria is becoming widely used as a substitute for clay in low-solids muds. Since the polymer is attacked readily by bacteria, a bactericide such as paraformaldehyde or a chlorinated phenol also must be used with the biopolymer. As little as 0.5 lbm/bbl biopolymer can cause a large increase in the apparent viscosity observed at shear rates representative of annular flow conditions. At higher shear rates representative of conditions in the drillstring, a much lower apparent viscosity is observed. This shear thinning permits a better hydraulic cleaning action beneath the bit and allows the solids control equipment to operate more efficiently. The biopolymer is not affected greatly by salinity and can be used in saturated salt water.

A fine grade of asbestos also is used as a clay substitute. The asbestos fibers build a tangled structure much like attapulgite, but relatively smaller amounts are required. Concentrations as low as 4 lbm/bbl commonly are used. Since asbestos produces viscosity primarily by mechanical entanglement, it is not affected by salinity. Asbestos muds have shear thinning characteristics giving a low viscosity at high shear rates. Like most low solids muds, a high API filtration rate is obtained unless a filtration control additive such as starch or CMC is used. Asbestos muds tend to settle but are remixed easily by pumping or stirring. However, it must be noted that certain species of asbestos are regarded as carcinogenic and are not used.

2.3.12 Density Control Additives.

Barium sulfate is the primary additive used to increase the density of clay/water muds. Densities ranging from 9 to 19 lbm/gal can be obtained using mixtures of barium sulfate, clay, and water. The specific gravity of pure barium sulfate is 4.5, but the commercial grade used in drilling fluids (API barite) has an average specific gravity of about 4.2.

Recently, alternative density-control agents such as hematite (Fe_2O_3) with specific gravity ranging from 4.9 to 5.3 and ilmenite ($FeO \cdot TiO_2$), with specific gravity ranging from 4.5 to 5.11 have been introduced. These materials are heavier and about twice as hard as conventional barite. The increased densities of these materials will decrease the weight/volume requirements in the mud. Since they are harder than for barite, the particle-size attrition rate will be significantly less than barite. However, the increased hardness raises the question of how abrasive these materials would be in the circulating system.

Galena (PbS), a dense mineral having a specific gravity of 7.5, can be used for special drilling problems that require a mud density greater than 19 lbm/gal. However, the formation pore pressure never requires a mud weight in excess of 19 lbm/gal during normal drilling operations since this is the average weight of the minerals and fluids present in the earth's crust. The hardness of galena is about the same as for barite.

The ideal mixing calculations presented in Sec. 2.2 accurately describe the addition of weight material to clay/water muds. These calculations allow the determination of the amount of mud and weight material required to obtain a specific volume of mud having a specified density. When excess storage capacity is not available, the density increase will require discarding a portion of the mud. In this case, the proper volume of old mud should be discarded before adding weight material. For ideal mixing the volume of mud, V_1, and weight material, V_B, must sum to the desired new volume, V_2:

$$V_2 = V_1 + V_B = V_1 + \frac{m_B}{\rho_B}.$$

Likewise, the total mass of mud and weight material must sum to the desired density-volume product:

$$\rho_2 V_2 = \rho_1 V_1 + m_B.$$

Solving these simultaneous equations for unknowns V_1 and m_B yields

$$V_1 = V_2 \frac{(\rho_B - \rho_2)}{(\rho_B - \rho_1)}, \quad \dots \dots \dots \dots \dots (2.16)$$

and

$$m_B = (V_2 - V_1)\rho_B. \quad \dots \dots \dots \dots \dots (2.17)$$

When the final volume of mud is not limited, the final volume can be calculated from the initial volume by rearranging Eq. 2.16:

$$V_2 = V_1 \frac{(\rho_B - \rho_1)}{(\rho_B - \rho_2)}. \quad \dots \dots \dots \dots \dots (2.18)$$

Example 2.16. It is desired to increase the density of 200 bbl of 11-lbm/gal mud to 11.5 lbm/gal using API barite. The final volume is not limited. Compute the weight of API barite required.

Solution. From Table 2.7 the density of API barite is 35.0 lbm/gal. Using Eq. 2.18, the final volume V_2 is given by

$$V_2 = V_1 \frac{(\rho_B - \rho_1)}{(\rho_B - \rho_2)} = 200 \frac{(35.0 - 11.0)}{(35.0 - 11.5)}$$
$$= 204.255 \text{ bbl}.$$

Using Eq. 2.17, the weight of API barite required is given by

$$m_B = (V_2 - V_1)\rho_B = (204.255 - 200)(42)(35)$$
$$= 6,255 \text{ lbm}.$$

The addition of large amounts of API barite to the drilling fluid can cause the drilling fluid to become quite viscous. The finely divided API barite has an extremely large surface area and can adsorb a significant amount of free water in the drilling fluid. This problem can be overcome by adding water with the weight material to make up for the water adsorbed on the surface of the finely divided particles. However, this solution has the disadvantage of requiring additional weight material to achieve a given increase in mud density because the additional water tends to lower the density of the mixture. Thus, it is often desirable to add only the minimum water required to wet the surface of the weight material. The addition of approximately 1 gal of water per 100 lbm of API barite is usually sufficient to prevent an unacceptable increase in fluid viscosity. Including a required water volume per unit mass of barite, V_{wB}, in the expression for total volume yields

$$V_2 = V_1 + V_B + V_w = V_1 + \frac{m_B}{\rho_B} + m_B V_{wB}.$$

Likewise, including the mass of water in the mass balance expression gives

$$\rho_2 V_2 = \rho_1 V_1 + m_B + \rho_w m_B V_{wB}.$$

Solving these simultaneous equations for unknowns V_1 and m_B yields

$$V_1 = V_2 \left[\frac{\rho_B \left(\frac{1 + \rho_w V_{wB}}{1 + \rho_B V_{wB}} \right) - \rho_2}{\rho_B \left(\frac{1 + \rho_w V_{wB}}{1 + \rho_B V_{wB}} \right) - \rho_1} \right], \quad \dots (2.19)$$

and

$$m_B = \frac{\rho_B}{1 + \rho_B V_{wB}} (V_2 - V_1). \quad \dots \dots \dots (2.20)$$

Example 2.17. It is desired to increase the density

of 800 bbl of 12-lbm/gal mud to 14 lbm/gal. One gallon of water will be added with each 100-lbm sack of API barite to prevent excessive thickening of the mud. A final mud volume of 800 bbl is desired. Compute the volume of old mud that should be discarded and the mass of API barite to be added.

Solution. The water requirement for barite, V_{wB}, is 0.01 gal/lbm. The initial volume of 12-lbm/gal mud needed can be computed using Eq. 2.19. For a final volume of 800 bbl, V_1 is given by

$$V_1 = 800 \left\{ \frac{35.0 \left[\frac{1 + 8.33(0.01)}{1 + 35.0(0.01)} \right] - 14.0}{35.0 \left[\frac{1 + 8.33(0.01)}{1 + 35.0(0.01)} \right] - 12.0} \right\}$$

$$= 700.53 \text{ bbl.}$$

Thus, 99.47 bbl of mud should be discarded before adding any API barite. Using Eq. 2.20, the mass of API barite needed is given by

$$m_B = \frac{35.0}{1 + 35.0(0.01)} (800 - 700.53)(42)$$

$$= 108,312 \text{ lbm.}$$

The volume of water to be added with the barite is

$$0.01 \, m_B = 1,083 \text{ gal or } 25.79 \text{ bbl.}$$

Considerable savings can be achieved in mud treatment costs for a weighted mud system if (1) the total mud volume is held as low as possible and (2) the initial conversion from an unweighted system to a weighted system is done with the low-specific-gravity solids at the minimum possible concentration. The reduction in solids concentration by dilution is much less expensive before the conversion is made because the mud discarded during dilution does not contain any API barite. When mud dilution just precedes weighting operations, the proper volume of old mud, V_1, dilution water, V_w, and mass of weight material, m_B, that should be combined to obtain a desired volume of new mud, V_2, can be determined using the ideal mixing calculations. For ideal mixing, the total new volume is given by

$$V_2 = V_1 + V_w + \frac{m_B}{\rho_B}$$

Similarly, a mass balance on the mixing procedure requires

$$\rho_2 V_2 = \rho_1 V_1 + \rho_w V_w + m_B.$$

A third equation can be obtained by performing a volume balance on the low-specific-gravity solids. The volume of low-specific-gravity solids can be expressed in terms of the present volume fraction, f_{c1} and the desired new volume fraction, f_{c2}:

$$f_{c2} V_2 = f_{c1} V_1$$

Solving these three equations simultaneously yields

$$V_1 = V_2 \frac{f_{c2}}{f_{c1}}, \quad \dots\dots\dots\dots\dots (2.21)$$

$$V_w = \frac{(\rho_B - \rho_2)V_2 - (\rho_B - \rho_1)V_1}{(\rho_B - \rho_w)}, \quad \dots\dots (2.22)$$

and

$$m_B = (V_2 - V_1 - V_w)\rho_B. \quad \dots\dots\dots\dots (2.23)$$

Example 2.18. After cementing casing in the well, it is desirable to increase the density of the 9.5-lbm/gal mud to 14 lbm/gal before resuming drilling operations. It also is desired to reduce the volume fraction of low-specific-gravity solids from 0.05 to 0.03 by dilution with water. The present mud volume is 1,000 bbl, but a final mud volume of 800 bbl is considered adequate. Compute the amount of original mud that should be discarded and the amount of water and API barite that should be added.

Solution. The initial volume of 9.5-lbm/gal mud needed can be computed using Eq. 2.21. For a final volume of 800 bbl, V_1 is given by

$$V_1 = 800 \frac{0.03}{0.05} = 480 \text{ bbl.}$$

Thus, 520 bbl of the initial 1,000 bbl should be discarded before adding any water or barium sulfate. Using Eq. 2.22, the volume of water needed is

$$V_w = \frac{(35.0 - 14.0) \, 800 - (35.0 - 9.5) \, 480}{(35.0 - 8.33)}$$

$$= 171 \text{ bbl.}$$

Using Eq. 2.23, the mass of API barite needed is given by

$$m_B = (800 - 480 - 171)(42)(35.0) = 219,000 \text{ lbm.}$$

2.3.13 Solids Control for Weighted Muds. The addition of solids for increasing density lowers the amount of inert formation solids that can be tolerated. The ideal composition of a weighted clay/water mud is (1) water, (2) active clay, and (3) inert weight material. Thus, every possible effort should be made to remove the undesirable low-gravity solids by screening before their particle size is reduced within the size range of the API barite particles. Hydrocyclones cannot be used alone on weighted systems because their cut point falls in the particle-size range of the API barite as shown in Fig. 2.26. However, they sometimes are used in conjunction with a shaker screen to increase the flow rate capacity of the solids removal equipment. A series arrangement of a hydrocyclone and a shaker screen is called a *mud cleaner*. It is suited best for muds of moderate density (below 15 lbm/gal). The fine solids that pass through the screen can be handled by dilution and deflocculation.

At higher densities, the mud cleaners are much less efficient. Much of the coarse solids in the mud remain in the liquid stream exiting the top of the unit and, thus, bypass the screen. Also, dilution requires

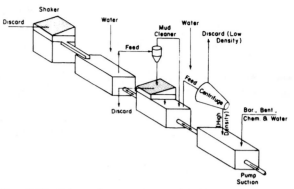

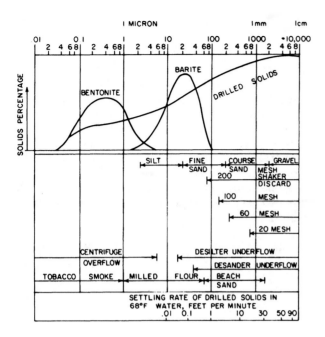

Fig. 2.26—Particle size range for common solids found in weighted water-base muds.

Fig. 2.27—Schematic arrangement of solids-control equipment for weighted mud systems (after Ref. 5).

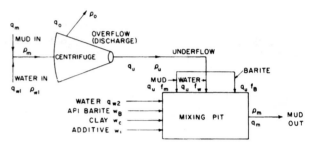

Fig. 2.28—Flow diagram of a centrifuge.

discarding a large volume of API barite with a portion of the old mud and the cost of dilution can become quite high. In this situation, centrifuges often are employed to separate the particles having sizes that fall in the API barite range from the liquid and extremely fine solids. In this manner the mud stream is divided into (1) a low-density overflow slurry (approximately 9.5 lbm/gal) and (2) a high-density slurry (approximately 23.0 lbm/gal). The high-density slurry is returned to the active mud system, and the low-density slurry usually is discarded. An example solids removal system for weighted clay/water muds is shown in Fig. 2.27.

About three-fourths of the bentonite and chemical content of the mud is discarded with the fine solids when the centrifuge is used. New bentonite and chemicals must be added to prevent depleting the mud. Also, since some of the API barite and drilled solids are discarded in the overflow, the volume of mud reclaimed from the underflow will be less than the volume of mud processed. A small additional volume of new mud must be built in order to maintain the total mud volume constant. A material balance calculation can be made to determine the proper amounts of API barite, clay, chemicals, and water required to reconstruct a barrel of mud that has been processed with a centrifuge.

2.3.13.1 Centrifuge Analysis.
Consider the flow diagram for a centrifuge shown in Fig. 2.28. Dilution water and mud enter the centrifuge and a high-density slurry (ρ_u) containing the API barite exits the underflow while a low-density slurry (ρ_o) containing the low-specific-gravity solids and most of the water

and chemicals exits the overflow.

Thus, the flow rate of the overflow is the sum of the mud flow rate and water flow rate into the centrifuge less the underflow rate:

$$q_o = q_m + q_{w1} - q_u. \quad \dots \dots \dots \dots \dots (2.24)$$

The total mass rate into the centrifuge is given by

$$\text{mass rate in} = q_{w1}\rho_w + q_m\rho_m.$$

Similarly, the total mass rate out of the centrifuge is given by

$$\text{mass rate out} = q_u\rho_u + q_o\rho_o.$$

For continuous operation, the mass rate into the centrifuge must equal the mass rate out of the centrifuge. Equating the expressions for mass rate in and mass rate out gives

$$q_{w1}\rho_w + q_m\rho_m = q_u\rho_u + q_o\rho_o. \quad \dots \dots (2.25)$$

Substitution of Eq. 2.24 for the overflow rate in Eq. 2.25 and solving for the underflow rate, q_u, gives the following equation.

$$q_u = \frac{q_m(\rho_m - \rho_o) - q_{w1}(\rho_o - \rho_w)}{(\rho_u - \rho_o)}. \quad \dots \dots (2.26)$$

This equation allows the calculation of the underflow rate from a knowledge of (1) water flow rate and mud flow rate into the centrifuge, (2) the densities of the water and mud entering the centrifuge and (3) the densities of the underflow and overflow slurries exiting the centrifuge.

The reconstruction of the mud from the centrifuge underflow occurs in a pit downstream of the centrifuge. It is desired to obtain a final mud flow rate from the mixing pit equal to the mud feed rate into the centrifuge. In addition, the final mud density should be equal to the density of the feed mud. A knowledge of volume fraction of feed mud, dilution water, and API barite in the underflow stream simplifies the calculations required for reconstruction of a mud with the desired properties.

Consider the underflow stream to be composed of (1) old mud, (2) dilution water, and (3) additional barite stripped from the discarded mud. The density of the underflow can be expressed by

$$\rho_u = \rho_m f_{um} + \rho_w f_{uw} + \rho_B f_{uB}, \quad \ldots\ldots\ldots (2.27)$$

where f_{um}, f_{uw}, and f_{uB} are the volume fraction of mud, dilution water, and API barite in the underflow stream. Furthermore, if we assume perfect mixing of the feed mud and dilution water in the centrifuge, then the diluted feed mud in the underflow should consist of feed mud and dilution water in the same ratio as they existed going into the centrifuge. From this assumption we obtain

$$f_{uw} = f_{um} \frac{q_{w1}}{q_m} . \quad \ldots\ldots\ldots\ldots\ldots (2.28)$$

Also, since all the volume fractions must sum to one, then f_{uB} can be expressed by

$$f_{uB} = 1 - f_{um} - f_{uw} = 1 - f_{um} - f_{um} \frac{q_{w1}}{q_m} . \quad \ldots (2.29)$$

Substituting these expressions for f_{uw} and f_{uB} into Eq. 2.27 and solving for f_{um} gives

$$f_{um} = \frac{(\rho_B - \rho_u)}{\rho_B - \rho_m + \frac{q_{w1}}{q_m}(\rho_B - \rho_w)} . \quad \ldots\ldots (2.30)$$

The flow rate of old mud, dilution water, and API barite to the pit from the centrifuge underflow are given by $q_u f_{um}$, $q_u f_{uw}$, and $q_u f_{uB}$, respectively.

The fraction of the underflow stream composed of old mud already has Wyoming bentonite, deflocculants, filtration control additives, etc., at the desired concentration. Thus, only sufficient additives to treat the remaining portion of the mud stream from the mixing pit are required. The fraction of the mud stream from the mixing pit composed of old mud is given by

$$f_m = \frac{q_u f_{um}}{q_m} .$$

If c_i is the desired concentration in pounds per barrel of a given additive in the mud stream, then this additive must be added to the mixing pit at the following mass rate.

$$w_i = c_i q_m (1 - f_m) = c_i (q_m - q_u f_{um}), \quad \ldots (2.31)$$

where q_m and q_u are expressed in barrels per unit time. This expression also can be used to obtain the mass rate at which commercial clay should be added.

The volume of water and API barite needed to maintain the density of the mud leaving the mixing pit at the same density as the mud entering the centrifuge can be determined through a balance of the material entering

and exiting the mixing pit. The total flow rate exiting the mixing pit can be expressed by

$$q_m = q_u + q_{w2} + \frac{w_B}{\rho_B} + \frac{w_c}{\rho_c} + \sum_{i=1}^{n} \frac{w_i}{\rho_i} .$$

Similarly, the mass rate of material exiting the mixing pit is given by

$$q_m \rho_m = q_u \rho_u + q_{w2} \rho_w + w_B + w_c + \sum_{i=1}^{n} w_i .$$

Solving these simultaneous equations for unknowns q_{w2} and w_B yields

$$q_{w2} = \left[q_m (\rho_B - \rho_m) - q_u (\rho_B - \rho_u) \right.$$
$$\left. - w_c \left(\frac{\rho_B}{\rho_c} - 1 \right) - \sum_{i=1}^{n} w_i \left(\frac{\rho_B}{\rho_i} - 1 \right) \right]$$
$$\div (\rho_B - \rho_w), \quad \ldots\ldots\ldots\ldots\ldots (2.32)$$

and

$$w_B = \left(q_m - q_u - q_{w2} - \frac{w_c}{\rho_c} - \sum_{i=1}^{n} \frac{w_i}{\rho_i} \right) \rho_B .$$
$$\ldots\ldots\ldots\ldots\ldots\ldots\ldots\ldots\ldots (2.33)$$

Example 2.19. A 16.2-lbm/gal mud is entering a centrifuge at a rate of 16.53 gal/min along with 8.33-lbm/gal dilution water which enters the centrifuge at a rate of 10.57 gal/min. The density of the centrifuge underflow is 23.4 lbm/gal while the density of the centrifuge overflow is 9.3 lbm/gal. The mud contains 22.5 lbm/bbl Wyoming bentonite and 6 lbm/bbl deflocculant. Compute the rate at which Wyoming bentonite, deflocculant, water, and API barite should be added to the mixing pit to maintain the mud properties constant.

Solution. The flow rate of the centrifuge underflow is given by Eq. 2.26:

$$q_u = \frac{16.53(16.2 - 9.3) - 10.57(9.3 - 8.33)}{(23.4 - 9.3)}$$
$$= 7.36 \text{ gal/min}.$$

The fraction of old mud in the centrifuge underflow is given by Eq. 2.30:

$$f_{um} = \frac{(35.0 - 23.4)}{35.0 - 16.2 + \frac{10.57}{16.53}(35.0 - 8.33)} = 0.324.$$

The mass rate of clay and deflocculant can be computed using Eq. 2.31:

$$w_c = \frac{22.5[16.53 - 7.36(0.324)]}{42 \text{ gal/bbl}}$$
$$= 22.5(0.337) = 7.58 \text{ lbm/min clay}.$$

$$w_d = 6 \times 0.337 = 2.02 \text{ lbm/min deflocculant}.$$

The water flow rate into the mixing pit is given by Eq. 2.32. Assuming a density of 21.7 lbm/gal for

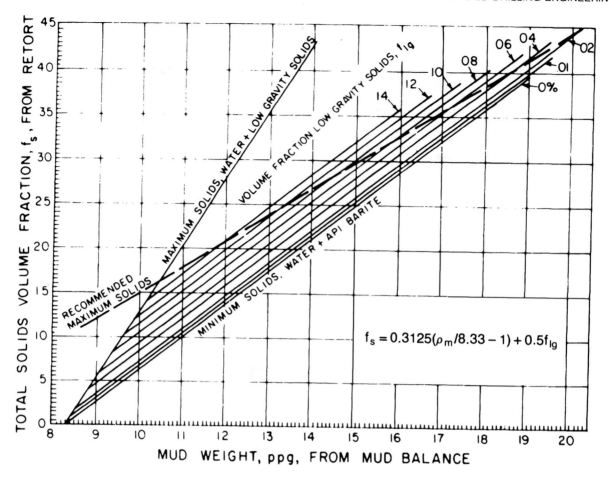

Fig. 2.29—Solids content vs. mud weight for freshwater muds.

both the clay and deflocculant and a density of 35 lbm/gal for the weight material gives

$$q_{w2} = \Big[16.53(35.0-16.2)-7.36(35.0-23.4)$$

$$-7.58\Big(\frac{35.0}{21.7}-1\Big)-2.02\Big(\frac{35.0}{21.7}-1\Big)\Big]$$

$$\div(35.0-8.33)=8.23 \text{ gal/min water.}$$

The mass rate of API barite into the mixing pit is given by Eq. 2.33:

$$w_B = \Big(16.53-7.36-8.23-\frac{7.58}{21.7}-\frac{2.02}{21.7}\Big)35.0$$

$$=17.4 \text{ lbm/min API barite.}$$

2.3.13.2 Solids Content Determination. Empirically established guidelines on the recommended minimum and maximum solids content of freshwater muds are given in Fig. 2.29. Within the limits shown, the optimal amount of solids is difficult to determine. Lowering the concentration of inert low-specific-gravity solids improves hole cleaning and penetration rate but at the same time requires discarding large quantities of expensive mud additives. Also, the use of deflocculants would permit a higher solids content

to be maintained and still have reasonable mud properties. Mud treatment usually is prescribed to achieve a given yield point, plastic viscosity, and filtration rate at the lowest possible mud-treatment cost. Typical ranges of acceptable plastic viscosities and yield points are shown in Figs. 2.30 and 2.31. The maximum yield point decreases with mud density because of the larger quantity of API barite that must be suspended to obtain the higher mud densities. The maximum suggested plastic viscosity increases with density because addition of API barite causes an increase in plastic viscosity and the cost of reducing viscosity increases with density.

The determination of both the low-specific-gravity and the high-specific-gravity solids content of the mud is an essential part of the proper interpretation of the API diagnostic tests. A solids content analysis generally is based on mud retort data and material balance considerations. From a mass balance we have

$$\rho_m = \rho_w f_w + \rho_{lg} f_{lg} + \rho_B f_B + \rho_o f_o, \quad \dots \dots (2.34)$$

where ρ_m is mud density, ρ_w is water density, f_w is the fraction of water present, ρ_{lg} is the density of the low-gravity solids, f_{lg} is the volume fraction of low-gravity solids, ρ_B is the density of the API barite, f_B is the volume fraction of API barite in the mud, ρ_o is the density of the oil in the mud, and f_o is the volume fraction of oil in the mud.

Using the relationship $f_B = 1 - f_w - f_{lg} - f_o$, Eq.

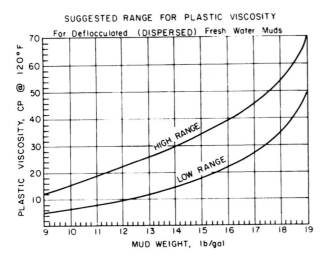

Fig. 2.30—Typical range of acceptable viscosities for clay/water muds.[5]

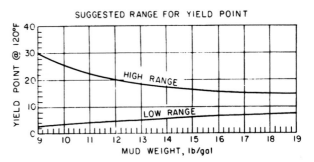

Fig. 2.31—Typical range of acceptable yield points for clay/water muds.[5]

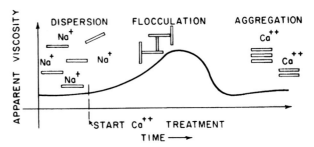

Fig. 2.32—Behavior of clay/water mud during lime treatment.

2.34 can be written

$$f_{lg} = \frac{\rho_w f_w + (1 - f_o - f_w)\rho_B + \rho_o f_o - \rho_m}{\rho_B - \rho_{lg}}$$

$$\dots\dots\dots\dots\dots\dots\dots\dots (2.35)$$

Example 2.20. Calculate the fraction of low-gravity solids and API barite for the 12-lbm/gal mud described in Example 2.8. Assume the oil is diesel oil. Recall that the water fraction, corrected for the effect of dissolved salt, is 0.773.

Solution. From Tables 2.3 and 2.7, the water specific gravity is 1.0857 and the oil specific gravity is 0.86. Using Eq. 2.35,

$$f_{lg} = \left[(1.0857)(0.773) + (1 - 0.06 - 0.773)4.2 \right.$$
$$\left. + (0.86)(0.06) - \frac{12}{8.33} \right] / (4.2 - 2.6) = 0.095,$$

and

$$f_B = 1 - f_{lg} - f_w - f_o$$
$$= 1 - 0.095 - 0.773 - 0.06$$
$$= 0.072.$$

For fresh water muds that do not contain any oil it is possible to determine the volume fraction of low-gravity solids using Fig. 2.29, after the total solids are determined using the mud retort and the mud weight is determined with the mud balance. The equation in Fig. 2.29 is determined using Eq. 2.34 and the relationships

$$f_w = 1 - f_s,$$

and

$$f_B = f_s - f_{lg},$$

where f_s is the volume fraction of all solids in the mud. Thus, using Eq. 2.34 gives

$$f_s = \frac{\rho_m + f_{lg}(\rho_B - \rho_{lg}) - \rho_w}{\rho_B - \rho_w}. \dots\dots\dots (2.36)$$

For the mud additives densities and metric units listed in Table 2.7, Eq. 2.36 becomes

$$f_s = \frac{\rho_m/8.33 + f_{lg}(4.2 - 2.6) - 1}{4.2 - 1}$$
$$= 0.3125(\rho_m/8.33 - 1) + 0.5 f_{lg}.$$

Also shown in Fig. 2.29 is the "recommended maximum solids" line. This line, specified by field experience, provides the maximum amount of solids to be tolerated in freshwater muds before major solids removal methods are undertaken.

Example 2.21. A 14-lbm/gal freshwater mud is reported to contain no oil and to have a total solids content of 28%, a plastic viscosity measured at 120°F of 32 cp, and a yield point measured at 120°F of 15 lbf/100 sq ft. Is this a good mud? If not, what steps should be taken?

Solution. From Fig. 2.29 we see that the maximum recommended solids for a 14-lbm/gal mud is 26.5%, well below the 28% reported. From Fig. 2.30 we see that the plastic viscosity for a 14-lbm/gal mud should not exceed 30 cp. The high plastic viscosity reported confirms the presence of excess solids. Steps should be taken to reduce the total solids content to 26.5%.

2.3.13.3 Quality of Low-Gravity Solids. Once the volume fraction of low-specific-gravity solids has been determined, it is desirable to know what approximate fraction of these low-gravity solids is active clay and what fraction is inert drilled solids. In Sec. 2.1.12, it was noted that the methylene blue test can be used to assist this determination. Recall that the cation exchange capacity is reported in

TABLE 2.14—SPECIFICATIONS FOR BARITES

Property	Specified API	OCMA Values
Minimum specific gravity	4.2	4.2
Wet screen analysis, wt%		
Maximum residue on No. 200 sieve	3	3
Minimum residue on No. 350 sieve	5	5
Maximum residue on No. 350 sieve	—	15
Maximum soluble metals or calcium, ppm	250	—
Maximum soluble solids, wt%	—	0.1
Performance of 2.5-g/mL barite, cp (apparent)	—	125
Water suspension after contamination with gypsum		viscosity at 600 rpm

TABLE 2.15—THEORETICAL SOLUBILITY OF Ca^{2+} AS A FUNCTION OF pH

pH	$[OH^-]$ (mol/L)	$[OH^-]^2$	$[Ca^{2+}]$ (mol/L)
11.0	0.0010	1.0×10^{-6}	1.3000
11.5	0.0032	1.0×10^{-5}	0.1300
12.0	0.0100	1.0×10^{-4}	0.0130
12.5	0.0316	1.0×10^{-3}	0.0013
13.0	0.1000	1.0×10^{-2}	0.0001

milliequivalent weights (meq) of methylene blue per 100 mL of mud and that the equivalent montmorillonite content of the mud in pounds mass per barrel is approximately five times the cation exchange capacity. While application of this simple rule of thumb is often sufficient when performing a mud solids analysis, it is sometimes desirable to refine the use of the cation exchange capacity data. This is accomplished by performing methylene blue tests on a sample of the bentonite being used and a sample of the drilled solids being encountered. These cation exchange capacity (Z_V) results are reported in meq of methylene blue per 100 g of solids. Typical results for several common solids are shown in Table 2.6.

The cation exchange capacity of the mud, $\overline{Z_{Vm}}$, reported in meq/100 mL of mud sample, can be predicted from a knowledge of the cation exchange capacity of the bentonite clay $\overline{Z_{Vc}}$, the bentonite clay fraction, f_c, the cation exchange capacity of the drilled solids, $\overline{Z_{Vds}}$, and the drill solids fraction, f_{ds}, using the following equation.

$$\overline{Z_{Vm}} = 100 \text{ mL} \left(f_c \rho_c \frac{\overline{Z_{Vc}}}{100} + f_{ds} \rho_{ds} \frac{\overline{Z_{Vds}}}{100} \right).$$

Since both ρ_c and ρ_{ds} are approximately 2.6 g/mL this equation reduces to

$$\overline{Z_{Vm}} = 2.6(f_c \overline{Z_{Vc}} + f_{ds} \overline{Z_{Vds}})$$

Substituting the identity

$$f_{ds} = f_{lg} - f_c$$

and solving for f_c yields

$$f_c = \frac{\overline{Z_{Vm}} - 2.6 f_{lg} \overline{Z_{Vds}}}{2.6(\overline{Z_{Vc}} - \overline{Z_{Vds}})}. \quad \ldots \ldots \ldots \ldots (2.37)$$

Eq. 2.37 allows the direct determination of the bentonite clay fraction from the methylene blue titration results on samples of mud, bentonite clay, and drilled solids and a knowledge of the total fraction of low-gravity solids.

Example 2.22. A mud retort analysis of a 15-lbm/gal

freshwater mud indicates a solids content of 29% and an oil content of zero. Methylene blue titrations of samples of mud, bentonite clay, and drilled solids indicates a $\overline{Z_{Vm}}$ of 7.8 meq/100 mL, a $\overline{Z_{Vc}}$ of 91 meq/100 g, and a $\overline{Z_{Vds}}$ of 10 meq/100 g. The mud has a plastic viscosity of 32 cp and yield point of 20 lbm/100 sq ft when measured at 120°F. Determine (1) the total fraction of low-gravity solids, (2) the volume fraction of bentonite, (3) the volume fraction of drilled solids, and (4) whether there is sufficient bentonite in the mud.

Solution.
1. From Fig. 2.29, $f_{lg} = 0.08$.
2. Using this value in Eq. 2.37 yields

$$f_c = \frac{7.8 - 2.6(0.08)(10)}{2.6(91 - 10)}$$

$$= 0.027.$$

3. $f_{ds} = f_{lg} - f_c = 0.08 - 0.027 = 0.053$.
4. For a volume fraction of 0.027, the bentonite content in lbm/bbl is

$$910(0.027) = 24.6 \text{ lbm/bbl},$$

which is a little high for a 15-lbm/gal mud. Figure 2.31 suggests such a claim since the yield point is slightly above the maximum recommended value.

2.4 Inhibitive Water-Base Muds

An inhibitive water-base mud is one in which the ability of active clays to hydrate has been reduced greatly. Inhibitive muds prevent formation solids from readily disintegrating into extremely small particles and entering the mud. They also partially stabilize the portions of the wellbore drilled through easily hydrated clays and shales. Inhibitive water-base muds are formed by the controlled contamination of a clay/water mud by electrolytes and/or deflocculants. Thus, they also are highly resistant to subsequent contaminants encountered during the drilling operation.

2.4.1 Calcium-Treated Muds. Lime-treated muds were among the first inhibitive muds. The beneficial effects of calcium-treated muds at a high pH were discovered largely by accident as a result of cement contamination.

The properties of a calcium-treated mud depend to a great extent on the amount of calcium that enters into solution. When Ca^{2+} and OH^- ions are introduced into an aqueous solution, they react to form $Ca(OH)_2$.

The solubility product, K_{sp}, for this reaction has a value of about 1.3×10^{-6}:

$$Ca^{2+} + 2OH^- \rightleftharpoons Ca(OH)_2.$$

$$K_{sp} = [Ca^{2+}][OH^-]^2 = 1.3 \times 10^{-6} \text{ (at } 70°F).$$

The *solubility principle* states that the product $[Ca^{2+}] \cdot [OH^-]^2$ must remain constant. If we represent the solubility of the Ca^{2+} ion by S, then

$$[Ca^{2+}] = S,$$
$$[OH^-] = 2S, \text{ and}$$
$$S \cdot (2S)^2 = 1.3 \times 10^{-6}.$$

Solving for the solubility, S, gives

$$S = \sqrt[3]{\frac{1.3 \times 10^{-6}}{4}} = 6.89 \times 10^{-3} \text{ mol/L.}$$

The concentration of OH^- ions in a saturated solution of lime is $2S = 1.38 \times 10^{-2}$. This corresponds to a pH given by

$$pH = -\log\left(\frac{1 \times 10^{-14}}{1.38 \times 10^{-2}}\right) = 12.14.$$

The solubility of Ca^{2+} can be altered greatly by changing the pH of the solution. For conditions at which the lime solubility reaction is predominant, the solubility product principle implies

$$[Ca^{2+}] = \frac{1.3 \times 10^{-6}}{[OH]^2} = \frac{1.3 \times 10^{-6}}{[\text{antilog (pH} - 14)]^2}$$

Using this relation, the solubility of Ca^{2+} as a function of pH has been tabulated in Table 2.15.

The chemistry of a lime-treated mud is more complicated than that of an aqueous solution of lime because of the presence of clays. However, the solubility of Ca^{2+} in a lime-treated mud still remains a function of the pH of the mud. Thus, the solubility of Ca^{2+} in a lime mud can be controlled by the addition of caustic (NaOH).

When Ca^{2+} enters a clay/water mud, the mud begins to flocculate and thicken. At the same time, a clay cation exchange reaction begins in which the Na^+ cations of the clay are replaced by Ca^{2+} cations. Calcium montmorillonite does not hydrate as extensively as sodium montmorillonite, and the clay platelets begin to stack closer together as the cation exchange reaction proceeds. This process is called aggregation and is marked by a thinning of the mud. The magnitude of the change in viscosity associated with flocculation and subsequent aggregation depends on the temperature and the concentration of solids and deflocculants in the mud.

The conversion to a lime mud generally is made in the wellbore during drilling operations. The high temperature and pressure environment of the well greatly aids the chemical conversion. Lime, caustic, and a deflocculant are added to the mud at a rate that will accomplish the conversion in about one circulation. A variety of deflocculants can be used. Quebracho, chrome lignosulfonate, calcium lignosulfonate, and processed lignite are all applicable. A water-soluble polymer usually is added during the second circulation to improve the filtration properties and stabilize the mud.

The amount of lime, caustic, and deflocculant added to make the conversion usually is determined using pilot tests. The average conversion is made using about 8 lbm/bbl lime, 3 lbm/bbl caustic, 3 lbm/bbl deflocculant, and 1 lbm/bbl polymer. An excess amount of lime present in the mud is desirable to prevent the slow depletion of Ca^{2+} concentration in solution by formation solids and mud dilution. A frequent practice is to maintain the lime content of the mud in pounds per barrel numerically equal to the P_f.

Calcium-treated muds sometimes are formed using gypsum ($CaSO_4 \cdot 2H_2O$) rather than lime [$Ca(OH)_2$]. Gypsum-treated muds are particularly well-suited for drilling massive gypsum or anhydrite formations. The solubility reaction for gypsum is given by

$$Ca^{2+} + SO_4^{2-} \rightleftharpoons CaSO_4.$$

$$K_{sp} = [Ca^{2+}][SO_4^{2-}] = 2.4 \times 10^{-5} \text{ (at } 70°F).$$

If we represent the solubility of Ca^{2+} by S,

$$[Ca^{2+}] = [SO_4^{2-}] = S,$$
$$S \cdot S = 2.4 \times 10^{-5}, \text{ and}$$
$$S = \sqrt{2.4 \times 10^{-5}} = 4.9 \times 10^{-3} \text{ mol/L.}$$

As shown in Table 2.15, a pH of approximately 12.3 would be required to reduce $[Ca^{2+}]$ to this value in a lime solution. A gypsum-treated mud offers the advantage of low $[Ca^{2+}]$ at lower values of pH. However, calcium solubility is not altered easily in a gypsum-treated mud. Chrome lignosulfonate is the deflocculant most often used, and the pH usually is maintained between 9.5 and 10.5.

High-lime muds cannot be used at temperatures above about 275°F. At high temperatures, a progressive gelation is experienced that requires high annular pressures to begin circulation after a trip. In extreme cases, a hard rigid cement is formed by a reaction between the caustic, calcium, and clays in the mud. In addition, the water-soluble polymers degrade rapidly at temperatures above 275°F and good filtration properties cannot be maintained.

2.4.2 Lignosulfonate-Treated Muds. Inhibitive water-base muds also can be formed by adding a large amount of deflocculant (thinner) to a clay/water mud. Deflocculant concentrations as high as 12 lbm/bbl are common in this type of mud. Chrome or ferrochrome lignosulfonate are the primary deflocculants used for this purpose because of effectiveness and relatively low cost. Chrome lignosulfonate has a high tolerance for both sodium chloride and calcium and will deflocculate at temperatures in excess of 300°F. However, temperature degradation quickly reduces the concentration of chrome lignosulfonate at temperatures above 300°F. Lignite often is used with the chrome lignosulfonate in

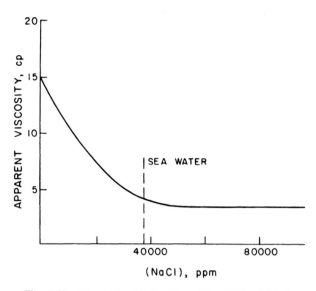

Fig. 2.33 – Viscosity obtained by adding 20 lbm/bbl of Wyoming bentonite to saline water.

TABLE 2.16—IONS PRESENT IN SEA WATER

Ion	Concentration (ppm)
Cations:	
Na^+	16,550
Mg^{2+}	1,270
Ca^{2+}	400
K^+	300
Anions:	
Cl^-	18,970
SO_4^{2-}	2,650
Br^-	65

inhibitive muds because of its ability to chelate calcium and magnesium ions and improve the mud filtration characteristics.

The conversion to a lignosulfonate mud is accomplished simply by increasing the concentration of chrome lignosulfonate and lignite to the desired level. In many areas, a fully inhibitive mud is not justified economically. The concentration of the deflocculants used depends on the drilling conditions encountered. The concentration of deflocculants can be increased easily if mud contamination or borehole instability becomes a problem.

Chrome or ferrochrome lignosulfonate and lignite are emulsifiers of oil and the oil content of the mud can be maintained as high as 15% in unweighted muds or 5% in weighted muds. The resulting oil-in-water emulsion may decrease the adverse effect of the mud on the formations. However, the presence of oil in the mud also requires more stringent pollution control procedures and can reduce the effectiveness of some formation evaluation techniques.

As environmental regulations become more stringent throughout the world, the use of chrome or ferrochrome lignosulfates may be restricted in the future. Other lignosulfates, such as potassium, iron, or titanium lignosulfates, are on the market. However, they are not as effective as chrome or ferrochrome lignosulfates. Their use may be appropriate in low-solids, low-weight muds at temperatures not greater than 250°F.

2.4.3 High-Salinity Muds. Salt water presents inherent problems for treatment and maintenance as a result of contaminants in the make-up water. Severe mud problems can arise from the combined effects of high temperature, salts, and hardness. Inhibitive muds having a concentration of NaCl in excess of 1% by weight are called *saltwater muds*. Saltwater muds usually are formed from another type of inhibitive mud capable of tolerating salt contamination. Calcium-treated muds, lignite-treated muds, and lignosulfonate-treated muds all

are used commonly to form saltwater muds. Seawater muds often are used in marine areas where a source of fresh water is not readily available. Saturated saltwater muds are used primarily to permit drilling a relatively in-gauge hole through salt formations. After conversion, saltwater muds are not affected greatly by subsequent increases in salinity through contamination. Also, the presence of salt further reduces the ability of active clays to hydrate. However, introducing salt into the mud increases the chemical treatments required to maintain acceptable mud properties and reduces the number of well-logging tools that can be used effectively.

Bentonite is added to mud for viscosity, gel strength, and fluid loss control. The presence of bentonite is beneficial for cuttings-carrying-capacity and filter cake characteristics. Dry bentonite does not hydrate significantly in water of even moderate salinity. Shown in Fig. 2.33 are the viscosities obtained by adding 20 lbm/bbl Wyoming bentonite to water of various salinities. Note that for salinities greater than that of seawater, Wyoming bentonite behaves essentially like an inert solid. However, an entirely different behavior is observed if the Wyoming bentonite is prehydrated using fresh water and treated with a deflocculant before the salinity is increased. In this case, the clay retains a moderate degree of hydration and improves the viscous characteristics and filtration characteristics of the mud. The amount of deflocculant (thinner) required increases with salt concentration and can be determined using pilot tests. At extremely high salt concentrations, saltwater clays such as attapulgite and sepiolite clay may be required to obtain acceptable mud properties.

The salinity of seawater usually is expressed as equivalent NaCl concentration and is determined stoichiometrically from the titration of seawater with $AgNO_3$. The equivalent NaCl concentration of seawater is about 35,000 ppm. However, a more detailed analysis of seawater shows many salts other than NaCl to be present in significant concentrations. A typical analysis of seawater is shown in Table 2.16. The Ca^{2+} and Mg^{2+} ions present in seawater often are as detrimental to the mud as the NaCl. The method used to reduce the detrimental effect of the Ca^{2+} and Mg^{2+} ions depends on the type of mud used to form the saltwater mud. They can be tolerated in a calcium-treated mud, removed through sequestering by treating with a phosphate, or reduced by chelation through treating with a lignite. In addition, precipitation of Mg^{2+} and Ca^{2+} from the

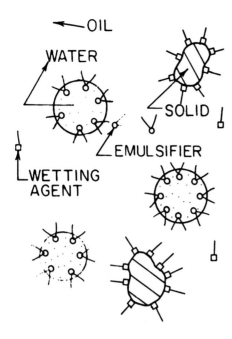

Fig. 2.34—Schematic of an oil mud.

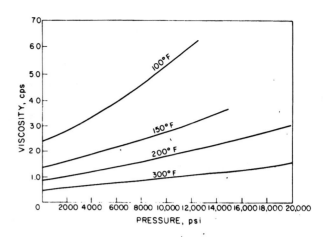

Fig. 2.35—Effect of temperature and pressure on the viscosity of diesel oil.[5]

seawater can be accomplished before mixing the mud by treating with a strong base such as NaOH and a soluble carbonate such as Na_2CO_3.

Potassium chloride is more effective than sodium chloride in suppressing the hydration and swelling of active clay minerals. The presence of K^+ causes a cation exchange reaction to occur in which sodium montmorillonite is converted to potassium montmorillonite. The presence of K^+ reduces the ability of the active clays to hydrate almost as much as a divalent ion. The effectiveness of the K^+ ion is thought to be a result of its smaller size, which permits it to enter the silicate lattice of the montmorillonite structure. Prehydrated bentonite can be used with KCl muds in the same manner as with NaCl muds. However, since KCl is a stronger contaminant, a higher concentration of deflocculant will be required. Lignite that has been neutralized with KOH often is used as a deflocculant and filtration control agent for KCl muds. When KCl is added to seawater muds, KOH and K_2CO_3 sometimes are substituted for NaOH and Na_2CO_3 in any water-softening treatments.

2.5 Oil Muds

Drilling fluids are called *oil muds* if the continuous phase is composed of a liquid hydrocarbon. No. 2 diesel usually is used for the oil phase because of its viscosity characteristics, low flammability, and low solvency for rubber. As shown in Fig. 2.34, water present in an oil mud is in the form of an emulsion. A chemical *emulsifier* must be added to prevent the water droplets from coalescing and settling out of the emulsion. The emulsifier also permits water originally present in the rock destroyed by the bit to emulsify easily. A chemical *wettability* reversal agent is added to make the solids in the mud preferentially wet by oil rather than water. Otherwise, the solids will be absorbed by the water droplets and cause high viscosities and eventual settling of barite.

The advantages and disadvantages of using oil muds are summarized in Table 2.17. Because of the higher initial cost and pollution control problems associated with oil muds, they are used much less frequently than water-base muds. The most common application for oil muds include (1) drilling deep hot formations (temperatures > 300°F), (2) drilling salt, anhydrite, carnallite, potash, or active shale formations or formations containing H_2S or CO_2, (3) drilling producing formations easily damaged by water-base muds, (4) corrosion control, (5) drilling directional or slim holes where high torque is a problem, (6) preventing or freeing stuck pipe, and (7) drilling weak formations of subnormal pore pressure. For these applications, the use of oil muds often can reduce the overall drilling costs enough to offset the higher initial mud cost and cuttings disposal costs. Many suppliers often buy back the used oil muds at a discount rate.

In recent years, a newly modified oil mud known as *relaxed fluid-loss oil mud* or *low-colloid oil mud* has been used successfully in the field. This mud has proven to improve the drilling rate compared with conventional oil mud in the U.S. gulf coast area and

TABLE 2.17—ADVANTAGES AND DISADVANTAGES OF USING OIL MUDS

Advantages	Disadvantages
1. Good rheological properties at temperatures as high as 500°F	1. Higher initial cost
2. More inhibitive than inhibitive water base muds	2. Requires more stringent pollution-control procedures
3. Effective against all types of corrosion	3. Reduced effectiveness of some logging tools
4. Superior lubricating characteristics	4. Remedial treatment for lost circulation is more difficult
5. Permits mud densities as low as 7.5 lbm/gal	5. Detection of gas kicks is more difficult because of gas solubility in diesel oil

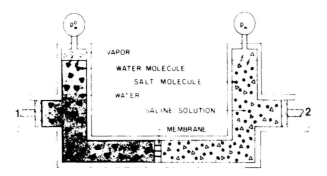

Fig. 2.36—Osmosis of water through a semipermeable membrane.

Texas. Since the oil mud formulation has been modified by reducing the concentration of the colloidal material, the thermal stability also is compromised.

2.5.1 Oil Phase. In addition to the commonly used No. 2 diesel oil, weathered crude oils and various refined oils have been used as the oil phase for oil muds. Recently, several mineral oils have been developed that have a lower toxicity than No. 2 diesel oil. These oils were developed to help solve the potential pollution problems associated with use of oil muds in a marine environment. Safety requires that the oil phase selected has a relatively low flammability. Oils having an open cup *fire point* above 200°F are considered safe. The flash point is the minimum temperature at which the vapors above the oil can be ignited by a flame. The fire point is the minimum temperature at which sustained combustion of the vapors above the oil can be maintained. In addition to the flammability requirements, the oil should be relatively free of aromatic hydrocarbons, which have a tendency to soften the rubber parts of the blowout preventers and other drilling equipment. Oils having an *aniline point* above 140°F are considered acceptable. The aniline point is the temperature below which oil containing 50% by volume of aniline ($C_6H_5 - NH_2$) becomes cloudy. The solvent powers for many other materials (such as rubber) are related to the solvent power for aniline. The oil selected also should exhibit an acceptable viscosity over the entire range of temperatures and pressures to be encountered in the well. The effects of temperature and pressure on the viscosity of No. 2 diesel oil are shown in Fig. 2.35.

2.5.2 Water Phase. The emulsified water of an oil mud tends to increase the viscosity of the mud in the same manner as inert solids. It also causes a slight increase in fluid density. Since water is much less expensive than oil, it also decreases the total cost of an oil mud. Water contents as high as 50% of the mud volume have been used in oil muds. However, as mud density is increased, it is necessary to decrease the water content to prevent excessive mud viscosity. A highly weighted mud usually has a water content less than 12%. In some applications, it is desirable to maintain the water content as low as possible. However, even when no water is added to the mud at the surface, the water content gradually increases during drilling operations. Water contents in the

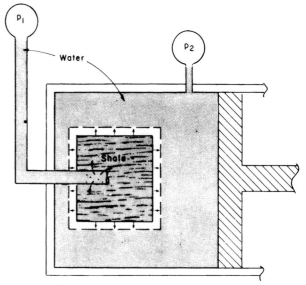

Fig. 2.37—Adsorption of water by shale.

range of 3 to 5% frequently are tolerated even when a zero water content is desired because of the high cost of dilution with oil.

The emulsified water phase of oil muds retains the ability to contact the subsurface formations and also to influence subsurface corrosion. Thus the chemical composition of the water phase is an important factor affecting the inhibitive properties of oil muds as well as water-base muds. When active shales are not a problem, fresh water or seawater can be used for the water phase. However, troublesome shale sections often require an increase in the electrolyte concentration of the water. Techniques have been developed for determining the exact electrolyte concentration at which a given shale will neither swell nor dehydrate.[6-8] These techniques involve adding sufficient NaCl or $CaCl_2$ to the water phase of the mud so that the chemical potential of the water in the mud is equal to the chemical potential of the water in the shale. These muds are called *balanced-activity* oil muds.

2.5.3 Balanced-Activity Oil Muds. The free energy, G, of a given system is a function of the temperature, T, the pressure, p, and moles, n_i, of each component present:

$$G = f(T,p,n_1, n_2, \ldots, n_i, \ldots) .$$

The change in free energy with T, p, and n_i is given by

$$dG = \frac{\partial G}{\partial T}dT + \frac{\partial G}{\partial p}dp + \sum \frac{\partial G}{\partial n_i}dn_i. \quad \ldots\ldots (2.38)$$

In this thermodynamic expression, the partial derivatives have the following meanings.

$$\frac{\partial G}{\partial T} = -S \text{ (entropy)}, \frac{\partial G}{\partial p} = V \text{ (volume)},$$

and

$$\frac{\partial G}{\partial n_i} \doteq \mu_i \text{ (chemical potential)}.$$

At Points 1 and 2 in a closed system, the condition for thermal equilibrium is $T_1 = T_2$ and the condition for mechanical equilibrium is $p_1 = p_2$. Similarly, the condition for chemical equilibrium of component i is $\mu_{i_1} = \mu_{i_2}$.

The hydration of shale is somewhat similar in mechanism to the osmosis of water through a semipermeable membrane. Consider the system shown in Fig. 2.36 in which pure water is separated from a saline solution by a membrane permeable to water alone. If the water is in equilibrium with its vapor, then the chemical potential of the water in the liquid and vapor phase must be the same. The change in chemical potential of the water with pressure is given by

$$d\mu_w = \frac{\partial \mu_w}{\partial p} dp = \frac{\partial^2 G}{\partial n_w \partial p} dp = \frac{\partial V_w}{\partial n_w} dp$$

$$= \bar{V}_w dp, \qquad \qquad (2.39)$$

where $\bar{V}_w$ is the partial molar volume of water. For the vapor phase, the partial molar volume can be expressed in terms of vapor pressure using the ideal gas law·

$$\bar{V}_w = \frac{RT}{p_w}, \qquad \qquad (2.40)$$

where R is the universal gas constant and p_w is the partial pressure of the water vapor. The chemical potential, μ_w, of the water in the vapor above the saline solution relative to the chemical potential, μ_w^o, of the vapor phase above the pure water can be obtained by substituting Eq. 2.40 into Eq. 2.39:

$$\int_{\mu_w^o}^{\mu_w} d\mu_w = \int_{p_w^o}^{p_w} \frac{RT}{p_w} dp_w,$$

and

$$\mu_w = \mu_w^o + RT \ln \frac{p_w}{p_w^o}, \qquad \qquad (2.41)$$

where p_w^o is the vapor pressure of pure water at the given temperature. Since the liquid on each side of the osmosis cell is in equilibrium with its vapor, Eq. 2.41 also expresses the chemical potential of the liquid phases.

For an ideal solution, the escaping tendency of each component is proportional to the mole fraction of that component in the solution. Since there are fewer water molecules per unit volume in the saline solution than in the pure water, the vapor pressure above the saline solution is less than the vapor pressure above the pure water. Note that for $p_w < p_w^o$, Eq. 2.41 shows that $\mu_w < \mu_w^o$. Thus, there is a tendency for water to move through the semipermeable membrane to dilute the saline solution and increase the vapor pressure above the saline solution. This would be true even if no vapor was physically present. For example, if the positions of Pistons 1 and 2 in Fig. 2.36 were altered in such a manner as to liquify the vapor without changing the fluid pressure, the chemical potential of the liquids would remain unchanged. Water still would tend to move from low salinity to high salinity.

The pressure required to prevent water from moving through a semipermeable membrane from a solution of low salinity (high vapor pressure) to a solution of high salinity (low vapor pressure) is called the *osmotic pressure*, Π. For example, if sufficient hydraulic pressure were applied using Piston 2 in Fig. 2.36 to prevent water movement through the membrane, the pressure difference existing across the membrane would be the osmotic pressure. For this to occur, the chemical potential of the liquid phases on either side of the membrane must be equal. This requires that the decrease in chemical potential due to salinity be offset exactly by the increase in chemical potential caused by the imposed hydraulic pressure. The effect of external pressure on chemical potential is given by Eq. 2.39. For an incompressible liquid, we have

$$\Delta \mu_w = \int_{p_1}^{p_2} \bar{V}_w dp = \bar{V}_w \Delta p = -\Pi \bar{V}_w.$$

For $\Pi \bar{V}_w$ to counterbalance the change in chemical potential due to salinity requires

$$-\Pi \bar{V}_w = RT \ln \frac{p_w}{p_w^o}.$$

Solving for the osmotic pressure, Π, yields

$$\Pi = -\frac{RT}{\bar{V}_w} \ln \frac{p_w}{p_w^o}, \qquad \qquad (2.42)$$

where the potential molal volume, V_w, of the liquid phase can be evaluated as molecular weight divided by density.

When shale is exposed to a source of fresh water, there is a tendency for the shale to adsorb water and swell. The adsorptive pressure of shales is analogous to the osmotic pressure of semipermeable membranes. The attraction of the shale platelets for water lowers the escaping tendency (vapor pressure) of the water in the shale, causing the chemical potential of the water in the shale to be less than the chemical potential of pure water. Eq. 2.41 can be applied to water adsorption into shale if the term p_w is considered to be the vapor pressure of water in the shale. Similarly, Eq. 2.42 also can be applied to water adsorption into shale if the term Π is considered to be the adsorptive pressure. For the arrangement shown in Fig. 2.37, the adsorptive pressure causes the confining pressure to exceed the fluid pressure. The adsorptive pressure is equal to the pressure difference, $p_2 - p_1$.

Eq. 2.42 was derived for an ideal system. To apply the same concepts to nonideal systems, the term *fugacity* was introduced. Basically, the fugacity, f_i, of component i in solution is the imaginary value of vapor pressure that would have to be used in Eq. 2.42 to make it apply accurately to a nonideal system. Thus, for nonideal solutions, the ratio p_w/p_w^o is replaced by f_w/f_w^o. The ratio f_i/f_i^o also is called the *activity*, a_i, of component i. Thus, the chemical potential of water in solution or in shale compared to the chemical potential of pure water is given by

$$\mu_w = \mu_w^o + RT \ln a_w. \qquad \qquad (2.43)$$

For water not to enter and swell a shale formation in which the confining pressure has been relieved by drilling, the activity of the water in the mud and in the shale must be equal.

The activity of the water in the mud and in shale

TABLE 2.18—SATURATED SOLUTIONS FOR CALIBRATING ELECTROHYGROMETER

Salt	Activity
$ZnCl_2$	0.10
$CaCl_2$	0.30
$MgCl_2$	0.33
$Ca(NO_3)_2$	0.51
$NaCl$	0.75
$(NH_4)_2SO_4$	0.80
Pure water	1.00

Fig. 2.38—Electrohygrometer apparatus.[3]

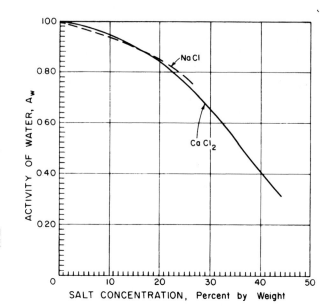

Fig. 2.39—Water activity in calcium chloride and sodium chloride at room temperature.

cuttings is measured in the field using an *electrohygrometer* (Fig. 2.38). The probe of the electrohygrometer is placed in the equilibrium vapor over the sample being tested. The electrical resistance of the probe is sensitive to the amount of water vapor present. Since the test always is conducted at atmospheric pressure, the water vapor pressure is directly proportional to the volume fraction of water in the air/water vapor mixture. The instrument normally is calibrated using the saturated solutions of known activity shown in Table 2.18.

Sodium chloride and calcium chloride are the salts generally used to alter the activity of the water in the mud. Calcium chloride is quite soluble, allowing the activity to be varied over a wide range. In addition, it is a relatively inexpensive additive. The resulting water activity for various concentrations of $NaCl$ and $CaCl_2$ are shown in Fig. 2.39.

Example 2.23. The activity of a sample of shale cuttings drilled with an oil mud is determined to be 0.69 by an electrohygrometer. Determine the concentration of calcium chloride needed in the water phase of the mud in order to have the activity of the mud equal to the activity of the shale.

Solution. Using Fig. 2.39, a calcium chloride concentration of 28.2% by weight is needed to give an activity of 0.69. Converting this concentration to lbm/bbl gives

$$0.282 = \frac{x}{350 + x}.$$

$$x = 137.5 \text{ lbm/bbl.}$$

2.5.4 Emulsifiers.

A calcium or magnesium fatty acid soap frequently is used as an emulsifier for oil muds. Fatty acids are organic acids present in naturally occurring fats and oils that have a structure that can be represented by

$$CH_3 - CH_2 - (CH_2)_n - \underset{\underset{OH}{|}}{C} = O.$$

The unbranched acids with 12, 14, 16, or 18 carbon atoms are especially common in animal and vegetable fats. Fatty acid soaps are the salts formed by the reaction of fatty acids with a base. For example, the reaction of a fatty acid with caustic yields a sodium fatty acid salt:

$$CH_3 - CH_2 - (CH_2)_n - \underset{\underset{OH}{|}}{C} = O + NaOH \rightarrow$$

$$\underset{\text{(fatty acid soap)}}{CH_3 - CH_2 - (CH_2)_n - \underset{\underset{O - Na^+}{|}}{C} = O} + \underset{\text{(water)}}{HOH}$$

The long hydrocarbon chain portion of the soap molecule tends to be soluble in oil and the ionic portion of the molecule tends to be soluble in water. When soap is introduced to a mixture of oil and water, the soap molecule will accumulate at the oil/water interfaces with the water-soluble end

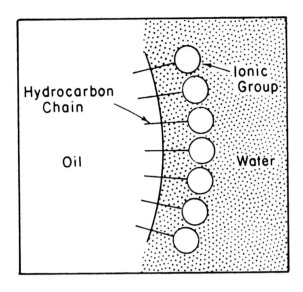

(a) MONOVALENT CATION

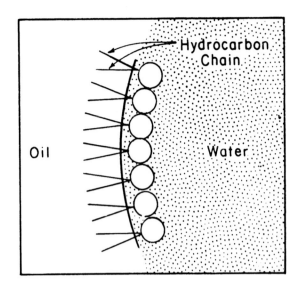

(b) DIVALENT CATION

Fig. 2.40—Orientation of fatty acid soap molecules at oil/water interface.

residing in the water and the oil-soluble end residing in the oil. This greatly reduces the surface energy of the interface and permits the formation of a stable emulsion. Fatty acid soaps formed from monovalent ions have a single hydrocarbon chain. As shown in Fig. 2.40, the packing of this type of soap molecules at an oil/water interface tends to form a concave oil surface and favors an oil-in-water emulsion. Fatty acid soaps formed from divalent ions such as Ca^{2+} or Mg^{2+} have two hydrocarbon chains. The packing of this type of soap molecules at an oil/water interface tends to form a convex oil surface and favors a water-in-oil emulsion. Of course, the relative amounts of oil and water present also influence the type of emulsion formed.

While the fatty acid soaps are the most common type of emulsifier used in oil muds, almost any type of oil-soluble soap can be used. Calcium naphthenic acid soaps and soaps made from rosin (pine tree sap) also are common organic acid-type soaps. Napthenic acid soaps can be formed economically from coal tar. They have aromatic ring structures rather than the straight hydrocarbon chains of the fatty acids. The rosin soaps are produced economically by treating components of pine tree sap. Rosin primarily contains branched hydrocarbon chains and ring structures. In addition, soaps formed from organic amines rather than organic acids are used.

The effectiveness of a given oil mud emulsifier depends upon the alkalinity and electrolytes present in the water phase. Also, some emulsifiers tend to degrade at high temperatures. However, the suitability of a particular emulsifier for a given oil mud application can be determined using pilot tests if previous test data are not available.

2.5.5 Wettability Control. When a drop of liquid is placed on the surface of a solid, it may spread to cover the solid surface or it may remain as a stable drop. The shape that the drop assumes depends upon the strength of the adhesive forces between molecules of the liquid and solid phases. The *wettability* of a given solid surface to a given liquid is defined in terms of the contact angle, θ, shown in Fig. 2.41. A liquid that exhibits a small contact angle has a strong wetting tendency. If the contact angle is equal to 180°, the liquid is said to be completely nonwetting.

When two liquids are brought simultaneously in contact with a solid and with each other, one of the liquids will preferentially wet the solid. For example, while both oil and water tend to wet a silica surface, the silica is preferentially wet by water. Thus, the water tends to spread under the oil and occupy the position in contact with the silica surface. The contact angle, θ, is less than 90° measured through the liquid that preferentially wets the surface.

Most natural minerals are preferentially wet by water. When water-wet solids are introduced to a water-in-oil emulsion, the solids tend to agglomerate with the water, causing high viscosities and settling. Water-wet solids also tend to cause the formation of an oil-in-water emulsion rather than a water-in-oil emulsion. To overcome these problems, wettability control agents are added to the oil phase of the mud. The wetting agents are surfactants similar to the emulsifiers. One end of the molecule tends to be soluble in the oil phase while the other end has a high affinity for the solid surface. The molecules accumulate at the oil/solid interfaces with the oil-soluble end pointing toward the oil phase. This effectively changes the solids from being preferentially wet by water to preferentially wet by oil.

The soaps added to serve as emulsifiers also function to some extent as wetting agents. However, they usually do not act fast enough to handle a large influx of water-wet solids during fast drilling or mud weighting operations. Several special surfactants are available for more effective oil wetting.

An electrical stability test is used to indicate emulsion stability. This test indicates the voltage at which the mud will conduct current in the test apparatus. A loose emulsion often is due to the

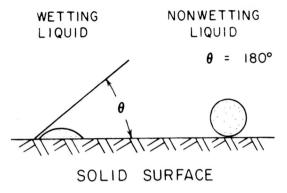

Fig. 2.41—Contact angle, Θ.

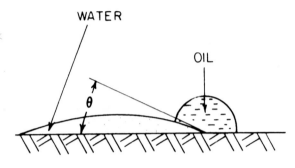

Fig. 2.42—Relative wettability.

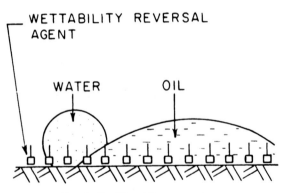

Fig. 2.43—Wettability reversal.

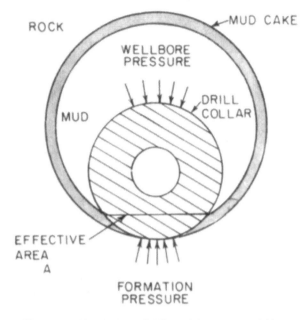

Fig. 2.44—Mechanism of differential pressure sticking.

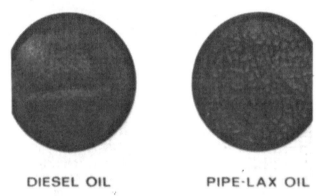

Fig. 2.45—Cracking of mud cake by oil mud.

presence of water-wet solids or free water. A gradual decrease in emulsion stability with time indicates the need for more emulsifier. Visual observations also are useful in determining the presence of water-wet solids. The surface of an oil mud will become less shiny and have less apparent dispersion rings or swirls when water-wet solids are present. The cuttings also tend to adhere to each other or to the shale shaker screen and may have a gummy feeling.

2.5.6 Viscosity Control. The emulsified water tends to increase the mud viscosity as well as lower the total mud cost. To a lesser extent, the soaps added to the oil also tend to increase viscosity. Further increases in viscosity can be achieved by adding solids to the mud. Asphalts and amine-treated

bentonite are the main viscosity control additives. Some of the heavy hydrocarbons present in asphalt go into solution in the mud. The less soluble components are carried as colloidal solids. High-molecular-weight polar molecules present in the asphalt probably act to make other solids preferentially wet by oil. The amine-treated bentonite easily disperses in oil muds to form a colloid.

2.5.7 Filtration Control. Since oil is the continuous phase in an oil mud, only the oil phase is free to form a filtrate. This property makes oil muds especially suitable for drilling formations easily damaged by water invasion. In addition, oil muds usually have excellent filtration properties and rarely require filtration control additives. However, when additional fluid loss control is desired, asphalt, polymers, manganese oxide, and amine-treated lignite can be used.

2.5.8 Density Control.
API barite is the main density control additive used in oil muds as well as water-base muds. Calcium carbonate also is used sometimes when a relatively low mud density is required. Settling of API barite is more severe in oil muds because of the lower gel strengths. Also, if the API barite is not converted completely to an oil-wet condition, the API barite particles will aggregate, greatly increasing their tendency to settle.

2.5.9 Alkalinity Control.
Lime is used to maintain the alkalinity of oil muds at an acceptable level. A high pH (8.5-10.0) is needed to control corrosion and to obtain the best performance from the emulsifiers. When formation gases such as CO_2 or H_2S that form acids upon ionization are expected, an even higher alkalinity is used. The ability to contain a large reserve of undissolved lime makes an oil mud superior to a water mud when drilling hydrogen sulfide or carbon dioxide bearing zones. The usual range of the methyl orange alkalinity of the mud is 0.5 to 1.0 cm^3. However, a value of 2.0 cm^3 may be desirable when H_2S or CO_2 is anticipated.

2.5.10 Control of Solids and Water Content.
Hydrocyclones and centrifuges cannot be used economically on oil muds since a significant volume of the expensive liquid phase would be discarded by these devices. Dilution is also quite expensive. Screening is the only economical means of solids control of oil muds. Since oil muds are inhibitive, cutting disintegration is limited and screens are very effective. Using several screens in series, it usually is possible to screen the returning mud stream as fine as 200 mesh. When the desired solids level cannot be maintained by screening, dilution will be required.

The water content of oil muds also must be maintained within limits. When the mud temperature is high, water evaporation will be significant. Evaporation losses must be replaced to prevent changing the salinity and activity of the mud. In addition, if the saline solution becomes saturated, the precipitation of salts can cause a decrease in the emulsion stability. The water content will have to be decreased when increasing mud density to prevent excessive viscosity. This is accomplished by dilution with oil. It usually is more economical to discard a portion of the oil mud when diluting rather than continually increasing total mud volume.

2.5.11 Oil Muds for Freeing Stuck Pipe.
A frequent application of oil muds is for freeing a drillstring held against the mud cake by hydrostatic pressure in the wellbore. This problem is illustrated in Fig. 2.44 and is called *differential pressure sticking* to distinguish it from other causes of stuck pipe such as insufficient cutting removal or borehole collapse. Several oil mud formulations designed specifically for freeing stuck pipe are available. The technique involves displacing a volume of oil mud sufficient to fill the annular region

where the pipe is stuck and then alternately applying compression, tension, and torque until the pipe is free. The use of an oil mud of equal density with the water-base mud in the well will prevent premature migration of the oil mud up the annulus. In some cases, a formation tester is used to hydraulically isolate and lower the wellbore pressure opposite the stuck pipe.

The force required to free differentially stuck pipe is given by

$$F_{st} = \Delta p A f, \dots\dots\dots\dots\dots\dots\dots (2.44)$$

where F_{st} is the freeing force, Δp is the pressure differential between the wellbore and the permeable formation, A is the effective area of contact with the mud cake, and f is the coefficient of friction between the pipe and mud cake.

The effective area of contact, A, used in Eq. 2.44 is the chord length of the imbedded portion of the drill collars (see Fig. 2.44) multiplied by the thickness, h_f, of the low pressure, permeable formation against which the drill collars are held. It can be shown that for an in-gauge borehole, A is expressed by

$$A = 2h_f \sqrt{ \left(\frac{d_2}{2} - h_{mc} \right)^2 - \left(\frac{d_2}{2} - h_{mc} \frac{d_2 - h_{mc}}{d_2 - d_1} \right)^2 }$$
$$\dots\dots\dots\dots\dots\dots\dots\dots\dots\dots\dots (2.45)$$

for

$$h_{mc} \le \frac{d_1}{2} \le \frac{d_2 - h_{mc}}{2}$$

and where h_{mc} is the thickness of the mud cake, d_1 is the outer diameter of the drill collars, and d_2 is the diameter of the borehole.

Eq. 2.44 indicates that these following factors tend to increase the sticking force: (1) high wellbore pressure caused by unnecessarily high mud density, (2) low formation pore pressure in permeable zone (e.g., a depleted oil or gas sand), (3) thick, permeable formation, which causes a greater effective area, (4) thick mud cake, which causes a greater effective area, (5) large pipe diameter, which causes a greater effective area, and (6) a mud cake with high coefficient of friction. Thus, muds having a low density, a low water loss, and a thin, slick mud cake are best for preventing differential pressure sticking. Also, pipe shape is an important factor, and several drill collar configurations have been developed to decrease the sticking tendency. These include (1) drill collars with spiral grooves, (2) square drill collars, (3) drill collars with external upsets, and (4) drill collars with upsets in the middle and on each end. All of these designs reduce the effective area of contact.

The oil mud soaking technique for freeing stuck pipe is thought to work by cracking the mud cake as shown in Fig. 2.45. Cracking the mud cake allows the pressure differential to equalize. Undoubtedly, the better lubricating characteristics of the oil mud also help.

2.5.12 Oil Muds for Lost Circulation.
A mixture of diesel oil and bentonite or diesel oil, bentonite, and cement sometimes is used to seal off a fractured

formation to which drilling fluid is being lost. Bentonite concentrations as high as 300 lbm/bbl commonly are used. The technique involves pumping the diesel oil slurry down the drillstring while mud is pumped down the annulus. The diesel slurry and the mud come together in the formation and set to a stiff consistency. About two parts slurry to one part mud is required. The displacements are accomplished using two cementing pump trucks (one each for the slurry and the mud).

Exercises

2.1 Discuss the relation between the mud properties determined in the API diagnostic tests and the functions of the drilling fluid.

2.2 Determine the concentration of H^+ and OH^- in moles per liter in an aqueous solution having a pH of 11.6. *Answer*: 2.51×10^{-12}, 0.00398.

2.3 The solubility product, K_{sp}, for $Mg(OH)_2$ is 8.9×10^{-12}. Determine the following.
a. The solubility of Mg^{++} in moles per liter for a pH of 11. *Answer*: 1.3×10^{-4}.
b. The pH of a saturated solution of $Mg(OH)_2$. *Answer*: 10.4.
c. The solubility of Mg^+ in moles per liter for a pH of 11.0. *Answer*: 8.9×10^{-6}.

2.4 A 15-in. hole is drilled to a depth of 4,000 ft. The API water loss of the mud is 10 mL. Approximately 30% of the lithology is permeable sandstone and the rest is impermeable shale.
a. Construct a plot of estimated filtration loss in barrels vs. time in hours (0 to 24 hours) that would occur if the hole were drilled instantaneously. Assume porosity is 0.25. *Answer*: 42.4 bbl after 24 hours.
b. Compute the radius of the invaded zone for Part a in inches after 24 hours. *Answer*: 9.62 in.
c. Repeat Part a assuming a drilling rate of 200 ft/hr. *Answer*: 31.6 bbl after 24 hours.
d. Do you feel the API water loss test is representative of conditions in the well during drilling operations? (Hint: Find an article on "dynamic filtration.")

2.5 A filtrate volume of 5 cm^3 is collected in 10 min in a filter press having an area of 90 cm^2. A spurt loss of 0.5 cm^3 was observed. Compute the API water loss. *Answer*: 4.15 cm^3.

2.6 A saline solution contains 175.5 g of NaCl per liter of solution. Using a water density of 0.9982 g/cm^3, express the concentration of NaCl in terms of (1) molality, (2) molarity, (3) normality, (4) parts per million, (5) milligrams per liter, (6) weight percent, and (7) pounds per barrel of water. *Answer*: 3.198; 3.00; 3.00; 157,440; 175,500; 15.7; 65.4.

2.7 What is the theoretical phenolphthalein alkalinity of a saturated solution of $Ca(OH)_2$? *Answer*: 0.69 cm^3.

2.8 Discuss the difference between these alkalinity values: (1) P_m and P_f, and (2) P_f and M_f.

2.9 One liter of solution contains 3.0 g of NaOH and 8.3 g of Na_2CO_3. Compute the theoretical values of P_f and M_f. *Answer*: 7.7 cm^3; 11.6 cm^3.

2.10 Alkalinity tests on a mud give a P_m value of 5.0 and a P_f value of 0.7. Determine the approximate amount of undissolved lime in the mud. The volume fraction of water in the mud is 80%. *Answer*: 1.154 lbm/bbl.

2.11 A volume of 20 mL of 0.0282 N $AgNO_3$ was required to titrate 1 mL of saline water in the API test for salinity. Determine the concentration of Cl^- and NaCl in the solution in mg/L assuming only NaCl was present. *Answer*: 20,000 and 33,000.

2.12 A 1,000-mL solution contains 5.55 g of $CaCl_2$ and 4.77 g of $MgCl_2$.
a. How many milliliters of 0.02 N standard versenate solution would be required in the API titration for total hardness? *Answer*: 10 mL per 1-mL sample.
b. Express the concentrations of Ca^{2+} and Mg^{2+} in parts per million. *Answer*: 5,550 and 4,770 total salt (approximate for $\rho_s = \rho_w$).

2.13 Titrations for the total hardness of the mud and mud filtrate require 5.0 and 0.5 mL of 0.02 N versenate solution, respectively. If the volume fraction of water in the mud is 0.85, determine the equivalent free $CaSO_4$ concentration in pounds per barrel. *Answer*: 2.18.

2.14 An 11.4-lbm/gal freshwater mud is found to have a solids content of 16.2 vol%.
a. Compute the volume fraction of API barite and low-specific-gravity solids. *Answer*: 0.068 and 0.094.
b. Compute the weight fraction of API barite and low-specific-gravity solids in the mud. *Answer*: 0.209 and 0.179.
c. Compute the API barite and low-specific-gravity solids content in pounds per barrel of mud. *Answer*: 100 and 85.5 lbm/bbl.

2.15 A freshwater mud has a methylene blue capacity of 5 meq/100 mL of mud. Determine the approximate sodium montmorillonite content of the mud. *Answer*: 25 lbm/bbl.

2.16 A titration test has shown that a drilling mud contains 150 mg/L of calcium. The mud engineer plans to add enough SAPP $(Na_2H_2P_2O_7)$ to his 1,000-bbl system to reduce the calcium concentration to 30 mg/L. Determine the amount of SAPP that must be added to the mud system. *Answer*: 116.5 lbm.

2.17 Compute the density of a mud mixed by adding 30 lbm/bbl of clay and 200 lbm of API barite to 1 bbl of water. *Answer*: 11.8 lbm/gal.

2.18 Determine the density of a brine mixed by adding 150 lbm of $CaCl_2$ to 1 bbl of water. *Answer*: 10.7 lbm/gal.

2.19 Discuss the desirable and undesirable aspects of a high mud viscosity.

2.20 Compute the yield of a clay that requires addition of 35 lbm/bbl of clay to 1 bbl of water to raise the apparent viscosity of water to 15 cp (measured in a Fann viscometer at 600 rpm). *Answer*: 59.3 bbl/ton.

2.21 A mud cup is placed under one cone of a

hydrocyclone unit being used to process an unweighted mud. Twenty seconds were required to collect 1 qt of ejected slurry having a density of 20 lbm/gal. Compute the mass of solids and water being ejected by the cone per hour. *Answer:* 852 lbm/hr and 47.6 lbm/hr.

2.22 The only available source of water for the drilling fluid has a $[Ca^{2+}]$ of 900 ppm and a $[Mg^{2+}]$ of 400 ppm. Determine the concentration of caustic and soda ash that would be required to remove the Ca^{2+} and Mg^{2+} by precipitation. Would any other undesirable ions still be present? *Answer:* 0.46 lbm/bbl of NaOH and 0.84 lbm/bbl of Na_2CO_2. Yes, NaCl.

2.23 A 1,000-bbl unweighted freshwater mud system has a density of 9.5 lbm/gal. What mud treatment would be required to reduce the solids content to 4% by volume? The total mud volume must be maintained at 1,000 bbl and the minimum allowable mud density is 8.8 lbm/gal. *Answer:* Discard 544 bbl, add 544 bbl of water.

2.24 Name the three common causes of flocculation. Also name four types of mud additives used to control flocculation.

2.25 The density of 600 bbl of 12-lbm/gal mud must be increased to 14 lbm/gal using API barite. One gallon of water per sack of barite will be added to maintain an acceptable mud consistency. The final volume is not limited. How much barite is required? *Answer:* 92,800 lbm.

2.26 The density of 800 bbl of 14-lbm/gal mud must be increased to 14.5 lbm/gal using API barite. The total mud volume is limited to 800 bbl. Compute the volume of old mud that should be discarded and the weight of API barite required. *Answer:* Discard 19.05 bbl, add 28,000 lbm of barite.

2.27 The density of 900 bbl of a 16-lbm/gal mud must be increased to 17 lbm/gal. The volume fraction of low-specific-gravity solids also must be reduced from 0.055 to 0.030 by dilution with water. A final mud volume of 900 bbl is desired. Compute the volume of original mud that must be discarded and the amount of water and API barite that should be added. *Answer:* Discard 409 bbl, add 257.6 bbl of water and 222,500 lbm of barite.

2.28 Assuming a clay and chemical cost of $10.00/bbl of mud discarded and a barium sulfate cost of $0.10/lbm, compute the value of the mud discarded in Problem 2.27. If an error of +0.01% is made in determining the original volume fraction of low-specific-gravity solids in the mud, how much mud was unnecessarily discarded? *Answer:* $16,697; 191 bbl.

2.29 Derive expressions for determining the amounts of barite and water that should be added to increase the density of 100 bbl of mud from ρ_1 to ρ_2. Also derive an expression for the increase in mud volume expected upon adding the barite and the water. Assume a water requirement of 1 gal per sack of barite. *Answer:* $M_B = 109,000 \ (\rho_2 - \rho_1)/(28.08 - \rho_2);$ $V_w = M_B/4,200;$ $V = 0.0091 \ M_B.$

2.30 A 16.5-lbm/gal mud is entering a centrifuge at a rate of 20 gal/min along with 8.34 lbm/gal of dilution water, which enters the centrifuge at a rate of 10 gal/min. The density of the centrifuge underflow is 23.8 lbm/gal while the density of the overflow is 9.5 lbm/gal. The mud contains 25 lbm/bbl bentonite and 10 lbm/bbl deflocculant. Compute the rate at which bentonite, deflocculant, water, and API barite should be added downstream of the centrifuge to maintain the mud properties constant. *Answer:* 10.02 lbm/min of clay, 4.01 lbm/min of deflocculant, 9.78 gal/min of water, and 20.81 lbm/min of barite.

2.31 A well is being drilled and a mud weight of 17.5 lbm/gal is predicted. Intermediate casing has just been set in 15 lbm/gal freshwater mud that has a solids content of 29%, a plastic viscosity of 32 cp, and a yield point of 20 lbf/100 sq ft (measured at 120°F). What treatment is recommended upon increasing the mud weight to 17.5 lbm/gal?

2.32 A mud retort analysis of a 16-lbm/gal freshwater mud indicates a solids content of 32.5% and an oil content of zero. Methylene blue titrations of samples of mud, bentonite clay, and drilled solids indicates a CEC_m of 6 meq/100 mL, a CEC_c of 75 meq/100 g, and a CEC_{ds} of 15 meq/100 g. Determine (1) the total volume fraction of low-gravity solids, (2) the volume fraction of bentonite, and (3) the volume fraction of drilled solids. *Answer:* 0.0745; 0.0198; and 0.0547.

2.33 Define an inhibitive mud. Name three types of inhibitive water-base muds.

2.34 Discuss why prehydrated bentonite is used in high-salinity muds.

2.35 Discuss the advantages and disadvantages of using oil muds.

2.36 Compute the osmotic pressure developed across the membrane shown in Fig. 2.36 if the saline water has a weight fraction of $CaCl_2$ of (1) 0.10, (2) 0.30, or (3) 0.44 (assume $T = 70°F$). *Answer:* 1,000; 8,500; and 22,500 psi.

2.37 Compute the adsorptive pressure developed by a shale having an activity of 0.5 in contact with an oil mud containing emulsified fresh water (assume $T = 70°F$). *Answer:* 13,665 psi.

2.38 Compute the pounds per barrel of $CaCl_2$ that should be added to the water phase of an oil mud to inhibit hydration of a shale having an activity of 0.8. If the oil mud will contain 30% water by volume, how much $CaCl_2$ per barrel of mud will be required? *Answer:* 98.7 lbm/bbl of water and 29.6 lbm/bbl of mud.

2.39 Define these terms: (1) emulsifier, (2) wetting agent, (3) preferentially oil wet, (4) fatty acid soap, and (5) balanced activity mud.

2.40 A 6.125-in. hole is being drilled through a 100-ft depleted gas sand. The pressure in the wellbore is 2,000 psi greater than the formation pressure of the depleted sand. The mud cake has a thickness of 0.5 in. and a coefficient of friction of 0.10. If the 4.75-in. collars become differentially stuck over the entire sand interval, what force would be required to pull the collars free? *Answer*: 1,129,000 lbf.

References

1. "Standard Procedure for Testing Drilling Fluids," API R.P. 13B, Dallas (1974).
2. *Drilling Mud Data Book*, NL Baroid, Houston (1954).
3. *Drilling Fluid Engineering Manual*, Magcobar Div., Dresser Industries Inc., Houston (1972).
4. Grim, R.E.: *Clay Mineralogy*, McGraw-Hill Book Co., New York City (1968).
5. Annis, M.R.: *Drilling Fluids Technology*, Exxon Co. U.S.A., Houston (1974).
6. Chenevert, M.E.: "Shale Control With Balanced-Activity Oil-Continuous Muds," *J. Pet. Tech.* (Oct. 1970) 1309-1316; *Trans.*, AIME, **249**.
7. Darley, H.C.H.: "A Laboratory Investigation of Borehole Stability," *J. Pet. Tech.* (July 1969) 883-892; *Trans.*, AIME, **246**.
8. Mondshine, T.C.: "New Technique Determines Oil-Mud Salinity Needs in Shale Drilling," *Oil and Gas J.* (July 14, 1969) 70.
9. "Specifications for Oil-Well Drilling-Fluid Materials," API Spec. 13A, Dallas (1979).

Nomenclature

a = activity
A = area; treating agent
C = contaminant
C_i = concentration of ith member of alkaline series (e.g., C_1 is used for methane, C_2 for ethane, etc.)
C_f = correction factor used on water fraction to account for loss of salt during retorting
$\overline{CEC}$ = cation exchange capacity
d = diameter
D = depth
F = force
f = fractional volume; fugacity; coefficient of friction
G = free energy
h = thickness
k = permeability
K_{sp} = solubility product constant
K_w = ion product constant of water
m = mass
M = methyl orange alkalinity
n = moles present
N = number of revolutions per minute
P = phenolphthalein alkalinity
p = pressure; vapor pressure; partial pressure
Δp = pressure differential
q = flow rate
R = gas constant
S = solubility
t = time
T = temperature
V = volume
$\overline{V}$ = molar volume
θ = contact angle
θ_N = dial reading on Fann viscometer at rotor speed N
μ = viscosity; chemical potential

μ_a = apparent viscosity
μ_p = plastic viscosity
Π = osmotic pressure; shale adsorption pressure
ρ = density
τ_y = yield point
ϕ = porosity

Superscript

o = signifies pure component

Subscripts

a = agent
B = API barite
c = bentonite clay; contaminant
ds = drilled solids
f = filtrate; formation
i = component i in mixture
lg = low specific gravity
m = mud
mc = mud cake
mt = total for mixture
o = oil; overflow
r = rock
s = solids
sc = solids in mud cake
sm = solids in mud
sp = spurt loss
st = stuck pipe
t = time; titration
tf = titration volume per unit volume of filtrate
tm = titration volume per unit volume of mud
u = underflow
um = mud in underflow
uw = water in underflow
uB = API barite in underflow
w = water

SI Metric Conversion Factors

bbl ×	1.589 873	E−01	= m³
cp ×	1.0*	E−03	= Pa·s
cu ft ×	2.831 685	E−02	= m³
cu in. ×	1.638 706	E+01	= cm³
°F	(°F−32) 1.8		= °C
gal ×	3.785 412	E−03	= m³
in. ×	2.54*	E+00	= cm
lbf/100 sq ft ×	4.788 026	E−01	= Pa
lbm ×	4.535 924	E−01	= kg
lbm/bbl ×	2.853 010	E+00	= kg/m³
lbm/gal ×	1.198 264	E+02	= kg/m³
mol/L ×	1.0*	E−03	= kmol/L
qt ×	9.463 529	E−01	= dm³
sq ft ×	9.290 304*	E−02	= m²
sq in. ×	6.451 6*	E+00	= cm²
ton ×	1.0*	E+00	= Mg

*Conversion factor is exact.

Chapter 3
Cements

The purposes of this chapter are to present (1) the primary objectives of cementing, (2) the test procedures used to determine if the cement slurry and set cement have suitable properties for meeting these objectives, (3) the common additives used to obtain the desirable properties under various well conditions, and (4) the techniques used to place the cement at the desired location in the well. The mathematical modeling of the flow behavior of the cement slurry is not discussed in this chapter but is presented in detail in Chap. 4.

Cement is used in the drilling operation to (1) protect and support the casing, (2) prevent the movement of fluid through the annular space outside the casing, (3) stop the movement of fluid into vugular or fractured formations, and (4) close an abandoned portion of the well. A cement slurry is placed in the well by mixing powdered cement and water at the surface and pumping it by hydraulic displacement to the desired location. Thus, the hardened, or reacted, cement slurry becomes "set" cement, a rigid solid that exhibits favorable strength characteristics.

The drilling engineer is concerned with the selection of the best cement composition and placement technique for each required application. A deep well that encounters abnormally high formation pressure may require several casing strings to be cemented properly in place before the well can be drilled and completed successfully. The cement composition and placement technique for each job must be chosen so that the cement will achieve an adequate strength soon after being placed in the desired location. This minimizes the waiting period after cementing. However, the cement must remain pumpable long enough to allow placement to the desired location. Also, each cement job must be designed so that the density and length of the unset cement column results in sufficient subsurface pressure to control the movement of pore fluid while not causing formation fracture. Consideration must be given to the composition of subsurface contaminating fluids to which the cement will be exposed.

The main ingredient in almost all drilling cements is *portland cement,* an artificial cement made by burning a blend of limestone and clay. This is the same basic type of cement used in making concrete. A slurry of portland cement in water is ideal for use in wells because it can be pumped easily and hardens readily in an underwater environment. The name "portland cement" was chosen by its inventor, Joseph Aspdin, because he thought the produced solid resembled a stone quarried on the Isle of Portland off the coast of England.

3.1 Composition of Portland Cement

A schematic representation of the manufacturing process for portland cement is shown in Fig. 3.1. The oxides of Ca, Al, Fe, and Si react in the extreme temperature of the kiln (2600 to 2800°F), resulting in balls of cement *clinker* upon cooling. After aging in storage, the seasoned clinker is taken to the grinding mills where gypsum ($CaSO_4 \cdot 2H_2O$) is added to retard setting time and increase ultimate strength. The unit sold by the cement company is the *barrel,* which contains 376 lbm or four 94-lbm sacks.

Cement chemists feel that there are four crystalline compounds in the clinker that hydrate to form or aid in the formation of a rigid structure. These are (1) tricalcium silicate ($3CaO \cdot SiO_2$ or "C_3S"), (2) dicalcium silicate ($2CaO \cdot SiO_2$ or "C_2S"), (3) tricalcium aluminate ($3CaO \cdot Al_2O_3$ or "C_3A"), and (4) tetracalcium aluminoferrite ($4CaO \cdot Al_2O_3 \cdot Fe_2O_3$ or "C_4AF"). The hydration reaction is exothermic and generates a considerable quantity of heat, especially the hydration of C_3A.

The chemical equations representing the hydration

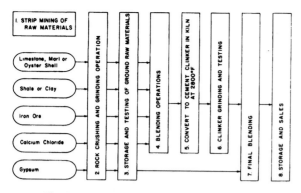

Fig. 3.1 – Manufacture of Portland cement.

of the cement compounds when they are mixed with water are as follows.

$$2(3CaO \cdot SiO_2) + 6H_2O \rightarrow$$
$$3CaO \cdot 2SiO_2 \cdot 3H_2O + 3Ca(OH)_2.$$

$$2(2CaO \cdot SiO_2) + 4H_2O \rightarrow$$
$$(slow)3CaO \cdot 2SiO_2 \cdot 3H_2O + Ca(OH)_2.$$

$$4CaO \cdot Al_2O_3 \cdot Fe_2O_3 + 10H_2O + 2Ca(OH)_2 \rightarrow$$
$$(slow)6CaO \cdot Al_2O_3 \cdot Fe_2O_3 \cdot 12H_2O.$$

$$3CaO \cdot Al_2O_3 + 12H_2O + Ca(OH)_2 \rightarrow$$
$$(fast)3CaO \cdot Al_2O_3 \cdot Ca(OH)_2 \cdot 12H_2O.$$

$$3CaO \cdot Al_2O_3 + 10H_2O + CaSO_4 \cdot 2H_2O \rightarrow$$
$$3CaO \cdot Al_2O_3 \cdot CaSO_4 \cdot 12H_2O.$$

The main cementing compound in the reaction products is $3CaO \cdot 2SiO_2 \cdot 3H_2O$, which is called tobermorite gel. The gel has an extremely fine particle size and, thus, a large surface area. Strong surface attractive forces causes the gel to adsorb on all crystals and particles and bind them together. Excess water that is not hydrated reduces cement strength and makes the cement more porous and permeable.

C_3S is thought to be the major contributor to strength, especially during the first 28 days of curing. C_2S hydrates very slowly and contributes mainly to the long term strength. C_3A hydrates very rapidly and produces most of the heat of hydration observed during the first few days. The gypsum added to the clinker before grinding controls the rapid hydration of C_3A. The C_3A portion of the cement also is attacked readily by water containing sulfates. C_4AF has only minor effects on the physical properties of the cement.

The chemical composition of portland cement generally is given in terms of oxide analysis. The relative amounts of the four crystalline compounds present are computed from the oxide analysis. API[1] uses the following equations for calculating the weight percent of the crystalline compounds from the weight percent of the oxides present.

$$C_3S = 4.07C - 7.6S - 6.72A - 1.43F - 2.85SO_3.$$
$$\dots\dots\dots\dots\dots\dots\dots\dots\dots\dots (3.1)$$
$$C_2S = 2.87S - 0.754 \, C_3S. \dots\dots\dots (3.2)$$
$$C_3A = 2.65A - 1.69 \, F. \dots\dots\dots (3.3)$$
$$C_4AF = 3.04 \, F. \dots\dots\dots\dots (3.4)$$

These equations are valid as long as the weight ratio of Al_2O_3 to Fe_2O_3 present is greater than 0.64.

Example 3.1. Calculate the percentages of C_3S, C_2S, C_3A, and C_4AF from the following oxide analysis of a standard portland cement.

Oxide	Weight Percent
Lime (CaO or C)	65.6
Silica (SiO$_2$ or S)	22.2
Alumina (Al$_2$O$_3$ or A)	5.8
Ferric oxide (Fe$_2$O$_3$ or F)	2.8
Magnesia (MgO)	1.9
Sulfur trioxide (SO$_3$)	1.8
Ignition loss	0.7

Solution. The A/F ratio is $5.8/2.8 = 2.07$. Thus, using Eqs. 3.1 through 3.4 yields

$$C_3S = 4.07(65.6) - 7.6(22.2) - 6.72(5.8)$$
$$- 1.43(2.8) - 2.85(1.8)$$
$$= 50.16\%.$$
$$C_2S = 2.87(22.2) - 0.754(50.16)$$
$$= 25.89\%.$$
$$C_3A = 2.65(5.8) - 1.69(2.8)$$
$$= 10.64\%.$$
$$C_4AF = 3.04(2.8)$$
$$= 8.51\%.$$

3.2 Cement Testing

API[1] presents a recommended procedure for testing drilling cements. These tests were devised to help drilling personnel determine if a given cement composition will be suitable for the given well conditions. Cement specifications almost always are stated in terms of these standard tests. The test equipment needed to perform the API tests includes: (1) a *mud balance* for determining the slurry density, (2) a *filter press* for determining the filtration rate of the slurry, (3) a *rotational viscometer* for determining the rheological properties of the slurry, (4) a *consistometer* for determining the thickening rate characteristics of the slurry, (5) a *cement permeameter* for determining the permeability of the set cement, (6) *specimen molds* and *strength testing machines* for determining the tensile and compressive strength of the cement, (7) an *autoclave* for determining the soundness of the cement, and (8) a *turbidimeter* for determining the fineness of the cement. Unlike drilling fluid testing, routine testing of the cement slurry normally is not done at the rig site. However, it is imperative for the drilling engineer to understand the nature of these tests if he is to interpret cement specifications and reported test results properly.

The mud balance, filter press, and rotational viscometer used for cement testing are basically the same equipment described in Chap. 2 for testing drilling fluids. However, when measuring the density of cement slurries, entrained air in the sample is more difficult to remove. The pressurized mud balance

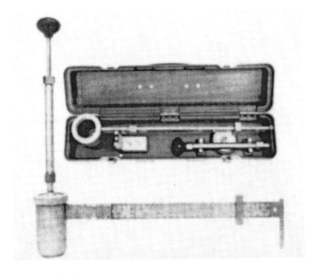

Fig. 3.2 — Pressurized mud balance.

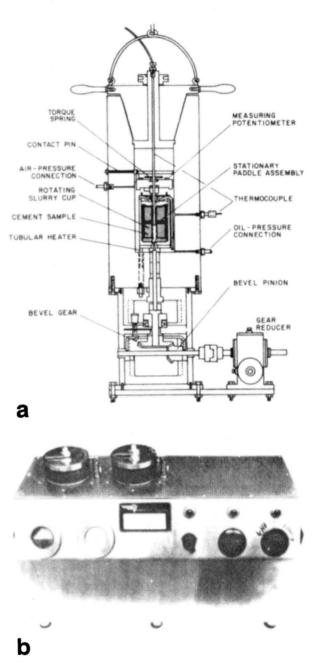

a

b

Fig. 3.3 — Cement consistometer: (a) schematic of high-pressure consistometer, (b) atmospheric-pressure consistometer.

shown in Fig. 3.2 can be used to minimize the effect of the entrained air.

3.2.1 Cement Consistometer. The pressurized and atmospheric-pressure consistometers used in testing cement are shown in Fig. 3.3. The apparatus consists essentially of a rotating cylindrical slurry container equipped with a stationary paddle assembly, all enclosed in a pressure chamber capable of withstanding temperatures and pressures encountered in well cementing operations. The cylindrical slurry chamber is rotated at 150 rpm during the test. The slurry consistency is defined in terms of the torque exerted on the paddle by the cement slurry. The relation between torque and slurry consistency is given by

$$B_c = \frac{T - 78.2}{20.02}, \quad \dots\dots\dots\dots\dots\dots\dots (3.5)$$

where T=the torque on the paddle in g-cm and B_c=the slurry consistency in API consistency units designated by B_c. The thickening time of the slurry is defined as the time required to reach a consistency of 100 B_c. This value is felt to be representative of the upper limit of pumpability. The temperature and pressure schedule followed during the test must be given with the thickening time for the test results to be meaningful. API periodically reviews field data concerning the temperatures and pressures encountered during various types of cementing operations and publishes recommended schedules for use with the consistometer. At present, 31 published schedules are available for simulating various cementing operations. Schedule 6, designed to simulate the average conditions encountered during the cementing of casing at a depth of 10,000 ft, is shown in Table 3.1.

The atmospheric-pressure consistometer is frequently used to simulate a given history of slurry pumping before making certain tests on the slurry. For example, the rheological properties of cement slurries are time dependent since the cement thickens with time. The history of shear rate, temperature, and pressure before measuring the cement rheological properties using a rotational viscometer can be specified in terms of a schedule followed using the consistometer. The consistometer also is used to determine the maximum, minimum, normal, and free water content of the slurry. In these tests, the sample is placed first in the consistometer and stirred for a period of 20 minutes at 80°F and atmospheric pressure. The *minimum water content* is the amount of mixing water per sack of cement that will result in a consistency of 30 B_c at the end of this period. The *normal water content* is the amount of mixing water per sack of cement that will result in a consistency of 11 B_c at the end of the test. The *free water content* is determined by pouring a 250-mL sample from the consistometer into a

TABLE 3.1 – EXAMPLE CONSISTOMETER SCHEDULE[1]
(Schedule 6 – 10,000-ft (3050 m) casing cement specification test)

Surface temperature, °F (°C)	80 (27)
Surface pressure, psi (kg/cm²)	1,250 (88)
Mud density	
lbm/gal (kg/L)	12 (1.4)
lbm/cu ft	89.8
psi/Mft (kg/cm³/m)	623 (0.144)
Bottomhole temperature, °F (°C)	144 (62)
Bottomhole pressure, psi (kg/cm²)	7,480 (526)
Time to reach bottom, minutes	36

Time (minutes)	Pressure (psi)	Pressure (kg/cm²)	Temperature (°F)	Temperature (°C)
0	1,250	88	80	27
2	1,600	113	84	29
4	1,900	134	87	31
6	2,300	162	91	33
8	2,600	183	94	34
10	3,000	211	98	37
12	3,300	232	101	38
14	3,700	260	105	41
16	4,000	281	108	42
18	4,400	309	112	44
20	4,700	330	116	47
22	5,100	359	119	48
24	5,400	380	123	51
26	5,700	401	126	52
28	6,100	429	130	54
30	6,400	451	133	56
32	6,800	478	137	58
34	7,100	499	140	60
36	7,480	526	144	62

Final temperature and pressure should be held constant to completion of test, within ±2°F (±1°C) and ±100 psi (±7 kg/cm²), respectively.

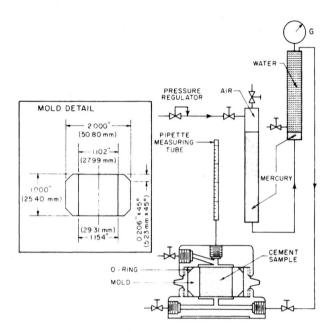

Fig. 3.4 – Cement permeameter.

glass graduated cylinder and noting the amount of free supernatant water that separates from the slurry over a 2-hour period. The *maximum water content* is defined as the amount of water per sack of cement that will result in 3.5 mL of free water. A consistometer designed to operate only at atmospheric pressure is frequently used in conjunction with the determination of the slurry rheological properties and water content.

Example 3.2. The torque required to hold the paddle assembly stationary in a cement consistometer rotating at 150 rpm is 520 g-cm. Compute the slurry consistency.

$$B_c = \frac{T - 78.2}{20.02} = \frac{520 - 78.2}{20.02}$$

$$= 22 \text{ consistency units.}$$

3.2.2 Cement Permeameter. A schematic of the permeameter used in the cement permeability test recommended by API is shown in Fig. 3.4.[1] The permeability of a set cement core to water is determined by measuring the flow rate through the core at

a given pressure differential across the length of the core. The permeability then is computed using an appropriate form of Darcy's law:

$$k = 14,700 \frac{q\mu L}{A\Delta p}, \quad \ldots \ldots \ldots \ldots \ldots \ldots (3.6)$$

where

- k = permeability, md,
- q = flow rate, mL/s,
- μ = water viscosity, cp,
- L = sample length, cm,
- A = sample cross-sectional area, cm², and
- Δp = pressure differential, psi.

The curing time, temperature, and pressure of the sample usually are reported with the cement permeability.

Example 3.3. A Class E cement core having a length of 2.54 cm and a diameter of 2.865 cm allows a water flow rate of 0.0345 mL/s when placed under a pressure differential of 20 psi. A second core containing 40% silica cured in a similar manner allows only 0.00345 mL/s of water to flow under a pressure differential of 200 psi. Compute the permeability of the two cement samples.

Solution. Using Darcy's law for linear flow of liquids as defined by Eq. 3.6 gives

TABLE 3.2 – WELL SIMULATION TEST SCHEDULES FOR CURING STRENGTH SPECIMENS

Schedule Number	Depth (ft)	Depth (m)	Pressure* (psi)	Pressure* (kg/cm²)	Temperature, °F (°C) — Elapsed Time from First Application of Heat and Pressure									
					0:30	0:45	1:00	1:15	1:30	2:00	2:30	3:00	3:30	4:00**
1S	1,000	310	800	56	82 (28)	83 (28)	84 (29)	85 (29)	86 (30)	87 (31)	89 (31)	91 (33)	93 (34)	95 (35)
2S	2,000	610	1,600	113	93 (34)	94 (34)	96 (36)	97 (36)	98 (37)	100 (38)	103 (39)	105 (41)	108 (42)	110 (43)
3S	4,000	1220	3,000	211	106 (41)	108 (42)	110 (43)	113 (45)	115 (46)	120 (49)	125 (52)	130 (54)	135 (57)	140 (60)
4S	6,000	1830	3,000	211	116 (47)	120 (49)	124 (51)	128 (53)	131 (55)	139 (59)	147 (64)	155 (68)	162 (72)	170 (77)
5S	8,000	2440	3,000	211	126 (52)	131 (55)	136 (58)	142 (61)	147 (64)	158 (70)	168 (76)	179 (82)	190 (88)	200 (93)
6S	10,000	3050	3,000	211	133 (56)	148 (64)	154 (68)	161 (72)	167 (75)	180 (82)	192 (89)	205 (96)	218 (103)	230 (110)
7S	12,000	3660	3,000	211	143 (62)	173 (78)	179 (82)	186 (86)	193 (89)	206 (97)	220 (104)	233 (112)	246 (119)	260 (127)
8S	14,000	4270	3,000	211	153 (67)	189 (87)	210 (99)	216 (103)	223 (106)	236 (113)	250 (121)	263 (128)	277 (136)	290 (143)
9S	16,000	4880	3,000	211	164 (73)	206 (97)	248 (120)	254 (123)	260 (127)	272 (133)	284 (140)	296 (147)	308 (153)	320 (160)
10S	18,000	5490	3,000	211	179 (82)	227 (108)	277 (136)	302 (150)	307 (153)	315 (157)	324 (162)	333 (167)	341 (172)	350 (177)
11S	20,000	6100	3,000	211	184 (84)	236 (113)	288 (142)	340 (171)	344 (173)	351 (177)	358 (181)	366 (186)	373 (189)	380 (193)

*The test pressure shall be applied as soon as specimens are placed in the pressure vessel and maintained at the given pressure within the following limits for the duration of the curing period:

- Schedule 1S . . . . 800 ± 100 psi (56 ± 7 kg/cm²)
- Schedule 2S . . . . 1600 ± 200 psi (113 ± 14 kg/cm²)
- Schedules 3S through 11S . . . . 3000 ± 500 psi (211 ± 35 kg/cm²)

**Final temperature (Col. 13) shall be maintained ± 3°F (± 2°C) throughout the remainder of the curing period.

$$k_1 = 14,700 \frac{0.0345(1.0)(2.54)}{\frac{\pi}{4}(2.865)^2(20)} = 10 \text{ md.}$$

The permeability observed after the addition of silica is given by

$$k_2 = 14,700 \frac{0.00345(1.0)(2.54)}{\frac{\pi}{4}(2.865)^2(200)} = 0.1 \text{ md.}$$

3.2.3 Strength, Soundness and Fineness Tests.

The standard tests for cement compressive strength, tensile strength, soundness, and fineness published in the latest ASTM C 190, C 109, C 151, and C 115 also are used by API for drilling cements. The compressive strength of the set cement is the compressional force required to crush the cement divided by the cross-sectional area of the sample. Test schedules for curing strength test specimens are recommended by API. These schedules are based on average conditions encountered in different types of cementing operations and are updated periodically on the basis of current field data. The test schedules published in RP 10B[1] in Jan. 1982 are given in Table 3.2. *The compressive strength of the cement is usually about 12 times greater than the tensile strength* at any given curing time. Thus, frequently only the compressive strength is reported.

The *soundness* of the cement is the percent linear expansion or contraction observed after curing in an autoclave under saturated steam at a pressure of 295 psig for 3 hours. A cement that changes dimensions upon curing may tend to bond poorly to casing or to form cracks.

The *fineness* of the cement is a measure of the size of the cement particle achieved during grinding. The fineness is expressed in terms of a calculated total particle surface area per gram of cement. The fineness is calculated from the rate of settlement of cement particles suspended in kerosene in a Wagner turbidimeter. The finer the cement is ground during manufacture, the greater the surface area available for contact with water, and the more rapid is the hydration process.

3.3 Standardization of Drilling Cements

API has defined eight standard classes and three standard types of cement for use in wells. The eight classes specified are designated Class A to Class H. The intended meanings of the various classes are defined in Table 3.3. The three types specified are (1) ordinary "O," (2) moderate sulfate-resistant "MSR," and (3) high sulfate-resistant "HSR." The chemical requirement for the various types and classes are given in Table 3.4 and the physical requirements are given in Table 3.5.

The physical requirements of the various classes of cement given in Table 3.5 apply to cement samples prepared according to API specifications. To provide uniformity in testing, it is necessary to specify the amount of water to be mixed with each type of cement. These water-content ratios, shown in Table 3.6, often are referred to as the normal water content or "API water" of the cement class. As will be discussed in the next section, Wyoming bentonite sometimes is added to the cement slurry to reduce the slurry density, or barium sulfate is added to increase the slurry density. API specifies that the water content be increased 5.3 wt% for each weight percent of bentonite added and 0.2 wt% for each weight percent of barium sulfate added.

The relation between well depth and cementing time used in the specifications for the various cement classes is shown in Fig. 3.5. The relation shown assumes a cement mixing time of 20 cu ft/min and a displacement rate after mixing of 50 cu ft/min. Also, a 7.0-in.-OD casing having a cross-sectional area of 33.57 sq in. is assumed. For these conditions, which are felt to be representative of current field practice, the time required to mix and displace various volumes of cement has been plotted as a function of depth. Also plotted are the cement volumes used in determining the recommended minimum thickening time.

3.3.1 Construction Industry Cement Designations.

The majority of the cement produced in this country is used in construction with only about 5% being

TABLE 3.3—STANDARD CEMENT CLASSES DESIGNATED BY API[1]

Class A: Intended for use from surface to 6,000-ft (1830-m) depth, when special properties are not required. Available only in ordinary type (similar to ASTM C 150 Type I).

Class B: Intended for use from surface to 6,000-ft (1830-m) depth, when conditions require moderate to high sulfate-resistance. Available in both moderate (similar to ASTM C 150, Type II) and high sulfate-resistant types.

Class C: Intended for use from surface to 6,000-ft (1830-m) depth, when conditions require high early strength. Available in ordinary and moderate (similar to ASTM C 150, Type III) and high sulfate-resistant types.

Class D: Intended for use from 6,000- to 10,000-ft depth (1830- to 3050-m) depth, under conditions of moderately high temperatures and pressures. Available in both moderate and high sulfate-resistant types.

Class E: Intended for use from 10,000- to 14,000-ft (3050- to 4270-m) depth, under conditions of high temperatures and pressures. Available in both moderate and high sulfate-resistant types.

Class F: Intended for use from 10,000- to 16,000-ft (3050- to 4880-m) depth, under conditions of extremely high temperatures and pressures. Available in both moderate and high sulfate-resistant types.

Class G: Intended for use as a basic cement from surface to 8,000-ft (2400-m) depth as manufactured, or can be used with accelerators and retarders to cover a wide range of well depths and temperatures. No additions other than calcium sulfate or water, or both, shall be interground or blended with the clinker during manufacture of Class G cement. Available in moderate and high sulfate-resistant types.

Class H: Intended for use as a basic cement from surface to 8,000-ft (2440-m) depth as manufactured, and can be used with accelerators and retarders to cover a wide range of well depths and temperatures. No additions other than calcium sulfate or water, or both, shall be interground or blended with the clinker during manufacture of Class H cement. Available only in moderate sulfate-resistant type.

used in oil and gas wells. In some cases, it may be necessary to use cement products normally marketed for the construction industry. This is especially true when working in foreign countries. Five basic types of portland cements are used commonly in the construction industry. The ASTM classifications and international designations for these five cements are shown in Table 3.7. Note that ASTM Type I, called normal, ordinary, or common cement, is similar to API Class A cement. Likewise, ASTM Type II, which is modified for moderate sulfate resistance is similar to API Class B cement. ASTM Type III, called high early strength cement, is similar to API Class C cement.

3.4 Cement Additives

Today more than 40 chemical additives are used with various API classes of cement to provide acceptable

slurry characteristics for almost any subsurface environment. Essentially all of these additives are free-flowing powders that either can be dry blended with the cement before transporting it to the well or can be dispersed in the mixing water at the job site. At present, the cement Classes G and H can be modified easily through the use of additives to meet almost any job specifications economically. The use of a modified Class H cement has become extremely popular.

The cement additives available can be subdivided into these functional groups: (1) density control additives, (2) setting time control additives, (3) lost circulation additives, (4) filtration control additives, (5) viscosity control additives, and (6) special additives for unusual problems. The first two categories are perhaps the most important because they receive consideration on almost every cement job. Some additives serve more than one purpose and, thus, would fit under more than one of the classifications shown above.

The nomenclature used by the petroleum industry to express the concentration of cement additives often is confusing to the student. However, most of the confusion can be cleared up by pointing out that the reference basis of cement mixtures is a unit weight of cement. When the concentration of an additive is expressed as a "weight percent" or just "percent," the intended meaning is usually that the weight of the additive put in the cement mixture is computed by multiplying the weight of cement in the mixture by the weight percent given by 100%. The concentration of liquid additives sometimes is expressed as gallons per sack of cement. A sack of cement contains 94 lbm unless the cement product is a blend of cement and some other material. The water content of the slurry sometimes is expressed as water cement ratio in gallons per sack and sometimes expressed as a weight percent. The term "percent mix" is used for water content expressed as a weight percent. Thus,

$$\text{percent mix} = \frac{\text{water weight}}{\text{cement weight}} \times 100.$$

The theoretical volume of the slurry mixture is calculated using the same procedure outlined in Sec. 2.2 of Chap. 2 for drilling fluids. Ideal mixing can be assumed unless one or more of the components are dissolved in the water phase of the cement. Many components are used in low concentration and have very minor effects on slurry volume. Physical properties of cement components needed to perform the ideal mixing calculations are given in Table 3.8.

The volume of slurry obtained per sack of cement used is called the *yield* of the cement. This term should not be confused with the yield of a clay or the yield point of a fluid as discussed in Chap. 2.

Self study HW

Example 3.4. It is desired to mix a slurry of Class A cement containing 3% bentonite, using the normal mixing water as specified by API (Table 3.6). Determine the weight of bentonite and volume of

TABLE 3.4—CHEMICAL REQUIREMENTS OF API CEMENT TYPES[1]

			Cement Class			
Ordinary Type (O)	A	B	C	D,E,F	G	H
Magnesium oxide (MgO), maximum, %	5.00	—	5.00	—	—	—
Sulfur trioxide (SO_3), maximum, %	3.50	—	4.50	—	—	—
Loss on ignition, maximum, %	3.00	—	3.00	—	—	—
Insoluble residue, maximum, %	0.75	—	0.75	—	—	—
Tricalcium aluminate ($3CaO \cdot Al_2O_3$), maximum, %	—	—	15.00	—	—	—
Moderate Sulfate-Resistant Type (MSR)						
Magnesium oxide (MgO), maximum, %	—	5.00	5.00	5.00	5.00	5.00
Sulfur trioxide (SO_3), maximum, %	—	3.00	3.50	2.50	2.50	2.50
Loss on ignition, maximum, %	—	3.00	3.00	3.00	3.00	3.00
Insoluble residue, maximum, %	—	0.75	0.75	0.75	0.75	0.75
Tricalcium silicate ($3CaO \cdot SiO_2$), %						
maximum	—	—	—	—	58.00	58.00
maximum	—	—	—	—	48.00	48.00
Tricalcium aluminate ($3CaO \cdot Al_2O_3$), maximum, %	—	8.00	8.00	8.00	8.00	8.00
Total alkali content expressed as sodium oxide (Na_2O) equivalent, maximum, %	—	—	—	—	0.60	0.60
High Sulfate-Resistant Type (HSR)						
Magnesium oxide (MgO), maximum, %	—	5.00	5.00	5.00	5.00	—
Sulfur trioxide (SO_3), maximum %	—	3.00	3.50	2.50	2.50	—
Loss on ignition, maximum, %	—	3.00	3.00	3.00	3.00	—
Insoluble residue, maximum, %	—	0.75	0.75	0.75	0.75	—
Tricalcium silicate ($3CaO \cdot SiO_2$), %						
maximum	—	—	—	—	65.00	—
maximum	—	—	—	—	48.00	—
Tricalcium aluminate ($3CaO \cdot Al_2O_3$), maximum, %	—	3.00	3.00	3.00	3.00	—
Tetracalcium aluminoferrite ($4CaO \cdot Al_2O_3 \cdot Fe_2O_3$) plus twice the tricalcium aluminate ($3CaO \cdot Al_2O_3$), maximum, %	—	24.00	24.00	24.00	24.00	—
Total alkali content expressed as sodium oxide (Na_2O) equivalent, maximum, %[3]			0.60			

TABLE 3.5—PHYSICAL REQUIREMENTS OF API CEMENT TYPES[1]

			Cement Case					
	A	B	C	D	E	F	G	H
Soundness (autoclave expansion) maximum, %	0.80	0.80	0.80	0.80	0.80	0.80	0.80	0.80
Fineness* (specific surface), minimum, cm^2/g	1,500	1,600	2,200	—	—	—	3.5**	2.5**
Free water content, maximum, mL								

Compressive Strength Test 8-Hour Curing Time	Schedule Number, Table 6.1 RP10B	Curing Temperature (°F)	(°C)	Curing Pressure (psi)	(kg/cm^2)	Minimum Compressive Strength, psi (kg/cm^2)											
	—	100	38	Atmos.		250	(18)	200	(14)	300	(21)	—	—	—	—	1,500	(106)	1,500	(106)
	1S	95	35	800	56	—	—	—	—	—	—	—	—	—	—			
	3S	140	60	3.000	211	—	—	—	—	—	—	—	—	—	—			
	6S	230	110	3.000	211	—	—	—	—	—	500	(35)	—	—	—	—		
	8S	290	143	2.000	211	—	—	—	—	—	—	500	(35)	—	—	—		
	9S	330	2.000	211	—	—	—	—	—	—	—	—	500	(35)	—	—	—	—

Compressive Strength Test 24-Hour Curing Time	Schedule Number, Table 6.1 RP10B	Curing Temperature (°F)	(°C)	Curing Pressure (psi)	(kg/cm^2)	Minimum Compressive Strength, psi (kg/cm^2)												
	—	100	38	Atmos.		1.800	(127)	1.500	(106)	2.000	(141)	—	—	—	—	—	—	—
	4S	170	77	2.000	211	—	—	—	—	—	1.000	(70)	1.000	(70)	—	—	—	—
	6S	230	110	3.000	211	—	—	—	—	—	2.000	(141)	—	—	1.000	(70)	—	—
	8S	290	143	3.000	211	—	—	—	—	—	—	2.000	(141)	—	—	—	—	—
	9S	320	160	2.000	211	—	—	—	—	—	—	—	—	1.000	(70)	—	—	—

Pressure Temperature Thickening Time Test	Well Simulation Test Schedule Number, Table 7.2 RP10B	Simulated Well Depth (ft)	(m)	Maximum Consistency 15 to 30-Minute Stirring Period U_c†	Minimum Thickening Time (minutes)***							
	1	1.000	310	30	90	90	90	—	—	—	—	—
	4	6.000	1830	30	90	90	90	90	—	—	—	—
	5	8.000	2440	30	—	—	—	—	—	—	90	90
	6	10.000	3050	30	—	—	—	100	100	100	—	—
	8	14.000	4270	30	—	—	—	—	154	—	—	—
	9	16.000	4880	30	—	—	—	—	—	190	—	—

*Determined by Wagner turbidimeter apparatus described in ASTM C 115 *Fineness of Portland Cement by the Turbidimeter,* current edition of *ASTM Book of Standards,* Part 9

**Based on 250 mL volume, percentage equivalent of 3.5 mL is 1.4%

***Thickening time requirements are based on 75 percentile values of the total cementing times observed in the casing survey, plus a 25% safety factor

†Units of slurry consistency (U) formerly referred to as "poises"

Maximum thickening-time requirement for Schedule 5 is 120 minutes

TABLE 3.6—NORMAL WATER CONTENT OF CEMENT RECOMMENDED BY API

API Class Cement	Water (%) by Weight of Cement	Water gal per sack	Water L per sack
A and B	46	5.19	19.6
C	56	6.32	23.9
D, E, F, and H	38	4.29	16.2
G	44	4.97	18.8
J (tentative)	*	*	*

*As recommended by the manufacturer.

Note 1: The addition of bentonite to cement requires that the amount of water be increased. It is recommended, for testing purposes, that 5.3% water be added for each 1% bentonite in all API classes of cement. For example, a Class A cement slurry having a water/cement ratio of 0.46, to which is added 3% bentonite, will require an increase in water/cement ratio to 0.619.

Note 2: The addition of barite to cement generally requires that the amount of water be increased. It is recommended, for testing purposes, that 0.2% water be added for each 1% barite. For example, a cement slurry having a normal water/cement ratio of 0.38 and weighted to 18 lbm/gal (134.6 lbm/cu ft) (2.2 kg/L) by addition of 60% barite, will require an increase in water/cement ratio to 0.50.

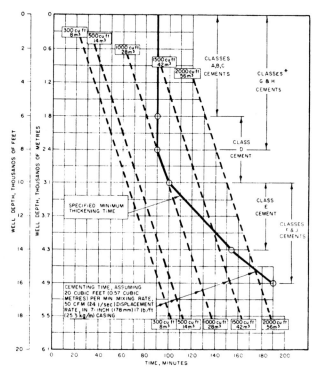

Fig. 3.5 – Well depth and cementing time relationship used in definition of API cement classes.[1]

water to be mixed with one 94-lbm sack of cement. Also, compute the percent mix, yield, and density of the slurry.

Solution. The weight of bentonite to be blended with one sack of Class A cement is

$$0.03(94) = 2.82 \text{ lbm}.$$

The normal water content for Class A cement is 46% (Table 3.6). However, 5.3% water must be added for each percent bentonite. Thus, the percent mix is

$$46 + 3(5.3) = 61.9\%.$$

The water volume to be added per sack of Class A cement is given by

$$\frac{0.619(94 \text{ lbm/sack})}{8.33 \text{ lbm/gal}} = 6.98 \text{ gal}.$$

The specific gravities of cement and bentonite are 3.14 and 2.65, respectively (Table 3.8). The volume of the slurry is given by

$$\frac{94 \text{ lbm}}{3.14(8.33)\text{lbm/gal}} + \frac{2.82 \text{ lbm}}{2.65(8.33)\text{lbm/gal}}$$
$$+ 6.98 \text{ gal/sack} = 10.7 \text{ gal/sack}.$$

The yield of the slurry is

$$\frac{10.7 \text{ gal/sack}}{7.48 \text{ gal/cu ft}} = 1.43 \text{ cu ft/sack}.$$

The density of the slurry is the total mass divided by the total volume or

$$\frac{94 + 2.82 + 8.33(6.98)}{10.7} = 14.48 \text{ lbm/gal}.$$

3.4.1 Density Control. The density of the cement slurry must be high enough to prevent the higher-pressured formations from flowing into the well during cementing operations, yet not so high as to cause fracture of the weaker formations. In most cases, the density of the cement slurry obtained by mixing cement with the normal amount of water will be too great for the formation fracture strength, and it will be desirable to lower the slurry density.

TABLE 3.7—BASIC ASTM CEMENT TYPES[2]

ASTM Type	International Designation	API Class	Common Name	Typical Composition C_3S	C_2S	C_3A	C_4AF
I	OC	A	normal, ordinary, or common	53	24	8	8
II		B	modified	47	32	3	12
III	RHC	C	high early strength	58	16	8	8
IV	LHC		low heat	26	54	2	12
V	SRC		sulfate-resisting	—	—	—	—

Reducing the cement density also tends to reduce the overall cost of the cement slurry. Slurry density is reduced by using a higher water/cement ratio or adding low-specific-gravity solids, or both. Also, nonstandard cements having a lower specific gravity are available. Trinity Lite-wate[TM] cement, which has a specific gravity of 2.8, is a very popular low-density cement.

The low-specific-gravity solids commonly used to reduce slurry density include (1) bentonite (sodium montmorillonite), (2) diatomaceous earth, (3) solid hydrocarbons, (4) expanded perlite, and (5) pozzolan. When extremely weak formations are present, it may not be possible to reduce slurry density sufficiently to prevent fracture. In this case, the mud column in front of the cement slurry can be aerated

TABLE 3.8—PHYSICAL PROPERTIES OF CEMENTING MATERIALS[5]

Material	Bulk Weight (lbm/cu ft)	Specific Gravity	Weight 3.6[a] Absolute Gal	Absolute Volume gal/lbm	Absolute Volume cu ft/lbm
API cements	94	3.14	94	0.0382	0.0051
Ciment Fondu	90	3.23	97	0.0371	0.0050
Lumnite cement	90	3.20	96	0.0375	0.0050
Trinity Lite-Wate	75	2.80	75.0[e]	0.0429	0.0057
Activated charcoal	14	1.57	47.1	0.0765	0.0102
Barite	135	4.23	126.9	0.0284	0.0038
Bentonite (gel)	60	2.65	79.5	0.0453	0.0060
Calcium chloride, flake[b]	56.4	1.96	58.8	0.0612	0.0082
Calcium chloride, powder[b]	50.5	1.96	58.8	0.0612	0.0082
Cal-Seal, gypsum cement	75	2.70	81.0	0.0444	0.0059
CFR-1[b]	40.3	1.63	48.9	0.0736	0.0098
CFR-2[b]	43.0	1.30	39.0	0.0688	0.0092
DETA (liquid)	59.5	0.95	28.5	0.1258	0.0168
Diacel A[b]	60.3	2.62	78.6	0.0458	0.0061
Diatomaceous earth	16.7	2.10	63.0	0.0572	0.0076
Diacel LWL[b]	29.0	1.36	40.8	0.0882	0.0118
Diesel Oil No. 1 (liquid)	51.1	0.82	24.7	0.1457	0.0195
Diesel Oil No. 2 (liquid)	53.0	0.85	25.5	0.1411	0.0188
Gilsonite	50	1.07	32	0.1122	0.0150
HALDAD®-9[b]	37.2	1.22	36.6	0.0984	0.0131
HALDAD®-14[b]	39.5	1.31	39.3	0.0916	0.0122
Hematite	193	5.02	150.5	0.0239	0.0032
HR-4[b]	35	1.56	46.8	0.0760	0.0103
HR-7[b]	30	1.30	39	0.0923	0.0123
HR-12[b]	23.2	1.22	36.6	0.0984	0.0131
HR-L (liquid)[b]	76.6	1.23	36.9	0.0976	0.0130
Hydrated lime	31	2.20	66	0.0545	0.0073
Hydromite	68	2.15	64.5	0.0538	0.0072
LA-2 Latex (liquid)	68.5	1.10	33	0.1087	0.0145
LAP-1 Latex[b]	50	1.25	37.5	0.0960	0.0128
LR-11 Resin (liquid)	79.1	1.27	38.1	0.0945	0.0126
NF-1 (liquid)[b]	61.1	0.98	29.4	0.1225	0.0164
NF-P[b]	40	1.30	39.0	0.0923	0.0123
Perlite regular	8[c]	2.20	66.0	0.0546	0.0073
Perlite Six	38[d]	—	—	0.0499	0.0067
Pozmix® A	74	2.46	74	0.0487	0.0065
Pozmix® D	47	2.50	73.6	0.0489	0.0065
Salt (dry NaCl)	71	2.17	65.1	0.0553	0.0074
Salt (in solution at 77°F with fresh water)					
6%, 0.5 lbm/gal	—	—	—	0.0384	0.0051
12%, 1.0 lbm/gal	—	—	—	0.0399	0.0053
18%, 1.5 lbm/gal	—	—	—	0.0412	0.0055
24%, 2.0 lbm/gal	—	—	—	0.0424	0.0057
Saturated, 3.1 lbm/gal	—	—	—	0.0445	0.0059
Salt (in solution at 140°F with fresh water)					
saturated, 3.1 lbm/gal	—	—	—	0.0458	0.0061
Sand (Ottawa)	100	2.63	78.9	0.0456	0.0061
Silica flour (SSA-1)	70	2.63	78.9	0.0456	0.0061
Coarse silica (SSA-2)	100	2.63	78.9	0.0456	0.0061
Tuf Additive No. 1	—	1.23	36.9	0.0976	0.0130
Tuf-Plug	48	1.28	38.4	0.0938	0.0125
Water	62.4	1.00	30.0	0.1200	0.0160

[a] Equivalent to one 94-lbm sack of cement in volume.
[b] When less than 5% is used, these chemicals may be omitted from calculations without significant error.
[c] For 8 lbm of Perlite regular use a volume of 1.43 gal at zero pressure.
[d] For 38 lbm of Perlite Six use a volume of 2.89 gal at zero pressure.
[e] 75 lbm = 3.22 absolute gal.

TABLE 3.9—WATER REQUIREMENTS OF CEMENTING MATERIALS[4]

Material	Water Requirements
API Class A and B cements	5.2 gal (0.70 cu ft)/94-lbm sack
API Class C cement (Hi Early)	6.3 gal (0.84 cu ft)/94-lbm sack
API Class D and E cements (retarded)	4.3 gal (0.58 cu ft)/94-lbm sack
API Class G cement	5.0 gal (0.67 cu ft)/94-lbm sack
API Class H cement	4.3 to 5.2 gal/94-lbm sack
Chem Comp cement	6.3 gal (0.84 cu ft)/94-lbm sack
Ciment Fondu	4.5 gal (0.60 cu ft)/94-lbm sack
Lumnite cement	4.5 gal (0.60 cu ft)/94-lbm sack
HLC	7.7 to 10.9 gal/87-lbm sack
Trinity Lite-Wate cement	7.7 gal (1.03 cu ft)/75-lbm sack (maximum)
Activated charcoal	none at 1 lbm/sack of cement
Barite	2.4 gal (0.32 cu ft)/100-lbm sack
Bentonite (gel)	1.3 gal (0.174 cu ft)/2% in cement
Calcium chloride	none
Gypsum hemihydrate	4.8 gal (0.64 cu ft)/100-lbm sack
CFR-1	none
CFR-2	none
Diacel A	none
Diacel D	3.3 to 7.2 gal/10% in cement (see Lt. Wt. Cement)
Diacel LWL	none (up to 0.7%)
	0.8 to 1.0 gal/1% in cement (except gel or Diacel D slurries)
Gilsonite	2.0 gal (0.267 cu ft)/50 lbm/cu ft
HALAD -9	none (up to 0.5%)
	0.4 to 0.5 gal/sack of cement at over 0.5%
HALAD -14	none
Hematite	0.36 gal (0.048 cu ft)/100-lbm sack
HR-4	none
HR-7	none
HR-12	none
HR-20	none
Hydrated lime	0.153 gal (0.020 cu ft)/lbm
Hydromite	3.0 gal (0.40 cu ft)/100-lbm sack
LA-2 Latex	0 to 0.8 gal/sack of cement
LAP-1 powdered latex	1.7 gal (0.227 cu ft)/1% in cement
NF-P	none
Perlite regular	4.0 gal (0.535 cu ft)/8 lbm/cu ft
Perlite Six	6.0 gal (0.80 cu ft)/38 lbm/cu ft
Pozmix A	3.6 gal (0.48 cu ft)/74 lbm/cu ft
Salt (NaCl)	none
Sand, Ottawa	none
Silica flour (SSA-1)	1.5 gal (0.20 cu ft)/35% in cement (32.9 lbm)
Coarse silica (SSA-2)	none
Tuf Additive No. 1	none
Tuf Plug	none

with nitrogen to reduce hydrostatic pressure further.

In areas where the formation pore pressure is extremely high, it may be necessary to increase the slurry density. Slurry density usually is increased by using a lower water content or adding high-specific-gravity solids. The high-specific-gravity solids commonly used to increase slurry density include (1) hematite, (2) ilmenite, (3) barite (barium sulfate), and (4) sand. The specific gravity of selected cement additives are shown in Table 3.8. The water requirements for the various additives are shown in Table 3.9.

3.4.2 Bentonite. The use of bentonite (sodium montmorillonite) clay for building drilling fluid viscosity has been discussed previously in Chap. 2. This same clay mineral is used extensively as an additive for lowering cement density. However, bentonite marketed for use in drilling fluid sometimes is treated with an organic polymer that is

undesirable for use in cement slurries since it tends to increase slurry viscosity. The addition of bentonite lowers the slurry density because of its lower specific gravity and because its ability to hydrate permits the use of much higher water concentrations. Bentonite concentrations as high as 25% by weight of cement have been used. The bentonite usually is blended dry with the cement before mixing with water, but it can be prehydrated in the mixing water. Much higher increases in water content can be obtained for each percent bentonite added when the bentonite is prehydrated in the mixing water. The ratio of bentonite dry blended to bentonite prehydrated is about 3.6:1 for comparable slurry properties.

In addition to lowering slurry density, the addition of bentonite lowers slurry cost. However, a high percentage of bentonite in cement also will cause a reduction in cement strength and thickening time. Also, the higher water content lowers the resistance to sulfate attack and increases the permeability of the set cement. At temperatures above 230°F, the use of bentonite promotes retrogression of strength in cements with time. Typical data showing the effect of bentonite concentration on various properties of Class A cement are shown in Table 3.10. However, test results have been found to vary significantly from batch to batch. When exact data are needed, tests should be conducted using the same materials and mixing water that will be used in the cementing operations.

3.4.3 Diatomaceous Earth. A special grade of diatomaceous earth also is used in portland cements to reduce slurry density. The diatomaceous earth has lower specific gravity than bentonite (Table 3.8) and permits higher water/cement ratios without resulting in free water. In addition, the silica contained in the diatomite reacts chemically with the calcium hydroxide released as portland cement sets, and it produces a gel that becomes cementitious with age and temperature. Diatomaceous earth concentrations as high as 40% by weight of cement have been used. As in the case of bentonite, the thickening time and strength of the cement are increased as the diatomaceous earth concentration is increased. Example slurry properties obtained with various concentrations of diatomaceous earth are shown in Table 3.11.

3.4.4 Solid Hydrocarbons. Gilsonite (an asphaltite) and sometimes coal are used as extremely low-specific-gravity solids for reducing slurry density without greatly increasing the water content. The addition of gilsonite has almost no effect on slurry thickening time, and low-density cements obtained using gilsonite have much higher compressive strengths than other types of low-density cements. Example properties obtained with gilsonite and Class A cements are shown in Table 3.12.

3.4.5 Expanded Perlite. Perlite is volcanic glass containing a small amount of combined water. The raw ore is expanded by introduction into a kiln, where the temperature is raised to the fusion point of

TABLE 3.10—CLASS A CEMENT WITH BENTONITE[4]

Bentonite (%)	Maximum Water Requirements		Slurry Weight		Slurry Volume (cu ft/sk)
	(gal/sk)	(cu ft/sk)	(lbm/gal)	(lbm/cu ft)	
0	5.2	0.70	15.6	117	1.18
2	6.5	0.87	14.7	110	1.36
4	7.8	1.04	14.1	105	1.55
6	9.1	1.22	13.5	101	1.73
8	10.4	1.39	13.1	98	1.92

Thickening Time—Hours: Minutes
(Pressure-Temperature Thickening-Time Test)

Bentonite (%)	API Casing Tests			API Squeeze Tests		
	4,000 ft	6,000 ft	8,000 ft	2,000 ft	4,000 ft	6,000 ft
0	3:00+	2:25	1:40	2:14	1:32	1:01
2	2:25	1:48	1:34	2:25	1:29	0:56
4	2:34	1:57	1:32	2:26	1:18	0:58
6	2:35	1:45	1:22	2:16	1:26	0:56
8	2:44	1:50	1:24	2:31	1:28	0:58

Compressive Strengths—psi
Atmospheric Pressure

Bentonite (%)	60°F	80°F	100°F	120°F	Bentonite (%)	60°F	80°F	100°F	120°F
12 Hours					24 Hours				
0	80	580	1,035	1,905	0	615	1,905	2,610	3,595
2	55	455	635	1,280	2	365	1,090	1,520	2,040
4	20	220	375	780	4	225	750	1,015	1,380
6	15	85	245	500	6	85	360	730	925
8	15	50	155	310	8	60	265	510	610

TABLE 3.11 – CLASS A CEMENT WITH DIATOMACEOUS EARTH[4]

Diacel D (%)	Water (gal/sk)	Slurry Weight (lbm/gal)	Slurry Volume (cu ft/sk)
0	5.2	15.6	1.18
10	10.2	13.2	1.92
20	13.5	12.4	2.42
30	18.2	11.7	3.12
40	25.6	11.0	4.19

Thickening Time – Hours:Minutes
(Pressure-Temperature Thickening-Time Test)

Diacel D (%)	API Casing Tests		
	4,000 ft	6,000 ft	8,000 ft
0	3:36	2:41	1:59
10	4:00+	3:00+	2:14
20	4:00+	3:00+	2:38
30	4:00+	3:00+	3:00+
40	4:00+	3:00+	3:00+

Compressive Strength (psi)

Diacel D (%)	Temperature			
	80°F	100°F	120°F	140°F
24 Hours				
0	875	2,305	3,850	4,690
10	280	620	1,125	1,435
20	140	295	660	1,215
40	20	45	190	635
72 Hours				
0	3,535	5,965	7,285	7,090
10	1,010	1,430	2,140	2,210
20	530	905	1,895	1,860
40	95	245	820	795

TABLE 3.12 – CLASS A CEMENT WITH GILSONITE[3]

Gilsonite (lbm/sk)	Water Ratio		Slurry Weight		Slurry Weight (cu ft/sk)
	(gal/sk)	(cu ft/sk)	(lbm/gal)	(lbm/cu ft)	
0	5.2	0.70	15.6	117	1.18
5	5.4	0.72	15.1	113	1.27
10	5.4	0.72	14.7	110	1.36
12.5	5.6	0.75	14.4	108	1.42
15	5.7	0.76	14.3	107	1.47
20	5.7	0.76	14.0	105	1.55
25	6.0	0.80	13.6	102	1.66
50	7.0	0.94	12.5	94	2.17
100	9.0	1.20	11.3	85	3.18

the ore. At this temperature, the combined water in the ore expands, producing a cellular, thin-walled structure. Mixing water will enter this cellular structure under high pressure when perlite is used in cements. Laboratory tests have shown that approximately 4.5 gallons of water are required to completely saturate 1 cu ft of expanded perlite under pressure. If this additional water is not included in the slurry, the loss of slurry water to the perlite will cause the slurry to become too viscous to pump.

Expanded perlite is marketed in a variety of blends. The term "regular" usually is applied to the unblended material. Blends of 13 lbm expanded perlite mixed with 30 lbm of waste volcanic glass fines and blends of 8 lbm of expanded perlite with 30 lbm of pozzolanic material also are common.

TABLE 3.13 – CLASS A CEMENT WITH EXPANDED PERLITE[3]

Bentonite (%)	Water (gal/sk)	Atmospheric Pressure		Pressure, 3,000 psi	
		Slurry Weight (lbm/gal)	Slurry Volume (cu ft/sk)	Slurry Weight (lbm/gal)	Slurry Volume (cu ft/sk)
1 Sack Cement, 1/2 cu ft Oil Patch Regular					
2	8.5	13.01	1.78	13.84	1.68
4	9.7	12.70	1.95	13.43	1.84
6	10.8	12.44	2.11	13.10	2.00
8	11.7	12.29	2.24	12.90	2.14
1 Sack Cement, 1 cu ft Oil Patch Regular					
2	10.5	11.91	2.20	13.20	1.99
4	11.7	11.74	2.28	12.90	2.16
6	12.8	11.60	2.54	12.83	2.32
8	13.7	11.50	2.66	12.50	2.45

TABLE 3.14 – CLASS A CEMENT WITH POZZOLAN[3]

Mixture (%)		Weight (lbm/sk)	Water Solids Ratio	Water Ratio		Slurry Weight		Slurry Volume (cu ft/sk)
Pozmix "A"	Portland Cement			gals/sk of Mix	cu ft/sk of Mix	(lbm/gal)	(lbm cu ft)	
0	100	94	0.46	5.20	0.70	15.60	117	1.18
25	75	89	0.56	5.98	0.80	14.55	109	1.29
*50	50	84	0.57	5.75	0.77	14.15	106	1.26
60	40	82	0.58	5.71	0.76	14.00	105	1.25
75	25	79	0.63	5.97	0.80	13.55	101	1.29

*Most commonly recommended blend.

TABLE 3.15 – CLASS E CEMENT WITH HEMATITE[3]

Hematite Hi-Dense No. 3 (lbm/sk cement)	Water (gal/sk)	Slurry Volume (cu ft/sk)	Slurry Weight (lbm/gal)	24-Hour Compressive Strength (psi), Curing Pressure 3,000 psi	
				260°F	290°F
0	4.5	1.08	16.25	6,965	4,125
12	4.55	1.12	17.0	6,425	4,090
28	4.6	1.18	18.0	6,290	4,275
46	4.65	1.24	19.0	5,915	5,575

Bentonite generally is used with expanded perlite to minimize water separation at the surface and to raise the viscosity. Without bentonite, the perlite tends to separate and float above the slurry because of its low density. The approximate water requirements for various concentrations of perlite and water are shown in Table 3.13. Also shown are the slurry densities obtained at atmospheric pressure and at a pressure of 3,000 psi. The density increase at the elevated pressure results from water in the slurry being forced into the cellular structure of the expanded perlite.

3.4.6 Pozzolan. Pozzolans are siliceous and aluminous mineral substances that will react with calcium hydroxide formed in the hydration of portland cement to form calcium silicates that possess cementitious properties. Diatomaceous earth, which has been discussed previously, is an example of a pozzolan. However, the term pozzolan as used in marketing cement additives usually refers to finely ground pumice or fly ash (flue dust) produced in coal-burning power plants. The specific gravity of pozzolans is only slightly less than the specific gravity of portland cement, and the water requirement of pozzolans is about the same as for portland cements. Thus, only slight reductions in density can be achieved with this material. The range of slurry densities possible using various concentrations of one type of pozzolan is shown in Table 3.14. Because of this relatively low cost, considerable cost savings can be achieved through the use of pozzolans.

3.4.7 Hematite. Hematite is reddish iron oxide ore (Fe_2O_3) having a specific gravity of approximately 5.02. Hematite can be used to increase the density of a cement slurry to as high as 19 lbm/gal. Metallic powders having a higher specific gravity than hematite have been tried but were found to settle out of the slurry rapidly unless they were ground extremely fine. When ground fine enough to prevent settling, the

TABLE 3.16 – CLASS A, B, G, or H CEMENT WITH ILMENITE[3]

Ilmenite Hi-Dense No. 2 (lbm/sk cement)	Water (gal/sk)	Slurry Volume (cu ft/sk)	Slurry Weight (lbm/gal)
0	5.2	1.18	15.6
7	5.2	1.20	16.0
22	5.2	1.25	17.0
39	5.2	1.31	18.0

TABLE 3.17 – CLASS E CEMENT WITH BARITE[3]

Barite (lbm/sk)	Water (gal/sk)	Slurry Volume (cu ft/sk)	Slurry Weight (lbm/gal)	Thickening Time. 14.000 ft. Casing Schedule (hours:minutes)	24-Hour Compressive Strength (psi) Curing Pressure. 3.000 psi 260 F	290 F
0	4.5	1.08	16.25	3:00 +	6.965	4.125
22	5.1	1.24	17.0	3:00 +	4.440	4.225
55	5.8	1.46	18.0	2:28	4.315	4.000
108	7.1	1.83	19.0	1:45	3.515	2.650

TABLE 3.18 – RETARDED CEMENT WITH OTTAWA SAND[3]

Ottawa Sand (20-40) (lbm/sk cement)	Water (gal/sk)	Slurry Volume (cu ft/sk)	Slurry Weight (lbm/gal)
0	4.5	1.08	16.25
10	4.5	1.14	16.50
28	4.5	1.25	17.00
51	4.5	1.39	17.50
79	4.5	1.56	18.00

increased water requirement results in slurry densities below that possible with hematite. The water requirement for hematite is approximately 0.36 gal/100 lbm hematite. The effect of hematite on the thickening time and compressive strength of the cement has been found to be minimal at the concentrations of hematite generally used. The range of slurry densities possible using various concentrations of hematite is shown in Table 3.15.

3.4.8 Ilmenite. Ilmenite is a black mineral composed of iron, titanium, and oxygen that has a specific gravity of approximately 4.67. Although ilmenite has a slightly lower specific gravity than hematite, it requires no additional water and provides about the same slurry density increase as hematite at comparable concentrations. Like hematite, ilmenite has little effect on thickening time or compressive strength. The range of slurry densities possible using various concentrations of ilmenite is shown in Table 3.16.

3.4.9 Barite. The use of barite, or barium sulfate, for increasing the density of drilling fluids has been discussed previously in Chap. 2. This mineral also is used extensively for increasing the density of a cement slurry. The water requirements for barite are considerably higher than for hematite or ilmenite,

TABLE 3.19 – CLASS A CEMENT WITH CALCIUM CHLORIDE[4]

Bentonite (%)	Water Requirement (gal/sk)	Slurry Weight (lbm/gal)	Slurry Volume (cu ft/sk)
0	5.2	15.6	1.18
2	6.5	14.7	1.36
4	7.8	14.1	1.55

Thickening Time (hours:minutes)
(pressure-temperature thickening-time test)

API Casing Tests

Calcium Chloride (%)	0% Bentonite 2,000 ft	4,000 ft	2% Bentonite 2,000 ft	4,000 ft	4% Bentonite 2,000 ft	4,000 ft
0	3:36	2:25	3:20	2:25	3:46	2:34
2	1:30	1:04	2:00	1:30	2:41	2:03
4	0:47	0:41	0:56	1:10	1:52	2:00

API Compressive Strength (psi)
0% Bentonite

Calcium Chloride (%)	Curing Temperature and Pressure 60°F, 0 psi	80°F, 0 psi	95°F, 800 psi	110°F, 1,600 psi	140°F, 3,000 psi
6 Hours					
0	Not Set	115	260	585	2,060
2	190	685	945	1,220	2,680
4	355	960	1,195	1,600	3,060
8 Hours					
0	20	265	445	730	2,890
2	300	1,230	1,250	1,750	3,380
4	450	1,490	1,650	2,350	2,950
12 Hours					
0	80	580	800	1,120	3,170
2	555	1,675	2,310	2,680	3,545
4	705	2,010	2,500	3,725	4,060
18 Hours					
0	375	1,405	1,725	2,525	3,890
2	970	2,520	3,000	4,140	5,890
4	1,105	2,800	3,640	4,690	4,850
24 Hours					
0	615	1,905	2,085	2,925	5,050
2	1,450	3,125	3,750	5,015	6,110
4	1,695	3,080	4,375	4,600	5,410

requiring about 2.4 gal/100 lbm of barite. The large amount of water required decreases the compressive strength of the cement and dilutes the other chemical additives. The range of slurry densities possible using various concentrations of barite is shown in Table 3.17.

3.4.10 Sand. Ottawa sand, even though it has a relatively low specific gravity of about 2.63, sometimes is used to increase slurry density. This is possible since the sand requires no additional water to be added to the slurry. Sand has little effect on the strength or pumpability of the cement, but causes the cement surface to be relatively hard. Because of the tendency to form a hard cement, sand often is used to form a plug in an open hole as a base for setting a whipstock tool used to change the direction of the hole. The range of slurry densities possible using various concentrations of sand is shown in Table 3.18.

gulf sandy

Example 3.5. It is desired to increase the density of a Class H cement slurry to 17.5 lbm/gal. Compute the amount of hematite that should be blended with each sack of cement. The water requirements are 4.5 gal/94 lbm Class H cement and 0.36 gal/100 lbm hematite.

Solution. Let x represent the pounds of hematite per sack of cement. The total water requirement of the slurry then is given by $4.5 + 0.0036x$. Expressing the slurry density in terms of x yields

$$\rho = \frac{\text{total mass, lbm}}{\text{total volume, gal}},$$

$$17.5 = \frac{94 + x + 8.34(4.5 + 0.0036x)}{\left[\dfrac{94}{3.14(8.34)} + \dfrac{x}{5.02(8.34)} + (4.5 + 0.0036x)\right]}$$

Solving this expression yields $x = 18.3$ lbm hematite/94 lbm cement.

3.4.11 Setting-Time Control.

The cement must set and develop sufficient strength to support the casing and seal off fluid movement behind the casing before drilling or completion activities can be resumed. The exact amounts of compressive strength needed is difficult to determine, but a value of 500 psi commonly is used in field practice. Experimental work by Farris[6] has shown that a tensile strength of only a few psi was sufficient to support the weight of the casing under laboratory conditions. However, some consideration also must be given to the shock loading imposed by the rotating drillstring during subsequent drilling operations. It is possible for the drillstring to knock off the lower joint of casing and junk the hole if a good bond is not obtained. The cement strength required to prevent significant fluid movement behind the casing was investigated by Clark.[7] His data show that tensile strengths as low as 40 psi are acceptable with maximum bonding being reached at a value of about 100 psi. Since the ratio of compressive strength to tensile strengths usually is about 12:1, 40- and 100-psi tensile strengths correspond to compressive strengths of 480 and 1,200 psi.

When cementing shallow, low-temperature wells, it may be necessary to accelerate the cement hydration so that the waiting period after cementing is minimized. The commonly used cement accelerators are (1) calcium chloride, (2) sodium chloride, (3) hemihydrate form of gypsum, and (4) sodium silicate. Cement setting time also is a function of the cement composition, fineness, and water content. For example, API Class C cement is ground finer and has a higher C_3A content to promote rapid hydration. When low water/cement ratios are used to reduce setting time, friction-reducing agents (dispersants) sometimes are used to control rheological properties. However, the dispersant must be chosen with care since many dispersants tend to retard the setting of the cement. Organic dispersants such as tannins and lignins already may be present in the water available for mixing cement, especially in swampy locations. Thus, it often is important to measure cement thickening time using a water sample taken from the location.

3.4.12 Calcium Chloride.

Calcium chloride in concentrations up to 4% by weight commonly is used as a cement accelerator in wells having bottomhole temperatures of less than 125°F. It is available in a regular grade (77% calcium chloride) and an anhydrous grade (96% calcium chloride). The anhydrous grade is in more general use because it absorbs moisture less readily and is easier to maintain in storage. The effect of calcium chloride on the compressive strength of API Class A cement is shown in Table 3.19.

3.4.13 Sodium Chloride.

Sodium chloride is an accelerator when used in low concentrations. Maximum acceleration occurs at a concentration of about 5% (by weight of mixing water) for cements containing no bentonite. At concentrations above 5%, the effectiveness of sodium chloride as an accelerator is reduced. Saturated sodium chloride solutions tend to act as a retarder rather than an accelerator. Saturated sodium chloride cements are used primarily for cementing through salt formations and through shale formations that are highly sensitive to fresh water. Potassium chloride is more effective than sodium chloride for inhibiting shale hydration and can be used for this purpose when the additional cost is justified. The effect of sodium chloride on the compressive strength of API Class A cement is shown in Table 3.20.

Seawater often is used for mixing cement when drilling offshore. The sodium, magnesium, and calcium chlorides at the concentrations present in the seawater all act as cement accelerators. Typical effects of seawater on cement slurry properties as compared with fresh water are shown in Table 3.21.

This thickening time obtained with seawater usually is adequate for cement placement where bottomhole temperatures do not exceed 160°F. Cement retarders can be used to counteract the effect of the seawater at higher temperatures, but laboratory tests always should be made before this type of application.

3.4.14 Gypsum.

Special grades of gypsum hemihydrate cement can be blended with portland cement to produce a cement with a low thickening time at low temperatures. These materials should not be used at high temperatures, because the gypsum hydrates may not form a stable set. The maximum working temperature depends on the grade of gypsum cement used, varying from 140°F for the regular grade to 180°F for the high-temperature grade. A full range of blends, from as little as 1 sack gypsum/20 sacks cement to pure gypsum, have been used for various applications. The water requirement of gypsum hemihydrate is about 4.8 gal/100-lbm sack.

TABLE 3.20—CLASS A CEMENT WITH SODIUM CHLORIDE[4]

Water Requirements		% Salt by Weight of Water	Weight of Dry Salt (lbm/sk cement)	Slurry Weight		Slurry Volume (cu ft/sk)
(gal/sk)	(cu ft/sk)			(lbm/gal)	(lbm/cu ft)	
5.2	0.70	0	0	15.6	117	1.18
		5	2.17	15.7	117	1.19
		10	4.33	15.8	118	1.20
		15	6.50	15.9	119	1.21
		20	8.66	16.0	120	1.22
		sat (140°F)	16.12	16.1	120	1.27

Thickening Time and Compressive Strength

Salt (%)	Calcium Chloride (%)	Thickening Time, 2,000 ft, Casing Test (hours:minutes)	Compressive Strength (psi)			
			8 Hours		24 Hours	
			95 F, 800 psi	110 F, 1,600 psi	95 F 800 psi	110 F 1,600 psi
0	0	4:15	305	925	2,240	3,230
0	2	1:40	1,365	2,000	3,920	4,815
5	0	2:30	1,050	2,060	3,990	4,350
5	2	1:49	1,630	2,515	4,530	5,465
10	0	2:30	965	1,925	4,150	4,730
10	2	1:48	1,235	2,200	3,775	4,650
15	0	3:01	700	1,735	4,015	4,480
15	2	2:31	945	1,605	3,075	3,820
20	0	3:00	380	1,140	3,175	3,495
20	2	3:13	490	1,065	2,390	3,155
sat	0	7:15+	not set	15	930	1,955
sat	2	5:00+	50	290	1,570	2,450

API Class A Cement With HR-4 Retarder, Water, 5.2 gal/sk

Thickening Time (hours:minutes)

HR-4 (%)	API Casing Cementing				API Squeeze Cementing		
	4,000 ft	6,000 ft	8,000 ft	10,000 ft	4,000 ft	6,000 ft	8,000 ft
0% Salt							
0.0	3:36	2:25	1:59	1:14	1:32	1:01	0:44
10% Salt Water							
0.0	1:53	1:30	1:10	0:50	1:08	1:00	0:30
0.2	2:29	2:05	1:33	1:18	1:43	1:15	0:34
0.4	3:00+	3:00+	3:08	3:14	2:59	2:45	—
15% Salt Water							
0.0	2:05	1:33	1:25	1:10	1:35	0:55	0:21
0.2	3:00+	2:17	2:02	1:48	2:19	2:07	0:22
0.4	3:00+	3:00+	3:00+	3:00	—	—	0:15
20% Salt Water							
0.0	2:00+	2:00	1:50	1:13	1:47	1:23	0:42
0.2	3:00+	3:00+	2:36	1:36	3:00+	2:12	—

For very shallow wells and surface applications at low temperatures where an extremely short setting time combined with rapid strength development is desired, a small amount of sodium chloride can be used with a gypsum cement blend. For example, a laboratory blend of 90 lbm of gypsum hemihydrate, 10 lbm of Class A portland cement, and 2 lbm of salt when mixed with 4.8 gal of water will develop over 1,000 psi of compression strength when cured at only 50°F for 30 minutes.

3.4.15 Sodium Silicate. Sodium silicate is used as an accelerator for cements containing diatomaceous earth. It is used in concentrations up to about 7% by weight.

3.4.16 Cement Retarders. Most of the organic compounds discussed in Chap. 2 for use as drilling fluid deflocculants tend to retard the setting of portland cement slurries. These materials also are called thinners or dispersants. Calcium lignosulfonate, one of the common mud deflocculants, has been found to be very effective as a cement retarder at very low concentrations. Laboratory data on the thickening time of Class A and Class H cements at various concentrations of calcium lignosulfonate are shown in Table 3.22.

TABLE 3.21 – TYPICAL EFFECT OF SEAWATER ON THICKENING TIME
(water ratio: 5.2 gal/sk)

	Thickening Time Hours:Minutes		Compressive Strength (psi at 24 hours)		
	6,000 ft*	8,000 ft*	50°F	110°F, 1,600 psi	140°F, 3,000 psi
API Class A Cement					
Fresh water	2:25	1:59	435	3,230	4,025
Seawater	1:33	1:17	520	4,105	4,670
API Class H Cement					
Fresh water	2:59	2:16	–	1,410	2,575
Seawater	1:47	1:20	–	2,500	3,085

*API RP 10B Casing Schedule.

The addition of an organic acid to the calcium lignosulfonate (Halliburton HR-12™) has been found to give excellent retarding characteristics at extremely high temperatures. It also improves the rheological properties of the slurry to a greater extent than calcium lignosulfonate alone. When the addition of the organic acid increases the effectiveness of the retarder to the extent that less than 0.3% would be used, it may be difficult to obtain a uniform blend. In this case, the use of calcium lignosulfonate is best. The addition of the organic acid also has been found to be effective in Class A cement.

Calcium-sodium lignosulfonate has been found to be superior to a calcium lignosulfonate when high concentrations of bentonite are used in the cement. The use of calcium-sodium lignosulfonate has been found to produce a slurry having a lower viscosity during mixing and helps to reduce air entrainment.

Sodium tetraborate decahydrate (borax) can be used to enhance the effectiveness of the organic deflocculants as retarders, especially in deep, high-temperature wells where a large increase in thickening time is needed. The optimum borax concentration is thought to be about one-third of the concentration of deflocculant used. Laboratory tests have shown that in addition to increasing the pumping time, the borax reduces the detrimental effect of the deflocculant on the early compressive strength of the cement.

Carboxymethyl hydroxyethyl cellulose (CMHEC) commonly is used both for cement retardation and for fluid loss control. It is used more commonly with cements containing diatomaceous earth, but it is effective as a retarder in essentially all portland cements. Laboratory data on the thickening time and compressive strength of Class H cement at various concentrations of CMHEC are shown in Table 3.23.

TABLE 3.22 – CLASS A AND CLASS H CEMENT
WITH CALCIUM LIGNOSULFONATE
(Halliburton HR-4)[3]

API Class A Cement or Pozmix Cement

Depth (ft)	Temperature (°F) Static	Circulating	Calcium Lignosulfate	Approximate Thickening Time (hours)
Casing Cementing (Primary)				
2,000 to 6,000	110 to 170	91 to 113	0.0	2 to 4
6,000 to 10,000	170 to 230	113 to 144	0.0 to 0.5	2 to 4
10,000 to 14,000	230 to 290	144 to 206	0.5 to 1.0	3 to 4
Squeeze Cementing				
2,000 to 4,000	110 to 140	98 to 116	0.0	2 to 4
4,000 to 8,000	140 to 200	116 to 159	0.0 to 0.5	2 to 4
8,000 to 12,000	200 to 260	159 to 213	0.5 to 1.0	3 to 4

API Class H Cement

Depth (ft)	Static	Circulating	Calcium Lignosulfate	Approximate Thickening Time (hours)
Casing Cementing				
4,000 to 6,000	140 to 170	103 to 113	0.0	3 to 4
6,000 to 10,000	170 to 230	113 to 144	0.0 to 0.3	3 to 4
10,000 to 14,000	230 to 290	144 to 206	0.3 to 0.6	2 to 4
14,000 to 18,000	290 to 350	206 to 300	0.6 to 1.0	–
Squeeze Cementing				
2,000 to 4,000	110 to 140	98 to 116	0.0	3 to 4
4,000 to 6,000	140 to 170	116 to 136	0.0 to 0.3	2 to 4
6,000 to 10,000	170 to 230	136 to 186	0.3 to 0.5	3 to 4
10,000 to 14,000	230 to 290	186 to 242	0.5 to 1.0	2 to 4

3.4.17 Lost-Circulation Additives. *Lost circulation* is defined as the loss of drilling fluid or cement from the well to subsurface formations. This condition is detected at the surface when the flow rate out of the annulus is less than the pump rate into the well. Lost circulation occurs when (1) extremely high-permeability formations are encountered, such as a gravel bed, oyster bed, or vugular limestone, or (2) a fractured formation is encountered or created because of excessive wellbore pressure.

Lost circulation usually occurs while drilling and can be overcome by adding lost-circulation material to the drilling fluid or reducing the drilling fluid density. In some cases, however, lost-circulation material is added to the cement slurry to minimize the loss of cement to a troublesome formation during cementing and, thus, to ensure placing the cement in the desired location. The lost-circulation additives are classified as (1) fibrous, (2) granular, or (3) lamellated. In laboratory experiments, fibrous and granular additives are effective in high-permeability gravel beds. In simulated fractures, granular and lamellated additives are found to be effective. The commonly used granular additives include gilsonite, expanded perlite, plastics, and crushed walnut shells. Fibrous materials used include nylon fibers, shredded wood bark, sawdust, and hay. However, the use of wood products can cause cement retardation because they contain tannins. Lamellated materials include cellophane and mica flakes.

Semi-solid and flash-setting slurries are available for stopping severe lost-circulation problems encountered while drilling that cannot be remedied by adding lost-circulation additives to the mud. The placement of these cements in the lost-circulation zone requires a special operation, since they are not merely additives to the fluid being circulated while drilling. The slurries most often used for this purpose include (1) the gypsum hemihydrate cements, (2) mixtures of bentonite and diesel oil (gunk), and (3) mixtures of cement, bentonite, and diesel oil (bengum).

3.4.18 Filtration-Control Additives. Cement filtration-control additives serve the same function as the mud filtration-control additives discussed in Chap. 2. However, cement slurries containing no filtration control additives have much higher filtration rates than clay/water muds. An untreated slurry of Class H cement has a 30-minute API filter loss in excess of 1,000 cm^3. It is desirable to limit the loss of water filtrate from the slurry to permeable formations to (1) minimize the hydration of formations containing water-sensitive shales, (2) prevent increases in slurry viscosity during cement placement, (3) prevent the formation of annular bridges, which can act as a packer and remove hydrostatic pressure holding back potentially dangerous high-pressure zones, and (4) reduce the rate of cement dehydration while pumping cement into abandoned perforated intervals and, thus, allow plugging longer perforated intervals in a single operation.

The commonly used filtration-control additives include (1) latex, (2) bentonite with a dispersant, (3) CMHEC, and (4) various organic polymers. The

TABLE 3.23—CLASS H CEMENT WITH CMHEC[4]

Water Requirement		Slurry Density		Slurry Volume
(gal/sk)	(cu ft/sk)	(lbm/gal)	(lbm/cu ft)	(cu ft/sk)
4.3	0.58	16.4	123	1.06
5.2	0.70	15.6	117	1.18

Thickening Time (hours:minutes)
(Pressure-Temperature Thickening-Time Tests)

Diacel LWL %	Well Depth (ft), Water, 4.3 gal/sk		
	10,000	12,000	14,000
0.05	1:34	1:28	0:53
0.10	3:52	2:24	2:07
0.15	4:00 +	3:37	2:24
0.20	4:00 +	4:00 +	3:04

Compressive Strength (psi)

Diacel LWL (%)	Curing Time (hours)	Curing Temperature (°F, 3,000 psi), Water, 4.3 gal/sk			
		200	230	260	290
0.00	8	5,450	6,350	6,950	6,375
	24	8,400	7,950	8,525	7,700
0.05	8	3,500	5,850	6,850	6,750
	24	8,350	8,125	8,800	7,200
0.10	8	2,025	5,100	6,700	7,050
	24	8,200	8,300	9,000	6,700
0.15	8	850	3,600	6,400	7,250
	24	7,200	8,450	9,150	6,200
0.20	8	0	1,750	6,100	7,400
	24	4,925	8,600	9,200	5,650

Thickening Time (hours:minutes)
(Pressure-Temperature Thickening-Time Tests)

Diacel LWL %	Well Depth (ft), Water, 5.2 gal/sk		
	10,000	12,000	14,000
0.05	2:32	1:40	1:04
0.10	4:12	3:28	2:30
0.15	4:00 +	4:23	2:35
0.20	4:00 +	4:00 +	3:32

filtration rate in cm^3/30 min through a 325-mesh screen caused by a 1,000-psi pressure differential for various concentrations of one of the organic polymers (Halliburton HALAD-9[TM]) in Class H cement is shown in Table 3.24.

3.4.19 Viscosity-Control Additives. Untreated cement slurries have a high effective viscosity at the shear rates present during cement placement. It is desirable to reduce the effective viscosity of the slurry so that (1) less pump horsepower will be required for cement placement, (2) there will be reduction in the annular frictional pressure gradient and, thus, a smaller chance of formation fracture, and (3) the slurry can be placed in turbulent flow at a lower pumping rate. Some evidence indicates that the drilling fluid is displaced with less mixing and, thus, less cement contamination when the flow pattern is turbulent. Shown in Table 3.25 is a comparison of the velocity required to achieve turbulence of a Class H cement with and without a viscosity-control additive. The commonly used viscosity-control additives include (1) the organic deflocculants such as

TABLE 3.24 – CLASS H CEMENT WITH A FILTRATION CONTROL ADDITIVE[3]

HALAD-9 (%)	Salt (%)	Water Requirement		Slurry Weight		Slurry Volume (cu ft/sk)
		(gal/sk)	(cu ft/sk)	(lbm/gal)	(lbm/cu ft)	
0.0	0	5.20	0.70	15.60	117	1.18
0.6 to 1.2	0	5.64	0.75	15.26	114	1.24

Fluid Loss Tests
(cm^3/30 min; 325-mesh screen; 1,000-psi pressure)

HALAD-9 (%)	0% Salt	10% Salt
0.6	72	192
0.8	52	84
1.0	38	62
1.2	24	36

calcium lignosulfonate, (2) sodium chloride, and (3) certain long-chain polymers. Deflocculants reduce cement viscosity in the same manner as discussed previously in Chap. 2 for drilling fluids. However, it should be remembered that deflocculants act as retarders as well as thinners. Certain organic polymers are available that will act as thinners without accelerating or retarding the cement.

3.4.20 Other Additives. Miscellaneous additives and slurries not discussed in the previous categories given include (1) paraformaldehyde and sodium chromate, which are used to counteract the effect of cement contamination by organic deflocculants from the drilling mud, (2) silica flour, which is used to form a stronger, more stable, and less permeable cement for high-temperature applications, (3) hydrazine, an oxygen scavenger used to control corrosion, (4) radioactive tracers used to determine where the cement has been placed, (5) special fibers such as nylon to make the cement more impact resistant, and (6) special compounds which slowly evolve small gas bubbles as the cement begins to harden.

The formation of small gas bubbles in the cement is thought to be desirable when there is danger of gas flow occurring in the newly cemented borehole when the cement begins to harden. There have been several cases of gas flow through the annulus several hours after cementing operations were completed. A gas flow outside the casing can be particularly difficult to stop because conventional well control procedures cannot be used easily.

The mechanism by which gas blowouts occur shortly after cementing is not fully understood. However, it is known that the formation of a semirigid gel structure begins as soon as cement placement is completed. Initially, the formation of the static gel structure is similar to that occurring in a drilling fluid when fluid movement is stopped. Later, as the cement begins to set, the cement gel becomes much more progressive than that of a drilling fluid. As the cement slurry goes through a transition from a liquid to a solid, it begins to lose the ability to transmit hydrostatic pressure to the lower part of the cemented annulus. (Equations applicable to this phenomenon are developed later in Sec. 12 of Chap. 4.) If this loss in ability to transmit hydrostatic pressure is accompanied by a cement slurry volume reduction, the wellbore pressure can fall sufficiently to permit gas from a permeable high-pressure formation to enter the annulus. The semirigid slurry may not be able to withstand the higher stresses created when gas begins to flow. Gas flow may increase and communicate with a more shallow formation. In an extreme case, gas flow may reach the surface.

TABLE 3.25 – CRITICAL FLOW RATES FOR TURBULENCE[3]

API Class H Cement, Water Ratio:5.2 gal/sk
Slurry Weight:15.6 lbm/gal

Slurry Properties	Neat	Dispersant, 1.0%
n', flow behavior	0.30	0.67
K', consistency index	0.195	0.004

Pipe Size (in.)	Hole Size (in.)	Critical Velocity (ft/sec)		Critical Flow Rate (bbl/min)	
		Without Dispersant	With Dispersant	Without Dispersant	With Dispersant
4½	6¾	9.2	2.61	13.6	3.85
4½	7⅞	8.6	2.13	20.9	5.18
5½	7⅞	9.1	2.54	16.9	4.70
5½	8¾	8.6	2.17	23.3	5.85
7	8¾	9.6	2.96	15.5	4.76
8⅝	11	9.1	2.54	24.4	6.90
8⅝	12¼	8.5	2.05	37.0	9.05
9⅝	12¼	9.0	2.41	29.6	8.08

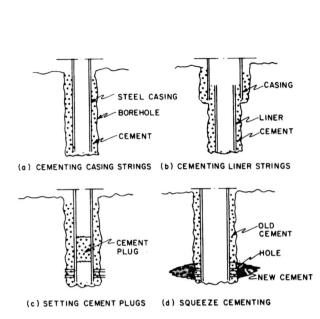

Fig. 3.6 – Common cement placement requirements.

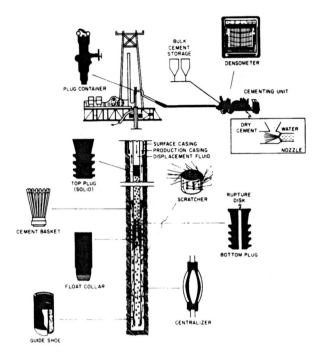

Fig. 3.7 – Conventional placement technique used for cementing casing.[5]

Volume reductions occurring while the cement is making the transition from a liquid slurry to a rigid solid can be traced to two sources. A small volume reduction, as measured in the soundness test, occurs due to the cement hydration reaction. For most cements used in current practice, this volume reduction is small, generally being on the order of 0.1 to 0.3%.[10] Much larger volume reductions are thought to be possible due to the loss of water filtrate to the borehole walls.

The magnitude of the pressure loss per unit volume of filtrate loss is controlled primarily by the cement compressibility during the early stages of the hardening process. A cement with a high compressibility is desirable because it will give a small pressure loss per unit volume of filtrate loss. The introduction of a compound that will react slowly to form small gas bubbles as the cement begins to harden will greatly increase the compressibility of the cement. Cement compressibility also can be increased by blending small volumes of nitrogen gas with the cement slurry during mixing.

A high-compressibility cement permits much larger volumes of water filtrate to be lost without greatly increasing the potential for gas flow into the well. Other methods for reducing the potential for gas flow after cementing include use of (1) filtration control additives to reduce the volume of filtrate loss, (2) shorter cement columns to reduce the effectiveness of the gel strength in blocking the transmission of hydrostatic pressure, and (3) cements that build gel strength quickly after pumping is stopped and harden more rapidly.

3.5. Cement Placement Techniques

Different cementing equipment and placement techniques are used for (1) cementing *casing* strings, (2) cementing *liner* strings, (3) setting cement *plugs*, and (4) *squeeze* cementing. These different types of cementing applications are illustrated in Fig. 3.6. A casing string differs from a liner in that casing extends to the surface, while the top of a liner is attached to subsurface casing previously cemented in place. Cement plugs are placed in open hole or in casing before abandoning the lower portion of the well. Cement is squeezed into lost-circulation zones, abandoned casing perforations, or a leaking cemented zone to stop undesired fluid movement.

3.5.1 Cement Casing. The conventional method for cementing casing is described in Fig. 3.7. Cement of the desired composition is blended at a bulk blending station where cement is moved by aerating the fine powder and blowing it between pressurized vessels under 30 to 40 psi air pressure. The blended cement is transported to the job in bulk transport units. When the casing string is ready to be cemented, cement is mixed with water in a special cementing unit. The cementing unit usually is truck-mounted for land jobs and skid-mounted for offshore operations. The unit mixes the slurry by pumping water under high pressure through a nozzle and loading dry cement through a hopper just downstream of the nozzle. Dry cement is moved pneumatically from the bulk units. The cement slurry enters a pump on the cementing unit and is pumped to a special cementing head or plug container screwed into the top joint of casing.

When the cementing operation begins, the bottom rubber wiper plug is released from the plug container ahead of the cement slurry. This plug wipes the mud from the casing ahead of the slurry to minimize the contamination of the cement with the mud. In some cases a slug of liquid called a *mud preflush* is pumped into the casing before cementing to assist in isolating the cement from the drilling fluid. After the desired volume of slurry has been mixed and pumped into the casing, a top wiper plug is released from the plug container. The top plug differs from the bottom plug

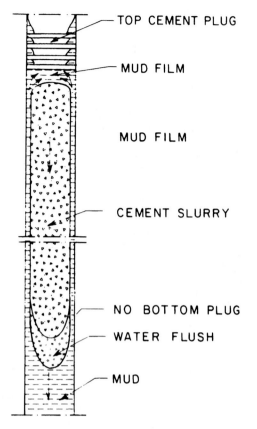

Fig. 3.8 – Cement contamination without bottom plug.

in that the top plug is solid rubber while the bottom plug contains a thin rupture diaphragm. The cement slug is displaced down the casing by pumping drilling fluid or completion fluid into the casing behind the top plug. When the bottom plug reaches the float collar, the diaphragm in the plug ruptures, allowing the cement slurry to be displaced through the guide shoe and into the annulus. When the top plug reaches the bottom plug, the pressure increase at the surface signifies the end of the displacement operation. By counting pump strokes, the approximate position of the top plug can be determined at all times and the pump can be slowed down at the end of the displacement to prevent an excessive pressure increase.

The float collar can act as a check valve to prevent cement from backing up into the casing. Top plugs that have pressure seals and latch in place can be used in addition to the float collar. Fluid movement also can be prevented by holding pressure on the top plug with the displacing fluid. However, it is not good practice to hold excessive pressure in the casing while the cement sets because when the pressure is released, the casing may change diameter sufficiently to break the bond with the cement and form a small annular channel.

Not all operators use a bottom plug when cementing. Indeed, there have been cases when the solid plug mistakenly was placed below the cement slurry. Also, with some of the first plug designs, the float collar was stopped up by plug fragments before completing the cement displacement. When a bottom plug is not used, however, the cement does not wipe all

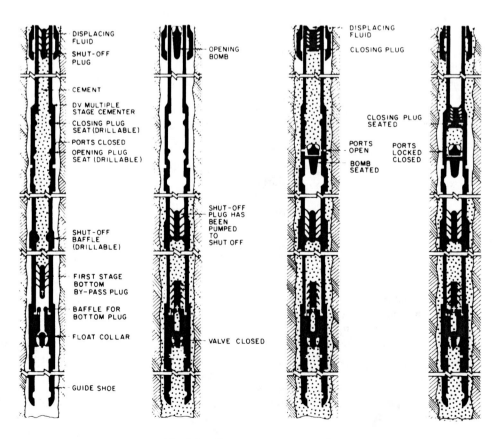

Fig. 3.9 – Two-stage cementing.[5]

the mud from the wall of the casing. This results in a contaminated zone being built up in front of the top plug as shown in Fig. 3.8.

Subsurface equipment that often is used in the conventional casing cementing operation includes one or more joints of casing below the float collar, called the *shoe* joints, a *guide shoe* or *float shoe* at the bottom of the casing, and *centralizers, scratchers,* and *baskets* on the outside of the casing. The shoe joints allow for entrapment of contaminated mud or cement, which may result from the wiping action of the top cementing plug. The simplest guide shoe design is the open-ended collar type with or without a molded nose. The guide shoe simply guides the casing past irregularities in the borehole wall. Circulation is established through the open end of the guide shoe or through side ports designed to create more agitation as the cement slurry is circulated up the annulus. Should the casing be resting on bottom, circulation can be achieved more easily through side port openings in the guide shoe. A float shoe serves the function of both a guide shoe and a float collar when no shoe joints are desired. Float shoes and float collars containing a packer also are available for isolating the lower portion of the hole from the cemented zone. These devices have side ports for slurry exit above the packer. Centralizers are placed on the outside of the casing to help hold the casing in the center of the hole. Scratchers are used to help remove mudcake from the borehole walls. Some scratchers are designed for cleaning by reciprocating the casing, while others are designed for cleaning by rotating the casing. Cement baskets are used to help

support the weight of the cement slurry at points where porous or weak formations are exposed.

In addition to the conventional placement method for cementing casings, there are several modified techniques used in special situations. These include: (1) stage cementing, (2) inner-string cementing, (3) annular cementing through tubing, (4) multiple-string cementing, (5) reverse-circulation cementing, and (6) delayed-setting cementing.

3.5.2 Stage Cementing. Stage cementing is one of the procedures developed to permit using a cement column height in the annulus that normally would cause fracture of one or more subsurface formations. It also can be used to reduce the potential for gas flow after cementing. The first stage of the cementing operation is conducted in the conventional manner. After the slurry hardens, a bomb is dropped (Fig. 3.9) to open a side port in a staging tool placed in the casing string. The second-stage cement then is pumped through this side port and into the annulus above the set first-stage cement. Equipment also is available for three-stage cementing.

3.5.3 Inner-String Cementing. Inner-string cementing was developed to reduce the cementing time and the amount of cement left in the shoe joint of extremely large-diameter casing. The technique uses a float collar or shoe modified with sealing adapters, which permits tubing or drillpipe to be landed and hydraulically sealed. The cement then is displaced down the inner drillpipe or tubing string rather than the casing. When the float collar or shoe is equipped with a backpressure valve or latch-down

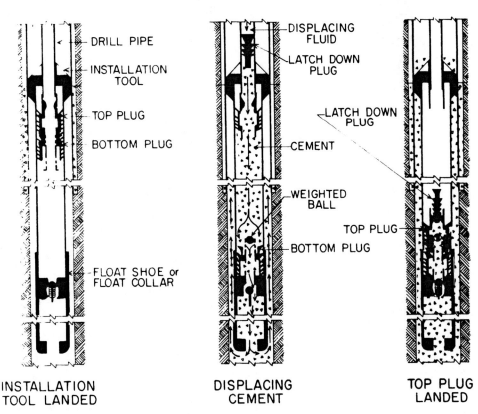

Fig. 3.10—Conventional placement techniques for cementing a liner.

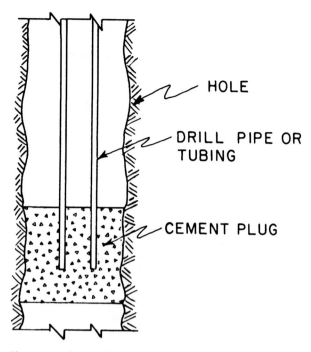

Fig. 3.11 – Placement technique used for setting cement plug.

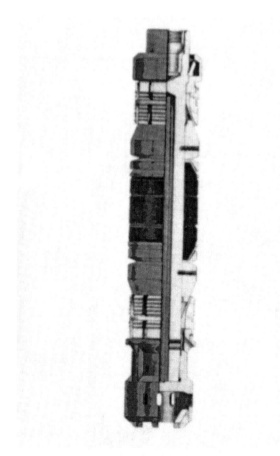

Fig. 3.12 – Bridge plug.

plug, the inner string can be withdrawn immediately after displacing the cement.

3.5.4 Annular Cementing Through Tubing. This technique consists of pumping cement through tubing run in the annulus between two casing strings or between the casing and the open hole. It usually is used to bring the top of the previously placed cement to the surface or to repair casing.

3.5.5 Multiple-String Cementing. Multiple-string cementing is a multiple completion method that involves cementing several strings of tubing in the hole without the use of an outer casing string. This type of completion is an alternative to the more conventional multiple-completion method in which the tubing strings are set using packers inside a larger-diameter casing.

3.5.6 Reverse-Circulation Cementing. Reverse-circulation cementing consists of pumping the slurry down the annulus and displacing the mud back through the casing. This method has been used in some instances when extremely low-strength formations were present near the bottom of the hole. A special float collar or shoe as well as a special wellhead assembly is required to use this method. Since wiper plugs cannot be used, it is difficult to detect the end of the cement displacement. The cement usually is overdisplaced at least 300 ft into the bottom of the casing to enhance the probability of a good cement job at the shoe.

3.5.7 Delayed-Setting Cementing. Delayed-setting cementing is used to obtain a more uniform mud displacement from the annulus than is possible using the conventional cement placement technique. This method consists of placing a retarded cement slurry having good filtration properties in the wellbore before running the casing. Cement placement is accomplished down the drillpipe and up the annulus. The drillpipe then is removed from the well and casing is lowered into the unset cement slurry. This method also can be modified for multiple-string cementing with cement displacement being made through the lower string. The delayed-setting cementing method requires a drilling engineer with a strong nervous system.

3.5.8 Cementing Liners. The conventional method of liner cementing is illustrated in Fig. 3.10. The liner is attached to drillpipe using a special liner-setting tool. The liner is lowered to a position with several hundred feet of overlap between the top of the liner and the bottom of the casing. The liner-setting tool then is actuated so that the liner is attached mechanically to and supported by the casing without hydraulically sealing the passage between the liner and casing. The cement is pumped down the drillpipe, separated from the displacing fluid by a latch-down plug. This latch-down plug actuates a special wiper plug in the liner-setting tool after the top of the cement column reaches the liner. When the wiper plug reaches the float collar, a pressure increase at the surface signifies the end of the cement displacement. The drillstring then must be released from the liner-setting tool and withdrawn before the cement hardens.

Usually, a volume of cement sufficient to extend past the top of the liner is displaced. When the drillstring is withdrawn, this cement collects at the top of the liner but generally does not fall to bottom. This cement can be washed out using the drillpipe or drilled out after the cement sets. When cement is not displaced to the top of the liner, cement must be forced into this area using a squeeze-cementing method, which is discussed in Sec. 3.5.10.

A large variety of liner-setting tools is available. Most are set with either a mechanical or hydraulic device. The hydraulically set devices are actuated by drillpipe rotation or by dropping a ball or plug and then set by applying pump pressure. The mechanically set devices are actuated by drillpipe rotation and set by lowering the drillpipe. A liner-setting tool must be selected on the basis of the liner weight, annular dimensions, cement displacement rate, and liner-cementing procedure. There is a trend toward not setting the liner until after cement displacement. This allows the liner to be moved while cementing to improve mud removal in the annulus. This practice imposes some limitations on the liner-setting tool selected.

After drilling the well, it may be desirable to extend the liner back to the surface if a commercial hydrocarbon deposit is found and the well is to be completed. This type of casing is called a *tieback liner*. If casing is used to extend the liner up the hole, but not all the way to the surface, it is called a *stub liner*. Stub liners are used primarily to repair the top of a leaking liner.

3.5.9 Plug Cementing. Cement plugs can be set in open hole or in casing. Plugs are set to prevent fluid communication between an abandoned lower portion of the well and the upper part of the well. Plugs also are set to provide a seat for directional drilling tools used to sidetrack the well.

Cement plugs are placed using drillpipe or tubing as shown in Fig. 3.11. When a plug is placed off bottom in an open hole, a caliper log can be used to locate an in-gauge portion of the hole. Centralizers and scratchers can be placed on the bottom section of the pipe that will be opposite the section of hole to be plugged. Cement is pumped down the drillpipe or tubing and into the hole. The cement slurry has a natural tendency to form a bridge below the tubing, causing the slurry to move up the annular space opposite the drillpipe or tubing. When cementing in casing, a bridge plug (Fig. 3.12) sometimes is placed below the cement plug to assist in forming a good hydraulic seal.

When the drillpipe or tubing is open-ended, cement displacement is continued until the fluid columns are *balanced* − i.e., have the same height of slurry inside the pipe and annulus. As shown in Fig. 3.13, the pump pressure during the displacement provides a good indication of when the fluid columns are balanced. The drillstring or tubing then is pulled slowly from the cement slurry.

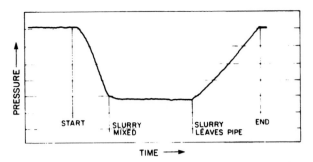

Fig. 3.13 − Idealized pressure/time chart for balanced pressure plug.

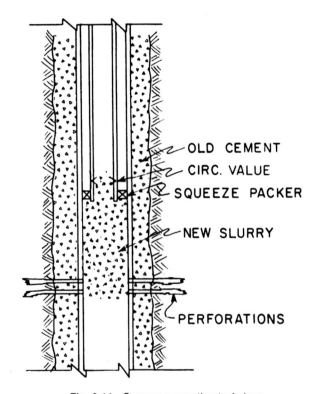

Fig. 3.14—Squeeze-cementing technique.

An improved plug placement technique has been described by Doherty[8] and Goins[9] as an alternative to the balanced-column method. Mud contamination is minimized by placing a slug of water in front of and behind the cement slurry. The cement is displaced almost completely from the drillpipe or tubing. To prevent backflow while the pipe is being pulled, a special backpressure valve or plug catcher can be used at the bottom of the pipe string. Cement plugs also can be set using a wireline device called a *dump bailer*. After setting a bridge plug at the desired depth, the dump bailer is filled with cement and lowered into the well on a wireline. The dump bailer is designed to empty the slurry into the hole when the bailer reaches bottom. After the first batch of cement takes an initial set, a second batch can be placed. The dump bailer seldom is used during drilling because of the time required to set a long plug.

3.5.10 Squeeze Cementing. Squeeze cementing consists of forcing a cement slurry into an area of the well or formation by means of an applied hydraulic pressure. The purpose of squeeze cementing is to form a hydraulic seal between the wellbore and the zone squeezed. If the slurry is placed using sufficient pressure to fracture the formation, the process is called a *high-pressure squeeze*. The pressure required to fracture the formation and allow rapid slurry placement is called the *breakdown pressure*. In the high-pressure squeeze, whole cement slurry is forced into the fractured formation. If the slurry is placed using less than the breakdown pressure, the process is called a *low-pressure squeeze*. The low-pressure squeeze causes a cement cake to plate out against the formation as filtrate is lost to the formation. Common applications of squeeze cementing are (1) plugging abandoned casing perforations, (2) plugging severe lost-circulation zones, and (3) repairing annular leaks in previously cemented casing, caused by failure of either the casing or previously placed cement.

The conventional method for squeeze cementing is shown in Fig. 3.14. A squeeze packer and a circulating valve (Fig. 3.15) are placed above the perforations and lowered into the well on drillpipe or tubing. A slug of water or preflush is placed opposite the zone to be squeezed, and then the squeeze packer is set just above the zone of interest. The circulating valve is opened above the packer, and the cement slurry is displaced down the drillpipe. A slug of water is used before and after the cement slurry to prevent mud contamination of the slurry and mud plugging of the zone of interest. The circulating valve usually is closed after most of the leading water slug has been pumped into the annulus. The cement then is pumped into the zone of interest until the final desired squeeze pressure is obtained. With the low-pressure squeeze, it is common practice to stop pumping periodically, or *hesitate*, during the squeeze process. This assists buildup of filter cake nodes against the formation or perforations. The circulating valve then is opened, and the excess cement is pumped up the drillpipe to the surface. This process is called *reversing out*. If the desired squeeze pressure is not obtained, the operation is repeated after waiting for the first batch of cement to take an initial set.

In addition to the conventional squeeze-cementing method, several modified techniques have been developed. In certain instances, it may be necessary to isolate the section below the perforation by placing a bridge plug below the perforations. In other cases, neither a squeeze packer nor a bridge plug is used and pressure is held against the well by closing the blowout preventers. This method is called the *bradenhead squeeze*. A *block squeeze* consists of isolating a production zone by perforating above and below the producing interval and squeezing through both perforated intervals in separate steps.

3.5.11 Cement Volume Requirements. In addition to selecting the cement composition and placement technique, the drilling engineer must determine the volume of cement slurry needed for the job. The volume required usually is based on past experience

Fig. 3.15 — Squeeze packer assembly.[5]

CIRCULATING VALVE ASSEMBLY

UPPER SLIPS

PACKER ELEMENTS

LOWER SLIPS

JUNK CATCHER

and regulatory requirements in the area. As little as 300 ft of fillup has been used behind relatively deep casing strings. However, in some cases, the entire annulus is cemented. It usually is necessary to include considerably more slurry than indicated by the theoretical hole size because of hole enlargement while drilling. Thus, an *excess factor*, based on prior experience in the area, usually is applied to the theoretical cement volume. When hole size or *caliper logs* are available, a more accurate slurry volume can be determined. Example 3.6 illustrates one method of estimating slurry volume requirements for conventional casing cementing operation.

Example 3.6. Casing having an OD of 13.375 in. and an ID of 12.415 in. is to be cemented at a depth of 2,500 ft. A 40-ft shoe joint will be used between the float collar and the guide shoe. It is desired to place a 500-ft column of high-strength slurry at the bottom of the casing. The high-strength slurry is composed of Class A cement mixed using a 2% calcium chloride flake (by weight of cement) and a water/cement ratio of 5.2 gal/sack. The upper 2,000 ft of the annulus is to be filled with a low-density slurry of Class A cement mixed with 16% bentonite and 5% sodium chloride (by weight of cement) and a water/cement ratio of 13 gal/sack. Compute the slurry volume requirements if the excess factor in the annulus is 1.75. The bit size used to drill the hole is 17 in.

Solution. The specific gravity of Class A cement is 3.14, and the specific gravity of bentonite is 2.65 (Table 3.8). The specific gravity of the brines must be determined using Tables 2.3 and 2.4 in Chap. 2. The NaCl brine is made by adding 4.7 lbm [0.05 (94)] of NaCl to 108.4 lbm [13 (8.34)] of water. Interpolating in Table 2.3 for a weight fraction of 0.0415 [or 4.7/(108.4 + 4.7)] gives a NaCl brine specific gravity of 1.0279. The $CaCl_2$ brine is made by adding 1.88 lbm [0.02 (94)] of $CaCl_2$ to 43.4 lbm [5.2 (8.34)] of water. Interpolating in Table 2.4 for a weight fraction of 0.0415 [or 1.88/(43.4 + 1.88)] gives a $CaCl_2$ brine specific gravity of 1.0329.

The volume of each component in the low-density lead slurry per sack of cement is given by:

Component	Volume (cu ft)	
Cement	$\dfrac{94}{3.14\,(62.4)}$	= 0.4797
Bentonite	$\dfrac{0.16\,(94)}{2.65\,(62.4)}$	= 0.0910
Salt water	$\dfrac{108.4 + 4.7}{1.0279\,(62.4)}$	= 1.7633
Yield		= 2.334 cu ft/sack

The annular capacity for the 17-in. hole and 13.375-in. casing is given by

$$A_a = \frac{\pi}{4}\,(17^2 - 13.375^2)\,\frac{\text{sq ft}}{144\ \text{sq in.}}$$

$$= 0.6006 \text{ sq ft}.$$

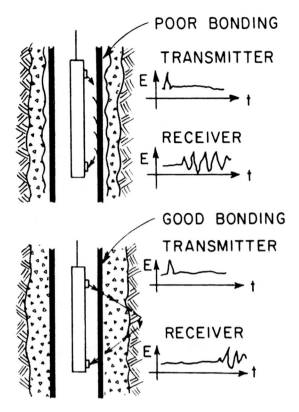

Fig. 3.16 – Acoustic energy travel in cased wells.

Using a column length of 2,000 ft and an excess factor of 1.75, the slurry volume required is

$$0.6006\,(2,000)\,(1.75) = 2,102 \text{ cu ft}.$$

This slurry volume will require mixing

$$\frac{2,102 \text{ cu ft}}{2.334 \text{ cu ft/sack}} = 901 \text{ sacks}.$$

The volume of each component in the high-strength tail slurry is given by:

Component	Volume (cu ft)	
Cement	$\dfrac{94}{3.14\,(62.4)}$	= 0.4797
$CaCl_2$ brine	$\dfrac{(43.4 + 1.88)}{1.0329\,(62.4)}$	= 0.7025
Yield		= 1.182 cu ft/sack

The cement required for a 500-ft annular column and a 40-ft column in the shoe joint is given by

$$0.6006\,(500)\,(1.75) + \frac{\pi}{4}\,(12.415)^2\,\frac{40}{144}$$

$$= 559.2 \text{ cu ft}.$$

This slurry volume will require mixing

$$\frac{559.2 \text{ cu ft}}{1.182 \text{ cu ft/sack}} = 473 \text{ sacks of cement}.$$

Total slurry volume then will be

$$2,102 \text{ cu ft} + 559.2 \text{ cu ft} = 2,661.2 \text{ cu ft},$$

and the total sacks of cement required for the job will be

$$901 + 473 = 1,374 \text{ sacks}.$$

3.5.12 Cementing Time Requirements. As discussed previously, the time required to place the cement slurry is one of the more important variables in the engineering design of the slurry properties. The relationships between well depth and cementing time used by the API in the specifications for the various cement classes (see Fig. 3.5) represent average well conditions and may not be applicable in all cases. A more accurate estimation of cementing time can be made based on the actual slurry volume and pumping rates to be used. Also, it is always prudent to allow some extra cementing time for unforeseen operational problems. Example 3.7 illustrates one method of estimating cementing time requirements for a conventional casing cementing operation.

Example 3.7. Estimate the cementing time for the cementing operation described in Example 3.6 if one cementing truck having a mixing capacity of approximately 20 cu ft/min is used. The rig pump will be operated at 60 strokes/min and has a pump factor of 0.9674 cu ft/stroke.

Solution. The volume of cement to be mixed is 901 sacks for the lead slurry and 473 sacks for the tail slurry. Each sack of cement has an approximate bulk volume of 1 cu ft. However, the lead slurry will be blended with 16% bentonite and 5% salt, and the tail slurry will be blended with 2% calcium chloride. The bulk weight of bentonite, salt, and calcium chloride (flake) are 60, 71, and 56.4 lbm/cu ft, respectively (Table 3.8). Thus, the additional volume of solids to be mixed is given by

$$\frac{901(94)(0.16)}{60} + \frac{901(94)(0.05)}{71}$$
$$+ \frac{473(94)(0.02)}{56.4} = 301 \text{ cu ft}.$$

The total mixing time is given by

$$\frac{1,374 + 301}{20} = 83.8 \text{ minutes}.$$

The mixing time accounts for the time period before placing the top plug at the top of the casing. The time required to displace the top plug from the surface to the float collar is given by

$$\frac{\pi}{4} (12.415)^2 \frac{(2,500 - 40)}{144}$$
$$\times \frac{1}{60(0.9674)} = 35.6 \text{ minutes}.$$

Thus, the total cementing time is $83.8 + 35.6 = 119.4$ minutes or 2 hours.

3.5.13 Cementing Evaluation. After cementing operations are complete and the cement left in the wellbore, along with portions of the subsurface cementing equipment, have been drilled out, the cement job usually is evaluated to ensure that the cementing objectives have been accomplished. It always is a good practice to pressure-test cemented casing to the maximum pressure anticipated in subsequent drilling operations. The top of the cement can be located by making a temperature survey of the well from 6 to 10 hours after completing the cement displacement. When cement is present behind the pipe, heat liberated due to the exothermic hydration reaction will cause an increase in temperature. In addition, acoustic logging tools are available for evaluating the bond between the cement and the pipe. When the cement is not bonded acoustically to both the pipe and the formation, a strong early sound reflection will be received by the acoustic logging device, indicating sound travel primarily through the casing (Fig. 3.16).

Exercises

3.1 List four steps in the manufacture of portland cement. What is the approximate weight and bulk volume of portland cement sold in 1 bbl and in one sack?

3.2 Name four crystalline phases in portland cement that hydrate when mixed with water. Show the chemical equation for the hydration of each phase. Identify the reaction product that is the main cementing ingredient in portland cement. Identify the crystalline phase that forms reaction products readily attacked by water-containing sulfates.

3.3 The oxide analysis of a cement is as follows.

Oxide	Weight Percent
Lime (CaO or C)	66.5
Silica (SiO_2 or S)	21.1
Alumina (Al_2O_3 or A)	4.8
Ferric oxide (Fe_2O_3 or F)	2.6
Magnesia (MgO)	1.2
Sulfur trioxide (SO_3)	2.7
Ignition loss	0.9

Calculate the theoretical weight fraction of C_3S, C_2S, C_3A, and C_4AF present in the cement. *Answer:* 0.666, 0.103, 0.083, 0.079.

3.4 List the equipment needed to perform the standard API tests for drilling cements.

3.5 Define the following: minimum water content, normal water content, free water content, and maximum water content.

3.6 The torque required to hold the paddle assembly stationary in a cement consistometer rotating at 150 rpm is 800 g-cm. What is the slurry consistency? *Answer:* 36 Bc.

3.7 A Class H cement core sample having a length of 2.54 cm and a diameter of 2.865 cm allows a water flow rate of 0.05 mL/s when placed under a pressure differential of 20 psi. Compute the permeability of the cement. *Answer:* 14.5 md.

3.8 The cement tensile strength required to support the weight of a string of casing is estimated to be 8 psi. If the cement is known to have a compressive strength of 200 psi, do you think the casing could be supported by the cement? Why? *Answer*: Yes, tensile strength $\cong 17$ psi.

3.9 List the eight standard classes and three standard types of API cement. Compare these classes and types with the ASTM cement types used in the construction industry.

3.10 How is the composition of a high-sulfate-resistant cement different from a standard portland cement?

3.11 List the normal (API) water content of each class of cement used in slurry preparation. *Answer*: 5.19 gal/sack for Classes A and B, 6.32 gal/sack for Class C, 4.29 gal/sack for Classes D through H, 4.97 gal/sack for Class G.

3.12 Compute the yield and density of each class of cement when mixed with the normal amount of water as defined by API. *Answer*: 1.17 cf/sack and 15.6 lbm/gal for Classes A and B.

3.13 It is desired to reduce the density of a Class A cement to 12.8 lbm/gal by adding bentonite. Using the water requirements for Class A cement and bentonite given in Table 3.9, compute the weight of bentonite that should be blended with each sack of cement. Compute the yield of the slurry. What is the "percent mix" of the slurry? *Answer*: 9.3 lbm/sack, 2.09 cf/sack, 102.9%.

3.14 Repeat Exercise 3.13 using diatomaceous earth instead of bentonite as the low-specific-gravity solids and a water requirement of 3.3 gal/10% diatomaceous earth. *Answer*: 11.5 lbm/sack, 1.75 cf/sack, 84.2%.

3.15 Identify the following cement additives: gilsonite, expanded perlite, and pozzolan.

3.16 It is desired to increase the density of a Class H cement to 17.5 lbm/gal using barite. Compute the weight of barite that should be blended with each sack of cement. Use the water requirements for Class H cement (maximum strength) and barite given in Table 3.9. Compute the yield of the slurry. What is the "percent mix" of the slurry? *Answer*: 29.4 lbm/sack, 1.26 cf/sack, 44.3%.

3.17 Repeat Problem 3.16 using sand instead of barite. *Answer*: 40.6 lbm/sack, 1.30 cf/sack, 38.2%.

3.18 List two common cement accelerators for Class A cement.

3.19 List two common cement retarders for Class H cement.

3.20 List the three types of lost-circulation additives used in cement, and give one example of each type.

3.21 List four common filtration-control additives.

3.22 Describe in your own words the conventional cement placement techniques used for cementing a casing string, cementing a liner string, setting a cement plug, and squeeze cementing.

3.23 What is the purpose of a shoe joint? Why is a shoe joint especially important in a deep cementing job or when no bottom wiper plug is used?

3.24 Casing having an OD of 9.625 in. and an ID of 8.535 in. is to be cemented at a depth of 13,300 ft in a 12.25-in. borehole. A 40-ft shoe joint will be used between the float collar and the guide shoe. It is desired to place 2,500 ft of cement in the annulus. Each sack of Class H cement will be mixed with 4.3

gal of water to which is added 18% salt (by weight of water). A small quantity of dispersant will be blended with the cement, but this additive has no significant effect on the slurry yield or density. Compute the density of the slurry, the yield of the slurry, the number of sacks of cement required, and the cementing time. Assume that cement can be mixed at a rate of 20 sacks/min and displaced at a rate of 9 bbl/min and use an excess factor of 1.5. The salt will be added to the water phase and thus does not blend with the dry cement. *Answer*: 16.7 lbm/gal; 1.09 cf/sack; 1,092 sacks; 159 minutes.

3.25 A 7.0-in. liner having an ID of 6.276 in. is to be cemented at a depth of 15,300 ft in an 8.5-in. hole. Casing is set at 13,300 ft as described in Problem 3.24, and a 300-ft overlap between the casing and liner is desired. A 40-ft shoe joint will be used between the float collar and the guide shoe. It is desired to use 1,000 ft of preflush (in annulus) and then fill the total annular space opposite the liner with cement. Class H cement containing 35% silica flour, and 1.2% HALAD-22A will be mixed with 5.8 gal of water containing 18% salt (by weight of water). Compute the slurry density, the slurry yield, the volume of cement slurry required if the caliper log shows an average washout of 2.0-in. increase in hole diameter, and the number of sacks of Class H cement needed. Neglect the effect of the HALAD-22A on the density and volume of the slurry. *Answer*: 16.3 lbm/gal, 1.61 cf/sack, 445 sacks.

Nomenclature

A = area
B_c = slurry consistency
k = permeability
L = length
p = pressure
q = flow rate
T = torque
μ = viscosity
ρ = slurry density

References

1. *Specification for Materials and Testing for Well Cements*, API Spec. 10, Dallas (Jan. 1982).
2. *ASTM Standards on Cement Manual of Cement Testing*, Philadelphia (1975) Part 13.
3. *Technical Data on Cementing*, Halliburton Services, Duncan, OK.
4. *Halliburton Cementing Tables*, Halliburton Services, Duncan, OK.
5. *Sales and Service Catalog*, Halliburton Services, Duncan, OK.
6. Farris, R.F.: "Method for Determining Minimum Waiting-on-Cement Time," *Trans.*, AIME (1946) **165**, 175-188.
7. Clark, R.C.: "Requirements of Casing Cement for Segregating Fluid Bearing Formation," *Oil and Gas J.* (1953) **51**, 173.
8. Doherty, W.T.: "Oil-Well Cementing in the Gulf Coast Area," *Production Bull.*, API, Sec. 4, **60** (1933).
9. Goins, W.C. Jr.: "Open-Hole Plugback Operations," *Oil-Well Cementing Practices in the United States*, API, New York City (1959) 193-197.
10. Sutton, D.L., Sabins, F., and Faul, R.: "Preventing Annular Gas Flow After Cementing," *Oil and Gas J.* (Dec. 10 and Dec. 17, 1984) 84 and 92.

SI Metric Conversion Factors

$$\text{bbl} \times 1.589\ 873 \qquad \text{E}-01 = \text{m}^3$$
$$\text{cp} \times 1.0^* \qquad \text{E}-03 = \text{Pa·s}$$
$$\text{cu ft} \times 2.831\ 685 \qquad \text{E}-02 = \text{m}^3$$
$$\text{cu ft/lbm} \times 6.242\ 796 \qquad \text{E}-02 = \text{m}^3/\text{kg}$$
$$\text{ft} \times 3.048^* \qquad \text{E}-01 = \text{m}$$
$$°\text{F} \quad (°\text{F}-32)\ 1.8 \qquad \qquad = °\text{C}$$
$$\text{gal} \times 3.785\ 412 \qquad \text{E}-03 = \text{m}^3$$
$$\text{gal/cu ft} \times 1.336\ 806 \qquad \text{E}+02 = \text{dm}^3/\text{m}^3$$

$$\text{gal/lbm} \times 8.345\ 404 \qquad \text{E}+00 = \text{dm}^3/\text{kg}$$
$$\text{in.} \times 2.54^* \qquad \text{E}+00 = \text{cm}$$
$$\text{lbm} \times 4.535\ 924 \qquad \text{E}-01 = \text{kg}$$
$$\text{lbm/cu ft} \times 1.601\ 846 \qquad \text{E}+01 = \text{kg}/\text{m}^3$$
$$\text{lbm/gal} \times 1.198\ 264 \qquad \text{E}+02 = \text{kg}/\text{m}^3$$
$$\text{psi} \times 6.894\ 757 \qquad \text{E}+00 = \text{kPa}$$
$$\text{sq ft} \times 9.290\ 304 \qquad \text{E}-02 = \text{m}^2$$
$$\text{sq in.} \times 6.451\ 6^* \qquad \text{E}+00 = \text{cm}^2$$

*Conversion factor is exact.

Chapter 4
Drilling Hydraulics

Chaps. 2 and 3 provided information about the composition and properties of drilling fluids and cement slurries. In this chapter, the relation between the fluid properties and the subsurface hydraulic forces present in the well will be developed.

The science of fluid mechanics is very important to the drilling engineer. Extremely large fluid pressures are created in the long slender wellbore and tubular pipe strings by the presence of drilling mud or cement. The presence of these subsurface pressures must be considered in almost every well problem encountered. In this chapter, the relations needed to determine the subsurface fluid pressures will be developed for three common well conditions. These well conditions include (1) a static condition in which both the well fluid and the central pipe string are at rest, (2) a circulating operation in which the fluids are being pumped down the central pipe string and up the annulus, and (3) a tripping operation in which a central pipe string is being moved up or down through the fluid. The second and third conditions listed are complicated by the non-Newtonian behavior of drilling muds and cements. Also included in this chapter are the relations governing the transport of rock fragments and immiscible formation fluids to the surface by the drilling fluid.

While it was not feasible to illustrate fully all the drilling applications of the fundamental concepts developed in this chapter, several of the more important applications are presented in detail. These applications include (1) calculation of subsurface hydrostatic pressures tending to burst or collapse the well tubulars or fracture exposed formations, (2) several aspects of blowout prevention, (3) displacement of cement slurries, (4) bit nozzle size selection, (5) surge pressures due to a vertical pipe movement, and (6) carrying capacity of drilling fluids.

The order in which the applications are presented parallels the development of the fundamental fluid mechanics concepts given in the chapter. This approach best serves the needs of the beginning students of drilling engineering.

4.1 Hydrostatic Pressure in Liquid Columns

Subsurface well pressures are determined most easily for static well conditions. The variation of pressure with depth in a fluid column can be obtained by considering the free-body diagram (Fig. 4.1) for the vertical forces acting on an element of fluid at a depth D in a hole of cross-sectional area A. The downward force on the fluid element exerted by the fluid above is given by the pressure p times the cross-sectional area of the element A:

$$F_1 = pA.$$

Likewise, there is an upward force on the element exerted by the fluid below given by

$$F_2 = \left(p + \frac{dp}{dD}\Delta D\right)A.$$

In addition, the weight of the fluid element is exerting a downward force given by

$$F_3 = F_{wv}A\Delta D,$$

where F_{wv} is the specific weight of the fluid. Since the fluid is at rest, no shear forces exist and the three forces shown must be in equilibrium:

$$pA - \left(p + \frac{dp}{dD}\Delta D\right)A + F_{wv}A\Delta D = 0.$$

Expansion of the second term and division by the element volume $A\Delta D$ gives

$$dp = F_{wv}dD. \quad \dots\dots\dots\dots\dots\dots\dots\dots (4.1)$$

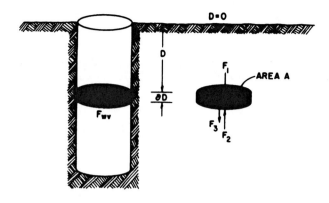

Fig. 4.1—Forces acting on a fluid element.

If we are dealing with a liquid such as drilling mud or salt water, fluid compressibility is negligible, and specific weight can be considered constant with depth. Integration of Eq. 4.1 for an incompressible liquid gives

$$p = F_{wv}D + p_0, \quad \dots\dots\dots\dots\dots\dots\dots (4.2a)$$

where p_0, the constant of integration, is equal to the surface pressure ($D=0$). Normally the static surface pressure p_0 is zero unless the blowout preventer of the well is closed and the well is trying to flow. The specific weight of the liquid in field units is given by

$$F_{wv} = 0.052\rho, \quad \dots\dots\dots\dots\dots\dots\dots (4.3)$$

$$\overset{ppg \quad ft}{\underset{p = .052 \cdot mw \cdot D}{\downarrow \quad \downarrow}}$$

where F_{wv} is the specific weight in pounds per square inch per foot and ρ is the fluid density in pounds mass per gallon. Thus Eq. 4.2a, in field units, is given by

$$p = 0.052\rho D + p_0. \quad \dots\dots\dots\dots\dots\dots (4.2b)$$

An important application of the hydrostatic pressure equation is the determination of the proper drilling fluid density. The fluid column in the well must be of sufficient density to cause the pressure in the well opposite each permeable stratum to be greater than the pore pressure of the formation fluid in the permeable stratum. This problem is illustrated in the schematic drawing shown in Fig. 4.2. However, the density of the fluid column must not be sufficient to cause any of the formations exposed to the drilling fluid to fracture. A fractured formation would allow some of the drilling fluid above the fracture depth to leak rapidly from the well into the fractured formation.

Example 4.1. Calculate the static mud density required to prevent flow from a permeable stratum at 12,200 ft if the pore pressure of the formation fluid is 8,500 psig.

Solution. Using Eq. 4.2b,

$$\rho = \frac{p}{0.052D} = \frac{8,500}{0.052(12,200)} = 13.4 \text{ lbm/gal.}$$

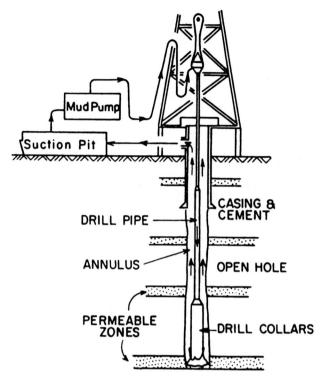

Fig. 4.2—The well fluid system.

Thus, the mud density must be at least 13.4 lbm/gal to prevent the flow of formation fluid into the wellbore when the well is open to the atmosphere ($p_0 = 0$ psig) and there is no mud circulation.

4.2 Hydrostatic Pressure in Gas Columns

In many drilling and completion operations, a gas is present in at least a portion of the well. In some cases, gas is injected in the well from the surface while in other cases gas may enter the well from a subsurface formation. The variation of pressure with depth in a static gas column is more complicated than in a static liquid column because the gas density changes with changing pressure.

The gas behavior can be described using the real gas equation defined by

$$pV = znRT = z\frac{m}{M}RT, \quad \dots\dots\dots\dots\dots (4.4)$$

where

p = absolute pressure,
V = gas volume,
n = moles of gas,
R = universal gas constant,
T = absolute temperature,
m = mass of gas,
M = gas molecular weight, and
z = gas deviation factor.

The *gas deviation factor z* is a measure of how much the gas behavior deviates from that of an *ideal gas*. An ideal gas is one in which there are no attractive forces between gas molecules. Gas deviation factors for natural gases have been determined experimentally as a function of temperature and pressure and are readily available in the petroleum literature.[1-3] In this chapter the simplifying assumption of ideal gas behavior will generally be made to assist the student in focusing more easily on the drilling hydraulics concepts being developed.

The gas density can be expressed as a function of pressure by rearranging Eq. 4.4. Solving this equation for gas density ρ yields

$$\rho = \frac{m}{V} = \frac{pM}{zRT}. \quad \dots\dots\dots\dots\dots\dots (4.5a)$$

Changing units from consistent units to common field units gives

$$\rho = \frac{pM}{80.3zT}, \quad \dots\dots\dots\dots\dots\dots (4.5b)$$

where ρ is expressed in pounds mass per gallon, p is in pounds per square inch absolute, and T is in degrees Rankine.

When the length of the gas column is not great and the gas pressure is above 1,000 psia, the hydrostatic equation for incompressible liquids given by Eq. 4.2b can be used together with Eq. 4.5b without much loss in accuracy. However, when the gas column is not short or highly pressured, the variation of gas density with depth within the gas column should be taken into account. Using Eqs. 4.1, 4.3, and 4.5a we obtain

$$dp = \frac{0.052\,pM}{80.3\ zT}dD.$$

If the variation in z within the gas column is not too great, we can treat z as a constant, $\bar{z}$. Separating variables in the above equation yields

$$\int_{P_0}^{P} \frac{1}{p}dp = \frac{M}{1,544\ zT}\int_{D_0}^{D} dD.$$

Integration of this equation gives

Pressure in the tubing @ certain depth.

$$p = p_0 e^{\frac{M(D-D_0)}{1,544\ zT}} \quad \dots\dots\dots\dots\dots\dots (4.6)$$

Example 4.2. A well contains tubing filled with methane gas (molecular weight = 16) to a vertical depth of 10,000 ft. The annular space is filled with a 9.0-lbm/gal brine. Assuming ideal gas behavior, compute the amount by which the exterior pressure on the tubing exceeds the interior tubing pressure at 10,000 ft if the surface tubing pressure is 1,000 psia and the mean gas temperature is 140°F. If the collapse resistance of the tubing is 8,330 psi, will the tubing collapse due to the high external pressure?

Solution. The pressure in the annulus at a depth of 10,000 ft is given by Eq. 4.2b.

$$p_2 = 0.052(9.0)(10,000) + 14.7 = 4,695 \text{ psia}.$$

The pressure in the tubing at a depth of 10,000 ft is given by Eq. 4.6.

$$p_1 = 1,000e^{\frac{16(10,000)}{1,544(1)(460+140)}} = 1,188 \text{ psia}.$$

Thus, the pressure difference is given by

$$p_2 - p_1 = 4,695 - 1,188 = 3,507 \text{ psi},$$

which is considerably below the collapse pressure of the tubing.

The density of the gas in the tubing at the surface could be approximated using Eq. 4.5b as follows.

$$\rho = \frac{1,000(16)}{80.3(1)(600)} = 0.331 \text{ lbm/gal}.$$

It is interesting to note that the use of this density in Eq. 4.2b gives

$$p_1 = 0.052(0.331)(10,000) + 1,000 = 1,172 \text{ psia},$$

which is within 16 psi of the answer obtained using the more complex Eq. 4.6.

4.3 Hydrostatic Pressure in Complex Fluid Columns

During many drilling operations, the well fluid column contains several sections of different fluid densities. The variation of pressure with depth in this type of complex fluid column must be determined by separating the effect of each fluid segment. For example, consider the complex liquid column shown in Fig. 4.3. If the pressure at the top of Section 1 is known to be p_0, then the pressure at the bottom of Section 1 can be computed from Eq. 4.2b:

$$p_1 = 0.052\,\rho_1(D_1 - D_0) + p_0.$$

The pressure at the bottom of Section 1 is essentially equal to the pressure at the top of Section 2. Even if an interface is present, the capillary pressure would be negligible for any reasonable wellbore geometry. Thus, the pressure at the bottom of Section 2 can be expressed in terms of the pressure at the top of Section 2:

$$p_2 = 0.052\rho_2(D_2 - D_1) + 0.052\rho_1(D_1 - D_0) + p_0.$$

In general, the pressure p at any vertical distance depth D can be expressed by

$$p = p_0' + 0.052\sum_{i=1}^{n} \rho_i(D_i - D_{i-1}). \quad \dots\dots\dots (4.7)$$

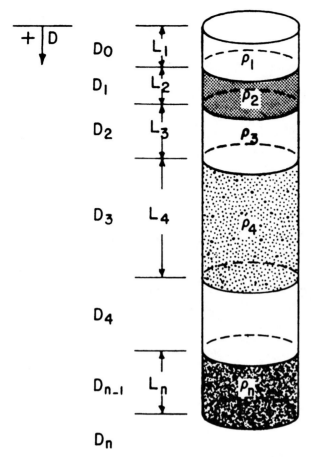

Fig. 4.3—A complex liquid column.

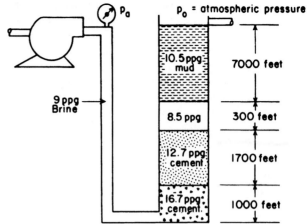

Fig. 4.4—Viewing the well as a manometer.

and the change in hydrostatic pressure is added to the known pressure; conversely, when moving up through a section, $(D_{i+1} - D_i)$ is negative and the change in hydrostatic pressure is subtracted from the known pressure.

$$p_a = p_0 + 0.052 \, [10.5(7,000) + 8.5(300)$$

$$+ 12.7(1,700) + 16.7(1,000) - 9.0(10,000)].$$

Since the known pressure p_0 is 0 psig, then

$$p_a = 1,266 \text{ psig.}$$

4.3.1 Equivalent Density Concept

Field experience in a given area often allows guidelines to be developed for the maximum mud density that formations at a given depth will withstand without fracturing during normal drilling operations. It is sometimes helpful to compare a complex well fluid column to an equivalent single-fluid column that is open to the atmosphere. This is accomplished by calculating the equivalent mud density ρ_e, which is defined by

$$\rho_e = \frac{p}{0.052D}. \quad\quad\quad\quad\quad\quad\quad\quad\quad (4.8)$$

The equivalent mud density always should be referenced at a specified depth.

Example 4.4. Calculate the equivalent density at a depth of 10,000 ft for Example 4.3 for static well conditions after the cement has been displaced completely from the casing.

Solution. At a depth of 10,000 ft,

$$p = 0.052(9.0)(10,000) + 1,266 = 5,946 \text{ psig.}$$

It is frequently desirable to view the well fluid system shown in Fig. 4.3 as a manometer when solving for the pressure at a given point in the well. The drillstring interior usually is represented by the left side of the manometer, and the annulus usually is represented by the right side of the manometer. A hydrostatic pressure balance can then be written in terms of a known pressure and the unknown pressure using Eq. 4.7.

Example 4.3. An intermediate casing string is to be cemented in place at a depth of 10,000 ft. The well contains 10.5-lbm/gal mud when the casing string is placed on bottom. The cementing operation is designed so that the 10.5-lbm/gal mud will be displaced from the annulus by (1) 300 ft of 8.5-lbm/gal mud flush, (2) 1,700 ft of 12.7-lbm/gal filler cement, and (3) 1,000 ft of 16.7-lbm/gal high-strength cement. The high-strength cement will be displaced from the casing with 9-lbm/gal brine. Calculate the pump pressure required to completely displace the cement from the casing.

Solution. The complex well fluid system is understood more easily if viewed as a manometer (Fig. 4.4). The hydrostatic pressure balance is written by starting at the known pressure and moving through the various fluid sections to the point of the unknown pressure. When moving down through a section, $(D_{i+1} - D_i)$ is positive

Using Eq. 4.8,

$$\rho_e = \frac{5,946}{0.052\,(10,000)} = 11.4 \text{ lbm/gal.}$$

4.3.2 Effect of Entrained Solids and Gases in Drilling Fluid

Drilling engineers seldom deal with pure liquids or gases. For example, both drilling fluids and cements are primarily a mixture of water and finely divided solids. The drilling mud in the annulus also contains the drilled solids from the rock broken up by the bit and the formation fluids that were contained in the rock. As long as the foreign materials are suspended by the fluid or settling through the fluid at their terminal velocity, the effect of the foreign materials on hydrostatic pressure can be computed by replacing the fluid density in Eq. 4.2b with the density of the mixture. However, particles that have settled out of the fluid and are supported by grain-to-grain contact do not influence hydrostatic pressure.

The density of an ideal mixture can be computed using the calculation procedure discussed in Sec. 2 of Chap. 2. The average density of a mixture of several components is given by

$$\rho = \frac{\sum_{i=1}^{n} m_i}{\sum_{i=1}^{n} V_i} = \frac{\sum_{i=1}^{n} \rho_i V_i}{\sum_{i=1}^{n} V_i} = \sum_{i=1}^{n} \rho_i f_i, \quad \ldots \ldots (4.9)$$

where m_i, V_i, ρ_i, and f_i are the mass, volume, density, and volume fraction of component i, respectively. As long as the components are liquids and solids, the component density is essentially constant throughout the entire length of the column. Thus, the average density of the mixture also will be essentially constant.

If one component is a finely divided gas, the density of the gas component does not remain constant but decreases with the decreasing pressure. A drilling fluid that is measured to have a low density due to the presence of gas bubbles is said to be *gas cut*.

The determination of hydrostatic pressure at a given depth in a gas cut mud can be made through use of the *real gas* equation. If N_v moles of gas are dispersed in (or associated with) 1 gal of drilling fluid, the volume fraction of gas at a given point in the column is given by

$$f_g = \frac{\dfrac{zN_vRT}{p}}{1+\dfrac{zN_vRT}{p}}. \quad \ldots \ldots (4.10)$$

In addition, the gas density ρ_g at that point is defined by Eq. 4.5a. Thus, the effective density of the mixture is given by

$$\rho = \rho_f(1-f_g)+\rho_g f_g = \frac{(\rho_f+MN_v)p}{p+zN_vRT}, \quad \ldots \ldots (4.11)$$

where M is the average molecular weight of the gas.

For common field units, substitution of this expression for mean density in Eq. 4.3 and combining with Eq. 4.1 yields

$$\int_{D_1}^{D_2} dD = \int_{p_1}^{p_2} \frac{p+zN_vRT}{0.052(\rho_f+MN_v)}\frac{dp}{p}. \quad \ldots \ldots (4.12)$$

If the variation of z and T is not too great over the column length of interest, they can be treated as constants of mean values $\bar{z}$ and $\bar{T}$. Integration of Eq. 4.12 gives

$$D_2-D_1 = \frac{p_2-p_1}{a}+\frac{b}{a}\ln\frac{p_2}{p_1}, \quad \ldots \ldots (4.13)$$

where

$$a = 0.052\,(\rho_f+MN_v), \quad \ldots \ldots (4.14)$$

and

$$b = \bar{z}N_vR\bar{T}. \quad \ldots \ldots (4.15)$$

It is unfortunate that the pressure p_2 appears within the logarithmic term in Eq. 4.13. This means that an iterative calculation procedure must be used for the determination of the change in pressure with elevation for a gas-cut fluid column. However, if the gas/liquid mixture is highly pressured and not very long, the variation of gas density with pressure can be ignored. In this case, the mixture density $\bar{\rho}$ given by Eq. 4.11 can be assumed constant and the change in hydrostatic pressure can be computed using Eq. 4.2b.

Example 4.5. A massive low-permeability sandstone having a porosity of 0.20, a water saturation of 0.3, and a methane gas saturation of 0.7 is being drilled at a rate of 50 ft/hr with a 9.875-in. bit at a depth of 12,000 ft. A 14-lbm/gal drilling fluid is being circulated at a rate of 350 gal/min while drilling. Calculate the change in hydrostatic pressure caused by the drilled formation material entering the mud. Assume that the mean mud temperature is 620°R and that the formation water has a density of 9.0 lbm/gal. Also assume that the gas behavior is ideal and that both the gas and the rock cuttings move at the same annular velocity as the mud. The density of the drilled solids is 21.9 lbm/gal.

Solution. The hydrostatic head exerted by 12,000 ft of 14-lbm/gal mud would be

$$p = 14.7+0.052(14)(12,000) = 8,751 \text{ psia.}$$

The formation is being drilled at a rate of

$$50\left[\frac{\pi(9.875)^2}{4(144)}\right]\left(\frac{7.48}{60}\right) = 3.31 \text{ gal/min.}$$

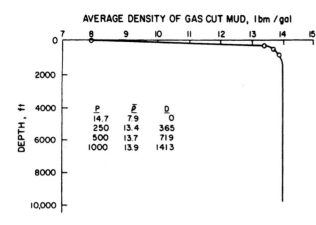

Fig. 4.5—Annular density plot for Example 4.5.

Drilled solids are being added to the drilling fluid at a rate of

$$3.31(1-0.2)=2.65 \text{ gal/min.}$$

Formation water is being added to the drilling fluid at a rate of

$$3.31(0.2)(0.3)=0.2 \text{ gal/min.}$$

The density of the drilling fluid after the addition of the water and drilled solids would be

$$\bar{\rho}=\frac{14(350)+21.9(2.65)+9(0.2)}{350+2.65+0.2}$$

$$=14.057 \text{ lbm/gal.}$$

Methane gas is being added to the drilling fluid at a rate of

$$3.31(0.2)(0.7)=0.464 \text{ gal/min.}$$

Assuming the gas is ideal and the formation pressure is approximately 8,751 psia, the gas density given by Eq. 4.5b is

$$\rho_g=\frac{(8,751)(16)}{80.3(1.00)(620)}=2.8 \text{ lbm/gal.}$$

Thus, the gas mass rate entering the well is given by

$$\frac{(2.8)(0.464)}{16}=0.081 \text{ mol/min.}$$

Since the mud is being circulated at a rate of 350 gal/min, the moles of gas per gallon of mud is given by

$$N_v=\frac{0.081}{350}=0.000231 \text{ mol/gal.}$$

Using Eqs. 4.14 and 4.15 gives

$$a=0.052[14.057+16(0.000231)]=0.7312,$$

and

$$b=(1)(0.000231)(80.3)(620)=11.5.$$

Since the well is open to the atmosphere, the surface pressure p_1 is 14.7 psia. The bottomhole pressure p_2 must be estimated from Eq. 4.13 in an iterative manner. As shown in the table below, various values for p_2 were assumed until the calculated (D_2-D_1) was equal to the well depth of 12,000 ft.

p_2 (psia)	$\dfrac{p_2-p_1}{0.7312}$	$15.72 \ln\dfrac{p_2}{p_1}$	D_2-D_1
8,750	11,946	100.43	12,046
8,700	11,878	100.34	11,978
8,716	11,900	100.37	12,000

Thus, the change in hydrostatic head due to the drilled formation material entering the mud is given by

$$\Delta p=8,716-8,751=-35 \text{ psi.}$$

Example 4.5 indicates that the loss in hydrostatic head due to normal contamination of the drilling fluid is usually negligible. In the past this was not understood by many drilling personnel. The confusion was caused mainly by a severe lowering of density of the drilling fluid leaving the well at the surface. This lowering of density was due to the rapidly expanding entrained gas resulting from the decrease in hydrostatic pressure on the drilling fluid as it approached the surface. The theoretical surface mud density that would be seen in Example 4.5 is given by Eq. 4.11 as

$$\rho=\frac{[14.057+16(0.000231)]14.7}{14.7+1(0.000231)(80.3)(620)}$$

$$=7.9 \text{ lbm/gal.}$$

As one driller remarked, this amount of loss in mud density "would cause even a monkey to get excited." In the past, it was common practice to increase the density of the drilling fluid when gas-cut mud was observed on the surface because of a fear of a potential blowout. However, Example 4.5 clearly shows that this should not be done unless the well will flow with the pump off. As shown in Fig. 4.5, significant decreases in annular mud density occur only in the relatively shallow part of the annulus. The rapid increase in annular density with depth occurs because the gas volume decreases by a factor of two when the hydrostatic pressure doubles. For example, increasing the hydrostatic pressure at the surface from 14.7 to 117.6 psia causes a unit volume of gas to decrease to one-eighth of its original size.

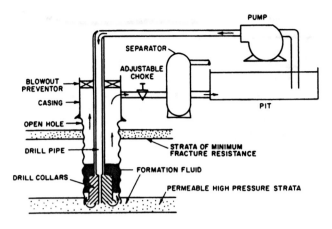

Fig. 4.6—Schematic of well control operations.

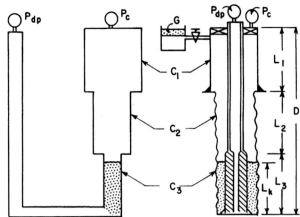

Fig. 4.7—Schematic of initial well conditions during well control operations.

4.4 Annular Pressures During Well Control Operations

One of the more important applications of the hydrostatic pressure relationships is the determination of annular pressures during *well control operations*. Well control operations refer to the emergency procedures followed when formation fluids begin flowing into the well and displacing the drilling fluid. The flow of formation fluids into the well is called a *kick*. A schematic illustrating the hydraulic flow paths during well-control operations is shown in Fig. 4.6. Formation fluids that have flowed into the wellbore generally must be removed by circulating the well through an adjustable choke at the surface. The bottomhole pressure of the well at all times must remain above the pore pressure of the formation to prevent additional influx of formation fluid. However, a complicating factor is the danger of fracturing a weaker stratum that also is exposed to the hydraulic pressure. Fracturing of an exposed stratum often results in an *underground blowout* in which an uncontrolled flow of formation fluids from the high-pressure stratum to the fractured stratum occurs. Thus, the proper well control strategy is to adjust a surface choke so that the bottomhole pressure of the well is maintained just slightly above the formation pressure.

A plot of the surface annular pressure vs. the volume of drilling fluid circulated is called an annular pressure profile. Although annular pressure calculations are not required for well control procedure used by most operators today, a prior knowledge of kick behavior helps in the preparation of appropriate contingency plans. Since annular frictional pressure losses are generally small at the circulating rates used in well control operations, the calculations can be made using the hydrostatic pressure equations.[5-9]

4.4.1 Kick Identification

The annular pressure profile that will be observed during well control operations depends to a large extent on the composition of the kick fluids. In general, a gas kick causes higher annular pressures than a liquid kick. This is true because a gas kick (1) has a lower density than a liquid kick and (2) must be allowed to expand as it is pumped to the surface. Both of these factors result in a lower hydrostatic pressure in the annulus. Thus, to main-

tain a constant bottomhole pressure, a higher surface annular pressure must be maintained using the adjustable choke.

Kick composition must be specified for annular pressure calculations made for the purpose of well planning. Kick composition generally is not known during actual well control operations. However, the density of the kick fluid can be estimated from the observed drillpipe pressure, annular casing pressure, and pit gain. The density calculation often will determine if the kick is predominantly gas or liquid.

The density of the kick fluid is estimated most easily by assuming that the kick fluid entered the annulus as a slug. A schematic of initial well conditions after closing the blowout preventer on a kick is shown in Fig. 4.7. The volume of kick fluid present must be ascertained from the volume of drilling fluid expelled from the annulus into the pit before closing the blowout preventer. The pit gain G usually is recorded by pit volume monitoring equipment. If the kick volume is smaller than the total capacity of the annulus opposite the drill collars, the length of the kick zone, L_k, can be expressed in terms of the kick volume, V_k, and the annular capacity, C_3. Assuming the well diameter is approximately constant, we obtain

$$L_k = V_k C_3, \qquad \qquad (4.16)$$

where L_k is the length of the kick zone, V_3 is the volume of the kick zone, and C_3 is the annular capacity of the hole opposite the drill collars expressed as length per unit volume.

If the kick volume is larger than the total capacity of the annulus opposite the drill collars, then the length of the kick zone, L_k, is given by

$$L_k = L_3 + \left(V_k - \frac{L_3}{C_3} \right) C_2, \qquad (4.17)$$

where L_3 is the total length of the drill collars and C_2 is the annular capacity of the hole opposite the drill pipe expressed as length per unit volume. A pressure balance on the initial well system for a uniform mud density, ρ_m, is given by

$$p_c + 0.052[\rho_m(D-L_k) + \rho_k L_k - \rho_m D] = p_{dp}.$$

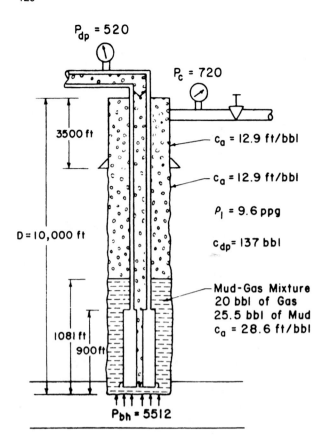

3500 ft

c$_a$ = 12.9 ft/bbl

c$_a$ = 12.9 ft/bbl

ρ_l = 9.6 ppg

c$_{dp}$= 137 bbl

D = 10,000 ft

Mud-Gas Mixture
20 bbl of Gas
25.5 bbl of Mud
c$_a$ = 28.6 ft/bbl

1081 ft

900ft

P$_{bh}$ = 5512

Fig. 4.8—Schematic for Example 4.6.

Solving this expression for the density of the kick, ρ_k, yields

$$\rho_k = \rho_m - \frac{p_c - p_{dp}}{0.052 \, L_k}. \qquad \dots \dots \dots \dots \dots (4.18)$$

A kick density less than about 4 lbm/gal should indicate that the kick fluid is predominantly gas, and a kick density greater than about 8 lbm/gal should indicate that the kick fluid is predominantly liquid.

Several factors can cause large errors in the calculation of kick fluid density when the kick volume is small. Hole washout can make the determination of kick length difficult. In addition, the pressure gauges often do not read accurately at low pressures. Also, the effective annular mud density may be slightly greater than the mud density in the drillpipe because of entrained drilled solids. Furthermore, the kick fluid is mixed with a significant quantity of mud and often cannot be represented accurately as a slug. Thus, the kick density computed using Eq. 4.18 should be viewed as only a rough estimate.

Some improvement in the accuracy of the kick density calculation can be achieved if the volume of mud mixed with the formation fluids is known. The minimum mud volume that was mixed with the kick fluids can be estimated using the expression

$$V_m = q t_d, \qquad \dots \dots \dots \dots \dots \dots \dots (4.19)$$

where q is the flow rate of the pumps, and t_d is the kick detection time before stopping the pump and closing the

blowout preventer. The volume of the kick-contaminated zone can be estimated using

$$V_k = G + q t_d, \qquad \dots \dots \dots \dots \dots \dots (4.20)$$

thus allowing the mean density of the kick-contaminated zone to be computed using Eq. 4.18. The mean density $\bar{\rho}_k$ of the mixed zone then can be related to the density of the kick fluid using the mixture equations. Since a significant amount of natural mixing occurs (even if the pump is not operating when formation gas enters the well), Eq. 4.20 tends to predict a mixture volume that is too low.

Example 4.6. A well is being drilled at a vertical depth of 10,000 ft while circulating a 9.6-lbm/gal mud at a rate of 8.5 bbl/min when the well begins to flow. Twenty barrels of mud are gained in the pit over a 3-minute period before the pump is stopped and the blowout preventers are closed. After the pressures stabilized, an initial drillpipe pressure of 520 psig and an initial casing pressure of 720 psig are recorded. The annular capacity of the casing opposite the drillpipe is 12.9 ft/bbl. The annular capacity opposite the 900 ft of drill collars is 28.6 ft/bbl. Compute the density of the kick fluid. The total capacity of the drillstring is 130 bbl.

Solution. A schematic illustrating the geometry of this example is given in Fig. 4.8. The total capacity opposite the 900 ft of drill collars is

$$\frac{900 \text{ ft}}{28.6 \text{ ft/bbl}} = 31.5 \text{ bbl.}$$

If it is assumed that the kick fluids entered as a slug, then the volume of kick fluid is less than the total annular capacity opposite the drill collars. Thus,

$$L_k = 20 \text{ bbl } (28.6 \text{ ft/bbl}) = 572 \text{ ft.}$$

Using Eq. 4.18, the density of the kick fluid is given by

$$\rho_k = 9.6 - \frac{720 - 520}{0.052(572)} = 2.9 \text{ lbm/gal.}$$

The results should be interpreted as an indication of low-density kick fluid—i.e., a gas.

If it is assumed that the kick fluids are mixed with the mud pumped while the well was flowing,

$$V_k = 20 \text{ bbl} + 8.5 \text{ bbl/min}(3 \text{ min}) = 45.5 \text{ bbl.}$$

The length of mixed zone is given by Eq. 4.17 as

$$L_k = 900 + (45.5 - 31.5)12.9 = 1081 \text{ ft.}$$

Using Eq. 4.18, the mean density of the mixed zone is given by

$$\bar{\rho}_k = 9.6 - \frac{720 - 520}{(0.052)(1,081)} = 6.04 \text{ lbm/gal.}$$

Since the column of mixed zone is only 1,081 ft long and under high pressure, the mean density can be related to the kick fluid density using equations for the effective density of incompressible mixtures.

$$6.04 = \frac{20\rho_k + 25.5(9.6)}{45.5}.$$

Solving this equation for the kick fluid density yields

$$\rho_k = \frac{6.04(45.5) - 9.6(25.5)}{20} = 1.5 \text{ lbm/gal},$$

which also indicates that the kick fluid is a gas.

___

4.4.2 Annular Pressure Prediction

The same hydrostatic pressure balance approach used to identify the kick fluids also can be used to estimate the pressure at any point in the annulus for various well conditions. During well control operations, the bottomhole pressure will be maintained constant at a value slightly above the formation pressure through the operation of an adjustable choke. Thus, it is usually convenient to express the pressure at the desired point in the annulus in terms of the known bottomhole pressure. This requires a knowledge of only the length and density of each fluid region between the bottom of the hole and the point of interest. When a gas kick is involved, the length of the gas region must be determined using the real gas equation. For simplicity, it usually is assumed that the kick region remains as a continuous slug that does not slip relative to the mud.

___

Example 4.7. Again consider the kick described in Example 4.6. Compute the mud density required to kill the well and the equivalent density at the casing seat for the shut-in well conditions. Also compute the equivalent density that would occur at the casing seat after pumping 300 bbl of kill mud while maintaining the bottomhole pressure 50 psi higher than the formation pressure through use of an adjustable surface choke. Assume the kick is methane gas at a constant temperature of 140°F and that the gas behaves as an ideal gas.

Solution. The formation pressure is given by

$$p_{bh} = 520 + 0.052(9.6)(10,000) = 5,512 \text{ psig}.$$

The mud density required to overcome this formation pressure is given by

$$\rho_2 = \frac{520 + 0.052(9.6)(10,000)}{0.052(10,000)} = \frac{520}{0.052(10,000)}$$

$$+ 9.6 = 10.6 \text{ lbm/gal}.$$

Similarly, the equivalent density at the casing seat depth of 3,500 ft is given by

$$\rho_e = \frac{720}{0.052(3,500)} + 9.6 = 13.56 \text{ lbm/gal}.$$

After pumping 300 bbl of 10.6-lbm/gal mud, the volume of 10.6-lbm/gal mud in the annulus at the bottom of the hole is

$$300 - 130 = 170 \text{ bbl}.$$

This is true since the total drillstring capacity is 130 bbl. The length of this region is given by

$$L_1 = 900 + (170 - 31.5)12.9 = 2,687 \text{ ft}.$$

The region above the new mud in the annulus will contain 130 bbl of 9.6-lbm/gal mud that was displaced from the drillpipe. The length of this region is given by

$$L_2 = 130(12.9) = 1,677 \text{ ft}.$$

The region above the 9.6-lbm/gal mud will contain methane gas. The approximate pressure of the gas is needed to compute the gas volume. The pressure at the bottom of the gas region can be computed from the known bottomhole pressure, which is 50 psi higher than the formation pressure.

$$p_g = 5,512 + 50 - 0.052(10.6)(2,687)$$

$$-0.052(9.6)(1,677) = 3,244 \text{ psig}.$$

This pressure occurs at a depth of

$$10,000 - 2,687 - 1,677 = 5,636 \text{ ft}.$$

For ideal gas behavior, the volume of the gas is given by

$$V_g = 20 \text{ bbl} \frac{(5,512 + 14.7)}{(3,244 + 14.7)} = 33.9 \text{ bbl},$$

and the length of the gas region is given by

$$L_g = 33.9(12.9) = 437 \text{ ft}.$$

The density of the gas is given by Eq. 4.5b.

$$\rho_g = \frac{pM}{80.3zT} = \frac{(3,244 + 14.7)16}{80.3(1)(460 + 140)}$$

$$= 1.08 \text{ lbm/gal}.$$

The region above the gas contains 9.6-lbm/gal mud. Thus, we can compute the pressure at 3,500 ft from the known pressure of 3,244 psig at 5,636 ft using

$$p = 3,244 - 0.052(1.08)(437)$$

$$-0.052(9.6)(5,636 - 437 - 3,500) = 2,371 \text{ psig}.$$

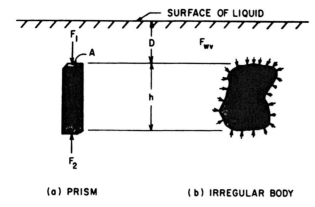

Fig. 4.9—Hydraulic forces acting on a foreign body.

The equivalent density corresponding to this pressure is given by

$$\rho_e = \frac{2,371}{0.052(3,500)} = 13.0 \text{ lbm/gal.}$$

4.5 Buoyancy

In the previous sections of this chapter, we have focused our attention on the calculation of hydrostatic pressure at given points in the well. In this section we will turn our attention to the forces on the subsurface well equipment and pipe strings that are due to hydrostatic pressure. In some cases only the resultant force or bending moment created by hydrostatic pressure is needed, while in others the axial tension or compression at a given point in the submerged equipment is desired. The determination of the resultant force on a submerged body will be considered first.

The net effect of hydraulic pressure acting on a foreign material immersed in the well fluid is called buoyancy. Buoyancy is understood most easily for a vertical prism such as the one shown in Fig. 4.9. Hydraulic pressure acting on a side of the prism at any given depth is balanced by an equal hydraulic pressure acting on the opposite side of the prism. Thus, the net force exerted by the fluid on the prism is the resultant of force F_1 acting at the top and force F_2 acting at the bottom of the prism. The downward force F_1 is given by the hydraulic pressure at depth D times the cross-sectional area A.

$$F_1 = p_1 A = F_{wv} D A.$$

Similarly, the upward force on the prism is given by the pressure at a depth of $(D+h)$ times the cross-sectional area A.

$$F_2 = p_2 A = F_{wv}(D+h)A.$$

Thus, the resultant buoyant force F_{bo} exerted by the fluid on the prism is given by

$$F_{bo} = F_2 - F_1$$

$$= F_{wv} A(D+h) - F_{wv} A D = F_{wv} A h,$$

which indicates that the *upward buoyant force F_{bo} is equal to the weight of the displaced fluid.* This relation was first used by Archimedes about 250 B.C.

Archimedes' relation is valid for a foreign body immersed in a fluid regardless of its shape. For an understanding of the more general case of an irregular body immersed in a fluid, consider that the fluid pressures at the surface of the body would be unchanged if the body were not present and this surface were considered an imaginary surface drawn in the liquid. Since the fluid element contained by the imaginary surface is at rest, the sum of the vertical forces must be zero and the weight of the contained fluid must be equal to the buoyant force. The surrounding fluid acts with the same system of forces on the foreign body and this body also experiences a net upward force equal to the weight of the fluid region occupied by the body—i.e., equal to the weight of the displaced fluid. Note that this same argument could be applied even if the foreign body were immersed only partially in the fluid.

The *effective weight, W_e,* of well equipment immersed in a fluid is defined by

$$W_e = W - F_{bo}, \quad \ldots\ldots\ldots\ldots\ldots\ldots\ldots (4.21)$$

where W is the weight of the well equipment in air and F_{bo} is the buoyant force.

Using Archimedes' relation, the buoyant force is given by

$$F_{bo} = \rho_f V = \rho_f \frac{W}{\rho_s}, \quad \ldots\ldots\ldots\ldots\ldots\ldots (4.22)$$

where ρ_f is the fluid density, ρ_s is the density of steel, and V is the fluid volume displaced. Substitution of Eq. 4.22 into Eq. 4.21 yields

$$W_e = W\left(1 - \frac{\rho_m}{\rho_s}\right). \quad \ldots\ldots\ldots\ldots\ldots\ldots (4.23)$$

The density of steel is approximately 490 lbm/cu ft or 65.5 lbm/gal.

Example 4.8. Ten thousand feet of 19.5-lbm/ft drillpipe and 600 ft of 147-lbm/ft drill collars are suspended off bottom in a 15-lbm/gal mud. Calculate the effective hook load that must be supported by the derrick.

Solution. The weight of the drillstring in air is given by

$$W = 19.5(10,000) + 147(600) = 283,200 \text{ lbm.}$$

The effective weight of the drillstring in mud can be computed using Eq. 4.23.

$$W_e = W\left(1 - \frac{\rho_m}{\rho_s}\right) = 283,200\left(1 - \frac{15}{65.5}\right)$$

$$= 218,300.$$

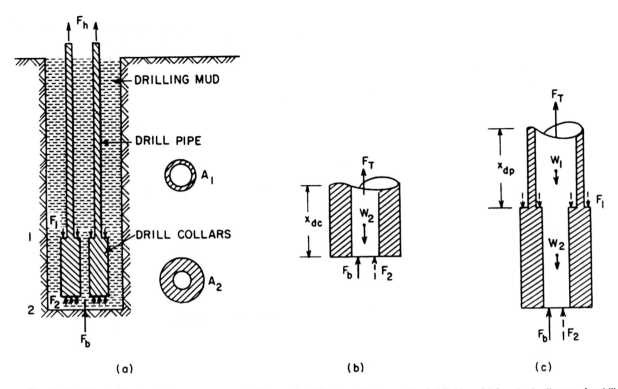

Fig. 4.10—Effect of hydrostatic pressure on axial forces in drillstring: (a) schematic of drillstring, (b) free body diagram for drill collars, and (c) free body diagram for drillpipe.

4.5.1 Determination of Axial Stress

In Example 4.8, only the net effect of hydrostatic pressure on the pipe string was required. However, in some cases it may be necessary to compute the axial stress at a given point in the pipe string. The axial stress is the axial tension in the pipe string divided by the cross-sectional area of steel. When axial stress must be determined, the effective points of application of the hydrostatic pressure must be considered and Archimedes' relation cannot be used.

Consider the idealized schematic of a drillstring suspended in a well (Fig. 4.10a). The lower portion of the drillstring is composed of drill collars, while the upper portion is composed of drillpipe. To apply a downward force, F_b, on the bit, the drillstring is lowered until a portion of the weight of the pipe string is supported by the bottom of the hole. The cross-sectional area of the drill collars, A_2, is much greater than the cross-sectional area of the drillpipe, A_1. Note that hydrostatic pressure is applied to the bottom of the drill collars against a cross-sectional area, A_2, and at the top of the drill collars against the cross section, $A_2 - A_1$. To determine the axial tension F_T in the drill collars, consider a free body diagram of the lower portion of the drillstring (Fig. 4.10b). For the system to be in equilibrium,

$$F_T = W_2 - F_2 - F_b = w_{dc}x_{dc} - p_2 A_2 - F_b,$$

$$\dots\dots\dots\dots\dots\dots\dots\dots\dots\dots\dots\dots (4.24a)$$

where

w_{dc} = weight per unit length of drill collars in air,

x_{dc} = distance from the bottom of the drill collars to the point of interest,

p_2 = hydrostatic pressure at Point 2, and

F_b = force applied to the bit.

To determine the axial tension in the drillpipe, consider a free-body diagram of the upper portion of the drillstring (Fig. 4.10c). As before, system equilibrium requires that

$$F_T = W_1 + W_2 + F_1 - F_2 - F_b$$

$$F_T = w_{dp}x_{dp} + W_2 + p_1(A_2 - A_1)$$

$$-p_2 A_2 - F_b, \dots\dots\dots\dots\dots\dots (4.24b)$$

where

w_{dp} = weight per foot of drillpipe in air,

x_{dp} = distance from the bottom of the drillpipe (top of drill collars) to the point of interest, and

p_1 = hydrostatic pressure at Point 1.

Axial stress is obtained by dividing the axial tension by the cross-sectional area of steel.

Example 4.9. Prepare a graph of axial tension vs. depth for the drillstring described in Example 4.8. Also develop expressions for determining axial stress in the drillstring.

Solution. The hydrostatic pressure p at the top of the drill collars is given by

$$p_1 = 0.052(15)(10,000) = 7,800 \text{ psig.}$$

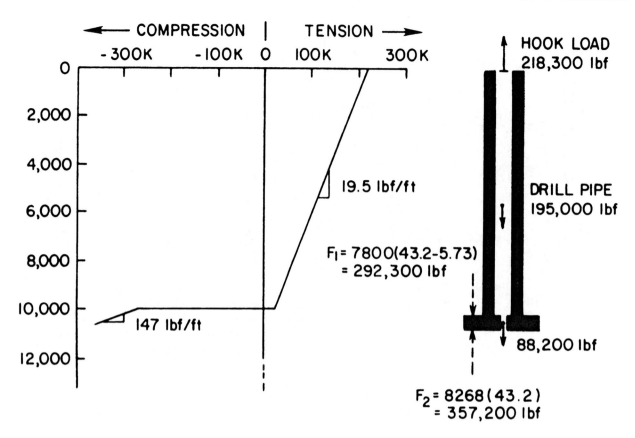

Fig. 4.11—Axial tensions as a function of depth for Example 4.9.

Similarly, the hydrostatic pressure p_2 at the bottom of the drill collars is given by

$$p_2 = 0.052(15)(10,600) = 8,268 \text{ psig.}$$

The cross-sectional area of 19.5-lbm/ft drillpipe is given approximately by

$$A_1 = \frac{19.5 \text{ lbm/ft}}{490 \text{ lbm/cu ft}} \times 144 \text{ sq in./sq ft} = 5.73 \text{ sq in.}$$

Similarly, the cross-sectional area of 147-lbm/ft drill collars is given by

$$A_2 = \frac{147 \text{ lbm/ft}}{490 \text{ lbm/cu ft}} \times 144 \text{ sq in./sq ft} = 43.2 \text{ sq in.}$$

The tension in the drillpipe as a function of depth is given by Eq. 4.24b.

$$F_T = 19.5(10,000 - D) + 147(600)$$

$$+ 7,800(43.2 - 5.73) - 8,268(43.2) - 0.0.$$

After simplifying this equation, we obtain

$$F_T = 218,300 - 19.5D$$

for the $0 < D < 10,000$ ft range. Note that this is the equation of a straight line with a slope of -19.5 lbf/ft

and an intercept of 218,300 lbf (at surface). Note that this is the same result obtained in Example 4.8.

At depths below 10,000 ft, Eq. 4.24a can be used.

$$F_T = 147(10,600 - D) - 8,268(43.2) - 0.$$

After simplifying this equation, we obtain

$$F_T = 1,201,000 \text{ lbf} - 147D$$

for the $10,000 < D < 10,600$ ft range. Using the equations for F_T as a function of depth, the graph shown in Fig. 4.11 was obtained.

Axial stress in the pipe string is obtained by dividing the axial tension by the cross-sectional area of steel. Thus, axial stress σ_z at any point in the drillpipe is given by

$$\sigma_z = \frac{T}{A_s} = \frac{218,300 - 19.5\,D}{5.73}$$

$$= 38,098 \text{ psi} - 3.403\,D$$

for the $0 < D < 10,000$-ft range. Similarly, the axial stress at any point in the drill collars is given by

$$\sigma_z = \frac{1,201,000 - 147\,D}{43.2}$$

$$= 27,800 \text{ psi} - 3.403\,D$$

for the $10,000 < D < 10,600$-ft range.

DRILLING HYDRAULICS   125

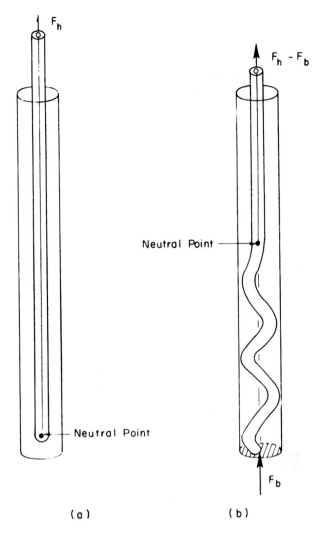

Fig. 4.12—Helical buckling of slender pipe in well: (a) slender pipe suspended and (b) partially buckled slender pipe.

The preparation of a graph of axial stress vs. depth is left as a student exercise.

4.5.2 Effect of Buoyancy on Buckling

Long slender columns such as drillpipe have a low resistance to any applied bending moments and tend to fail by *buckling* when subjected to a vertical compressional load. As shown in Fig. 4.12, if long, slender drillpipe that is confined by a wellbore or casing is subjected to a compressional load on bottom that is less than the hook load, helical buckling can occur in the lower portion of the pipe. Buckling forces are resisted by the moment of inertia of the pipe. The moment of inertia of circular pipe is given by

$$I = \pi/64(d_n{}^4 - d^4),$$

where d_n is the nominal or outside diameter and d is the inside diameter. For drill collars, the moment of inertia is large and generally is assumed to be great enough to prevent buckling. However, the moment of inertia of drillpipe is small and generally is assumed to be negligi-

ble. Thus, if a buckling tendency exists above the drill collars, helical buckling may occur in the drillpipe, as shown in Fig. 4.13.

If drillpipe is rotated in a buckled condition, the tool joints will fatigue quickly and fail. It is common practice to use enough heavy walled drill collars in the lower section of the drillstring so that the desired weight may be applied to the bit without creating a tendency for the drillpipe to buckle. The point above which there is no tendency to buckle is sometimes referred to as the *neutral point*. At the neutral point, the axial stress is equal to the average of the radial and tangential stresses (Fig. 4.14). Current design practice is to maintain the neutral point below the drillpipe during drilling operations.

If the drilling fluid is air and the torque required to rotate the bit is low, the radial and tangential stress in the drillpipe may be negligible. For these simplified conditions, the neutral point is the point of zero axial stress. The length of drill collars then can be chosen such that the weight of the collars is equal to the desired weight to be applied to the bit. In this case, the minimum length of drill collars, L_{dc}, is given by

$$L_{dc} = \frac{F_b}{w_{dc}} \text{(in air)}, \quad\dots\dots\dots\dots\dots\dots (4.25a)$$

where F_b is the maximum force to be applied to the bit during drilling operations, and w_{dc} is the weight per foot of the drill collars. Note that the use of this length of drill collars under these simplified well conditions would result in the neutral point occurring at the junction between the drillpipe and drill collars. No portion of the slender drillpipe is subjected to axial compression.

The effect of buoyancy on buckling should not be ignored if a liquid drilling fluid is used. However, there has been a great amount of confusion in the past about the proper procedure for including buoyancy into the analysis. Many people have reasoned that the net vertical compressional force due to buoyancy simply should be added to the compressional loading, F_b, when computing the minimum length of drill collars by Eq. 4.25a. However, this approach ignores the effect of hydrostatic pressure on the radial and tangential stresses present at the neutral point and thus may be overly conservative. For example, this approach would predict the need for a considerable length of drill collars even for a bit loading, F_b, of zero.

One of the easiest and most general approaches to including the effect of buoyancy on buckling was proposed by Goins.[4] Goins introduces a *stability force* due to fluid pressure p_i inside the pipe, and pressure p_o outside the pipe. The stability force is defined by

$$F_s = A_i p_i - A_o p_o,$$

where A_i is the cross-sectional area computed using the inside pipe diameter, d, and A_o is the cross-sectional area computed using the outside diameter of the pipe, d_n. The stability force can be plotted on a tension/compression diagram such as the one shown in Fig. 4.11. The neutral point then can be determined from the intersection of the axial compression force and the stability force.

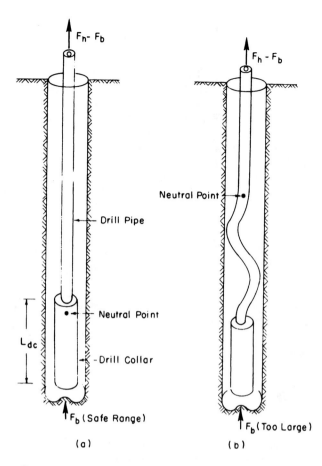

Fig. 4.13—Helical buckling of drillpipe above drill collars: (a) desired condition and (b) undesired buckled condition.

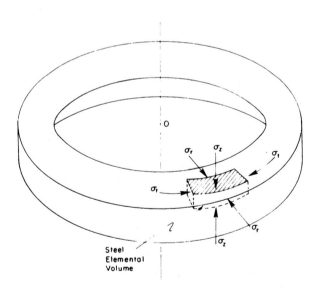

Fig. 4.14—Stress state in steel at neutral point.

For simplified conditions, when the fluid pressures are due to the hydrostatic pressure of a drilling fluid of uniform density, a corollary of Archimedes' law can be applied. Recall that in developing Archimedes' relation it was pointed out that the fluid pressure acting on the surface of any foreign body would be unchanged if the body were not present and that this surface was merely an imaginary surface drawn in the liquid. It was reasoned that if the fluid element contained by the surface is at rest, the sum of the vertical forces must be zero and the weight of the contained fluid must be equal in magnitude but opposite in direction to the buoyant force. However, it is also true that for the fluid element contained by the surface to be at rest, the sum of the moments acting on the fluid element must be zero. *Thus, the moment caused by the hydrostatic forces acting on the fluid element must be equal in magnitude but opposite in direction to the moment caused by the weight of the contained fluid, regardless of the shape of the surface.* The weight of the contained fluid in the imaginary surface and the weight of the foreign body both are distributed loads and have the same moment arm with respect to a given point. This means that for a long slender column immersed in fluid, the *effective weight* of the rod in the fluid should be used instead of the weight of the rod in air when computing bending moments. Thus, the proper length of drill collars, L_{dc}, required to eliminate a tendency for the drillpipe to buckle is given by

$$L_{dc} = \frac{F_b}{w_{dc}\left(1 - \dfrac{\rho_f}{\rho_s}\right)} \quad \text{(in liquid)} \dots \dots \dots (4.25b)$$

It can be shown that the use of a drill collar length predicted by Eq. 4.25b for hydrostatic conditions will result in an intersection of the stability force line and the axial compression line at the junction between the drillpipe and the drill collars.

Note that when Eq. 4.25b is used, only hydrostatic pressures were considered and the pressures due to fluid circulation are neglected. Also neglected are the effects of the torque needed for drillpipe rotation. These two factors can have a significant effect on the radial, tangential, and axial stresses in the pipe wall and thus can cause a significant shift in the neutral point. Also, wall friction makes it difficult to determine the bit loading, F_b, from the observed hook load. Thus, when Eq. 4.25b is used, it is advisable to include a safety factor of at least 1.3.

Example 4.10. A maximum bit weight of 68,000 lbf is anticipated in the next section of hole, which starts at a depth of 10,600 ft. The drill collars have an internal diameter of 3.0 in. and an external diameter of 8.0 in. and the mud density is 15 lbm/gal. Compute the minimum length of drill collars required to prevent a buckling tendency from occurring in the 19.5-lbf/ft drillpipe. The drillpipe has an internal diameter of 4.206 in. and an external diameter of 5.0 in. Also show that the intersection of a plot of axial compressive stress and a plot of the stability force occurs at the top of the drill collars.

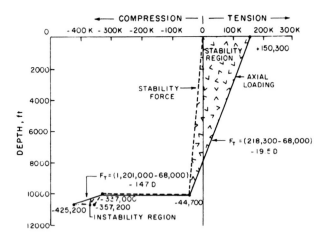

Fig. 4.15—Stability analysis plot for Example 4.10.

Solution. The weight per foot of the drill collars is given by

$$w_{dc} = \frac{\pi(8^2 - 3^2)(490)}{4(144)} = 147 \text{ lbf/ft.}$$

The effective weight per foot in mud is given by

$$w_{dce} = 147\left(1 - \frac{15}{65.5}\right) = 113.3 \text{ lbf/ft.}$$

Thus, the minimum length of drill collars required for a bit weight of 68,000 lbf is given by

$$L_{dc} = \frac{68,000 \text{ lbf}}{113.3 \text{ lbf/ft}} = 600 \text{ ft.}$$

Note that the well conditions are identical to those of Example 4.9, except that the load on the bit is 68,000 lbf. The stability force at the bottom of the collars is given by

$$F_s = A_i p_i - A_o p_o$$

$$= \frac{\pi}{4}(3)^2(8,268) - \frac{\pi}{4}(8)^2(8,268)$$

$$= -357,200 \text{ lbf,}$$

and the stability force at the top of the collars is given by

$$F_s = \frac{\pi}{4}(3)^2(7,800) - \frac{\pi}{4}(8)^2(7,800)$$

$$= -337,000 \text{ lbf.}$$

Similarly, the stability force at the bottom of the drillpipe is

$$F_s = \frac{\pi}{4}(4,206)^2(7,800) - \frac{\pi}{4}(5)^2(7,800)$$

$$= -44,700 \text{ lbf,}$$

and the stability force at the surface is zero. These points have been used to plot stability force as a function of depth in Fig. 4.15. Note that the intersection of the axial compression and stability force lines occur at the junction between the drillpipe and drill collars.

The actual length of drill collars used in practice should be increased by applying a safety factor. Using a safety factor of 1.3 gives

$$L_{dc} = 600(1.3) = 780 \text{ ft.}$$

4.6 Nonstatic Well Conditions

The determination of pressure at various points in the well can be quite complex when either the drilling mud or the drillstring is moving. Frictional forces in the well system can be difficult to describe mathematically. However, in spite of the complexity of the system, the effect of these frictional forces must be determined for the calculation of (1) the flowing bottomhole pressure or equivalent circulating density during drilling or cementing operations, (2) the bottomhole pressure or equivalent circulating density during tripping operations, (3) the optimum pump pressure, flow rate, and bit nozzle sizes during drilling operations, (4) the cuttings-carrying capacity of the mud, and (5) the surface and downhole pressures that will occur in the drillstring during well control operations for various mud flow rates.

The basic physical laws commonly applied to the movement of fluids are (1) conservation of mass, (2) conservation of energy, and (3) conservation of momentum. All of the equations describing fluid flow are obtained by application of these physical laws using an assumed rheological model and an equation of state. Example rheological models used by drilling engineers are the *Newtonian* model, the *Bingham plastic* model, and the *power-law* model. Example equations of state are the incompressible fluid model, the slightly compressible fluid model, the ideal gas equation, and the real gas equation.

4.6.1 Mass Balance

The law of conservation of mass states that the net mass rate into any volume V is equal to the time rate of increase of mass within the volume. The drilling engineer normally considers only *steady-state* conditions in which the mass concentration or fluid density at any point in the well remains constant. Also, with the exception of air or gas drilling, the drilling fluid can be considered incompressible—i.e., the fluid density is essentially the same at all points in the well system. In the absence of any accumulation or leakage of well fluid in the surface equipment or underground formations, the flow rate of an incompressible well fluid must be the same at all points in the well.

The mean velocity at a given point is defined as the flow per unit area at that point. Because of nonuniform flow geometry, the mean velocity at various points in the well may be different even though the flow rate at all points in the well is the same. A knowledge of the mean velocity at a given point in the well often is desired. For example, the drilling engineer frequently will compute the mean upward flow velocity in the annulus to ensure

TABLE 4.1

Pipe	Annulus
$v = \dfrac{17.16 \ (q, \ \text{bbl/min})}{d^2}$	$v = \dfrac{17.16 \ (q, \ \text{bbl/min})}{d_2^2 - d_1^2}$
$v = \dfrac{3.056 \ (q, \ \text{cu ft/min})}{d^2}$	$v = \dfrac{3.056 \ (q, \ \text{cu ft/min})}{d_2^2 - d_1^2}$
$v = \dfrac{(q, \ \text{gal/min})}{2.448 \ d^2}$	$v = \dfrac{(q, \ \text{gal/min})}{2.448 \ (d_2^2 - d_1^2)}$

where

v = average velocity, ft/s,
d = internal diameter of pipe, in.,
d_2 = internal diameter of outer pipe or borehole, in., and
d_1 = external diamter of inner pipe, in.

that it is adequate for rock-cutting removal. Shown in Table 4.1 are convenient forms of q/A for units frequently used in the field.

Sclf study i-32

Example 4.11. A 12-lbm/gal mud is being circulated at 400 gal/min. The 5.0-in. drillpipe has an internal diameter of 4.33 in., and the drill collars have an internal diameter of 2.5 in. The bit has a diameter of 9.875 in. Calculate the average velocity in the (1) drillpipe, (2) drill collars, and (3) annulus opposite the drillpipe.

Solution. Using the expression given in Table 4.1 for units of gallons per minute, inches, and feet per second gives:

1. $v_{dp} = \dfrac{400}{2.448(4.33)^2} = 8.715$ ft/s .

2. $v_{dc} = \dfrac{400}{2.448(2.5)^2} = 26.143$ ft/s .

3. $v_{dpa} = \dfrac{400}{2.448(9.875^2 - 5^2)} = 2.253$ ft/s .

4.6.2 Energy Balance

The law of conservation of energy states that the net energy rate out of a system is equal to the time rate of work done within the system. Consider the generalized flow system shown in Fig. 4.16. The energy entering the system is the sum of

$E_1 + p_1 \bar{V}_1$ = enthalpy per unit mass of the fluid entering the system at Point 1,

$-gD_1$ = potential energy per unit mass of the fluid entering the system at Point 1,

$\bar{v}_1^2/2$ = kinetic energy per unit mass of the fluid entering the system at Point 1, and

Q = heat per unit mass of fluid entering the system.

The energy leaving the system is the sum of

$E_2 + p_2 \bar{V}_2$ = enthalpy per unit mass of the fluid leaving the system at Point 2,

$-gD_2$ = potential energy per unit mass of the fluid leaving the system at Point 2, and

$\bar{v}_2^2/2$ = kinetic energy per unit mass leaving the system at Point 2.

The work done by the fluid is equal to the energy per unit mass of fluid given by the fluid to a fluid engine (or equal to minus the work done by a pump on the fluid). Thus, the law of conservation of energy yields

$$(E_2 - E_1) + (p_2 \bar{V}_2 - p_1 \bar{V}_1) - g(D_2 - D_1)$$

$$+ \frac{1}{2}(\bar{v}_2^2 - \bar{v}_1^2) = W + Q.$$

Simplifying this expression using differential notations yields

$$\Delta E - g\Delta D + \frac{\Delta v^2}{2} + \Delta(p\bar{V}) = W + Q. \qquad \ldots \ldots (4.26)$$

Eq. 4.26 is the first law of thermodynamics applied to a steady flow process. This equation is best suited for flow systems that involve either heat transfer or adiabatic processes involving fluids whose thermodynamic properties have been tabulated previously. This form of the equation seldom has been applied by drilling engineers. The change in internal energy of the fluid and the heat gained by the fluid usually is considered using a *friction loss* term, which can be defined in terms of Eq. 4.26 using the following expression.

$$F = \Delta E + \int_1^2 p \, d\bar{V} - Q. \qquad \ldots \ldots \ldots \ldots \ldots (4.27)$$

The frictional loss term can be used conveniently to account for the lost work or energy wasted by the viscous forces within the flowing fluid. Substitution of Eq. 4.27 into Eq. 4.26 yields

$$\int_1^2 \bar{V} dp - g\Delta D + \frac{\Delta \bar{v}^2}{2} = W - F. \qquad \ldots \ldots \ldots (4.28)$$

Eq. 4.28 often is called the mechanical energy balance equation. This equation was in use even before heat flow was recognized as a form of energy transfer by Carnot and Joule and is a completely general expression containing no limiting assumptions other than the exclusion of phase boundaries and magnetic, electrical, and chemical effects. The effect of heat flow in the system is included in the friction loss term F.

The first term in Eq. 4.28,

$$\int_1^2 \bar{V} dp,$$

may be difficult to evaluate if the fluid is compressible unless the exact path of compression or expansion is known. Fortunately, drilling engineers deal primarily with essentially incompressible fluids having a constant specific volume $\bar{V}$.

Since for incompressible fluids, the term

$$\int_1^2 \bar{V}dp$$

is given by

$$\int_1^2 \bar{V}dp = \frac{\Delta p}{\rho},$$

Eq. 4.28 also can be expressed by

$$\Delta p - \rho g\,\Delta D + \rho \frac{\Delta \bar{v}^2}{2} = \rho W - \rho F.$$

Expressing this equation in practical field units of pounds per square inch, pounds per gallons, feet per second, and feet gives

$$p_1 + 0.052\rho(D_2 - D_1) - 8.074$$

$$\underbrace{8 \sim \times 10^{-4}\rho(\bar{v}_2^2 - \bar{v}_1^2)}_{\Delta P_b} + \Delta p_p - \Delta p_f = p_2. \quad \ldots\ldots (4.29)$$

$$\Delta P_f = \Delta P_p - \Delta P_b$$

Example 4.12. Determine the pressure at the bottom of the drillstring if the frictional pressure loss in the drillstring is 1,400 psi, the flow rate is 400 gals/min, the mud density is 12 lbm/gal, and the well depth is 10,000 ft. The internal diameter of the drill collars at the bottom of the drillstring is 2.5 in. and the pressure increase developed by the pump is 3,000 psi.

Solution. The average velocity in the drill collars is

$$v_{dc} = \frac{400}{2.448(2.5)^2} = 26.14 \text{ ft/s.}$$

The average velocity in the mud pits is essentially zero.

$$p_2 = 0 + 0.052(12)(10,000) - 8.074$$

$$\times 10^{-4}(12)(26.14)^2 + 3,000 - 1,400$$

$$= 0 + 6,240 - 6.6 + 3,000 - 1,400$$

$$= 7,833 \text{ psi.}$$

Example 4.12 illustrates the minor effect of the kinetic energy term of Eq. 4.29 in this drilling application. In general, the change in kinetic energy caused by fluid acceleration can be ignored, except for the flow of drilling fluid through the bit nozzles.

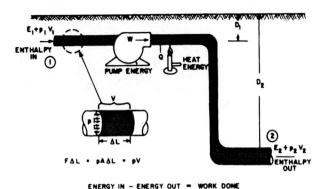

$$F\Delta L = \rho A \Delta L = \rho V$$

ENERGY IN − ENERGY OUT = WORK DONE

Fig. 4.16—Generalized flow system.

4.7 Flow Through Jet Bits

A schematic of incompressible flow through a short constriction, such as a bit nozzle, is shown in Fig. 4.17. In practice, it generally is assumed that (1) the change in pressure due to a change in elevation is negligible, (2) the velocity v_o upstream of the nozzle is negligible, compared with the nozzle velocity v_n, and (3) the frictional pressure loss across the nozzle is negligible. Thus, Eq. 4.29 reduces to

$$p_1 - 8.074 \times 10^{-4}\rho v_n^2 = p_2.$$

Substituting the symbol Δp_b for the pressure drop $(p_1 - p_2)$ and solving this equation for the nozzle velocity v_n yields

$$v_n = \sqrt{\frac{\Delta p_b}{8.074 \times 10^{-4}\rho}}. \quad \ldots\ldots\ldots\ldots (4.30)$$

Unfortunately, the exit velocity predicted by Eq. 4.30 for a given pressure drop across the bit, Δp_b, never is realized. The actual velocity is always smaller than the velocity computed using Eq. 4.30 primarily because the assumption of frictionless flow is not strictly true. To compensate for this difference, a correction factor or discharge coefficient C_d usually is introduced so that the modified equation,

$$v_n = C_d\sqrt{\frac{\Delta p_b}{8.074 \times 10^{-4}\rho}}, \quad \ldots\ldots\ldots\ldots (4.31)$$

will result in the observed value for nozzle velocity. The discharge coefficient has been determined experimentally for bit nozzles by Eckel and Bielstein.[5] These authors indicated that the discharge coefficient may be as high as 0.98 but recommended a value of 0.95 as a more practical limit.

A rock bit has more than one nozzle, usually having the same number of nozzles and cones. When more than one nozzle is present, the pressure drop applied across all of the nozzles must be the same (Fig. 4.18). According to Eq. 4.31, if the pressure drop is the same for each nozzle, the velocities through all nozzles are equal.

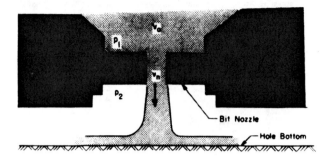

Fig. 4.17—Flow through a bit nozzle.

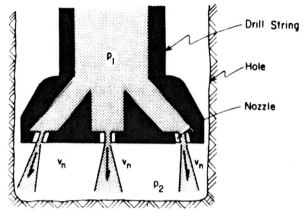

Fig. 4.18—Flow through parallel nozzles.

Therefore, if the nozzles are of different areas, the flow rate q through each nozzle must adjust so that the ratio q/A is the same for each nozzle. If three nozzles are present,

$$\bar{v} = \frac{q_1}{A_1} = \frac{q_2}{A_2} = \frac{q_3}{A_3}.$$

Note also that the total flow rate of the pump, q, is given by

$$q = q_1 + q_2 + q_3 = \bar{v}_n A_1 + \bar{v}_n A_2 + \bar{v}_n A_3.$$

Simplifying this expression yields

$$q = \bar{v}_n (A_1 + A_2 + A_3) = \bar{v}_n A_t.$$

Thus, the velocity of flow through each nozzle is also equal to the total flow rate divided by the total nozzle area.

$$\frac{q}{A_t} = \frac{q_1}{A_1} = \frac{q_2}{A_2} = \dots = \frac{q_i}{A_i}. \quad \dots \dots \dots \dots (4.32)$$

In field units, the nozzle velocity v_n is given by

$$v_n = \frac{q}{3.117\, A_t}, \quad \dots \dots \dots \dots \dots \dots \dots (4.33)$$

where v_n has units of feet per second, q has units of gallons per minute, and A_t has units of square inches. Combining Eqs. 4.31 and 4.33 and solving for the pressure drop across the bit, Δp_b, yields

$$\Delta p_b = \frac{8.311 \times 10^{-5} \rho q^2}{C_d^2 A_t^2}. \quad \dots \dots \dots \dots \dots (4.34)$$

Since the viscous frictional effects are essentially negligible for flow through short nozzles, Eq. 4.34 is valid for both Newtonian and non-Newtonian liquids.

Bit nozzle diameters often are expressed in 32nds of an inch. For example, if the bit nozzles are described as "12-13-13," this denotes that the bit contains one nozzle having a diameter of $^{12}\!/_{32}$ in. and two nozzles having a diameter of $^{13}\!/_{32}$ in.

Example 4.13. A 12.0-lbm/gal drilling fluid is flowing through a bit containing three $^{13}\!/_{32}$-in. nozzles at a rate of 400 gal/min. Calculate the pressure drop across the bit.

Solution. The total area of the three nozzles is given by

$$A_t = \frac{\pi}{4(32)^2} (13^2 + 13^2 + 13^2)$$

$$= 7.67 \times 10^{-4}(169 + 169 + 169)$$

$$= 0.3889 \text{ sq in.}$$

Using Eq. 4.34, the pressure drop across the bit is given by

$$\Delta p_b = \frac{8.311 \times 10^{-5}(12)(400)^2}{(0.95)^2(0.3889)^2}$$

$$= 1,169 \text{ psi.}$$

4.7.1 Hydraulic Power

Since power is the rate of doing work, pump energy W can be converted to hydraulic power P_H by multiplying W by the mass flow rate ρq. Thus,

$$P_H = \rho W q = \Delta p_p q.$$

If the flow rate q is expressed in gallons per minute and the pump pressure Δp_p is expressed in pounds per square inch,

$$P_H = \frac{\Delta p_p q}{1,714}, \quad \dots \dots \dots \dots \dots \dots \dots (4.35)$$

where P_H is expressed in hydraulic horsepower. Likewise, other terms in Eq. 4.29, the pressure balance equation, can be expressed as hydraulic horsepower by multiplying the pressure term by $q/1,714$.

Example 4.14. Determine the hydraulic horsepower being developed by the pump discussed in Example 4.12. How much of this power is being lost due to the viscous forces in the drillstring.

Solution. The pump power being used is given by Eq. 4.35.

$$P_H = \frac{\Delta p_p q}{1,714} = \frac{3,000(400)}{1,714} = 700 \text{ hp.}$$

The power consumed due to "friction" in the drillstring is

$$P_f = \frac{\Delta p_f q}{1,714} = \frac{1,400(400)}{1,714} = 327 \text{ hp.}$$

4.7.2 Hydraulic Impact Force

The purpose of the jet nozzles is to improve the cleaning action of the drilling fluid at the bottom of the hole. Before jet bits were introduced, rock chips were not removed efficiently and much of the bit life was consumed regrinding the rock fragments. While the cleaning action of the jet is not well-understood, several investigators have concluded that the cleaning action is maximized by maximizing the total hydraulic impact force of the jetted fluid against the hole bottom. If it is assumed that the jet stream impacts the bottom of the hole in the manner shown in Fig. 4.17, all of the fluid momentum is transferred to the hole bottom. Since the fluid is traveling at a vertical velocity v_n before striking the hole bottom and is traveling at zero vertical velocity after striking the hole bottom, the time rate of change of momentum (in field units) is given by

$$F_j = \frac{\Delta(m\bar{v})}{\Delta t} \cong \left(\frac{m}{\Delta t}\right) \Delta\bar{v} = \frac{(\rho q)\bar{v}_n}{32.17(60)}, \quad \ldots \ldots (4.36)$$

where (ρq) is the mass rate of the fluid. Combining Eqs. 4.31 and 4.36 yields

$$F_j = 0.01823 C_d q \sqrt{\rho \Delta p_b}, \quad \ldots \ldots \ldots (4.37)$$

where F_j is given in pounds.

Example 4.15. Compute the impact force developed by the bit discussed in Example 4.13.

Solution. Using Eq. 4.37,

$$F_j = 0.01823(0.95)(400)\sqrt{(12)(1,169)}$$

$$= 820 \text{ lbf.}$$

4.8 Rheological Models

The frictional pressure loss term in the pressure balance equation given as Eq. 4.29 is the most difficult to evaluate. However, this term can be quite important since extremely large viscous forces must be overcome to move drilling fluid through the long slender conduits used in the rotary drilling process. A mathematical description of the viscous forces present in a fluid is required for the development of friction loss equations. The rheological models generally used by drilling engineers to approximate fluid behavior are (1) the Newtonian model, (2) the Bingham plastic model, and (3) the power-law model.

4.8.1 Newtonian Model

The viscous forces present in a simple Newtonian fluid are characterized by the fluid viscosity. Examples of Newtonian fluids are water, gases, and high gravity oils. To understand the nature of viscosity, consider a fluid contained between two large parallel plates of area A, which are separated by a small distance L (Fig. 4.19). The upper plate, which is initially at rest, is set in motion in the x direction at a constant velocity v. After sufficient time has passed for steady motion to be achieved, a constant force F is required to keep the upper plate moving at constant velocity. The magnitude of the force F was found experimentally to be given by

$$\frac{F}{A} = \mu \frac{V}{L}.$$

The term F/A is called the *shear stress* exerted on the fluid. Thus, shear stress is defined by

$$\tau = \frac{F}{A}. \quad \ldots \ldots \ldots \ldots \ldots \ldots \ldots (4.38)$$

Note that the area of the plate A is the area in contact with the fluid. The velocity gradient v/L is an expression of the *shear rate*.

$$\dot{\gamma} = \frac{v}{L} = \frac{dv}{dL}. \quad \ldots \ldots \ldots \ldots \ldots \ldots (4.39)$$

Thus, the Newtonian model states that the shear stress τ is directly proportional to the shear rate $\dot{\gamma}$.

$$\tau = \mu \dot{\gamma}, \quad \ldots \ldots \ldots \ldots \ldots \ldots \ldots (4.40)$$

where μ, the constant of proportionality, is known as the viscosity of the fluid (Fig. 4.20). In terms of the moving plate this means that if the force F is doubled, the plate velocity v also will double.

Viscosity is expressed in poises. A poise is 1 dyne-s/cm^2 or 1 g/cm·s. In the drilling industry,

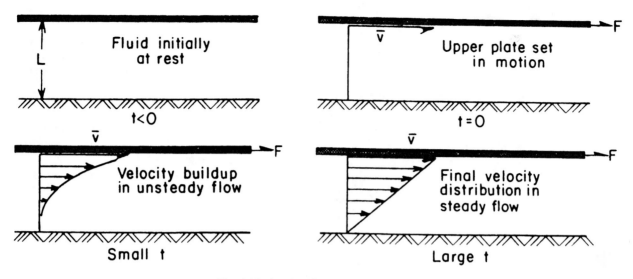

Fig. 4.19—Laminar flow of Newtonian fluids.

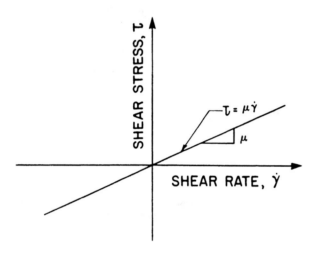

Fig. 4.20—Shear stress vs. shear rate for a Newtonian fluid.

viscosity generally is expressed in centipoises, where 1 cp = 0.01 poise. Occasionally, viscosity is expressed in units of lbf-s/sq ft. The units of viscosity can be related (at sea level) by

$$\frac{\text{lbf-s}}{\text{ft}^2} \times \frac{(454 \text{ g/lbf})(980 \text{ cm/s}^2)}{(30.48 \text{ cm/ft})^2}$$

$$= 479 \text{ dyne-s/cm}^2$$

$$= 479 \text{ poise} = 47{,}900 \text{ cp}.$$

Example 4.16. An upper plate of 20-cm^2 area is spaced 1 cm above a stationary plate. Compute the viscosity in centipoise of a fluid between the plates if a force of 100 dyne is required to move the upper plate at a constant velocity of 10 cm/s.

Solution. The shear stress τ is given by

$$\tau = \frac{100 \text{ dyne}}{20 \text{ cm}^2} = 5 \text{ dyne/cm}^2.$$

The shear rate $\dot{\gamma}$ is given by

$$\dot{\gamma} = \frac{10 \text{ cm/s}}{1 \text{ cm}} = 10 \text{ seconds}^{-1}.$$

Using Eq. 4.40,

$$\mu = \tau/\dot{\gamma} = \frac{5 \text{ dyne/cm}^2}{10 \text{ seconds}^{-1}} = 0.5 \text{ dyne·s/cm}^2,$$

or

$$\mu = 50 \text{ cp}.$$

The linear relation between shear stress and shear rate described by Eq. 4.40 is valid only as long as the fluid moves in layers or laminae. A fluid that flows in this manner is said to be in *laminar flow*. This is true only at relatively low rates of shear. At high rates of shear, the flow pattern changes from laminar flow to *turbulent flow*, in which the fluid particles move downstream in a tumbling chaotic motion so that vortices and eddies are formed in the fluid. Dye injected into the flow stream thus would be dispersed quickly throughout the entire cross section of the fluid. The turbulent flow of fluids has not been described mathematically. Thus, when turbulent flow occurs, frictional pressure drops must be determined by empirical correlations.

4.8.2 Non-Newtonian Models

Most drilling fluids are too complex to be characterized by a single value for viscosity. The apparent viscosity measured depends on the shear rate at which the measurement is made and the prior shear rate history of the fluid. Fluids that do not exhibit a direct proportionality between shear stress and shear rate are classified as *non-Newtonian*. Non-Newtonian fluids that are shear-rate dependent (Fig. 4.21) are *pseudoplastic* if the apparent viscosity decreases with increasing shear

$$\tau = \frac{F}{A}$$

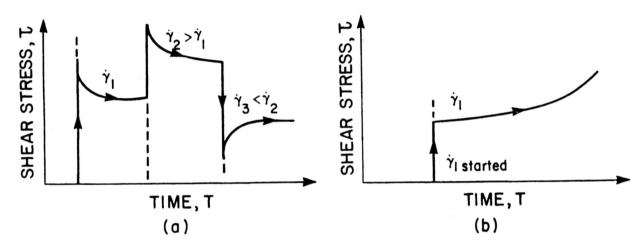

Fig. 4.21—Shear stress vs. shear rate for pseudoplastic and dilatant fluids: (a) pseudoplastic behavior, $\mu_{a2} < \mu_{a1}$, and (b) dilatant behavior, $\mu_{a2} > \mu_{a1}$.

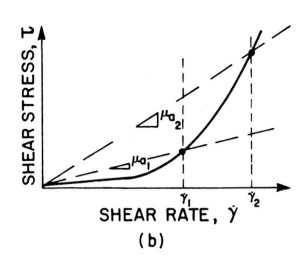

Fig. 4.22—Shear stress vs. time for thixotropic and rheopectic fluids: (a) thixotropic behavior and (b) rheopectic behavior.

rate and are *dilatant* if the apparent viscosity increases with increasing shear rate. Drilling fluids and cement slurries are generally pseudoplastic in nature.

Non-Newtonian fluids that are shear-time-dependent (Fig. 4.22) are *thixotropic* if the apparent viscosity decreases with time after the shear rate is increased to a new constant value and are *rheopectic* if the apparent viscosity increases with time after the shear rate is increased to a new constant value. Drilling fluids and cement slurries are generally thixotropic.

The Bingham plastic and power-law rheological models are used to approximate the pseudoplastic behavior of drilling fluids and cement slurries. At present, the thixotropic behavior of drilling fluids and cement slurries is not modeled mathematically. However, drilling fluids and cement slurries generally are stirred before measuring the apparent viscosities at various shear rates so that steady-state conditions are obtained. Not accounting for thixotropy is satisfactory for most cases, but significant errors can result when a large number of direction changes and diameter changes are present in the flow system.

4.8.3 Bingham Plastic Model

The Bingham plastic model is defined by

$$\tau = \mu_p \, \dot{\gamma} + \tau_y \; ; \; \tau > \tau_y \longrightarrow \mu_p = \frac{\tau - \tau_y}{\dot{\gamma}} \quad \dots \, (4.41a) \quad \tau = \frac{F}{A}$$

$$\dot{\gamma} = 0 \; ; \; -\tau_y \le \tau \le +\tau_y , \quad \dots \dots \dots \dots (4.41b)$$

and

$$\tau = \mu_p \, \dot{\gamma} - \tau_y \; ; \; \tau < -\tau_y . \quad \dots \dots \dots \dots (4.41c)$$

A graphical representation of this behavior is shown in Fig. 4.23.

A Bingham plastic will not flow until the applied shear stress τ exceeds a certain minimum value τ_y known as the yield point. After the yield point has been exceeded, changes in shear stress are proportional to changes in shear rate and the constant of proportionality is called the plastic viscosity, μ_p. Eqs. 4.41a through 4.41c are valid only for laminar flow. Note that the units of plastic viscosity are the same as the units of Newtonian or "apparent" viscosity. To be consistent, the units of the yield

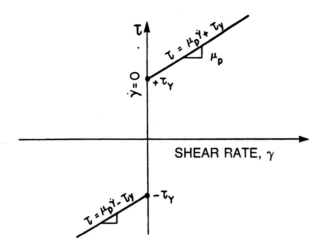

Fig. 4.23—Shear stress vs. shear rate for a Bingham plastic fluid.

point τ_y must be the same as the units for shear stress τ. Thus, the yield point has consistent units of dynes per square centimeter. However, yield point usually is expressed in field units of pounds per 100 sq ft. The two units can be related (at sea level) by

$$\frac{1 \text{ lbf}}{100 \text{ sq ft}} \times \frac{(454 \text{ g/lbf})(980 \text{ cm/s}^2)}{(30.48 \text{ cm/ft})^2}$$

$$= 4.79 \text{ dyne/cm}^2.$$

Example 4.17. An upper plate of 20-cm^2 area is spaced 1 cm above a stationary plate. Compute the yield point and plastic viscosity of a fluid between the plates if a force of 200 dynes is required to cause any movement of the upper plate and a force of 400 dynes is required to move the upper plate at a constant velocity of 10 cm/s.

Solution. The yield point τ_y is given by Eq. 4.41a with $\dot{\gamma} = 0$.

$$\tau_y = \tau = \frac{200 \text{ dyne}}{20 \text{ cm}^2} = 10 \text{ dyne/cm}^2.$$

In field units,

$$\tau_y = \frac{10}{4.79} = 2.09 \text{ lbf/100 sq ft}.$$

The plastic viscosity μ_p is given by Eq. 4.41a with $\dot{\gamma}$ given by

$$\dot{\gamma} = \frac{10 \text{ cm/s}}{\text{cm}} = 10 \text{ seconds}^{-1}.$$

Thus, μ_p is given by

$$\mu_p = \frac{400/20 - 10}{10} = 1.0 \text{ dyne-s/cm}^2,$$

or

$$\mu_p = 100 \text{ cp}.$$

4.8.4 Power-Law Model

The power-law model is defined by

$$\tau = K |\dot{\gamma}|^{n-1} \dot{\gamma}. \quad \ldots\ldots\ldots\ldots\ldots\ldots\ldots (4.42)$$

A graphical representation of the model is shown in Fig. 4.24. Like the Bingham plastic model, the power-law model requires two parameters for fluid characterization. However, the power-law model can be used to represent a pseudoplastic fluid ($n < 1$), a Newtonian fluid ($n = 1$), or a dilatant fluid ($n > 1$). Eq. 4.42 is valid only for laminar flow.

The parameter K usually is called the *consistency index* of the fluid, and the parameter n usually is called either the power-law exponent or the *flow-behavior index*. The deviation of the dimensionless flow-behavior index from unity characterizes the degree to which the fluid behavior is non-Newtonian. The units of the consistency index K depend on the value of n. K has units of dyne-s^n/cm^2 or g/cm·s^{2-n}. In this text, a unit called the equivalent centipoise (eq cp) will be used to represent 0.01 dyne-s^n/cm^2. Occasionally, the consistency index is expressed in units of lbf-s^n/sq ft. The two units of consistency index can be related (at sea level) by

$$\frac{1 \text{ lbf-s}^n}{\text{sq ft}} \times \frac{(454 \text{ g/lbf})(980 \text{ cm/s}^2)}{(30.48 \text{ cm/ft})^2}$$

$$= 479 \text{ dyne-s}^n/\text{cm}^2$$

$$= 47,900 \text{ eq cp}.$$

Example 4.18. An upper plate of 20 cm^2 is spaced 1 cm above a stationary plate. Compute the consistency index and flow-behavior index if a force of 50 dyne is required to move the upper plate at a constant velocity of 4 cm/s and a force of 100 dyne is required to move the upper plate at a constant velocity of 10 cm/s.

Solution. Application of Eq. 4.42 at the two rates of shear observed yields

$$\frac{50}{20} = K \left(\frac{4}{1}\right)^n,$$

and

$$\frac{100}{20} = K \left(\frac{10}{1}\right)^n.$$

Dividing the second equation by the first gives

$$\frac{100}{50} = \left(\frac{10}{4}\right)^n.$$

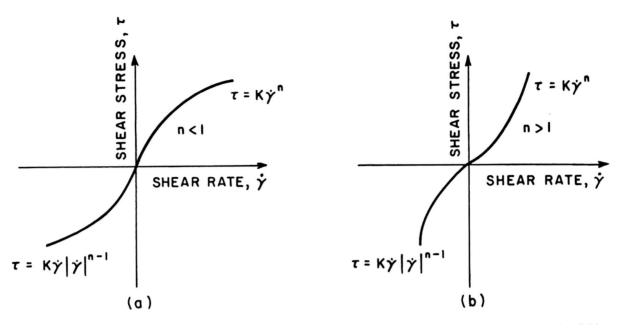

Fig. 4.24—Shear stress vs. shear rate for a power-law fluid: (a) pseudoplastic power-law fluid and (b) dilatant power-law fluid.

Taking the log of both sides and solving for n yields

$$n = \frac{\log (100/50)}{\log (10/4)} = 0.756.$$

Substituting this value of n in the first equation above yields

$$K = \frac{50}{20(4)^{0.756}} = 0.8765 \ \frac{\text{dyne-s}^{0.756}}{\text{cm}^2}$$

$$= 87.65 \text{ eq cp.}$$

4.9 Rotational Viscometer

Examples 4.16 through 4.18 illustrate the physical meaning of the Newtonian, Bingham plastic, and power-law parameters. Unfortunately, it would be extremely difficult to build a viscometer based on the relative movement of two flat parallel plates. However, as shown in Fig. 4.25, the rotation of an outer sleeve about a concentric cylinder is somewhat similar to the relative movement of parallel plates. The viscometer described in the standard API diagnostic tests for drilling fluids (see Chap. 2) is a rotational viscometer.

Rotation of the outer sleeve instead of the inner bob has been found to extend the transition from laminar to turbulent flow to higher shear rates. Since only the laminar flow regime can be described analytically, all fluid characterization measurements must be made in laminar flow. In practice, the torque exerted by the fluid on the stationary bob usually is measured by a torsion spring attached to the bob. The rotor and bob dimensions available on the rotational viscometer are shown in Table 4.2. The $(r_1)_1/(r_2)_1$ rotor/bob combination is the standard combination used for field testing of drilling fluid.

The fluid shear rate between stationary and moving parallel plates was assumed to be constant in Examples 4.16 through 4.18. However, the fluid shear rate in a rotational viscometer is a function of the radius r. The fluid velocity v at a given radius is related to the angular velocity ω by

$$v = r\omega . \quad\dotfill (4.43)$$

Thus, the change in velocity v with radius r is given by

$$\frac{dv}{dr} = r\frac{d\omega}{dr} + \omega .$$

If the fluid layers were not slipping past one another but moving together as a solid plug, the change in velocity with radius would be given by

$$\frac{dv}{dr} = \omega \text{ (no slip).}$$

Thus, the shear rate due to slippage between fluid layers is given by

$$\dot{\gamma} = r\frac{d\omega}{dr} . \quad\dotfill (4.44)$$

When the rotor is rotating at a constant angular velocity ω_2 and the bob is held motionless ($\omega_1 = 0$), the torque applied by the torsion spring to the bob must be equal but opposite in direction to the torque applied to the rotor by the motor. The torque is transmitted between the rotor and the bob by the viscous drag between successive layers of fluid. If there is no slip at the rotor wall, the layer of fluid immediately adjacent to the rotor also is moving at an angular velocity ω_2. Successive layers of fluid between r_2 and r_1 are moving at successively lower velocities. If there is no slip at the bob wall, the layer of fluid immediately adjacent to the bob is motionless. If the small end effect at the bottom of the bob

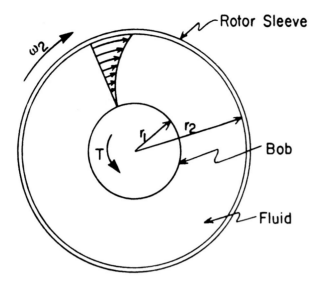

Fig. 4.25—Bottom view of rotational viscometer.

TABLE 4.2—ROTOR AND BOB DIMENSIONS
FOR ROTATIONAL VISCOMETERS

	Bob Dimensions			Rotor Dimensions	
Type	Radius (cm)	Length (cm)	Type		Radius (cm)
$(r_1)_1$	1.7245	3.80	$(r_2)_1$		1.8415
$(r_1)_2$	1.2276	3.80	$(r_2)_2$		1.7589
$(r_1)_3$	0.86225	3.80	$(r_3)_3$		2.5867
$(r_1)_4$	0.86225	1.89			

Assuming that no slip occurs at the walls of the viscometer, the angular velocity is zero at r_1 and ω_2 at r_2. Thus, separating variables in Eq. 4.46 gives

$$\int_0^{\omega_2} d\omega = \frac{360.5\,\theta}{2\pi h\mu} \int_{r_1}^{r_2} \frac{dr}{r^3}. \quad \dots \dots \dots \dots (4.47)$$

Integrating and solving for viscosity μ yields

$$\mu = \frac{360.5\,\theta}{4\pi h\omega_2}\left(\frac{1}{r_1^2} - \frac{1}{r_2^2}\right). \quad \dots \dots \dots \dots (4.48a)$$

Substituting the value $2\pi N/60$ for ω_2, the values of r_1, r_2, and h shown in Table 4.2, and changing viscosity units to centipoise simplifies this equation to the following.

$$\mu = 300\,\frac{\theta_N}{N}, \quad \dots \dots \dots \dots \dots \dots (4.48b)$$

where

μ = fluid viscosity, cp,
θ_N = dial reading of the rotational viscometer, and
N = speed of rotation of the outer cylinder, rpm.

Note that if the rotational viscometer is operated at 300 rpm, the dial reading of the viscometer is numerically equal to the viscosity in centipoise. Most viscometers used in the field will operate at either 300 or 600 rpm. However, in some cases, a multispeed viscometer that will operate at 3, 6, 100, 200, 300, and 600 rpm is used.

It is often desirable to know the shear rates present in a rotational viscometer for a given speed of rotation. Eq. 4.48a can be rearranged to give

$$\frac{360.5\,\theta}{\pi h\mu} = \frac{4\omega_2}{\left(\dfrac{1}{r_1^2} - \dfrac{1}{r_2^2}\right)}.$$

Substituting this expression with Eq. 4.46 yields

$$\frac{d\omega}{dr} = \frac{4\omega_2}{2r^3\left(\dfrac{1}{r_1^2} - \dfrac{1}{r_2^2}\right)} = \frac{4\pi N/60}{r^3(0.04137)}.$$

is ignored, then the torque T can be related to the shear stress in the fluid at any radius r between the bob radius r_1 and the rotor radius r_2 using the following equation.

$$T = \tau\,(2\pi rh)\,r.$$

The spring constant of the torsion spring generally used in testing drilling fluids is chosen such that

$$T = 360.5\,\theta,$$

where θ is the dial reading of the Fann measured in degrees of angular displacement. Equating the two expressions for torque and solving for shear stress yields

$$\tau = \frac{360.5\,\theta}{2\pi hr^2}. \quad \dots \dots \dots \dots \dots \dots (4.45)$$

Eq. 4.45 indicates that the shear stress present in the fluid varies inversely with the square of the radius. This relation is a consequence of the geometry of the viscometer and does not depend on the nature of the fluid. The shear rate can be related to shear stress using the defining equation for the Newtonian, Bingham plastic, or power-law fluid models discussed in the previous sections.

4.9.1 Newtonian Model

If the fluid can be described by the Newtonian fluid model, then the shear stress at any point in the fluid is given by

$$\tau = \mu\dot{\gamma} = \mu r\,\frac{d\omega}{dr}.$$

Combining this equation with Eq. 4.45 yields

$$\frac{d\omega}{dr} = \frac{360.5\,\theta}{2\pi hr^3\mu}. \quad \dots \dots \dots \dots (4.46)$$

Thus, the shear rate $\dot{\gamma}$ is given by

$$\dot{\gamma} = r\frac{d\omega}{dr} = \frac{5.066\,N}{r^2}. \qquad \ldots\ldots\ldots\ldots\ldots (4.49)$$

Example 4.19. A Fann viscometer normally is operated at 300 and 600 rpm in the standard API diagnostic test. Compute the shear rate that would occur at the bob radius of 1.7245 cm for these two rotor speeds if the viscometer contained a Newtonian fluid.

Solution. Using Eq. 4.49, we obtain

$$\dot{\gamma} = \frac{5.066N}{(1.7245)^2} = 1.703\ N.$$

At a rotor speed of 300 rpm, the shear rate at the bob is given by

$$\dot{\gamma} = 1.703(300) = 511 \text{ seconds}^{-1}.$$

Similarly, at a rotor speed of 600 rpm, the shear rate at the bob is given by

$$\dot{\gamma} = 1.703(600) = 1{,}022 \text{ seconds}^{-1}.$$

4.9.2 Non-Newtonian Models

The rotational viscometer can also be used to determine the flow parameters of the Bingham plastic and power-law fluid models. The equations needed for the calculations of these flow parameters can be derived by following the same steps used in the derivation of the equations for the Newtonian model. These derivations are presented in Appendix A and the final equations are summarized in Table 4.3. Since two flow parameters must be calculated for both the Bingham plastic and power-law models, two readings must be made with a rotational viscometer at different rotor speeds. Normally the 300 and 600 rpm readings are used in the computation. However, when it is desirable to characterize the fluid at lower shear rates, the flow parameters can be computed using readings taken at lower rotor speeds.

Example 4.20. A rotational viscometer containing a non-Newtonian fluid gives a dial reading of 12 at a rotor speed of 300 rpm and a dial reading of 20 at a rotor speed of 600 rpm. Compute the consistency index and flow-behavior index of the power-law model for this fluid.

Solution. Using Table 4.3, the flow-behavior index and consistency index are given by

$$n = 3.32 \log(20/12) = 0.737,$$

$$K = \frac{510(12)}{(511)^{0.737}} = 61.8 \text{ eq cp},$$

or

$$K = 0.618 \text{ dyne-s}^{0.737}/\text{cm}^2.$$

4.10 Laminar Flow in Pipes and Annuli

The drilling engineer deals primarily with the flow of drilling fluids and cements down the circular bore of the drillstring and up the circular annular space between the drillstring and the casing or open hole. If the pump rate is low enough for the flow to be laminar, the Newtonian, Bingham plastic, or power-law model can be employed to develop the mathematical relation between flow rate and frictional pressure drop. In this development, these simplifying assumptions are made: (1) the drillstring is placed concentrically in the casing or open hole, (2) the drillstring is not being rotated, (3) sections of open hole are circular in shape and of known diameter, (4) the drilling fluid is incompressible, and (5) the flow is isothermal.

In reality, none of these assumptions are completely valid, and the resulting system of equations will not describe perfectly the laminar flow of drilling fluids in the well. In addition, the student should keep in mind that the Newtonian, Bingham plastic, and power-law fluid rheological models do not take into account the thixotropic nature of drilling mud and only approximate the actual laminar flow fluid behavior. Some research has been conducted on the effect of pipe eccentricity,[6] pipe rotation, and temperature and pressure variations[7,8] on flowing pressure gradients. However, the additional computational complexity required to remove the assumptions listed above seldom is justified in practice.

Fluid flowing in a pipe or a concentric annulus does not have a uniform velocity. If the flow pattern is laminar, the fluid velocity immediately adjacent to the pipe walls will be zero, and the fluid velocity in the region most distant from the pipe walls will be a maximum. Typical flow velocity profiles for a laminar flow pattern are shown in Fig. 4.26. As shown in this figure, concentric rings of fluid laminae are telescoping down the conduit at different velocities. Pipe flow can be considered as a limiting case of annular flow in which the inner radius of the pipe, r_1, has a value of zero.

A relation between radius r, shear stress τ, and frictional pressure gradient dp_f/dL can be obtained from a consideration of Newton's law of motion for a shell of fluid at radius r. Shown in Fig. 4.27 is a free-body diagram of a shell of fluid of length ΔL and of thickness Δr. The sign convention used in Fig. 4.27 is such that the direction of flow is from left to right and that the velocity of flow is decreasing with increasing radius. Thus, the next shell of fluid *enclosed by* the fluid element of interest is moving faster than the fluid element of interest. Furthermore, the next shell of fluid *enclosing* the element of interest is moving slower than the element of interest. The force F_1 applied by the fluid pressure at Point 1 is given by

$$F_1 = p(2\pi r\Delta r).$$

TABLE 4.3—SUMMARY OF EQUATIONS FOR ROTATIONAL VISCOMETER

Newtonian Model $\quad\quad \mu_a = \dfrac{300}{N}\theta_N$ $\quad\quad\quad\quad\quad\quad \dot{\gamma} = \dfrac{5.066}{r^2}N$

Bingham Plastic Model $\quad \mu_p = \theta_{600} - \theta_{300}$

or

$\mu_p = \dfrac{300}{N_2 - N_1}(\theta_{N_2} - \theta_{N_1})$ $\quad\quad \dot{\gamma} = \dfrac{5.066}{r^2}N + \dfrac{479\tau_y}{\mu_p}\left(\dfrac{3.174}{r^2} - 1\right)$

$\tau_y = \theta_{300} - \mu_p$

or

$\tau_y = \theta_{N_1} - \mu_p\dfrac{N_1}{300}$

$\tau_g = \theta_{max}$ at 3 rpm

Power-Law Model $\;\;$ *flow behav index* $\; n = 3.322 \log\left(\dfrac{\theta_{600}}{\theta_{300}}\right)$

or

$n = \dfrac{\log\left(\dfrac{\theta_{N_2}}{\theta_{N_1}}\right)}{\log\left(\dfrac{N_2}{N_1}\right)}$

consist. index $\quad K = \dfrac{510\,\theta_{300}}{(511)^n}$ $\quad\quad\quad \dot{\gamma} = 0.2094N\;\dfrac{\dfrac{1}{r^{2/n}}}{n\left[\dfrac{1}{r_1^{\,2/n}} - \dfrac{1}{r_2^{\,2/n}}\right]}$

or

$K = \dfrac{510\,\theta_N}{(1.703\,N)^n}$

Likewise, the force F_2 applied by the fluid pressure at Point w is given by

$$F_2 = p_2(2\pi r\Delta r) = \left(p - \dfrac{dp_f}{dL}\Delta L\right)(2\pi r\Delta r).$$

The negative sign for the dp_f/dL term is required because the frictional pressure change Δp_f is used to represent $(p_1 - p_2)$ rather than $(p_2 - p_1)$.

The frictional force exerted by the adjacent shell of fluid enclosed by the fluid element of interest is given by

$$F_3 = \tau(2\pi r\Delta L).$$

Similarly, the frictional force exerted by the adjacent shell of fluid that encloses the fluid element of interest is given by

$$F_4 = \tau_{r+\Delta r}\left[2\pi(r+\Delta r)\Delta L\right]$$

$$= \left(\tau + \dfrac{d\tau}{dr}\Delta r\right)\left[2\pi(r+\Delta r)\Delta L\right].$$

If the fluid element is moving at a constant velocity, the sum of the forces acting on the elements must equal zero. Summing forces, we obtain

$$(2\pi r\Delta r)p - (2\pi r\Delta r)\left(p - \dfrac{dp_f}{dL}\Delta L\right)$$

$$+ (2\pi r\Delta L)\tau - 2\pi\Delta L(r+\Delta r)\left(\tau + \dfrac{d\tau}{dr}\Delta r\right) = 0.$$

Expanding this equation, dividing through by $(2\pi r\Delta r\Delta L)$, and taking the limit as $\Delta r \to 0$ yields

$$\dfrac{dp_f}{dL} - \dfrac{1}{r}\dfrac{d(\tau r)}{dr} = 0. \quad\quad\quad\quad\quad\quad (4.50)$$

Since dp/dL is not a function of r, Eq. 4.50 can be integrated with respect to r. Separating variables yields

$$\int d(\tau r) = \dfrac{dp_f}{dL}\int r\,dr.$$

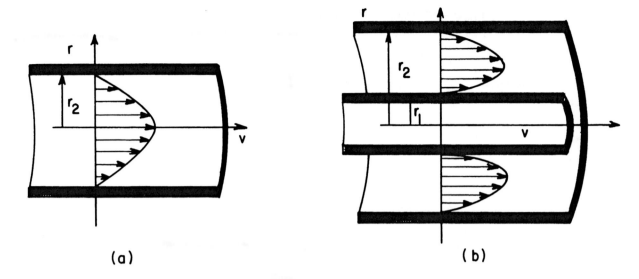

Fig. 4.26—Velocity profiles for laminar flow: (a) pipe flow and (b) annular flow.

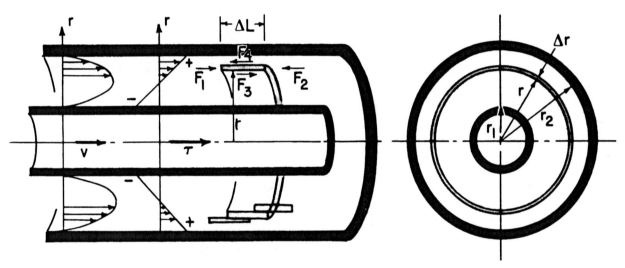

Fig. 4.27—Free body diagram for annular fluid element.

Upon integration we obtain

$$\tau = \frac{r}{2}\frac{dp_f}{dL} + \frac{C_1}{r}, \quad \dots\dots\dots\dots\dots(4.51)$$

where C_1 is the constant of integration. Note that for the special case of pipe flow, the constant C_1 must be zero if the shear stress is not to be infinite at $r = 0$. Eq. 4.51, which relates shear stress and frictional pressure gradient at a given radius, is a consequence of the geometry of the system and does not require the assumption of a fluid rheological model.

The shear rate $\dot\gamma$ for the sign convention used in the derivation is given by

$$\dot\gamma = -\frac{dv}{dr}. \quad \dots\dots\dots\dots\dots\dots\dots(4.52)$$

The shear rate can be related to shear stress using the defining equation for the Newtonian, Bingham plastic, or power-law fluid model.

4.10.1 Newtonian Model

If the fluid can be described with the Newtonian fluid model, the shear stress at any point in the fluid is given by

$$\tau = \mu\,\dot\gamma = -\mu\frac{dv}{dr}.$$

Combining this equation with Eq. 4.51 gives

$$-\mu\frac{dv}{dr} = \frac{r}{2}\frac{dp_f}{dL} + \frac{C_1}{r}.$$

After separating variables and integrating, we obtain

$$v = -\frac{r^2}{4\mu}\frac{dp_f}{dL} - \frac{C_1}{\mu}\ln r + C_2, \quad \dots\dots\dots(4.53a)$$

where C_2 is the second constant of integration. Since the drilling fluid wets the pipe walls, the fluid layers immediately adjacent to the pipe walls (at $r=r_1$ and at $r=r_2$) have a velocity of zero. Using these boundary conditions in Eq. 4.53a yields

$$0 = -\frac{r_1^2}{4\mu}\frac{dp_f}{dL} - \frac{C_1}{\mu}\ln r_1 + C_2,$$

and

$$0 = -\frac{r_2^2}{4\mu}\frac{dp_f}{dL} - \frac{C_1}{\mu}\ln r_2 + C_2.$$

Solving these two equations simultaneously for C_1 and C_2 gives

$$C_1 = -\frac{1}{4}\frac{dp_f}{dL}\frac{(r_2^2 - r_1^2)}{\ln r_2/r_1},$$

and

$$C_2 = -\frac{1}{4\mu}\frac{dp_f}{dL}\frac{(r_2^2 \ln r_1 - r_1^2 \ln r_2)}{\ln r_2/r_1}.$$

Substituting these expressions for C_1 and C_2 in Eq. 4.53a yields

$$v = \frac{1}{4\mu}\frac{dp_f}{dL}\left[(r_2^2 - r^2) - (r_2^2 - r_1^2)\frac{\ln r_2/r}{\ln r_2/r_1}\right].$$

$$\dotfill (4.53b)$$

Note that in the limit as $r_1 \rightarrow 0$, the second term in the brackets of Eq. 4.53b also approaches zero. Thus, for pipe flow,

$$v = \frac{1}{4\mu}\frac{dp_f}{dL}(r_2^2 - r^2). \dotfill (4.53c)$$

If the pressure gradient and viscosity are known, Eqs. 4.53b and 4.53c can be used to determine the velocity distribution within an annulus or pipe, respectively. However, a relation between pressure gradient and total flow rate is needed for most engineering applications. The total flow rate can be obtained by summing the flow contained in each concentric shell of fluid. Thus,

$$q = \int v(2\pi r)\, dr$$

$$= \frac{2\pi}{4\mu}\frac{dp_f}{dL}\int_{r_1}^{r_2}\left[(r_2^2 r - r^3)\right.$$

$$\left. - (r_2^2 r - r_1^2 r)\frac{\ln r_2/r}{\ln r_2/r_1}\right]dr.$$

Upon integration, this equation becomes

$$q = \frac{\pi}{8\mu}\frac{dp_f}{dL}\left[r_2^4 - r_1^4 - \frac{(r_2^2 - r_1^2)^2}{\ln r_2/r_1}\right]. \dotfill (4.54a)$$

This relation was developed first by Lamb.[9]

For annular flow, the flow rate q is the mean velocity v multiplied by the annular cross-sectional area.

$$q = \pi(r_2^2 - r_1^2)\bar{v}.$$

Substituting this expression for q in Eq. 4.54a and solving for the frictional pressure gradient dp_f/dL gives

$$\frac{dp_f}{dL} = \frac{8\mu\bar{v}}{\left(r_2^2 + r_1^2 - \dfrac{r_2^2 - r_1^2}{\ln r_2/r_1}\right)}. \dotfill (4.54b)$$

Converting from consistent units to more convenient field units, Lamb's equation becomes

$$\frac{dp_f}{dL} = \frac{\mu\bar{v}}{1,500\left(d_2^2 + d_1^2 - \dfrac{d_2^2 - d_1^2}{\ln d_2/d_1}\right)}, \dotfill (4.54c)$$

where

$\bar{v}$ = mean flow velocity, ft/s,

μ = viscosity, cp,

dp_f/dL = frictional pressure gradient, psi/ft,

d_1 = outside diameter of the inner pipe, in., and

d_2 = inside diameter of the outer pipe, in.

Note that in the limit as $d_1 \rightarrow 0$, Eq. 4.54c becomes

$$\frac{dp_f}{dL} = \frac{\mu\bar{v}}{1,500\, d^2}. \dotfill (4.54d)$$

This equation is the familiar Hagen-Poiseuille law for circular pipe in field units.

Example 4.21. A 9-lbm/gal Newtonian fluid having a viscosity of 15 cp is being circulated in a 10,000-ft well containing a 7-in.-ID casing and a 5-in.-OD drillstring at a rate of 80 gal/min. Compute the static and circulating bottomhole pressure by assuming that a laminar flow pattern exists.

Solution. The static bottomhole pressure is given by Eq. 4.2b. Since the annulus is open to the atmosphere at the surface,

$$p = 0.052(9)(10,000) + 0 = 4,680 \text{ psig}.$$

If fluid acceleration effects are neglected, the circulating bottomhole pressure is the sum of the hydrostatic pressure plus the frictional pressure loss in the annulus. As shown in Table 4.1, the mean annular velocity is given by

$$\bar{v} = \frac{q}{2.448(d_2^2 - d_1^2)} = \frac{80}{2.448(7^2 - 5^2)} = 1.362 \text{ ft/s}.$$

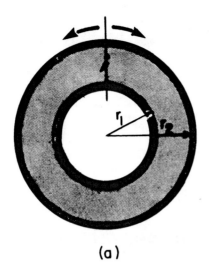
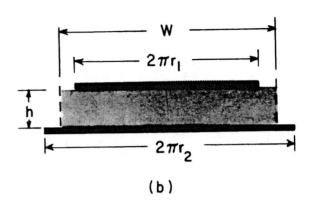

(a)   **(b)**

Fig. 4.28—Representing the annulus as a slot: (a) annular and (b) equivalent slot.

The frictional pressure gradient is determined using Eq. 4.54c.

$$\frac{dp_f}{dL} = \frac{15(1.362)}{1500\left(7^2 + 5^2 - \dfrac{7^2 - 5^2}{\ln 7/5}\right)}$$

$$= 0.0051 \text{ psi/ft.}$$

Since the total frictional pressure loss in the annulus is

$$\Delta p_f = 0.0051 \text{ psi/ft (10,000 ft)} = 51 \text{ psi,}$$

the circulating bottomhole pressure is given by

$$p_c = 4680 + 51 = 4{,}731 \text{ psig.}$$

4.10.2 Representing the Annulus As a Slot

Annular flow also can be approximated using equations developed for flow through rectangular slots. The slot flow equations are much simpler to use and are reasonably accurate as long as the ratio $d_1/d_2 > 0.3$. This minimum ratio almost always is exceeded in rotary drilling applications. As shown in Fig. 4.28, an annular space can be represented as a narrow slot having an area A and height h, given by

$$A = Wh = \pi(r_2^2 - r_1^2), \quad \dots\dots\dots\dots\dots\dots (4.55a)$$

and

$$h = r_2 - r_1. \quad \dots\dots\dots\dots\dots\dots\dots\dots (4.55b)$$

The relation between shear stress and frictional pressure gradient for a slot can be obtained from a consideration of the pressure and viscous forces acting on an element of fluid in the slot (Fig. 4.29). If we consider an element of fluid having width W and thickness Δy, the force F_1 applied by the fluid pressure at Point 1 is given by

$$F_1 = p \ W \ \Delta y.$$

Likewise, the force F_2 applied by the fluid pressure at Point 2 is given by

$$F_2 = p_2 \ W \Delta y = \left(p - \frac{dp_f}{dL} \Delta L\right) W \Delta y.$$

The frictional force exerted by the adjacent layer of fluid below the fluid element of interest is given by

$$F_3 = \tau \ W \Delta L.$$

Similarly, the frictional force exerted by the adjacent layer of fluid above the fluid element of interest is given by

$$F_4 = \tau_{y+\Delta y} W \Delta L = \left(\tau + \frac{d\tau}{dy} \Delta y\right) W \Delta L.$$

If the flow is steady, the sum of the forces acting on the fluid element must be equal to zero. Summing forces, we obtain

$$F_1 - F_2 + F_3 - F_4 = 0$$

and

$$p W \Delta y - \left(p - \frac{dp_f}{dL} \Delta L\right) W \Delta y + \tau W \Delta L$$

$$- \left(\tau + \frac{d\tau}{dy} \Delta y\right) W \Delta L = 0.$$

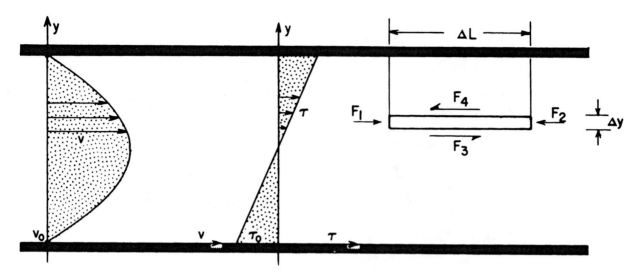

Fig. 4.29—Free body diagram for fluid element in a narrow slot.

Expanding this equation and dividing through by $(W\Delta L\Delta y)$ yields

$$\frac{dp_f}{dL} - \frac{d\tau}{dy} = 0. \qquad\dots\dots\dots\dots\dots\dots (4.56)$$

Since dp_f/dL is not a function of y, Eq. 4.56 can be integrated with respect to y. Separating variables and integrating gives

$$\tau = y\frac{dp_f}{dL} + \tau_0, \qquad\dots\dots\dots\dots\dots (4.57)$$

where τ_0 is the constant of integration that corresponds to the shear stress at $y=0$. For the sign convention used, the shear rate $\dot\gamma$ is given by

$$\dot\gamma = -\frac{dv}{dy}. \qquad\dots\dots\dots\dots\dots\dots\dots (4.58)$$

Thus, for the Newtonian model, we obtain

$$\tau = \mu\dot\gamma = -\mu\frac{dv}{dy} = y\frac{dp_f}{dL} + \tau_0.$$

Separating variables and integrating gives

$$v = -\frac{y^2}{2\mu}\frac{dp_f}{dL} - \frac{\tau_0 y}{\mu} + v_0, \qquad\dots\dots\dots (4.59a)$$

where v_0 is the second constant of integration, which corresponds to the fluid velocity at $y=0$. Since the fluid wets the pipe walls, the velocity v_0 is zero for $y=0$ and for $y=h$. Applying these boundary conditions to Eq. 4.59a yields

$$0 = 0 - 0 + v_0,$$

and

$$0 = -\frac{h^2}{2\mu}\frac{dp_f}{dL} - \frac{\tau_0 h}{\mu} + v_0.$$

Thus, the constants of integration τ_0 and v_0 are given by

$$\tau_0 = -\frac{h}{2}\frac{dp_f}{dL},$$

and

$$v_0 = 0.$$

Substituting these values for τ_0 and v_0 in Eq. 4.59a yields

$$v = \frac{1}{2\mu}\frac{dp_f}{dL}(hy - y^2). \qquad\dots\dots\dots\dots (4.59b)$$

The flow rate q is given by

$$q = \int_0^h v\,dA = \int_0^h vW\,dy = \frac{W}{2\mu}\frac{dp_f}{dL}\int_0^h (hy - y^2)\,dy.$$

Integrating this equation yields

$$q = \frac{Wh^3}{12\mu}\frac{dp_f}{dL}. \qquad\dots\dots\dots\dots\dots (4.60a)$$

Substituting the expressions for (Wh) and h (given by Eqs. 4.55a and 4.55b) in Eq. 4.60a gives

$$q = \frac{\pi}{12\mu}\frac{dp_f}{dL}(r_2^2 - r_1^2)(r_2 - r_1)^2. \qquad\dots\dots (4.60b)$$

Expressing the flow rate in terms of the mean flow velocity $\bar{v}$ and solving for the frictional pressure gradient dp_f/dL gives

$$\frac{dp_f}{dL} = \frac{12\mu\bar{v}}{(r_2 - r_1)^2}. \quad \ldots\ldots\ldots\ldots\ldots (4.60c)$$

Converting from consistent units to more convenient field units of pounds per square inch, centipoise, feet per second, and inches, we obtain

$$\frac{dp_f}{dL} = \frac{\mu\bar{v}}{1,000(d_2 - d_1)^2}. \quad \ldots\ldots\ldots (4.60d)$$

Example 4.22. Compute the frictional pressure loss for the annulus discussed in Example 4.21 using a slot flow representation of the annulus. Assume that the flow pattern is laminar.

Solution. The ratio d_1/d_2 has a value of 0.714. Since this ratio is greater than 0.3, Eq. 4.60d can be applied.

$$\Delta p_f = \frac{dp}{dL} D = \frac{(1.362)(15)(10,000)}{1,000(2^2)}$$

$$= 51 \text{ psi.}$$

Note that this is the same value for frictional pressure loss that was obtained using Eq. 4.54c.

4.10.3 Determination of Shear Rate

A knowledge of the shear rate present in the well sometimes can lead to improved accuracy in the pressure loss determination. Care can be taken to measure the apparent fluid viscosity at values of shear rates near those present in the well. If this is done, good accuracy sometimes can be achieved using flow equations for Newtonian fluids even if the well fluid does not follow closely the Newtonian model over a wide range of shear rates. The maximum value of shear rate will occur at the pipe walls. For circular pipe, the shear stress is given by Eq. 4.51 with $C_1 = 0$. Thus, the shear stress at the wall where $r = r_w$ is given by

$$\tau_w = \frac{r_w}{2} \frac{dp_f}{dL} \quad \text{(circular pipe).} \quad \ldots\ldots\ldots (4.61)$$

Substituting the expression for the frictional pressure gradient dp_f/dL for a circular pipe into Eq. 4.61 yields

$$\tau_w = \frac{r_w}{2}\left(\frac{8\mu\bar{v}}{r_w^2}\right) = 4\frac{\mu\bar{v}}{r_w}.$$

The shear rate at the pipe wall can be obtained from the shear stress at the pipe wall using the defining equation of the Newtonian model.

$$\dot{\gamma}_w = \tau_w/\mu = \frac{4\bar{v}}{r_w}. \quad \ldots\ldots\ldots\ldots\ldots (4.62a)$$

Changing from consistent units to field units, we obtain

$$\dot{\gamma} = \frac{96\bar{v}}{d} \quad \text{(circular pipe),} \quad \ldots\ldots\ldots\ldots (4.62b)$$

where the mean velocity $\bar{v}$ has units of feet per second, the internal diameter of the pipe has units of inches, and the shear rate has units of seconds^{-1}.

The shear stress for an annulus (slot flow approximation) is given by Eq. 4.57. Thus, the shear stress at the wall where $y = h$ is given by

$$\tau_w = \frac{h}{2}\frac{dp_f}{dL} = \frac{(r_2 - r_1)}{2}\frac{dp_f}{dL}. \quad \ldots\ldots\ldots (4.63)$$

Substituting the expression for the frictional pressure gradient (given by Eq. 4.60c) in Eq. 4.63 yields

$$\tau_w = \frac{(r_2 - r_1)}{2}\left[\frac{12\mu\bar{v}}{(r_2 - r_1)^2}\right].$$

Thus, for laminar flow of Newtonian fluids, the shear rate at the pipe wall is given by

$$\dot{\gamma}_w = \tau_w/\mu = \frac{6\bar{v}}{(r_2 - r_1)}. \quad \ldots\ldots\ldots\ldots (4.64a)$$

In field units, this equation becomes

$$\dot{\gamma}_w = \frac{144\bar{v}}{(d_2 - d_1)} \quad \text{(annulus).} \quad \ldots\ldots\ldots (4.64b)$$

Example 4.23. Compute the shear rate at the wall for the annulus discussed in Example 4.21. Assume that the flow pattern is laminar.

Solution. The shear rate at the wall is given by Eq. 4.64b.

$$\dot{\gamma}_w = \frac{144(1.362)}{7 - 5} = 98 \text{ seconds}^{-1}.$$

Thus, for improved accuracy, the apparent viscosity of the well fluid should be measured at a shear rate near 98 seconds^{-1}.

$$N_{RE} < 2100$$

TABLE 4.4—SUMMARY OF LAMINAR FLOW EQUATIONS FOR PIPES AND ANNULI

	Frictional Pressure Loss	Shear Rate At Pipe Wall
Newtonian	**Pipe** $$\frac{dp_f}{dL} = \frac{\mu \bar{v}}{1{,}500\, d^2}$$	**Pipe** $$\dot{\gamma}_w = \frac{96\bar{v}}{d}$$
	Annulus $$\frac{dp_f}{dL} = \frac{\mu \bar{v}}{1{,}000\, (d_2 - d_1)^2}$$	**Annulus** $$\dot{\gamma}_w = \frac{144\bar{v}}{(d_2 - d_1)}$$
Bingham Plastic	**Pipe** $$\frac{dp_f}{dL} = \frac{\mu_p \bar{v}}{1{,}500\, d^2} + \frac{\tau_y}{225\, d}$$	**Pipe** $$\dot{\gamma}_w = \frac{96\bar{v}}{d} + 159.7\frac{\tau_y}{\mu_p}$$
	Annulus $$\frac{dp_f}{dL} = \frac{\mu_p \bar{v}}{1{,}000\, (d_2 - d_1)^2} + \frac{\tau_y}{200(d_2 - d_1)}$$	**Annulus** $$\dot{\gamma}_w = \frac{144\bar{v}}{(d_2 - d_1)} + 239.5\frac{\tau_y}{\mu_p}$$
Power-Law	**Pipe** $$\frac{dp_f}{dL} = \frac{K\bar{v}^n}{144{,}000\, d^{1+n}}\left(\frac{3 + 1/n}{0.0416}\right)^n$$	**Pipe** $$\dot{\gamma}_w = \frac{24\bar{v}}{d}(3 + 1/n)$$
	Annulus $$\frac{dp_f}{dL} = \frac{K\bar{v}^n}{144{,}000\, (d_2 - d_1)^{1+n}}\left(\frac{2 + 1/n}{0.0208}\right)^n$$	**Annulus** $$\dot{\gamma}_w = \frac{48\bar{v}}{d_2 - d_1}(2 + 1/n)$$

4.10.4 Non-Newtonian Models

Analytical expressions for the isothermal, laminar flow of non-Newtonian fluids can be derived by following essentially the same steps used for Newtonian fluids. The reader is referred to the work of Laird[10] and Fredrickson and Bird[11] for a discussion of the development of the annular flow equations for Bingham plastic fluids. However, as in the case of Newtonian fluids, annular flow can be modeled accurately for the usual geometry of interest to drilling engineers through use of the less-complex flow equations for a narrow slot. The derivations of the laminar flow equations for the Bingham plastic and power-law fluid model are given in Appendix B. The annulus was represented as a narrow slot in these derivations. The flow equations derived are summarized in Table 4.4.

Example 4.24. A cement slurry that has a flow-behavior index of 0.3 and a consistency index of 9400 eq cp is being pumped in an 8.097×4.5 in. annulus at a rate of 200 gal/min. Assuming the flow pattern is laminar, compute the frictional pressure loss per 1,000 ft of annulus. Also estimate the shear rate at the pipe wall.

Solution. Using Table 4.1, the average velocity in the annulus is given by

$$\bar{v} = \frac{200}{2.448(8.097^2 - 4.5^2)} = 1.803 \text{ ft/s.}$$

Using Table 4.4, the frictional pressure loss predicted by the power-law model is given by

$$\frac{dp_f}{dL} = \frac{9{,}400(1.803)^{0.3}}{144{,}000(8.097 - 4.5)^{1+0.3}}\left(\frac{2 + 1/0.3}{0.0208}\right)^{0.3}$$

$$= 0.0779 \text{ psi/ft or } 77.9 \text{ psi/1,000 ft.}$$

Also using Table 4.4, the approximate shear rate at the pipe wall is given by

$$\dot{\gamma}_w = \frac{48(1.803)}{(8.097 - 4.5)}(2 + 1/0.3)$$

$$= 128 \text{ seconds}^{-1}.$$

4.11 Turbulent Flow in Pipes and Annuli

In many drilling operations, the drilling fluid is pumped at too high a rate for laminar flow to be maintained. The fluid laminae become unstable and break into a chaotic diffused flow pattern. The transfer of momentum caused by this chaotic fluid movement causes the velocity distribution to become more uniform across the center portion of the conduit than for laminar flow. However, a thin boundary layer of fluid near the pipe walls generally remains in laminar flow. A schematic representation of laminar and turbulent pipe flow is shown in Fig. 4.30.

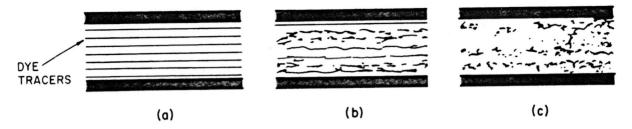

Fig. 4.30—Laminar and turbulent flow patterns in a circular pipe: (a) laminar flow, (b) transition between laminar and turbulent flow, and (c) turbulent flow.

A mathematical development of flow equations for turbulent flow has not been possible to date. However, a large amount of experimental work has been done in straight sections of circular pipe, and the factors influencing the onset of turbulence and the frictional pressure losses due to turbulent flow have been identified. By applying the method of dimensional analysis, these factors have been grouped so that the empirical data could be expressed in terms of dimensionless numbers.

4.11.1 Newtonian Fluids

The experimental work of Osborne Reynolds[12] has shown that the onset of turbulence in the flow of Newtonian fluids through pipes depends on (1) pipe diameter d, (2) density of fluid ρ, (3) viscosity of fluid μ, (4) average flow velocity $\bar{v}$. In terms of the primary units of mass M, length L, and time T, these variables have the following dimensions.

Parameter:	d	ρ	μ	$\bar{v}$
Units:	L	m/L^3	m/(Lt)	L/t

The *Buckingham π theorem* of dimensional analysis states that the number of independent dimensionless groups N that can be obtained from n parameters is given by

$$N = n - m,$$

where m is the number of primary units involved. Since all three primary units (m, L, and t) are used in at least one of the four parameters shown previously,

$$N = 4 - 3 = 1,$$

and only one *independent* dimensionless group is possible. The dimensionless grouping commonly used is expressed in consistent units by

$$N_{Re} = \frac{\rho \bar{v} d}{\mu}, \qquad \dots \dots \dots \dots (4.65a)$$

where N_{Re} is the Reynolds number. In field units, this equation becomes

$$N_{Re} = \frac{928 \, \rho \bar{v} d}{\mu}, \qquad \dots \dots \dots (4.65b)$$

where

ρ = fluid density, lbm/gal
$\bar{v}$ = mean fluid velocity, ft/s
d = pipe diameter, in., and
μ = fluid viscosity, cp.

For engineering purposes, flow of a Newtonian fluid in pipes usually is considered to be laminar if the Reynolds number is less than 2,100 and turbulent if the Reynolds number is greater than 2,100. However, for Reynolds numbers of about 2,000 to 4,000, the flow is actually in a transition region between laminar flow and fully developed turbulent flow. Also, careful experimentation has shown that the laminar region may be made to terminate at a Reynolds number as low as 1,200 by artificially introducing energy into the system—e.g., hitting the pipe with a hammer. Likewise, the laminar flow region can be extended to Reynolds numbers as high as 40,000 by using extremely smooth, straight pipes that are insulated from vibrations. However, these conditions generally are not realized in rotary drilling situations.

Example 4.25. A 9.0-lbm/gal brine having a viscosity of 1.0 cp is being circulated in a well at a rate of 600 gal/min. Determine whether the fluid in the drillpipe is in laminar or turbulent flow if the internal diameter of the drillpipe is 4.276 in.

Solution. The average velocity in the drillpipe is given by

$$\bar{v} = \frac{q}{2.448 \, d^2} = \frac{600}{2.448(4.276)^2} = 13.4 \text{ ft/s}.$$

Using Eq. 4.65b, the Reynolds number is given by

$$N_{Re} = \frac{928 \, \rho \bar{v} d}{\mu} = \frac{928(9.0)(13.4)(4.276)}{(1)}$$

$$= 478,556.$$

Since the Reynolds number is well above 2,100, the fluid in the drillpipe is in turbulent flow.

Once it has been established that the flow pattern is turbulent, the determination of the frictional pressure loss must be based on empirical correlations. The most

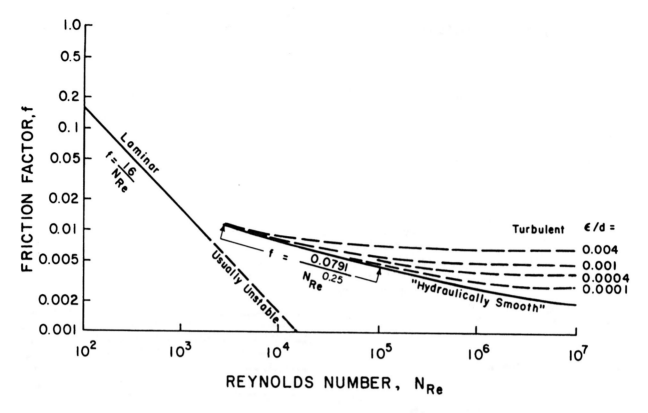

Fig. 4.31—Stanton chart showing fanning friction factors for turbulent flow in circular pipe.

widely used correlations are based on a dimensionless quantity known as the *friction factor*. The friction factor is defined by

$$f = \frac{F_k}{AE_k}, \quad \dots\dots\dots\dots\dots\dots\dots\dots (4.66a)$$

where

F_k = force exerted on the conduit walls due to fluid movement,

A = characteristic area of the conduit, and

E_k = kinetic energy per unit volume of fluid.

For pipe flow, the shear stress on the conduit walls is given by Eq. 4.61.

$$\tau_w = \frac{r_w}{2} \frac{dp_f}{dL} = \frac{d}{4} \frac{dp_f}{dL}.$$

The force F_k exerted at the pipe wall due to fluid motion is given by

$$F_k = (2\pi r_w \Delta L)\tau_w = \frac{\pi d^2}{4} \frac{dp_f}{dL} \Delta L.$$

The kinetic energy per unit volume of fluid is given by

$$E_k = \tfrac{1}{2}\rho v^2.$$

Substituting these expressions for F_k and E_k into Eq. 4.66a yields

$$f = \frac{\pi d^2 \dfrac{dp_f}{dL} \Delta L}{2\rho v^2 A}. \quad \dots\dots\dots\dots\dots\dots (4.66b)$$

If the characteristic area A is chosen to be $2\pi r_w \Delta L$, Eq. 4.66b reduces to

$$f = \frac{d}{2\rho v^2} \frac{dp_f}{dL}. \quad \dots\dots\dots\dots\dots\dots\dots (4.66c)$$

Eq. 4.66c is known as the *Fanning equation*, and the friction factor defined by this equation is called the *Fanning friction factor*. The friction factor f is a function of the Reynolds Number N_{Re} and a term called the *relative roughness*, ϵ/d. The relative roughness is defined as the ratio of the *absolute roughness*, ϵ, to the pipe diameter where the absolute roughness represents the average depth of pipe-wall irregularities. An empirical correlation for the determination of friction factors for fully developed turbulent flow in circular pipe has been presented by Colebrook.[13] The Colebrook function is given by

$$\frac{1}{\sqrt{f}} = -4 \log\left(0.269 \, \epsilon/d + \frac{1.255}{N_{Re}\sqrt{f}}\right). \quad \dots (4.67a)$$

TABLE 4.5—ABSOLUTE PIPE ROUGHNESS FOR SEVERAL TYPES OF CIRCULAR PIPES (after Streeter[14])

Type of Pipe	Absolute Roughness, ϵ (in.)
Riveted steel	0.00025 to 0.0025
Concrete	0.000083 to 0.00083
Cast iron	0.000071
Galvanized iron	0.000042
Asphalted cast iron	0.000033
Commercial steel	0.000013
Drawn tubing	0.0000004

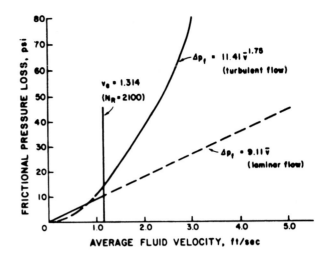

Fig. 4.32—Comparison of laminar and turbulent pressure-loss equations of Example 4.26.

The friction factor f appears both inside and outside the log term of Colebrook's equation requiring an iterative solution technique. This difficulty can be avoided by a graphical representation of the Colebrook function. A plot of friction factor against Reynolds number on log-log paper is called a *Stanton chart*. A Stanton chart for the Colebrook function is shown in Fig. 4.31. However, the solution of Eq. 4.67a using an electronic calculator is not difficult and yields more precise results than is possible using the graphical solution.

The selection of an appropriate absolute roughness ϵ for a given application is often difficult. Shown in Table 4.5[14] are average roughness values determined empirically for several types of conduits. Also, Cullender and Smith[15] in a study of published data obtained in clean steel pipes in gas well and pipeline service found an average pipe roughness of 0.00065 in. to apply to most of the data. Fortunately, in rotary drilling applications involving the use of relatively viscous drilling fluids, the Reynolds number seldom exceeds 100,000. Also for most wellbore geometries, the relative roughness is usually less than 0.0004 in all sections. For these conditions, the friction factors for smooth pipe (zero roughness) can be applied for most engineering calculations. For smooth pipe, Eq. 4.67a reduces to

$$\frac{1}{\sqrt{f}} = 4\log\left(N_{Re}\sqrt{f}\right) - 0.395. \qquad (4.67b)$$

In addition, for smooth pipe and a Reynolds number range of 2,100 to 100,000, a straight-line approximation (on a log-log plot) of the Colebrook function is possible. This approximation, first presented by Blasius,[16] is given by

$$f = \frac{0.0791}{N_{Re}^{0.25}}, \qquad (4.67c)$$

where $2,100 \le N_{Re} \le ,100,000$ and $\epsilon/d = 0$. The Blasius formula allows the construction of simplified hydraulic nomographs and special hydraulic slide rules widely used in the past by field personnel in the drilling industry.

The Fanning equation can be rearranged for the calculation of frictional pressure drop due to turbulent flow in circular pipe. Rearranging Eq. 4.66c and converting to field units gives

$$\frac{dp_f}{dL} = \frac{f\rho\bar{v}^2}{25.8d}. \qquad (4.66d)$$

In addition, the Fanning equation can be extended to the laminar flow region if the friction factor for the laminar region is defined by

$$f = \frac{16}{N_{Re}}. \qquad (4.67d)$$

The proof of this relation is left as a student exercise.

A simplified turbulent flow equation can be developed for smooth pipe and moderate Reynolds numbers by substituting Eq. 4.67c into Eq. 4.66d.

$$\frac{dp_f}{dL} = \frac{0.0791\dfrac{\rho\bar{v}^2}{25.8d}}{\left(\dfrac{928\rho\bar{v}d}{\mu}\right)^{0.25}}.$$

Simplifying this expression yields

$$\frac{dp_f}{dL} = \frac{\rho^{0.75}\bar{v}^{1.75}\mu^{0.25}}{1,800d^{1.25}} = \frac{\rho^{0.75}q^{1.75}\mu^{0.25}}{8,624d^{4.75}},$$

$$\qquad (4.66e)$$

for circular pipe where $\epsilon/d = 0$ and N_{Re} is between 2,100 and 100,000. Eq. 4.66e is in a form that readily identifies the relative importance of the various hydraulic parameters on turbulent frictional pressure loss. For example, it can be shown that changing from 4.5-in. to 5-in. drillpipe would reduce the pressure loss in the drillpipe by about a factor of two.

Example 4.26. Determine the frictional pressure drop in 10,000 ft of 4.5-in. drillpipe having an internal diameter of 3.826 in. if a 20-cp Newtonian fluid having a density of 9 lbm/gal is pumped through the drillpipe at a rate of 400 gal/min.

Solution. The mean fluid velocity is given by

$$\bar{v} = \frac{q}{2.448d^2} = \frac{400}{2.448(3.826)^2} = 11.16 \text{ ft/s.}$$

The Reynolds number is given by

$$N_{Re} = \frac{928\rho\bar{v}d}{\mu} = \frac{928(9)(11.16)(3.826)}{20} = 17,831.$$

Since the Reynolds number is greater than 2,100, the flow pattern is turbulent. For commercial steel, the absolute roughness is given in Table 4.5 as 0.000013. Thus, the relative roughness is

$$\frac{\epsilon}{d} = \frac{0.000013}{3.826} = 0.0000034.$$

Note that this corresponds closely to the smooth pipeline on Fig. 4.31 at a Reynolds number of 17,831. Solving Eq. 4.67b by trial and error, the Fanning friction factor is 0.00666. Thus, the frictional pressure loss is given by

$$\Delta p_f = \frac{dp_f}{dL}\Delta L = \frac{f\rho\bar{v}^2}{25.8d}\Delta L$$

$$= \frac{0.00666(9)(11.16)^2(10,000)}{25.8(3.826)} = 756 \text{ psi.}$$

It is interesting to note that the use of the simplified turbulent flow equation defined by Eq. 4.66e gives

$$\Delta p_f = \frac{\rho^{0.75}\bar{v}^{1.75}\mu^{0.25}}{1,800d^{1.25}}\Delta L$$

$$= \frac{(9)^{0.75}(11.16)^{1.75}(20)^{0.25}10,000}{1,800(3.826)^{1.25}}$$

$$= 777 \text{ psi.}$$

The student should be warned that the Fanning friction factor presented in this text and commonly used in the drilling industry may be different from the friction factor used in other texts. A common friction factor used in many engineering texts is the *Moody friction factor*. The Moody friction factor is four times larger than the Fanning friction factor. Thus, a friction factor read from a Moody chart must be divided by four before being used with the equations presented in this text.

4.11.2 Alternate Turbulence Criteria

In some design problems, it is desirable to determine the frictional pressure losses associated with a wide range of fluid velocities. As discussed in the preceding sections, the frictional pressure loss associated with the pipe flow

of a Newtonian fluid must be determined using a different equation when the flow pattern is turbulent than when the flow pattern is laminar. However, neither equation may predict accurately the pressure loss in the transition region between laminar and turbulent flow. Furthermore, the use of a Reynolds number 2,100 as the criteria for changing from the laminar flow equation to the turbulent flow equation often causes an artificial discontinuity in the relation between pressure loss and mean flow velocity which generally is not observed experimentally. This problem can be illustrated using the pipe geometry and fluid properties given in Example 4.26.

Consider the data given in Example 4.26. At low fluid velocities, the flow pattern is laminar and the frictional pressure loss equation shown in Table 4.4 gives

$$\Delta p_f = \frac{\mu\bar{v}\Delta L}{1,500d^2} = \frac{20\bar{v}(10,000)}{1,500(3.826)^2} = 9.11\bar{v}.$$

This equation has been plotted in Fig. 4.32 for a wide range of mean fluid velocities. At higher fluid velocities, the flow pattern is fully turbulent and the frictional pressure loss can be approximated using Eq. 4.66c:

$$p_f = \frac{\rho^{0.75}\bar{v}^{1.75}\mu^{0.25}\Delta L}{1,800d^{1.25}}$$

$$= \frac{(9)^{0.75}\bar{v}^{1.75}(20)^{0.25}(10,000)}{1,800(3.826)^{1.25}}.$$

$$\Delta p_f = 11.41\bar{v}^{1.75}.$$

This equation also has been plotted in Fig. 4.32. The Reynolds number for the data of Example 4.26 is given by

$$N_{Re} = \frac{928\rho\bar{v}d}{\mu} = \frac{928(9.0)\bar{v}(3.826)}{20}$$

$$= 1,598\bar{v}.$$

If we assume that the flow pattern changes from laminar to turbulent at a Reynolds number of 2,100, the *critical velocity* at which the change in flow pattern occurs is given by

$$v_c = \frac{2,100}{1,598} = 1.314 \text{ ft/s.}$$

It can be seen from Fig. 4.32 that there would be a discontinuity in computed frictional pressure loss if the transition from laminar to turbulent flow is assumed to occur at this fluid velocity. This fictitious discontinuity is caused by the assumption that the flow pattern suddenly changes from laminar to fully developed turbulent flow at a discrete Reynolds number of 2,100 rather than over a range of Reynolds numbers between 2,000 and 4,000.

To avoid the discontinuity in the relation between frictional pressure loss and mean fluid velocity, it sometimes is assumed that the flow pattern changes from laminar to turbulent flow where the laminar and turbulent flow equations yield the same value of frictional pressure loss—e.g., where the two equations cross in Fig. 4.32. When this procedure is used, it is necessary to compute the frictional pressure loss using both the laminar and turbulent flow equations and then select the result that is numerically the highest. This method is well suited to numerical solution techniques performed using a computer. This is especially true for root-finding techniques that require the use of a continuous relation between flow rate and pressure.

4.11.3 Extension of Pipe Flow Equations to Annular Geometry

A large amount of experimental work has been done in circular pipe. Unfortunately, this is not true for flow conduits of other shapes. When noncircular flow conduits are encountered, a common practice is to calculate an effective conduit diameter such that the flow behavior in a circular pipe of that diameter would be roughly equivalent to the flow behavior in the noncircular conduit. One criterion often used in determining an equivalent circular diameter for a noncircular conduit is the ratio of the cross-sectional area to the wetted perimeter of the flow channel. This ratio is called the *hydraulic radius*. For the case of an annulus, the hydraulic radius is given by

$$r_H = \frac{\pi(r_2{}^2 - r_1{}^2)}{2\pi(r_{1} + r_2)} = \frac{r_2 - r_1}{2} = \frac{d_2 - d_1}{4}.$$

The equivalent circular diameter is equal to four times the hydraulic radius.

$$d_e = 4r_H = d_2 - d_1. \quad \dots \dots \dots \dots \dots (4.68a)$$

Note that for $d_1 = 0$ (no inner pipe), the equivalent diameter correctly reduces to the diameter of the outer pipe.

A second criterion used to obtain an equivalent circular radius is the geometry term in the pressure-loss equation for the laminar flow region. Consider the pressure loss equations for pipe flow and concentric annular flow of Newtonian fluids given as Eqs. 4.54c and 4.54d. Comparing the geometry terms in these two equations yields

$$d^2 \sim d_2{}^2 + d_1{}^2 - \frac{d_2{}^2 - d_1{}^2}{\ln(d_2/d_1)}.$$

Thus, the equivalent circular diameter of an annulus obtained using these criteria is given by

$$d_e = \sqrt{d_2{}^2 + d_1{}^2 - \frac{d_2{}^2 - d_1{}^2}{\ln(d_2/d_1)}}. \quad \dots \dots \dots (4.68b)$$

A third expression for the equivalent diameter of an annulus can be obtained by comparing Eqs. 4.54c and

4.60d, the slot flow approximation for an annulus. Comparing the denominator of these two equations yields

$$1{,}500d^2 \sim 1{,}000(d_2 - d_1)^2.$$

Thus, the equivalent circular diameter of a slot representation of an annulus is given by

$$d_e = 0.816(d_2 - d_1). \quad \dots \dots \dots \dots \dots (4.68c)$$

For most annular geometries encountered in drilling operations, $d_1/d_2 > 0.3$, and Eqs. 4.68b and 4.68c give almost identical results.

A fourth expression for the equivalent diameter of an annulus was developed empirically by Crittendon[17] from a study of about 100 hydraulic fracture treatments of producing wells in which lease crude was used as a fracturing fluid. Expressed in terms of d_1 and d_2, Crittendon's equivalent diameter is given by

$$d_e = \frac{\sqrt[4]{d_2{}^4 - d_1{}^4 - \frac{(d_2{}^2 - d_1{}^2)^2}{\ln(d_2/d_1)}} + \sqrt{d_2{}^2 - d_1{}^2}}{2}.$$

$$\dots \dots \dots \dots \dots \dots \dots (4.68d)$$

When using Crittendon's empirical correlation, a fictitious average velocity also must be used in describing the flow system. The fictitious average velocity is computed using the cross-sectional area of the equivalent circular pipe rather than the true cross-sectional area. This *is not true* when using Eqs. 4.68a, 4.68b, and 4.68c. The true average velocity is used when employing these equations.

All four expressions for equivalent diameter shown above have been used in practice to represent annular flow. Eq. 4.68a is probably the most widely used in the petroleum industry. However, this is probably due to the simplicity of the method rather than a superior accuracy.

Example 4.27. A 9.0 lbm/gal brine having a viscosity of 1.0 cp is being circulated in a well at a rate of 200 gal/min. Apply the four criteria for computing equivalent diameter given by Eqs. 4.68a through 4.68d to the annulus opposite the drillpipe to determine the flow pattern and frictional pressure gradient. The drillpipe has an external diameter of 5.0 in. and the hole has a diameter of 10.0 in.

Solution. The equivalent diameters given by Eqs. 4.68a through 4.68d are as follows.

$$d_e = d_2 - d_1 = 10.0 - 5.0 = 5.0 \text{ in.} \quad \dots \dots \dots (4.68a)$$

$$d_e = \sqrt{d_2{}^2 + d_1{}^2 - \frac{d_2{}^2 - d_1{}^2}{\ln d_2/d_1}}$$

$$= \sqrt{10^2 + 5^2 - \frac{10^2 - 5^2}{\ln 2}} = 4.099 \text{ in.} \quad \dots (4.68b)$$

$d_e = 0.816(d_2 - d_1) = 0.816(10-5) = 4.080$ in.

$$\dots\dots\dots\dots\dots\dots\dots (4.68c)$$

$$d_e = \frac{\sqrt[4]{d_2{}^4 - d_1{}^4 - \dfrac{(d_2{}^2 - d_1{}^2)^2}{\ln d_2/d_1}} + \sqrt{d_2{}^2 - d_1{}^2}}{2}.$$

$$= \frac{\sqrt[4]{10^4 - 5^4 - \dfrac{(10^2 - 5^2)^2}{\ln 2}} + \sqrt{10^2 - 5^2}}{2}.$$

$$= 7.309 \text{ in.} \quad \dots\dots\dots\dots\dots (4.68d)$$

The true average velocity is given by

$$\bar{v} = \frac{q}{2.448(d_2{}^2 - d_1{}^2)} = \frac{200}{2.448(10^2 - 5^2)}$$

$$= 1.089 \text{ ft/s}.$$

The fictitious equivalent velocity needed to apply Crittendon's criterion is given by

$$\bar{v}_e = \frac{q}{2.448 d_e{}^2} = \frac{200}{2.448(7.309)^2}$$

$$= 1.529 \text{ ft/s}.$$

Expressing the Reynolds number in terms of $\bar{v}$ and d_e yields

$$N_{Re} = \frac{928 \rho \bar{v} d_e}{\mu} = \frac{928(9.0)}{(1.0)} \bar{v} d_e = 8352 \bar{v} d_e.$$

Expressing the frictional pressure gradient given by Eq. 4.66e in terms of $\bar{v}$ and d_e yields

$$\frac{dp_f}{dL} = \frac{\rho^{0.75} \bar{v}^{1.75} \mu^{0.25}}{1,800 d_e{}^{1.25}}$$

$$= \frac{(9)^{0.75}(1)^{0.25}}{1,800} \frac{\bar{v}^{1.75}}{d_e{}^{1.25}}$$

$$= 0.002887 \frac{\bar{v}^{1.75}}{d_e{}^{1.25}}.$$

The results obtained for each of the four methods are summarized as follows.

Eq.	d_e	$\bar{v}$	N_{Re}	$\dfrac{dp_f}{dL}$
4.68a	5.000	1.089	45,476	4.48×10^{-4}
4.68b	4.099	1.089	37,282	5.75×10^{-4}
4.68c	4.080	1.089	37,109	5.78×10^{-4}
4.68d	7.309	1.529	93,337	4.97×10^{-4}

For this problem, all four criteria indicate that the fluid in the annulus is in turbulent flow.

Note the close agreement in Example 4.27 between the results obtained with Eqs. 4.68b and 4.68c. This should be expected since $d_1/d_2 > 0.3$.

4.11.4 Bingham Plastic Model

The frictional pressure loss associated with the turbulent flow of a Bingham plastic fluid is affected primarily by density and plastic viscosity. While the yield point of the fluid affects both the frictional pressure loss in laminar flow and fluid velocity at which turbulence begins, at higher shear rates corresponding to a fully turbulent flow pattern, the yield point is no longer a highly significant parameter. It has been found empirically that the frictional pressure loss associated with the turbulent flow of a Bingham plastic fluid can be predicted using the equations developed for Newtonian fluids if the plastic viscosity is substituted for the Newtonian viscosity. This substitution can be made in the Reynolds number used in the Colebrook function defined by Eq. 4.67b or in the simplified turbulent flow equation given by Eq. 4.66e.

Accurately predicting the onset of turbulent flow is even more difficult for fluids that follow the Bingham plastic model than for fluids that follow the Newtonian model. When only the frictional pressure loss is desired, this problem can be avoided by calculating the frictional pressure loss using both the laminar and turbulent flow equations and then selecting the result that is numerically the highest. The pressure loss computed in this manner will be reasonably accurate even though the incorrect flow pattern may be assumed in some cases. However, in some design problems, it may be necessary to establish the actual flow rate at which turbulence begins. For example, many engineers feel that cement slurry and preflush solutions should be pumped in turbulent flow for more efficient mud removal during cementing operations. In this type problem, the use of more accurate turbulence criteria is required.

The most commonly used turbulence criterion involves the calculation of a representative apparent viscosity that can be used in the Reynolds number criterion developed for Newtonian fluids. The apparent viscosity most often used is obtained by comparing the laminar flow equations for Newtonian and Bingham plastic fluids. For example, combining the pipe flow equations for the Newtonian and Bingham plastic model given in Table 4.4 yields

$$\frac{\mu_a \bar{v}}{1,500 d^2} = \frac{\mu_p \bar{v}}{1,500 d^2} + \frac{\tau_y}{225 d}.$$

Solving for μ_a, the apparent Newtonian viscosity gives

$$\mu_a = \mu_p + \frac{6.66 \tau_y d}{\bar{v}}. \quad \dots\dots\dots\dots\dots (4.69a)$$

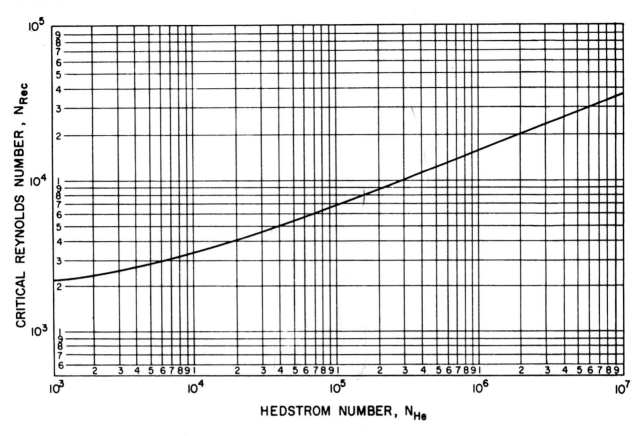

Fig. 4.33—Critical Reynolds numbers for Bingham plastic fluids.

A similar comparison of the laminar flow equations (given in Table 4.4 for Newtonian and Bingham fluids in an annulus) yields

$$\mu_a = \mu_p + \frac{5\tau_y(d_2 - d_1)}{\bar{v}}. \quad \ldots\ldots\ldots\ldots\ldots (4.69b)$$

These apparent viscosities can be used in place of the Newtonian viscosity in the Reynolds number formula. As in the case of Newtonian fluids, a Reynolds number greater than 2,100 is taken as an indication that the flow pattern is turbulent.

A promising new turbulence criterion for fluids that follow the Bingham plastic model has been presented recently by Hanks.[18] If the yield point and plastic viscosity are included in the dimensional analysis previously presented for Newtonian fluids, we have the following.

Parameter: ρ $\bar{v}$ d μ_p τ_y
Units: m/L^3 L/t L m/Lt m/Lt^2

Since all three fundamental units (m, L, and t) are included among the five parameters, two independent dimensionless groups are possible (5−3=2). As shown previously, one possible group is the Reynolds number.

$$\frac{\rho \bar{v} d}{\mu_p} \sim \frac{(m/L^3)(L/t)L}{m/Lt}.$$

However, an additional group, called the *Hedstrom number* also is possible.

$$\frac{\rho \tau_y d^2}{\mu_p{}^2} \sim \frac{(m/L^3)(m/Lt^2)L^2}{(m/Lt)^2}.$$

In field units, the Hedstrom number is given by

$$N_{He} = \frac{37,100 \rho \tau_y d^2}{\mu_p{}^2}. \quad \ldots\ldots\ldots\ldots\ldots (4.70)$$

Hanks has found that the Hedstrom number could be correlated with the *critical Reynolds number*, $(N_{Re})_c$—i.e., the Reynolds number above which the flow pattern is turbulent. The correlation has been presented graphically in Fig. 4.33. It is based on the simultaneous solution of the following two equations.

$$\frac{\left(\dfrac{\tau_y}{\tau_w}\right)}{\left(1 - \dfrac{\tau_y}{\tau_w}\right)^3} = \frac{N_{He}}{16,800}. \quad \ldots\ldots\ldots\ldots (4.71)$$

$$(N_{Re})_c = \frac{1 - \dfrac{4}{3}\left(\dfrac{\tau_y}{\tau_w}\right) + \dfrac{1}{3}\left(\dfrac{\tau_y}{\tau_w}\right)^4}{8\left(\dfrac{\tau_y}{\tau_w}\right)} N_{He}.$$

$$\ldots\ldots\ldots\ldots\ldots\ldots (4.72)$$

If the flow pattern is turbulent, the Reynolds number can be used in the Colebrook function to determine the friction factor.

Example 4.28. A 10-lbm/gal mud having a plastic viscosity of 40 cp and a yield point of 15 lbf/100 sq ft is being circulated at a rate of 600 gal/min. Estimate the frictional pressure loss in the annulus opposite the drill collars if the drill collars are in a 6.5-in. hole, have a length of 1,000 ft, and a 4.5-in. OD. Check for turbulence using both the apparent viscosity concept and the Hedstrom number approach. Use an equivalent diameter given by Eq. 4.68c to represent the annular geometry.

Solution. The average velocity is given by

$$\bar{v} = \frac{q}{2.448(d_2^2 - d_1^2)} = \frac{600}{2.448(6.5^2 - 4.5^2)}$$

$$= 11.14 \text{ ft/s}.$$

The apparent viscosity at this mean velocity is given by Eq. 4.69b.

$$\mu_a = \mu_p + \frac{5\tau_y(d_2 - d_1)}{\bar{v}}$$

$$= 40 + \frac{5(15)(2)}{11.14} = 53.5 \text{ cp}.$$

Computing an equivalent diameter using Eq. 4.68c yields

$$d_e = 0.816(d_2 - d_1) = 0.816(2) = 1.632 \text{ in.}$$

Thus, the Reynolds number for an apparent viscosity of 53.5 cp is given by

$$N_{Re} = \frac{928 \rho \bar{v} d_e}{\mu_a} = \frac{928(10)(11.14)(1.632)}{53.5} = 3,154.$$

Since $N_{Re} > 2,100$, a turbulent flow pattern is indicated.

$$N_{He} = \frac{37,100 \rho \tau_y d_e^2}{\mu_p^2}$$

$$= \frac{37,100(10)(15)(1.632)^2}{(40)^2} = 9,263.$$

Using Fig. 4.33, a critical Reynolds number of 3,300 is indicated. The Reynolds number for a plastic viscosity of 40 cp is given by

$$N_{Re} = \frac{928 \rho \bar{v} d_e}{\mu_p} = \frac{928(10)(11.14)(1.632)}{40}$$

$$= 4,218.$$

Since 4,218 is greater than $(N_{Re})_c = 3,300$, a turbulent flow pattern again is indicated.

The Colebrook function for smooth pipe gives a friction factor of 0.0098 for a Reynolds number of 4,218 (see Eq. 4.67b). Thus, the pressure drop is given by

$$\Delta p_f = \frac{dp_f}{dL} \Delta L = \frac{f\rho \bar{v}^2}{25.8 d_e} \Delta L$$

$$= \frac{0.0098(10)(11.14)^2(1,000)}{25.8(1.632)} = 289 \text{ psi.}$$

It is interesting to note that the simplified flow equation given by Eq. 4.66e gives

$$\Delta p_f = \frac{\rho^{0.75} \bar{v}^{1.75} \mu_p^{0.25} \Delta L}{1,800 d_e^{1.25}}$$

$$= \frac{(10)^{0.75}(11.14)^{1.75}(40)^{0.25}(1,000)}{1,800(1.632)^{1.25}}$$

$$= 289 \text{ psi.}$$

It is also interesting to note that the use of the laminar flow equation gives

$$\Delta p_f = \frac{\mu_p \bar{v} \Delta L}{1,000(d_2 - d_1)^2} + \frac{\tau_y \Delta L}{200(d_2 - d_1)}$$

$$= \frac{40(11.14)(1,000)}{1,000(6.5 - 4.5)^2} + \frac{15(1,000)}{200(6.5 - 4.5)}$$

$$= 149 \text{ psi,}$$

which is less than the value predicted by the turbulent flow relations. Thus, the flow pattern giving the greatest frictional pressure loss is turbulent flow.

4.11.5 Power-Law Model

Dodge and Metzner[19] have published a turbulent flow correlation for fluids that follow the power-law model. Their correlation has gained widespread acceptance in the petroleum industry. An apparent viscosity for use in the Reynolds number criterion is obtained by comparing the laminar flow equations for Newtonian and power-law fluids. For example, combining the Newtonian and power-law equations for laminar flow given in Table 4.4 yields

$$\frac{\mu_a \bar{v}}{1,500 d^2} = \frac{K \bar{v}^n}{144,000 d^{(1+n)}} \left(\frac{3 + 1/n}{0.0416}\right)^n .$$

Solving for μ_a, the apparent Newtonian viscosity, yields

$$\mu_a = \frac{K d^{(1-n)}}{96 \bar{v}^{(1-n)}} \left(\frac{3 + 1/n}{0.0416}\right)^n .$$

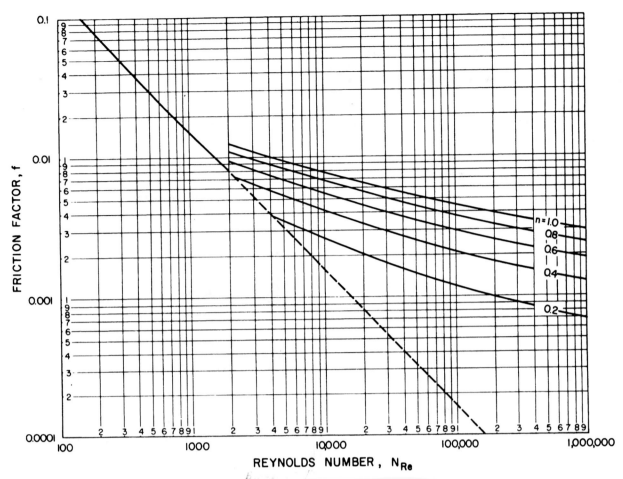

Fig. 4.34—Friction factors for power-law fluid model.

Substituting the apparent viscosity in the Reynolds number equation gives

$$N_{Re} = \frac{89,100 \rho \bar{v}^{(2-n)}}{K} \left(\frac{0.0416d}{3+1/n} \right)^n . \quad \ldots \ldots \ldots (4.73)$$

As in the case of the Bingham plastic model, the use of the apparent viscosity concept in the calculation of Reynolds number does not yield accurate friction factors when used with the Colebrook function. However, Dodge and Metzner developed a new empirical friction factor correlation for use with the Reynolds number given by Eq. 4.73. The friction factor correlation is given by

$$\sqrt{1/f} = \frac{4.0}{n^{0.75}} \log(N_{Re} f^{1-n/2}) - \frac{0.395}{n^{1.2}} . \quad \ldots \ldots (4.74)$$

The correlation was developed only for smooth pipe. However, this is not a severe limitation for most drilling fluid applications. A graphical representation of Eq. 4.74 is shown in Fig. 4.34. The upper line on this graph is for $n = 1$ and is identical to the smooth pipeline on Fig. 4.31.

The critical Reynolds number, above which the flow pattern is turbulent, is a function of the flow-behavior index n. It is recommended that the critical Reynolds number for a given n value be taken from Fig. 4.34 as the starting point of the turbulent flow line for the given

n value. For example, the critical Reynolds number for an n value of 0.2 is 4,200.

The Dodge and Metzner correlation can be applied to annular flow by the development of an apparent viscosity from a comparison of the laminar annular flow equations for Newtonian and power-law fluids given in Table 4.4.

$$\frac{\mu \bar{v}}{1,000(d_2 - d_1)^2}$$

$$= \frac{K \bar{v}^n}{144,000(d_2 - d_1)^{1+n}} \left(\frac{2+1/n}{0.0208} \right)^n .$$

Solving for μ_a, the apparent Newtonian viscosity gives

$$\mu_a = \frac{K(d_2 - d_1)^{1-n}}{144 \bar{v}^{(1-n)}} \left(\frac{2+1/n}{0.0208} \right)^n .$$

Substituting this apparent viscosity in the Reynolds number equation and using Eq. 4.68c for equivalent diameter gives

$$N_{Re} = \frac{109,000 \rho \bar{v}^{(2-n)}}{K} \left[\frac{0.0208(d_2 - d_1)}{2+1/n} \right]^n .$$

$$\ldots \ldots \ldots \ldots \ldots \ldots \ldots \ldots \ldots (4.75)$$

Example 4.29. A 15.6-lbm/gal cement slurry having a consistency index of 335 eq cp and a flow-behavior index of 0.67 is being pumped at a rate of 672 gal/min between a 9.625-in. hole and a 7.0-in. casing. Determine the frictional pressure loss per 100 ft of slurry. Use Eq. 4.68c to obtain the equivalent diameter.

Solution. The mean fluid velocity is given by

$$\bar{v} = \frac{q}{2.448(d_2{}^2 - d_1{}^2)} = \frac{672}{2.448(9.625^2 - 7^2)}$$

.29 ft/s.

The Reynolds number is given by Eq. 4.75.

$$N_{Re} = \frac{109{,}000(15.6)(6.29)^{(2-0.67)}}{335}$$

$$\cdot \left[\frac{0.0208(9.625 - 7)}{2 + 1/0.67} \right]^{0.67} = 3{,}612.$$

From Fig. 4.34, the critical Reynolds number for $n=0.67$ is 2,100. Thus, the flow pattern is turbulent. Also, from Fig. 4.34, the friction factor for $n=0.67$ and $N_{Re}=3{,}612$ is 0.00815. This can be verified using Eq. 4.74. Thus, the frictional pressure drop per 100 ft is given by

$$\Delta p = \frac{f \rho \bar{v}^2}{25.8 d_e} \Delta L$$

$$= \frac{0.00815(15.6)(6.29)^2(100)}{25.8[0.816(9.625 - 7)]} = 9.1 \text{ psi.}$$

4.11.6 Summary of Frictional Pressure-Loss Equations

It is unfortunate that the development of the frictional pressure-loss equations for circular pipes and annuli is a rather lengthy endeavor. The student reader should not be overly concerned if he feels somewhat bewildered at this point. Many of those who have gone before you have had the same feeling. To facilitate the application of the various frictional pressure-loss equations, they have been summarized in Table 4.6 in practical field units.

4.12. Initiating Circulation of the Well

Drilling muds usually will exhibit a thixotropic behavior at the time circulation is started. Since the frictional pressure-loss equations presented in Sections 4.10 and 4.11 neglect the thixotropic characteristics of the mud, they should be applied only after the mud has been sheared for some time. The entire annulus may have to be displaced before a steady frictional pressure loss is observed. In some cases, the pressure required to initiate circulation in the annulus may be a good deal higher than the pressure required to sustain circulation at the desired flow rate.

The pressure gradient required to start circulation can be computed if the gel strength of the drilling fluid is known. Since the shear stress is greatest at the pipe wall, this is where the initial fluid movement will occur. Equating the wall shear stress to the gel strength yields

$$\tau_g = \frac{r_w}{2} \frac{dp_f}{dL} \quad \text{(pipe)}, \quad \ldots\ldots\ldots\ldots\ldots(4.76a)$$

and

$$\tau_g = \frac{(r_2 - r_1)}{2} \frac{dp_f}{dL} \quad \text{(annulus)}. \quad \ldots\ldots\ldots(4.76b)$$

Changing from consistent units to field units and solving for the pressure gradient dp_f/dL gives

$$\frac{dp_f}{dL} = \frac{\tau_g}{300d} \quad \text{(pipe)}, \quad \ldots\ldots\ldots\ldots\ldots(4.77)$$

and

$$\frac{dp_f}{dL} = \frac{\tau_g}{300(d_2 - d_1)} \quad \text{(annulus)}, \quad \ldots\ldots\ldots(4.78)$$

where τ_g has units of pounds per 100 sq ft.

At extremely low annular velocities, the wall shear stress and the yield point may be approximately equal and the frictional pressure-loss equation for the Bingham plastic model will give too high a value. This occurs because the wall shear stress is not at least twice the yield point and the assumption made in the derivation of the pressure-loss equation for Bingham plastic fluids will not be valid (see Appendix B). When this condition occurs, it usually is better to approximate the frictional pressure gradient using Eq. 4.78 with the yield point substituted in place of the gel strength.

HW: Prob

Example 4.30. Compute the equivalent density below the casing seat at 3,000 ft when a mud having a density of 9.0 lbm/gal and a gel strength of 50 lbf/100 sq ft just begins to flow. The casing has an internal diameter of 10.0 in., and the drillpipe has an external diameter of 5.0 in.

Solution. Using Eq. 4.78, the pressure gradient required to break the gel at the pipe wall is given by

$$\frac{dp_f}{dL} = \frac{\tau_g}{300(d_2 - d_1)} = \frac{50}{300(10 - 5)} = 0.0333 \text{ psi/ft.}$$

The pressure at the casing seat when the gel begins to break is given by

$$p = 0.052 \rho D + \frac{dp_f}{dL} D$$

$$= 0.052(9)(3{,}000) + 0.0333(3{,}000)$$

$$= 1{,}404 + 100$$

$$= 1{,}504 \text{ psi.}$$

$N_{Re} > 2180$

TABLE 4.6—SUMMARY OF FRICTIONAL PRESSURE LOSS EQUATIONS*

	Newtonian Model	Bingham Plastic Model	Power-Law Model

Mean Velocity, $\bar{v}$

Pipe

$$\bar{v} = \frac{q}{2.448d^2} \qquad \bar{v} = \frac{q}{2.448d^2} \qquad \bar{v} = \frac{q}{2.448d^2}$$

Annulus

$$\bar{v} = \frac{q}{2.448(d_2^2 - d_1^2)} \qquad \bar{v} = \frac{q}{2.448(d_2^2 - d_1^2)} \qquad \bar{v} = \frac{q}{2.448(d_2^2 - d_1^2)}$$

Flow Behavior Parameters

$$\mu = \theta_{300} \qquad\qquad \mu_p = \theta_{600} - \theta_{300} \qquad\qquad n = 3.32 \log\frac{\theta_{600}}{\theta_{300}}$$

$$\tau_y = \theta_{300} - \mu_p \qquad\qquad K = \frac{510\,\theta_{300}}{511^n}$$

Turbulence Criteria

Pipe

$$N_{Rec} = 2,100 \qquad\qquad N_{He} = \frac{37,100\,\rho\,\tau_y\,d^2}{\mu_p^2} \qquad\qquad N_{Rec} \text{ from Fig. 4.34}$$

$$N_{Re} = \frac{928\,\rho\,\bar{v}\,d}{\mu} \qquad\qquad N_{Rec} \text{ from Fig. 4.33} \qquad\qquad N_{Re} = \frac{89,100\,\rho\bar{v}^{2-n}}{K}\left(\frac{0.0416d}{3+1/n}\right)^n$$

$$N_{Re} = \frac{928\,\rho\,\bar{v}\,d}{\mu_p}$$

Annulus

$$N_{Rec} = 2,100 \qquad\qquad N_{He} = \frac{24,700\,\rho\,\tau_y\,(d_2 - d_1)^2}{\mu_p^2} \qquad\qquad N_{Rec} \text{ from Fig. 4.34}$$

$$N_{Re} = \frac{757\,\rho\bar{v}(d_2 - d_1)}{\mu} \quad N_{Re} = \frac{757\,\rho\bar{v}(d_2 - d_1)}{\mu_p} \quad N_{Re} = \frac{109,000\,\rho(\bar{v})^{2-n}}{K}\left[\frac{0.0208(d_2 - d_1)}{2+1/n}\right]^n$$

Laminar Flow Frictional Pressure Loss

Pipe

$$\frac{dp_f}{dL} = \frac{\mu\bar{v}}{1,500\,d^2} \qquad \frac{dp_f}{dL} = \frac{\mu_p\bar{v}}{1,500\,d^2} + \frac{\tau_y}{225\,d} \qquad \frac{dp_f}{dL} = \frac{K\bar{v}^n\left(\frac{3+1/n}{0.0416}\right)^n}{144,000\,d^{1+n}}$$

Annulus

$$\frac{dp_f}{dL} = \frac{\mu\bar{v}}{1,000\,(d_2 - d_1)^2} \quad \frac{dp_f}{dL} = \frac{\mu_p\bar{v}}{1,000\,(d_2 - d_1)^2} + \frac{\tau_y}{200\,(d_2 - d_1)} \quad \frac{dp_f}{dL} = \frac{K\bar{v}^n\left(\frac{2+1/n}{0.0208}\right)^n}{144,000\,(d_2 - d_1)^{1+n}}$$

Turbulent Flow Frictional Pressure Loss

Pipe

$$\frac{dp_f}{dL} = \frac{f\rho\bar{v}^2}{25.8\,d} \qquad \frac{dp_f}{dL} = \frac{f\rho\bar{v}^2}{25.8\,d} \qquad \frac{dp_f}{dL} = \frac{f\rho\bar{v}^2}{25.8\,d}$$

or

or

$$\frac{dp_f}{dL} = \frac{\rho^{0.75}\bar{v}^{1.75}\mu^{0.25}}{1,800\,d^{1.25}} \qquad \frac{dp_f}{dL} = \frac{\rho^{0.75}\bar{v}^{1.75}\mu_p^{0.25}}{1,800\,d^{1.25}}$$

Annulus

$$\frac{dp_f}{dL} = \frac{f\rho\bar{v}^2}{21.1\,(d_2 - d_1)} \qquad \frac{dp_f}{dL} = \frac{f\rho\bar{v}^2}{21.1\,(d_2 - d_1)} \qquad \frac{dp_f}{dL} = \frac{f\rho\bar{v}^2}{21.1\,(d_2 - d_1)}$$

or

or

$$\frac{dp_f}{dL} = \frac{\rho^{0.75}\bar{v}^{1.75}\mu^{0.25}}{1,396\,(d_2 - d_1)^{1.25}} \qquad \frac{dp_f}{dL} = \frac{\rho^{0.75}\bar{v}^{1.75}\mu_p^{0.25}}{1,396\,(d_2 - d_1)^{1.25}}$$

*Alternate turbulence criteria are to assume the flow pattern which gives the greatest frictional pressure loss.

The equivalent density is given by

$$\rho_e = \frac{p}{0.052D} = \frac{1,504}{0.052(3,000)}$$

$$= 9.64 \text{ lbm/gal.}$$

When the drilling fluid becomes severely gelled in annuli of small clearance, excessive pressures may be required to break circulation. In some cases, the pressure required at a given depth to initiate circulation may exceed the fracture pressure of an exposed formation at that depth. To reduce the pressure requirements, the drillstring can be rotated before the pump is started. In addition, the pump speed can be increased very slowly while the drillstring is being rotated. If the rig is equipped with a direct drive power system and the pump speed cannot be increased gradually, a bypass line can be installed between the pump discharge and the pit. This allows the flow into the well to be increased gradually by starting the pump with the bypass line open and then gradually closing the bypass.

4.13. Jet Bit Nozzle Size Selection

The determination of the proper jet bit nozzle sizes is one of the more frequent applications of the frictional pressure-loss equations by drilling personnel. Significant increases in penetration rate can be achieved through the proper choice of bit nozzles. In relatively competent formations, the penetration rate increase is felt to be due mainly to improved cleaning action at the hole bottom. Wasteful regrinding of cuttings is prevented if the fluid circulated through the bit removes the cuttings as rapidly as they are made. In soft formations, the jetted fluid also may aid in the destruction of the hole bottom.

The true optimization of jet bit hydraulics cannot be achieved yet. Before this can be done, accurate mathematical relations must be developed that define the effect of the level of hydraulics on (1) penetration rate, (2) operational costs, (3) bit wear, (4) potential hole problems such as hole washout, and (5) drilling-fluid carrying capacity. At present, there is still disagreement as to what hydraulic parameter should be used to indicate the level of the hydraulic cleaning action. The most commonly used hydraulic design parameters are (1) bit nozzle velocity, (2) bit hydraulic horsepower, and (3) jet impact force. Current field practice involves the selection of the bit nozzle sizes that will cause one of these parameters to be a maximum.

4.13.1 Maximum Nozzle Velocity

Before jet bits were introduced, rig pumps usually were operated at the flow rate corresponding to the estimated minimum annular velocity that would lift the cuttings. To some extent, this practice continues even today. If the jet nozzles are sized so that the surface pressure at this flow rate is equal to the maximum allowable surface pressure, then the fluid velocity in the bit nozzles will be the maximum that can be achieved and still lift the cuttings. This can be proved using Eq. 4.31, the nozzle velocity equation. As shown in this equation, nozzle velocity is directly proportional to the square root of the pressure drop across the bit.

$$\bar{v}_n \propto \sqrt{\Delta p_b}.$$

Thus, the nozzle velocity is a maximum when the pressure drop available at the bit is a maximum. The pressure drop available at the bit is a maximum when the pump pressure is a maximum and the frictional pressure loss in the drillstring and annulus is a minimum. The frictional pressure loss is a minimum when the flow rate is a minimum.

4.13.2 Maximum Bit Hydraulic Horsepower

In 1958, Speer[20] published a paper that pointed out that the effectiveness of jet bits could be improved by increasing the hydraulic power of the pump. Speer reasoned that penetration rate would increase with hydraulic horsepower until the cuttings were removed as fast as they were generated. After this "perfect cleaning" level was achieved, there should be no further increase in penetration rate with hydraulic power. Shortly after Speer published his paper, several authors pointed out that due to the frictional pressure loss in the drillstring and annulus, the hydraulic power developed at the bottom of the hole was different from the hydraulic power developed by the pump. They concluded that bit horsepower rather than pump horsepower was the important parameter. Furthermore, it was concluded that bit horsepower was not necessarily maximized by operating the pump at the maximum possible horsepower. The conditions for maximum bit horsepower were derived by Kendall and Goins.[21]

The pump pressure is expended by (1) frictional pressure losses in the surface equipment, Δp_s, (2) frictional pressure losses in the drillpipe, Δp_{dp}, and drill collars, Δp_{dc}, (3) pressure losses caused by accelerating the drilling fluid through the nozzle, and (4) frictional pressure losses in the drill collar annulus, Δp_{dca}, and drillpipe annulus, Δp_{dpa}. Stated mathematically,

$$p_p = \Delta p_s + \Delta p_{dp} + \Delta p_{dc} + \Delta p_b + \Delta p_{dca} + \Delta p_{dpa}.$$

$$\dots \dots \dots \dots \dots \dots \dots \dots \dots \dots \dots (4.79)$$

If the total frictional pressure loss to and from the bit is called the *parasitic pressure loss* Δp_d, then

$$\Delta p_d = \Delta p_s + \Delta p_{dp} + \Delta p_{dc} + \Delta p_{dca} + \Delta p_{dpa},$$

$$\dots \dots \dots \dots \dots \dots \dots \dots \dots (4.80a)$$

and

$$p_p = \Delta p_b + \Delta p_d. \dots \dots \dots \dots \dots \dots (4.80b)$$

Since each term of the parasitic pressure loss can be computed for the usual case of turbulent flow using Eq. 4.66e,

$$\Delta p_f \propto q^{1.75},$$

and we can represent the total parasitic pressure loss using

$$\Delta p_d \propto q^m = cq^m, \dots \dots \dots \dots \dots \dots (4.81)$$

where m is a constant that theoretically has a value near 1.75, and c is a constant that depends on the mud properties and wellbore geometry. Substitution of this expression for Δp_d into Eq. 4.80b and solving for Δp_b yields

$$\Delta p_b = \Delta p_p - cq^m.$$

Since the bit hydraulic horsepower P_{Hb} is given by Eq. 4.35,

$$P_{Hb} = \frac{\Delta p_b q}{1{,}714} = \frac{\Delta p_p q - cq^{m+1}}{1{,}714}.$$

Using calculus to determine the flow rate at which the bit horsepower is a maximum gives

$$\frac{dP_{Hb}}{dq} = \frac{\Delta p_p - (m+1)cq^m}{1{,}714} = 0.$$

Solving for the root of this equation yields

$$\Delta p_p = (m+1)cq^m = (m+1)\Delta p_d, \quad \dots \dots (4.82a)$$

or

$$\Delta p_d = \frac{\Delta p_p}{(m+1)}. \quad \dots \dots \dots \dots \dots (4.82b)$$

Since $(d^2 P_{Hb})/(dq^2)$ is less than zero for this root, the root corresponds to a maximum. Thus, bit hydraulic horsepower is a maximum when the parasitic pressure loss is $[1/(m+1)]$ times the pump pressure.

From a practical standpoint, it is not always desirable to maintain the optimum $\Delta p_d/\Delta p_p$ ratio. It is usually convenient to select a pump liner size that will be suitable for the entire well rather than periodically reducing the liner size as the well depth increases to achieve the theoretical maximum. Thus, in the shallow part of the well, the flow rate usually is held constant at the maximum rate that can be achieved with the convenient liner size. For a given pump horsepower rating P_{HP} this maximum rate is given by

$$q_{max} = \frac{1{,}714\, P_{HP} E}{p_{max}}, \quad \dots \dots \dots \dots (4.83)$$

where E is the overall pump efficiency, and p_{max} is the maximum allowable pump pressure set by contractor. This flow rate is used until a depth is reached at which $\Delta p_d/\Delta p_p$ is at the optimum value. The flow rate then is decreased with subsequent increases in depth to maintain $\Delta p_d/\Delta p_p$ at the optimum value. However, the flow rate never is reduced below the minimum flow rate to lift the cuttings.

4.13.3 Maximum Jet Impact Force

Some rig operators prefer to select bit nozzle sizes so that the jet impact force is a maximum rather than bit hydraulic horsepower. McLean[22] concluded from experimental work that the velocity of the flow across the

bottom of the hole was a maximum for the maximum impact force. Eckel,[23] working with small bits in the laboratory, found that the penetration rate could be correlated to a bit Reynolds number group so that

$$\frac{dD}{dt} \propto \left(\frac{\rho \bar{v}_n d_n}{\mu_a} \right)^{a_8},$$

where

$$\frac{dD}{dt} = \text{penetration rate},$$

ρ = fluid density,
$\bar{v}_n$ = nozzle velocity,
d_n = nozzle diameter,
μ_a = apparent viscosity of the fluid at a shear rate of 10,000 seconds^{-1}, and
a_8 = constant.

It can be shown that when nozzle sizes are selected so that jet impact force is a maximum, the Reynolds number group defined by Eckel is also a maximum. (The proof of this is left as a student exercise.) The derivation of the proper conditions for maximum jet impact was published first by Kendall and Goins.[21]

The jet impact force is given by Eq. 4.37.

$$F_j = 0.01823 C_d q \sqrt{\rho \Delta p_b}$$
$$= 0.01823 C_d q \sqrt{\rho(\Delta p_p - \Delta p_d)}.$$

Since the parasitic pressure loss is given by Eq. 4.81,

$$F_j = 0.01823 C_d (\rho \Delta p_p q^2 - \rho c q^{m+2})^{0.5}.$$

Using calculus to determine the flow rate at which the bit impact force is a maximum gives

$$\frac{dF_j}{dq} = \left\{ 0.009115 C_d \left[2\rho \Delta p_p q - (m+2) \cdot \rho c q^{m+1} \right] \right\}$$

$$\left/ \left[\left(\rho \Delta p_p q^2 - \rho c q^{m+2} \right)^{0.5} \right] = 0. \right.$$

Solving for the root of this equation yields

$$2\rho \Delta p_p q - (m+2)\rho c q^{m+1} = 0,$$

$$\rho q [2\Delta p_p - (m+2)\Delta p_d] = 0,$$

or

$$\Delta p_d = \frac{2\Delta p_p}{(m+2)}. \quad \dots \dots \dots \dots \dots (4.84)$$

Since $(d^2 F_j)/dq^2$ is less than zero for this root, the root corresponds to a maximum. Thus, the jet impact force is a maximum when the parasitic pressure loss is $[2/(m+2)]$ times the pump pressure.

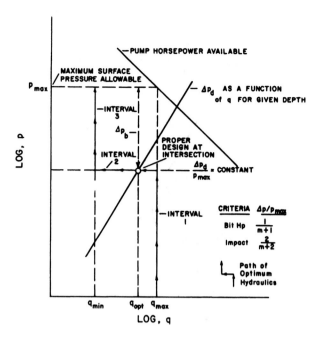

Fig. 4.35—Use of log-log plot for selection of proper pump operation and bit nozzle sizes.

4.13.4 Known Cleaning Needs

When large high-pressure pumps are available and the parasitic pressure loss is low because of a large-diameter drillstring and a low-viscosity drilling fluid, it may be possible to achieve a higher level of bit hydraulics than is needed to clean the hole bottom adequately. If the hole cleaning needs can be established from penetration rate data taken in similar lithology under conditions of varying bit hydraulics, it is wasteful to provide a higher level of bit hydraulics than needed. Under those conditions, the pump energy input should be reduced by decreasing the flow rate until the desired level of bit hydraulics just can be obtained if the pump is operated at the maximum allowable pressure. This same logic could be applied using either hydraulic horsepower or impact force as the hydraulic parameter.

The student should not be overly concerned about which parameter, bit power, or impact force is best. Perhaps the main reason neither criteria has been proved superior in all cases is because there is not a great amount of difference in the application of the two procedures. If hydraulic horsepower is a maximum, the jet impact force will be within 90% of the maximum and vice versa.

4.13.5 Graphical Analysis

The selection of bit nozzle sizes can be simplified somewhat through use of a graphical solution technique involving the use of log-log paper. Since for the usual case of turbulent flow, the parasitic pressure loss is approximated using Eq. 4.66e, a straight line representation of parasitic pressure loss is possible on log-log paper. Since Δp_d is proportional to $q^{1.75}$, a plot of $\log(\Delta p_d)$ vs. $\log(q)$ theoretically has a slope m of 1.75. It also can be shown from Eq. 4.35 that lines of constant hydraulic horsepower plot as a straight line with a slope of -1.0 on a graph of $\log(p_p)$ vs. $\log(q)$.

Shown in Fig. 4.35 is a summary of the conditions for the selection of bit nozzle sizes using the various hydraulic parameters. The conditions for proper pump operation and bit nozzle selection occur at the intersection of the line representing the parasitic pressure loss and the path of optimum hydraulics. The path of optimum hydraulics has three straight-line segments labeled Intervals 1, 2, and 3. Interval 1, defined by $q = q_{max}$, corresponds to the shallow portion of the well where the pump is operated at the maximum allowable pressure and the maximum possible flow rate for the convenient pump liner size and pump horsepower rating. Interval 2, defined by constant Δp_d, corresponds to the intermediate portion of the well where the flow rate is reduced gradually to maintain $\Delta p_d / p_{max}$ at the proper value for maximum bit hydraulic horsepower or impact force. Interval 3, defined by $q = q_{min}$, corresponds to the deep portion of the well where the flow rate has been reduced to the minimum value that efficiently will lift the cuttings to the surface. In Fig. 4.35, the intersection of the parasitic pressure-loss line and path of optimum hydraulics occurs in Interval 2. This corresponds to a bit run at an intermediate depth. Since parasitic pressure loss increases with depth, a shallow bit run would tend to intersect at Interval 1 and a deep bit run would tend to intersect at Interval 3. Once the intersection point is obtained the proper flow rate, q_{opt}, can be read from the graph. In addition, the proper pressure drop across the bit, $(\Delta p_b)_{opt}$, corresponds to $(p_{max} - \Delta p_d)$ on the graph at the intersection point. The proper nozzle area, $(A_t)_{opt}$, then can be computed by rearranging Eq. 4.34.

$$(A_t)_{opt} = \sqrt{\frac{8.311 \times 10^{-5} \rho q^2}{C_d^2 (\Delta p_b)_{opt}}}. \quad \ldots \ldots \ldots \ldots (4.85)$$

Three nozzles then are selected so that the total area of the nozzles is near $(A_t)_{opt}$.

The easiest and perhaps the most accurate method for determining the total parasitic loss at a given depth is by direct measurement of pump pressure. It is customary to measure the pump pressure for at least two flow rates periodically during drilling operations. As will be discussed in the next section, these pump pressures are useful during well control operations in the event the well "kicks." Since the total nozzle area of the bit currently in use is generally known, the pressure drop across the bit can be computed easily at the given flow rates using Eq. 4.34. The parasitic pressure then can be obtained at these flow rates as the difference between the pump pressure and the pressure drop across the bit.

Example 4.31. Determine the proper pump operating conditions and bit nozzle sizes for maximum jet impact force for the next bit run. The bit currently in use has three 12/32-in. nozzles. The driller has recorded that when the 9.6-lbm/gal mud is pumped at a rate of 485 gal/min, a pump pressure of 2,800 psig is observed and when the pump is slowed to a rate of 247 gal/min, a pump pressure of 900 psig is observed. The pump is rated at 1,250 hp and has an efficiency of 0.91. The minimum flow rate to lift the cuttings is 225 gal/min.

The maximum allowable surface pressure is 3,000 psig. The mud density will remain unchanged in the next bit run.

Solution. Using Eq. 4.34, the pressure drop through the bit at flow rates of 485 and 247 gal/min are as follows.

$$\Delta p_b = \frac{8.311 \times 10^{-5} \, \rho q^2}{C_d^2 \, A_t^2}.$$

$$\Delta p_{b1} = \frac{8.311 \times 10^{-5} \, (9.6)(485)^2}{(0.95)^2 \left[\frac{3\pi}{4} \left(\frac{12}{32} \right)^2 \right]^2} = 1,894 \text{ psig}.$$

$$\Delta p_{b2} = \frac{8.311 \times 10^{-5} \, (9.6)(247)^2}{(0.95)^2 \left[\frac{3\pi}{4} \left(\frac{12}{32} \right)^2 \right]^2} = 491 \text{ psig}.$$

Thus, the parasitic pressure loss at flow rates 485 and 247 gal/min are as follows.

$$\Delta p_d = \Delta p_p - \Delta p_b.$$

$$\Delta p_{d1} = 2,800 - 1,894 = 906 \text{ psig}.$$

$$\Delta p_{d2} = 900 - 491 = 409 \text{ psig}.$$

These two points are plotted in Fig. 4.36 to establish the parasitic pressure loss/flow rate relation for the given well geometry and mud properties. The slope of the line is determined graphically to have a value *m* of 1.2. Alternatively, *m* could be determined using a two-point method:

$$m = \frac{\log(906/409)}{\log(485/247)} = 1.18.$$

The path of optimum hydraulics is determined as follows.

Interval 1

$$q_{max} = \frac{1,714 P_{HP} E}{p_{max}} = \frac{1,714(1,250)(0.91)}{3,000}$$

$$= 650 \text{ gal/min}.$$

Interval 2

$$\Delta p_d = \frac{2}{m+2} p_{max} = \frac{2}{1.2+2}(3,000) = 1,875 \text{ psig}.$$

Interval 3

$$q_{min} = 225 \text{ gal/min}.$$

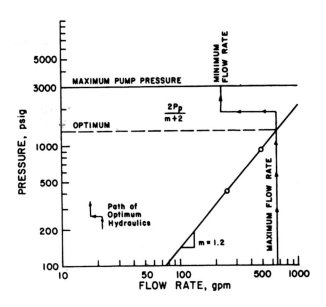

Fig. 4.36—Application of graphical analysis techniques for selection of bit nozzle sizes.

As shown in Fig. 4.36, the intersection of the parasitic pressure-loss line and the path of optimum hydraulics occurs in Interval 1 at

$$q_{opt} = 650 \text{ gal/min}, \quad \Delta p_d = 1,300 \text{ psi}.$$

Thus,

$$\Delta p_b = 3,000 - 1,300 = 1,700 \text{ psig}.$$

Thus, the proper total nozzle area is given by

$$(A_t)_{opt} = \sqrt{\frac{8.311 \times 10^{-5} \, \rho q^2}{C_d^2 \, \Delta p_{bopt}}}$$

$$= \sqrt{\frac{8.311 \times 10^{-5}(9.6)(650)^2}{(0.95)^2(1,700)}} = 0.47 \text{ sq in.}$$

In Example 4.31, the slope of the parasitic pressure loss/flow rate line was found to have a value of 1.2 rather than the theoretical value of 1.75. Reported values of *m* determined from field data are frequently much lower than the theoretical value. Thus, it is generally best to determine *m* from field data rather than assume a value of 1.75. This requires pump pressure data for at least two flow rates. Also, since the flow pattern of different portions of the well are subject to change at different flow rates, it is best not to extrapolate too far from the current operating conditions with a constant slope.

It is sometimes desirable to calculate the proper pump operating conditions and nozzle sizes during the well planning phase. This is done primarily when engineering personnel will not be available at the rig site. In this case, it is often desirable to attach the recommended flow rates and bit nozzle sizes to the well plan furnished the field personnel. The calculation of the parasitic pressure loss at various depths during the well planning

TABLE 4.7—TURBULENT FLOW RESISTANCE OF SURFACE CONNECTIONS

	Typical Combinations							
	No. 1		No. 2		No. 3		No. 4	
Components	ID (in.)	L (ft)	ID (in.)	L (ft)	ID (in.)	L (ft)	ID (in.)	L (ft)
Standpipe	3	40	3½	40	4	45	4	45
Drilling hose	2	45	2½	55	3	55	3	55
Swivel washpipe and gooseneck	2	4	2½	5	2½	5	3	6
Kelly	2¼	40	3¼	40	3¼	40	4	40

Drillpipe		Equivalent Length of Surface Connections in Feet of Drillpipe			
OD (in.)	Weight (lbm/ft)				
3½	13.3	437	161		
4½	16.6		761	479	340
5	19.5			816	579

phase can be accomplished using the frictional pressure-loss equations developed in the preceding sections. The mud properties and hole geometry must be planned as a function of depth before the nozzle size computation can be made. The frictional pressure loss in the surface connections usually is estimated by representing the surface equipment as an equivalent length of drillpipe. Shown in Table 4.7 are equivalent lengths for several typical combination of surface connections. The equivalent lengths shown were obtained by considering the losses in the standpipes, drilling hose, swivel, and kelly.

Since optimum hydraulics calculations for an entire well are quite lengthy, they are best accomplished using a computer. However, if a suitable computer program is not available, an approximate solution can be achieved by assuming an arbitrary flow rate to perform the pressure loss calculations and then using a graphical method to extrapolate to other flow rates.

Example 4.32. It is desired to estimate the proper pump operating conditions and bit nozzle sizes for maximum bit horsepower at 1,000-ft increments for an interval of the well between surface casing at 4,000 ft and intermediate casing at 9,000 ft. The well plan calls for the following conditions.

Pump: 3,423 psi maximum surface pressure
         1,600 hp maximum input
         0.85 pump efficiency

Drillstring: 4.5-in., 16.6-lbm/ft drillpipe (3.826-in. I.D.)
                600 ft of 7.5-in.-O.D. × 2.75-in.-I.D. drill collars

Surface Equipment: Equivalent to 340 ft of drillpipe

Hole Size: 9.857 in. washed out to 10.05 in.
              10.05-in.-I.D. casing

Minimum Annular Velocity: 120 ft/min

Mud Program:

Depth (ft)	Mud Density (lbm/gal)	Plastic Viscosity (cp)	Yield Point (lbf/100 sq ft)
5,000	9.5	15	5
6,000	9.5	15	5
7,000	9.5	15	5
8,000	12.0	25	9
9,000	13.0	30	12

Solution. The path of optimum hydraulics is as follows.

Interval 1

$$q_{max} = \frac{1,714 P_{HP} E}{p_{max}} = \frac{1,714(1,600)(0.85)}{3,423}$$

$$= 681 \text{ gal/min.}$$

Interval 2

Since measured pump pressure data are not available and a simplified solution technique is desired, a theoretical m value of 1.75 is used. For maximum bit horsepower,

$$\Delta p_d = \left(\frac{1}{m+1}\right) p_{max} = \left(\frac{1}{1.75+1}\right)(3,423)$$

$$= 1,245 \text{ psia.}$$

Interval 3

For a minimum annular velocity of 120 ft/min opposite the drillpipe,

$$q_{min} = 2.448\left(10.05^2 - 4.5^2\right)\left(\frac{120}{60}\right)$$

$$= 395 \text{ gal/min.}$$

Parasitic Pressure Losses

The parasitic pressure losses can be computed at any convenient flow rate. A flow rate of 500 gal/min will be used in this example. The frictional pressure loss in each section of drillpipe and annulus will be calculated using the equations for the Bingham plastic model outlined in Table 4.6.

The mean velocity in the drillpipe is given by

$$\bar{v}_{dp} = \frac{q}{2.448d^2} = \frac{500}{2.448(3.826)^2} = 13.95 \text{ ft/s.}$$

The Hedstrom number is given by

$$N_{He} = \frac{37,100\rho\tau_y d^2}{\mu_p^{\,2}}$$

$$= \frac{37,100(9.5)(5)(3.826)^2}{15^2} = 114,650.$$

From Fig. 4.33, the critical Reynolds number is 7,200. The Reynolds number is given by

$$N_{Re} = \frac{928\rho\bar{v}d}{\mu_p} = \frac{928(9.5)(13.95)(3.826)}{15}$$

$$= 31,370.$$

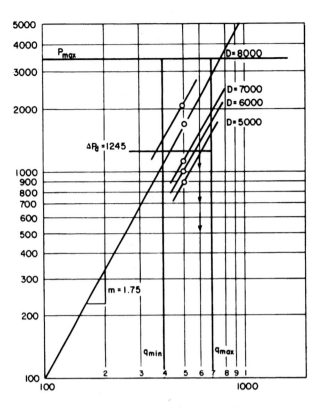

Fig. 4.37—Hydraulics plot for Example 4.33.

Thus, the flow pattern is turbulent. The frictional pressure loss in the drillpipe is given by

$$\Delta p_{dp} = \frac{\rho^{0.75}\bar{v}^{1.75}\mu_p^{0.25}}{1,800d^{1.25}}\Delta L$$

$$= \frac{(9.5)^{0.75}(13.95)^{1.75}(15)^{0.25}}{1,800(3.826)^{1.25}}(5,000-600)$$

$$= 490 \text{ psi.}$$

The pressure loss in surface equipment is equal to the pressure loss in 340 ft of drillpipe; thus,

$$\Delta p_s = 490\frac{340}{4,400} = 38 \text{ psi.}$$

The frictional pressure loss in other sections is computed following a procedure similar to that outlined above for the section of drillpipe. The entire procedure then can be repeated to determine the total parasitic losses at depths of 6,000, 7,000, 8,000, and 9,000 ft. The results of these computations are summarized in the following table.

Depth	Δp_s	Δp_{dp}	Δp_{dc}	Δp_{dca}	Δp_{dpa}	Δp_d
5,000	38	490	320	20	20	888
6,000	38	601	320	20	25	1,004
7,000	38	713	320	20	29	1,120
8,000	51	1,116	433	28	75*	1,703
9,000	57	1,407	482	27*	111*	2,084

*Laminar flow pattern indicated by Hedstrom number criteria.

The parasitic pressure losses expected at each depth assumed are plotted in Fig. 4.37 at the assumed flow rate of 500 gal/min. The parasitic pressure losses expected at other flow rates can be estimated by drawing a line with a slope of 1.75 through the computed points. The proper hydraulic conditions are obtained as in the previous example from the intersection of the parasitic pressure-loss lines and the path of optimum hydraulics. Note that all intersections, other than that for $D=9,000$ ft, occur in Interval 2 where $\Delta p_d = 1,245$ psi. The proper pump operating conditions and nozzle areas, are as follows.

(1) Depth (ft)	(2) Flow Rate (gal/min)	(3) Δp_d (psi)	(4) Δp_b (psi)	(5) A_t (sq in.)
5,000	600	1,245	2,178	0.380
6,000	570	1,245	2,178	0.361
7,000	533	1,245	2,178	0.338
8,000	420	1,245	2,178	0.299
9,000	395	1,370	2,053	0.302

The first three columns were read directly from Fig. 4.37. Col. 4 was obtained by subtracting Δp_d shown in Col. 3 from the maximum pump pressure of 3,423 psi. Col. 5 was obtained using Eq. 4.85.

The results of the hydraulics calculations indicate only the total nozzle area of the bit. Since jet bits have three nozzles, there are a large number of nozzle size combinations that will approximate closely the correct total nozzle area. Sutko,[24] in experimental work using a physical model of a rock fragment, found that the force on a rock fragment beneath a bit is increased when unequal nozzle sizes are used. For example, flow only from one nozzle was found to move across the hole bottom and exit on the opposite side of the bit from the nozzle. This flow pattern created larger forces on the rock

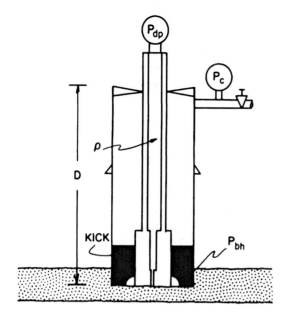

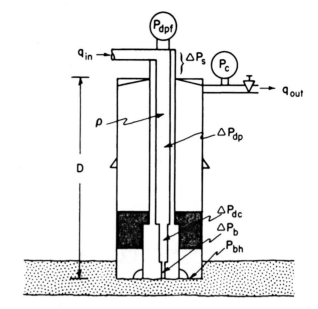

$$P_{bh} = P_{dp} + 0.052\,\rho D$$

(a)

$$P_{bh} = P_{dpf} + 0.052\rho D - \triangle P_s - \triangle P_{dp} - \triangle P_{dc} - \triangle P_b$$

(b)

Fig. 4.38—Relation between BHP and surface drillpipe pressure during well control operations: (a) shut-in conditions and (b) circulating conditions.

fragment than if the flow path had to turn and exit near the nozzle as in the case of three equal nozzle sizes. However, many engineers still prefer to divide the flow as evenly as possible among the three nozzles because of reports of uneven bit cooling and bit cone cleaning when using only one or two nozzles. Additional work is needed in this area to resolve this controversy.

4.14 Pump Pressure Schedules for Well Control Operations

During well control operations, the bottomhole pressure must be maintained at a value slightly above the formation pressure while the formation fluids are circulated from the well and kill mud is circulated into the well. This is accomplished by maintaining a backpressure on the annulus through the use of an adjustable choke. Unfortunately, a direct measurement of bottomhole pressure is not possible at present. Thus, it is necessary to infer the bottomhole pressure from surface pressure measurements while the well is being circulated. In Section 4.4, the relation between bottomhole pressure and surface annular pressure during well control operations was developed from hydrostatic considerations. This is possible because annular frictional pressure losses are generally small. However, the calculation of a meaningful annular pressure profile requires an accurate knowledge of the composition of the kick fluids and their distribution in the annulus. Since this information is generally not available at the time of the kick, annular pressure profiles cannot be used for accurate maintenance of constant bottomhole pressure during well control operations. A more accurate bottomhole pressure control is possible through use of the surface pressure in

the drillpipe since the drilling fluid in the drillstring generally is not contaminated with formation fluids. Thus, most modern well control procedures involve the use of drillpipe pressure schedules designed to maintain the bottomhole pressure at the proper value. Unfortunately, frictional pressure losses in the drillstring are not negligible as they are in the annulus.

The determination of the proper drillpipe pressure schedule for a given well control operation can be achieved by considering a flowing pressure balance of the well. Consider the shut-in well shown in Fig. 4.38A. The relation between the bottomhole pressure and surface drillpipe pressure for shut-in conditions is given by

$$p_{dp} + 0.052\rho D = p_{bh}.$$

However, after circulation of the well is initiated, the frictional pressure drops must be considered and the relation changed to

$$p_{dpf} + 0.052\rho D - \Delta p_s - \Delta p_{dp} - \Delta p_{dc} - \Delta p_b$$

$$= p_{bh}. \quad\ldots\ldots\ldots\ldots\ldots\ldots\ldots(4.86a)$$

It is usually convenient to choose a circulating drillpipe pressure as the sum of the static drillpipe pressure, p_{dp}, and a routinely measured circulating pump pressure, Δp_p, measured at the selected pump speed.

$$p_{dpf} = p_{dp} + \Delta p_p. \quad\ldots\ldots\ldots\ldots\ldots(4.86b)$$

While this routinely measured pump pressure includes the annular friction loss as well as the pressure loss in the drillstring and through the bit (Eq. 4.79), the annular friction loss is usually only a small fraction of the total pressure change. Thus, the use of Eq. 4.86b results in the selection of a circulating bottomhole pressure only slightly higher than the shut-in value. When using Eq. 4.86b, the circulating bottomhole pressure exceeds the shut-in bottomhole pressure by an amount equal to the frictional pressure loss in the annulus. This can be seen by substituting Eq. 4.79 for Δp_p in Eq. 4.86b, and then substituting Eq. 4.86b for p_{dpf} in Eq. 4.86a. This small excess bottomhole pressure is desirable since it is difficult to maintain the exact choke setting required to keep the drillpipe pressure at the value intended. Also, since a large portion of the annular frictional pressure losses occurs below the casing seat, the adverse effect of the excess bottomhole pressure is minimal.

It is important to measure the circulating pump pressure Δp_p frequently enough so that an accurate value will be available in the event a kick is taken. Most well operators require the measurement of a circulating pressure at a rate suitable for well control operations at least once a tour. Additional measurements may be required if the drilling fluid properties, bit nozzle sizes, or drillstring dimensions are changed. It is often desirable to measure the circulating pump pressure at several pump speeds so that the most suitable flow rate can be selected when a kick is taken.

An alternative method of determining the proper initial circulating drillpipe pressure is available if the well is shut in with the kick fluids confined to the lower portion of the well. The annular pressure required for a constant bottomhole pressure does not change rapidly, even for a gas kick, until the top of the kick reaches the upper portion of the annulus. Thus, if the surface casing is held constant at the shut-in value while the pump speed is first brought up to the desired constant value, the bottomhole pressure will increase only by an amount equal to the frictional pressure loss in the annulus, and the circulating drillpipe pressure will stabilize at a value of $p_{dp}+\Delta p_p$ for the flow rate used. The normal circulating pump pressure then can be taken as the difference between the drillpipe pressure observed after kick circulation is initiated and the stabilized drillpipe pressure observed when the well was shut in. This alternative method of determining Δp_p is quite useful when accurate circulating pump pressure data are not available at the time the kick was taken. Unfortunately, it is not well-suited to rigs that have high frictional pressure losses in the choke lines, such as a floating drilling vessel with underwater blowout preventers. When using this alternative procedure, the frictional pressure losses in the choke line are included in the annular frictional pressure losses used as excess bottomhole pressure. However, the frictional pressure losses in the choke lines are applied above the casing seat and may cause fracture of an unprotected formation.

As long as the average mud density in the drillstring remains constant, the bottomhole pressure can be held at the proper value by maintaining the pump speed constant and the circulating drillpipe pressure at a value of $(p_{dp}+\Delta p_p)$ for the given pump speed. However, when the average mud density in the drillstring changes significantly, both the hydrostatic pressure and the frictional pressure losses in the drillstring are altered. Thus, the circulating drillpipe pressure must be varied to just offset the change in the hydrostatic and frictional

pressure loss terms if the bottomhole pressure is to remain constant. The change in hydrostatic pressure due to a change in mud density is given by

$$\Delta p_h = 0.052(\rho_2 - \rho_1)D.$$

The change in pressure drop through the bit varies linearly with mud density. Also, the frictional pressure loss in the drillstring for the usual case of turbulent flow is proportional to mud density raised to the 0.75 power. Thus, for reasonable mud density increase, a linear relation between mud density and Δp_f can be assumed without introducing a large error. Since the annular pressure losses are small, the increase in frictional pressure loss due to a change in mud density in the drillstring can be approximated using

$$\Delta p_f = \Delta p_p \frac{(\rho_2 - \rho_1)}{\rho_1}.$$

The net decrease in circulating drillpipe pressure required to offset the increase in hydrostatic pressure and pressure loss between the surface and the bit is given by

$$\Delta p_{dpf} = \Delta p_h - \Delta p_f = (\rho_2 - \rho_1)\left(0.052D - \frac{\Delta p_p}{\rho_1}\right).$$

$$\dots\dots\dots\dots\dots\dots\dots(4.87)$$

Since this relation is linear with respect to mud density increase, it is usually convenient to calculate only the final circulating drillpipe pressure, corresponding to the final mud density reaching the bit. Intermediate drillpipe pressures then are determined by means of graphical or tabular interpolations.

Example 4.33. A 20-bbl kick is taken at a depth of 10,000 ft (Example 4.6). After the pressures stabilized, an initial drillpipe pressure of 520 psig and an initial casing pressure of 720 psig were recorded. The internal capacity of the 9,100-ft drillpipe is 0.01422 bbl/ft, and the internal capacity of the 900-ft drill collars is 0.0073 bbl/ft. A pump pressure of 800 psig was recorded previously at a reduced rate of 20 strokes/min. The pump factor is 0.2 bbl/stroke. Compute the drillpipe pressure schedule required to keep the bottomhole pressure constant as the mean mud density in the drillstring increases from an initial value of 9.6 lbm/gal to the final kill mud density.

Solution. The initial drillpipe pressure required after the pump speed is stabilized at 20 strokes/min is given by

$$p_{dpf} = p_{dp} + \Delta p_p = 520 + 800 = 1,320 \text{ psig.}$$

The kill mud density is given by

$$\rho_2 = \rho_1 + \frac{p_{dp}}{0.052D} = 9.6 + \frac{520}{0.052(10,000)}$$

$$= 10.6 \text{ lbm/gal.}$$

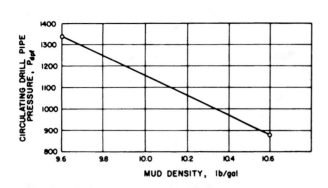

Fig. 4.39A—Drillpipe pressure schedule for the circulate-and-weight method.

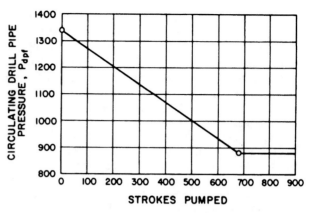

Fig. 4.39B—Drillpipe pressure schedule for the wait-and-weight method.

The total drillpipe pressure change required to maintain the bottomhole pressure constant as the mud density in the drillstring increases from 9.6 to 10.6 lbm/gal is given by Eq. 4.87:

$$\Delta p_{dpf} = (\rho_2 - \rho_1)\left(0.052D - \frac{\Delta p_p}{\rho_1}\right)$$

$$= (10.6 - 9.6)\left[0.052(10,000) - \frac{800}{9.6}\right]$$

$$= 437 \text{ psig.}$$

Thus, the final circulating drillpipe pressure is given by

$$1,320 - 437 = 883 \text{ psig.}$$

Because the relation between mean mud density in the drillstring and circulating drillpipe pressure is linear, intermediate values of circulating drillpipe pressure can be obtained graphically. If the mud density is increased slowly while circulating (circulate-and-weight method), a record of the mud density pumped into the drillstring must be maintained so that the mean mud density in the drillstring is known at all times. A frequent simplification of this technique is obtained by assuming that the mean mud density in the drillstring is the mud density at a point in the drillstring that has an equal volume of mud above and below. Half the total volume of the drillstring is given by

$$\frac{1}{2}[9,100(0.01422) + 900(0.0073)] = 68 \text{ bbl.}$$

The number of pump strokes required to pump 68 bbl is given by

$$\frac{68 \text{ bbl}}{0.2 \text{ bbl/strokes}} = 340 \text{ strokes.}$$

Since the pump rate is 20 strokes/min, it will take

$$\frac{340 \text{ strokes}}{20 \text{ strokes/min}} = 17 \text{ minutes}$$

for the mud to move from the surface to this point. Thus, the proper drillpipe pressure at a given point in time can be obtained by entering Fig. 4.39A for the mud density that was measured at the pump suction 17 minutes earlier.

If the mud density is increased to the final value of 10.6 lbm/gal before circulation of the kick is initiated (wait-and-weight method), it is usually convenient to express the mean mud density in the drillstring as a function of the number of strokes pumped. For simplicity, a linear relation usually is assumed. Since the internal diameter of the drillstring is not uniform, this assumption will give some error, but the error is small for all common drillstrings. The total number of strokes required to pump the kill mud to the bit is given by

$$\frac{[9,100(0.01422) + 900(0.0073)]}{0.2} = 680 \text{ strokes.}$$

Thus, the mean mud density in the drillstring increases 1 lbm/gal after pumping 680 strokes. Assuming a linear relation, intermediate values of drillpipe pressure can be obtained graphically as shown in Fig. 4.39B.

4.15 Surge Pressures Due to Vertical Pipe Movement

In Sections 4.10 and 4.11, equations that described the movement of fluids through conduits were developed. However, when running casing or making a trip, a slightly different situation is encountered in that the conduit is moved through the fluid rather than the fluid through the conduit. If a casing string or drillstring is lowered into the well, the pressure at a given point in the well increases. A pressure increase due to a downward pipe movement commonly is called a *surge* pressure. Similarly, if a casing string or drillstring is pulled from the well, the pressure at a given point in the well decreases. A pressure decrease due to an upward pipe movement commonly is called a *swab* pressure.

As pipe is moved downward in a well, the drilling fluid must move upward to exit the region being entered by the new volume of the extending pipe. Likewise, an upward pipe movement requires a downward fluid movement. The flow pattern of the moving fluid can be either laminar or turbulent depending on the velocity at which the pipe is moved. It is possible to derive mathematical equations for surge and swab pressures only for the laminar flow pattern. Empirical correlations must be used if the flow pattern is turbulent.

4.15.1 Laminar Flow

The basic differential equations derived in Section 4.10 to describe laminar flow in circular pipes and annuli apply to conduit movement through the fluid as well as fluid movement through the conduit. Only the boundary conditions are different.

A typical velocity profile for laminar flow caused by pulling pipe out of the hole at velocity, $-v_p$, is shown in Fig. 4.40. Note that the velocity profile inside the inner pipe caused by a vertical pipe movement is identical to the velocity profile caused by pumping fluid down the inner pipe. If the mean fluid velocity in the pipe is expressed relative to the pipe wall, the pipe flow equation developed in Section 4.10 can be applied. Substituting the term $[\bar{v}_i - (-v_p)]$ for $\bar{v}$ in Eq. 4.54d yields

$$\frac{\mathrm{d}p_f}{\mathrm{d}L} = \frac{\mu(\bar{v}_i + v_p)}{1{,}500d^2}. \qquad (4.88)$$

The velocity profile in the annulus caused by vertical pipe movement differs from the velocity profile caused by pumping fluid through the annulus in that the velocity at the wall of the inner pipe is not zero. However, Eq. 4.53a developed for annular flow and Eq. 4.59a developed for slot flow are still applicable if the proper boundary conditions are used to evaluate the constants of integration. The slot flow representation of the annular geometry usually is preferred because of its relative simplicity. Eq. 4.59a for slot flow is given by

$$v = -\frac{y^2}{2\mu}\frac{\mathrm{d}p_f}{\mathrm{d}L} - \tau_0\frac{y}{\mu} + v_0.$$

The constants of integrations, τ_0 and v_0, can be evaluated using the boundary conditions

$$v = -v_p \text{ at } y = 0,$$

and

$$v = 0 \text{ at } y = h.$$

Applying these boundary conditions yields

$$-v_p = 0 - 0 + v_0,$$

and

$$0 = \frac{-h^2}{2\mu}\frac{\mathrm{d}p_f}{\mathrm{d}L} - \tau_0\frac{h}{\mu} + v_0.$$

Solving these two equations simultaneously for τ_0 and v_0 gives

$$\tau_0 = -\frac{h}{2}\frac{\mathrm{d}p_f}{\mathrm{d}L} - \frac{v_p\mu}{h},$$

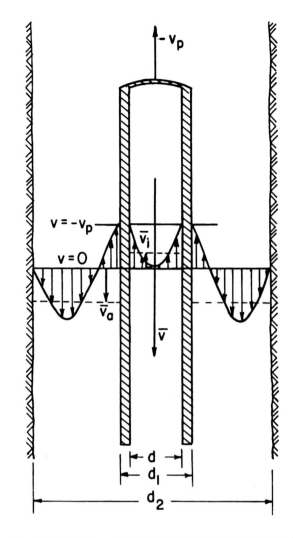

Fig. 4.40—Velocity profiles for laminar flow pattern when pipe is pulled out of hole.

and

$$v_0 = -v_p.$$

Substituting these values for τ_0 and v_0 in Eq. 4.59a yields

$$v = \frac{1}{2\mu}\frac{\mathrm{d}p_f}{\mathrm{d}L}(yh - y^2) - v_p\left(1 - \frac{y}{h}\right). \qquad (4.89)$$

The flow rate q is given by

$$q = \int_0^h v\,\mathrm{d}A = \int_0^h vW\mathrm{d}y$$

$$= \frac{W}{2\mu}\frac{\mathrm{d}p_f}{\mathrm{d}L}\int_0^h (hy - y^2)\mathrm{d}y - v_pW\int_0^h\left(1 - \frac{y}{h}\right)\mathrm{d}y.$$

Integrating this equation yields

$$q = \frac{Wh^3}{12\mu}\frac{\mathrm{d}p_f}{\mathrm{d}L} - \frac{v_pWh}{2}. \qquad (4.90a)$$

In terms of annular geometry,

$$A = Wh = \pi(r_2{}^2 - r_1{}^2),$$

and

$$h = r_2 - r_1.$$

Substitution of these equivalent relations for Wh and h in Eq. 4.90a gives

$$q = \frac{\pi}{12\mu} \frac{dp_f}{dL}(r_2{}^2 - r_1{}^2)(r_2 - r_1)^2$$

$$- \frac{\pi v_p}{2}(r_2{}^2 - r_1{}^2). \quad \ldots\ldots\ldots\ldots (4.90b)$$

Expressing the flow rate in terms of the mean flow velocity in the annulus, $\bar{v}_a$, and solving for the frictional pressure gradient dp_f/dL gives

$$\frac{dp_f}{dL} = \frac{12\mu\left(\bar{v}_a + \dfrac{v_p}{2}\right)}{(r_2 - r_1)^2}. \quad \ldots\ldots\ldots\ldots (4.90c)$$

Converting from consistent units to more convenient field units, we obtain

$$\frac{dp_f}{dL} = \frac{\mu\left(\bar{v}_a + \dfrac{v_p}{2}\right)}{1,000(d_2 - d_1)^2}. \quad \ldots\ldots\ldots\ldots (4.90d)$$

The total upward flow rate of drilling fluid caused by lowering pipe into the hole must be equal in magnitude but opposite in direction to the rate at which fluid is displaced from the bottom of the hole by the elongating pipe section. Likewise, the total downward flow rate of drilling fluid caused by pulling pipe out of the hole must be equal in magnitude but opposite in direction to the volume rate at which pipe is being withdrawn. If the bottom of the pipe is closed, the magnitude of the total flow rate is given by

$$q_t = v_p \frac{(\pi d_1{}^2)}{4},$$

where d_1 is the outer diameter of the inner pipe. Since the pipe is closed, all the flow is in the annulus, and the magnitude of the mean annular velocity $\bar{v}_a$ is given by

$$\bar{v}_a = \frac{4q_t}{\pi(d_2{}^2 - d_1{}^2)} = \frac{d_1{}^2 v_p}{(d_2{}^2 - d_1{}^2)}. \quad \ldots\ldots\ldots (4.91)$$

If the bottom of the pipe is open, vertical pipe movement will cause fluid flow both in the pipe and in the annulus. However, since the pipe and annulus have a common pressure at both the top and the bottom, the total frictional pressure loss must be the same inside the pipe and the annulus. For uniform annular geometry, as in the case of running casing, the frictional pressure gradient is also uniform, and a direct combination of Eqs. 4.88 and 4.90d is possible such that

$$\frac{\mu(\bar{v}_i + v_p)}{1,500d^2} = \frac{\mu\left(\bar{v}_a + \dfrac{v_p}{2}\right)}{1,000(d_2 - d_1)^2}. \quad \ldots\ldots\ldots (4.92)$$

Also, since the combined flow in the pipe and annulus must be equal in magnitude to the rate at which steel is moved into or out of the well,

$$q_t = q_i + q_a,$$

or

$$v_p \pi(d_1{}^2 - d^2) = \bar{v}_i \pi d^2 + \bar{v}_a \pi(d_2{}^2 - d_1{}^2). \quad \ldots (4.93)$$

Solving Eqs. 4.92 and 4.93 simultaneously yields the following expression for *laminar flow* due to the vertical movement of an open-ended pipe of uniform cross-sectional area in a Newtonian fluid.

$$\bar{v}_a = v_p \frac{3d^4 - 4d_1{}^2(d_2 - d_1)^2}{-6d^4 - 4(d_2 - d_1)^2(d_2{}^2 - d_1{}^2)}. \quad \ldots (4.94)$$

When using this equation, recall that the direction of fluid flow was assumed opposite to the direction of pipe movement in the equation derivations. When pipe movement is upward, the computed frictional pressure change at the bottom of the pipe will cause the pressure at this point to decrease. Similarly, when pipe movement is downward, the frictional pressure change at the bottom of the pipe will cause the pressure at this point to increase.

Example 4.34. Calculate the equivalent density below the bottom joint of 4,000 ft of 10.75-in. casing (having a 10.0-in. ID) if the casing is being lowered at a rate of 1.0 ft/s in a 12-in. hole containing 9.0-lbm/gal brine having a viscosity of 2.0 cp. Perform the calculation for (1) casing that is open and (2) casing with a closed bottom end. Assume that the flow pattern is laminar.

Solution. If the bottom of the casing is open, the mean annular velocity is given by Eq. 4.94:

$$\bar{v}_a = (1.0)\frac{3(10)^4 - 4(10.75)^2(12 - 10.75)^2}{-6(10)^4 - 4(12 - 10.75)^2(12^2 - 10.75^2)}$$

$$= -0.4865 \text{ ft/s}.$$

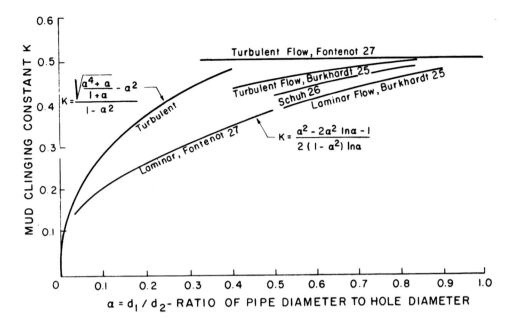

Fig. 4.41—Mud clinging constant, K, for computing surge-and-swab pressure.

The viscous pressure gradient is given by Eq. 4.90d:

$$\frac{dp_f}{dL} = \frac{\mu\left(\bar{v}_a + \frac{v_p}{2}\right)}{1,000(d_2 - d_1)^2}$$

$$= \frac{2(-0.4865 + \frac{1.0}{2})}{1,000(12 - 10.75)^2} = 0.0000175 \text{ psi/ft}.$$

Thus, the surge pressure is negligible when the large-diameter casing is open.

If the bottom of the casing is closed, the mean annular velocity is given by Eq. 4.91:

$$\bar{v}_a = \frac{d_1{}^2 v_p}{(d_2{}^2 - d_1{}^2)} = \frac{10.75^2(1.0)}{(12.0^2 - 10.75^2)}$$

$$= 4.06 \text{ ft/s}.$$

The viscous pressure gradient is given by Eq. 4.90:

$$\frac{dp_f}{dL} = \frac{\mu\left(\bar{v}_a + \frac{v_p}{2}\right)}{1,000(d_2 - d_1)^2}$$

$$= \frac{2\left(4.06 + \frac{1.0}{2}\right)}{1,000(12.0 - 10.75)^2} = 0.00584 \text{ psi/ft}.$$

Thus, the total surge pressure below 4,000 ft of closed casing is given by

$$\Delta p_f = \frac{dp_f}{dL}\Delta L = 0.00584(4,000) = 23 \text{ psi}.$$

The equivalent density at a depth of 4,000 ft for this surge pressure is given by

$$\rho_e = \rho + \frac{\Delta p_f}{0.052D} = 9.0 + \frac{23}{0.052(4,000)}$$

$$= 9.11 \text{ lbm/gal}.$$

The use of Eq. 4.94 is not directly applicable to nonuniform pipe such as drillstring containing drillpipe, drill collars, and a jet bit. It also cannot be applied when a turbulent flow pattern is possible. A complex problem is best solved using an iterative technique. The usual approach is to start at the bottom of the drillstring with an assumed flow split between the pipe interior, q_p, and well annulus, q_a, such that

$$q_t = q_a + q_p.$$

For simplicity, the bottom of the drillstring can be represented by a drill collar with a restricted exit (Fig. 4.42). The restricted exit represents the nozzles in the bit and has an area equal to the total area of the nozzles. The total flow rate displaced by the bottom of the simplified drillstring is given by

$$(q_t)_1 = v_p\left[\frac{\pi}{4}(d_1)_1{}^2 - A_j\right]. \quad \dots\dots\dots\dots (4.95)$$

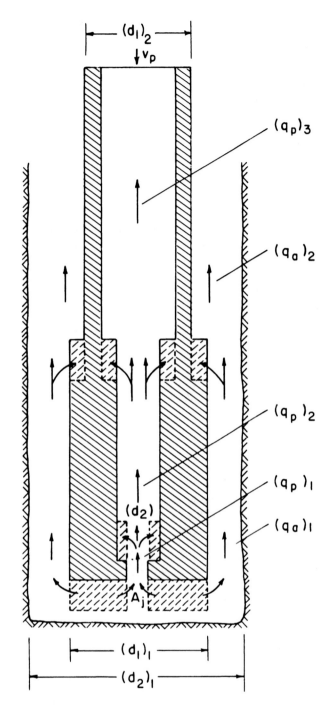

Fig. 4.42—Simplified hydraulic representation of the lower part of a drillstring.

If a value for f_a is assumed such that

$$f_a = \frac{q_a}{q_t},$$

then the annular flow rate opposite the bottom portion of the drillstring, $(q_a)_1$, can be computed using

$$(q_a)_1 = f_a(q_t)_1$$

$$= f_a\, v_p \left[\frac{\pi}{4}(d_1)_1^2 - A_j \right]. \quad \dotsc\dotsc\dotsc\dotsc (4.96a)$$

At any point in the drillstring where the outer diameter, d_1, changes (such as at the juncture between the drillpipe and drill collars), the annular flow rate changes according to

$$(q_a)_i = (q_a)_{i-1} - \frac{\pi}{4} v_p \left[(d_1)_{i-1}^2 - (d_1)_i^2 \right]. \quad (4.96b)$$

Similarly, the interior flow rate in the pipe exit, q_{p1}, can be computed using

$$(q_p)_1 = (1 - f_a)(q_t)_1$$

$$= (1 - f_a)v_p \left[\frac{\pi}{4}(d_1)_1^2 - A_j \right]. \quad \dotsc\dotsc\dotsc (4.96c)$$

Likewise, at any point in the drillstring where the inner diameter, d, changes (such as above the restricted exit and at the juncture between the drillpipe and drill collars), the interior flow rate changes to

$$(q_p)_i = (q_p)_{i-1} - \frac{\pi}{4} v_p \left[(d)_i^2 - (d)_{i-1}^2 \right]. \quad \dotsc (4.96d)$$

The total flow rate at any elevation in the well is the sum of the annular flow rate and the flow rate in the pipe interior:

$$q_t = q_a + q_p.$$

It can be shown from combination of Eqs. 4.96b and 4.96d that the total flow rate at any elevation can be expressed by

$$q_t = \frac{\pi}{4} v_p (d_1^2 - d^2). \quad \dotsc\dotsc\dotsc\dotsc\dotsc\dotsc (4.96e)$$

However, it is possible that the flow rate inside the pipe may be in a different direction than the flow rate in the annulus. If this is true, one must take care to keep track of the direction (sign) as well as the magnitude of the flow rates.

To determine the correct surge or swab pressure, the value of f_a must be chosen so that the sum of the frictional pressure changes through all sections of the annulus is equal to the sum of frictional pressure changes through all sections of the drillstring interior. If the flow rate in the annulus is in a different direction than the flow rate in the pipe interior, root values of f_a that are greater than one or less than zero are possible. Values of f_a which are greater than one tend to occur when the internal area is very small in comparison to the annular area. Values of f_a which are negative tend to occur when the annular area is very small in comparison to the internal area.

In the case of a closed-end pipe, no flow through the pipe interior is possible and f_a is known to be equal to one. For this simplified situation, an iterative calculation procedure is not required. Total flow is based on the outer diameter of the pipe and annular velocity is given by Eq. 4.91. Examples of this situation included running casing or drillpipe in the well with a check valve (float) present in the pipe string.

4.15.2 Non-Newtonian Fluid Models

It is possible to derive laminar flow surge pressure equations using non-Newtonian fluid models such as the Bingham plastic and power-law models. This can be accomplished by changing the boundary conditions at the pipe wall from $v=0$ to $v=-v_p$ in the annular flow derivations for the Bingham plastic model and power-law model given in Appendix B. However, the resulting surge pressure equations are far too complex for field application.

A simplified technique for computing surge pressures was presented by Burkhardt[25] in 1961. The simplified method is based on the use of an effective fluid velocity in the annular flow equations. The suitability of the annular flow equations for predicting surge pressure is suggested by the similarity of the annular flow and surge pressure equations for the Newtonian fluid model. For example, the Newtonian surge pressure equation given by Eq. 4.90d is obtained if an effective mean annular velocity $\bar{v}_{ae}$ defined by

$$\bar{v}_{ae} = \bar{v} + 0.5 v_p$$

is used in the slot flow equation given by Eq. 4.60d. Burkhardt suggested using an effective mean annular velocity given by

$$\bar{v}_{ae} = \bar{v} + K v_p,$$

where the constant K, called the *mud clinging constant*, is obtained for a given annular geometry using Fig. 4.41. Burkhardt obtained the correlation for K using complex equations derived for the Bingham plastic model using a slot flow representation of the annulus. Note that for small annular clearances, where surge and swab pressures will be most significant, the value of K approaches 0.5.

Clinging constant values also can be obtained from the work of Schuh[26] for the power-law fluid model and a slot approximation of annular geometry. The resulting curve is used irrespective of flow pattern and falls between Burkhardt's curves for laminar and turbulent flow.

4.15.3 Turbulent Flow

Empirical correlations have not been developed specifically for the calculations of surge or swab pressures. However, Burkhardt[25] and Schuh[26] have presented (Fig. 4.41) a correlation for the clinging constant K required to apply the turbulent flow equations commonly used for annular geometry. Recall that these annular flow equations, in turn, are based on an empirical correlation developed for circular pipes. Unfortunately, no published criteria for establishing the onset of turbulence are available. The usual procedure is to calculate surge or swab pressures for both the laminar and turbulent flow patterns and then to use the larger value.

Example 4.35. Using the Bingham plastic model, calculate the swab pressure in a 7.875-in. hole below a 15,000-ft drillstring composed of 14,300 ft of 4.5-in. drillpipe (3.826-in. ID), 700 ft of 6.25-in. drill collars (2.75-in. ID), and a bit containing three 11/32-in. nozzles (0.2784 sq in.). The drillstring is moving through a 10-lbm/gal drilling fluid having a plastic viscosity of 60 cp and a yield point of 10 lbf/100 sq ft at a maximum rate of 4.0 ft/s.

Solution: The nomenclature shown in Fig. 4.42 has been adopted in the solution of this problem. Pipe and annular sections are numbered starting at the bottom of the drillstring. For simplicity, the effect of the outer shape of the bit and the effect of tool joints are assumed to be negligible and are ignored. In addition, it is assumed that the hole is kept full and that the fluid level in the pipe and annulus are maintained approximately equal. The total flow rate near the bottom of the drillstring is given by Eq. 4.95:

$$(q_t)_1 = v_p \left[\frac{\pi}{4}(d_1)^2 - A_j \right]$$

$$= 4 \text{ ft/s} \left[\frac{\pi}{4}(6.25)^2 - 0.2784 \right] \text{sq in.}$$

$$\times \left(\frac{\text{sq ft}}{144 \text{ sq in.}} \right)$$

$$= 0.8445 \text{ cu ft/s.}$$

The flow rate through the bit jets is given by

$$(q_p)_1 = (1 - f_a)(q_t)_1$$

$$= [0.8445(1 - f_a)] \text{cu ft/s.}$$

The flow rates in Sections 2 and 3 of the drill collar and drill pipe interior are given by Eq. 4.96d.

$$(q_p)_2 = 0.8445(1 - f_a) - 4 \text{ ft/s}$$

$$\left[\frac{\pi}{4}(2.75)^2 - 0.2784 \right] \text{sq in.} \left(\frac{\text{sq ft}}{144 \text{ sq in.}} \right)$$

$$= [0.8445(1 - f_a) - 0.1573] \text{cu ft/s.}$$

TABLE 4.8—SUMMARY OF SWAB PRESSURE CALCULATION FOR EXAMPLE 4.35

Variable				
$f_a = (q_a/q_t)_1$	0.5	0.75	0.70	0.692
$(q_p)_1$, cu ft/s	0.422	0.211	0.251	0.260
$(q_p)_2$, cu ft/s	0.265	0.054	0.093	0.103
$(q_p)_3$, cu ft/s	0.111	−0.101	−0.061	−0.052
Δp_b, psig	442	115	160	171
Δp_{dc}, psig	104	33	44	46
Δp_{dp}, psig	449	273	293	297
Total Δp_i, psig	995	421	497	514
$(q_a)_1$, cu ft/s	0.422	0.633	0.594	0.585
$(q_a)_2$, cu ft/s	0.012	0.223	0.183	0.174
Δp_{dca}, psig	104	139	128	126
Δp_{dpa}, psig	335	405	392	389
Total Δp_a, psig	439	544	520	515

$$(q_p)_3 = (q_p)_2 - 4\left[\frac{\pi}{4}(3.826)^2 - \frac{\pi}{4}(2.75)^2\right]\frac{1}{144}$$

$$= 0.8445(1-f_a) - 0.1573 - 0.1544$$

$$= [0.8445(1-f_a) - 0.3117] \text{ cu ft/s}.$$

The mean fluid velocity (with respect to an observer at the surface) through the bit jets is given by

$$(v_i)_1 = (q_p)_1/(A_p)_1 = \frac{0.8445(1-f_a)}{0.2784/144}$$

$$= [436.8(1-f_a)] \text{ ft/s}.$$

Similarly, the fluid velocity in the drill collar and drill pipe interior is given by

$$(v_i)_2 = (q_p)_2/(A_p)_2 = \frac{0.8445(1-f_a)-0.1573}{\frac{\pi}{4}(2.75)^2\frac{1}{144}}$$

$$= [20.47(1-f_a) - 3.814] \text{ ft/s}.$$

$$(v_i)_3 = (q_p)_3/(A_p)_3 = \frac{0.8445(1-f_a)-0.3117}{\frac{\pi}{4}(3.826)^2\frac{1}{144}}$$

$$= [10.58(1-f_a) - 3.904] \text{ ft/s}.$$

The effective fluid velocity (with respect to the nozzle wall) through the bit jets is given by

$$(v_{ie})_1 = (v_i)_1 + v_p = [436.8(1-f_a)+4] \text{ ft/s}.$$

Similarly, the effective fluid velocity (with respect to the pipe wall) in the drill collar and drill pipe is given by

$$(v_{ie})_2 = (v_i)_2 + v_p = [20.47(1-f_a)+0.186] \text{ ft/s}.$$

$$(v_{ie})_3 = (v_i)_3 + v_p = [10.58(1-f_a)+0.096] \text{ ft/s}.$$

The pressure loss through the bit jets, Δp_b, is calculated through rearrangement of Eq. 4.31 and the use of $(v_{ie})_1$ for v_n. The frictional pressure losses through the drill collars and drill pipe are calculated using the flow equations given in Table 4.6 with v replaced by $(v_{ie})_2$ and $(v_{ie})_3$, respectively. The frictional pressure losses for each section are calculated for both laminar and turbulent flow, and then the larger of the two results is selected as the correct value. Once the pressure loss for each section is determined, they are summed to give the total pressure losses in the drillstring interior. The results of these calculations are given in Table 4.8 for various assumed values of f_a. The flow rates, velocities, effective velocities, and frictional pressure losses in the annulus are determined using a procedure similar to that used for the pipe interior. The flow rates in Sections 1 and 2 of the drill collar and drill pipe annulus are computed using Eqs. 4.96a and 4.96b.

$$(q_a)_1 = f_a(q_t)_1 = (0.8445\, f_a)\text{ft}^3/\text{s}.$$

$$(q_a)_2 = (q_a)_1 - \frac{\pi}{4}v_p[(d_1)_1^2 - (d_1)_2^2]$$

$$= 0.8445 f_a - 4 \text{ ft/s}\left[\frac{\pi}{4}(6.25^2 - 4.5^2)\right]$$

$$\text{sq in.}\frac{\text{sq ft}}{144 \text{ sq in.}}$$

$$= (0.8445\, f_a - 0.4104) \text{ cu ft/s}.$$

The mean annular fluid velocity (with respect to an observer at the surface) is given by

$$(v_a)_1 = (q_a)_1/(A_a)_1 = \frac{0.8445 f_a}{\frac{\pi}{4}(7.875^2 - 6.25^2)\frac{1}{144}}$$

$$= (6.746 f_a) \text{ ft/s}.$$

$$(v_a)_2 = (q_a)_2/(A_a)_2 = \frac{0.8445 f_a - 0.4104}{\frac{\pi}{4}(7.875^2 - 4.5^2)\frac{1}{144}}$$

$$= 3.707 f_a - 1.802 \text{ ft/s}.$$

The effective annular fluid velocity is given by $(\bar{v}_a + K v_p)$ where the mud clinging constant, K, is defined by Fig. 4.41. The ratio d_1/d_2 is 0.571 opposite the drillpipe and 0.794 opposite the drill collar. The mud clinging constants obtained for these values from Fig. 4.41 are 0.410 and 0.460, respectively, for laminar flow, and 0.465 and 0.486, respectively, for turbulent flow. Thus, the effective annular velocity opposite the drillpipe is

$$(v_{ae})_1 = (v_a)_1 + K v_p = \begin{cases} 6.746 f_a + 0.410 \,(4) \\ 6.746 f_a + 0.465 \,(4). \end{cases}$$

$$(v_{ae})_1 = \begin{cases} 6.746 f_a + 1.64 \text{ for laminar flow} \\ 6.746 f_a + 1.86 \text{ for turbulent flow}. \end{cases}$$

Similarly, the effective annular velocity opposite the drill collars is

$$(v_{ae})_2 = (v_a)_2 + Kv_p = \begin{cases} 3.707\, f_a - 1.802 + 0.46\ (4) \\ 3.707\, f_a - 1.802 + 0.486\ (4) \end{cases}$$

$$(v_{ae})_2 = \begin{cases} 3.707\, f_a - 0.038 \ \text{for laminar flow} \\ 3.707\, f_a + 0.142 \ \text{for turbulent flow.} \end{cases}$$

The frictional pressure losses in the drill collar and drill-pipe annulus are calculated using the flow equations given in Table 4.6 with v replaced by $(v_{ae})_1$ and $(v_{ae})_2$, respectively. The frictional pressure losses for each section are calculated for both laminar and turbulent flow, and then the larger of the two results is selected as the correct value. Once the pressure loss for each annular section is determined, they are summed to give the total pressure losses in the annulus. Values of f_a are tried until the total frictional pressure losses in the pipe interior and in the annulus are equal, thus yielding a final solution. The results of these calculations are summarized in Table 4.8. The results indicate that the total loss in pressure beneath the drillstring due to upward pipe movement was 515 psig. The results also indicate the 69.2% of the flow at the bottom of the drillstring was from the annulus and 30.8% of the flow was from the interior of the drillstring.

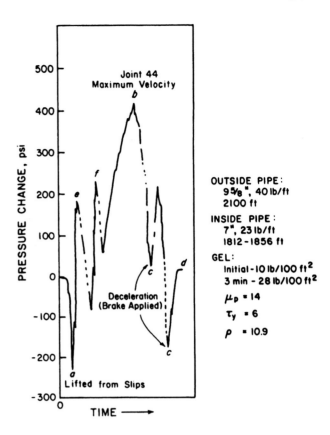

Fig. 4.43—Typical pressure surge pattern measured as a joint of casing was lowered into the wellbore (after Ref. 25).

In Example 4.35, both the fluid velocity in the annulus and the fluid velocity in the pipe were computed with respect to a stationary observer at the surface. The total flow rate or fluid flux (Eq. 4.96e) was based on the area of steel. An alternate equivalent approach used by some authors is always to compute total flow rate or fluid flux based on a closed-end pipe. This can be done using Eq. 4.96e with d equal to zero. Flow up the pipe interior is considered to be a leak that subtracts from the annular flow rate. The fluid velocity inside the pipe is computed with respect to an observer riding along on the pipe. In this case, the interior flow rate or fluid flux is a constant value over the entire pipe length. The effective velocity inside the pipe is equal to this constant value, divided by the interior pipe area at the point of interest. It is not necessary to add the pipe velocity to the average flow velocity. However, the amount of the flow traveling up the pipe interior still must be determined by trial and error. An excellent description of this equivalent alternative approach is given by Fontenot and Clark.[27]

Fontenot and Clark also compared computed values of surge pressures to experimentally measured values in two wells. Best results were achieved for the power-law fluid model with downhole fluid rheological properties corrected for downhole temperature and pressure. The well was broken into 10 sections to allow fluid rheological properties to be easily varied with downhole temperature and pressure. This required the use of a computer program.

In addition to the pressure surges caused by the viscous effects of vertical pipe movements, pressure surges also occur in the wellbore due to (1) the pressure required to break the gel and (2) fluid acceleration and deceleration. Shown in Fig. 4.43 is a record of changes in bottomhole pressure that occurred in a well as single joint of casing was added to the casing string and lowered into the well. Note that in Fig. 4.43 bottomhole pressure peaks are caused by (a) breaking the gel when the casing was lifted from the slips, (b) the viscous drag when the pipe was lowered at maximum velocity, and (c) the inertial effects of rapid deceleration when the brakes were applied. The effect of breaking the gel by pipe movement is identical to the effect of breaking the gel by pumping fluid. Thus, the equations developed in Section 4.12 also can be used to calculate the effect of breaking the gel by vertical pipe movement. In general, the pressure surges caused by inertial effects tend to be less than those caused by viscous drag.

4.15.4 Inertial Effects

The pressure change in a fluid due to inertial effects can be estimated from consideration of the vertical forces acting on an element of the fluid at a depth D in a hole of cross-sectional area A. Consider for example the fluid element discussed in Sec. 4.1 (Fig. 4.1) not to be at rest but accelerating in a downward direction. In this case, the sum of the downward forces must be equal to the mass of the fluid element times its acceleration:

$$\Sigma F = Ma$$

$$pA - \left(p + \frac{dp}{dD}\Delta D \right) A + F_{wv}A\Delta D = \rho A \Delta D\, a.$$

TABLE 4.9—
SPHERICITIES FOR
VARIOUS PARTICLE
SHAPES

Shape	Sphericity
Sphere	1.0
Octahedron	0.85
Cube	0.81
Prism	
$\ell \cdot \ell \cdot 2\ell$	0.77
$\ell \cdot 2\ell \cdot 2\ell$	0.76
$\ell \cdot 2\ell \cdot 3\ell$	0.73
Cylinders	
$h = r/15$	0.25
$h = r/10$	0.32
$h = r/3$	0.59
$h = r$	0.83
$h = 2r$	0.87
$h = 3r$	0.96
$h = 10r$	0.69
$h = 20r$	0.58

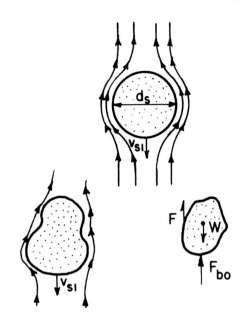

Fig. 4.44—Streamlines of fluid movement about a settling particle.

Expansion of the second term and division by the element volume $A\Delta D$ gives

$$\frac{dp}{dD} = F_{wv} - \rho a.$$

This expression says that the change in pressure with depth in a downwardly accelerating fluid is equal to the specific weight of the fluid (hydrostatic gradient) less the inertial effect, which is equal to the fluid density times the fluid acceleration. If we are interested in only the magnitude of the pressure gradient due to fluid acceleration, then

$$\frac{dp_a}{dL} = \rho a. \quad \dots \dots \dots \dots \dots \dots \dots \dots \dots (4.97)$$

For a pipe with a closed end in an incompressible fluid, the relation between fluid acceleration and pipe acceleration is given by

$$a = a_p \frac{d^2_1}{d^2_2 - d^2_1}. \quad \dots \dots \dots \dots \dots \dots \dots (4.98)$$

Substituting Eq. 4.98 into 4.97 and converting to field units yields the following expression for a closed-end pipe.

$$\frac{dp_a}{dL} = \frac{0.00162\rho a_p d^2_1}{d^2_2 - d^2_1}. \quad \dots \dots \dots \dots \dots (4.99)$$

For a pipe with an open end, the fluid acceleration can occur inside and outside the pipe at different rates. This situation is too complex for a convenient approximate solution to be easily developed. However, for this case, the pressure surge caused by inertial effects are generally small and of no practical interest.

When applying Eq. 4.99, one should recognize that the assumption of an incompressible fluid may not be realistic, especially when the change in pipe velocity is very rapid. In addition, the elastic characteristics of the hole or casing wall can also dampen the inertial pressure surge seen. A discussion of elastic theory and water hammer effects is beyond the scope of this book. However, elastic theory predicts that, for an instantaneous change in fluid velocity, Δv, the magnitude of the pressure surge is given by

$$\Delta p_a = 0.00162 \, \rho v_{wave} \, \Delta v,$$

where v_{wave} is the effective velocity of the pressure wave through the fluid. For a completely rigid (inelastic) pipe or borehole, the effective velocity of the pressure wave (ft/sec) is given by

$$v_{wave} = \sqrt{\frac{619}{c_e \rho}},$$

where c_e is the effective compressibility of the fluid (psi^{-1}) and ρ is the fluid density (lbm/gal).

Example 4.36. Compute the surge pressure due to inertial effects caused by downward 0.5-ft/s^2 acceleration of 10,000 ft of 10.75-in. casing with a closed end through a 12.25-in. borehole containing 10-lbm/gal mud.

Solution. The pressure surge is given by Eq. 4.99:

$$\Delta p_a = \frac{0.00162\rho a_p d^2_1}{d^2_2 - d^2_1} \Delta L$$

$$= \frac{0.00162(10)(0.5)(10.75)^2}{12.25^2 - 10.75^2}(10,000) = 271 \text{ psi}.$$

4.16 Particle Slip Velocity

The rate at which solid particles will settle out of the well fluids is often of concern to the drilling engineer. As discussed in Chap. 2, the removal of rock cuttings from the well is one of the primary functions of the drilling fluids. Unfortunately, because of the complex geometry and boundary conditions involved, analytical expressions describing particle slip velocity have been obtained only for very idealized conditions. Again the engineer is forced to depend primarily on empirical correlations and direct observations for most applications.

4.16.1 Newtonian Fluids

For a particle of foreign material falling through a fluid at its terminal velocity, the sum of the vertical forces acting on the particle must be zero—i.e., the downward force W due to gravity is exactly counterbalanced by the sum of the buoyant force F_{bo} and viscous drag F caused by the fluid. The weight W of a particle of density ρ_s and the volume V_s can be expressed by

$$W = \rho_s V_s g,$$

where g is the acceleration of gravity. The buoyant force F_{bo} can be expressed in terms of the weight of displaced liquid by

$$F_{bo} = \rho_f V_s g.$$

Summing the vertical forces we obtain

$$F = W - F_{bo} = (\rho_s - \rho_f) g V_s.$$

For a spherical particle, the volume of the particle is given by $\pi d_s^3 / 6$, and the viscous force F can be expressed in terms of the particle diameter.

$$F = (\rho_s - \rho_f) g (\pi d_s^3 / 6). \quad \dots \dots \dots \dots (4.100)$$

Stokes[28] has shown that for *creeping flow* (i.e., the streamlines of fluid movement pass smoothly about the spherical particle and there is no eddying downstream of the particle) the viscous drag F is related to the slip velocity v_s of the sphere through the fluid by

$$F = 3\pi d_s \mu v_{s1}. \quad \dots \dots \dots \dots \dots (4.101)$$

Equating Eqs. 4.100 and 4.101 and solving for the particle slip velocity yields

$$v_{s1} = \frac{1}{18} \frac{d_s^2}{\mu} (\rho_s - \rho_f) g, \quad \dots \dots \dots (4.102a)$$

which is known as Stokes' law. Converting from consistent units to convenient field units gives

$$v_{s1} = \frac{138(\rho_s - \rho_f) d_s^2}{\mu}. \quad \dots \dots \dots (4.102b)$$

Stokes' law can be used to determine the slip velocity of spherical particles through Newtonian liquids as long as turbulent eddies are not present in the wake of the particle. The onset of turbulence can be corrected to a parti-cle Reynolds number given in field units by

$$N_{Re} = \frac{928 \rho_f v_{s1} d_s}{\mu}. \quad \dots \dots \dots \dots (4.103)$$

Stokes' law is found to give acceptable accuracy for Reynolds number below 0.1. For Reynolds numbers greater than 0.1, empirically determined friction factors must be used. The friction factor in this case is defined by

$$f = \frac{F}{A E_K}, \quad \dots \dots \dots \dots \dots (4.104a)$$

where

F = force exerted on the particle due to viscous drag,
A = characteristic area of the particle, and
E_K = kinetic energy per unit volume.

The force F is defined by Eq. 4.100, and the kinetic energy per unit volume is given by

$$E_K = \frac{1}{2} \rho_f v_{s1}^2.$$

If the characteristic area A is chosen to be $(\pi d_s^2 / 4)$, then Eq. 4.104a reduces to

$$f = \frac{4}{3} g \frac{d_s}{v_{s1}^2} \frac{\rho_s - \rho_f}{\rho_f}. \quad \dots \dots \dots (4.104b)$$

The friction factor f is a function of the Reynolds number and, in the case of nonspherical shapes, a term called the *sphericity*, Ψ. Sphericity is defined as the surface area of a sphere containing the same volume as the particle divided by the surface area of the particle. A list of some shapes and their sphericity are shown in Table 4.9. The friction-factor/Reynolds-number relation is shown in Fig. 4.45. The slanted lines in Fig. 4.45 are provided to facilitate a noniterative graphical solution technique (Example 4.37).

The friction factor equation can be rearranged for the calculation of particle slip velocity. Converting Eq. 4.104b from consistent units to field units gives

$$f = 3.57 \frac{d_s}{v_{s1}^2} \left(\frac{\rho_s - \rho_f}{\rho_f} \right). \quad \dots \dots \dots (4.104c)$$

Solving this equation for the particle slip velocity yields

$$v_{s1} = 1.89 \sqrt{\frac{d_s}{f} \left(\frac{\rho_s - \rho_f}{\rho_f} \right)}. \quad \dots \dots \dots (4.104d)$$

This equation can be extended to Reynolds numbers below 0.1 if the friction factor at low Reynolds number is defined by

$$f = \frac{24}{N_{Re}}. \quad \dots \dots \dots \dots \dots (4.105)$$

The proof of this relation is left as an exercise.

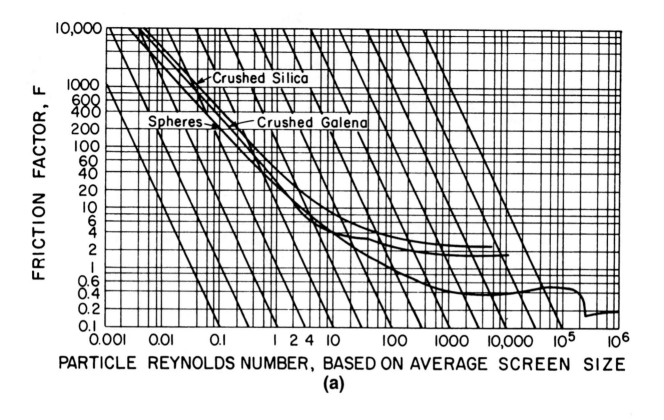

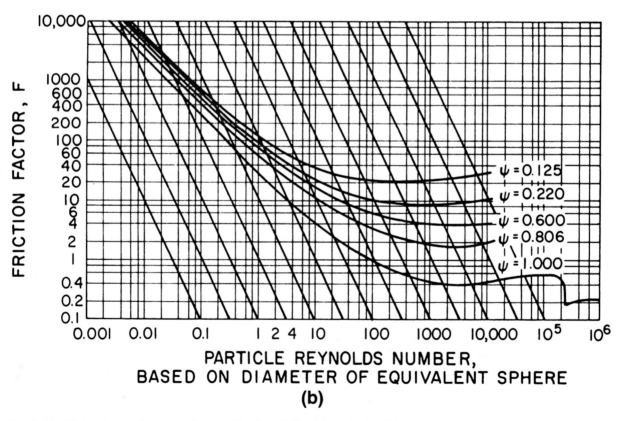

Fig. 4.45—Friction factors for computing particle-slip velocity: (a) crushed solids and spheres and (b) particles of different sphericities.

Example 4.37. How much sand having a mean diameter of 0.025 and a sphericity of 0.81 will settle to the bottom of the hole if circulation is stopped for 30 minutes. The drilling fluid is 8.33-lbm/gal water having a viscosity of 1 cp and containing about 1% sand by volume. The specific gravity of the sand is 2.6.

Solution. A slip velocity first must be assumed to establish a point on the friction factor plot shown in Fig. 4.45. Assuming Stokes' law is applicable,

$$v_{sl} = \frac{138(\rho_s - \rho_f)d_s{}^2}{\mu}$$

$$= \frac{138[2.6(8.33) - 8.33](0.025)^2}{1}$$

$$= 1.15 \text{ ft/s}.$$

This slip velocity corresponds to a friction factor and Reynolds number given by

$$f = \frac{3.57(\rho_s - \rho_f)d_s}{\rho_f v_{sl}{}^2}$$

$$= \frac{3.57[2.6(8.33) - 8.33]0.025}{8.33(1.15)^2}$$

$$= 0.108,$$

and

$$N_{Re} = \frac{928\rho_f v_{sl} d_s}{\mu} = \frac{928(8.33)(1.15)(0.025)}{1}$$

$$= 222.$$

Entering Fig. 4.45a at the point ($f = 0.108$, $N_{Re} = 222$) and moving parallel to the slant lines to the curve for $\Psi = 0.81$ yields an intersection point ($f = 5$, $N_{Re} = 40$). Thus, the slip velocity is given by

$$v_{sl} = 1.89\sqrt{\frac{d_s}{f}\left(\frac{\rho_s - \rho_f}{\rho_f}\right)}$$

$$= 1.89\sqrt{\frac{0.025}{5}\frac{[2.6(8.33) - (8.33)]}{8.33}}$$

$$= 0.17 \text{ ft/s or } 10.1 \text{ ft/min.}$$

If circulation is stopped for 30 minutes, the sand will settle from approximately

$$30(10.1) = 303 \text{ ft}$$

of the bottom portion of the hole. If the sand packs with a

porosity of 0.40, the fill on bottom is approximately

$$303\frac{(0.01)}{(1 - 0.4)} = 5 \text{ ft.}$$

4.16.2 Non-Newtonian Fluids

Particles will not settle through a static non-Newtonian fluid unless the net force on the particle due to gravity and buoyancy is sufficient to overcome the gel strength of the fluid. For a sphere, the surface area is $\pi d_s{}^2$, and the force required to break the gel is equal to $\pi d_s{}^2 \tau_g$. Equating this force to the net force given by Eq. 4.100 gives

$$\pi d_s{}^2 \tau_g = (\rho_s - \rho_f)g\left(\frac{1}{6}\pi d_s{}^3\right).$$

Thus, the gel strength τ_g needed to suspend a particle of diameter d_s is given by

$$\tau_g = \frac{d_s}{6}(\rho_s - \rho_f), \quad \dots\dots\dots\dots\dots\dots (4.106a)$$

Converting this equation from consistent units to field units gives

$$\tau_g = 10.4d_s(\rho_s - \rho_f), \quad \dots\dots\dots\dots\dots (4.106b)$$

or, conversely, the particle diameter must exceed

$$d_s = \frac{\tau_g}{10.4(\rho_s - \rho_f)} \quad \dots\dots\dots\dots\dots\dots (4.106c)$$

to settle through a fluid having gel strength τ_g.

Particles having a diameter slightly greater than that given by Eq. 4.106c will settle slowly such that the flow pattern around the sphere corresponds to creeping flow. An analytical solution for creeping flow has not been developed for non-Newtonian fluids.

Example 4.38. Compute the maximum-diameter sand particle having a specific gravity of 2.6 that can be suspended by a mud that has a density of 9 lbm/gal and a gel strength of 5 lbm/100 sq ft.

Solution. The maximum diameter of a spherical sand grain is given by Eq. 4.106c:

$$d_s = \frac{\tau_g}{10.4(\rho_s - \rho_f)} = \frac{5}{10.4[2.6(8.33) - 9.0]}$$

$$= 0.038 \text{ in.}$$

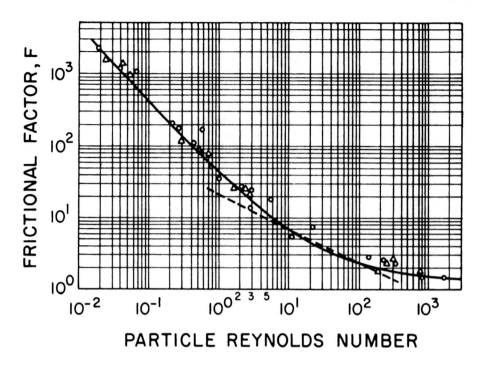

Fig. 4.46—Particle-slip velocity correlation of Moore.

4.16.3 Carrying Capacity of a Drilling Fluid

In rotary drilling operations, both the fluid and the rock fragments are moving. The situation is complicated further by the fact that the fluid velocity varies from zero at the wall to a maximum at the center of pipe. In addition, the rotation of the drillpipe imparts centrifugal force on the rock fragments, which affects their relative location in the annulus. Because of the extreme complexity of this flow behavior, drilling personnel have relied primarily on observation and experience for determining the lifting ability of the drilling fluid. In practice, either the flow rate or effective viscosity of the fluid is increased if problems related to inefficient cuttings removal are encountered. This has resulted in a natural tendency toward thick muds and high annular velocities. However, as pointed out in Section 4.13, increasing the mud viscosity or flow rate can be detrimental to the cleaning action beneath the bit and cause a reduction in the penetration rate. Thus, there may be a considerable economic penalty associated with the use of a higher flow rate or mud viscosity than necessary.

Experimental studies of drilling-fluid carrying capacity have been conducted by several authors. Williams and Bruce[29] were among the first to recognize the need for establishing the minimum annular velocity required to lift the cuttings. In 1951, they reported the results of extensive laboratory and field measurements on mud-carrying capacity. Before their work, the minimum annular velocity generally used in practice was about 200 ft/min. As a result of their work, a value of about 100 ft/min gradually was accepted. More recent experimental work by Sifferman et al.[30] indicates that while 100 ft/min may be required when the drilling fluid is water, a minimum annular velocity of 50 ft/min should provide satisfactory cutting transport for a typical drilling mud.

Several investigators have proposed empirical correlations for estimating the cutting slip velocity experienced during rotary drilling operations. While these correlations should not be expected to give extremely accurate results for such a complex flow behavior, they do provide valuable insight in the selection of drilling fluid properties and pump operating conditions. The correlations of Moore,[31] Chien,[32] and Walker and Mayes[33] have achieved the most widespread acceptance.

4.16.4 Moore Correlation

Moore[31] has proposed a procedure for applying the slip velocity equation for static fluids (Eq. 4.104d) to the average flowing condition experienced during drilling operations. The method is based on the computation of an apparent Newtonian viscosity using the method proposed by Dodge and Metzner[19] and presented in Sec. 4.11. This method involves equating the annular frictional pressure-loss expressions for the power-law and Newtonian-fluid models and then solving for the apparent Newtonian viscosity. The apparent Newtonian viscosity obtained in this manner is given by

$$\mu_a = \frac{K}{144} \left(\frac{d_2 - d_1}{\bar{v}_a} \right)^{1-n}$$

$$\left(\frac{2 + \frac{1}{n}}{0.0208} \right)^n . \quad\ldots\ldots\ldots\ldots\ldots\ldots(4.107)$$

This apparent viscosity is used in place of the Newtonian viscosity in computing the particle Reynolds number defined by Eq. 4.103. The Reynolds number computed in this manner then can be used with the friction factor correlation given in Fig. 4.46. The data shown in Fig. 4.46 was obtained using limestone and shale cuttings

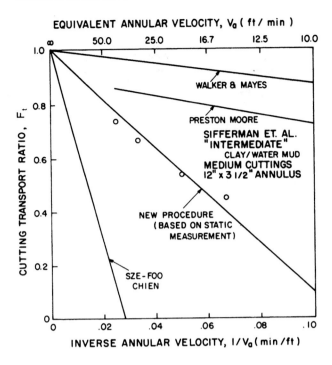

Fig. 4.47—Comparison of various methods of predicting cuttings-transport ratio.

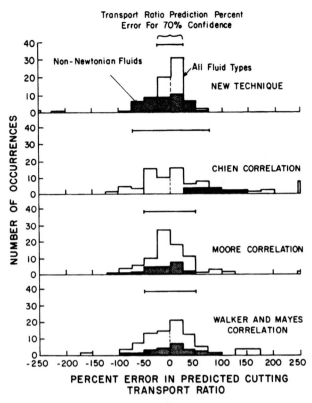

Fig. 4.48—Histograms of error in cuttings-transport ratio predictions by various methods.

from field drilling operations. For Reynolds numbers greater than 300, the flow around the particle is fully turbulent and the friction factor becomes essentially constant at a value of about 1.5. For this condition, the slip velocity (Eq. 4.104d) reduces to

$$\bar{v}_{sl} = 1.54 \sqrt{d_s \frac{\rho_s - \rho_f}{\rho_f}} \ . \ \ldots\ldots\ldots\ldots\ldots\ (4.108a)$$

For particle Reynolds numbers of 3 or less, the flow pattern is considered to be laminar and the friction factor plots as a staight line such that

$$f = \frac{40}{N_{Re}} \ .$$

For this condition, the slip velocity equation reduces to

$$\bar{v}_{sl} = 82.87 \frac{d_s^2}{\mu_a} (\rho_s - \rho_f). \ \ \ldots\ldots\ldots\ldots\ (4.108b)$$

For intermediate Reynolds numbers, the dashed line approximation shown in Fig. 4.46 is given by

$$f = \frac{22}{\sqrt{N_{Re}}} \ .$$

For this relation, the slip velocity equation reduces to

$$\bar{v}_{sl} = \frac{2.90 d_s (\rho_s - \rho_f)^{0.667}}{\rho_f^{0.333} \mu_a^{0.333}} \ . \ \ \ldots\ldots\ldots\ (4.108c)$$

This corresponds to a transitional flow pattern between laminar flow and fully developed turbulent flow.

4.16.5 Chien Correlation

The Chien correlation[32] is similar to the Moore correlation in that it involves the computation of an apparent Newtonian viscosity for use in the particle Reynolds number determination. For polymer-type drilling fluids, Chien recommends computing the apparent viscosity using

$$\mu_a = \mu_p + 5 \frac{\tau_y d_s}{\bar{v}_a} . \ \ \ldots\ldots\ldots\ldots\ldots\ (4.109)$$

However, for suspensions of bentonite in water, it is recommended that the plastic viscosity be used for the apparent viscosity. For particle Reynolds numbers above 100, Chien recommends the use of 1.72 for the friction factor. This is only slightly higher than the value of 1.5 recommended by Moore. For lower particle Reynolds numbers, the following correlation was presented:

$$\bar{v}_{sl} = 0.0075 \left(\frac{\mu_a}{\rho_f d_s} \right)$$

$$\left[\sqrt{ \frac{36,800 d_s}{\left(\frac{\mu_a}{\rho_f d_s} \right)^2} \left(\frac{\rho_s - \rho_f}{\rho_f} \right) + 1 } - 1 \right]$$

$$\ldots\ldots\ldots\ldots\ldots\ldots\ (4.110)$$

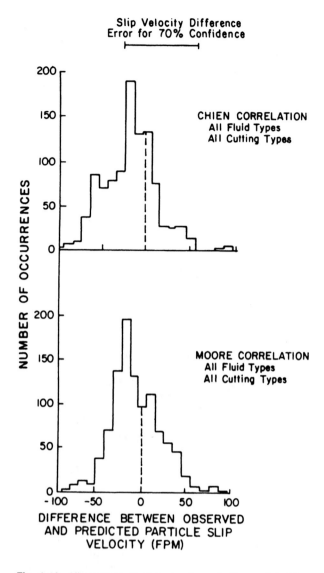

Fig. 4.49—Histograms of error in slip velocity predicted by Moore and Chien correlation.

4.16.6 Walker and Mayes Correlation

The correlation proposed by Walker and Mayes[33] uses a friction factor defined for a circular disk in flat fall (flat face horizontal) rather than for a sphere. For this particle configuration,

$$f = \frac{2gh}{\bar{v}_{sl}^2}\left(\frac{\rho_s - \rho_f}{\rho_f}\right), \quad \dots\dots\dots\dots\dots (4.111)$$

where h is the thickness of the disk. For particle Reynolds number greater than 100, the flow pattern is considered turbulent and f is assumed constant at a value of 1.12. Substituting this value of f in Eq. 4.111 and converting to field units gives

$$\bar{v}_{sl} = 2.19\sqrt{h\left(\frac{\rho_s - \rho_f}{\rho_f}\right)} \quad \dots\dots\dots (4.112)$$

This equation is also equivalent to a slip velocity equation first presented by Williams and Bruce.[29]

For the calculation of particle Reynolds numbers, Walker and Mayes developed an empirical relation for the shear stress due to particle slip. The shear stress relation is given in field units by

$$\tau_s = 7.9\sqrt{h(\rho_s - \rho_f)}. \quad \dots\dots\dots\dots (4.113)$$

The shear rate $\dot{\gamma}_s$ corresponding to the shear stress τ_s then is determined using a plot of shear stress (dial reading $\times$ 1.066) vs. shear rate (rotor speed $\times$ 1.703) obtained using a standard rotational viscometer. The apparent viscosity for use in the particle Reynolds number determination then is obtained using

$$\mu_a = 479\frac{\tau_s}{\dot{\gamma}_s}. \quad \dots\dots\dots\dots\dots (4.114)$$

If the Reynolds number is greater than 100, the slip velocity is computed using Eq. 4.112. The following correlation is provided in field units for Reynolds numbers less than 100.

$$\bar{v}_{sl} = 0.0203\,\tau_s\sqrt{\frac{d_s\dot{\gamma}}{\sqrt{\rho_f}}}, \quad \dots\dots\dots (4.115)$$

where d_s is the diameter of the disk. It should be pointed out that this apparent viscosity is based on the relative shear rate of the particle to the fluid and does not take into account fluid shear due to the liquid velocity in the annulus. Thus, the slip velocity predicted by the Walker and Mayes correlation is independent of annular velocity.

For field applications, a representative particle diameter d_s and thickness h must be estimated from a sample of rock cuttings. The diameter is chosen to represent an equivalent disk diameter, and the thickness is chosen to represent an equivalent disk thickness. For odd shapes, the diameter can be chosen as four times the projected sectional area divided by perimeter around the projected area. The thickness can be obtained by measuring the settling velocity of the cuttings in water and solving Eq. 4.112 for h.

4.16.7 Cutting Transport Ratio

Rock cuttings advance toward the surface at a rate equal to the difference between the fluid velocity and the particle slip velocity. The particle velocity relative to the surface is called the *transport velocity*.

$$v_T = \bar{v}_a - \bar{v}_{sl}.$$

The *transport ratio* is defined as the transport velocity divided by the mean annular velocity:

$$F_T = \frac{\bar{v}_T}{\bar{v}_a} = 1 - \frac{v_{sl}}{v_a}. \quad \dots\dots\dots\dots (4.116)$$

For positive cutting transport ratios, the cuttings will be transported to the surface. For a particle slip velocity of zero, the mean cutting velocity is equal to the mean annular velocity and the cutting transport ratio is unity. As

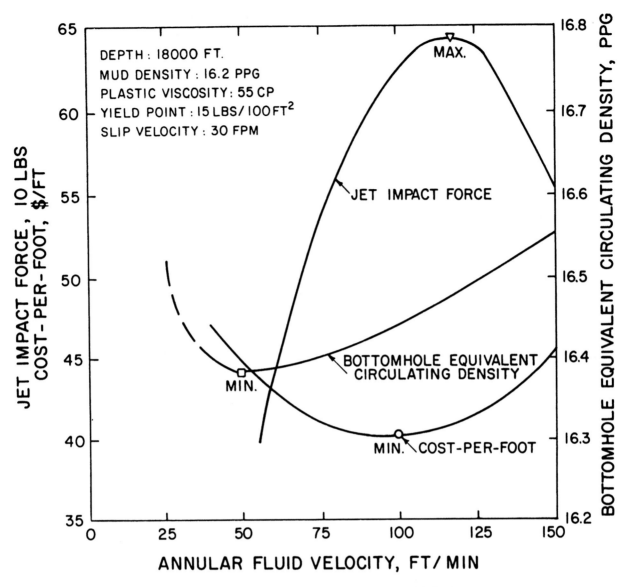

DEPTH : 18000 FT.
MUD DENSITY : 16.2 PPG
PLASTIC VISCOSITY: 55 CP
YIELD POINT : 15 LBS/100 FT^2
SLIP VELOCITY : 30 FPM

Fig. 4.50—Example of results of cuttings-transport optimization.

the slip velocity increases, the transport ratio decreases and the concentration of cuttings in the annulus en route to the surface increases. Cutting transport ratio is, thus, an excellent measure of the carrying capacity of a particular drilling fluid.

In recent work by Sample and Bourgoyne,[34] a plot of transport ratio vs. the reciprocal of the annular velocity was found to be an extremely convenient graphical technique. As can be seen from Eq. 4.116, if slip velocity is independent of annular velocity, a straight-line plot should result. The slope of the line is numerically equal to the particle slip velocity, and the x intercept is equal to the reciprocal of the particle slip velocity. The y intercept, which corresponds to an infinite annular velocity, must be equal to a transport ratio of one. Sample and Bourgoyne found that for annular velocities below about 120 ft/min, the slip velocity was essentially independent of annular velocity. Thus, by making an experimental determination of slip velocity in a static column and then drawing a line from the y intercept of 1.0 to the x intercept of $1/v_s$, an approximate representation of cutting

transport ratio could be obtained. This procedure was applied using data obtained in full-scale experiments by Sifferman *et al.*, and the results are shown in Fig. 4.47. The correlations of Moore, Chien, and Walker and Mayes also were applied, and the results are plotted in Fig. 4.47 for comparison. Note that the method of Sample and Bourgoyne gave the best results for this example. Note also that the experimental data as well as the computed results obtained with the various correlations gave essentially a straight-line plot.

Sample and Bourgoyne compiled a computer data file containing all the available published experimental data on cuttings slip velocity in flowing fluids. The data file consists of measurements obtained for different fluid types (water, polymer, and clay muds) using a variety of particle types and sizes (spheres, disks, rectangular prisms, and actual rock cuttings). The data file was used to evaluate the accuracy of the various methods for predicting cuttings transport ratio. The error histograms are shown in Fig. 4.48. The most accurate approach was found to be the use of an experimental slip velocity

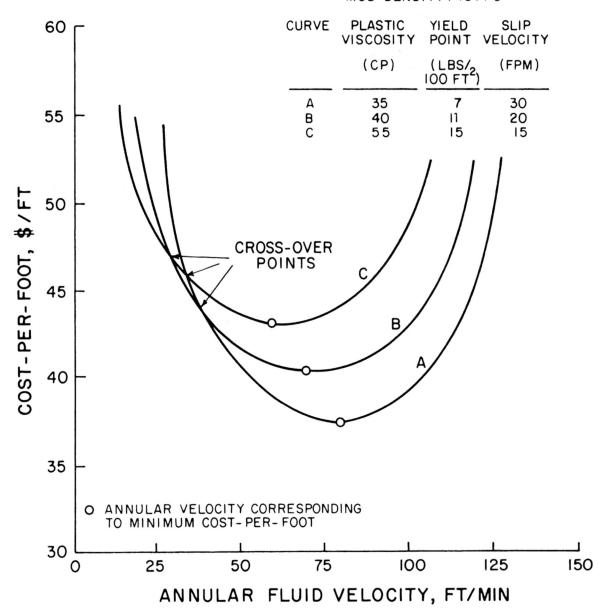

Fig. 4.51—Example effect of mud viscosity on optimal cuttings-transport and minimum cost per foot.

measurement made in a recently stirred static sample of the drilling fluid. The most accurate empirical correlation was found to be that of Moore.[31] A slip velocity error histogram for the Moore correlation is shown in Fig. 4.49.

Sample and Bourgoyne[34] also developed a computer model for estimating the optimal cuttings transport ratio for a given set of field conditions. The computer model predicts the cost per foot using Eq. 1.16 presented in Sec. 1.10 of Chap. 1. The penetration rate was predicted using the drilling model defined by Eq. 5.28 presented in Sec. 5.7 of Chap. 5. The penetration rate equation assumes an exponential decline in penetration rate with increasing bottomhole pressure. It also assumes that penetration rate increases proportionally to the jet impact force raised to the 0.3 power.

For low values of cuttings transport ratio, the concentration of cuttings in the annulus is high, causing a high effective mud density. This in turn causes a high circulating bottomhole pressure and a low penetration rate.

The volume fraction of cuttings in the mud can be determined by considering the feed rate, q_s, of cuttings at the bit, and the cuttings transport ratio, F_T. For a given bit penetration rate, (dD/dt), the feed rate of cuttings is

$$q_s = A_b \frac{dD}{dt} , \dots\dots\dots\dots\dots\dots\dots (4.117)$$

where A_b is the area cut by the bit. This equation assumes that the cuttings do not disintegrate to the size of individual grains. Otherwise, a factor of $(1-\phi)$ must be applied, where ϕ is the rock porosity. The transport velocity of the cuttings, v_T, and fluid, v_a, in a borehole annulus of area A_a is given by

$$\bar{v}_T = \frac{q_s}{A_a f_s},$$

$$\bar{v}_a = \frac{q_m}{A_a(1-f_s)},$$

where f_s is the volume fraction of cuttings in the mud and q_m is the flow rate of the mud.

Since the transport ratio, F_T, is defined as $\bar{v}_T/\bar{v}_a$, then

$$F_T = \frac{\dfrac{q_s}{A_a f_s}}{\dfrac{q_m}{A_a(1-f_s)}}.$$

Solving this expression for the volume fraction gives

$$f_s = \frac{q_s}{q_s + F_T q_m}. \quad \ldots\ldots\ldots\ldots\ldots (4.118)$$

Once the volume fraction of cuttings is known, the effective annular mud density, $\bar{\rho}$, can be computed using

$$\bar{\rho} = \rho_m(1-f_s) + \rho_s f_s, \quad \ldots\ldots\ldots\ldots (4.119)$$

where ρ_s is the average density of the cuttings.

The average density of the mud in the annulus can be decreased by increasing the mud flow rate and thus increasing the transport ratio. However, as the mud flow rate is increased, a point is reached at which bottomhole pressure begins increasing with increasing flow rate due to excessive frictional pressure losses in the annulus. Also the jet impact force available at the bit begins to decrease with increasing flow rate due to the excessive frictional pressure losses in the drillstring. Thus, there exists an optimal flow rate which results in a minimum theoretical cost per foot.

Typical results obtained using the computer model of Sample and Bourgoyne are shown in Fig. 4.50 for a 16.2-lbm/gal mud having a plastic viscosity of 40 cp, a yield point of 15, and a cuttings slip velocity of 30 ft/min while drilling an 8.5-in. hole at 18,000 ft. Note that the minimum theoretical cost per foot occurred at an annular velocity of 100 ft/min. The maximum impact force occurred at an annular velocity of 120 ft/min and the minimum equivalent circulating density occurred at an annular velocity of 50 ft/min. For most drilling conditions studied, the maximum impact force criteria tended to yield only slightly higher costs per foot than the true optimum.

Cuttings transport ratio can be increased by increasing the annular fluid velocity or by adjusting the fluid properties such as viscosity or density. In most cases, a lower theoretical cost per foot could be achieved through the use of low-viscosity fluids. Example results computed for muds of varying viscosity are shown in Fig. 4.51.

Exceptions to this rule are seen primarily for shallow, large-diameter boreholes, where the theoretical minimum cost per foot occurs to the left of the crossover region shown in Fig. 4.51. Of course, the drilling fluid must serve many other functions not considered in the cost per foot formula, and the drilling fluid viscosity used is often a result of many considerations. For example, it may be impossible to achieve a low-viscosity drilling fluid and simultaneously maintain a high fluid density which may be required to prevent a blowout.

Example 4.39. Compute the transport ratio of a 0.25-in. cutting having a specific gravity of 2.6 (21.6 lbm/gal) in a 9.0-lbm/gal clay/water mud being pumped at an annular velocity of 120 ft/min (2.0 ft/s) in a 10×5 in. annulus. Apply the correlations of Moore, Chien, and Walker and Mayes. The following data were obtained for the drilling fluid using a rotational viscometer.

Rotor Speed (rpm)	Dial Reading (degree)
3	2.0
6	3.3
100	13
200	22
300	30
600	50

Assume both the diameter and thickness of the cuttings are approximately 0.25 in.

Solution.

Moore Correlation

The consistency index K and flow behavior index n based on the 300- and 600-rpm reading normally are used with the Moore correlation.

$$n = 3.32 \log\left(\frac{50}{30}\right) = 0.74.$$

$$K = \frac{510\,(30)}{511^{0.74}} = 152 \text{ eq cp.}$$

The apparent Newtonian viscosity at an annular velocity of 120 ft/min (2 ft/s) is given by Eq. 4.107

$$\mu_a = \frac{K}{144}\left(\frac{d_2-d_1}{\bar{v}_a}\right)^{1-n}\left(\frac{2+1/n}{0.0208}\right)^n$$

$$= \frac{152}{144}\left(\frac{10-5}{2}\right)^{0.26}\left(\frac{2+1/0.74}{0.0208}\right)^{0.74}$$

$$= 58 \text{ cp.}$$

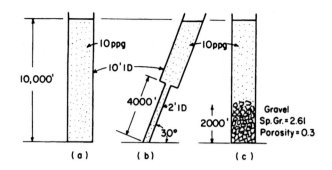

Fig. 4.52—Example fluid columns for Exercise 4.1.

Assuming an intermediate particle Reynolds number (transitional flow pattern), the slip velocity is given by Eq. 4.108c:

$$v_{sl} = \frac{2.90 d_s (\rho_s - \rho_f)^{0.667}}{\rho_f^{0.333} \mu_a^{0.333}}$$

$$= \frac{2.9(0.25)(21.6 - 9.0)^{0.667}}{9^{0.333} 58^{0.333}} = 0.49 \text{ ft/s}.$$

The particle Reynolds number is given by

$$N_{Re} = \frac{928 \rho_f \bar{v}_{sl} d_s}{\mu_a} = \frac{928(9)(0.49)(0.25)}{58} = 18.$$

Since this Reynolds number is between 3 and 300, the use of Eq. 4.108c is justified. The transport ratio for an annular velocity of 2.0 ft/s and a cutting slip velocity of 0.49 ft/s is given by

$$F_T = \frac{2.0 - 0.49}{2.0} = 0.755 \text{ or } 75.5\%.$$

Chien Correlation

For clay/water muds, Chien recommends the use of the plastic viscosity as the apparent viscosity. Using the 300- and 600-rpm readings on the Fann gives

$$\mu_p = 50 - 30 = 20 \text{ cp}.$$

Assuming a transitional flow pattern, the slip velocity is given by Eq. 4.110:

$$\bar{v}_{sl} = 0.0075 \left(\frac{\mu_a}{\rho_f d_s} \right)$$

$$\cdot \left[\sqrt{\frac{36,800 d_s}{\left(\frac{\mu_a}{\rho_f d_s} \right)^2} \left(\frac{\rho_s - \rho_f}{\rho_f} \right) + 1} - 1 \right].$$

Since

$$\left(\frac{\mu_a}{\rho_f d_s} \right) = \frac{20}{9(0.25)} = 8.89,$$

then

$$\bar{v}_{sl} = 0.0075(8.89)$$

$$\cdot \left[\sqrt{\frac{36,800(0.25)}{(8.89)^2} \left(\frac{21.6 - 9}{9} \right) + 1} - 1 \right]$$

$$= 0.79 \text{ ft/s}.$$

The particle Reynolds number for this slip velocity is given by

$$N_{Re} = \frac{928(9)(0.25)}{20} = 82.5.$$

Since this Reynolds number is less than 100, the use of Eq. 4.110 is justified. The transport ratio is given by

$$F_T = \frac{2.0 - 0.79}{2.0} = 0.605 \text{ or } 60.5\%.$$

Walker and Mayes Correlation

The shear stress of the particle falling in a static liquid is estimated using Eq. 4.113.

$$\tau_s = 7.9 \sqrt{h(\rho_s - \rho_f)}$$

$$= 7.9 \sqrt{0.25(21.6 - 9.0)}$$

$$= 14.0 \text{ lbm/100 sq ft}.$$

A shear stress of 14 lbm/100 sq ft corresponds to a Fann dial reading given by

$$\frac{14}{1.065} = 13.1.$$

This, in turn, corresponds roughly to a rotor speed of 100 rpm for the data given. The shear rate at the bob in the Fann viscometer is 1.703 times the rotor speed. Thus, we have

$$\mu_a = 479 \frac{14}{(1.703)(100)} = 39 \text{ cp}.$$

Assuming a transitional flow pattern, the slip velocity is given by Eq. 4.115.

$$v_s = 0.0203 \, \tau_s \sqrt{\frac{d_s \, \dot{\gamma}}{\sqrt{\rho}}}$$

$$= 0.0203(14) \sqrt{\frac{0.25(1.703)(100)}{\sqrt{9}}} = 1.07 \text{ ft/s}.$$

The particle Reynolds number for this slip velocity is given by

$$N_{Re} = \frac{928(9)(1.07)(0.25)}{39} = 57.$$

Since this Reynolds number is less than 100, the use of Eq. 4.115 is justified. The transport ratio is given by

$$F_T = \frac{2.0 - 1.07}{2.0} = 0.465 \text{ or } 46.5\%.$$

Exercises

4.1 Calculate the hydrostatic pressure at the bottom of the fluid column for each case shown in Fig. 4.52. *Answer*: 5,200 psig for Case A.

4.2 Calculate the mud density required to fracture a stratum at 5,000 ft if the fracture pressure is 3,800 psig. *Answer*: 14.6 lbm/gal.

4.3 An ideal gas has an average molecular weight of 20. What is the density of the gas at 2,000 psia and 600°R? *Answer*: 0.8 lbm/gal.

4.4 The mud density of a well is being increased from 10 to 12 lbm/gal. If the pump is stopped when the interface between the two muds is at depth of 8,000 ft in the drillstring, what pressure must be held at the surface by the annular blowout preventers to stop the well from flowing? What is the equivalent density in annulus at 4,000 ft after the blowout preventers are closed? *Answer*: 832 psig, 14 lbm/gal.

4.5 A well contains methane gas occupying the upper 6,000 ft of annulus. The mean gas temperature is 170°F and the surface pressure is 4,000 psia.
   a. Estimate the pressure exerted against a sand below the bottom of the surface casing at a depth of 5,500 ft. Assume ideal gas behavior. *Answer*: 4,378 psia.
   b. Calculate the equivalent density at a depth of 5,500 ft. *Answer*: 15.3 lbm/gal.

4.6 A casing string is to be cemented in place at a depth of 10,000 ft. The well contains 10-lbm/gal mud when the casing string is placed on bottom. The cementing operation is designed so that the 10-lbm/gal mud will be displaced from the annulus by (1) 500 ft of 8.5-lbm/gal mud flush, (2) 2,000 ft of 12.7-lbm/gal filler cement, and (3) 1,500 ft of 16.7-lbm/gal high-strength cement. The high-strength cement will be displaced from the casing by 9-lbm/gal brine. Calculate the minimum pump pressure required to completely displace the casing. Assume no shoe joints are used. *Answer*: 1,284 psig.

4.7 A well is being drilled at 12,000 ft using an 11-lbm/gal mud when a permeable formation having a fluid pressure of 7,000 psig is cut by the bit.
   a. If fluid circulation is stopped, what hydrostatic pressure will be exerted by the mud against the permeable formation? *Answer*: 6,864 psig.
   b. Will the well flow if the blowout preventers are left open? *Answer*: yes.
   c. Calculate the surface drillpipe pressure if the blowout preventers are closed. *Answer*: 136 psig.

   d. If formation brine of specific gravity 1.1 enters the annulus up to a depth of 11,000 ft before the blowout preventers are closed, what will be the surface annular pressure after the well is shut in? *Answer*: 231 psig.

4.8 The penetration rate of the rotary drilling process can be increased greatly by lowering the hydrostatic pressure exerted against the hole bottom. In areas where formation pressures are controlled easily, the effective hydrostatic pressure sometimes is reduced by injecting gas with the well fluids. Calculate the volume of methane gas per volume of water (standard cubic feet per gallon) that must be injected at 5,000 ft to lower the effective hydrostatic gradient of fresh water to 6.5 lbm/gal. Assume ideal gas behavior and an average gas temperature of 174°F. Neglect the slip velocity of the gas relative to the water velocity. *Answer*: 0.764 scf/gal.

4.9 A massive gas sand at 10,000 ft having a porosity of 0.30 and a water saturation of 0.35 is being drilled at a rate of 80 ft/hr using a 9.875-in. bit. The drilling mud has a density of 12 lbm/gal and is being circulated at a rate of 400 gal/min. The annular capacity is 2.8 gal/ft. The mean temperature of the well is 600°R. Ignore the slip velocity of the gas bubbles and rock cuttings.
   a. After steady-state conditions are reached, what is the effective bottomhole pressure? Assume that the gas is pure methane and behaves as an ideal gas. *Answer*: 6,205 psia.
   b. What is the equivalent mud weight in the annulus? *Answer*: 11.9 lbm/gal.
   c. What is the mud density of the mud leaving the annulus at the surface at atmospheric pressure? *Answer*: 5.8 lbm/gal.
   d. Make a plot of the density of the drilling fluid in the annulus vs. depth.
   e. Can the gas-cut mud at the surface be eliminated completely by increasing the mud density? *Answer*: no.

4.10 A well is being drilled at a vertical depth of 12,200 ft while circulating a 12-lbm/gal mud at a rate of 9 bbl/min when the well begins to flow. Fifteen barrels of mud are gained in the pit over a 5-minute period before the pump is stopped and the blowout preventers are closed. After the pressures stabilized, an initial drillpipe pressure of 400 psia and an initial casing pressure of 550 psia were recorded. The annular capacity opposite the 5-in., 19.5-lbf/ft drillpipe is 0.0775 bbl/ft. The annular capacity opposite the 600 ft of 3-in. ID drill collars is 0.035 bbl/ft. Assume $M = 16$ and $T = 600°R$.
   a. Compute the density of the kick material assuming the kick entered as a slug. *Answer*: 5.28 lbm/gal.
   b. Compute the density of the kick material assuming the kick mixed with the mud pumped during the detection time. *Answer*: 1.54 lbm/gal.
   c. Do you think that the kick is a liquid or a gas?
   d. Compute the pressure that will be observed at

the casing depth of 4,000 ft when the top of the kick zone reaches the casing if the kick is circulated from the well before increasing the mud density. *Answer*: 3,198 psia.

e. Compute the annular pressure that will be observed at the surface when the top of the kick zone reaches the surface if the kick is circulated to the surface before increasing the mud density. The annular capacity inside the casing is also 0.0775 bbl/ft. *Answer*: 1,204 psia.

f. Compute the surface annular pressure that would be observed at the surface when the top of the kick zone reaches the surface if the mud density is increased to the kill mud density before circulation of the well. *Answer*: 1,029 psia.

4.11 Using the data from Exercise 4.10, compute the pit gain that will be observed when the kick reaches the surface if the kick is circulated to the surface before increasing the mud density. Assume that the kick remains as a slug and that any gas present behaves as an ideal gas. *Answer*: 99.8 bbl.

4.12 A 20-bbl influx of 9.0-lbm/gal salt water enters a 10,000-ft well containing 10-lbm/gal mud. The annular capacity is 0.0775 bbl/ft opposite the drillpipe and 0.0500 bbl/ft opposite the 600 ft of drill collars. The capacity factor inside the drillpipe is 0.01776 bbl/ft, and the capacity factor inside the drill collars is 0.008 bbl/ft. The formation pressure is 6,000 psia. Assume $T=140°F$.

a. Compute the shut-in drillpipe and casing pressure that would be observed after the kick entered the well. *Answer*: 785 psig; 806 psig.

b. Compute the surface annular pressure that would be observed when the top of the saltwater kick reaches the surface if the mud density is increased to the kill mud density before circulation of the well. *Answer*: 208 psig.

c. Compute the total pit gain that would be observed when the top of the kick reaches the surface. *Answer*: 20 bbl.

d. Compute the surface annular pressure that would be observed if the kick was methane gas instead of brine. *Answer*: 1,040 psig.

e. Compute the surface annular pressure that would be observed if the kick was methane gas and the annular capacity was 0.1667 bbl/ft instead of 0.0775 bbl/ft. Assume the gas density is negligible. *Answer*: 684 psig.

4.13 A derrick is capable of supporting 500,000 lbf. How many feet of 13⅜-in., 72-lbm/ft casing could be supported by the derrick if the casing was run open-ended through the 12-lbm/gal mud and not "floated in"? *Answer*: 8,500 ft.

4.14 The lost circulation alarm is sounded by the well monitoring equipment, and the driller notices that the hook load slowly increases. A total of 4,000 lbm of hook load was gained before the hook load again stabilized. Calculate the depth of the fluid level in the well. The well contained 14.5-lbm/gal mud, and the drillstring consisted of 9,400 ft of 5-in. drillpipe and 600 ft of 7-in. drill collars. The drillpipe weighs 19.5 lbm/ft in air, and the drill collars weigh 100 lbm/ft in air. Casing having an internal diameter of 12.347 in. is set at 4,000 ft. The open hole having a diameter of 9,875 in., extends from the casing seat to the total well depth of 10,000 ft. Steel has a density of 490 lbm/cu ft. *Answer*: 927 ft.

4.15 A 20,000-ft drillstring is composed of 19,400 ft of 5-in., 19.5-lbm/ft drillpipe and 600 ft of 8.0- × 3.0-in. drill collars. The mud density is 18 lbm/gal. Assume the drill string is suspended off bottom.

a. Construct a graph of axial tension (or compression) vs. depth.

b. Determine the maximum weight that can be applied to the bit without causing a bending moment in the drillstring. *Answer*: 64 K-lbf.

c. Assume that the driller slacks off the hook load so that the computed weight is applied to the bit. Construct a graph of axial tension (or compression) vs. depth.

d. Lubinski[4] defines the *neutral point of buckling* as the point at which the axial compressive stress is equal to the external hydrostatic pressure. Is this true for this example? *Answer*: yes; $p=\sigma=18,150$ psi.

4.16 The maximum weight to be applied to the bit during the next interval of hole is 60,000 lbm. The drillstring will be composed of 5-in., 19.5-lbm/ft drillpipe and 2.75×8.0 in. drill collars and the maximum mud density anticipated is 13 lbm/gal. Compute the minimum length of drill collars required to prevent a buckling tendency in the drillpipe. *Answer*: 496 ft.

4.17 A 10-lbm/gal mud is flowing at a steady rate of 160 gal/min down a drillpipe having an internal diameter of 4.33 in. and an external diameter of 5 in. The diameter of the hole is 10 in.

a. Compute the average flow velocity in the drillpipe. *Answer*: 3.49 ft/s.

b. Compute the average flow velocity in the annulus opposite the drillpipe. *Answer*: 0.871 ft/s.

4.18 Determine the pressure at the bottom of the drill collars if the frictional loss in the drillstring is 900 psi, the flow rate is 350 gal/min, the mud density is 10-lbm/gal, and the well depth is 8,000 ft. The internal diameter of the drill collars is 2.75 in., and the pressure developed by the pump is 2,600 psig. *Answer*: 5,860 psig.

4.19 A 10-lbm/gal mud is being circulated at a rate of 600 gal/min. If the bit contains two ¹⁵⁄₃₂-in. nozzles and one ¹⁷⁄₃₂-in. nozzle and the pump pressure is 3,000 psi, what is the total frictional pressure loss in the well system? *Answer*: 1,969 psig.

4.20 Compute the pump pressure required to pump a 9-lbm/gal fluid from sea level to an elevation of 1,000 ft. Assume that inertial and viscous (frictional) pressure changes are negligible. *Answer*: 468 psig.

4.21 A pump is being operated at a rate of 800 gal/min and a pressure of 3,000 psig. The density of the drilling fluid is 15 lbm/gal, and the total nozzle area of the bit is 0.589 sq in.

a. Compute the power developed by the pump. *Answer*: 1,400 hp.

b. Compute the power loss to viscous effects. What happens to this energy? *Answer*: 210 hp.

c. Compute the impact force of the jets of fluid against the bottom of the hole. *Answer*: 2,709 lbf.

4.22 Define (1) Newtonian, (2) non-Newtonian, (3) shear stress, (4) shear rate, (5) pseudoplastic, (6) dilatant, (7) thixotropic, and (8) rheopectic.

4.23 A shear stress of 5 dynes/cm^2 is measured in a fluid for a shear rate of 20 seconds^{-1}. Compute the Newtonian or apparent viscosity in centipoise. *Answer*: 25 cp.

4.24 The following shear/stress-rate behavior was observed.

Shear Rate (seconds^{-1})	Shear Stress (dynes/cm^2)
20	11.0
30	15.2
40	19.1
50	22.9
60	26.5

a. Make a plot of shear stress (ordinate) vs. shear rate (abscissa) on Cartesian paper. Make a plot of shear stress (ordinate) vs. shear rate (abscissa) on log-log paper.

b. Can the fluid behavior be accurately modeled by the Newtonian, Bingham plastic, or power-law model?

c. Compute the apparent viscosity for each shear rate. *Answer*: 55 cp at 20 seconds^{-1}.

d. Compute the yield point and plastic viscosity using data taken at shear rates of 20 and 60 seconds^{-1}. *Answer*: $\tau_y = 3.3$ dynes/cm^2; $\mu_p = 39$ cp.

e. Compute the consistency index and flow-behavior index using data taken at shear rates of 20 and 60 seconds^{-1}. *Answer*: $n = 0.8$; $K = 100$ eq cp.

4.25 A rotational viscometer contains a fluid that gives a dial reading of 20 at a rotor speed of 300 rpm and a dial reading of 40 at a rotor speed of 600 rpm.

a. Is the fluid in the Fann viscometer a Newtonian fluid? Why?

b. What is the apparent viscosity at 300 rpm in poise, centipoise, and lbf-s/sq ft? *Answer*: 20 cp; 4.17×10^{-4} lbf-s/sq ft.

4.26 A rotational viscometer contains a fluid that has an apparent viscosity of 10 cp at a rotor speed of 600 rpm.

a. Make a plot of fluid velocity vs. radius for radii between the bob radius r_1 and the rotor radius r_2. Is the velocity profile approximately linear as in

the case of the stationary and moving plates discussed in the previous section?

b. Make a plot of shear rate $\dot{\gamma} = r(d\omega/dr)$ vs. radius for radii between the bob radius and the rotor radius r_2. Is shear rate approximately constant between r_1 and r_2?

c. Make a plot of shear stress τ vs. radius for radii between the bob radius r_1 and the rotor radius r_2.

4.27 Determine the shear rate at the bob of a rotational viscometer for rotor speeds of 3, 6, 100, 200, 300, and 600 rpm for Newtonian fluids. *Answer*: 5.11 seconds^{-1} at 3 rpm.

4.28 A rotational viscometer containing a Newtonian fluid gives a dial reading of five at a rotor speed of 6 rpm. Compute the viscosity of the fluid in centipoise. *Answer*: 250 cp.

4.29 A rotational viscometer contains a Bingham plastic fluid that gives a dial reading of 22 at a rotor speed of 300 rpm and a dial reading of 39 at a rotor speed of 600 rpm. Compute the plastic viscosity and yield point of the fluid. *Answer*: 17 cp and 5 lbf/100 sq ft.

4.30 Derive equations for obtaining the plastic viscosity and yield point from rotational dial readings obtained at (a) 3 and 6 rpm and (b) 100 and 200 rpm.

4.31 A rotational viscometer contains a power-law fluid that gives a dial reading of 22 at a rotor speed of 300 rpm and a dial reading of 39 at a rotor speed of 600 rpm. Compute the consistency index and flow-behavior index of the fluid. *Answer*: $K = 65$ eq cp; $n = 0.825$.

4.32 Derive equations for obtaining the consistency index and flow-behavior index from rotational dial readings obtained at (a) 3 to 6 rpm and (b) 100 and 200 rpm.

4.33 A fluid is reported as having a plastic viscosity of 40 cp and a yield point of 7 lbf/100 sq ft. Compute a consistency index and flow-behavior index for this fluid. *Answer*: $n = 0.888$; $K = 94$ eq cp.

4.34 A 40-cp oil is flowing through 9,000 ft of 3-in. tubing at a rate of 2,500 B/D. Compute the frictional pressure loss in the tubing. Assume that the flow pattern is laminar. *Answer*: 88 psig.

4.35 A 9.2-lbm/gal Newtonian fluid having a viscosity of 30 cp is being circulated at a rate of 100 gal/min in a vertical well containing a 6-in.-ID casing and a 4.5-in.-OD drillstring. Compute the static and circulating pressure in the annulus of 15,000 ft. Assume that the flow pattern is laminar. *Answer*: 7,176 psig; 7,694 psig.

4.36 A Bingham plastic fluid has a plastic viscosity of 50 cp and a yield point of 12 lbf/100 sq ft. Assuming that the flow pattern is laminar, compute the frictional pressure gradient resulting from (a) a flow rate of 50 gal/min through a drillstring having a 3.826-in. ID and (b) a flow rate of 90 gal/min through a 10- × 7-in. annulus. *Answer:* (a) 0.0171 psi/ft, (b) 0.0240 psi/ft.

4.37 The following readings were taken with a rotational viscometer.

N	θ_N
3	4.5
6	5.5
100	9.0
200	14.5
300	19.0
600	32.5

a. Using the power-law model, compute the frictional pressure gradient resulting from a 50-gal/min flow rate in a 2.9-in.-ID drillpipe. Assume the flow pattern is laminar. *Answer:* 0.00586 psi/ft.

b. Compute the shear rate at the wall of the drillpipe. What two rotational readings would you expect to give the best n and K factors for the pressure gradient computation. *Answer:* 86 seconds^{-1}.

c. Make a plot of flow velocity vs. pipe radius and a plot of shear stress vs. pipe radius.

4.38 A power-law fluid has a consistency index of 90 eq cp and a flow behavior index of 0.9. Compute the frictional pressure gradient resulting from a 120-gal/min flow rate in a 9- × 11-in. annulus. Assume the flow pattern is laminar. *Answer:* 0.0182 psi/ft.

4.39 Compute the equivalent density below the casing seat at 4,000 ft when a mud having a density of 10 lbm/gal and a gel strength of 70 lbm/100 sq ft just begins to flow. The casing has an internal diameter of 7.825 in., and the drillpipe has an external diameter of 5 in. Discuss how the pressure required to start circulation can be reduced. *Answer:* 11.6 lbm/gal.

4.40 A well is being drilled at a depth of 5,000 ft using water having a density of 8.33 lbm/gal and a viscosity of 1 cp as the drilling fluid. The drillpipe has an external diameter of 4.5 in. and an internal diameter of 3.826 in. The diameter of the hole is 6.5 in. The drilling fluid is being circulated at a rate of 500 gal/min. Assume a relative roughness of zero.

a. Determine the flow pattern in the drillpipe. *Answer:* turbulent.

b. Determine the frictional pressure loss per 1,000 ft of drillpipe. *Answer:* 51.3 psi/1,000 ft.

c. Determine the flow pattern in the annular opposite the drillpipe. *Answer:* turbulent.

d. Determine the frictional pressure loss per 1,000 ft of annulus. *Answer:* 72.9 psi/1,000 ft.

4.41 Work Exercise 4.40 for a Bingham plastic fluid having a density of 10 lbm/gal, a plastic viscosity of 25 cp, and a yield point of 5 lbf/100 sq ft. *Answer:* turbulent; 132 psi/1,000 ft; 185 psi/1,000 ft.

4.42 Work Exercise 4.40 for a power-law fluid having a density of 12 lbm/gal, a flow behavior index of 0.75, and a consistency index of 200 eq cp. *Answer:* turbulent; 144 psi/1,000 ft; turbulent; 205 psi/1,000 ft.

4.43 Show that Eq. 4.66d can be extended to the laminar flow region by the substitution of Eq. 4.67d for the friction factor f.

4.44 A 15-lbm/gal cement slurry has a flow-behavior index of 0.3 and a consistency index of 9,000 eq cp. Compute the flow rate required for turbulent flow in an 8.097-×4.5-in. annulus. Also estimate the frictional pressure loss and the shear rate at the wall for this flow rate. *Answer:* For $N_{Rec} = 3,200$ from Fig. 4.34, $\bar{v} = 11.5$ ft/s, and $q = 1,294$ gal/min; 130 psi/1,000 ft; $\dot{\gamma} = 817$.

4.45 Show that the conditions required to maximize the bit Reynolds number given by $(\rho \bar{v}_n d_n)/\mu_a$ are identical to the conditions required to maximize the hydraulic impact force.

4.46 The bit currently in use has three $^{12}\!/_{32}$-in. nozzles. The driller has recorded that when the 10-lbm/gal mud is pumped at a rate of 500 gal/min, a pump pressure of 3,000 psig is observed, and when the pump is slowed to a rate of 250 gal/min, a pump pressure of 800 psi is observed. The pump is rated at 1,000 hp and has an overall efficiency of 0.9. The minimum flow rate to lift the cuttings is 240 gal/min. The maximum allowable surface pressure is 3,000 psi.

a. Determine the proper pump operating conditions and bit nozzle sizes for maximum bit horsepower for the next bit run. *Answer:* $q = 514$ gal/min; $A_t = 0.345$.

b. What bit horsepower will be obtained at the conditions selected? *Answer:* 611 hp.

c. What impact force will be obtained at the conditions selected? *Answer:* 1,270 lbf.

d. What nozzle velocity will be obtained at the conditions selected? *Answer:* 477 ft/s.

4.47 Work Exercise 4.46 for maximum impact force rather than maximum bit horsepower. *Answer:* Answers are the same for conditions given because $q = q_{max}$ for both cases.

4.48 Work Exercise 4.46 for maximum nozzle velocity rather than maximum bit horsepower. *Answer:* $q = 240$ gal/min; $A_t = 0.136$; $v_n = 553$ ft/s.

4.49 The following well conditions are given.

Well depth, ft	15,000
Drillpipe OD, in.	5
lbm/ft	19.5
in. effective ID	4.33
Drill collars (8-in.-OD×3-in.-ID), ft	600
Casing (J-55) set at 4,500 ft, in.	10.75
lbm/ft	40.5

Bit diameter, in.   9.875
  (assume average hole size of 10 in.)
Nozzles (three), in. $^{13}/_{32}$
Surface equipment
  Double-acting duplex pump, hp   1,500
  Liner, in.   6.5
  Stroke, in.   18
  Rods, in.   2.5
  Overall efficiency, %   85
  Maximum surface pressure, psi   3,500
  Surface piping equivalent to
    200 ft of drillpipe
Mud properties
  Density, lbm/gal   12

Rotational speed	3	6	100	200	300	600
Reading	1.8	3	10	15	20	36

Minimum annular velocity, ft/min   120

a. Compute the equivalent circulating density at depths of 4,000, 14,000, and 15,000 ft corresponding to an annular velocity opposite the drillpipe of 120 ft/min using the Bingham plastic model. *Answer*: 12.10 lbm/gal; 12.10 lbm/gal; 12.13 lbm/gal.

b. Compute the total nozzle area of the bit needed for maximum bit horsepower. *Answer*: 0.324 sq in.

4.50 It is desired to estimate the proper pump operating conditions and bit nozzle sizes for maximum jet impact force at 1,000-ft increments for an interval of the well between surface casing at 4,000 ft and intermediate casing at 10,000 ft. The well plan calls for the following conditions.

Depth (ft)	Mud Density (lbm/gal)	Plastic Viscosity (cp)	Yield Point (lbf/100 sq ft)
5,000	10.0	25	5.0
6,000	10.0	25	5.0
7,000	10.0	25	5.0
8,000	11.0	30	6.0
9,000	12.0	35	8.0

Pump horsepower, hp   1,250
Pump efficiency   0.85
Maximum pump pressure, psig   3,200
Minimum annular velocity, ft/min   100
Drillpipe, in. OD × in. ID   4.5×3.826
Drill collars (7.5-in.-OD ×
    2.75-in.-ID), ft   600
Hole size (washed out), in.   10.05
Casing ID, in.   10.05
Surface equipment equivalent to 140 ft
    of drillpipe

4.51 A 50-bbl kick is taken at a depth of 14,000 ft while drilling with a 15-lbm/gal mud. After the pressures stabilized, an initial drillpipe pressure of 1,200 psig and an initial casing pressure of 2,600 psig were recorded. The drillstring is composed of 13,000 ft of drillpipe having an internal capacity of 0.01422 bbl/ft and 1,000 ft of drill collars having an internal capacity of 0.0087 bbl/ft. A pump pressure of 1,100 psig previously was recorded at a reduced rate of 25 strokes/min. The pump factor is 0.24 bbl/stroke.

a. Compute a drillpipe pressure schedule for the circulate-and-weight method of well control.

b. Compute a drillpipe pressure schedule for the wait-and-weight method of well control.

c. Compute the equivalent density at the casing seat of 13,000 ft just after the well was shut in. *Answer*: 18.9 lbm/gal.

4.52 Calculate the equivalent density below the bottom joint of 6,000 ft of 11.75-in. casing (having an 11-in. ID) if the casing is being lowered at a rate of 1 ft/s in a 12.25-in. hole containing 10-lbm/gal fluid with a viscosity of 3 cp. Perform the calculation assuming the flow pattern is laminar and the casing has a closed end. *Answer*: 12.8 lbm/gal.

4.53 Calculate the surge pressure in a 7.875-in. hole below a 12,000-ft drillstring composed of 11,400 ft of 5-in. drillpipe (4.3-in. ID), 600 ft of 6.25-in. drill collars (2.75-in. ID) and a bit containing three $^{12}/_{32}$-in. nozzles. The drillstring is being lowered through a 14-lbm/gal drilling fluid having a flow-behavior index of 0.8 and a consistency index of 250 eq cp at a maximum rate of 3.5 ft/s.

4.54 Calculate the equivalent density below the bottom joint of 10,000 ft of 7-in. casing being lowered at a rate of 3 ft/s in an 8.5-in. hole. Assume that the bottom of the casing string is closed and that the mud has a density of 13 lbm/gal, a plastic viscosity of 28 cp, and a yield point of 9 lbf/100 sq ft.

4.55 Compute the settling velocity of sand having a specific gravity of 2.6, a mean diameter of 0.018 in., and a sphericity of 0.81 through water having a density of 8.33 lbm/gal and a viscosity of 1 cp. *Answer*: 0.126 ft/s.

4.56 Compute the maximum diameter sand particle that can be suspended in a 12-lbm/gal fluid having a gel strength of 7 lbf/100 sq ft. *Answer*: 0.070 in.

4.57 Compute the transport ratio of a 0.375-in. cutting (both diameter and thickness) having a specific gravity of 2.5 in a 14-lbm/gal mud being pumped at an annular velocity of 90 ft/min in a 6.5- ×3.5-in. annulus. The following data were obtained for the drilling fluid using a rotational viscometer.

Rotor Speed (rpm)	Dial Reading (degrees)
3	4.0
6	6.6
100	26.0
200	44.0
300	60.0
600	100.0

(a) Compute the transport ratio using the Moore correlation.

(b) Compute the transport ratio using the Chien correlation.

(c) Compute the transport ratio using the Walker and Mayes correlation.

4.58 Compute the mud density in the annulus for the cutting transport ratios of 90, 50, and 5% if the mud density entering the well is 9 lbm/gal and the drilling rate is 100 ft/hr. Assume a flow rate of 600 gal/min, a bit size of 9.875 in., and a rock bulk density of 2.2 g/cm^3. *Answers*: 9.11, 9.20, 10.7 lbm/gal.

References

1. Burcik, E.J.: *Principles of Petroleum Reservoir Fluids*, John Wiley and Sons Inc., London (1957).
2. Standing, M.B. and Katz, D.L.: "Density of Natural Gases," *Trans.*, AIME (1942) **146**, 140–149.
3. Carr, N.L., Kobayashi, R., and Burrows, D.: "Viscosity of Hydrocarbon Gases under Pressure," *Trans.*, AIME (1954) **201**, 264–272.
4. Goins, W.C.: "Better Understanding Prevents Tubing Buckling Problems," *World Oil*, Part 1 (Jan. 1980) 101, Part 2 (Feb. 1980) 35.
5. Eckel, J.R. and Bielstein, W.J.: "Nozzle Design and Its Effect on Drilling Rate and Pump Operations," *Drill. and Prod. Prac.*, API (1951) 28–46.
6. Dodge, N.A.: "Friction Losses in Annular Flow," ASME Paper No. 63-WA-11 (1963).
7. Annis, M.R.: "High-Temperature Flow Properties of Water-Base Drilling Fluids," *J. Pet. Tech.* (Aug. 1967) 1074–80; *Trans.*, AIME, **240**.
8. Butts, H.B.: "The Effect of Temperature on Pressure Losses During Drilling," MS thesis, Louisiana State U., Baton Rouge (May 1972).
9. Lamb, H.: *Hydrodynamics*, sixth edition, Dover Publications, New York City (1945).
10. Laird, W.M.: "Slurry and Suspension Transport," *Ind. and Eng. Chem.* (1957) 138.
11. Fredrickson, A.G. and Bird, R.B.: "Non-Newtonian Flow in Annuli," *Ind. and Eng. Chem.* (1958) 347.
12. Reynolds, O.: "An Experimental Investigation of the Circumstances Which Determine Whether the Motion of Water Shall be Direct or Sinuous, and the Laws of Resistance in Parallel Channels," *Trans.*, Royal Soc. London (1883) **174**.
13. Colebrook, C.F.: "Turbulent Flow in Pipes, with Particular Reference to the Transition Region Between the Smooth and Rough Pipe Laws," *J. Inst. Civil Engs.*, London (1938–39).
14. Streeter, V.L.: *Fluid Mechanics*, McGraw-Hill Book Co., New York City (1962).
15. Cullender, M.H. and Smith, R.V.: "Practical Solution of Gas-Flow Equations for Wells and Pipelines with Large Temperature Gradients," *Trans.*, AIME (1956) **207**, 281–87.
16. Blasius, H.: "Das Aehnlichkeitsgesetz bei Reibungsvorgängen in Flüssigkeiten," VDL Forsch (1913) 131.
17. Crittendon, B.C.: "The Mechanics of Design and Interpretation of Hydraulic Fracture Treatment," *J. Pet. Tech.* (Oct. 1959) 21–29.
18. Hanks, R.W. and Pratt, D.R.: "On the Flow of Bingham Plastic Slurries in Pipes and Between Parallel Plates," *Soc. Pet. Eng. J.* (Dec. 1967) 342–46; *Trans.*, AIME, **240**.
19. Dodge, D.G. and Metzner, A.B.: "Turbulent Flow of Non-Newtonian Systems," *AIChE J.* (1959) **5**, 189.
20. Speer, J.W.: "A Method for Determining Optimum Drilling Techniques," *Drill. and Prod. Prac.*, API (1958) 130–147.
21. Kendal, W.A. and Goins, W.C.: "Design and Operation of Jet Bit Programs for Maximum Hydraulic Horsepower, Impact Force, or Jet Velocity," *Trans.*, AIME (1960) **219**, 238–247.
22. McLean, R.H.: "Velocities, Kinetic Energy and Shear in Crossflow Under Three-Cone Jet Bits," *J. Pet. Tech.* (Dec. 1965) 1443–48; *Trans.*, AIME, **234**.
23. Eckel, J.R.: "Microbit Studies of the Effect of Fluid Properties and Hydraulics on Drilling Rate," *Trans.*, AIME, **240**.
24. Sutko, A.: "Drilling Hydraulics—A Study of Chip Removal Force Under a Full-Size Jet Bit," *Soc. Pet. Eng. J.* (Aug. 1973) 233–238; *Trans.*, AIME, **255**.
25. Burkhardt, J.A.: "Wellbore Pressure Surges Produced by Pipe Movement," *J. Pet. Tech.* (June 1961) 595–605; *Trans.*, AIME, **222**.
26. Schuh, F.J.: "Computer Makes Surge Pressure Calculations Useful," *Oil and Gas J.* (Aug. 3, 1964) 96.
27. Fontenot, J.E. and Clark, R.E.: "An Improved Method for Calculating Swab and Surge Pressures and Circulating Pressures in a Drilling Well," *Soc. Pet. Eng. J.* (Oct. 1974) 451–62.
28. Stokes, G.G.: *Trans.*, Cambridge Phil. Soc. (1845, 1851) **8, 9**.
29. Williams, C.E. and Bruce, G.H.: "Carrying Capacity of Drilling Muds," *Trans.*, AIME (1951) **192**, 111–120.
30. Sifferman, T.R., Myers, G.M., Haden, E.L., and Wahl, H.A.: "Drill Cutting Transport in Full-Scale Vertical Annuli," *J. Pet. Tech.* (Nov. 1974) 1295–1302.
31. Moore, P.L.: *Drilling Practices Manual*, Petroleum Publishing Co., Tulsa (1974).
32. Chien, S.F.: "Annular Velocity for Rotary Drilling Operations," *Proc.*, SPE Fifth Conference on Drilling and Rock Mechanics, Austin, TX (Jan. 5-6, 1971) 5–16.
33. Walker, R.E. and Mayes, T.M.: "Design of Muds for Carrying Capacity," *J. Pet. Tech.* (July 1975) 893–900; *Trans.*, AIME, **259**.
34. Sample, K.J. and Bourgoyne, A.T.: "Development of Improved Laboratory and Field Procedures for Determining the Carrying Capacity of Drilling Fluids," paper SPE 7497 presented at the 1978 SPE Annual Technical Conference and Exhibition, Houston, Oct. 1-4.

Nomenclature

a = acceleration

a,b = coefficients of gas-cut mud equation

A = area

c = compressibility

C = capacity

C_d = discharge coefficient

d = diameter

D = depth

e = base of natural logarithm

E = internal energy

E_k = kinetic energy per unit volume

f = fractional volume

f_a = fractional flow in annulus, q_a/q_t

F = force

F_s = stability force

F_T = transport ratio; also tension force

F_{wv} = specific weight

g = acceleration of gravity

G = gain

I = moment of inertia

K = clinging constant; also consistency index

L = length

m = mass

M = molecular weight, also moment

n = number (of moles of gas, or of components or segments), also flow behavior index

N_{He} = Hedstrom number

N_{Re} = Reynolds number

N_{Rec} = critical Reynolds number

N_V = moles of gas dispersed in a unit volume (gal) of liquid

p = pressure

P_H = hydraulic power

q = flow rate

Q = heat energy per unit mass

r = radius

r_H = hydraulic radius

R = universal gas constant

t = time

t_d = time of kick detection

T = temperature; also torque

v = velocity

V = volume

$\bar{V}$ = specific volume

w = weight per foot

W = total weight, also work

x,y = spatial coordinates

z = gas deviation factor

α = d_1/d_2

$\dot{\gamma}$ = rate of shear

θ = dial reading of rotational viscometer

μ = viscosity

μ_a = apparent viscosity

μ_p = plastic viscosity

ρ = density

σ_r = radial stress

σ_t = tangential stress

σ_z = axial stress

τ = shear stress

τ_g = gel strength

τ_y = yield point

ϕ = porosity

ω = angular velocity

Subscripts

0, 1, 2, 3... = locations

a = annular, also acceleration

b = bit

bo = bouyancy

c = casing

d = parasitic

dc = drill collars

dca = drill collar annulus

dp = drillpipe

dpa = annulus opposite drillpipe

dpf = flowing or circulating drillpipe

e = effective, also equivalent

f = fluid, also fraction

g = gas

h = hook; also hydrostatic

j = jet

k = kick

m = mud

n = nozzle; also nominal

p = pump

s = steel; also solid; also surface equipment

sl = slip

T = transport

wave = pressure wave

$^-$ = average

SI Metric Conversion Factors

bbl	$\times$ 1.589 873	E$-$01	= m^3
bbl/ft	$\times$ 5.216 119	E$-$01	= m^3/m
cp	$\times$ 1.0*	E$-$03	= Pa·s
dyne/cm^2	$\times$ 1.0*	E$-$02	= mN/cm
ft	$\times$ 3.048*	E$-$01	= m
ft/bbl	$\times$ 1.917 134	E$+$00	= m/m^3
gal	$\times$ 3.785 412	E$-$03	= m^3
in.	$\times$ 2.54*	E$+$00	= cm
lbf/sq ft	$\times$ 4.788 026	E$-$02	= kPa
lbm	$\times$ 4.535 924	E$-$01	= kg
lbm/cu ft	$\times$ 1.601 846	E$+$01	= kg/m^3
lbm/ft	$\times$ 1.488 164	E$+$00	= kg/m
lbm/gal	$\times$ 1.198 264	E$+$02	= kg/m^3
lbm/sq ft	$\times$ 4.882 428	E$+$00	= kg/m^2
psi	$\times$ 6.894 757	E$+$00	= kPa
psi/ft	$\times$ 2.262 059	E$+$01	= kPa/m
sq ft	$\times$ 9.290 304*	E$-$02	= m^2
sq in.	$\times$ 6.451 6*	E$+$00	= cm^2

*Conversion factor is exact.

Chapter 5
Rotary Drilling Bits

The purpose of this chapter is to introduce the student to the selection and operation of drilling bits. Included in the chapter are discussions of (1) various bit types available, (2) criteria for selecting the best bit for a given situation, (3) standard methods for evaluating dull bits, (4) factors affecting bit wear and drilling speed, and (5) optimization of bit weight and rotary speed.

The process of drilling a hole in the ground requires the use of drilling bits. Indeed, the bit is the most basic tool used by the drilling engineer, and the selection of the best bit and bit operating conditions is one of the most basic problems that he faces. An extremely large variety of bits are manufactured for different situations encountered during rotary drilling operations. It is important for the drilling engineer to learn the fundamentals of bit design so he can understand fully the differences among the various bits available.

5.1 Bit Types Available

Rotary drilling bits usually are classified according to their design as either *drag bits* or *rolling cutter bits*. All drag bits consist of fixed cutter blades that are integral with the body of the bit and rotate as a unit with the drillstring. The use of this type of bit dates back to the introduction of the rotary drilling process in the 19th century. Rolling cutter bits have two or more cones containing the cutting elements, which rotate about the axis of the cone as the bit is rotated at the bottom of the hole. A two-cone rolling cutter bit was introduced in 1909.

5.1.1 Drag Bits

The design features of the drag bit include the number and shape of the cutting blades or stones, the size and location of the water courses, and the metallurgy of the bit and cutting elements. Drag bits drill by physically plowing cuttings from the bottom of the borehole much like a farmer's plow cuts a furrow in the soil. This type

of bit includes bits with steel cutters (Fig. 5.1), diamond bits (Fig. 5.2), and polycrystalline diamond (PCD) bits (Fig. 5.3). An advantage of drag bits over rolling cutting bits is that they do not have any rolling parts, which require strong, clean bearing surfaces. This is especially important in the small hole sizes, where space is not available for designing strength into both the bit cutter elements and the bearings needed for a rolling cutter. Also, since drag bits can be made from one solid piece of steel, there is less chance of bit breakage, which would leave *junk* in the bottom of the hole. Removing junk from a previous bit can lead to additional trips to the bottom and thus loss of considerable rig time.

Drag bits with steel cutter elements such as a fishtail bit perform best relative to other bit types in uniformly soft, unconsolidated formations. As the formations become harder and more abrasive, the rate of bit wear increases rapidly and the drilling rate decreases rapidly. This problem can be reduced by changing the shape of the cutter element and reducing the angle at which it intersects the bottom of the hole. Also, in soft formations that tend to be "gummy," the cuttings may stick to the blades of a drag bit and reduce their effectiveness. This problem can be reduced by placing a jet so that drilling fluid impinges on the upper surface of the blade. Because of the problems of rapid dulling in harder rocks and bit cleaning in gummy formations, drag bits with steel cutting elements largely have been displaced by other bit types in almost all areas.

Diamond bits perform best relative to other bit types in nonbrittle formations that have a plastic mode of failure for the stress conditions present at the bottom of the hole. The face or crown of the bit consists of many diamonds set in a tungsten carbide matrix. Under proper bit operation only the diamonds contact the hole bottom, leaving a small clearance between the matrix and the hole bottom. Fluid courses are provided in the matrix to direct the flow of drilling fluid over the face of the bit. These

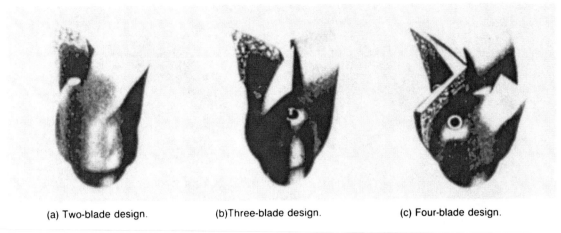

(a) Two-blade design.   (b)Three-blade design.   (c) Four-blade design.

Fig. 5.1—Example steel cutter drag bits dressed with tungsten carbide.

courses must be sized small enough so that some of the fluid is forced to flow between the matrix and the hole bottom, thereby cleaning and cooling the diamonds.

An important design feature of a diamond bit (Figs. 5.4A through 5.4C) is its shape or *crown profile* (Fig. 5.4C). A bit with a long taper assists in drilling a straight hole and allows the use of higher bit weights. On the other hand, a short taper is easier to clean because the available hydraulic energy can be concentrated over less surface area. A more concave bit face can be used in directional drilling applications to assist in increasing the angle of deviation of the borehole from vertical.

The size and number of diamonds used in a diamond bit depends on the hardness of the formation to be drilled. Bits for hard formations have many small (0.07- to 0.125-carat) stones, while bits for soft formations have a few large (0.75- to 2-carat) stones. Examples of diamond bits for both hard and soft formations are shown in Figs. 5.2A and 5.2B. If the diamonds used are too large for almost complete embedment in the formation, the unit loading on the diamond points will be excessive, resulting in localized heat generation and polishing of the cutting edge of the stones.

The design of the water-course pattern cut in the face of the bit and the *junk slots* cut in the side of the bit face controls cuttings removal and diamond cooling (Fig. 5.4A). Diamond bits are designed to be operated at a given flow rate and pressure drop across the face of the bit. Experiments conducted by bit manufacturers have indicated the need for approximately 2.0 to 2.5 hhp/sq in. of hole bottom with an approximate 500- to 1,000-psi pressure drop across the face of the bit to clean and cool the diamond adequately. The pressure drop across the face of the bit at a given flow rate can be established as the difference between the pump pressure measured with the bit off bottom and the pump pressure measured while drilling. The bit manufacturer usually will provide an estimate of the approximate circulating rate required to establish the needed pressure drop across the bit face.

5.1.2 Polycrystalline Diamond (PCD) Bits

Since the mid-1970's, a new family of drag bits has been made possible by the introduction of a sintered polycrystalline diamond *drill blank* as a bit cutter element. The drill blanks consist of a layer of synthetic

Courtesy of Norton Christensen

(a) Soft formation design.   (b) Hard formation design.

Fig. 5.2—Example diamond cutter drag bits.

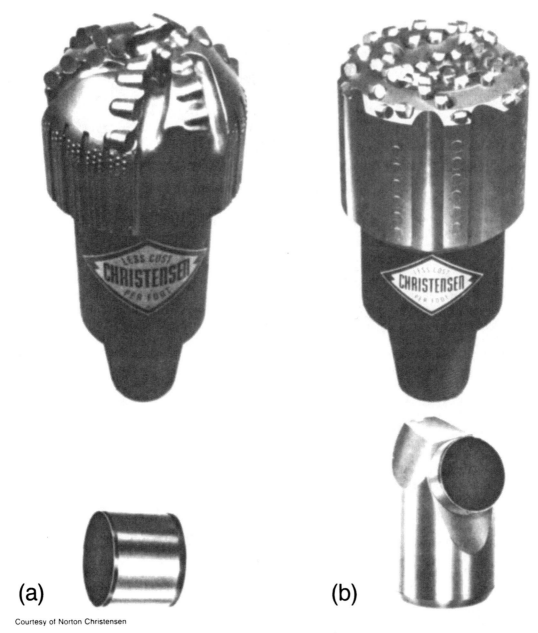

(a)   (b)

Fig. 5.3—Example polycrystalline diamond cutter bits.

polycrystalline diamond about $\frac{1}{64}$-in. thick that is bonded to a cemented tungsten carbide substrate in a high-pressure/high-temperature process. It contains many small diamond crystals bonded together. The cleavage planes of the diamond crystals have a random orientation that prevents any shock-induced breakage of an individual diamond crystal from easily propagating through the entire cutter. As shown in Fig. 5.3, the sintered polycrystalline diamond compact is bonded either to a tungsten carbide bit-body matrix or to a tungsten carbide stud that is mounted in a steel bit body.

The PCD bits are still evolving rapidly. They perform best in soft, firm, and medium-hard, nonabrasive formations that are not "gummy." Good results have been reported with PCD bits in drilling uniform sections of carbonates or evaporites that are not broken up with hard shale stringers. Successful use of these bits also has been

accomplished in sandstone, siltstone, and shale, although bit balling is a serious problem in very soft, gummy formations, and rapid cutter abrasion and breakage are serious problems in hard, abrasive formations. As in the case of the older, steel cutter drag bits, bit hydraulics can play an important role in reducing bit balling.

The bit shape or crown profile is also an important design feature of PCD bits. In addition to the double-cone profiles (Fig. 5.4C) used for diamond bits, single-cone profiles of various tapers and flat-bottom profiles are used for PCD bits. The hydraulic cleaning action is usually achieved primarily by using jets for steel-body PCD bits and by using water courses for matrix-body PCD bits.

Other important design features of a PCD bit include the size, shape, and number of cutters used and the angle

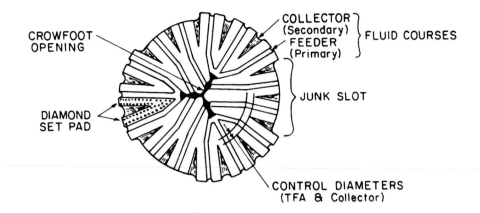

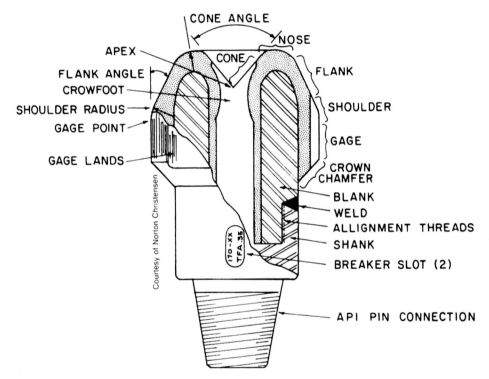

Fig. 5.4A—Diamond cutter drag bit—design nomenclature.

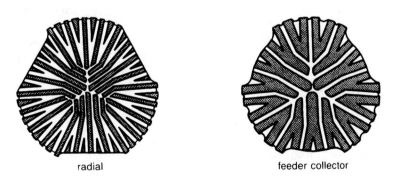

radial   feeder collector

Fig. 5.4B—Diamond cutter drag bit—example profiles and features.

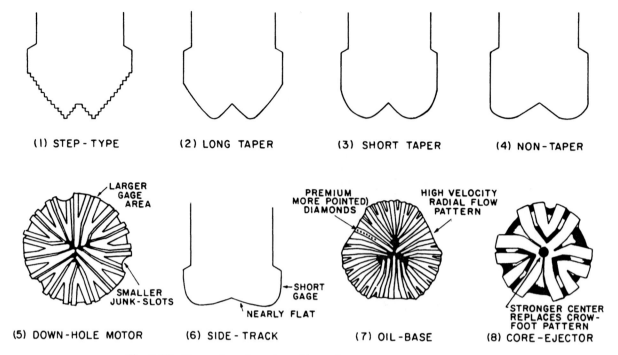

(1) STEP-TYPE   (2) LONG TAPER   (3) SHORT TAPER   (4) NON-TAPER

(5) DOWN-HOLE MOTOR   (6) SIDE-TRACK   (7) OIL-BASE   (8) CORE-EJECTOR

Fig. 5.4C—Diamond cutter drag bit—radial and feeder collectors.

of attack between the cutter and the surface of the exposed formation. Cutter orientation is defined in terms of back rake, side rake, and chip clearance or cutter exposure (Fig. 5.5).[1]

At present, a negative back-rake angle of 20° is standard on many steel-body PCD bits. However, smaller back-rake angles, which are better-suited for soft formations, are also available, especially in the matrix-body PCD bits. The side-rake assists in pushing the cuttings formed to the side of the hole, much like the action of a plow. The exposure of the cutter provides room for the cutting to peel off the hole bottom without impacting against the bit body and packing in front of the cutter.

Cutter orientation must be properly matched to the hardness of the formation being drilled. In soft, nonabrasive formations, where cutter wear is very slow, the orientation can be set to emphasize aggressive cutting. The high temperatures and wear rates caused by harder, more abrasive formations require a less-aggressive cutter orientation to prevent an excessive wear rate. The cutter orientation also depends on the expected cutter velocity, which in turn depends on the distance of the cutter location from the center of the hole.

5.1.3 Rolling Cutter Bits

The three-cone rolling cutter bit is by far the most common bit type currently used in rotary drilling operations. This general bit type is available with a large variety of tooth design and bearing types and, thus, is suited for a wide variety of formation characteristics. Fig. 5.6 is an example of a rolling cutter bit with the various parts labeled.[2] The three cones rotate about their axis as the bit is rotated on bottom.

The largest limitation a bit design engineer faces is that the bit must fit inside the borehole. The designer, thus, is required to make maximum use of a very limited amount of space. The size of every critical part can be increased

only at the expense of another critical part. This is especially true for the smaller bit sizes. Whereas most machines are designed to last for years, bits generally last at best only a few days.

The drilling action of a rolling cutter bit depends to some extent on the *offset* of the cones. As shown in Fig. 5.7, the offset of the bit is a measure of how much the cones are moved so that their axes do not intersect at a common point of the centerline of the hole. Offsetting causes the cone to stop rotating periodically as the bit is turned and scrape the hole bottom much like a drag bit. This action tends to increase drilling speed in most formation types. However, it also promotes faster tooth wear in abrasive formations. Cone offset is sometimes expressed as the angle the cone axis would have to be rotated to make it pass through the centerline of the hole. Cone offset angle varies from about 4° for bits used in soft formations to zero for bits used in extremely hard formations.

The shape of the bit teeth also has a large effect on the drilling action of a rolling cutter bit. Long, widely spaced, steel teeth are used for drilling soft formations. The long teeth easily penetrate the soft rock, and the scraping/twisting action provided by alternate rotation and plowing action of the offset cone removes the material penetrated. The action of this type bit often is compared with pushing a shovel into the ground and then leaning back on the handle to remove a large piece of earth. The wide spacing of the teeth on the cone promotes bit cleaning. Teeth cleaning action is provided by the intermeshing of teeth on different cones and by fluid jets between each of the three cones. As the rock type gets harder, the tooth length and cone offset must be reduced to prevent tooth breakage. The drilling action of a bit with zero cone offset is essentially a crushing action. The smaller teeth also allow more room for the construction of stronger bearings.

The metallurgy requirements of the bit teeth also depend on the formation characteristics. The two primary types used are (1) *milled tooth* cutters and (2) *tungsten carbide insert* cutters. The milled tooth cutters are manufactured by milling the teeth out of a steel cone, while the tungsten carbide insert bits are manufactured by pressing a tungsten carbide cylinder into accurately machined holes in the cone. The milled tooth bits designed for soft formations usually are faced with a wear-resistant material, such as tungsten carbide, on one side of the tooth. As shown in Fig. 5.8, the application of hard facing on only one side of the tooth allows more rapid wear on one side of the tooth than the other, and the tooth stays relatively sharp.

The milled tooth bits designed to drill harder formations are usually case hardened by special processing and heat treating the cutter during manufacturing. As shown in Fig. 5.8, this case-hardened steel should wear by chipping and tend to keep the bit tooth sharp.

The tungsten carbide teeth designed for drilling soft formations are long and have a chisel-shaped end. The inserts used in bits for hard formations are short and have a hemispherical end. These bits are sometimes called *button* bits. Examples of various insert bit tooth designs are shown in Fig. 5.9.

Considerable thought has gone into the position of the teeth on the cones of a rolling cutter bit. The inner rows of teeth are positioned on different cones so that they intermesh. This intermeshing (1) allows more room for a stronger bit design, (2) provides a self-cleaning action as the bit turns, and (3) allows maximum coverage of the hole bottom for a given number of teeth. The bottomhole coverage of most bits is about 70%. The outer row of teeth on each cone do not intermesh. This row of teeth, called the *heel teeth*, has by far the hardest job. Due to the circular geometry, more rock must be removed from the outermost annular ring of the hole bottom, and this rock is more difficult to remove because it tends to remain attached to the borehole wall.

Some of the heel teeth often are designed with *interruptions* or *identions* as shown in Fig. 5.6. These interruptions allow the heel teeth to generate a pattern on bottom having one-half the spacing of the cutter teeth. Thus, the cuttings are smaller than the space between the teeth and do not wedge between them readily. Because the heel teeth have a more difficult job, they may wear excessively, causing the bit to drill an out-of-gauge hole. This causes a gross misalignment of the load on the bearings and premature bit failure. Premature failure of the next bit is also likely if the hole remains undersized. Most bit manufacturers offer more than one heel tooth design with a given bit type so the drilling engineer may obtain the amount of gauge protection needed.

The common bearing assemblies used for rolling cutter bits are shown in Fig. 5.10. The standard or most inexpensive bearing assembly shown in Fig. 5.10a consists of (1) a roller-type outer bearing, (2) a ball-type intermediate bearing, and (3) a friction-type nose bearing. The roller-type outer bearing is the most heavily loaded member and usually tends to wear out first. The race that the roller bearing rolls over tends to spall and wear on the bottom side where the weight applied to the bit is transmitted from the pin to the cone. The intermediate ball bearings carry primarily axial or thrust loads on the

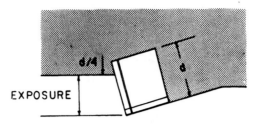

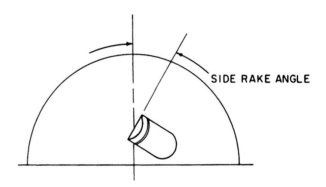

Fig. 5.5—Cutter orientation expressed in terms of exposure, back rake, and side rakes.

cones. They also serve to hold the cone in place on the bit. The nose bearings are designed to carry a portion of the axial or thrust loads after the ball bearings begin to wear. The nose bearing is a friction-type bearing in most bit sizes, but in the larger bit sizes another roller bearing is used. In the standard bearing design, all bearings are lubricated by the drilling fluid. When a gas is used as the drilling fluid, a modified bit is available with passageways permitting a portion of the gas to flow through the bearing assembly (Fig. 5.10b).

The intermediate-cost bearing assembly used in rolling cutter bits is the *sealed bearing* assembly. A cross section of a sealed bearing bit is shown in Fig. 5.10c. In this type bit, the bearings are maintained in a grease environment by grease seals, a grease reservoir, and a compensator plug that allows the grease pressure to be maintained equal to the hydrostatic fluid pressure at the bottom of the hole. While the grease seals require some space and, thus, a reduction in bearing capacity, the elimination of abrasive material from the bearings usually more than compensates for this disadvantage. As the

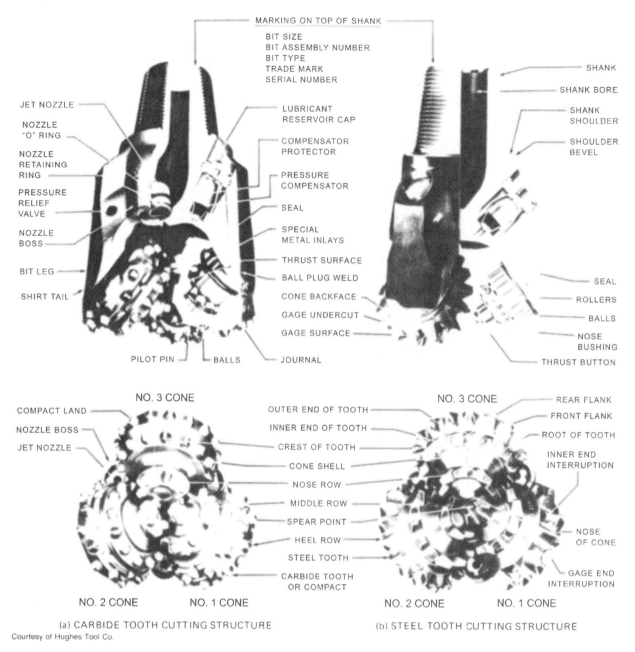

Fig. 5.6—Parts of a rolling cutter bit.

bit wears, the grease seals eventually fail and the drilling fluid can enter the bearings and accelerate the bearing wear.

Rolling cutter bits with the most advanced bearing assembly are the *journal bearing* bits (Fig. 5.10d). In this type bit, the roller bearings are eliminated and the cone rotates in contact with the journal bearing pin. This type bearing has the advantage of greatly increasing the contact area through which the weight on the bit is transmitted to the cone. Also, by eliminating one of the components (the rollers), additional space becomes available for strengthening the remaining components. Journal bearing bits require effective grease seals, special metallurgy, and extremely close tolerances during manufacture. Silver inlays in the journal help to minimize friction and prevent galling. While journal bearing bits are much more expensive than the standard

or sealed bearing bits, much longer bit runs can be obtained, thus eliminating some of the rig time spent on tripping operations.

5.1.4 Standard Classification of Bits

A large variety of bit designs are available from several manufacturers. The Intl. Assn. of Drilling Contractors (IADC) approved a standard classification system for identifying similar bit types available from various manufacturers.[3] The classification system adopted is the three-digit code.

The first digit in the bit classification scheme is called the bit series number. The letter "D" precedes the first digit if the bit is diamond or PCD drag bit. Series D1 through D5 are reserved for diamond bits and PCD bits in the soft, medium-soft, medium, medium-hard, and hard formation categories, respectively. Series D7

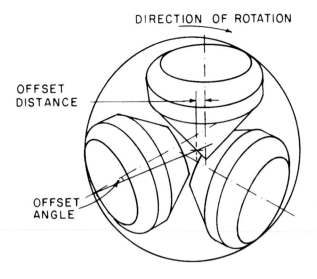

Fig. 5.7—Cone offset promotes cone slippage during rotation.

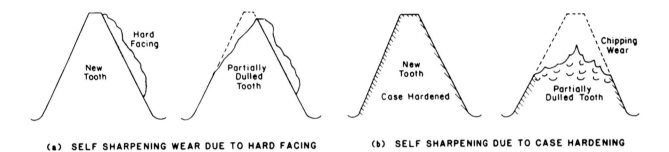

(a) SELF SHARPENING WEAR DUE TO HARD FACING   **(b) SELF SHARPENING DUE TO CASE HARDENING**

Fig. 5.8—Wear characteristics of milled-tooth bits.

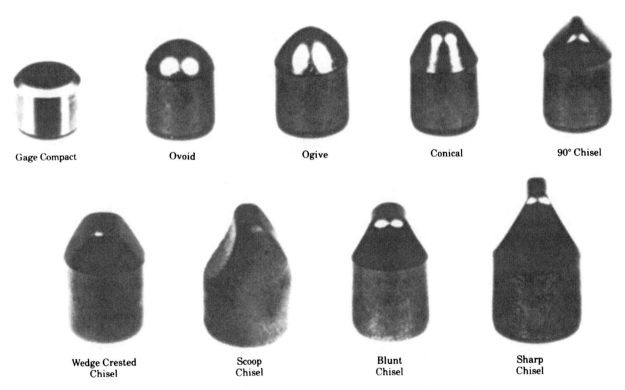

Fig. 5.9—Example tungsten carbide insert cutters used in rolling cutter bits.

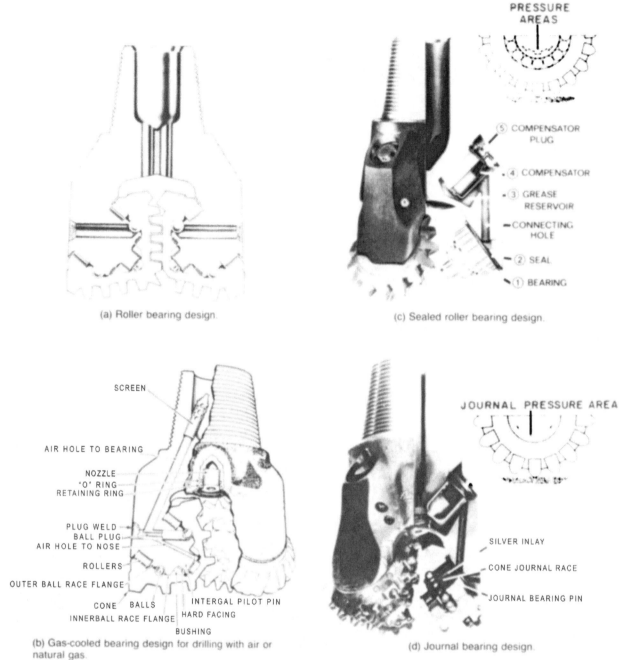

(a) Roller bearing design.

(c) Sealed roller bearing design.

PRESSURE AREAS

⑤ COMPENSATOR PLUG

④ COMPENSATOR

③ GREASE RESERVOIR

CONNECTING HOLE

② SEAL

① BEARING

SCREEN

AIR HOLE TO BEARING

NOZZLE
"O" RING
RETAINING RING

PLUG WELD
BALL PLUG
AIR HOLE TO NOSE

ROLLERS

OUTER BALL RACE FLANGE

CONE   BALLS

INNERBALL RACE FLANGE

INTERGAL PILOT PIN
HARD FACING

BUSHING

(b) Gas-cooled bearing design for drilling with air or natural gas.

JOURNAL PRESSURE AREA

SILVER INLAY

CONE JOURNAL RACE

JOURNAL BEARING PIN

(d) Journal bearing design.

Fig. 5.10—Common bearing assemblies used for rolling cutter bits.

through D9 are reserved for diamond core bits and PCD core bits in the soft-, medium-, and hard-formation categories. Series 1, 2, and 3 are reserved for milled tooth bits in the soft, medium, and hard formation categories, respectively. Series 5, 6, 7, and 8 are for insert bits in the soft, medium, hard, and extremely hard formation categories, respectively. Series 4 is reserved for future use with special categories such as a "universal" bit.

The second digit is called the type number. Type 0 is reserved for PCD drag bits. Types 1 through 4 designate a formation hardness subclassification from the softest to the hardest formation within each category.

The feature numbers are interpreted differently, depending on the general type of bit being described. Feature numbers are defined for (1) diamond and PCD drag bits, (2) diamond and PCD drag-type core-cutting bits, and (3) rolling cutter bits.

Eight standard diamond and PCD drag bit features are (1) step-type profile, (2) long-taper profile, (3) short-taper profile, (4) nontaper profile, (5) downhole-motor type, (6) sidetrack type, (7) oil-base type, and (8) core-ejector type. The remaining feature (No. 9) is reserved for special features selected by the bit manufacturer.

There are two standard feature numbers for diamond and PCD drag-type core-cutting bits. These bits are used to recover a length of formation sample cored from the central portion of the borehole. The two features are (1)

TABLE 5.1—IADC DIAMOND AND PCD DRILL BIT CLASSIFICATION CHART FOR FOUR MANUFACTURERS

MANUFACTURER: AMERICAN COLDSET (a)

FORMATION	IADC SERIES NUMBER	STEP TYPE 1	LONG TAPER 2	SHORT TAPER 3	NON-TAPER 4	DOWNHOLE MOTOR 5	SIDE TRACK 6	OIL BASE 7	CORE EJECTOR 8	OTHER 9
SOFT / D1	0			STRATACUT				STRATACUT		
	1		SHARKTOOTH			SHARKTOOTH T	ANGLE BUILDER			
	2		" TRIGG	EAGLE ARMADILLO		"	"	SHARKTOOTH	CORE CRUSHER	UD
	3		" "	"		"	"	"	"	"
	4		" "	"		"	"	"	"	"
MEDIUM SOFT / D2	0			STRATACUT				STRATACUT		
	1		SHARKTOOTH TRIGG	ARMADILLO		SHARKTOOTH T	ANGLE BUILDER	SHARKTOOTH	CORE CRUSHER	UD
	2		"	"		"	"	"	"	"
	3		"	"		"	"	"	"	"
	4		"	"		"	"	"	"	"
MEDIUM / D3	0			STRATACUT				STRATACUT		
	1		TRIGG	ARMADILLO		TBYY	ANGLE BUILDER		CORE CRUSHER	UD
	2		"	"		"	"		"	"
	3		"	SHREW		"	"		"	"
	4		"	"		"	"		"	"
MEDIUM HARD / D4	0									
	1		TRIGG	SHREW	WOLF	TBYY	ANGLE BUILDER		CORE CRUSHER	UD
	2		"	"		"	"		"	"
	3				"	"	"		"	"
	4				"	"	"		"	"
HARD / D5	0									
	1					TBYY	ANGLE BUILDER		CORE CRUSHER	UD
	2				WOLF RIBSET	"	"		"	MILL & DRILL
	3				"				"	"
	4									

MANUFACTURER: CHRISTENSEN (b)

FORMATION	IADC SERIES NUMBER	STEP TYPE 1	LONG TAPER 2	SHORT TAPER 3	NON-TAPER 4	DOWNHOLE MOTOR 5	SIDE TRACK 6	OIL BASE 7	CORE EJECTOR 8	OTHER 9
SOFT / D1	0		Rockut 1	Rockut 27		Rockut 11		Rockut 27		
	1	MD-18				MT128P				
	2		MD-38	MD-34		MT18P	MD-43ST	MD-38		
	3	MD-503	MD-341	MD-315						
	4	MD-197		MD-262						
MEDIUM SOFT / D2	0		Rockut 1	Rockut 26		Rockut 1		ROCKUT 26		
	1									
	2	MD-196	MD-261	MD-262		MT-51P	MD-43ST			
	3		MD-311	MD-34		MD-262		MD-38		
	4									
MEDIUM / D3	0		Rockut III	Rockut 26		Rockut 26		Rockut 26		
	1									
	2		MD-261	MD-331		MT-51P	MD-411ST	MD-262		
	3		MD-311			MD-262				MD-28
	4					MD-331				
MEDIUM HARD / D4	0									
	1							MD-331		
	2			MD-331		MT-54P	MD-411ST			MD-28
	3					MD-331				
	4			MD-262	MD-41	MD-262				
HARD / D5	0									
	1									
	2				MD-41	MT54P	MD-411ST		MD-37	
	3				MD-24				MD-23	
	4			MD-210	MD-240					

conventional core-barrel type and (2) face-discharge type. As in the previous case, feature No. 9 is reserved for special features selected by the bit manufacturer.

There are eight standard feature numbers for rolling-cutter bits. The standard feature numbers are (1) standard rolling cutter bit (jet bit or regular), (2) T-shaped heel teeth for gauge protection, (3) extra insert teeth for gauge protection, (4) sealed roller bearings, (5) combination of Nos. 3 and 4, (6) sealed friction bearing, and (7) com-

bination of Nos. 3 and 6. The remaining feature Nos. 8 and 9 were reserved for special features selected by the bit manufacturer. Feature No. 8 is often used to designate bits designed for directional drilling.

Tables 5.1 through 5.3 summarize many of the currently available bits by manufacturer and Intl. Assn. of Drilling Contractors (IADC) classification number for diamond bits, diamond core bits, and rolling-cutter bits, respectively. While bits from different manufacturers

MANUFACTURER: DOWDCO **(c)**

FORMATION	IADC SERIES NUMBER		STEP TYPE 1	LONG TAPER 2	SHORT TAPER 3	NON-TAPER 4	DOWNHOLE MOTOR 5	SIDE-TRACK 6	OIL BASE 7	CORE EJECTOR 8	OTHER 9
SOFT	D1	0									
		1	RFV-20	CHV-20	CHB-20	CH-20	RFT-20	ST-20	RFV-20		
		2	"	RFV-21	RFV-21	RF-21	RFT-21	"	RFV-21		
		3	"	"	"	"	"	"	"		
		4	"	"	"	"	"	"	"		
MEDIUM SOFT	D2	0									
		1	RFV-30	CHV-20	CHV-20	CH-20	RFT-20	ST-30	RFV-20		
		2	"	RFV-21	RFV-21	RF-21	RFT-21	"	RFV-21		
		3	"	CHV-30	CHV-30	CH-30	RFT-30	"	RFV-30		
		4	"	RFV-31	RFV-31	RF-31	RFT-31	"	RFV-31		
MEDIUM	D3	0									
		1	RFV-30	CHV-30	CHV-30	CH-30	RFT-30	ST-30	RFV-30		
		2	RFV-40	RFV-31	RFV-31	RF-31	RFT-31	"	RFV-31		
		3	"	"	"	"	"	"	"		
		4	"	"	"	"	"	"	"		
MEDIUM HARD	D4	0									
		1	RFV 40	RFV-40	RFV-40	RF-30	RFT-40	ST-40	RFV-40		
		2	"	RFV-41	RFV-41	RF-41	RFT-41	"	RFV-41		
		3	"	"	"	"	"	"	"		
		4	"	"	"	"	"	"	"		
HARD	D5	0									
		1		RFV-50	RFV-50	RF-50	RFT-50	ST-50	RFV-50	RF-50	
		2		RFV-51	RFV-51	RF-51	RFT-51	RST-50	RFV-51	RS-50	
		3		"	"	RS-50	"	"	"	"	
		4		"	"	CF-50	"	"	"	"	

MANUFACTURER: NL HYCALOG **(d)**

FORMATION	IADC SERIES NUMBER		STEP TYPE 1	LONG TAPER 2	SHORT TAPER 3	NON-TAPER 4	DOWNHOLE MOTOR 5	SIDE-TRACK 6	OIL BASE 7	CORE EJECTOR 8	OTHER 9
SOFT	D1	0		SAT III, SAT IV	JETPAX		TURBOPAX				
		1									
		2		901			MS1T				
		3		901			MS1T 901T	ST	901S	901CE	204 BI CNTR
		4		901			MS1T 901T	ST WM	901S	901CE	204 "
MEDIUM SOFT	D2	0		SAT III, SAT IV	JETPAX		TURBOPAX				
		1		901 MS1 730			MS1T 901 730T	ST	901S	901CE	204 "
		2		901 MS1 730			MS1T 901 730T	ST	901S	901CE	204 "
		3		901 MS1 730			MS1T 901 730T	ST	901S	901CE	204 "
		4		901 MS1 730			MS1T 901 730T	ST WM	901S	901CE	204 "
MEDIUM	D3	0									
		1		901 730			M1T 901T 730T	ST	901S	901CE 730CE	204 "
		2		901 730			M1T 901T 730T	ST	901S	901CE 730CE	204 "
		3		901 730			M1T 901T 730T	ST	901S	901CE 730CE	204 "
		4		901 730			M1T 901T 730T	ST WM	901S	901CE 730CE	204 "
MEDIUM HARD	D4	0									
		1		901 730		525 501	M1T 901T 730T	ST	901S	730CE	204 "
		2		901		525 501	M1T 901T 730T	ST	901S	730CE	204 "
		3		901 730		525 501	M1T 901T 730T	ST	901S	730CE	204 "
		4		901 730		525 501	M1T 901T 730T	ST WM	901S	730CE	204
HARD	D5	0									
		1		901 730		525 501	M1T 901T 730T	ST	901S	730CE	
		2		901 730		525 501	M1T 901T 730T	ST	901S	730CE	
		3		901 730		525 501	M1T 901T 730T			730CE	
		4				525 501	M1T	WM			

given the same classification number are not exactly equivalent, the classification scheme is often useful for grouping similar bits in drilling performance studies.

Some of the main design features of the various rolling-cutter bit types have been summarized by Estes.[4] Shown in Table 5.4 and Fig. 5.11 are some of the tooth design features of the various bit types and classes. Note that as the class number increases, the cone offset, tooth height, and amount of tooth hardfacing decreases while the number of teeth and amount of tooth case hardening increases. Shown in Fig. 5.12 are the relative bearing capacities for different bit classes. Note that an increase in bearing capacity is possible for the bits with a higher class number. This is possible due to the shorter length of bit teeth at higher bit class numbers.

5.2 Rock Failure Mechanisms

To operate a given bit properly, the drilling engineer needs to understand as much as possible about the basic mechanisms of rock removal that are at work, including (1) wedging, (2) scraping and grinding, (3) erosion by fluid jet action, (4) percussion or crushing, and (5) torsion or twisting. To some extent, these mechanisms are interrelated. While one may be dominant for a given bit

TABLE 5.2—IADC DIAMOND AND PCD CORE BIT CLASSIFICATION CHART FOR FOUR MANUFACTURERS

MANUFACTURER: AMERICAN COLDSET

(a)

FORMATION	IADC SERIES NUMBER		DESIGN FEATURES		
			CONVENTIONAL CORE BARREL 1	FACE DISCHARGE 2	OTHER 9
SOFT	D7	0		STRATACORE	
		1			
		2	SHARK HEAD	EH STAR FD	
		3	"	"	
		4	"	"	
MEDIUM	D8	0		STRATACORE	
		1	EH	SHARK HEAD FD	
		2	"	"	
		3	EH STAR	"	
		4	"	"	
HARD	D9	0			
		1	EH STAR	SHARK HEAD FD	
		2	"	"	
		3		"	
		4			

MANUFACTURER: CHRISTENSEN

(b)

FORMATION	IADC SERIES NUMBER		DESIGN FEATURES		
			CONVENTIONAL CORE BARREL 1	FACE DISCHARGE 2	OTHER 9
SOFT	D7	0	JC-2	JC-3	
		1			
		2	C-22		C-35
		3		C-18	C-28
		4			
MEDIUM	D8	0			
		1	C-22		
		2	C-20	C-18	C-25
		3	C-201		C-28
		4			
HARD	D9	0			
		1			
		2	C-23		C-28
		3	C-24		
		4	C-40		

MANUFACTURER: DOWDCO

(c)

FORMATION	IADC SERIES NUMBER		DESIGN FEATURES		
			CONVENTIONAL CORE BARREL 1	FACE DISCHARGE 2	OTHER 9
SOFT	D7	0	DOWDCOPAX-7C	DOWDCOPAX-7F	
		1	MS	FD-20	SV
		2	"	FD-30	"
		3	"	"	"
		4	"	"	"
MEDIUM	D8	0	DOWDCOPAX-8C	DOWDCOPAX-8F	
		1	M	FD-30	AWR
		2	"	FD-40	MV
		3	"	"	"
		4	"	"	"
HARD	D9	0	DOWDCOPAX-9C	DOWDCOPAX-9F	
		1	H	FD-50	RS
		2	"	"	"
		3	"	"	"
		4	"	"	"

MANUFACTURER: NL HYCALOG

(d)

FORMATION	IADC SERIES NUMBER		DESIGN FEATURES		
			CONVENTIONAL CORE BARREL 1	FACE DISCHARGE 2	OTHER 9
SOFT	D7	0	GeoPax	GeoPax	
		1	CS	CSFD	
		2	CS	CSFD	
		3	CS	CSFD	
		4	CS	CSFD	
MEDIUM	D8	0			
		1			
		2	CM		
		3	CMH		
		4	CH		
HARD	D9	0			
		1	CVH		
		2	CUH CUHR		
		3	CUH CUHR		
		4	CUH CUHP		Impregnated

TABLE 5.3—IADC ROLLING-CUTTER BIT CLASSIFICATION CHART FOR FOUR MANUFACTURERS

(a)

MANUFACTURER: HUGHES TOOL CO.

SERIES	FORMATIONS	TYPES	Standard Roller Bearing (1)	Roller Bearing Air (2)	Roller Brg Gage Protected (3)	Sealed Roller Bearing (4)	Sealed Roller Brg Gage Protected (5)	Sealed Friction Bearing (6)	Sealed Friction Brg Gage Protected (7)	Directional (8)	Other (9)
1	SOFT FORMATIONS WITH LOW COMPRESSIVE STRENGTH AND HIGH DRILLABILITY	1	ØSC3AJ			X3A		J1			MX3A
		2	ØSC3J	S				J2			
		3	ØSC1GJ					J3	JD3		MX1G
		4									
2	MEDIUM TO MEDIUM HARD FORMATIONS WITH HIGH COMPRESSIVE STRENGTH	1	ØWVJ/ØW4J	M				J4	JD4		
		2	WØ								
		3									
		4									
3	HARD SEMI ABRASIVE AND ABRASIVE FORMATIONS	1	W7J/W7C	H				J7			
		2	W7R2J								
		3		HR							
		4						J8	JD8		
4	SOFT FORMATIONS WITH LOW COMPRESSIVE STRENGTH AND HIGH DRILLABILITY	1									
		2									
		3					X11		J11		
		4									
5	SOFT TO MEDIUM FORMATIONS WITH LOW COMPRESSIVE STRENGTH	1					X22		J22		
		2									
		3	HH33				X33		J33		
		4									
6	MEDIUM HARD FORMATIONS WITH HIGH COMPRESSIVE STRENGTH	1	HH44				X44		J44		
		2							J44C/J55R		
		3	HH55						J55		
		4									
7	HARD SEMI ABRASIVE AND ABRASIVE FORMATIONS	1									
		2									
		3	HH77						J77		
		4									
8	EXTREMELY HARD & ABRASIVE FORMATIONS	1	HH88								
		2									
		3	HH99						J99		
		4									

MILLED TOOTH BITS = Series 1–3; INSERT BITS = Series 4–8.

(b)

MANUFACTURER: REED ROCK BIT CO.

SERIES	FORMATIONS	TYPES	Standard Roller Bearing (1)	Roller Bearing Air (2)	Roller Brg Gage Protected (3)	Sealed Roller Bearing (4)	Sealed Roller Brg Gage Protected (5)	Sealed Friction Bearing (6)	Sealed Friction Brg Gage Protected (7)	Directional (8)	Other (9)
1	SOFT FORMATIONS WITH LOW COMPRESSIVE STRENGTH AND HIGH DRILLABILITY	1	Y11			S11				Y11-JD	
		2	Y12		Y12T	S12		FP12			
		3	Y13		Y13T/Y13G	S13	S13G	FP13			
		4									
2	MEDIUM TO MEDIUM HARD FORMATIONS WITH HIGH COMPRESSIVE STRENGTH	1	Y21		Y21G	S21	S21G	FP21			
		2	Y22								
		3					S23G				
		4									
3	HARD SEMI-ABRASIVE AND ABRASIVE FORMATIONS	1	Y31		Y31G		S31G		FP31G		Y31-RAP
		2									
		3									
		4									
4	SOFT FORMATIONS WITH LOW COMPRESSIVE STRENGTH AND HIGH DRILLABILITY	1									
		2									
		3									
		4									
5	SOFT TO MEDIUM FORMATIONS WITH LOW COMPRESSIVE STRENGTH	1							HS51, FP51A		
		2					S52		HS51, FP52		
		3					S53		HPSM, FP53		
		4							HPSM, FP54		
6	MEDIUM HARD FORMATIONS WITH HIGH COMPRESSIVE STRENGTH	1							HPSM, FP62X		
		2	Y62-JA Y62B-JA		Y62 Y62B		S62		HPSM, FP62/FP62B		
		3	Y63-JA		Y63		S63		HPMH, FP63		
		4					S64		HPMH, FP64		
7	HARD SEMI-ABRASIVE AND ABRASIVE FORMATIONS	1							HPMH, HPH		
		2			Y72		S72		HPMH, HPH, FP72		
		3	Y73-JA		Y73				HPH, FP73		Y73-RAP
		4					S74		HPH, FP74		
8	EXTREMELY HARD & ABRASIVE FORMATIONS	1							HPH		
		2									
		3	Y83-JA		Y83		S83		FP83		
		4									

MILLED TOOTH BITS = Series 1–3; INSERT BITS = Series 4–8.

(c)

MANUFACTURER: SECURITY DIV.

FORMATIONS	SERIES		TYPES	STANDARD ROLLER BEARING (1)	ROLLER BEARING, AIR (2)	ROLLER BRG GAGE PROTECTED (3)	SEALED ROLLER BEARING (4)	SEALED ROLLER BRG GAGE PROTECTED (5)	SEALED FRICTION BEARING (6)	SEALED FRICTION BRG GAGE PROTECTED (7)	DIRECTIONAL (8)	OTHER (9)
		SOFT FORMATIONS WITH LOW COMPRESSIVE STRENGTH AND HIGH DRILLABILITY	1	S3S			S33S	S33SG	S33SF		S3SJD	
MILLED TOOTH BITS	1		2	S3		S3T	S33			S33F	S3JD	
			3	S4		S4T	S44	S44G				
			4								DS/DSS	
		MEDIUM TO MEDIUM HARD FORMATIONS WITH HIGH COMPRESSIVE STRENGTH	1	M4N			M44N	M44NG	M44NF			
	2		2	M4								
			3	M4L			M44L			M44LF	DM/DMM	
			4									
		HARD SEMI-ABRASIVE AND ABRASIVE FORMATIONS	1	H7		H7T	H77			H77F		
	3		2									
			3			H7SG		H77SG				
			4				H77C			H77CF		
		SOFT FORMATIONS WITH LOW COMPRESSIVE STRENGTH AND HIGH DRILLABILITY	1									
	4		2									
			3									
			4									
		SOFT TO MEDIUM FORMATIONS WITH LOW COMPRESSIVE STRENGTH	1					S84		S84F	DS84F	
INSERT BITS	5		2									
			3							S86F		
			4		S8JA			S88		S88F	DS88	
		MEDIUM HARD FORMATIONS WITH HIGH COMPRESSIVE STRENGTH	1							M84F		
	6		2		M8JA					M88F/M89TF		GM88
			3							M89F		
			4									
		HARD SEMI-ABRASIVE AND ABRASIVE FORMATIONS	1									
	7		2									
			3							H84F		
			4		H8JA			H88		H88F		
		EXTREMELY HARD & ABRASIVE FORMATIONS	1		H9JA			H99		H99F		
	8		2									
			3		H10JA			H100		H100F		
			4									

(d)

MANUFACTURER: SMITH TOOL

FORMATIONS	SERIES		TYPES	STANDARD ROLLER BEARING (1)	ROLLER BEARING, AIR (2)	ROLLER BRG GAGE PROTECTED (3)	SEALED ROLLER BEARING (4)	SEALED ROLLER BRG GAGE PROTECTED (5)	SEALED FRICTION BEARING (6)	SEALED FRICTION BRG GAGE PROTECTED (7)	DIRECTIONAL (8)	OTHER (9)
		SOFT FORMATIONS WITH LOW COMPRESSIVE STRENGTH AND HIGH DRILLABILITY	1	DS			SDS				DJ	
MILLED TOOTH BITS	1		2	DT		DTT	SDT		FDT		BHDJ	MSDT
			3	DG		DGT	SGT	SDGH	FDG			
			4									
		MEDIUM TO MEDIUM HARD FORMATIONS WITH HIGH COMPRESSIVE STRENGTH	1	V2		V2H	SV	SVH				
	2		2									
			3			T2H	ST2					
			4									
		HARD SEMI-ABRASIVE AND ABRASIVE FORMATIONS	1	L4		L4H	SL4	SL4H				
	3		2									
			3									
			4									
		SOFT FORMATIONS WITH LOW COMPRESSIVE STRENGTH AND HIGH DRILLABILITY	1									
	4		2									
			3									
			4									
		SOFT TO MEDIUM FORMATIONS WITH LOW COMPRESSIVE STRENGTH	1					2JS		F2		AI
INSERT BITS	5		2									
			3					3JS		F3		
			4									
		MEDIUM HARD FORMATIONS WITH HIGH COMPRESSIVE STRENGTH	1		4JA			4JS		F4/F45		4GA
	6		2		5JA			5JS		F5		5GA
			3							F47/F57		
			4									
		HARD SEMI-ABRASIVE AND ABRASIVE FORMATIONS	1		7JA					F6		
	7		2							F7		7GA
			3									
			4									
		EXTREMELY HARD & ABRASIVE FORMATIONS	1									
	8		2									
			3		9JA					F9		
			4									

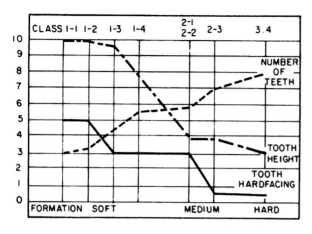

Fig. 5.11—Tooth design variations between bit types.[4]

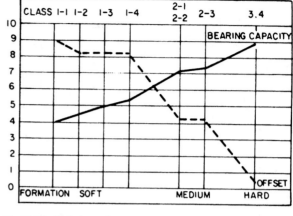

Fig. 5.12—Relative offset and bearing capacity between different bit types.[4]

design, more than one mechanism is usually present. In this discussion, only the two basic rotary drilling bit types will be discussed—i.e., the drag bit and the rolling cutter bits.

5.2.1 Failure Mechanisms of Drag Bits

Drag bits are designed to drill primarily by a wedging mechanism. If drag bits could be kept drilling by wedging, they would not dull so quickly. It is when they are dragging and, thus, scraping and grinding that they drill slowly and dull quickly. A twisting action also may contribute to rock removal from the center portion of the hole. A schematic illustrating the wedging action of a drag bit tooth just prior to cutting failure is shown in Fig. 5.13.[5] A vertical force is applied to the tooth as a result of applying drill collar weight to the bit, and a horizontal force is applied to the tooth as a result of applying the torque necessary to turn the bit. The result of these two forces defines the plane of thrust of the tooth or wedge. The cuttings are sheared off in a shear plane at an initial angle to the plane of thrust that is dependent on the properties of the rock.

The depth of the cut is controlled by the plane of thrust and is selected based on the strength of the rock and the radius to the cut. The depth of the cut is often expressed in terms of the bottom cutting angle, α. The angle α is a function of the desired cutter penetration per revolution L_p and radius r from the center of the hole. This relation can be defined by

$$\tan \alpha = \frac{L_p}{2\pi r}. \qquad (5.1)$$

The bottom clearance angle prevents the wedge from dragging the hole bottom while taking a chip and, thus, causing the bit to jump and chatter and to wear fast. The bottom clearance angle should not be too great, however, to prevent the bit from digging too deep and stalling the rotary whenever the weight-to-torque ratio is too great. A slight rake angle can help promote an efficient wedging mechanism, although a positive rake angle may not be necessary because of the downward slope of the hole bottom when the bit is operated properly. The bit tooth loses strength as the rake angle is increased.

TABLE 5.4—TOOTH DESIGN CHARACTERISTICS FOR ROLLING-CUTTER BITS
(after Estes[4])

Bit Type	Class	Formation Type	Tooth Description	Offset (degrees)
Steel-cutter milled-tooth	1–1, 1–2	very soft	hard-faced tip	3 to 4
	1–3, 1–4	soft	hard-faced side	2 to 3
	2–1, 2–2	medium	hard-faced side	1 to 2
	2–3	medium-hard	case hardened	1 to 2
	3	hard	case hardened	0
	4	very hard	case hardened, circumferential	0
Tungsten-carbide insert	5–2	soft	64° long blunt chisel	2 to 3
	5–3	medium-soft	65 to 80° long sharp chisel	2 to 3
	6–1	medium shales	65 to 80° medium chisel	1 to 2
	6–2	medium limes	60 to 70° medium projectile	1 to 2
	7–1	medium-hard	80 to 90° short chisel	0
	7–2	medium	60 to 70° short projectile	0
	8	hard chert	90° conical, or hemispherical	0
	9	very hard	120° conical, or hemispherical	0

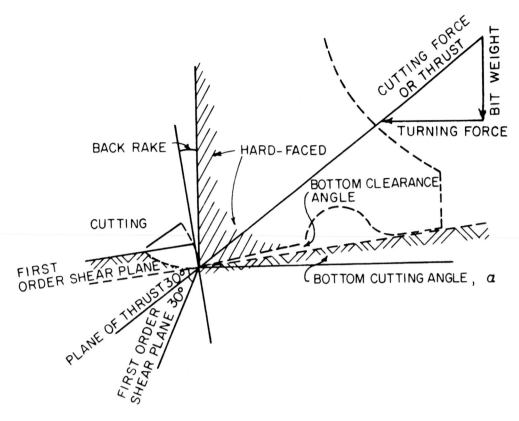

Fig. 5.13—Wedging action of drag bit cutter.

Diamond drag bits are designed to drill with a very small penetration into the formation. The diameter on the individual rock grains in a formation such as sandstone may not be much smaller than the depth of penetration of the diamonds. The drilling action of diamond drag bits in this situation is primarily a grinding action in which the cementaceous material holding the individual grains is broken by the diamonds.

Rock mechanics experts have applied several failure criteria in an attempt to relate rock strength measured in simple compression tests to the rotary drilling process. One such failure criterion often used is the Mohr theory of failure. The Mohr criterion states that yielding or fracturing should occur when the shear stress exceeds the sum of the cohesive resistance of the material c and the frictional resistance of the slip planes or fracture plane.

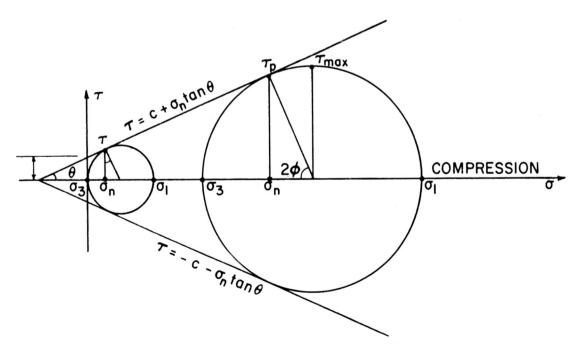

Fig. 5.14—Mohr's circle representation of Mohr failure criterion.

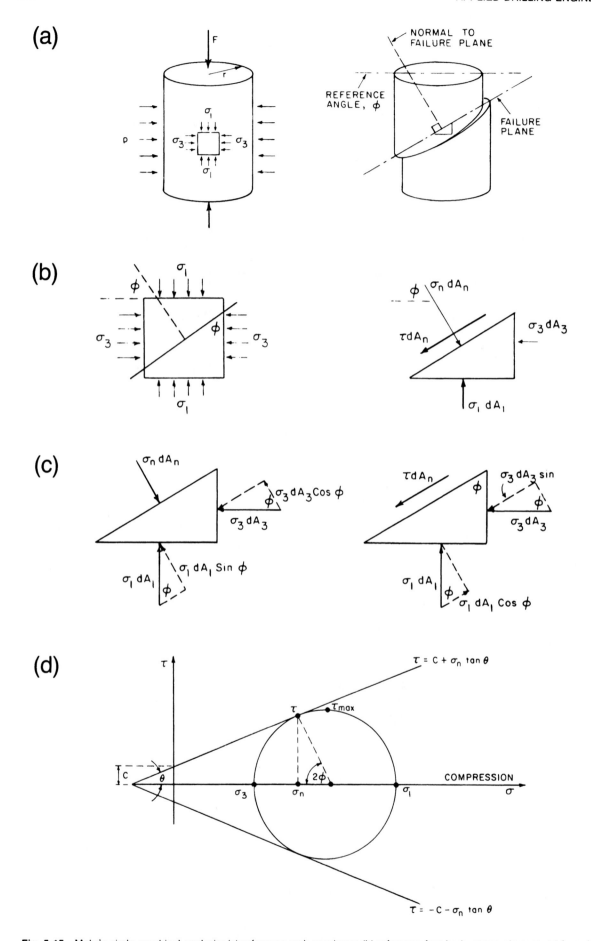

Fig. 5.15—Mohr's circle graphical analysis: (a) reference rock specimen; (b) reference free-body stress element; (c) force balance normal and parallel to failure plane, o; and (d) construction of Mohr's circle.

The Mohr criterion is stated mathematically by

$$\tau = \pm(c + \sigma_n \tan\theta), \quad\ldots\ldots\ldots\ldots\ldots\ldots (5.2)$$

where

τ = shear stress at failure,

c = cohesive resistance of the material,

σ_n = normal stress at the failure plane, and

θ = angle of internal friction.

As shown in Fig. 5.14, this is the equation of a line that is tangent to Mohr's circles drawn for at least two compression tests made at different levels of confining pressure.

To understand the use of the Mohr criterion, consider a rock sample to fail along a plane, as shown in Fig. 5.15, when loaded under a compressive force F and a confining pressure p. The compressive stress σ_1 is given by

$$\sigma_1 = \frac{F}{\pi r^2}.$$

The confining stress is given by

$$\sigma_3 = p.$$

If we examine a small element on any vertical plane bisecting the sample, the element is in the stress state given in Fig. 5.15b. Furthermore, we can examine the forces present along the failure plane at failure using the free-body elements shown in Fig. 5.15b. The orientation of the failure plane is defined by the angle ϕ between the normal-to-the-failure plane and a horizontal plane. It is also equal to the angle between the failure plane and the direction of the principal stress σ_1. Both a shear stress τ and a normal stress σ_n must be present to balance σ_1 and σ_3.

Summing forces normal to the fracture plane (Fig. 5.15c) gives

$$\sigma_n dA_n = \sigma_3 dA_3 \cos\phi + \sigma_1 dA_1 \sin\phi.$$

The unit area along the fracture plane dA_n is related to the unit areas dA_1 and dA_2 by

$$dA_3 = dA_n \cos\phi$$

and

$$dA_1 = dA_n \sin\phi.$$

Making these substitutions in the force balance equation gives

$$\sigma_n = \sigma_1 \sin^2\phi + \sigma_3 \cos^2\phi$$

$$= \tfrac{1}{2}(\sigma_1 + \sigma_3) - \tfrac{1}{2}(\sigma_1 - \sigma_3)\cos(2\phi). \quad\ldots\ldots (5.3a)$$

Summing forces parallel to the fracture plane gives

$$\tau dA_n = \sigma_1 dA_1 \cos\phi - \sigma_3 dA_3 \sin\phi.$$

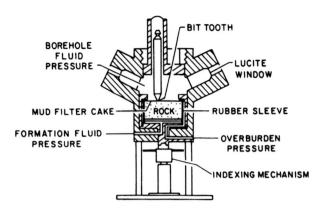

Fig. 5.16—Apparatus used for study of bit tooth penetration under simulated borehole conditions.[7]

Expressing all unit areas in terms of dA_n and simplifying yields

$$\tau = (\sigma_1 - \sigma_3) \sin\phi \cos\phi$$

$$= \tfrac{1}{2}(\sigma_1 - \sigma_3) \sin(2\phi). \quad\ldots\ldots\ldots\ldots (5.3b)$$

Note that Eqs. 5.3a and 5.3b are represented graphically by the Mohr's circle shown in Fig. 5.15d. Note also that the angle of internal friction, θ, and 2ϕ must sum to 90°. The angle of internal friction for the most rocks varies from about 30 to 40°.

The Mohr failure criterion can be used to predict the characteristic angle between the shear plane and the plane of thrust for a drag bit. Assuming an angle of internal friction of approximately 30° implies

$$2\phi = 90° - 30°$$

or

$$\phi = 30°.$$

This value of ϕ has been verified experimentally by Gray *et al.* in tests made at atmospheric pressure.[6]

Example 5.1. A rock sample under a 2,000-psi confining pressure fails when subjected to a compressional loading of 10,000 psi along a plane which makes an angle of 27° with the direction of the compressional load. Using the Mohr failure criterion, determine the angle of internal friction, the shear strength, and the cohesive resistance of the material.

Solution. The angles θ and 2ϕ must sum to 90°. Thus, the angle of internal friction is given by

$$\theta = 90 - 2(27) = 36°.$$

The shear strength is computed using Eq. 5.3b as follows.

$$\tau = \tfrac{1}{2}(\sigma_1 - \sigma_3) \sin(2\phi)$$

$$= \tfrac{1}{2}(10,000 - 2,000) \sin(54°) = 3,236.$$

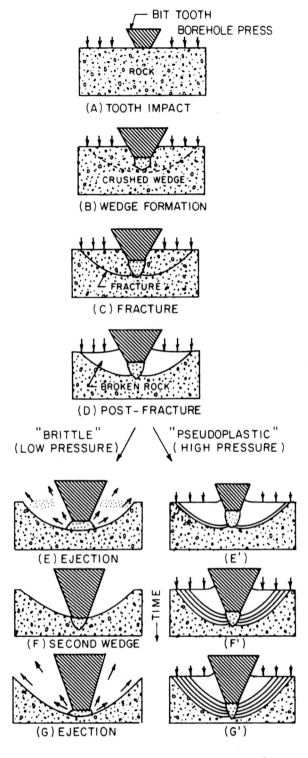

Fig. 5.17—Crater mechanism beneath a bit tooth.[7]

The stress normal to the fracture plane is computed using Eq. 5.3a as follows.

$$\sigma_n = \frac{1}{2}(\sigma_1 + \sigma_3) - \frac{1}{2}(\sigma_1 - \sigma_3) \cos(2\phi)$$

$$= \frac{1}{2}(10,000 + 2,000) - \frac{1}{2}(10,000$$

$$-2,000) \cos(54°) = 3,649 \text{ psi}.$$

The cohesive resistance can be computed by rearranging Eq. 5.2 as follows.

$$c = \tau - \sigma_n \tan \theta$$

$$= 3,236 - 3,649 \tan(36°) = 585 \text{ psi}.$$

5.2.2 Failure Mechanism of Rolling Cutter Bits

Rolling cutter bits designed with a large cone offset angle for drilling soft formations employ all of the basic mechanisms of rock removal. However, the percussion or crushing action is the predominant mechanism present for the IADC Series 3, 7, and 8 rolling cutter bits. Since these bit types are designed for use in hard, brittle formations in which penetration rates tend to be low and drilling costs tend to be high, the percussion mechanism is of considerable economic interest. Basic experimental tests conducted with an instrumented single tooth impacting on a rock sample have provided considerable insight into the basic mode of failure beneath the bit tooth. Maurer,[7] working with the apparatus shown in Fig. 5.16, studied bit tooth penetration under simulated borehole conditions. This apparatus, unlike those used prior to Maurer's work, allowed the borehole pressure, rock pore pressure, and rock confining pressure to be varied independently. The apparatus was equipped with a static loading device which used an air-actuated piston to simulate constant force impacts similar to those produced in rotary drilling. Strain gauges and a linear potentiometer were used to obtain force displacement curves on an x-y plotting oscilloscope.

Maurer found that the crater mechanism depended to some extent on the pressure differential between the borehole and the rock pore pressure. At low values of differential pressure, the crushed rock beneath the bit tooth was ejected from the crater, while at high values of differential pressure the crushed rock deformed in a plastic manner and was not ejected completely from the crater. The crater mechanism for both low and high differential fluid pressure is described in Fig. 5.17. The sequence of events shown in this figure is described by Maurer as follows.

As load is applied to a bit tooth (A), the constant pressure beneath the tooth increases until it exceeds the crushing strength of the rock and a wedge of finely powdered rock then is formed beneath the tooth (B). As the force on the tooth increases, the material in the wedge compresses and exerts high lateral forces on the solid rock surrounding the wedge until the shear stress τ exceeds the shear strength S of the solid rock and the rock fractures (C). These fractures propagate along a maximum shear surface, which intersect the direction of the principal stresses at a nearly constant angle as

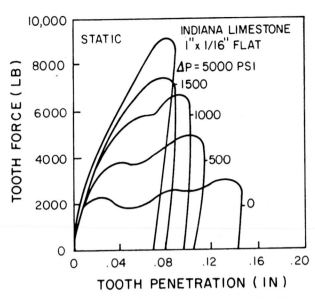

Fig. 5.18—Typical force displacement curves as a function of differential mud pressure,[7] ($\Delta p = p_{bh} - p_f$).

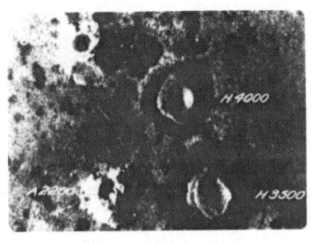

Courtesy of Hughes Tool Co.

Fig. 5.19—Example of craters formed in single-tooth impact apparatus.

predicted by the Mohr failure criteria. The force at which fracturing begins beneath the tooth is called the *threshold force*. As the force on the tooth increases above the threshold value, subsequent fracturing occurs in the region above the initial fracture, forming a zone of broken rock (D). At low differential pressure, the cuttings formed in the zone of broken rock are ejected easily from the crater (E). The bit tooth then moves forward until it reaches the bottom of the crater, and the process may be repeated (F,G). At high differential pressures, the downward pressure and frictional forces between the rock fragments prevent ejection of the fragments (E'). As the force on the tooth is increased, displacement takes place along fracture planes parallel to the initial fracture (F',G'). This gives the appearance of plastic deformation, and craters formed in this manner are called *pseudoplastic* craters. Typical force displacement curves for increasing values of differential pressure are shown in Fig. 5.18.

Examples of craters formed at both high and low values of differential pressure[2] are shown in Fig. 5.19. A 5-mm tungsten carbide penetrator was loaded to produce failure in a sample of Rush Springs sandstone. The sample was coated with plastic to simulate the buildup of a mudcake that would prevent the wellbore fluid from entering the pore space of the rock and equalizing the pressure differential. The two craters on the left were made with the formation at atmospheric pressure and with bit tooth loads of 1,600 and 2,200 lbf. The chips formed were removed easily. The two craters on the right were made at a pressure differential of 5,000 psi and bit tooth loads of 3,500 and 4,000 lbf. The material extruded from the craters that is characteristic of pseudoplastic crater formation was not removed easily, although it was weaker than the undisturbed formation.

High-speed movies[8] of full-scale bits drilling at atmospheric conditions with air as the circulating fluid have verified that the mechanisms of failure for rolling cutter bits with little or no offset is not too different from that observed in single bit-tooth impact experiments. This is shown in the sequence of photographs in Figs.

5.20a and 5.20b. Starting in Photograph 1, a tooth starts to apply pressure to the rock as the cone rolls forward. In Photograph 4, the threshold force has been transmitted to the rock and fracture is initiated. In Photograph 5, which is shown in more detail in Fig. 5.20b, ejection of the rock fragments from the crater proceeds in an explosive manner. Chips continue to be ejected in Photographs 6 through 9 as the tooth sinks further into the rock. Finally, in Photograph 10, particle discharge diminishes and the next tooth begins to apply force to the rock.

The drilling action of rolling cutter bits designed with a large offset for drilling soft, plastic formations is considerably more complex than the simple crushing action that results when no offset is used. Since each cone alternately rolls and drags, considerable wedging and twisting action is present. Shown in Fig. 5.21 is a comparison of the bottomhole patterns generated by a bit with no offset and a bit with considerable offset. The alternate rolling and dragging action of the high offset cones is evident from the bottomhole pattern of the bit teeth.

5.3 Bit Selection and Evaluation

Unfortunately, the selection of the best available bit for the job, like the selection of the best drilling fluid or drilling cement composition, can be determined only by trial and error. The most valid criterion for comparing the performance of various bits is the drilling cost per unit interval drilled. The cost-per-foot formula presented in Chap. 1 (Eq. 1.16) can be used for this purpose. Since no amount of arithmetic allows us to drill the same section of hole more than once, comparisons must be made between succeeding bits in a given well or between bits used to drill the same formations in different wells. The formations drilled with a given bit on a previous nearby well can be correlated to the well in progress using well logs and mud logging records.

The initial selection of bit type in a wildcat area can be made on the basis of what is known about the formation characteristics and drilling cost in an area. The terms

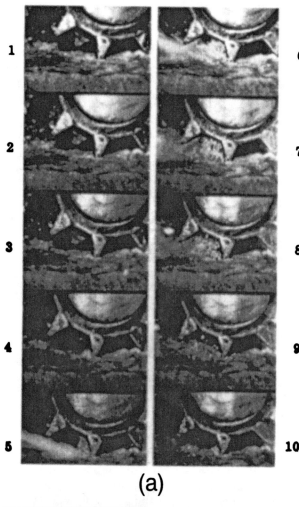

(a)

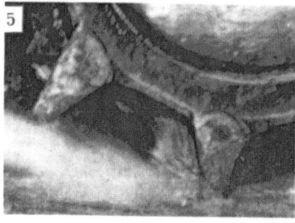

(b)

Fig. 5.20—Elastic rock failure beneath a rolling cutter bit.[8]: (a) high-speed photographic sequences and (b) enlargement of Sequence 5 showing ejection of crushed rock from crater.

usually used by drilling engineers to describe the formation characteristics are *drillability* and *abrasiveness*. The drillability of the formation is a measure of how easy the formation is to drill. It is inversely related to the compressive strength of the rock, although other factors are also important. Drillability generally tends to decrease with depth in a given area. The abrasiveness of the formation is a measure of how rapidly the teeth of a milled tooth bit will wear when drilling the formation. Although there are some exceptions, the abrasiveness tends to increase as the drillability decreases. Shown in Table 5.5 is a listing of bit types often used to drill various formation types. The formation types are listed approximately in order of the decreasing drillability and increasing abrasiveness.

In the absence of prior bit records, several rules of thumb often are used for initial bit selection. General rules for bit selection, like rules of grammar, are famous for the exception to the rules. Thus, the drilling cost per foot must eventually be the final criterion applied. However, the rules indicate certain tendencies shown to be common on the basis of past experience. Some of the rules of thumb used by many drilling engineers are as follows.

1. The IADC classification charts (Tables 5.1 through 5.3) provide an approximate listing of the bit types applicable in a given formation hardness.

2. The initial bit type and features selected should be governed by bit cost considerations. Premium rolling-cutter design features and high-cost diamond and PCD drag bits tend to be more applicable when the daily cost of the drilling operation is high. The cost of the bit probably should not exceed the rig cost per day.

3. Three-cone rolling-cutter bits are the most versatile bit type available and are a good initial choice for the shallow portion of the well.

4. When using a rolling-cutter bit:

a. Use the longest tooth size possible.

b. A *small amount* of tooth breakage should be tolerated rather than selecting a shorter tooth size.

c. When enough weight cannot be applied economically to a milled tooth bit to cause self-sharpening tooth wear, a longer tooth size should be used.

d. When the rate of tooth wear is much less than the rate of bearing wear, select a longer tooth size, a better bearing design, or apply more bit weight.

e. When the rate of bearing wear is much less than the rate of tooth wear, select a shorter tooth size, a more economical bearing design, or apply less bit weight.

5. Diamond drag bits perform best in nonbrittle formations having a plastic mode of failure, especially in the bottom portion of a deep well, where the high cost of tripping operations favors a long bit life, and a small hole size favors the simplicity of a drag bit design.

6. PCD drag bits perform best in uniform sections of carbonates or evaporites that are not broken up with hard shale stringers or other brittle rock types.

7. PCD drag bits should not be used in gummy formations, which have a strong tendency to stick to the bit cutters.

Since bit selection is done largely by trial and error, the importance of carefully evaluating a dull bit when it is removed from the well cannot be overstressed. It is

**TABLE 5.5—BIT TYPES OFTEN USED
IN VARIOUS FORMATION TYPES**

IADC Bit Classification	Formation
1–1 1–2 5–1 6–2	Soft formations having low compressive strength and high drillability (soft shales, clays, red beds, salt, soft limestone, unconsolidated formations, etc.)
1–3 6–1	Soft to medium formations or soft interspersed with harder streaks (firm, unconsolidated or sandy shales, red beds, salt, anhydrite, soft limestones, etc.)
2–1 6–2	Medium to medium hard formations (harder shales, sandy shales, shales alternating with streaks of sand and limestone, etc.)
2–3 6–2	Medium hard abrasive to hard formations (high compressive strength rock, dolomite, hard limestone, hard slaty shale, etc.)
3–1 7–2	Hard semiabrasive formations (hard sandy or chert bearing limestone, dolomite, granite, chert, etc.)
3–2 3–4 8–1	Hard abrasive formations (chert, quartzite, pyrite, granite, hard sand rock, etc.)

also important to maintain careful written records of the performance of each bit for future references. The IADC has adopted a numerical code for reporting the degree of bit wear relative to the (1) teeth, (2) bearings, and (3) bit diameter (gauge wear) structure. This code allows some of the more important aspects of bit wear to be quantified and logged quickly in the bit reports.

5.3.1 Grading Tooth Wear

The tooth wear of milled tooth bits is graded in terms of the fractional tooth height that has been worn away and is reported to the nearest eighth. For example, if half the original tooth height has been worn away, the bit will be graded as a T-4—i.e., the teeth are ⅘ worn. Unfortunately, it is sometimes difficult to characterize the tooth wear of an entire bit with a single number. Some teeth may be worn more than others, and some may be broken. Generally, the broken teeth are indicated by recording "BT" in a "remarks" column, and the average wear of the row of teeth with the most severe wear is reported. The best way to obtain the tooth wear is to measure the tooth height before and after the bit run. However, with experience, more rapid visual estimates of tooth condition can be made using a profile chart guide like the one shown in Fig. 5.22. Visual estimates are usually satisfactory when a single bit type is used in the well. Changes in the original tooth heights due to changing bit types can cause inaccurate visual estimates of tooth wear.

In some areas, unacceptably low penetration rates may occur before the tooth structure is completely worn. However, the penetration rate of the bit just before pull-

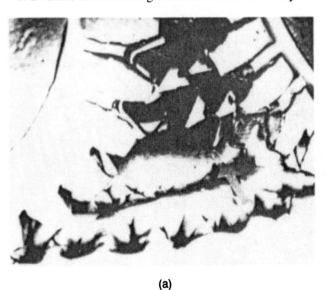

(a)

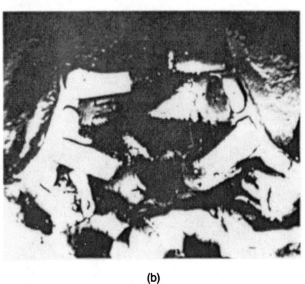

(b)

Fig. 5.21—Comparison of bottomhole patterns of hard and soft formation rolling cutter bits: (a) hard formation bit (zero cone offset) and (b) soft formation bit (maximum cone offset).

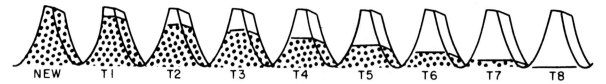

Fig. 5.22—Tooth wear guide chart for milled-tooth bits.

ing the bit should not influence the tooth wear evaluation. There are times when a T-3 will not drill, but this does not mean it should be reported as a T-8.

The cutting structures of insert bits generally are too hard to abrade as significantly as a milled steel tooth. The tooth inserts become broken or lost rather than worn. Thus, the tooth wear usually is reported as the fraction of the total number of inserts that have been broken or lost to the nearest eighth. Thus, an insert bit with half the inserts broken or lost would be graded a T-4—i.e., $\frac{4}{8}$ of the inserts are broken or lost.

5.3.2 Grading Bearing Wear

The field evaluation of bearing wear is very difficult. The bit would have to be disassembled to examine the condition of the bearings and journals. An examination of the dull bit will reveal only whether the bearings have failed or are still intact. Bearing failure usually results in (1) one or more "locked" cones so that they will no longer rotate or (2) one or more extremely loose cones so that the bearings have become exposed (Fig. 5.23).

A bearing failure is reported using the code B-8—i.e., the bearings are $\frac{8}{8}$ worn. A slightly loose cone usually is reported as a B-7. When bearing wear cannot be detected, it usually is estimated based on the number of hours of bearing life that the drilling engineer thought the bearings would last. Linear bearing wear with time is assumed in this estimate of bearing life. Thus, if a bit was pulled after 10 hours of operation and the drilling engineer felt the bearings should have lasted an additional 10 hours, the bearing wear will be reported as a B-4. A bearing grading chart such as the one shown in

Fig. 5.24 frequently is used in determining the proper bearing wear code.

5.3.3 Grading Gauge Wear

When the bit wears excessively in the base area of the rolling cones, the bit will drill an undersized hole. This can cause damage of the next bit run in the undersized hole. A ring gauge and a ruler must be used as shown in Fig. 5.25 to measure the amount of gauge wear. The loss of diameter is reported to the nearest eighth. Thus, a bit that has lost 0.5 in. of diameter is graded a G-O-4. The "O" indicates the bit is "out of gauge" and the "4" indicates the diameter has worn $\frac{4}{8}$ in. An "I" is used to indicate an "in-gauge" bit.

In addition to grading the bearings, teeth, and gauge of the bit, additional comments about the bit condition may be necessary. These remarks about the bit condition should enable those who subsequently will use the bit records to visualize readily the actual condition of the bit. Listed alphabetically in Table 5.6 are some common abbreviations used to describe the bit condition. Recall that the names used to describe various parts of a rolling cutter bit are given in Fig. 5.6.

Example 5.2. Describe the dull bit shown in Fig. 5.26. The use of a ring gauge indicated that the bit diameter has worn 1 in. from its initial value. The roller bearings have fallen out of the bit, and all the cones are very loose.

Courtesy of Smith Tool Co.

Fig. 5.23—Example of severe bearing wear.

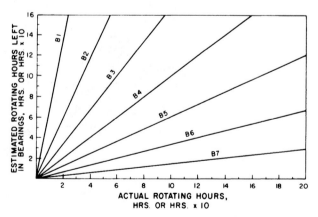

Fig. 5.24—Bearing grading guide for rolling cutter bits.

Solution. This bit should be graded using the code T-8, B-8, G-O-8 since the cutting structure is completely worn, the cones are very loose, and the bit is ⅝ in. out of gauge. In addition, "SD" should be placed in the remarks column to indicate that the shirttail is damaged.

The excessive tooth wear and gauge wear on this bit indicates a poor choice of drilling practices. The cost per foot for this bit run is probably unnecessarily high due to extremely low penetration rates of the end of the bit run. In addition, the undergauge hole drilled by this bit will reduce the efficiency of the next bit run by wasting bit life on reaming operations. A bit with additional gauge protection would be a better choice for this interval.

5.3.4 Abnormal Bit Wear

The ability to recognize the probable cause of the bit wear observed generally increases as experience is gained in evaluating dull bits run under various conditions. To provide at least a little of this experience, several bits that illustrate several types of abnormal bit wear or failure are shown in Figs. 5.27 through 5.30. Study each picture and attempt to identify the possible factors causing the type of wear shown before reading the discussion that follows.

The type of wear shown in Fig. 5.27 occurs when the cones are not free to rotate. This frequently is caused by bearing failure. However, in this case the bearings were in good condition and cone drag was caused by bit balling. Bit balling usually occurs in very soft, sticky formations when sufficient bit weight is applied to bury the teeth in the formation completely. The tendency for bit

CONDITION: 0 ⅜—BIT OUT OF GAUGE ⅜″

Courtesy of Security Rock Bits and Drill Tools

Fig. 5.25—Determination of gauge wear.

TABLE 5.6—COMMON ABBREVIATIONS USED IN DESCRIBING BIT CONDITION IN DULL BIT EVALUATION
(Courtesy of Hughes Tool Co.)

Location of Conditions		Formations (cont.)		Bit Body Conditions (cont.)		Journal Bearing Bits (cont.)	
Spearpoint	S	Hard sandy shale	HSSH	Broken circumferentially	BC	Broken spearpoint	BS
Nose	N	Sticky shale	STSH	Cracked	CC		
Middle row	M	Chert	CHE	Eroded cone shell	EC		
Heel	H	Chat	CHA				
Gauge	G	Granite	GRA	Bearing Conditions			
Cone or head number	1,2,3	Quartzite	Q				
		Pyrite	P	Bearing failure	BF		
Classification of Run		Chalk	CK	Broken bearing pin	BP		
				Broken rollers	BR		
Very good	Good +	Bit Body Conditions		Compensator plug damaged	CPD		
Good run	Good			Cone locked	CL		
Above average	Avg +	Bent legs	BL	Lost cone	LC		
Average run	Avg	Damaged bit	DB	Lost rollers	LR		
Below average	Avg –	Eroded nozzle	EN	Seal failure	SF		
Poor run	Poor	Lost nozzle	LN	Seals questionable	SQ		
Very poor run	Poor –	Plugged nozzle	PN	Seals effective	SE		
		Shirttail damaged	SD				
Formations				Journal Bearing Bits			
		Cone Teeth Conditions					
Sand	S			Seals effective	SE		
Lime	L	Broken teeth	BT	Seals questionable	SQ		
Sandy lime	SL	Balled up	BU	Seal failure	SF		
Dolomite	D	Cone dragged	CD	Bit rerunable	RR		
Sandy dolomite	SD	Cored	CR	Bit not rerunable	NR		
Anhydrite	A	Lost/loose compacts	LT	Bit was regreased	GR		
Gypsum	G	Off-center wear	OC	Pulled for torque	T		
Salt	SA	Rounded gauge	RG	Pulled on judgment			
Red beds	RB	Uniform wear	UW	(precaution)	J		
Shale	SH	Worn out of gauge	WG	Pulled for penetration rate	P		
Hard shale	HSH						
Sandy shale	SSH	Cone Shell Conditions					
		Broken axially	BA				

Courtesy of Hughes Tool Co.

Fig. 5.26—Example of a dull bit.

Courtesy of Smith Tool Co.

Fig. 5.27—Example of "cone dragged" bit wear.

balling can be reduced by applying less weight or by increasing the jet hydraulic cleaning action. A bit with a central nozzle often reduces the tendency for bit balling.

The type of wear shown in Fig. 5.28 occurs when the nose areas of the cones are worn away or lost. This frequently occurs because of excessive loads being applied to the cone tips. The cone tips break, allowing a "core" of rock to be cut in the center of the bottomhole pattern. The rock core causes subsequent abrading of inner cone metal. When this condition is detected, care must be taken on the next bit run to eliminate the formation core on bottom without breaking the cone tips of the new bit. This can be accomplished by breaking in the new bit using low bit weights and high rotary speeds.

The type of wear shown in Fig. 5.29 occurs when (1) the drilling fluid contains a high concentration of abrasive solids or (2) the circulation rate is extremely high. This problem is worse for regular bits than for jet bits since the fluid strikes directly on the cones of a regular bit. This problem usually can be eliminated through the operation of the drilling fluid desanders.

Off-center bit wear occurs (Fig. 5.30) when the bit does not rotate about the true center of the hole. This causes an oversized hole to be cut and circular ridges to develop on the bottom of the hole. These circular rings of rock wear away the cone shell area between the teeth as well as the front and back faces of the bit teeth. This problem usually indicates the need for a higher penetration rate, which could be achieved by using a bit with a longer tooth or perhaps by increasing the bit weight. Also, the bottomhole assembly could be altered to ensure that the bit is properly stabilized and centered in the borehole.

5.4 Factors Affecting Tooth Wear

One purpose for evaluating the condition of the dull bit is to provide insight about the selection of a more suitable time interval of bit use. If the dull bit evaluation indicates that the bit was pulled *green* (i.e., with considerable bit life remaining), expensive rig time may

have been wasted on unnecessary trip time. However, if the time interval of bit use is increased too much, the bit may break apart leaving *junk* in the hole. This will require an additional trip to *fish* the junk from the hole or may reduce greatly the efficiency of the next bit if an attempt is made to drill past the junk. Thus, a knowledge of the instantaneous rate of bit wear is needed to determine how much the time interval of bit use can be increased safely. Since drilling practices are not always the same for the new and old bit runs, a knowledge of how the various drilling parameters affect the instantaneous rate of bit wear also is needed. The rate of tooth wear depends primarily on (1) formation abrasiveness, (2) tooth geometry, (3) bit weight, (4) rotary speed, and (5) the cleaning and cooling action of the drilling fluid.

5.4.1 Effect of Tooth Height on Rate of Tooth Wear

Campbell and Mitchell[9] showed experimentally that the rate at which the height of a steel tooth can be abraded away by a grinding wheel is directly proportional to the area of the tooth in contact with the grinding wheel. The shape of steel bit teeth is generally triangular in cross section when viewed from either a front or side view. Thus, almost all milled tooth bits have teeth that can be described using the geometry shown in Fig. 5.31. The bit tooth initially has a contact area given by

$$A_i = w_{x1}\, w_{y1}.$$

After removal of tooth height, L_r, of the original tooth height, L_i, the bit tooth has a contact area given by

$$A = w_x w_y = \left[w_{x1} + \frac{L_r}{L_i}\left(w_{x2} - w_{x1}\right) \right]$$

$$\cdot \left[w_{y1} + \frac{L_r}{L_i}\left(w_{y2} - w_{y1}\right) \right].$$

Courtesy of Security Bit and Drill Tools
Fig. 5.28—Example of "cored" bit wear.

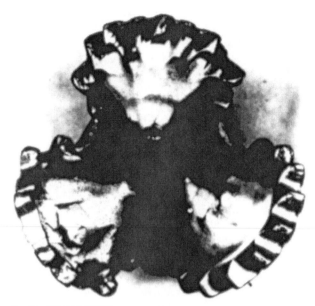

Courtesy of Hughes Tool Co.
Fig. 5.29—Example of "eroded teeth" bit wear.

The ratio L_r/L_i is defined as the fractional tooth wear h:

$$h = L_r/L_i. \qquad\qquad\qquad\qquad\qquad (5.4)$$

Expressing the contact area A in terms of fractional tooth wear h yields

$$A = \left[w_{x1} + h\left(w_{x2} - w_{x1} \right) \right]\left[w_{y1} + h\left(w_{y2} - w_{y1} \right) \right]$$
$$= \left(w_{x1} w_{y1} \right) + \left[w_{y1}\left(w_{x2} - w_{x1} \right) \right.$$
$$\left. + w_{x1}\left(w_{y2} - w_{y1} \right) \right] h$$
$$+ \left[\left(w_{x2} - w_{x1} \right)\left(w_{y2} - w_{y1} \right) \right] h^2.$$

If we define the geometry constants G_1 and G_2 by

$$G_1 = \left[w_{y1}\left(w_{x2} - w_{x1} \right) + w_{x1}\left(w_{y2} - w_{y1} \right) \right] A_i$$

and

$$G_2 = \left[\left(w_{x2} - w_{x1} \right)\left(w_{y2} - w_{y1} \right) \right] A_i,$$

the contact area A can be expressed by

$$A = A_i(1 + G_1 h + G_2 h^2).$$

Since the instantaneous wear rate dh/dt is proportional to the inverse of the contact area A,

$$\frac{dh}{dt} \propto \frac{1}{A_i(1 + G_1 h + G_2 h^2)}.$$

Courtesy of Hughes Tool Co.

Fig. 5.30—Example "off-center" bit wear.

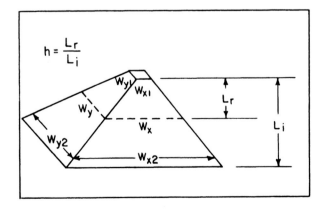

Fig. 5.31—Typical shape of a milled-tooth as a function of fractional tooth wear, h.

Fig. 5.32—Example cutter wear on a PCD drag bit. [10]

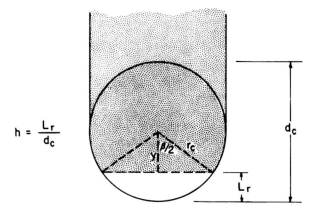

$$h = \frac{L_r}{d_c}$$

Fig. 5.33—PCD blank geometry as a function of fractional cutter wear, h, for a zero-back-rake angle.

The initial wear rate, when $h = 0$, is proportional to A_i. Thus, expressing dh/dt in terms of a standard initial wear rate $(dh/dt)_s$ gives

$$\frac{dh}{dt} \propto \left(\frac{dh}{dt}\right)_s \frac{1}{(1+G_1 h + G_2 h^2)} . \qquad \ldots \ldots \ldots (5.5a)$$

For most bit types, the dimension $(w_{x2} - w_{x1})$ will be small compared with $(w_{y2} - w_{y1})$. This allows a constant H_2 to be chosen such that the wear rate can be approximated using

$$\frac{dh}{dt} \propto \left(\frac{dh}{dt}\right)_s \frac{1}{(1+H_2 h)} . \qquad \ldots \ldots \ldots \ldots \ldots (5.5b)$$

The use of Eq. 5.5b in place of Eq. 5.5a greatly simplifies the calculation of tooth wear as a function of rotating time. A case-hardened bit tooth or a tooth with hard facing on one side often will have a self-sharpening type of tooth wear. Even though the mechanism of self-sharpening tooth wear is somewhat different than in the abrasive wear experiments of Campbell and Mitchell, a constant H_2 usually can be selected such that the instantaneous wear rate can be predicted using Eq. 5.5b.

Insert teeth used in rolling-cutter bits usually fail by fracturing of the brittle tungsten carbide. For this tooth type, fractional tooth wear, h, represents the fraction of the total number of bit teeth that have been broken. The wear rate (dh/dt) does not decrease with increasing fractional tooth wear, h. To the contrary, there is some evidence that the tooth breakage accelerates as the number of broken teeth beneath the bit increases. This type of behavior could be modeled with a negative value for H_2 in Eq. 5.5b. However, this phenomenon has not been studied in detail and in practice a value of zero is recommended for H_2 when using insert bits.

Diamond bits also wear by breakage or loss of the diamond cutter elements. The wear rate of diamond bits is thus not sensitive to the fractional cutter wear. The wear rate of diamond bits is far more sensitive to the amount of cooling provided by the flow of drilling fluid across the face of the bit.

PCD blanks tend to wear in a manner somewhat similar to a steel-tooth cutter due to the random orientation of the individual diamond crystals (Fig. 5.32). [10] However, the circular shape of the PCD blank provides a different relationship between fractional tooth wear, h, and cutter contact area. For a zero back-rake angle, the cutter contact area is proportional to the length of the chord, defined by the lower surface of the cutter remaining after removal of the cutter height, L_r (Fig. 5.33), since the fractional tooth wear, h, given by

$$h = \frac{L_r}{d_c}$$

and the dimension y shown in Fig. 5.33 is

$$y = r_c \left(\cos\frac{\beta}{2}\right).$$

Then

$$h = \frac{r_c - y}{d_c} = \frac{r_c - \left(r_c \cos\frac{\beta}{2}\right)}{2r_c} = \frac{1 - \cos\frac{\beta}{2}}{2}.$$

Solving this expression for the subtended angle, β, yields

$$\cos\frac{\beta}{2} = 1 - 2h. \quad\dots\dots\dots\dots\dots (5.5c)$$

Since the contact area is directly proportional to the chord length subtended by the angle β, then

$$A \propto 2\left(\frac{d_c}{2}\right)\sin\frac{\beta}{2},$$

and the wear rate (dh/dt) is inversely proportional to this contact area.

$$\frac{dh}{dt} \propto \left(\frac{dh}{dt}\right)_s \frac{1}{d_c \sin(\beta/2)}. \quad\dots\dots\dots (5.5d)$$

The wear rate (dh/dt) decreases with increasing fractional tooth wear, h, between 0 and 0.5. Above this range, the wear rate increases with increasing h.

For nonzero rake angles, the total contact area of both the PCD layer and the tungsten carbide substrate becomes more complex. However, the above analysis remains representative of the geometry of the thin PCD layer, which is believed to be the predominant contribution to the wear resistance of the PCD blank.

5.4.2 Effect of Bit Weight on Rate of Tooth Wear

Galle and Woods[11] published one of the first equations for predicting the effect of bit weight on the instantaneous rate of tooth wear. The relation assumed by Galle and Woods is given by

$$\frac{dh}{dt} \propto \frac{1}{1 - \log\left(\frac{W}{d_b}\right)}, \quad\dots\dots\dots\dots (5.6a)$$

where

W = bit weight in 1,000-lbm units,

d_b = bit diameter in inches, and

$W/d_b < 10.0$.

The wear rate at various bit weights can be expressed in terms of a standard wear rate that would occur for a bit weight of 4,000 lbf/in. Thus, the wear rate relative to this standard wear rate is given by

$$\frac{dh}{dt} \propto \frac{0.3979\left(\frac{dh}{dt}\right)_s}{1 - \log\left(\frac{W}{d_b}\right)}. \quad\dots\dots\dots (5.6b)$$

Note that dh/dt becomes infinite for $W/d_b = 10$. Thus, this equation predicts the teeth would fail instantaneously if 10,000 lbf/in. of bit diameter were applied. Later authors[11-15] used a simpler relation between the weight and tooth wear rate. Perhaps the most commonly used relation is given by

$$\frac{dh}{dt} \propto \frac{1}{\left(\frac{W}{d_b}\right)_m - \frac{W}{d_b}}, \quad\dots\dots\dots\dots (5.7a)$$

where $(W/d_b)_m$ is the maximum bit weight per inch of bit diameter at which the bit teeth would fail instantaneously and $W/d_b < (W/d_b)_m$. Expressing this relation in terms of a standard wear rate at 4,000 lbf/in. of bit diameter yields

$$\frac{dh}{dt} \propto \left(\frac{dh}{dt}\right)_s \left[\frac{\left(\frac{W}{d_b}\right)_m - 4}{\left(\frac{W}{d_b}\right)_m - \frac{W}{d_b}}\right]. \quad\dots\dots (5.7b)$$

Shown in Table 5.7 is a comparison of the relative wear rates predicted by Eqs. 5.6b and 5.7b assuming a maximum bit weight of 10,000 lbf/in. Since somewhat similar results are obtained over the range of conditions usually encountered in the field, the simpler relation given by Eq. 5.7b is more widely used. However, neither Eq. 5.6b nor Eq. 5.7b has been verified by published experimental data.

5.4.3 Effect of Rotary Speed on Rate of Tooth Wear

The first published relation between the instantaneous rate of tooth wear and the rotary speed also was presented by Galle and Woods for milled-tooth bits.[11] The Galle and Woods relation is given by

$$\frac{dh}{dt} \propto N + 4.34 \times 10^{-5} N^3. \quad\dots\dots\dots (5.8)$$

However, several more recent authors[11-15] have shown that essentially the same results can be obtained using the simpler relation:

$$\frac{dh}{dt} \propto (N)^{H_1}, \quad\dots\dots\dots\dots\dots (5.9a)$$

where H_1 is a constant. Also, H_1 was found to vary with the bit type used. The Galle and Woods relation applied only to milled-tooth bit types designed for use in soft formations. Expressing the tooth wear rate in terms of a standard wear rate that would occur at 60 rpm yields

$$\frac{dh}{dt} \propto \left(\frac{dh}{dt}\right)_s \left(\frac{N}{60}\right)^{H_1}. \quad\dots\dots\dots\dots (5.9b)$$

5.4.4 Effect of Hydraulics on Rate of Tooth Wear

The effect of the cooling and cleaning action of the drilling fluid on the cutter wear rate (dh/dt) is much more important for diamond and PCD drag bits than for rolling-cutter bits. Each diamond cutter must receive sufficient flow to prevent the buildup of excessive cutter

TABLE 5.7—COMPARISON OF EQUATIONS FOR MODELING THE EFFECT OF TOOTH HEIGHT ON TOOTH WEAR RATE

Bit Weight (lbf) per Inch (W/d_b)	Relative Wear Rate $\left[\dfrac{dh}{dt} \middle/ \left(\dfrac{dh}{dt}\right)_s \right]$	
	Eq. 5.6b	Eq. 5.7b
1	0.4	0.7
2	0.6	0.8
3	0.8	0.9
4	1.0	1.0
5	1.3	1.2
6	1.8	1.5

TABLE 5.8—RECOMMENDED TOOTH-WEAR PARAMETERS FOR ROLLING-CUTTER BITS

Bit Class	H_1	H_2	$(W/d)_{max}$
1–1 to 1–2	1.90	7	7.0
1–3 to 1–4	1.84	6	8.0
2–1 to 2–2	1.80	5	8.5
2–3	1.76	4	9.0
3–1	1.70	3	10.0
3–2	1.65	2	10.0
3–3	1.60	2	10.0
4–1	1.50	2	10.0

temperatures. The flow velocities must also be maintained high enough to prevent clogging of fluid passages with rock cuttings. The design of the fluid distribution passages in a diamond or PCD drag bit is extremely important and varies considerably among the various bits available. However, the manufacturer will generally specify the total flow area (TFA) of the fluid distribution system for each bit. In addition, the bit manufacturer will specify a recommended drilling-fluid flow rate or pressure drop across the bit face.

Mathematical models for estimating the effect of hydraulics on the rate of cutter wear have not yet been employed. The development of such models would be extremely difficult because of the wide variety of bit designs available. It is generally assumed that as long as the flow is present to clean and to cool the cutters, the effect of hydraulics on cutter wear rate can be ignored.

5.4.5 Tooth Wear Equation

A composite tooth wear equation can be obtained by combining the relations approximating the effect of tooth geometry, bit weight, and rotary speed on the rate of tooth wear. [15] Thus, the instantaneous rate of tooth wear is given by

$$\frac{dh}{dt} = \frac{1}{\tau_H} \left(\frac{N}{60}\right)^{H_1} \left[\frac{\left(\frac{W}{d_b}\right)_m - 4}{\left(\frac{W}{d_b}\right)_m - \left(\frac{W}{d_b}\right)} \right]$$

$$\cdot \left(\frac{1 + H_2/2}{1 + H_2 h}\right), \qquad \qquad (5.10)$$

where

h = fractional tooth height that has been worn away,
t = time, hours
$H_1, H_2,$
$(W/d_b)_m$ = constants,
W = bit weight, 1,000-lbf units,
N = rotary speed, rpm, and
τ_H = formation abrasiveness constant, hours.

The rock bit classification scheme shown in Table 5.3 can be used to characterize the many bit types available from the four major bit manufacturing companies. Recommended values of H_1, H_2, and $W/(d_b)_m$ are shown in Table 5.8 for the various rolling-cutter rock bit classes.

The tooth wear rate formula given by Eq. 5.10 has been normalized so that the abrasiveness constant τ_H is numerically equal to the time in hours required to completely dull the bit teeth of the given bit type when operated at a constant bit weight of 4,000 lbf/in. and a constant rotary speed of 60 rpm. The average formation abrasiveness encountered during a bit run can be evaluated using Eq. 5.10 and the final tooth wear h_f observed after pulling the bit. If we define a tooth wear parameter J_2 using

$$J_2 = \left[\frac{\left(\frac{W}{d_b}\right)_m - \left(\frac{W}{d_b}\right)}{\left(\frac{W}{d_b}\right)_m - 4} \right] \left(\frac{60}{N}\right)^{H_1} \left(\frac{1}{1 + H_2/2}\right),$$

$$\qquad \qquad (5.11)$$

Eq. 5.10 can be expressed by

$$\int_o^{t_b} dt = J_2 \tau_H \int_o^{h_f} (1 + H_2 h) \, dh. \qquad (5.12a)$$

Integration of this equation yields

$$t_b = J_2 \tau_H (h_f + H_2 h_f^2 / 2). \qquad (5.12b)$$

Solving for the abrasiveness constant τ_H gives

$$\tau_H = \frac{t_b}{J_2 (h_f + H_2 h_f^2 / 2)}. \qquad (5.13)$$

Although Eqs. 5.10 through 5.15 were developed for use in modeling the loss of tooth height of a milled tooth bit, they have also been applied with some degree of success to describe the loss of insert teeth by breakage. Insert bits are generally operated at lower rotary speeds than milled-tooth bits to reduce impact loading on the brittle tungsten carbide inserts. In hard formations, rotary speeds above about 50 rpm may quickly shatter the insert. [16]

Example 5.3. An 8.5-in. Class 1-3-1 bit drilled from a depth of 8,179 to 8,404 ft in 10.5 hours. The average bit weight and rotary speed used for the bit run was 45,000 lbf and 90 rpm, respectively. When the bit was pulled, it was graded T-5, B-4, G-1. Compute the average formation abrasiveness for this depth interval. Also estimate the time required to dull the teeth completely using the same bit weight and rotary speed.

Solution. Using Table 5.8 we obtain $H_1 = 1.84$, $H_2 = 6$, and $(W/d_b)_m = 8.0$. Using Eq. 5.11 we obtain

$$J_2 = \frac{8.0 - 45/8.5}{8.0 - 4.0} \left(\frac{60}{90}\right)^{1.84}$$

$$\cdot \frac{1}{1 + 6/2} = 0.08.$$

Solving Eq. 5.13 for the abrasiveness constant using a final fractional tooth dullness of $\frac{5}{8}$ or $0.625(T-5)$ gives

$$\tau_H = \frac{10.5 \text{ hours}}{0.080[0.625 + 6(0.625)^2/2]}$$

$$= 73.0 \text{ hours}.$$

The time required to dull the teeth completely ($hf = 1.0$) can be obtained from Eq. 5.12:

$$t_b = 0.08(73.0)\left[1 + 6(1)^2/2\right] = 23.4 \text{ hours}.$$

5.5 Factors Affecting Bearing Wear

The prediction of bearing wear is much more difficult than the prediction of tooth wear. Like tooth wear, the instantaneous rate of bearing wear depends on the current condition of the bit. After the bearing surfaces become damaged, the rate of bearing wear increases greatly. However, since the bearing surfaces cannot be examined readily during the dull bit evaluation, a linear rate of bearing wear usually is assumed. Also, bearing manufacturers have found that for a given applied force, the bearing life can be expressed in terms of total revolutions as long as the rotary speed is low enough to prevent an excessive temperature increase. Thus, bit bearing life usually is assumed to vary linearly with rotary speed.

The three main types of bearing assemblies used in rolling cutter bits are (1) nonsealed roller, (2) sealed roller, and (3) sealed journal. The price of the bit is lowest for the nonsealed roller and highest for the sealed journal.

The effect of bit weight on bearing life depends on the number and type of bearings used and whether or not the bearings are sealed. When the bearings are not sealed, bearing lubrication is accomplished with the drilling fluid, and the mud properties also affect bearing life.

The hydraulic action of the drilling fluid at the bit is also thought to have some effect on bearing life. As flow rate increases, the ability of the fluid to cool the bearings also increases. However, it is generally believed that flow rates sufficient to lift cuttings will also be sufficient to prevent excessive temperature buildup in the bearings.

TABLE 5.9—RECOMMENDED BEARING WEAR EXPONENT FOR ROLLING-CUTTER BITS

Bearing Type	Drilling Fluid Type	B_1	B_2
Nonsealed	barite mud	1.0	1.0
	sulfide mud	1.0	1.0
	water	1.0	1.2
	clay/water mud	1.0	1.5
	oil-base mud	1.0	2.0
Sealed roller bearings	—	0.70	0.85
Sealed journal bearings	—	1.6	1.00

Lummus[17] has indicated that too high a jet velocity can be detrimental to bearing life. Erosion of bit metal can occur, which leads to failure of the bearing grease seals. In the example discussed by Lummus, this phenomenon was important for bit hydraulic horsepower values above 4.5 hp/sq in. of hole bottom. However, a general model for predicting the effect of hydraulics on bearing wear was not presented.

A bearing wear formula[15] frequently used to estimate bearing life is given by

$$\frac{db}{dt} = \frac{1}{\tau_B}\left(\frac{N}{60}\right)^{B_1}\left(\frac{W}{4d_b}\right)^{B_2}, \quad \ldots \ldots \ldots (5.14)$$

where

b = fractional bearing life that has been consumed,
t = time, hours,
N = rotary speed, rpm,
W = bit weight, 1,000 lbf,
d_b = bit diameter, inches,
B_1, B_2 = bearing wear exponents, and
τ_B = bearing constant, hours.

Recommended values of the bearing wear exponent are given in Table 5.9. Note that the bearing wear formula given by Eq. 5.14 is normalized so that the bearing constant, τ_B, is numerically equal to the life of bearings if the bit is operated at 4,000 lbf/in. and 60 rpm. The bearing constant can be evaluated using Eq. 5.14 and the results of a dull bit evaluation. If we define a bearing wear parameter J_3 using

$$J_3 = \left(\frac{60}{N}\right)^{B_1}\left(\frac{4d_b}{W}\right)^{B_2}, \quad \ldots \ldots \ldots \ldots \ldots (5.15)$$

Eq. 5.14 can be expressed by

$$\int_o^{t_b} dt = J_3 \tau_B \int_o^{b_f} db,$$

where b_f is the final bearing wear observed after pulling the bit. Integration of this equation yields

$$t_b = J_3 \tau_B b_f. \quad \ldots \ldots \ldots \ldots \ldots \ldots \ldots (5.16)$$

Solving for the bearing constant τ_B gives

$$\tau_B = \frac{t_b}{J_3 b_f} . \qquad \qquad \text{...............(5.17)}$$

The bearing constant τ_B for an intermediate-size sealed roller-bearing-type bit is usually about 45 hours. An intermediate-size journal-type bit usually has a bearing constant of about 100 hours. However, bearing performance has a quite large statistical variation. This "statistical variation" is probably a function of accidental damage to the bearing seals (1) when the bit is run in the hole, (2) when the bit is placed on bottom and the new bottomhole pattern is established, or (3) when the bit is not stabilized properly. Also, the numerical value of the bearing constant depends on the selected value of the bearing wear exponents B_1 and B_2.

Example 5.4. Compute the bearing constant for a 7.875-in., Class 6-1-6 (sealed journal bearings) bit that was graded T-5, B-6, G-I after drilling 64 hours at 30,000 lbf and 70 rpm.

Solution. From Table 5.9, $B_1 = 1.6$ and $B_2 = 1.0$. Using Eq. 5.15, we obtain

$$J_3 = \left(\frac{60}{70}\right)^{1.6} \left[\frac{4(7.875)}{30}\right]^{1.0} = 0.820.$$

Solving Eq. 5.17 for the bearing constant using $b_f = \frac{5}{8}$ yields

$$\tau_B = \frac{64 \text{ hours}}{0.820(0.75)} = 104 \text{ hours.}$$

5.6 Terminating a Bit Run

There is almost always some uncertainty about the best time to terminate a bit run and begin tripping operations. The use of the tooth wear and bearing wear equations will provide, at best, a rough estimate of when the bit will be completely worn. In addition, it is helpful to monitor the rotary table torque. When the bearings become badly worn, one or more of the cones frequently will lock and cause a sudden increase or large fluctuation in the rotary torque needed to rotate the bit.

When the penetration rate decreases rapidly as bit wear progresses, it may be advisable to pull the bit before it is completely worn. If the lithology is somewhat uniform, the total drilling cost can be minimized by minimizing the cost of each bit run. In this case, the best time to terminate the bit run can be determined by keeping a current estimate of the cost per foot for the bit run, assuming that the bit would be pulled at the current depth. Even if significant bit life remains, the bit should be pulled when the computed cost per foot begins to increase. However, if the lithology is not uniform, this procedure will not always result in the minimum total well cost. In this case, an effective criterion for determining optimum bit

life can be established only after enough wells are drilled in the area to define the lithologic variations. For example, it is sometimes desirable to drill an abrasive formation with an already dull bit and then place a sharp bit in the next shale section. Alternatively, it may be best to terminate a bit run in order to place a hard formation bit in an extremely hard abrasive section where severe gauge problems are likely to develop.

Example 5.5. Determine the optimum bit life for the bit run described in the table below. The lithology is known to be essentially uniform in this area. The tooth wear parameter J_2 has a value of 0.4, the constant H_2 has a value of 6.0, and the bearing wear parameter J_3 has a value of 0.55. The formation abrasiveness constant τ_H has a value of 50 hours, and the bearing constant τ_B has a value of 30 hours. The bit cost is $800, the rig cost is $600/hr, and the trip time is 10 hours.

Footage ΔD (ft)	Drilling Time $t_b + t_c$ (hours)	Remarks
0	0	New bit
30	2.0	
50	4.0	
65	6.0	
77	8.0	
87	10.0	
96	12.0	
104	14.0	
111	16.0	Torque increased

Solution. The time required to wear out the teeth can be computed using Eq. 5.12:

$$t_b = (0.4)(50)\left[1 + 6(1)^2/2\right] = 80 \text{ hours.}$$

The time required to wear out the bearings can be computed using Eq. 5.16:

$$t_b = 0.55(30)(1) = 16.5 \text{ hours.}$$

The cost per foot of the bit run at various depths can be computed using Eq. 1.16. Thus, the overall cost per foot of the bit run that would result if the bit were pulled at the various depths shown are as follows.

Footage ΔD (ft)	Drilling Time $t_b + t_c$ (hours)	Drilling Cost C_f ($/ft)
0	0	0.0
30	2.0	266.66
50	4.0	184.00
65	6.0	160.00
77	8.0	150.65
87	10.0	147.12
96	12.0	145.83
104	14.0	146.15
111	16.0	147.75

Note that the lowest drilling cost would have resulted if the bit was pulled after 12 hours.

Alternate Solution. It usually is easier for field personnel to determine the point of minimum drilling cost using a graphical procedure. Eq. 1.16 can be rearranged to give

$$\frac{C_f}{C_r} = \frac{\dfrac{C_b}{C_r} + t_t + (t_b + t_c)}{\Delta D}.$$

Thus, if $C_b/C_r + t_t + (t_b + t_c)$ is plotted vs. ΔD, the cost per foot will be a minimum when the slope $\Delta D/(C_b/C_r + t_t + t_b + t_c)$ is a maximum.

$$\frac{C_b}{C_r} + t_t = \frac{800}{600} + 10 = 11.33 \text{ hours.}$$

This is plotted as a negative offset in the graphical construction shown in Fig. 5.34. Next, footage drilled is plotted vs. time and the maximum slope of a line starting at -11.33 hours is noted. The plot in Fig. 5.34 indicates a maximum slope after about 12 hours or 96 ft.

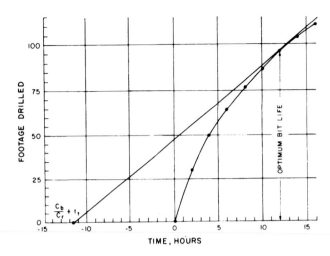

Fig. 5.34—Graphical determination of optimal bit life in uniform lithology.

5.7 Factors Affecting Penetration Rate

The rate of penetration achieved with the bit, as well as the rate of bit wear, has an obvious and direct bearing on the cost per foot drilled. The most important variables affecting penetration rate that have been identified and studied include (1) bit type, (2) formation characteristics, (3) drilling fluid properties, (4) bit operating conditions (bit weight and rotary speed), (5) bit tooth wear, and (6) bit hydraulics.

A considerable amount of experimental work has been done to study the effect of these variables on drilling rate. In most of this experimental work, the effect of a single variable was studied while holding the other variables constant.

5.7.1 Bit Type

The bit type selected has a large effect on penetration rate. For rolling cutter bits, the initial penetration rate is often highest in a given formation when using bits with long teeth and a large cone offset angle. However, these bits are practical only in soft formations because of a rapid tooth destruction and decline in penetration rate in hard formations. The lowest cost per foot drilled usually is obtained when using the longest tooth bit that will give a tooth life consistent with the bearing life at optimum bit operating conditions.

Drag bits are designed to obtain a given penetration rate. As discussed previously, drag bits give a wedging-type rock failure in which the bit penetration per revolution depends on the number of blades and the bottom cutting angle. The diamond and PCD bits are designed for a given penetration per revolution by the selection of the size and number of diamonds or PCD blanks. The width and number of cutters can be used to compute the effective number of blades. The length of the cutters projecting from the face of the bit (less the bottom clearance) limits the depth of the cut.

5.7.2 Formation Characteristics

The elastic limit and ultimate strength of the formation are the most important formation properties affecting penetration rate. The shear strength predicted by the Mohr failure criteria sometimes is used to characterize the strength of the formation. Maurer[7] has reported that the crater volume produced beneath a single tooth is inversely proportional to both the compressive strength of the rock and the shear strength of the rock. Bingham[18] found that the threshold force required to initiate drilling in a given rock at atmospheric pressure could be correlated to the shear strength of the rock as determined in a compression test at atmospheric pressure. To determine the shear strength from a single compression test, an average angle of internal friction of 35° was assumed. The angle of internal friction varies from about 30 to 40° for most rocks. Applying Eq. 5.3b for a standard compression test at atmospheric pressure ($\sigma_3 = 0$) gives

$$\tau_0 = \frac{1}{2}(\sigma_1 - 0)\sin(90 - \theta) = \frac{\sigma_1}{2}\cos\theta.$$

The threshold force or bit weight $(W/d)_t$ required to initiate drilling was obtained by plotting drilling rate as a function of bit weight per bit diameter and then extrapolating back to a zero drilling rate. The laboratory correlation obtained in this manner is shown in Fig. 5.35.

The permeability of the formation also has a significant effect on the penetration rate. In permeable rocks, the drilling fluid filtrate can move into the rock ahead of the bit and equalize the pressure differential acting on the chips formed beneath each tooth. This would tend to promote the more explosive elastic mode of crater formation described in Fig. 5.17. It also can be argued that the nature of the fluids contained in the pore spaces of the rock also affects this mechanism since more filtrate volume would be required to equalize the pressure in a rock containing gas than in a rock containing liquid.

The mineral composition of the rock also has some effect on penetration rate. Rocks containing hard, abrasive minerals can cause rapid dulling of the bit teeth. Rocks containing gummy clay minerals can cause the bit to ball up and drill in a very inefficient manner.

5.7.3 Drilling Fluid Properties

The properties of the drilling fluid reported to affect the penetration rate include (1) density, (2) rheological flow properties, (3) filtration characteristics, (4) solids content and size distribution, and (5) chemical composition.

Penetration rate tends to decrease with increasing fluid density, viscosity, and solids content, and tends to increase with increasing filtration rate. The density, solids content, and filtration characteristics of the mud control the pressure differential across the zone of crushed rock beneath the bit. The fluid viscosity controls the parasitic frictional losses in the drillstring and, thus, the hydraulic energy available at the bit jets for cleaning. There is also experimental evidence [16] that increasing viscosity reduces penetration rate even when the bit is perfectly clean. The chemical composition of the fluid has an effect on penetration rate in that the hydration rate and bit balling tendency of some clays are affected by the chemical composition of the fluid.

It has been reported [16] that the presence of colloid-size particles, which are less than 1 micron (1 μm), are an order of magnitude more detrimental to penetration rate than are particles coarser than about 30 μm. It is believed that the colloidal particles are much more efficient at plugging off the filtration beneath the bit. The development of the nondispersed, polymer muds discussed in Sections 2.3.9 through 2.3.11 of Chap. 2 was aimed at reducing the concentration of colloidal-size particles present.

The effect of drilling fluid density and the resulting bottomhole pressure on penetration rate has been studied by several authors. [7,19-24] The previously discussed experiments of Maurer, [7] which were conducted using a single bit tooth under simulated borehole conditions, have provided some insight into the mechanism by which an increase in drilling fluid density causes a decrease in penetration rate for rolling cutter bits. An increase in drilling fluid density causes an increase in the bottomhole pressure beneath the bit and, thus, an increase in the pressure differential between the borehole pressure and the formation fluid pressure. This pressure differential between the borehole pressure and formation

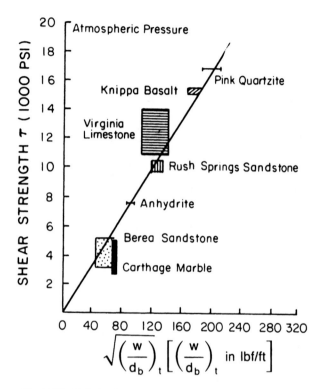

Fig. 5.35—Relation between rock shear strength and threshold bit weight at atmospheric pressure. [18]

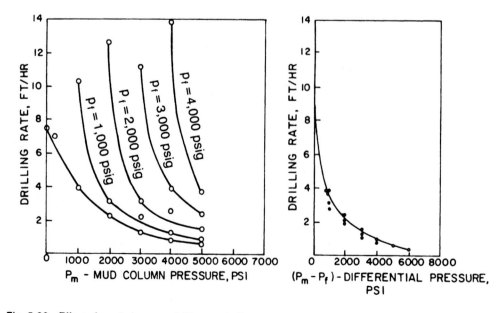

Fig. 5.36—Effect of overbalance on drilling rate in Berea sandstone for clay/water mud and 1.25-in. rolling cutter bit. [19]

fluid pressure often is called the *overbalance*. Recall the change in the crater formation mechanism with increasing overbalance described in Fig. 5.17. The Mohr failure criteria given by Eq. 5.3 predicts a similar effect of overbalance on drag bit performance. The normal stress at the failure plane σ_n for a wedging-type failure mechanism is directly related to overbalance.

Cunningham and Eenink,[19] working with a 1.25-in.-diameter rolling cutter bit in a laboratory drilling machine, studied the effect of overbalance on penetration rate for a wide range of rock permeabilities. Data obtained in Berea sandstone having a permeability in the range of 150 to 450 md are shown in Fig. 5.36 for a wide range of borehole and formation fluid pressures. Note that a good correlation is obtained when the data are replotted with drilling rate as a function of overbalance $(p_{bh} - p_f)$. Data obtained in Indiana limestone having a permeability of 8 to 10 md are shown in Fig. 5.37 and are similar to those obtained in the Berea sandstone, which had a much higher permeability. Apparently, formation damage beneath the bit caused by the deposition of a filter cake of mud and formation solids prevented a flow of mud filtrate ahead of the bit sufficient to equalize the pressure differential. Note that the effect of overbalance on penetration rate is more pronounced at low values of overbalance than at high values of overbalance. If the overbalance is quite large, additional increases in overbalance have essentially no effect on penetration rate.

Garnier and van Lingen[20] have published laboratory data obtained using both small drag bits and rolling cutter bits in a laboratory drilling apparatus. They concluded that the effective overbalance during chip removal by a drag bit often can be greater than the difference between the static borehole and rock pore pressures. When a chip is being lifted, a vacuum can be created under the chip unless sufficient liquid can be supplied to fill the

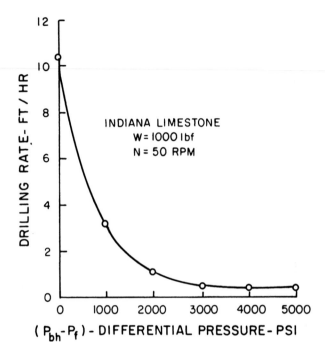

Fig. 5.37—Effect of overbalance on drilling rate in Indiana limestone for clay/water mud and 1.25-in. rolling cutter bit.

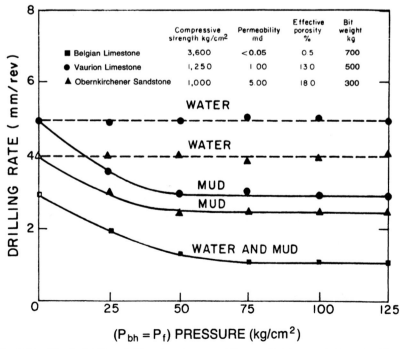

Fig. 5.38—Effect of drilling fluid and rock permeability on effective overbalance (at 32 rpm).[20]

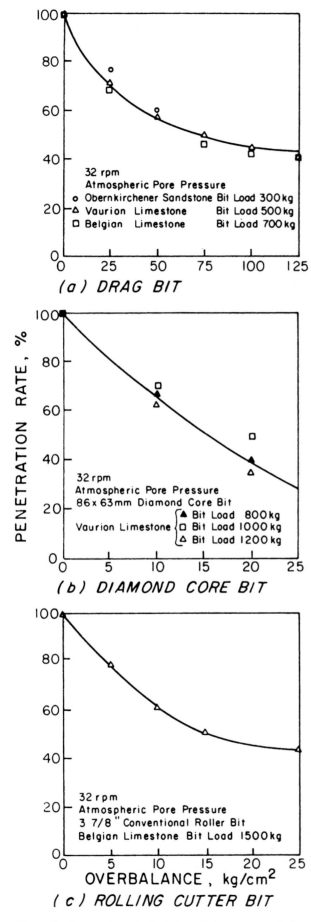

Fig. 5.39—Comparison of effect of overbalance on penetration rate.[20]

opening void space. The liquid can be supplied only by (1) drilling fluid flowing through the fracture, (2) drilling fluid filtrate flowing through the pores of the chip, and (3) formation fluid flowing into the void from the rock beneath the chip. When drilling a rock of low permeability with a clay/water mud, which readily forms a filter cake, the flow of liquid into the void beneath the chip was found to be too slow to prevent a pressure reduction beneath the chip. This was indicated by the data shown in Fig. 5.38, which were obtained with the static pore pressure and wellbore pressure maintained at the same value. Note that when mud is used as the drilling fluid, penetration rate decreased with increasing mud pressure, even though the static overbalance remained constant. This indicates that the effective dynamic overbalance during chip formation was greater than the static overbalance. When water was used as the drilling fluid, pressure equalization beneath the chip was more rapid for the rocks of moderate permeability, and penetration rate remained constant with increasing mud pressure.

To obtain the effect of overbalance on penetration rate for a drag bit, Garnier and van Lingen operated their drilling machine at various levels of borehole pressure while maintaining the pore pressure constant at atmospheric pressure. Since the pore pressure was already quite low, the dynamic and static overbalance were essentially equal. Shown in Fig. 5.39 are the results obtained using a 1.25-in. double-blade drag bit. Also shown for comparison are similar data obtained using a diamond core bit and a 3⅞-in. tricone rolling cutter bit.

Laboratory data on the effect of overbalance on penetration rate were recently obtained by using full-scale bits in a high-pressure wellbore simulator.[21] Typical experimental results obtained in Colton sandstone, which had an unconfined compressive strength of 7,600 psi and a permeability of 40 μd, are shown in Fig. 5.40. Note that these tests confirmed the behavior observed in the previous small-scale laboratory tests.

Some field data on the effect of overbalance on penetration rate are also available. The effect of overbalance on penetration rate in shale on seven wells drilled in south Louisiana was studied by Vidrine and Benit.[22] Data obtained on Well D at a depth of about 12,000 ft with an 8.5-in.-diameter rolling cutter bit are shown in Fig. 5.41. Note that the shape of the curve is quite similar to the laboratory data of Cunningham and Eenink. This type of behavior is accepted widely by field drilling personnel familiar with changes in penetration rate due to changes in mud density.

Bourgoyne and Young[15] observed that the relation between overpressure and penetration rate could be represented approximately by a straight line on semilog paper for the range of overbalance commonly used in field practice. In addition, they suggested normalizing the penetration rate data by dividing by the penetration rate corresponding to zero overbalance (borehole pressure equal to formation fluid pressure). The data given in Figs. 5.37 through 5.41 are presented as suggested by Bourgoyne and Young in Fig. 5.42. Note that a reasonably accurate straight-line representation of the data is possible for moderate values of overbalance. The equation for the straight line shown in Fig. 5.35 is given by

$$\log(R/R_0) = -m(p_{bh} - p_f), \quad \ldots \ldots \ldots \ldots (5.18a)$$

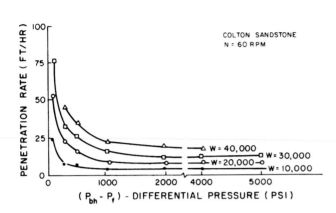

Fig. 5.40—Penetration rate as a function of pressure overbalance for Colton sandstone.[21]

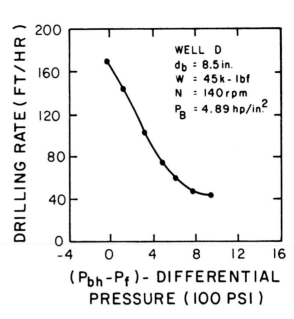

Fig. 5.41—Field measurement of the effect of overbalance on penetration rate in shale.[22]

where

R = penetrate rate,
R_0 = penetration rate at zero overbalance,
p_{bh} = bottomhole pressure in the borehole,
p_f = formation fluid pressure, and
m = the slope of the line.

If we express overbalance in terms of equivalent circulating mud density ρ_c and pore pressure gradient g_p, we obtain

$$(p_{bh} - p_f) = 0.052 \, D(\rho_c - g_p).$$

Substituting this expression for overbalance in Eq. 5.18a gives

$$\log(R/R_0) = -0.052 \, mD(\rho_c - g_p)$$

$$= 0.052 \, mD(g_p - \rho_c).$$

Bourgoyne and Young chose to replace the combination of constants $(-0.052m)$ by a single coefficient a_4.

$$\log(R/R_0) = a_4 D(g_p - \rho_c). \quad \ldots\ldots\ldots\ldots (5.18b)$$

This expression is useful for relating changes in mud density or pore pressure gradient to changes in penetration rate.

Example 5.6. The slope of the shale line in Fig. 5.42 has a value of -0.000666. Evaluate the coefficient a_4 for this value of m and estimate the change in penetration rate in this shale at 12,000 ft to be expected if the mud density is increased from 12 to 13 lbm/gal. The current penetration rate in shale is 20 ft/hr.

Solution. The coefficient a_4 is given by

$$a_4 = 0.052(0.000666) = 35 \times 10^{-6}.$$

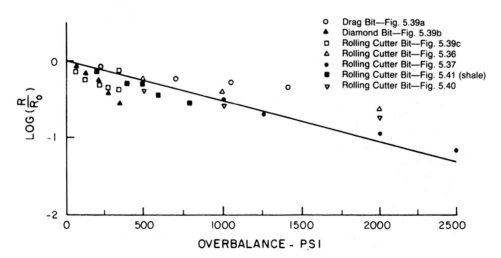

○ Drag Bit—Fig. 5.39a
▲ Diamond Bit—Fig. 5.39b
□ Rolling Cutter Bit—Fig. 5.39c
△ Rolling Cutter Bit—Fig. 5.36
● Rolling Cutter Bit—Fig. 5.37
■ Rolling Cutter Bit—Fig. 5.41 (shale)
▽ Rolling Cutter Bit—Fig. 5.40

Fig. 5.42—Exponential relation between penetration rate and overbalance for rolling cutter bits.

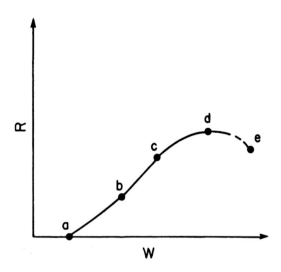

Fig. 5.43—Typical response of penetration rate to increasing bit weight.

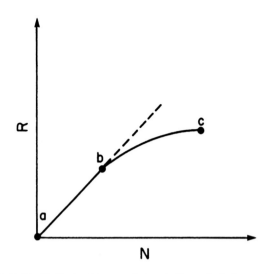

Fig. 5.44—Typical response of penetration rate to increasing rotary speed.

Eq. 5.18b can be rearranged using the definition of a common logarithm in terms of the initial penetration rate R_1 and mud density ρ_1 to give

$$R_1 = R_0 \times 10^{a_4 D(g_p - \rho_1)} = R_0 e^{2.303 a_4 D(g_p - \rho_1)}.$$

Similarly, for the final penetration rate R_2 and mud density ρ_2, we obtain

$$R_2 = R_0 \times 10^{a_4 D(g_p - \rho_2)} = R_0 e^{2.303 a_4 D(g_p - \rho_2)}.$$

Dividing the equation for R_2 by the equation for R_1 gives

$$\frac{R_2}{R_1} = e^{2.303 a_4 D(\rho_1 - \rho_2)}.$$

Solving for the final penetration rate R_2 yields

$$R_2 = R_1 \times e^{2.303 a_4 D(\rho_1 - \rho_2)}$$

$$= 20 \left[e^{2.303(35 \times 10^{-6})(12,000)(12-13)} \right] = 7.60 \text{ ft/hr.}$$

5.7.4 Operating Conditions

The effect of bit weight and rotary speed on penetration rate has been studied by numerous authors both in the laboratory and in the field. Typically, a plot of penetration rate vs. bit weight obtained experimentally with all other drilling variables held constant has the characteristic shape shown in Fig. 5.43. No significant penetration rate is obtained until the threshold bit weight is applied (Point a). Penetration rate then increases rapidly with increasing values of bit weight for moderate values of bit weight (Segment ab). A linear curve is often observed at moderate bit weights (Segment bc). However, at higher values of bit weight, subsequent increase in bit weight causes only slight improvements in penetration rate (Segment cd). In some cases, a decrease in penetration rate is observed at extremely high values

of bit weight (Segment de). This type of behavior often is called *bit floundering*. The poor response of penetration rate at high values of bit weight usually is attributed to less efficient bottomhole cleaning at higher rates of cuttings generation or to a complete penetration of the cutting element into the hole bottom.

A typical plot of penetration rate vs. rotary speed obtained with all other drilling variables held constant is shown in Fig. 5.44. Penetration rate usually increases linearly with rotary speed at low values of rotary speed. At higher values of rotary speed, the response of penetration rate to increasing rotary speed diminishes. The poor response of penetration rate at high values of rotary speed usually is also attributed to less efficient bottomhole cleaning.

Maurer[25] developed a theoretical equation for rolling cutter bits relating penetration rate to bit weight, rotary speed, bit size, and rock strength. The equation was derived from the following observation made in single-tooth impact experiments.

1. The crater volume is proportional to the square of the depth of cutter penetration.

2. The depth of cutter penetration is inversely proportional to the rock strength.

For these conditions, the penetration rate R is given by

$$R = \frac{K}{S^2} \left[\frac{W}{d_b} - \left(\frac{W}{d_b} \right)_t \right]^2 N, \quad \ldots\ldots\ldots\ldots (5.19)$$

where

K = constant of proportionality,
S = compressive strength of the rock,
W = bit weight,
W_0 = threshold bit weight,
d_b = bit diameter, and
N = rotary speed.

This theoretical relation assumes perfect bottomhole cleaning and incomplete bit tooth penetration.

The theoretical equation of Maurer can be verified using experimental data obtained at relatively low bit weight and rotary speeds corresponding to Segment ab in

Figs. 5.43 and 5.44. At moderate values of bit weight, the weight exponent usually is observed to be closer to a value of one than the value of two predicted by Eq. 5.19. At higher values of bit weight, a weight exponent of less than one usually is indicated. Bingham[18] suggested the following drilling equation on the basis of considerable laboratory and field data.

$$R = K\left(\frac{W}{d_b}\right)^{a_5} N, \quad \dots\dots\dots\dots\dots (5.20)$$

where K is the constant of proportionality that includes the effect of rock strength, and a_5 is the bit weight exponent.

In this equation the threshold bit weight was assumed to be negligible and the bit weight exponent must be determined experimentally for the prevailing conditions. However, a constant rotary speed exponent of one was used in the Bingham equation even though some of his data showed behavior similar to that described by Segment bc in Fig. 5.44.

More recently, several authors have proposed the determination of both a bit weight exponent and a rotary speed exponent using data representative of the prevailing conditions. Young[13] has pioneered the development of a computerized drilling control system in which both the bit weight and rotary speed could be varied systematically when a new formation type was encountered and the bit weight and rotary speed exponent automatically computed from the observed penetration rate response. Values of the bit weight exponent obtained from field data range from 0.6 to 2.0, while values of the rotary speed exponent range from 0.4 to 0.9.

Frequent changes in lithology with depth can make it difficult to evaluate the bit weight and rotary speed exponents from a series of penetration rate measurements made at various bit weights and rotary speeds. In many cases, the lithology may change before the tests are completed. To overcome this problem, a *drilloff test* can be performed. A drilloff test consists of applying a large weight to the bit and then locking the brake and monitoring the decrease in bit weight with time while maintaining a constant rotary speed. Hook's law of elasticity then can be applied to compute the amount the drillstring has stretched as the weight on the bit decreased and the hook load increased. In this manner, the response in penetration rate to changing bit weight can be determined over a very short depth interval.

Hook's law states that the change in stress is directly proportional to the change in strain.

$$\Delta\sigma = E\Delta\epsilon. \quad \dots\dots\dots\dots\dots\dots (5.21)$$

For the case of axial tension in a drillstring, the stress change is equal to the change in bit weight (axial tension) divided by the cross-sectional area of the drillpipe. The change in strain is equal to the change in drillpipe length per unit length. Thus, Hook's law becomes

$$\frac{\Delta W}{A_s} = E \frac{\Delta L}{L}.$$

Solving this expression for ΔL gives

$$\Delta L = \frac{L}{EA_s}\Delta W.$$

The average penetration rate observed for the change in bit weight ΔW can be obtained by dividing this equation by the time interval Δt required to drill off ΔW.

$$R = \frac{\Delta W}{\Delta t} = \frac{L}{EA_s}\frac{\Delta W}{\Delta t}.$$

Range 2 drillpipe has tool joint upsets over about 5% of its length that have a much greater cross-sectional area than the pipe body and essentially do not contribute to the length change observed. Replacing L by $0.95L$ gives

$$R = 0.95\frac{L}{EA_s}\frac{\Delta W}{\Delta t}. \quad \dots\dots\dots\dots (5.22)$$

The length change of the drill collars is also small and can be ignored.

Care must be taken to establish the bottomhole pattern of the bit at the initial bit weight of the test before performing the drilloff test. The following procedure was adapted from a Chevron U.S.A.[26] recommended practice.

1. Select a depth to run the drilloff test where a section of uniform lithology (usually shale) is expected.

2. While drilling with the bit weight currently in use, lock the brake and determine the time required to drill off 10% of this weight. This is called the characteristic time.

3. Increase the bit weight to the initial value of the drilloff test. This initial value should be at least a 20% increase in bit weight over the bit weight currently in use.

4. Drill at this bit weight long enough to establish the new bottomhole pattern of the bit. The time allowed is usually one characteristic time per 10% increase in bit weight—e.g., a time interval of twice the characteristic time would be used for a 20% increase in bit weight.

5. Lock the brake and maintain a constant rotary speed. Record the time each time the bit weight falls off 4,000 lbf. If the weight indicator is fluctuating, use the midpoint of the fluctuations as the bit weight. Continue the test until at least 50% of the initial bit weight has been drilled off.

6. Make a plot of Δt vs. W or R vs. W using log-log graph paper. A straight-line plot should result having a slope equal to the bit weight exponent. Deviation from straight-line behavior may occur at high bit weights if bit floundering occurs or is impending.

7. If time permits, repeat the test at a different rotary speed. If bit floundering (nonlinear behavior at high bit weights) was observed in the initial test, use a lower rotary speed in the second test. If no bit floundering occurred in the initial test, use a higher rotary speed in the second test.

The rotary speed exponent can be obtained using penetration rates obtained at two different rotary speeds but at the same bit weight.

TABLE 5.10—EXAMPLE DRILLOFF TEST ANALYSIS

Bit Weight (1,000 lbf)	Average Bit Weight (1,000 lbf)	N = 150			N = 100		
		Elapsed Time (seconds)	Δt (seconds	R (ft/hr)	Elapsed Time (seconds	Δt (seconds)	R (ft/hr)
76		0			0		
	74		52	16.6		54	16.6
72		52			54		
	70		53	16.6		60	14.4
68		105			114		
	66		55	15.7		66	13.1
64		160			180		
	62		58	14.9		73	11.8
60		218			253		
	58		63	13.7		81	10.7
56		281			334		
	54		71	12.2		90	9.6
52		352			424		
	50		80	10.8		101	8.6
48		432			525		
	46		90	9.6		116	7.4
44		522			641		
	42		104	8.3		132	6.5
40		626			773		
	38		120	7.2			
36		746					

Example 5.7. Using the following drilloff test data, evaluate the bit weight exponent and rotary speed exponent. The length of drillpipe at the time of the test was 10,000 ft, and the drillpipe has a cross-sectional area of 5.275 sq in. Young's modulus for steel is 30×10^6. Assume that the threshold bit weight is zero.

Test No. 1 (rotary speed = 150 rpm)

Bit Weight (1,000 lbm)	Elapsed Time (seconds)
76	0
72	52
68	105
64	160
60	218
56	281
52	352
48	432
44	522
40	626
36	746

Test No. 2 (rotary speed = 100 rpm)

Bit Weight (1,000 lbm)	Elapsed Time (seconds)
76	0
72	54
68	114
64	180
60	253
56	334
52	424
48	525
44	641
40	773

Solution. The penetration rate can be evaluated using Eq. 5.21:

$$R = 0.95 \frac{L}{EA_s} \frac{\Delta W}{\Delta t}$$

$$= 0.95 \frac{10,000}{30(10)^6 5.275} \frac{4,000}{\Delta t} = \frac{0.24}{\Delta t}.$$

If we express R in units of feet per hour and Δt in seconds, this expression becomes

$$R = \frac{0.24}{\Delta t} \left(\frac{3,600 \text{ seconds}}{1 \text{ hour}} \right) = \frac{864}{\Delta t}.$$

The drilloff test data have been evaluated using this expression in Table 5.10.

A plot of penetration rate vs. average bit weight can be constructed on log-log paper from the results of the drilloff test analysis. This has been done in Fig. 5.45. Graphical evaluation of the slope of the straight-line portion of either line on Fig. 5.45 yields a value of 1.6. Thus, the observed bit weight exponent is approximately 1.6 for values of bit weight below the flounder region. The rotary speed exponent can be evaluated from the spacing between the lines in the parallel region.

For example, a penetration rate of 13.7 ft/hr is observed for a bit weight of 58,000 lbf and a rotary speed of 150 rpm. Reducing the rotary speed to 100 rpm resulted in a penetration rate of 10.7 ft/hr at the same bit weight. Thus, we have

$$R = KN^{a_6},$$

$$13.7 = K(150)^{a_6},$$

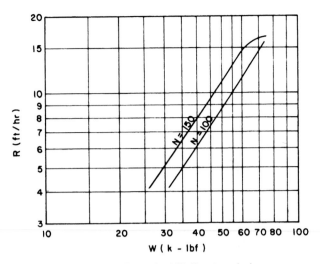

Fig. 5.45—Example drilloff test analysis.

Fig. 5.46—Cutaway view of extended-nozzle bit.[27]

and

$$10.7 = K(100)^{a_6}.$$

where K is the constant of proportionality, and a_6 is the rotary speed exponent.

Dividing the top equation by the bottom equation gives

$$\frac{13.7}{10.7} = \left(\frac{150}{100}\right)^{a_6}.$$

Taking the logarithm of both sides and solving for a_6 yields

$$a_6 = \frac{\log(13.7/10.7)}{\log(150/100)} = 0.6.$$

In this example, a good straight-line fit was obtained below the flounder region assuming the threshold bit weight was zero. When the threshold bit weight is not zero, it may be necessary to subtract the threshold bit weight from the bit-weight column before plotting the data. If the threshold bit weight is not known, it can be determined by trial and error as the value that gives the best straight-line fit.

5.7.5 Bit Tooth Wear

Most bits tend to drill slower as the bit run progresses because of tooth wear. The tooth length of milled tooth rolling cutter bits is reduced continually by abrasion and chipping. As previously discussed in Section 5.3, the teeth are altered by hardfacing or by case-hardening process to promote a self-sharpening type of tooth wear. However, while this tends to keep the tooth pointed, it does not compensate for the reduced tooth length. The teeth of tungsten carbide insert-type rolling cutter bits fail by breaking rather than by abrasion. Often, the entire tooth is lost when breakage occurs. Reductions in penetration rate due to bit wear usually are not as severe for insert bits as for milled tooth bits unless a large

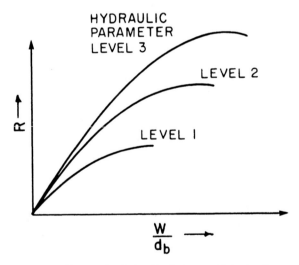

Fig. 5.47—Expected relationship between bit hydraulics and penetration rate.

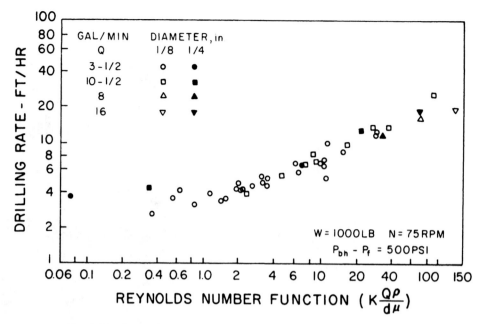

Fig. 5.48—Penetration rates as a function of bit Reynolds number.[28]

number of teeth are broken during the bit run. Diamond bits also fail from tooth breakage or loss of diamonds from the matrix.

Several authors have published mathematical models for computing the effect of tooth wear on penetration rate for rolling-cutter bits. Galle and Woods[11] published the following model in 1963.

$$R \propto \left(\frac{1}{0.928125h^2 + 6h + 1}\right)^{a_7} \quad \ldots\ldots (5.23)$$

where h is the fractional tooth height that has been worn away, and a_7 is an exponent.

A value of 0.5 was recommended for the exponent a_7 for self-sharpening wear of milled tooth bits, the primary bit type discussed in the publication. In a more recent work, Bourgoyne and Young[15] suggested a similar but less complex relationship given by

$$R \propto e^{-a_7h} \quad \ldots\ldots\ldots\ldots\ldots (5.24)$$

Bourgoyne and Young suggested that the exponent a_7 be determined based on the observed decline of penetration rate with tooth wear for previous bits run under similar conditions.

Example 5.8. An initial penetration rate of 20 ft/hr was observed in shale at the beginning of a bit run. The previous bit was identical to the current bit and was operated under the same conditions of bit weight, rotary speed, mud density, etc. However, a drilling rate of 12 ft/hr was observed in the same shale formation just before pulling the bit. If the previous bit was graded T-6, compute the approximate value of a_7.

Solution. The value of h for the previous bit just before the end of the bit run is ⅝ or 0.75. The value of h for the

new bit is zero. Thus, for the relation given as Eq. 5.24, we have

$$R = Ke^{-a_7h},$$

$$20 = Ke^{-a_7(0)},$$

and

$$12 = Ke^{-a_7(0.75)}.$$

Dividing the first equation by the second yields

$$\frac{20}{12} = e^{0.75a_7}$$

Taking the natural logarithm of both sides and solving for a_7 gives

$$a_7 = \frac{\ln(20/12)}{0.75} = 0.68.$$

5.7.6 Bit Hydraulics

The introduction of the jet-type rolling cutter bits in 1953 showed that significant improvements in penetration rate could be achieved through an improved jetting action at the bit. The improved jetting action promoted better cleaning of the bit teeth as well as the hole bottom. Some evidence has been presented[28] that the jetting action is most effective when using extended-nozzle bits in which the discharge ends of the jets are brought closer to the bottom of the hole (Fig. 5.46). A center jet must also be used with extended-nozzle bits to prevent bit balling in soft formations.

As discussed previously in Chap. 4, there is considerable uncertainty as to the best hydraulics parameter to use in characterizing the effect of hydraulics on penetration rate. Bit hydraulic horsepower, jet impact force, and nozzle velocity all are used commonly.

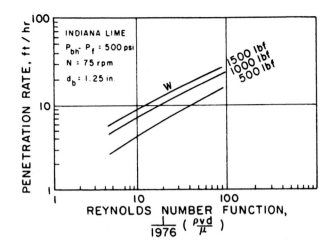

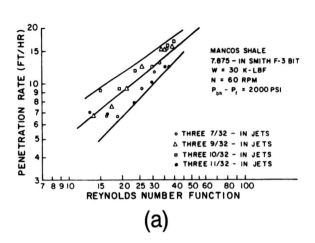

(a)

Fig. 5.49—Experimentally observed effect of bit weight and bit Reynolds number on penetration rate.[28]

The level of hydraulics achieved at the bit is thought by many to affect the flounder point of the bit. Shown in Fig. 5.47 is a hypothetical example of the type of behavior often reported. At low bit weights and penetration rates, the level of hydraulics required for hole cleaning is small. As more weight is applied to the bit and cuttings are generated faster, a flounder point is reached eventually where the cuttings are not removed as quickly as they are generated. If the level of hydraulics is increased, a higher bit weight and penetration rate will be reached before bit floundering occurs.

Eckel,[28] working with microbits in a laboratory drilling machine, has made the most extensive laboratory study to date of the relation between penetration rate and the level of hydraulics. Working at constant bit weight and rotary speed, Eckel found that penetration rate could be correlated to a Reynolds number group given by

$$N_{Re} = K \frac{\rho v d}{\mu_a}, \dots\dots\dots\dots\dots\dots (5.25)$$

where

K = a scaling constant,
ρ = drilling density,
v = flow rate,
d = nozzle diameter, and
μ_a = apparent viscosity of drilling fluid at 10,000 seconds^{-1}.

The shear rate of 10,000 seconds^{-1} was chosen as representative of shear rates present in the bit nozzle. The scaling constant, K, is somewhat arbitrary, but a constant value of 1/1,976 was used by Eckel to yield a convenient range of the Reynolds number group.

The results of Eckel's experiments are summarized in Figs. 5.48 and 5.49. Note that penetration rate was increased by increasing the Reynolds number function for the full range of Reynolds numbers studied. When the bit weight was increased, the correlation curve simply was shifted upward as shown in Fig. 5.49. The behavior at the flounder point was not studied by Eckel. It can be shown that, for a given drilling fluid, the Reynolds number function is a maximum when the jet impact force is a maximum (see Chap. 4).

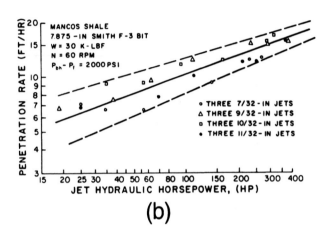

(b)

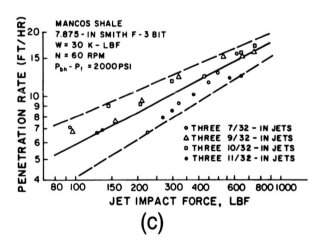

(c)

Fig. 5.50—Effect of hydraulics on penetration rate in Mancos shale under simulated borehole conditions: (a) observed correlation using Reynolds number function as hydraulics parameter, (b) observed correlations using bit hydraulic power as hydraulics parameter, and (c) observed correlation using jet impact force as hydraulics parameter.

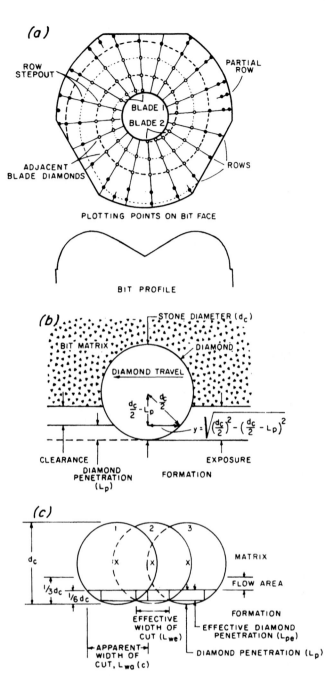

(a)

PLOTTING POINTS ON BIT FACE

BIT PROFILE

(b)

(c)

Fig. 5.51—Diamond bit stone layout assumed in penetration rate equation (after Ref. 30).

5.7.7 Penetration Rate Equation

The manner in which the important drilling variables that have been discussed affect penetration rate is quite complex and only partially understood. Thus, the development of an accurate mathematical model of the rotary drilling process is not yet possible. However, several mathematical models that attempt to combine the known relationships have been proposed. These models make it possible to apply formal optimization methods to the problem of selecting the best bit weight and rotary speed to achieve the minimum cost per foot. Many authors have reported[11-15] significant reductions in drilling cost through use of these approximate mathematical models.

Diamond bits, as well as other types of drag bits, are designed to achieve a given maximum penetration per revolution. Under ideal conditions, the bit weight and torque is such that the bit is kept feeding into the formation at the design cutting rate. The penetration rate of a drag bit for a given penetration of the cutting element into the formation is given by

$$R = L_{pe} n_{be} N, \quad\dots\dots\dots\dots\dots\dots\dots (5.26)$$

where

L_{pe} = effective penetration of each cutting element,
n_{be} = effective number of blades, and
N = rotary speed.

Peterson[30] has developed theoretical equations for the effective penetration L_{pe} and effective number of blades n_{be} for diamond bits. The equations were derived for a simplified model which assumed the following.

1. The bit has a flat face that is perpendicular to the axis of the hole.
2. Each blade is formed by diamonds laid out as a helix as shown in Fig. 5.51a.
3. The stones are spherical in shape as shown in Fig. 5.51b.
4. The diamonds are spaced so that the cross-sectional area removed per stone is a maximum for the design depth of penetration.
5. The bit is operated at the design depth of penetration.
6. The bit hydraulics are sufficient for perfect bottomhole cleaning.

For these conditions, the effective penetration L_{pe} and the effective number of blades n_{be} are given by

$$L_{pe} = 0.67 L_p \quad\dots\dots\dots\dots\dots\dots\dots (5.27a)$$

and

$$n_{be} = 1.92 \left(\frac{C_c}{s_d}\right) d_b \sqrt{d_c L_p - L_p^2}, \quad\dots\dots\dots (5.27b)$$

where

C_c = concentration of diamond cutters, carats/sq in.,
L_p = actual depth of penetration of each stone, in.,
d_b = bit diameter, in.,
d_c = average diameter of the face stone cutters, in., and
s_d = diamond size, carats/stone.

In spite of the convincing correlation presented in Figs. 5.48 and 5.49, Eckel's work has not been widely applied in practice. Hydraulic horsepower and jet impact force are more frequently used in the development of correlations between bit hydraulics and penetration rate. Recent data obtained in full-scale laboratory drilling experiments conducted under simulated borehole conditions[29] has shown that the jet Reynolds number group, hydraulic horsepower, and jet impact force all give similar results when used to correlate the effect of jet bit hydraulics on penetration rate. Fig. 5.50 shows correlations obtained for each of these parameters using data obtained in Mancos shale with a 7.875-in. Smith F3 bit.[29]

A formation property called the *formation resistance*, r_f, is used to compute the bit weight required to obtain the design penetration L_p. The formation resistance is the pressure needed to overcome the formation strength, allowing the stone to penetrate the rock.

$$r_f = \frac{W_e}{A_{dt}}, \quad \dots\dots\dots\dots\dots\dots\dots (5.27c)$$

where W_e is the effective weight applied to the bit after the hydraulic pumpoff forces have been taken into account, and A_{dt} is the total diamond area in contact with the formation. The formation resistance can be computed from an observed penetration rate after a bit is operated in the formation of interest.

For a spherical stone as shown in Fig. 5.51b, the contact area is given by

$$A_{dt} = \frac{\pi^2 d_b^2}{4}\left(\frac{C_c}{s_d}\right)\left(d_c L_p - L_p^2\right). \quad \dots\dots\dots (5.27d)$$

Example 5.9. An 8.625-in. diamond bit containing 270 0.23-in.-diameter stones of 1.00 carat is designed to operate at a depth of penetration of 0.01 in. Estimate the penetration rate that could be obtained with this bit if the formation characteristics are such that an acceptable bit weight and torque for this penetration could be maintained at a rotary speed of 200 rpm.

Solution. Ignoring the bit contouring required for proper hydraulic action and gauge protection, the bit is assumed to have a flat face that is perpendicular to the axis of the hole. Thus,

$$\frac{C_c}{s_d} = \frac{270}{\frac{\pi}{4}(8.625)^2} = 4.621 \text{ stones/sq in.}$$

The effective number of blades is given by Eq. 5.27b.

$$n_{be} = 1.92\left(\frac{C_c}{s_d}\right)d_b\sqrt{d_c L_p - L_p^2}$$

$$= 1.92(4.621)(8.625)\times\sqrt{0.23(0.01)-(0.01)^2}.$$

$$= 3.59.$$

The effective penetration is given by Eq. 5.27a.

$$L_{pe} = 0.67(0.01) = 0.0067 \text{ in.}$$

The penetration rate at a rotary speed of 200 rpm is given by

$$R = L_{pe}n_{be}N = \frac{0.0067}{12}(3.59)(60\times 200)$$

$$= 24 \text{ ft/hr.}$$

Penetration rate equations for rolling cutter bits have been proposed by various authors. The approach usually taken is to assume that the effects of bit weight, rotary speed, tooth wear, etc., on penetration rate are all independent of one another and that the composite effect can be computed using an equation of the form

$$R = (f_1)(f_2)(f_3)(f_4)\dots(f_n), \quad \dots\dots\dots (5.28a)$$

where f_1, f_2, f_3, f_4, etc., represent the functional relations between penetration rate and various drilling variables. The functional relations chosen usually are based on trends observed in either laboratory or field studies. Some authors have chosen to define the functional relation graphically, while others have used curve fitting techniques to obtain empirical mathematical expressions. Some relatively simple mathematical equations have been used that model only two or three of the drilling variables. An example is the Bingham model defined by Eq. 5.20.

Perhaps the most complete mathematical drilling model that has been used for rolling cutter bits is the model proposed by Bourgoyne and Young.[15] They proposed using eight functions to model the effect of most of the drilling variables discussed in the previous section. The Bourgoyne-Young drilling model can be defined by Eq. 5.28a with the following functional relations.

$$f_1 = e^{2.303a_1} = K. \quad \dots\dots\dots\dots\dots (5.28b)$$

$$f_2 = e^{2.303a_2(10,000-D)}. \quad \dots\dots\dots (5.28c)$$

$$f_3 = e^{2.303a_3 D^{0.69}(g_p - 9.0)}. \quad \dots\dots\dots (5.28d)$$

$$f_4 = e^{2.303a_4 D(g_p - \rho_c)}. \quad \dots\dots\dots (5.28e)$$

$$f_5 = \left[\frac{\left(\frac{W}{d_b}\right)-\left(\frac{W}{d_b}\right)_t}{4-\left(\frac{W}{d_b}\right)_t}\right]^{a_5} \quad \dots\dots\dots (5.28f)$$

$$f_6 = \left(\frac{N}{60}\right)^{a_6}. \quad \dots\dots\dots\dots (5.28g)$$

$$f_7 = e^{-a_7 h}. \quad \dots\dots\dots\dots\dots\dots (5.28h)$$

$$f_8 = \left(\frac{F_j}{1,000}\right)^{a_8}. \quad \dots\dots\dots\dots (5.28i)$$

In these equations,

D = true vertical well depth, ft,
g_p = pore pressure gradient, lbm/gal,
ρ_c = equivalent circulating density,
$(W/d_b)_t$ = threshold bit weight per inch of bit diameter at which the bit begins to drill, 1,000 lbf/in.,
h = fractional tooth dullness,
F_j = hydraulic impact force beneath the bit, lbf, and
a_1 to a_8 = constants that must be chosen based on local drilling conditions.

The constants a_1 through a_8 can be computed using prior drilling data obtained in the area when detailed drilling data are available. The drilling model can be used both for drilling optimization calculations and for the detection of changes in formation pore pressure.

The constants a_1 through a_8 can be computed using prior drilling data obtained in the area when detailed drilling data are available. The drilling model can be used both for drilling optimization calculations and for the detection of changes in formation pore pressure.

The function f_1 primarily represents the effects of formation strength and bit type on penetration rate. However, it also includes the effect of drilling variables such as mud type, solids content, etc., which are not included in the drilling model. The exponential expression for f_1 is useful when applying a multiple regression technique presented by Bourgoyne and Young[15] for computing the values of a_1 through a_8 from prior drilling data obtained in the area. The coefficient "2.303" allows the constant a_1 to be defined easily in terms of the common logarithm of an observed penetration rate. The utility of this will be demonstrated in Chap. 6, which includes a discussion of the determination of formation pore pressure from drilling data.

The functions f_2 and f_3 model the effect of compaction on penetration rate. The function f_2 accounts for the rock strength increase due to the normal compaction with depth, and the function f_3 models the effect of undercompaction experienced in abnormally pressured formations. Note that the $(f_2 f_3)$ product is equal to 1.0 for a pore pressure gradient equivalent to 9.0 lbm/gal and a depth of 10,000 ft.

The function f_4 models the effect of overbalance on penetration rate. This function has a value of 1.0 for zero overbalance—i.e., when the formation pore pressure is equal to the bottomhole pressure in the well.

The functions f_5 and f_6 model the effect of bit weight and rotary speed on penetration rate. Note that f_5 has a value of 1.0 when (W/d_b) has a value of 4,000 lbf/in. of bit diameter and f_6 has a value of 1.0 for a rotary speed of 60 rpm. This was chosen so that the $f_5 f_6$ product would have a value near 1.0 for common drilling conditions. The threshold bit weight is often quite small and can be neglected in areas such as the U.S. gulf coast where the formations are relatively soft. In more competent formations, the threshold bit weight can be estimated from drilloff tests terminated at very low bit weights. The function of f_5 has an upper limit corresponding to the bit flounder point, which must be established from drilloff tests. The constants a_5 and a_6 also can be determined from drilloff tests as in Example 5.7. Reported values of a_5 range from 0.5 to 2.0, and reported values of a_6 range from 0.4 to 1.0.

The function f_7 models the effect of tooth wear on penetration rate. The value of a_5 can be estimated from penetration rate measurements taken in similar formations at similar bit operating conditions at the beginning and end of a bit run as shown previously in Example 5.8. The term f_7 has a value of 1.0 for zero tooth wear. When tungsten carbide insert bits are used and operated at moderate bit weights and rotary speed, tooth wear is often insignificant and this term can be neglected. Typical values of a_7 for milled tooth bits range from 0.3 to 1.5.

The function f_8 models the effect of bit hydraulics on penetration rate. Jet impact force was chosen as the hydraulic parameter of interest, with a normalized value of 1.0 for f_8 at 1,000 lbf.

However, as shown in Fig. 5.50, the choice of impact force is arbitrary. Similar results could be obtained with bit hydraulic horsepower or nozzle Reynolds number as the hydraulic parameter affecting penetration rate. Typical values for a_8 range from 0.3 to 0.6.

In practice, it is prudent to select the best average values of a_2 through a_8 for the formation types in the depth interval of interest. However, the value of f_1 varies with the strength of the formation being drilled. The term f_1 is expressed in the same units as penetration rate and commonly is called the drillability of the formation. The drillability is numerically equal to the penetration rate that would be observed in the given formation type (under normal compaction) when operating with a new bit at zero overbalance, a bit weight of 4,000 lbf/in., a rotary speed of 60 rpm, and a depth of 10,000 ft. The drillability of the various formations can be computed using drilling data obtained from previous wells in the area.

Example 5.10. A 9.875-in. milled tooth bit operated at 40,000 lbf/in. and 80 rpm is drilling in a shale formation at a depth of 12,000 ft at a penetration rate of 15 ft/hr. The formation pore pressure gradient is equivalent to a 12.0 lbm/gal mud, and the equivalent mud density on bottom is 12.5 lbm/gal. The computed jet impact force beneath the bit is 1,200 lbf, and the computed fractional tooth wear is 0.3. Compute the apparent formation drillability f_1 using a threshold bit weight of zero and the following values of a_2 through a_8.

a_2	a_3	a_4	a_5	a_6	a_7	a_8
0.00007	0.000005	0.00003	1.0	0.5	0.5	0.5

Solution. The functional relations f_2 through f_8 are given by Eqs. 5.28a through 5.28h. The multiplier f_2 accounts for the normal decrease in penetration rate with depth from a reference depth of 10,000 ft.

$$f_2 = e^{2.303 a_2 (10,000 - D)}$$
$$= e^{2.303(0.00007)(10,000 - 12,000)} = 0.724.$$

The multiplier f_3 accounts for the increase in penetration rate due to undercompaction.

$$f_3 = e^{2.303 a_3 D^{0.69}(g_p - 9)}$$
$$= e^{2.303(0.000005)(12,000)^{0.69}(12 - 9)} = 1.023.$$

The multiplier f_4 accounts for the change in penetration rate with overbalance assuming a reference overbalance of zero.

$$f_4 = e^{2.303 a_4 D(g_p - \rho)}$$
$$= e^{2.303(0.00003)(12,000)(12.0 - 12.5)} = 0.6606.$$

The multiplier f_5 accounts for the change in penetration with bit weight assuming a reference bit weight of 4,000 lbf/in.

$$f_5 = \left[\frac{\left(\frac{W}{d_b}\right) - \left(\frac{W}{d_b}\right)_t}{4 - \left(\frac{W}{d_b}\right)_t} \right]^{a_5}$$

$$= \left[\frac{\left(\frac{40}{9.875}\right)}{4} \right]^{1.0} = 1.013.$$

The multiplier f_6 accounts for the change in penetration rate with rotary speed assuming a reference rotary speed of 60 rpm.

$$f_6 = \left(\frac{N}{60}\right)^{a_6} = \left(\frac{80}{60}\right)^{0.5} = 1.155.$$

The multiplier f_7 accounts for the change in penetration rate with tooth dullness using zero tooth wear as a reference.

$$f_7 = e^{-a_7 h} = e^{-0.5(0.3)} = 0.861.$$

The multiplier f_8 accounts for the change in penetration rate with jet impact force using an impact force of 1,000 lbf as a reference.

$$f_8 = \left(\frac{F_j}{1,000}\right)^{a_8} = \left(\frac{1,200}{1,000}\right)^{0.5} = 1.095.$$

Substitution of the computed values of f_2 to f_8 into Eq. 5.28 and solving for the formation drillability yields

$$R = f_1 \cdot f_2 \cdot f_3 \cdot \ldots \cdot f_8,$$

$$15 = f_1(0.724)(1.023)(0.6606)(1.013)$$

$$\cdot (1.155)(0.861)(1.095),$$

and

$$f_1 = \frac{15}{0.540} = 27.8 \text{ ft/hr.}$$

___

In Example 5.10, detailed drilling data were available at a given point in time. This requires the use of a modern well monitoring and data recording system. In many instances, data of this quality are not available and an average drillability for an entire bit run must be computed. For bits that show a significant tooth wear over the life of the bit, the change in the tooth wear function f_7 with time over the life of the bit must be taken into ac-

count. If we define a composite drilling variable J_1 using

$$J_1 = f_1 \cdot f_2 \cdot f_3 \cdot f_4 \cdot f_5 \cdot f_6 \cdot f_8, \quad \ldots \ldots (5.29)$$

Eq. 5.28 can be expressed by

$$R = \frac{dD}{dt} = J_1 f_7 = J_1 e^{-a_7 h}.$$

Separating variables in this equation yields

$$dD = J_1 e^{-a_7 h} dt. \quad \ldots \ldots \ldots \ldots \ldots \ldots (5.30)$$

The evaluation of this integral requires a relation between time t and tooth wear h. Recall that Eqs. 5.10 and 5.11 give

$$dt = J_2 \tau_H (1 + H_2 h) \, dh.$$

Substituting this expression into Eq. 5.30, we obtain

$$dD = J_1 J_2 \tau_H e^{-a_7 h} (1 + H_2 h) dh. \quad \ldots \ldots (5.31a)$$

Finally, integration of this equation leads to the following expression of bit footage in terms of the final tooth wear observed.

$$\Delta D = J_1 J_2 \tau_H \left[\frac{1 - e^{-a_7 h_f}}{a_7} \right.$$

$$\left. + \frac{H_2(1 - e^{-a_7 h_f} - a_7 h_f e^{-a_7 h_f})}{a_7^2} \right]. \quad . (5.31b)$$

This equation can be used to determine the footage corresponding to a given final tooth wear h_f and composite drilling parameter J_1. Conversely, it also can be used to compute an apparent or average value of J_1 for an observed footage ΔD and final tooth wear h_f. The formation drillability then can be computed from J_1 using Eq. 5.29.

In some cases, it is desirable to compute the footage drilled after a given time interval t_b of bit operation. To use Eq. 5.31 for this purpose, it is necessary to know the tooth dullness at the drilling time of interest. Recall that the time required to obtain a given tooth wear is given by Eq. 5.12b. Expressing this equation in terms of h_f, we obtain

$$\left(\frac{H_2 J_2 \tau_H}{2} \right) h_f^2 + (J_2 \tau_H) h_f - t_b = 0.$$

Solving this quadratic for h_f gives

$$h_f = \sqrt{\left(\frac{1}{H_2}\right)^2 + \left(\frac{2t_b}{H_2 J_2 \tau_H}\right)} - \left(\frac{1}{H_2}\right) \quad \ldots (5.32)$$

___

Example 5.11. Compute the average formation drillability for the bit run described in Example 5.3. Assume the

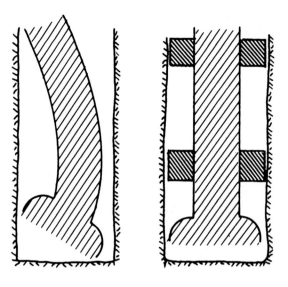

Fig. 5.52—Need for bit stabilizers.

average jet impact force was 1,000 lbf, the formation drilled was shale with a normal formation pressure gradient (equivalent to a 9.0 lbm/gal fluid), and the equivalent circulating density was 9.5 lbm/gal. Also, use the threshold bit weight and constants a_2 through a_8 given in Example 5.10.

Solution. Recall from Example 5.3 that H_2 had a value of 6, J_2 had a value of 0.080, and τ_H had a value of 73.0 hours. Also, the bit drilled from a depth of 8,179 to 8,404 ft in 10.5 hours and was graded as T-5 ($h_f = \frac{5}{8}$ or 0.625). The constant a_7 given in Example 5.10 had a value of 0.5. Substitution of these data into Eq. 5.31 yields

$$\Delta D = J_1 J_2 \tau_H \left[\frac{1-e^{-a_7 h_f}}{a_7} \right.$$

$$\left. + \frac{H_2(1-e^{-a_7 h_f} - a_7 h_f e^{-a_7 h_f})}{a_7^2} \right],$$

and

$$(8,404-8,179) = J_1(0.08)(73.0)$$

$$\times \left\{ \frac{1-e^{-0.5(0.625)}}{0.5} \right.$$

$$\left. + \frac{6\left[1-e^{-0.5(0.625)} - 0.5(0.625)e^{-0.5(0.625)}\right]}{(0.5)^2} \right\}.$$

Solving this equation for J_1 gives

$$J_1 = 25.8 \text{ ft/hr.}$$

The mean depth of the bit run is

$$\bar{D} = \frac{8,179+8,404}{2} = 8,292.$$

The multipliers f_2 through f_6 and f_8 can be obtained using Eqs. 5.28c through 5.28g and 5.28i.

$$f_2 = e^{2.303a_2(10,000-D)}$$

$$= e^{2.303(0.00007)(10,000-8,292)} = 1.32.$$

$$f_3 = e^{2.303a_3 D^{0.69}(g_p - \rho_c)} = 1.0 \text{ for } g_p = 9.0.$$

$$f_4 = e^{2.303a_4 D(g_p - \rho_c)}$$

$$= e^{2.303(0.00003)(8.292)(9.0-9.5)} = 0.751$$

$$f_5 = \left[\frac{\left(\frac{W}{d_b}\right) - \left(\frac{W}{d_b}\right)_t}{4 - \left(\frac{W}{d_b}\right)_t} \right]^{a_5}$$

$$= \left[\frac{\left(\frac{45}{8.5}\right) - 0}{4 - 0} \right]^{1.0} = 1.32.$$

$$f_6 = \left(\frac{N}{60}\right)^{a_6} = \left(\frac{90}{60}\right)^{0.5} = 1.225.$$

$$f_8 = \left(\frac{F_j}{1,000}\right)^{a_8} = 1.0 \text{ for } F_j = 1,000.$$

Substituting these values of f_2 through f_6 and f_8 into Eq. 5.29 gives

$$J_1 = f_1 \cdot f_2 \cdot f_3 \cdot f_4 \cdot f_5 \cdot f_6 \cdot f_8.$$

$$25.8 = f_1(1.32)(1.0)(0.751)(1.32)(1.225)(1.0).$$

Solving this equation for the formation drillability, we obtain $f_1 = 16.1$ ft/hr.

5.8 Bit Operation

In addition to selecting the best bit for the job, the drilling engineer must see that the bit selected is operated as efficiently as possible. Items of primary concern include (1) selection of bottomhole assembly, (2) prevention of accidental bit damage, (3) selection of bit weight and rotary speed, and (4) bit run termination.

Proper attention to all of these items must be given to approach a minimum-cost drilling operation.

5.8.1 Bottomhole Assembly

The bottomhole assembly used above the bit often has a significant effect on bit performance. The length of drill collars used should be adequate to prevent the development of bending moments in the drillpipe for the range of bit weight used. This can be accomplished through use of Eq. 4.25b as described in Chap. 4. Also, stabilizers should be used above the bit in the string of drill collars to prevent bending of the lower portion of the drill collars. A severe wobbling bit action results as the bit is rotated if the drill collars above the bit are not held in a concentric position in the borehole (Fig. 5.52). This can cause (1) severe shock loading on teeth, bearings, and grease seals of rolling cutter bits, (2) shock loading on diamond or PCD cutters and uneven fluid distribution beneath diamond bits, (3) a below-gauge borehole diameter, and (4) a crooked borehole.

The use of stabilizers having a diameter near the hole size can reduce the severity of these problems greatly. Special shock absorbing devices called *shock subs* also can be used above the bit to dampen the shock loads further. The additional cost of shock subs is justified more easily for the more expensive journal bearing bits, which have the potential of extremely long bit runs if the grease seals and bearing surfaces are not damaged.

5.8.2 Prevention of Accidental Bit Damage

Accidental bit damage before placing the bit in service at the bottom of the hole can reduce the life of the bit greatly. The bit should be tightened in the drillstring to the recommended torque using a special *breaker plate* designed for the bit type in use. Care also should be taken to see that the jet nozzles are installed properly using a shroud to minimize fluid erosion of the nozzle passages.

The bit is especially susceptible to damage during the tripping operations. The presence of tight spots observed when pulling the previous bit out of the hole should be noted in writing so that slower pipe velocities can be used at these points when running the new bit to bottom. Tight spots may be especially noticeable when running a fully stabilized bottomhole assembly after a bit that was observed to have significant gauge wear. When reaming is necessary, low bit weights should be used. The bit bearings are not designed for the inward thrust present during reaming operations. It is also possible to catch a bit cone on an irregular ledge in the borehole wall while running back to bottom. Plastic bit guides can be installed beneath the bit to minimize the risk of this type of damage.

Once the new bit reaches bottom, it should be "broken in" properly using a low bit weight and rotary speed for the first foot or two drilled. This allows any microscopic irregularities in the bearing surfaces to be smoothed and allows the bottomhole pattern of the new cutters to be established in the rock. The bit weight and rotary speed then can be increased slowly to the desired values. Also, it always is important to establish drilling fluid circulation before resuming drilling operations. Heat buildup can quickly damage the bit when fluid circulation stops during drilling operations.

5.8.3 Selection of Bit Weight and Rotary Speed

As discussed in the previous sections, the weight applied to the bit and the rotational speed of the drillstring have a major effect on both the penetration rate and the life of the bit. In addition, these parameters can be varied easily. Thus, the determination of the best bit weight and rotary speed for a given bit run is one of the routine problems faced by the drilling engineer. In selecting the bit weight and rotary speed to be used in drilling a given formation, consideration must be given to these items: (1) the effect of the selected operating conditions on the cost

TABLE 5.11—EXAMPLE COST-PER-FOOT TABLE[15]

Rotary Speed (rpm)	Bit Weight per Inch of Bit Diameter (1,000 lbf/in.)					
	2.0	3.0	4.0	5.0	6.0	7.0
20	$167.83	$103.51	$73.67	$56.88	$46.67	$42.82
40	114.94	71.45	51.48	40.61	34.92	38.55
60	95.19	59.84	43.84	35.56	32.36	42.65
80	85.77	54.61	40.77	34.08	32.80	49.76
100	81.15	52.37	39.85	34.30	34.70	58.52
120	79.25	51.83	40.17	35.51	37.48	68.36
140	79.07	52.38	41.29	37.37	40.86	79.01
160	80.08	53.68	42.96	39.69	44.68	90.29
180	81.97	55.54	45.06	42.37	48.85	102.11
200	84.52	57.83	47.48	45.32	53.29	114.39

per foot for the bit run in question and on subsequent bit runs, (2) the effect of the selected operating conditions on crooked hole problems, (3) the maximum desired penetration rate for the fluid circulating rates and mud processing rates available and for efficient kick detection, and (4) equipment limitations on the available bit weight and rotary speed.

In many instances, a wide range of bit weights and rotary speeds can be selected without creating crooked hole problems or exceeding equipment limitations. Also, penetration rates that can be achieved are usually less than the maximum desirable penetration rate in the deeper portions of the well. Under these conditions, the drilling engineer is free to select the bit weight and rotary speed that will result in the minimum cost per foot.

Several published methods for computing the optimum bit-weight/rotary-speed combinations for achieving minimum drilling costs are available.[11-17] All of these methods require the use of mathematical models to define the effect of bit weight and rotary speed on penetration rate and bit wear. Methods are available for computing both the best *variable* bit-weight/rotary-speed schedule and the best *constant* bit weight and rotary speed for the entire bit run. Galle and Woods[11] have reported that the simpler constant weight/speed methods result in only slightly higher costs per foot than the methods allowing the bit weights and rotary speeds to vary as the bit dulls or encounters different formation characteristics. Reed[14] indicated a difference of less than 3% in cost per foot between the variable and constant weight/speed schedules for the cases studied.

One straightforward technique that can be used to determine the best constant weight/speed schedule is to generate a cost-per-foot table. The cost per foot for various assumed bit weights and rotary speeds can be computed using the penetration rate and bit wear models and the results tabulated as shown in Table 5.11. The best combination of bit weight and rotary speed, the best bit weight for a given rotary speed, or the best rotary speed for a given desired bit weight then can be read from the table. The use of the best bit weight for a given rotary speed may be desirable when the rotary speed selection is limited by the rotary power transmission system. The best rotary speed for a given bit weight may be desirable when the bit weight is limited because of hole deviation problems.

Various algorithms can be used to evaluate the cost-per-foot table. When desired, a foot-by-foot analysis of the bit run can be made taking into account formation of different drillabilities that may be encountered during the bit run. However, when the use of a single average formation drillability is possible, the integrated forms of the penetration rate and tooth wear models can be used. This greatly reduces the number of calculation steps involved. For example, if the Bourgoyne-Young penetration rate and bit wear models are used, the following procedure could be used.

1. Assume a bit weight and rotary speed.
2. Compute the time required to wear out the bit teeth using Eqs. 5.11 and 5.12.
3. Compute the time required to wear out the bearings using Eqs. 5.15 and 5.16.
4. Using the smaller of the two computed times, compute the footage that would be drilled using Eqs. 5.29 and 5.31.
5. Compute the cost per foot using Eq. 1.16.

The procedure will give the cost per foot associated with complete bit wear. For a few cases where penetration rate decreases rapidly with tooth dullness, the minimum cost per foot can occur before complete bit wear. This situation can be determined by repeating Steps 4 and 5 using a drilling time slightly less than the bit life. If this results in a lower cost per foot, successively lower drilling times should be assumed until the optimum drilling time is determined.

Example 5.12. A Class 1-3 bit will be used to drill a formation at 7,000 ft having a drillability of 20 ft/hr. The abrasiveness constant τ_H has a value of 15.7 hours, the bearing constant τ_B has a value of 22 hours, and the bearing exponents B_1 and B_2 are equal to 1.0. The formation pore pressure gradient is equivalent to a 9.0-lbm/gal fluid, and the mud density is 10.0 lbm/gal. The bit costs $400, the operating cost of the drilling operation is $500/hr, the time required to trip for a new bit is 6.5 hours, and 3 minutes are required to make a connection. Using a threshold $(W/d_b)_t$ of 0.5 and the values of a_2 through a_8 as given, compute the cost per foot that would be observed for $(W/d_b) = 4.0$, $N = 60$ rpm, and a jet impact force of 900 lbf.

a_2	a_3	a_4	a_5	a_6	a_7	a_8
0.000087	0.000005	0.000017	1.2	0.6	0.9	0.4

Solution. Using Table 5.8 for a Class 1-3 bit, we obtain $H_1 = 1.84$, $H_2 = 6$, and $(W/d_b)_m = 8.0$. The value of J_2 as a function of bit weight and rotary speed is given by Eq. 5.11.

$$J_2 = \left[\frac{\left(\dfrac{W}{d_b}\right)_m - \dfrac{W}{d_b}}{\left(\dfrac{W}{d_b}\right) - 4} \right] \left(\frac{60}{N}\right)^{H_1} \left(\frac{1}{1 + \dfrac{H_2}{2}}\right)$$

$$= \left[\frac{8 - \left(\dfrac{W}{d_b}\right)}{4} \right] \left(\frac{60}{N}\right)^{1.84} \left(\frac{1}{1 + 6/2}\right)$$

$$= 0.250 \left(2 - \frac{W}{4d_b}\right)\left(\frac{60}{N}\right)^{1.84}.$$

For $(W/d) = 4$ and $N = 60$, J_2 has a value of 0.250. Using a final tooth dullness of 1.0, Eq. 5.12b gives

$$t_b = J_2 \tau_H \left[1 + 6(1)^2/2\right] = 4\tau_H J_2.$$

Substituting the values of τ_H and J_2 into this equation yields

$$t_b = 4(15.7)(0.25)\left(2 - \frac{W}{4d_b}\right)\left(\frac{60}{N}\right)^{1.84}$$

$$= 15.7\left(2 - \frac{W}{4d_b}\right)\left(\frac{60}{N}\right)^{1.84}.$$

For $(W/d_b) = 4$ and $N = 60$, the time required to reach a tooth dullness of 1.0 predicted by this equation is 15.7 hours.

The bearing life can be computed using Eqs. 5.15 and 5.16.

$$J_3 = \left(\frac{60}{N}\right)\left(\frac{4d_b}{W}\right)^{1.0}.$$

$$t_b = J_3 \tau_B (b_f) = J_3 (22)(1.0)$$

$$= 22\left(\frac{4d_b}{W}\right)\left(\frac{60}{N}\right).$$

For $(W/d_b) = 4$ and $N = 60$, the time required to completely wear the bearings predicted by this equation is 22 hours. Evaluation of the multipliers f_1 to f_4 and f_8 yields the following.

$$f_1 = 20.0.$$

$$f_2 = e^{2.303a_2(10,000 - D)}$$

$$\simeq e^{2.303(0.000087)(10,000 - 7,000)}$$

$$= 1.83.$$

$$f_3 = e^{2.303a_3 D^{0.69}(g_p - \rho_c)}$$

$$= 1.0 \text{ for } g_p = 9.0.$$

$$f_4 = e^{2.303a_4 D(g_p - \rho_c)}$$

$$= e^{2.303(0.000017)(7,000)(9 - 10)}$$

$$= 0.76.$$

$$f_8 = \left(\frac{F_j}{1,000}\right)^{a_8} = \left(\frac{900}{1,000}\right)^{0.4} = 0.959.$$

Substitution of these values into Eq. 5.29 gives

$$J_1 = f_1 \cdot f_2 \cdot f_3 \cdot f_4 \cdot f_5 \cdot f_6 \cdot f_8$$

$$= (20)(1.83)(1.0)(0.76)f_5 \cdot f_6 (0.959)$$

$$= 26.7 f_5 \cdot f_6.$$

For $(W/d_b) = 4$ and $N = 60$, both the weight function f_5 and the rotary function f_6 have a value of 1.0; thus, J_1 has a value of 26.7.

The footage drilled before tooth failure at 15.7 hours is given by Eq. 5.31.

$$\Delta D = J_1 J_2 \tau_H \left[\frac{1 - e^{-a_7 h_f}}{a_7} \right.$$
$$\left. + \frac{H_2 \left(1 - e^{-a_7 h_f} - a_7 h_f e^{-a_7 h_f} \right)}{a_7^2} \right].$$

Since the bit teeth will fail first, the final tooth dullness h_f is known to be 1.0. When the bearings fail first, it is necessary to compute h_f for the known value of t_b using Eq. 5.32. Solving the above equation for ΔD, we obtain

$$\Delta D = (26.7)(0.250)(15.7)$$
$$\times \left[\frac{1 - e^{-0.9}}{0.9} + \frac{6(1 - e^{-0.9} - 0.9 e^{-0.9})}{(0.9)^2} \right]$$

$$= 246 \text{ ft.}$$

This footage corresponds to approximately 8 joints of drillpipe at 3 minutes per connection. The total connection time is

$$t_c = \frac{3}{60}(8) = 0.4 \text{ hours.}$$

The cost per foot for the bit run is given by Eq. 1.16.

$$C_f = \frac{C_b + C_r(t_b + t_c + t_t)}{\Delta D}$$

$$= \frac{400 + 500(15.7 + 0.4 + 6.5)}{246} = \$47.56/\text{ft.}$$

This is the predicted cost per foot that corresponds to ending the bit run just before bit failure and is usually the minimum cost per foot for the bit weight and rotary speed assumed. However, to ensure that this is true, the cost per foot corresponding to a slightly shorter bit life should be checked. For example, if the bit was pulled after 15 hours, the final tooth dullness, as computed from Eq. 5.32, is given by

$$h_f = \sqrt{\left(\frac{1}{H_2} \right)^2 + \left(\frac{2 t_b}{H_2 J_2 \tau_H} \right)} - \frac{1}{H_2}$$

$$= \sqrt{\left(\frac{1}{6} \right)^2 + \frac{2(15)}{6(0.25)(15.7)}} - \frac{1}{6} = 0.974.$$

The footage drilled for this value of h_f would be

$$\Delta D = (26.7)(0.25)(15.7)$$
$$\times \left\{ \frac{1 - e^{-0.9(0.974)}}{0.9} + \left[6(1 - e^{-0.9(0.974)}) \right. \right.$$

$$\left. \left. -0.9(0.974) e^{-0.9(0.974)} \right] \middle/ (0.9)^2 \right\} = 238 \text{ ft.}$$

The cost per foot after 15 hours of drilling time is given by

$$C_f = \frac{400 + 500(15 + 0.4 + 6.5)}{238}$$

$$= \$47.69/\text{ft.}$$

Note that this cost per foot is slightly greater than the cost per foot corresponding to the maximum possible bit life.

Relatively simple analytical expressions for the best constant bit weight and rotary speed were derived by Bourgoyne and Young[15] for the case in which tooth wear limits bit life. Eq. 1.16, the cost-per-foot equation, can be rearranged to give

$$C_f = \frac{C_r}{\Delta D} \left(\frac{C_b}{C_r} + t_t + t_c + t_b \right).$$

Substituting Eq. 5.12a for t_b and Eq. 5.31a for ΔD in this cost-per-foot formula yields

$$C_f = \frac{C_r}{\int_0^{h_f} e^{-a_7 h}(1 + H_2 h)\,dh} \left[\frac{\frac{C_b}{C_r} + t_t + t_c}{J_1 J_2 \tau_H} \right.$$
$$\left. + \frac{\int_0^{h_f} (1 + H_2 h)\,dh}{J_1} \right].$$

Taking $(\partial C_f)/[\partial(W/d)] = 0$ and solving yields

$$\left(\frac{C_b}{C_r} + t_t + t_c \right) \left\{ a_5 - \left[\frac{\frac{W}{d_b} - \left(\frac{W}{d_b} \right)_t}{\left(\frac{W}{d_b} \right)_{max} - \left(\frac{W}{d_b} \right)} \right] \right\}$$

$$+ a_5 J_2 \tau_H \int (1 + H_2 h)\,dh = 0. \quad \ldots \ldots (5.33a)$$

Taking $(\partial C_f)/(\partial N) = 0$ and solving yields

$$\left(\frac{C_b}{C_r} + t_t + t_c \right) \left(1 - \frac{H_1}{a_6} \right) + J_2 \tau_H \int (1 + H_2 h)\,dh$$

$$= 0. \quad \ldots \ldots \ldots \ldots \ldots (5.33b)$$

Solving these two equations simultaneously for (W/d_b) gives the following expression for optimum bit weight.

$$\left(\frac{W}{d_b}\right)_{opt} = \frac{a_5 H_1 \left(\frac{W}{d_b}\right)_{max} + a_6 \left(\frac{W}{d_b}\right)_t}{a_5 H_1 + a_6}.$$

$$\dotfill (5.34)$$

If the optimum bit weight predicted by this equation is greater than the flounder bit weight, then the flounder bit weight must be used for the optimum. The optimum bit life is obtained by solving either Eq. 5.33a or Eq. 5.33b for $J_2 \tau_H \int (1 + H_2 h) dh$.

$$t_b = \left(\frac{C_b}{C_r} + t_t + t_c\right)\left(\frac{H_1}{a_6} - 1\right), \quad \dotfill (5.35)$$

The optimum rotary speed N_{opt} is obtained using the known value of t_b in Eq. 5.12b and solving for J_2. N_{opt} then can be obtained from J_2 using Eq. 5.11. This leads to the following expression for N_{opt}.

$$N_{opt} = 60 \left[\frac{\tau_H}{t_b} \frac{\left(\frac{W}{d_b}\right)_{max} - \left(\frac{W}{d_b}\right)_{opt}}{\left(\frac{W}{d_b}\right)_{max} - 4} \right]^{1/H_1}$$

$$\dotfill (5.36)$$

Unfortunately, for the case where bit life is limited by bearing wear or penetration rate, such simple expressions for the optimum conditions have not been found and the construction of a cost-per-foot table is the best approach. This type of calculation is most easily accomplished using a digital computer.

Example 5.13. Compute the optimum bit weight and rotary speed for the bit run described in Example 5.12. Bit floundering was observed to occur for bit weights above 6,700 lbf/in. at 60 rpm.

Solution. The optimum bit weight is computed using Eq. 5.34.

$$\left(\frac{W}{d_b}\right)_{opt} = \frac{a_5 H_1 \left(\frac{W}{d_b}\right)_{max} + a_6 \left(\frac{W}{d_b}\right)_t}{a_5 H_1 + a_6}$$

$$= \frac{1.2(1.84)(8.0) + 0.6(0.5)}{1.2(1.84) + 0.6} = 6.4.$$

Thus, the optimum bit weight is 6,400 lbf/in. of bit diameter. The optimum bit life is computed using Eq. 5.35.

$$t_b = \left(\frac{C_b}{C_r} + t_t + t_c\right)\left(\frac{H_1}{a_6} - 1\right).$$

Assuming a total trip time and connection time of about 7 hours, we obtain

$$t_b = \left(\frac{400}{500} + 7\right)\left(\frac{1.84}{0.6} - 1\right) = 16.1 \text{ hours.}$$

The optimum rotary speed can be calculated using Eq. 5.36.

$$N_{opt} = 60 \left[\frac{\tau_H}{t_b} \frac{\left(\frac{W}{d_b}\right)_{max} - \left(\frac{W}{d_b}\right)_{opt}}{\left(\frac{W}{d_b}\right)_{max} - 4} \right]^{1/H_1}$$

$$= 60 \left[\frac{15.7}{16.1} \frac{8 - 6.4}{8 - 4} \right]^{1/1.84} = 36 \text{ rpm.}$$

Since the computed optimum weight is below the flounder point, the use of 1.2 for a_5 is justified. If the computed optimum bit weight is above the flounder point, the weight at which floundering occurs should be used for the optimum bit weight.

Significant cost savings have been reported from the field use of mathematical methods for obtaining the optimum bit weight and rotary speed. However, these techniques should not be applied without engineering supervision on location. In many instances, the assumptions made in the bearing wear, tooth wear, and penetration rate equations yield inaccurate results and the computed optimums are not valid. When engineering supervision is present in the field, the progress of each bit run can be monitored to ensure that the deviation between the computed and observed results is acceptable. The bit manufacturers constantly are evaluating the performance of their bits in the various areas of drilling activity and can furnish guidelines for the driller when engineering supervision is not available. For example, the normal range of bit weights and rotary speeds recommended by one bit manufacturer for journal-bearing, insert-tooth, rolling-cutter bits is shown in Table 5.12. Using these guidelines, the driller can experiment using his own judgment and dull bit evaluation.

Exercises

5.1 List the two main types of bits in use today. Also, list two subclassifications of each basic bit type and discuss the conditions considered ideal for the application of each subclassification given.

5.2 Discuss how cone offset, tooth height, and number of teeth differ between soft- and hard-formation rolling cutter bits.

5.3 List five basic mechanisms of rock removal that are employed in the design of bits.

5.4 Discuss the primary mechanism of rock removal used in the design of drag bits.

5.5 Discuss the primary mechanism of rock removal used in the design of hard-formation rolling cutter bits.

TABLE 5.12—RANGE OF BIT WEIGHTS AND ROTARY SPEEDS RECOMMENDED BY ONE MANUFACTURER FOR JOURNAL-BEARING, INSERT-TOOTH, ROLLING CUTTER BITS

Bit Size Range (in.)	Bit Weight (lbf/in.)	Rotary Speed (rpm)	Bit Weight (lbf/in.)	Rotary Speed (rpm)	Bit Weight (lbf/in.)	Rotary Speed (rpm)
Class 5-1-7	**Soft Shales, Clays, and Salt**		**Unconsolidated Soft Shale With Sand Streaks**			
7⅞	1,500/3,500	55/90	2,500/4,700	40/60		
8½ to 8¾	1,500/3,600	55/90	2,500/4,800	40/60		
9⅞	1,500/3,700	55/90	2,500/4,800	40/60		
12¼	1,500/3,600	55/85	2,500/4,500	40/60		
Class 5-3-7	**Medium-Soft Shale**		**Soft and Medium-Soft Limestone and Dolomite**		**Unconsolidated Shale, Soft Lime, and Sand**	
6 to 6¾	2,400/3,200	55/70	3,200/4,300	45/65	3,200/4,000	40/55
7⅜ to 7⅞	2,900/4,000	55/70	3,900/5,000	45/65	3,900/4,500	40/55
8⅜ to 8¾	3,200/4,100	55/70	4,000/5,000	45/65	4,000/4,600	40/55
9½ to 9⅞	3,200/4,100	55/70	4,000/5,000	45/65	4,000/4,600	40/55
10⅝ to 11	3,000/4,000	55/70	3,900/4,500	45/65	3,900/4,500	40/55
12¼	2,800/3,800	55/70	3,800/4,300	45/65	3,800/4,400	40/55
14¾	2,700/3,700	55/65	3,700/4,300	45/60	3,700/4,300	40/55
17½	2,600/3,100	50/65	3,200/4,000	45/60	3,200/3,700	40/55
Class 6-1-7	**Medium-Hard Shale**		**Medium-Hard Lime and Shale Mixtures**		**Medium-Hard Sandy Lime and Shale**	
6 to 6¾	2,600/3,900	50/65	3,200/4,800	40/60	3,200/4,300	35/55
7⅜ to 7⅞	3,500/5,000	50/65	4,200/5,500	40/60	4,200/5,300	35/55
8⅜ to 8¾	3,600/5,100	50/65	4,400/5,600	40/60	4,400/5,400	35/55
9½ to 9⅞	3,600/5,100	50/65	4,400/5,500	40/60	4,400/5,400	35/55
10⅝ to 11	3,500/5,000	50/65	4,300/5,400	40/60	4,300/5,300	35/55
12¼	3,400/5,000	50/65	4,200/5,300	40/60	4,200/5,200	35/55
14¾	3,300/4,700	45/60	4,000/5,000	40/55	4,000/5,000	35/55
17½	3,100/4,100	45/60	3,500/4,700	40/55	3,500/4,700	35/50
Class 6-2-7	**Hard Shale With Lime**		**Hard Sandy Shale**			
6 to 6¾	3,300/4,500	40/60	3,300/4,400	35/50		
7⅜ to 7⅞	4,300/6,000	40/60	4,300/5,500	35/50		
8⅜ to 8¾	4,300/6,000	40/60	4,300/5,600	35/50		
9½ to 9⅞	4,300/6,000	40/60	4,300/5,600	35/50		
10⅝ to 11	4,200/5,800	40/60	4,200/5,500	35/50		
12¼	4,200/5,600	40/60	4,200/5,400	35/50		
Class 6-3-7	**Medium-Hard Limestone, Dolomite, and Brittle Shale**		**Medium-Hard Sandy Lime and Dolomite**			
6 to 6¾	3,200/4,500	35/60	3,000/4,300	35/50		
7⅞	4,500/6,300	35/60	4,500/6,000	35/50		
8½ to 8¾	4,500/6,300	35/60	4,500/6,000	35/50		
9½	4,400/6,300	35/60	4,300/6,000	35/50		
12¼	4,000/6,200	35/60	4,000/5,700	35/50		
Class 7-2-7	**Hard Lime and Dolomite**		**Hard Sandy Lime**			
6 to 6¾	2,800/4,600	35/55	2,700/4,300	35/45		
7⅜ to 7⅞	4,500/6,500	35/55	4,300/6,000	35/45		
8⅜ to 8¾	4,500/6,500	35/55	4,300/6,000	35/45		
Class 8-3-7	**Hardest Sandy Lime Chert, Basalt, Bromide, etc.**		**Hardest, Unconsolidated Abrasive Formations With Pyrite, Chert, etc.**			
6½	3,500/5,500	35/50	2,700/4,300	30/45		
7⅞	4,000/7,000	35/50	3,500/5,000	30/45		
8½ to 8¾	4,000/7,000	35/50	3,500/5,000	30/45		

5.6 A rock sample is placed in a strength-testing machine at atmospheric pressure and compressed axially to failure. A force of 12,000 lbf was required for rock failure, and the cross-sectional area of the sample was 2.0 sq in. The sample failed along a plane that makes a 35° angle with the direction of the compressional loading.

a. Construct Mohr's circle using the two principal stresses present.

b. Compute the shear stress present along the plane of failure. *Answer:* 2,819 psig.

c. Compute the normal stress to the plane of failure. *Answer:* 1,973 psig.

d. Compute the angle of interval friction. *Answer:* 20°.

e. Compute the cohesive resistance of the material. *Answer:* 2,100 psig.

f. Label the parameters computed in the previous four steps on the Mohr's circle construction. Using the Mohr criterion, compute the compressional force required for rock failure if the sample is placed under a 5,000-psi confining pressure. *Answer:* 16,200 psig.

5.7 Discuss how the mode of failure beneath the tooth of a rolling cutter bit changes as overbalance increases.

5.8 List seven rules of thumb of bit selection. What is the best basis of comparison when trying to choose between two different bit types?

5.9 The bit type currently used to drill a given formation consistently yields a drilling cost of about $50/ft. You are sending a new experimental bit type to the field for evaluation in this formation. The new bit is expected to have a bit life of about 150 hours as compared with the usual bit life of 15 hours. The new bit costs $10,000, and the operating cost of the drilling operation is $750/hr. Trip time is approximately 10 hours for the depth of interest. Prepare a graph that shows the *break-even* costs of $50/ft as a function of penetration rate and bit life to assist in the field evaluation of the new bit. Label the region of the graph that shows combinations of penetration rate and bit life which are not acceptable. If the initial penetration rate of the new bit during the first hour is 4 ft/hr, what would you recommend? *Answer:* Pull bit.

5.10 Grade the bit shown in the photograph below. *Answer:* T-8, B-8, G0.

5.11 Describe the difference between self-sharpening and abrasive tooth wear. Discuss what is done by the bit manufacturers to promote self-sharpening bit wear on milled-tooth, rolling-cutter bits.

5.12 A 9.875-in. Class 1-1-1 bit drilled from a depth of 12,000 to 12,200 ft in 12 hours. The average bit weight and rotary speed used for the bit run was 40,000 lbf and 90 rpm, respectively. When the bit was pulled, it was graded T-6, B-6. The drilling fluid was a barite-weighted clay/water mud having a density of 12 lbm/gal.

a. Compute the average formation abrasiveness constant for this depth interval. *Answer:* 43.7 hours.

b. Estimate the time required to completely dull the bit teeth using a bit weight of 45,000 lbf and a rotary speed of 100 rpm. *Answer:* 13.5 hours.

c. Compute the bearing constant for this depth interval. *Answer:* 24.3 hours.

d. Estimate the time required to completely dull the bearings using a bit weight of 45,000 lbf and a rotary speed of 100 rpm. *Answer:* 12.8 hours.

5.13 Compute the bearing constant, τ_B, for the bit of Example 5.3 if the drilling fluid were a weighted clay-water mud (barite mud). Use values of B_1 and B_2 recommended in Table 5.9. *Answer:* 41.7 hours.

5.14 Field data obtained on 7.875-in., Series 6, roller-bearing bits at a rotary speed of 60 rpm show an average bearing life of 32 hours for a bit weight of 5,700 lbf/in. and 45 hours for 3,800 lbf/in. Compute the apparent bearing weight exponent, B_2, and bearing constant, τ_B, for this bit type. *Answer:* 0.84 and 43.1 hours.

5.15 Field data observed on 7.875-in., Series 6, sealed journal-bearing bits at a rotary speed of 60 rpm shows an average bearing life of 67 hours at 5,700 lbf/in. and 100 hours at 3,800 lbf/in. Compute the apparent bearing weight exponent, B_2, and bearing constant, τ_B, for this bit type. *Answer:* 0.99 and 95 hours.

5.16 Field data observed on 7.875-in., Series 6, sealed journal-bearing bits operated with 4,000 to 5,000 lbf of bit weight per inch of bit diameter showed a medium bit life of 95 hours at a rotary speed of 60 rpm and 185 hours at 40 rpm. Using an assumed value of 1.0 for B_2, compute the apparent values for B_1 and τ_B from these observations. *Answer:* 1.64 and 107 hours.

5.17 Field data obtained using 7.875-in., Series 6, sealed roller-bearing insert bits operated at 4,000 lbf per inch of bit diameter indicated an average bit life of 42 hours at a rotary speed of 60 rpm and 55 hours at 40 rpm. Compute the apparent values of B_1 and τ_B. *Answer:* 0.67 and 42 hours.

5.18 Recommend values of B_2 and τ_B for 7.875-in., nonsealed, roller-bearing bits operated in oil muds, weighted clay-water muds (barite muds), and a clay-water mud containing H_2S (sulfide mud). The recommendation should be based on the laboratory

Fig. 5.53—Dull bit for Exercise 5.10.

TABLE 5.13—LABORATORY BEARING-LIFE DATA OBTAINED AT 60 rpm

Bit Weight/in. (K-lbf/in.)	Bearing Life (hours)		
	Sulfide Mud	Barite Mud	Oil Mud
3	14.0	48.0	—
6	7.5	17.5	80.0
9	—	—	25.0

bearing wear data shown in Table 5.13 and conducted at a rotary speed of 60 rpm. The bearing life was determined based on 0.1-in. wear in the bearing races. *Answer:* 2.87 and 256 hours; 1.46 and 31.5 hours; and 0.9 and 10.8 hours, respectively.

5.19 Field data obtained on 8.5-in. sealed roller-bearing insert bits are shown below. Use these data to obtain representative values of B_1, B_2, and τ_B for this bit type.

Bit Weight (1,000 lbf/in.)	Rotary Speed (rpm)	Bearing Life (hours)
4	60	41
6	60	30
4	40	81

Answer: 1.68, 0.77, and 41 hours.

5.20 Determine the optimum bit life for the bit run described in the following table. The lithology is known to be uniform for the depth range of interest. The tooth wear parameter J_2 has a value of 0.15, the constant H_2 has a value of 7.0, and the bearing wear parameter J_3 has a value of 0.56. The formation abrasiveness constant τ_H has a value of 40 hours, and the bearing constant τ_B has a value of 40 hours. The bit cost is $600, the rig operating cost is $1,000/hr, and the trip time is 6 hours.

Drilling Time t_b+t_c (hours)	Total Footage ΔD (ft)	Remarks
0	0	New bit
2	30	
4	54	
6	73	
8	88	
10	104	
12	117	
14	127	
16	135	
18	142	
20	147	
22	151	Torque increase

Answer: 12 hours.

5.21 List eight factors affecting penetration rate.

5.22 A penetration rate in shale of 20 ft/hr was obtained using a mud density of 12 lbm/gal at a depth of 10,000 ft. When the mud density was increased to 13 lbm/gal, the penetration rate was decreased to 9.5 ft/hr for similar drilling conditions. Compute the apparent value of the overbalance exponent a_4. *Answer:* 32×10^{-6}.

5.23 The penetration rate in shale is observed to increase from 12 to 18 ft/hr when the bit weight is increased from 30,000 to 50,000 lbf. Compute bit weight exponent a_5. *Answer:* 0.8.

5.24 Using the following drilloff test data, evaluate the bit weight exponent a_5 and rotary speed exponent a_6. The length of the 4.5-in., 16.6-lbm/ft drillpipe was 12,000 ft.

Test No. 1 (rotary speed = 120 rpm)

Bit Weight (1,000 lbf)	Elapsed Time (seconds)
80	0
76	104
72	210
68	320
64	436
60	562
56	704
52	864
48	1,045

Test No. 2 (rotary speed = 80 rpm)

Bit Weight (1,000 lbf)	Elapsed Time (seconds)
80	0
76	108
72	228
68	360
64	506
60	668
56	848
52	1,050

Answer: $a_5 = 1.6$ and $a_6 = 0.6$.

5.25 The average penetration rate in shale is observed to drop from 18 ft/hr for a new bit to 11 ft/hr at the end of the bit run. The bit was graded T-6, B-7. Assuming all variables other than tooth wear remained constant, evaluate the tooth wear exponent a_7. *Answer:* 0.7.

5.26 A bit contains three $\frac{14}{32}$-in. nozzles, and the mud, which has a density of 10 lbm/gal, is being circulated at a rate of 600 gal/min. The penetration rate is observed to decrease from 15 to 11 ft/hr when one of the two pumps is stopped temporarily, causing the circulation rate to fall from 600 to 400 gal/min. Compute the apparent hydraulics exponent a_8. *Answer:* 0.4.

5.27 A diamond bit with a total blade length of 5.585 in. contains 200 stones of 1.0 carats that have a width of 0.0848 in. for a penetration of 0.01 in. Compute the expected penetration rate if sufficient bit weight for a 0.01-in. depth of diamond penetration could be maintained at a rotary speed of 100 rpm. Assume that the diamonds are shaped and arrayed so that the penetration is two-thirds the maximum penetration depth. *Answer:* 8.0 ft/hr.

5.28 An 8.5-in. Class 1-1-1 bit operated at 35,000 lbf and 90 rpm is drilling in a shale formation at a depth of 9,000 ft at a penetration rate of 30 ft/hr. The formation pore pressure is equivalent to a 9.0-lbm/gal mud, and the equivalent mud density on bottom is 9.7 lbm/gal. The computed impact force beneath the bit is 1,300 lbf, and the computed fractional tooth wear is 0.4. Compute the apparent formation drillability f_1 for this bit type at 9,000 ft using a threshold bit weight of zero and the following values of a_2 through a_8.

a_2	a_3	a_4	a_5	a_6	a_7	a_8
0.00009	0.000004	0.00002	1.2	0.6	0.4	0.4

Answer: f_1 = 26.1 ft/hr.

5.29 A 9.875-in.-diameter Class 1-1-1 bit will be used to drill a formation at 9,000 ft that has a drillability of 40 ft/hr. The abrasiveness constant τ_H has a value of 38 hours, and the bearing constant τ_b has a value of 22 hours. The formation pore pressure gradient is equivalent to a 9.0 lbm/gal fluid gradient, and the weighted clay/water drilling fluid barite mud has a density of 9.7 lbm/gal. The bit cost is $600, the operating cost of the drilling operation is $800/hr, the time required to trip for a new bit is 7 hours, and 4 minutes are required to make a connection per 30-ft joint of drillpipe. Using a threshold bit weight per inch of 0.5 and the constants a_2 through a_8 given in Exercise 5.28, compute the cost per foot that would be observed for (W/d_b) = 4.5, N = 90 rpm, and a jet impact force of 1,100 lbf. *Answer:* 29.73 $/ft.

5.30 Write a FORTRAN program for determining a cost-per-foot table. Assume a constant bit weight and that rotary speed will be maintained throughout the bit run and that the formation drillability constant is a constant. Use the program to determine the optimum bit weight and rotary speed for Exercise 5.29.

5.31 Compute the optimum bit weight and rotary speed for the bit described in Exercise 5.29, assuming bit life is limited by tooth wear. Ignore the effect of connection time. The flounder bit weight is known to be 60,000 lbf/in. *Answer:* 5,650 lbf/in. and 60 rpm.

5.32 Use the program developed in Exercise 5.30 to determine the optimum bit weight and rotary speed for Exercise 5.29 if τ_B has a value of 95 instead of 22 hours.

References

1. Cerkovnik, J.: "Design, Application, and Future of Polycrystalline Diamond Compact Cutters in the Rocky Mountains," paper SPE 10893 presented at the 1982 SPE Rocky Mountain Regional Meeting, Billings, MT, May 19-21.
2. *Bit Handbook*, Hughes Tool Co., Houston (1976).
3. Durrett, E.: "Rock Bit Identification Simplified by IADC Action," *Oil and Gas J.* (May 22, 1972) 76.
4. Estes, J.C.: "Selecting the Proper Rotary Rock Bit," *J. Pet. Tech.* (Nov. 1971) 1359-67.
5. Hughes, R.V.: "Drag Bits Rate New Look in Light of Speed Drilling," *World Oil* (March 1965) 94.
6. Gray, K.E., Armstrong, F., and Gatlin, C.: "Two-Dimensional Study of Rock Breakage in Drag-Bit Drilling at Atmospheric Pressure," *J. Pet. Tech.* (Jan. 1962) 93-98; *Trans.*, AIME, **225**.
7. Maurer, W.C.: "Bit-Tooth Penetration Under Simulated Borehole Conditions," *J. Pet. Tech.* (Dec. 1965) 1433-42; *Trans.*, AIME, **234**.
8. Murray, A.S. and MacKay, S.P.: "Water Still Poses Tough Problem in Drilling with Air," *Oil and Gas J.* (June 10, 1957) 105.
9. Campbell, J.M. and Mitchell, B.J.: "Effect of Tooth Geometry on Tooth Wear Rate of Rotary Rock Bits," paper presented at API Mid-Continent Dist. Spring Meeting, March 1959.
10. Hoover, E.R. and Middleton, J.N.: "Laboratory Evaluation of PCD Drill Bits Under High Speed and High Wear Conditions," paper SPE 10326 presented at the 1981 SPE Annual Technical Conference and Exhibition, San Antonio, Oct. 4-7.
11. Galle, E.M. and Woods, A.B.: "Best Constant Weight and Rotary Speed for Rotary Rock Bits," *Drill. and Prod. Prac.*, API (1963) 48-73.
12. Edwards, J.H.: "Engineering Design of Drilling Operations," *Drill. and Prod. Prac.*, API (1964) 38-55.
13. Young, F.S. Jr.: "Computerized Drilling Control," *J. Pet. Tech.* (April 1969) 483-496; *Trans.*, AIME, **246**.
14. Reed, R.L.: "A Monte Carlo Approach to Optimal Drilling," *Soc. Pet. Eng. J.* (Oct. 1972) 423-438; *Trans.*, AIME, **253**.
15. Bourgoyne, A.T. and Young, F.S. Jr.: "A Multiple Regression Approach to Optimal Drilling and Abnormal Pressure Detection," *Soc. Pet. Eng. J.* (Aug. 1974) 371-384; *Trans.*, AIME, **257**.
16. Estes, J.C.: "Guidelines for Selecting Rotary Insert Rock Bits," *Pet. Eng.* (Sept. 1974).
17. Lummus, J.L.: "Analysis of Mud Hydraulics Interactions," *Pet. Eng.* (Feb. 1974).
18. Bingham, M.G.: "A New Approach to Interpreting Rock Drillability," reprinted from *Oil and Gas J.* series by Petroleum Publishing Co. (April 1965).
19. Cunningham, R.A. and Eenink, J.G.: "Laboratory Study of Effect of Overburden, Formation, and Mud Column Pressures on Drilling Rate of Permeable Formations," *Trans.*, AIME (1959) **216**, 9-17.
20. Garnier, A.J. and van Lingen, N.H.: "Phenomena Affecting Drilling Rates at Depth," *Trans.*, AIME (1959) **216**, 232-239.
21. Black, A.D. and Green, S.J.: "Laboratory Simulation of Deep Well Drilling," *Pet. Eng.* (March 1978).
22. Vidrine, D.J. and Benit, E.J.: "Field Verification of the Effect of Differential Pressure on Drilling Rate," *J. Pet. Te*
23. Murray, A.S. and Cunningham, R.A.: "Effect of Mud Column Pressure on Drilling Rates," *Trans.*, AIME (1955) **204**, 196-204.
24. Eckel, J.R.: "Effect of Pressure on Rock Drillability," *Trans.*, AIME (1957) **213**, 1-6.
25. Maurer, W.C.: "The 'Perfect-Cleaning' Theory of Rotary Drilling," *J. Pet. Tech.* (Nov. 1962) 1270-74; *Trans.*, AIME, **225**.
26. Vidder, A.: "Chevron Drill-Off Test (DOT)," Chevron Oil Co., New Orleans.
27. Pratt, C.A.: "Increased Penetration Rate Achieved With New Extended Nozzle Bit," *J. Pet. Tech.* (Aug. 1978) 1192-98.
28. Eckel, J.R.: "Microbit Studies of the Effect of Fluid Properties and Hydraulics on Drilling Rate, II," paper SPE 2244, SPE, Dallas (1968).
29. Tibbitts, G.A. *et al.*: "Effects of Bit Hydraulics on Full Scale Laboratory Drilled Shale," *J. Pet. Tech.* (July 1981) 1180-88.
30. Peterson, J.L.: "Diamond Drilling Model Verified in Field and Laboratory Tests," *J. Pet. Tech.* (Feb. 1976) 215-222; *Trans.*, AIME, **261**.

Nomenclature

a_1-a_8 = exponents in the penetration rate equation

A = area

A_b = area of bit

A_{dt} = total diamond area in contact with the formation

b_f = final bearing wear at end of bit run

B_1, B_2 = bearing wear exponents

c = cohesive resistance of material

c_f = cost per interval drilled

C_c = concentration of diamond cutters, carats/sq in.

C_r = fixed operating cost of rig per unit time

d = diameter

d_b = diameter of bit

d_c = diameter of cutter

D = depth

ΔD = depth interval drilled during bit run

E = Young's modulus of elasticity

f_1-f_8 = functions defining effect of various drilling variables

F = force

g_p = formation pore pressure gradient expressed as an equivalent fluid density

G = geometry constant for a given tooth design

h = fractional tooth wear

h_f = final tooth wear at end of bit run

H_1-H_3 = tooth geometry constants used to predict bit tooth wear

J_1-J_3 = composite functions of bit weight and rotary speed used in penetration rate, tooth wear, and bearing wear equations, respectively

K = constant

ℓ = tooth length

L = length

L_i = initial height

L_p = depth of penetration of drag bit cutter

L_r = height removed

L_w = width cut by an individual diamond for a penetration L_p

n_b = effective number of blades of drag bit

n_c = number of cutters

N = rotary speed

N_{Re} = Reynolds number

p = pressure

q = flow rate

r = radius

r_f = formation resistance

R = penetration rate

R_0 = penetration rate at zero overbalance

s_d = size of diamond, carats/stone

S = compressive strength of rock

t = time

t_b = bit life

t_c = nonrotating time during bit run (such as connection time)

t_t = time of tripping operations required to change bit

w = width

W = weight on bit

x,y = spatial coordinates

α = bottom cutting angle

β = angle subtended by wear surface on PCD blank

ϵ = axial strain or elongation per unit length

θ = angle of internal friction

μ = viscosity

μ_a = apparent viscosity

ρ = mud density

σ = normal stress

τ = shear stress

τ_B = bearing life constant

τ_H = formation abrasiveness constant

ϕ = angle between failure plane and direction of principal stress

log = logarithm, common base 10

ln = logarithm, natural, base e

Subscripts

a = apparent

b = bit

bh = bottomhole

c = cutter; also circulating

dp = drillpipe

e = effective

f = formation

i = initial

j = jet

max = maximum or destructive, also slope

n = normal to plane

opt = optimum

r = removed

s = steel, also standard or reference

t = threshold

x,y = spatial coordinates or directions

$\overline{}$ = average

SI Metric Conversion Factors

ft	× 3.048*	E−01	=	m
gal/min	× 3.785 412	E−03	=	m^3/min
in.	× 2.54*	E+00	=	cm
lbf	× 4.448 222	E+00	=	N
lbm	× 4.535 924	E−01	=	kg
lbm/gal	× 1.198 264	E+02	=	kg/m^3
psi	× 6.894 757	E+00	=	kPa
sq in.	× 6.451 6*	E−00	=	cm^2

*Conversion factor is exact.

Chapter 6
Formation Pore Pressure and Fracture Resistance

The objective of this chapter is to familiarize the student with commonly used methods of estimating the naturally occurring pressure of subsurface formation fluids and the maximum wellbore pressure that a given formation can withstand without fracture.

With the drilling of most deep wells, formations are penetrated that will flow naturally at a significant rate. In drilling these wells, safety dictates that the wellbore pressure (at any depth) be maintained between the naturally occurring pressure of the formation fluids and the maximum wellbore pressure that the formation can withstand without fracture. In Chap. 4, we focused on the determination of wellbore pressures during various types of drilling operations. In this chapter, the determination of formation fluid pressure and fracture pressure is discussed. Knowledge of how these two parameters vary with depth is extremely important in planning and drilling a deep well.

6.1 Formation Pore Pressure

To understand the forces responsible for subsurface fluid pressure in a given area, previous geologic processes must be considered. One of the simplest and most common subsurface pressure distributions occurs in the shallow sediments that were laid down slowly in a deltaic depositional environment (Fig. 6.1).

While detritus material, which is carried by river to the sea, is released from suspension and deposited, the sediments formed are initially unconsolidated and uncompacted and, thus, have a relatively high porosity and permeability. The seawater mixed with these sediments remains in fluid communication with the sea and is at hydrostatic pressure.

Once deposition has occurred, the weight of the solid particles is supported at grain-to-grain contact points and the settled solids have no influence on the hydrostatic fluid pressure below. Thus, hydrostatic pressure of the fluid contained within the pore spaces of the sediments depends only on the fluid density. With greater burial depth as deposition continues, the previously deposited rock grains are subjected to increased load through the grain-to-grain contact points. This causes realignment of the grains to a closer spacing, resulting in a more compacted, lower-porosity sediment.

As compaction occurs, water is expelled continually from the decreasing pore space. However, as long as there is a relatively permeable flow path to the surface, the upward flow potential gradient that is required to release the compaction water will be negligible and hydrostatic equilibrium will be maintained. Thus, the formation pore pressure can be computed by use of Eq. 4.2b in Chap. 4.

When formation pore pressure is approximately equal to theoretical hydrostatic pressure for the given vertical depth, formation pressure is said to be *normal*. Normal pore pressure for a given area usually is expressed in terms of the hydrostatic gradient. Table 6.1 lists the normal pressure gradient for several areas that have considerable drilling activity.

Example 6.1. Compute the normal formation pressure expected at a depth of 6,000 ft in the Louisiana gulf coast area.

Solution. The normal pressure gradient for the U.S. gulf coast area is listed in Table 6.1 as 0.465 psi/ft. Thus, the normal formation pore pressure expected at 6,000 ft is:

$$p_f = 0.465 \text{ psi/ft } (6,000 \text{ ft}) = 2,790 \text{ psi.}$$

6.1.1 Abnormal Formation Pressure

In many instances, formation pressure is encountered that is greater than the normal pressure for that depth. The term *abnormal formation pressure* is used to

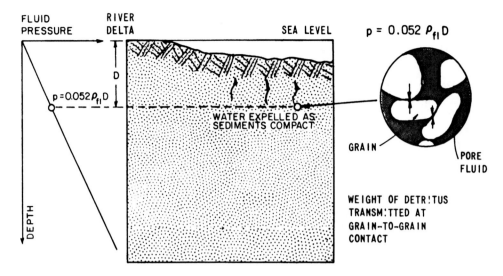

Fig. 6.1—Normal subsurface fluid pressure distribution in shallow deltaic sediments.

describe formation pressures that are greater than normal. Abnormally low formation pressures also are encountered, and the term *subnormal formation pressure* is used to describe these pressures.

Abnormal formation pressures are found in at least a portion of most of the sedimentary basins of the world. While the origin of abnormal formation pressure is not understood completely, several mechanisms that tend to cause abnormal formation pressure have been identified in sedimentary basins. These mechanisms can be classified generally as: (1) compaction effects, (2) diagenetic effects, (3) differential density effects, and (4) fluid migration effects.

6.1.2 Compaction Effects. Pore water expands with increasing burial depth and increased temperature, while the pore space is reduced by increasing geostatic load. Thus, normal formation pressure can be maintained only if a path of sufficient permeability exists to allow formation water to escape readily.

To illustrate this principle, a simple, one-dimensional, soil mechanics model is shown in Fig. 6.2. In the model, the rock grains are represented by pistons that contact one another through compressional springs. Connate water, which fills the space between the pistons, has a natural flow path to the surface. However, this path may become restricted (represented by closing the valve in the model). The pistons are loaded by the weight of the overburden, or geostatic load, σ_{ob}, at the given depth of burial. Resisting this load are (1) the support provided by the vertical grain-to-grain, or matrix, stress, σ_z, and (2) the pore fluid pressure, p. Thus, we have

$$\sigma_{ob} = \sigma_z + p. \qquad \qquad (6.1)$$

As long as pore water can escape as quickly as required by the natural compaction rate, the pore pressure will remain at hydrostatic pressure. The matrix stress will continue to increase as the pistons move closer together until the overburden stress is balanced.

TABLE 6.1—NORMAL FORMATION PRESSURE GRADIENTS FOR SEVERAL AREAS OF ACTIVE DRILLING

	Pressure Gradient (psi/ft)	Equivalent Water Density (kg/m³)
West Texas	0.433	1.000
Gulf of Mexico coastline	0.465	1.074
North Sea	0.452	1.044
Malaysia	0.442	1.021
Mackenzie Delta	0.442	1.021
West Africa	0.442	1.021
Anadarko Basin	0.433	1.000
Rocky Mountains	0.436	1.007
California	0.439	1.014

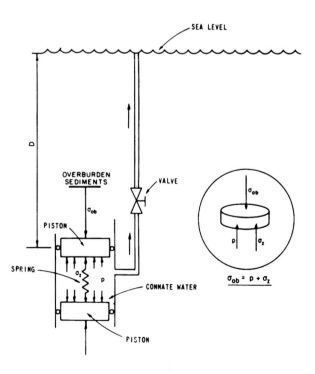

Fig. 6.2—One-dimensional sediment compaction model.

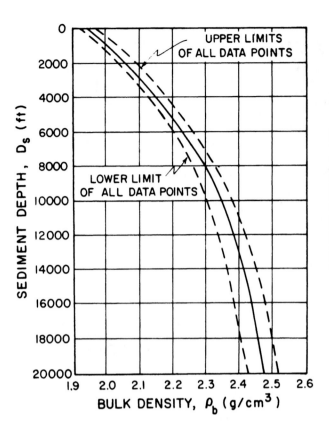

Fig. 6.3—Composite bulk density curve from density log data for the U.S. gulf coast.[1]

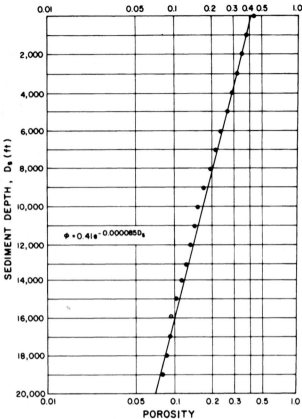

Fig. 6.4—Computed average porosity trend for U.S. gulf coast area.

TABLE 6.2—AVERAGE SEDIMENT POROSITY COMPUTATION FOR U.S. GULF COAST AREA

(1) Sediment Thickness D_s (ft)	(2) Bulk Density ρ_b (g/cm³)	(3) Average Porosity ϕ (frac.)
0	1.95	0.43
1,000	2.02	0.38
2,000	2.06	0.35
3,000	2.11	0.32
4,000	2.16	0.29
5,000	2.19	0.27
6,000	2.24	0.24
7,000	2.27	0.22
8,000	2.29	0.20
9,000	2.33	0.18
10,000	2.35	0.16
11,000	2.37	0.15
12,000	2.38	0.14
13,000	2.40	0.13
14,000	2.41	0.12
15,000	2.43	0.11
16,000	2.44	0.10
17,000	2.45	0.098
18,000	2.46	0.092
19,000	2.47	0.085
20,000	2.48	0.079

However, if the water flow path is blocked or severely restricted, the increasing overburden stress will cause pressurization of the pore water above hydrostatic pressure. The pore volume also will remain greater than normal for the given burial depth. The natural loss of permeability through compaction of fine-grained sediments, such as shale or evaporites, may create a seal that would permit abnormal pressures to develop.

The vertical overburden stress resulting from geostatic load at a sediment depth, D_s, for sediments having an average bulk density, ρ_b, is given by

$$\sigma_{ob} = \int_0^D \rho_b g dD, \quad \dots\dots\dots\dots\dots\dots \quad (6.2)$$

where g is the gravitational constant.

The bulk density at a given depth is related to the grain density, ρ_g, the pore fluid density, ρ_{fl}, and the porosity, ϕ, as follows.

$$\rho_b = \rho_g (1 - \phi) + \rho_{fl} \phi. \quad \dots\dots\dots\dots\dots \quad (6.3a)$$

In an area of significant drilling activity, the change in bulk density with depth usually is determined by conventional well logging methods. The effect of depth on average bulk density for sediments in the Texas and Louisiana gulf coast areas is shown in Fig. 6.3.[1]

The change in bulk density with burial depth is related primarily to the change in sediment porosity with compaction. Grain densities of the common minerals found in sedimentary deposits do not vary greatly and usually can be assumed constant at a representative average value. This is also true for pore fluid density.

In many areas, it is convenient to use the exponential relationship relating change in average sediment porosity to depth of burial when calculating the overburden stress, σ_{ob}, resulting from geostatic load at a given depth. To use this approach, the average bulk density data are expressed first in terms of average porosity. Solving Eq. 6.3a for porosity yields

$$\phi = \frac{\rho_g - \rho_b}{\rho_g - \rho_{fl}}. \qquad\qquad\qquad (6.3b)$$

This equation allows average bulk density data read from well logs to be expressed easily in terms of average porosity for any assumed grain density and fluid density. If these average porosity values are plotted vs. depth on semilog paper, a good straight-line trend usually is obtained. The equation of this line is given by

$$\phi = \phi_o e^{-KD_s}, \qquad\qquad\qquad (6.4)$$

where ϕ_o is the surface porosity, K is the porosity decline constant, and D_s is the depth below the surface of the sediments. The constants ϕ_o and K can be determined graphically or by the least-square method.

Example 6.2. Determine values for surface porosity, ϕ_o, and porosity decline constant, K, for the U.S. gulf coast area. Use the average bulk density data shown in Fig. 6.3, an average grain density of 2.60 g/cm^3, and an average pore fluid density of 1.074 g/cm^3.

Solution. The porosity calculations are summarized in Table 6.2. The bulk density given in Col. 2 was read from Fig. 6.3 at the depth given in Col. 1. The porosity values given in Col. 3 were computed using an average grain density of 2.60 and a fluid density of 1.074 g/cm^3 in Eq. 6.3b.

$$\phi = \frac{2.60 - \rho_b}{2.60 - 1.074} = \frac{2.60 - \rho_b}{1.526}.$$

The computed porosities are plotted in Fig. 6.4. A surface porosity, ϕ_o, of 0.41 is indicated on the trend line at zero depth. A porosity of 0.075 is read from the trend line at a depth of 20,000 ft. Thus, the porosity decline constant is

$$K = \frac{\ln \dfrac{\phi_o}{\phi}}{D_s} = \frac{\ln\left(\dfrac{0.41}{0.075}\right)}{20,000} = 0.000085 \text{ ft}^{-1}$$

and the average porosity can be computed using

$$\phi = 0.41 e^{-0.000085 D_s}.$$

The vertical overburden stress resulting from the geostatic load is computed easily at any depth once a convenient expression for the change in average sediment porosity with depth is obtained. Substitution of Eq. 6.3a into Eq. 6.2 gives

$$\sigma_{ob} = g \int_0^D [\rho_g(1-\phi) + \rho_{fl}\phi]dD. \qquad\qquad (6.5)$$

In offshore areas, Eq. 6.5 must be integrated in two parts. From the surface to the ocean bottom, the seawater density, ρ_{sw}, is equal to 8.5 lbm/gal and the porosity is 1. From the mudline to the depth of interest, the fluid density is assumed equal to the normal formation fluid density for the area and the porosity can be computed using Eq. 6.4. Thus, Eq. 6.5 becomes

$$\sigma_{ob} = g \int_0^{D_w} \rho_{sw}dD$$
$$+ g \int_{D_w}^D [\rho_g - (\rho_g - \rho_{fl})\phi_o e^{-KD}]dD.$$

Integration of this equation and substitution of D_s $(D - D_w)$, the depth below the surface of the sediments, yields

$$\sigma_{ob} = \rho_{sw}gD_w + \rho_g gD_s - \frac{(\rho_g - \rho_{fl})g\phi_o}{K}(1 - e^{-KD_s}).$$

$$\qquad\qquad\qquad\qquad (6.6)$$

Example 6.3. Compute the vertical overburden stress resulting from geostatic load near the Gulf of Mexico coastline at a depth of 10,000 ft. Use the porosity relationship determined in Example 6.2.

Solution. The vertical overburden stress resulting from geostatic load can be calculated using Eq. 6.6 with a water depth of zero. The grain density, surface porosity, and porosity decline constant determined in Example 6.2 were 2.60 g/cm^3, 0.41, and 0.000085 ft^{-1}, respectively. As shown in Table 6.1, the normal pore fluid density for the gulf coast area is 1.074 g/cm^3. Converting the density units to lbm/gal, using the conversion constant 0.052 to convert ρg to psi/ft, and inserting these values in Eq. 6.6, yields

$$\sigma_{ob} = 0.052(2.60)(8.33)(10,000)$$

$$- \frac{0.052(2.60 - 1.074)(8.33)(0.41)}{0.000085}$$

$$\cdot \left[1 - e^{-0.000085(10,000)}\right]$$

$$= 11,262 - 1,826 = 9,436 \text{ psi}.$$

The vertical overburden stress resulting from geostatic load often is assumed equal to 1.0 psi per foot of depth. This corresponds to the use of a constant value of bulk

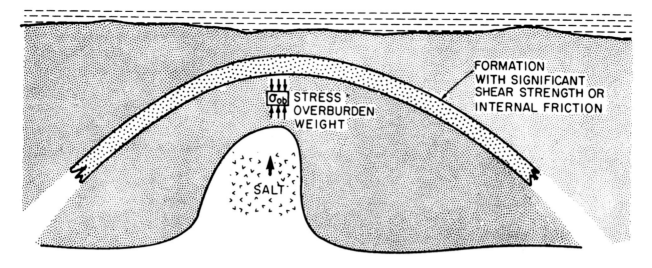

Fig. 6.5—Example of compressive stress in excess of geostatic load.

density for the entire sediment section. This simplifying assumption can lead to significant errors in the computation of overburden stress, especially for shallow sediments. Such an assumption should be made only when the change in bulk density with respect to depth is not known. Note that in Example 6.3, an average overburden stress gradient of 0.944 psi/ft was indicated.

The calculation of vertical overburden stress resulting from geostatic load does not always adequately describe the total stress state of the rock at the depth of interest. Compressive stresses resulting from geologic processes other than sedimentation may be present; these also tend to cause sediment compaction. For example, the upward movement of low-density salt or plastic shale domes is common in the U.S. gulf coast area. In the U.S. west coast area, continental drift is causing a collision of the North American and Pacific plates, which results in large lateral compressive stresses. If there are overlying rocks with significant shear resistance, the vertical stress state at depth may exceed the geostatic load. This is illustrated in Fig. 6.5. However, rocks generally fail readily when subjected to shear stress and faulting will occur, which tends to relieve the buildup of stresses above the geostatic load.

6.1.3 Diagenetic Effects

Diagenesis is a term that refers to the chemical alteration of rock minerals by geological processes. Shales and carbonates are thought to undergo changes in crystalline structure, which contributes to the cause of abnormal pressure. An often-cited example is the possible conversion of montmorillonite clays to illites, chlorites, and kaolinite clays during compaction in the presence of potassium ions. [2,3]

Water is present in clay deposits both as free pore water and as water of hydration, which is held more tightly within the shale outerlayer structure (see Fig. 6.6). Pore water is lost first during compaction of montmorillonite clays; water bonded within the shale interlayer structure tends to be retained longer. After reaching a burial depth at which a temperature of 200 to 300°F is present, dehydrated montmorillonite releases

the last water interlayers and becomes illite.

The water of hydration in the last interlayers has considerably greater density than free water, and, thus, undergoes a volume increase as it desorbs and becomes free water. When the permeability of the overlying sediments is sufficiently low, release of the last water interlayer can result in development of abnormal pressure. The last interlayer water to be released would be relatively free of dissolved salts. This is thought to explain the fresh water that sometimes is found at depth in abnormally pressured formations.

The chemical affinity for fresh water demonstrated by a clay such as montmorillonite is thought to cause shale formations to act in a manner somewhat analogous to a semipermeable membrane or a partial ion sieve. As discussed in Chap. 2, there are similarities between the osmotic pressure developed by a semipermeable membrane and the adsorptive pressure developed by a clay or shale. Water movement through shale may be controlled by a difference in chemical potential resulting from a salinity gradient as well as by a difference in darcy flow potential resulting from a pressure gradient.

For abnormal pressures to exist, an overlying pressure seal must be present. In some cases, a relatively thin section of dense caprock appears to form such a seal. A hypothesized mechanism, [4,5] by which a shale formation acts as a partial ion sieve to form such a caprock, is illustrated in Fig. 6.7.

In the absence of pressure, shales will absorb water only if the chemical potential or *activity* of the water is greater than that of the shale. However, shales will dehydrate or release water if the activity of the water is less than that of the shale. Since saline water has a lower activity than fresh water, there is less tendency for water molecules to leave a saline solution and enter the shale. However, if the saline water is abnormally pressured, the shale can be forced to accept water from a solution of lower activity. The higher the pressure, the greater the activity ratio that can be overcome. This reversal of the normal direction of water transfer sometimes is referred to as reverse osmosis. Ions that cannot enter the shale interlayers readily are left behind and become more concentrated, eventually forming precipitates. The

a. MONTMORILLONITE
BEFORE DIAGENESIS

b. LOSS OF SOME PORE WATER
AND INTERLAYER WATER

c. LOSS OF LAST INTERLAYER
CONVERTS MONTMORILLONITE
TO ILLITE

d. FINAL STAGE OF COMPACTION

Fig. 6.6—Clay diagenesis of montmorillonite to illite.[3]

precipitation of silica and carbonates would cause the upper part of the high-pressure zone to become relatively dense and impermeable.

Precipitation of minerals from solution also causes formation of permeability barriers in rock types other than shale. After loss of free water, gypsum ($CaSO_4 \cdot 2H_2O$) will give up water of hydration to become anhydrite ($CaSO_4$), an extremely impermeable evaporite. Evaporites are often nearly totally impermeable, resulting in abnormally pressured sediments below them. The pore water in carbonates tends to be saturated with the carbonate ion—i.e., the rate of solution is equal to the rate of recrystallization. However, when pressure is applied selectively at the grain contacts, the solubility is increased in these localized areas. Subsequent

recrystallization at adjacent sites can lead to a more compacted rock matrix. As in the case of shales, if a path does not exist to permit the pore water to escape as quickly as demanded by the natural rate of compaction, abnormal pore pressures result.

6.1.4 Differential Density Effects

When the pore fluid present in any nonhorizontal structure has a density significantly less than the normal pore fluid density for the area, abnormal pressures can be encountered in the updip portion of the structure. This situation is encountered frequently when a gas reservoir with a significant dip is drilled. Because of a failure to recognize this potential hazard, blowouts have occurred in familiar gas sands previously penetrated by other

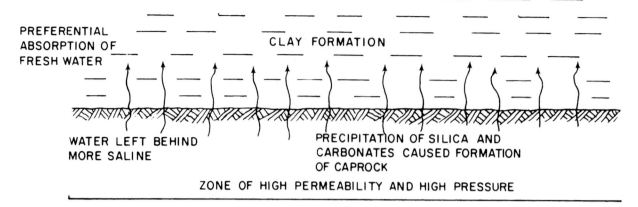

Fig. 6.7—Possible mechanism for formation of pressure seal above abnormal pressure zone.

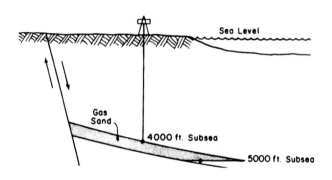

Fig. 6.8—Example illustrating origin of abnormal pressure caused by low-density pore fluid in a dipping formation.

wells. However, the magnitude of the abnormal pressure can be calculated easily by use of the hydrostatic pressure concepts presented in Chap. 4. A higher mud density is required to drill the gas zone safely near the top of the structure than is required to drill the zone near the gas/water contact.

Example 6.4. Consider the gas sand shown in Fig. 6.8, which was encountered in the U.S. gulf coast area. If the water-filled portion of the sand is pressured normally and the gas/water contact occurred at a depth of 5,000 ft, what mud weight would be required to drill through the top of the sand structure safely at a depth of 4,000 ft? Assume the gas has an average density of 0.8 lbm/gal.

Solution. The normal pore pressure gradient for the Gulf of Mexico area is given in Table 6.1 as 0.465 psi/ft, which corresponds to a normal water density of 8.94 lbm/gal. Thus, the pore pressure at the gas/water contact is

$$p=0.465(5,000)=2,325 \text{ psi.}$$

The pressure in the static gas zone at 4,000 ft is

$$p=2,325-0.052(0.8)(5,000-4,000)=2,283 \text{ psi.}$$

This corresponds to a gradient of

$$\frac{2,283}{4,000}=0.571 \text{ psi/ft.}$$

The mud density needed to balance this pressure while drilling would be

$$\rho=\frac{0.571}{0.052}=11 \text{ lbm/gal.}$$

In addition, an incremental mud density of about 0.3 lbm/gal would be needed to overcome pressure surges during tripping operations.

6.1.5 Fluid Migration Effects

The upward flow of fluids from a deep reservoir to a more shallow formation can result in the shallow formation becoming abnormally pressured. When this occurs, the shallow formation is said to be *charged*. As shown in Fig. 6.9, the flow path for this type of fluid migration can be natural or man-made. Even if the upward movement of fluid is stopped, considerable time may be required for the pressures in the charged zone to bleed off and return to normal. Many severe blowouts have occurred when a shallow charged formation was encountered unexpectedly. This situation is particularly common above old fields.

6.2 Methods for Estimating Pore Pressure

The fluid pressure within the formations to be drilled establishes one of the most critical parameters needed by the drilling engineer in planning and drilling a modern deep well. In well planning, the engineer must first determine whether abnormal pressures will be present. If they will be, the depth at which the fluid pressures depart from normal and the magnitude of the pressures must be estimated also. Many articles have appeared in the drilling literature over the past 25 years on the detection and estimation of abnormal pore pressures. The attention

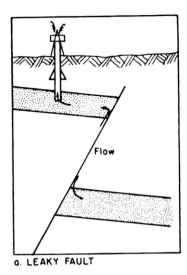 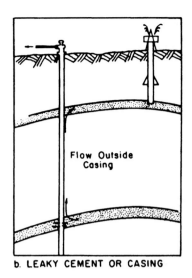 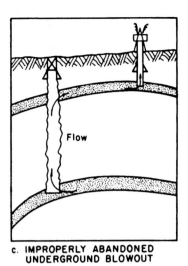

a. LEAKY FAULT   **b. LEAKY CEMENT OR CASING**   **c. IMPROPERLY ABANDONED UNDERGROUND BLOWOUT**

Fig. 6.9—Situations where upward fluid migration can lead to abnormally pressured shallow formations.

given to this problem is a reflection of both the importance of the information and the difficulties that have been experienced in establishing a method of accurately providing this information when it is needed most urgently.

For formation pore pressure data to have the greatest utility, they must be available as early as possible. However, direct measurement of formation pressure is very expensive and is possible only after the formation has been drilled. Such tests generally are made only to evaluate potential producing zones. Even if many previous wells had been drilled in an area, measured formation pressures would be available for only a limited number of them. Thus, the drilling engineer generally is forced to depend on indirect estimates of formation pressure.

Most methods for detecting and estimating abnormal formation pressure are based on the fact that formations with abnormal pressure also tend to be less compacted and have a higher porosity than similar formations with normal pressure at the same burial depth. Thus, any measurement that reflects changes in formation porosity also can be used to detect abnormal pressure. Generally, the porosity-dependent parameter is measured and plotted as a function of depth as shown in Fig. 6.10.

If formation pressures are normal, the porosity-dependent parameter should have an easily recognized trend because of the decreased porosity with increased depth of burial and compaction. A departure from the normal pressure trend signals a probable transition into abnormal pressure. The upper portion of the region of abnormal pressure is commonly called the *transition zone*. Detection of the depth at which this departure occurs is critical because casing must be set in the well before excessively pressured permeable zones can be drilled safely.

Two basic approaches are used to make a quantitative estimate of formation pressure from plots of a porosity-dependent parameter vs. depth. One approach is based on the assumption that similar formations having the same value of the porosity-dependent variable are under the same effective matrix stress, σ_z. Thus the matrix stress state, σ_z, of an abnormally pressured formation at

depth D is the same as the matrix stress state, σ_{zn}, of a more shallow normally pressured formation at depth D_n, which gives the same measured value of the porosity-dependent parameter. The depth, D_n, is obtained graphically (Fig. 6.10b) by entering the plot at the depth of interest, moving vertically from the abnormal pressure line at Point b to the normal trend line at Point c, and reading the depth corresponding to this point. Then the matrix stress state, σ_z, is computed at this depth by use of Eq. 6.1:

$$\sigma_z = \sigma_{zn} = \sigma_{obn} - p_n,$$

where σ_{obn} is evaluated first at depth D_n, as previously described in Example 6.3. The pore pressure at depth D is computed again through use of Eq. 6.1:

$$p = \sigma_{ob} - \sigma_z,$$

where σ_{ob} is evaluated at depth D.

The second approach for calculating formation pressure from plots of a porosity-dependent parameter vs. depth involves the use of empirical correlations. The empirical correlations are generally thought to be more accurate than the assumption of equivalent matrix stress at depths having equal values for the porosity-dependent parameter. However, considerable data must be available for the area of interest before an empirical correlation can be developed. When using an empirical correlation, values of the porosity-dependent parameter are read at the depth of interest both from the extrapolated normal trend line and from the actual plot. In Fig. 6.10b, values of X_n and X are read at Points a and b. The pore pressure gradient is related empirically to the observed departure from the normal trend line. Departure sometimes is expressed as a difference $(X - X_n)$ or a ratio (X_n/X). Empirical correlations also have been developed for normal trend lines.

Graphical overlays have been constructed that permit pressure gradients based on empirical correlations to be estimated quickly and conveniently from the basic plot of the porosity-dependent parameter vs. depth.

POROSITY DEPENDENT PARAMETER(X)   POROSITY DEPENDENT PARAMETER (X)

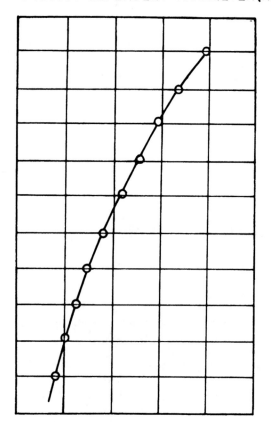

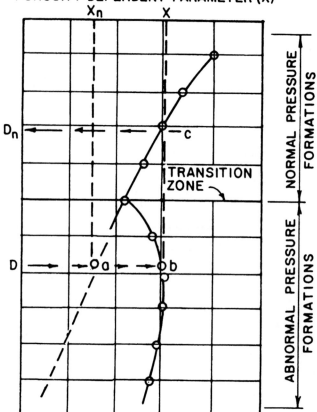

a. Normally Pressured Formations   **b. Abnormally Pressured Formations**

Fig. 6.10—Generalized example showing effect of abnormal pressure on a porosity-dependent parameter.

TABLE 6.3—REPRESENTATIVE INTERVAL TRANSIT TIMES FOR COMMON MATRIX MATERIALS AND PORE FLUIDS

Matrix Material	Matrix Transit Time (10^{-6} s/ft)
Dolomite	44
Calcite	46
Limestone	48
Anhydrite	50
Granite	50
Gypsum	53
Quartz	56
Shale	62 to 167
Salt	67
Sandstone	53 to 59
Pore Fluid	
Water (distilled)	218
100,000 ppm NaCl	208
200,000 ppm NaCl	189
Oil	240
Methane	626*
Air	910*

*Valid only near 14.7 psia and 60°F.

Techniques for detecting and estimating abnormal formation pressure often are classified as (1) predictive methods, (2) methods applicable while drilling, and (3) verification methods. Initial wildcat well planning must incorporate formation pressure information obtained by a predictive method. Those initial estimates are updated constantly during drilling. After drilling the target interval, the formation pressure estimates are checked again before casing is set, using various formation evaluation methods.

6.2.1 Prediction of Formation Pressure

Estimates of formation pore pressures made before drilling are based primarily on (1) correlation of available data from nearby wells and (2) seismic data. When planning development wells, emphasis is placed on data from previous drilling experiences in the area. For wildcat wells, only seismic data may be available.

To estimate formation pore pressure from seismic data, the average acoustic velocity as a function of depth must be determined. A geophysicist who specializes in computer-assisted analysis of seismic data usually performs this for the drilling engineer. For convenience, the reciprocal of velocity, or *interval transit time*, generally is displayed.

TABLE 6.4—AVERAGE INTERVAL TRANSIT TIME DATA COMPUTED FROM SEISMIC RECORDS OBTAINED IN NORMALLY PRESSURED SEDIMENTS IN UPPER MIOCENE TREND OF GULF COAST AREA[6]

Depth Interval (ft)	Average Interval Transit Time (10^{-6} s/ft)
1,500 to 2,500	153
2,500 to 3,500	140
3,500 to 4,500	132
4,500 to 5,500	126
5,500 to 6,500	118
6,500 to 7,500	120
7,500 to 8,500	112
8,500 to 9,500	106
9,500 to 10,500	102
10,500 to 11,500	103
11,500 to 12,500	93
12,500 to 13,500	96

TABLE 6.5—EXAMPLE CALCULATION OF APPARENT MATRIX TRANSIT TIME FROM SEISMIC DATA

Average Depth (ft)	Average Porosity (%)	Average Interval Transit Time (10^{-6} s/ft)	Apparent Matrix Transit Time (10^{-6} s/ft)
2,000	0.346	153	122
3,000	0.318	140	108
4,000	0.292	132	100
5,000	0.268	126	96
6,000	0.246	118	88
7,000	0.226	120	94
8,000	0.208	112	87
9,000	0.191	106	82
10,000	0.175	102	79
11,000	0.161	103	83
12,000	0.148	93	73
13,000	0.136	96	78

The observed interval transit time t is a porosity-dependent parameter that varies with porosity, ϕ, according to the following relation.

$$t = t_{ma}(1-\phi) + t_{fl}\phi, \dots\dots\dots\dots\dots(6.7)$$

where t_{ma} is the interval transit time in the rock matrix and t_{fl} is the interval transit time in the pore fluid. Interval transit times for common matrix materials and pore fluids are given in Table 6.3. Since transit times are greater for fluids than for solids, the observed transit time in rock increases with increasing porosity.

When plotting a porosity-dependent parameter vs. depth to estimate formation pore pressure, it is desirable to use a mathematical model to extrapolate a normal pressure trend (observed in shallow sediments) to deeper depths, where the formations are abnormally pressured. Often a linear, exponential, or power-law relationship is assumed so the normal pressure trend can be plotted as a straight line on cartesian, semilog, or log-log graph paper. In some cases, an acceptable straight-line trend will not be observed for any of these approaches, and a more complex model must be used.

A mathematical model of the normal compaction trend for interval transit time can be developed by substituting the exponential porosity expression defined by Eq. 6.4 for porosity in Eq. 6.7. After rearrangement of terms, this substitution yields

$$\ln\left[\frac{t}{\phi_o(t_{fl}-t_{ma})} - \frac{t_{ma}}{\phi_o(t_{fl}-t_{ma})}\right] = -KD. \dots(6.8)$$

This normal pressure relationship of average observed sediment travel time, t, and depth, D, is complicated by the fact that matrix transit time, t_{ma}, also varies with porosity. This variance results from compaction effects on shale matrix travel time. As shown in Table 6.3, t_{ma} for shales can vary from 167 μs/ft for uncompacted shales to 62 μs/ft for highly compacted shales. In addition, formation changes with depth also can cause changes in both matrix travel time and the normal compaction constants ϕ_o and K. These problems can be resolved only if sufficient normal pressure data are available.

Example 6.5. The average interval transit time data shown in Table 6.4 were computed from seismic records of normally pressured sediments occurring in the Upper Miocene trend of the Louisiana gulf coast. These sediments are known to consist mainly of sands and shales. Using these data and the values of K and ϕ_o computed previously for the U.S. gulf coast area in Example 6.2, compute apparent average matrix travel times for each depth interval given and curve fit the resulting values as a function of porosity. A water salinity of approximately 90,000 ppm is required to give a pressure gradient of 0.465 psi/ft.

Solution. The values of ϕ_o and K determined for the U.S. gulf coast area in Example 6.2 were 0.41 and 0.000085 ft^{-1}, respectively. From Table 6.3, a value of 209 is indicated for interval transit time in 90,000-ppm brine. Inserting these constants in Eqs. 6.4 and 6.7 gives

$$\phi = 0.41e^{-0.000085D}$$

and

$$t_{ma} = \frac{t-209\phi}{1-\phi}.$$

For the first data entry in Table 6.4, the mean interval depth is 2,000 ft and the observed travel time is 153 μs/ft. Using these values for D and t yields

$$\phi = 0.41e^{-0.000085(2,000)} = 0.346$$

and

$$t_{ma} = \frac{153-209(0.346)}{1-0.346} = 122 \ \mu\text{s/ft}.$$

Similar calculations for other depth intervals yield results shown in Table 6.5.

A plot of matrix transit time vs. porosity is shown in Fig. 6.11. From this plot, note that for the predominant

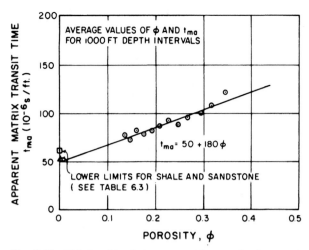

Fig. 6.11—Relationship between matrix transit time and porosity computed for sediments in the upper Miocene trend of the U.S. gulf coast area.

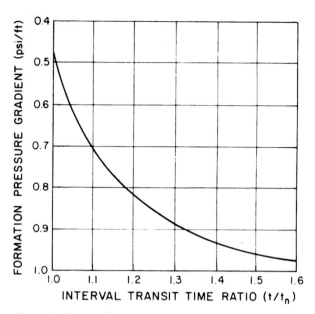

Fig. 6.13—Pennebaker relationship between formation pore pressure and seismic-derived interval transit time.[6]

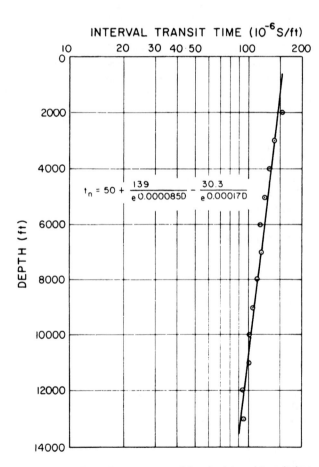

Fig. 6.12—Normal-pressure trend line for interval transit time computed from seismic data in upper Miocene trend of the U.S. gulf coast area.

shale lithology of the U.S. gulf coast area, the average matrix transit time can be estimated by

$$t_{ma} = 50 + 180\phi.$$

Use of this expression for t_{ma} and 209 for t_{fl} in Eq. 6.7 gives

$$t_n = 50 + 339\phi - 180\phi^2.$$

Substituting the expression defined by Eq. 6.4 for ϕ yields the following mathematical model for normally pressured Louisiana gulf coast sediments.

$$t_n = 50 + 339\phi_o e^{-0.000085D} - 180\phi_o^2 e^{-0.00017D}.$$

This relationship is plotted in Fig. 6.12 with surface porosity equal to 0.41. For comparison, the interval transit time data from Table 6.4 are shown also.

Other authors have assumed both a logarithmic (power-law) relationship[6-10] and an exponential relationship[11] between interval transit time and depth for normally pressured sediments. It can be shown that the mathematical model developed in Example 6.5 does not yield a straight-line extrapolation on either logarithmic or semilogarithmic plots, although a good straight-line fit could be made for a limited depth range using either approach. Significant departure from a straight line occurs below 15,000 ft at low porosity values.

The geologic age of sediments has been found to affect the normal pressure relationship between interval travel time and depth even within the same general type of lithology. Drilling older sediments that have had more time for compaction to occur produces an upward shift in the normal pressure trend line, in which a given interval transit time appears at a more shallow depth. Similarly, younger sediments produce a downward shift, in which a given interval transit time appears at a greater depth. In

TABLE 6.6—AVERAGE INTERVAL TRANSIT TIME DATA COMPUTED FROM SEISMIC RECORDS AT A WELL LOCATION IN THE SOUTH TEXAS FRIO TREND[6]

Depth Interval (ft)	Average Interval Transit Time (10^{-6} s/ft)
1,500 to 2,500	137
2,500 to 3,500	122
3,500 to 4,500	107
4,500 to 5,500	104
5,500 to 6,500	98
6,500 to 7,500	95
7,500 to 8,500	93
8,500 to 9,500	125
9,500 to 10,500	132
10,500 to 11,500	130
11,500 to 12,500	126

TABLE 6.7—EXAMPLE CALCULATION OF SURFACE POROSITY CONSTANT

Depth Interval (ft)	$\bar{D}$ (ft)	t_n (10^{-6} s/ft)	ϕ_o
1,500 to 2,500	2,000	137	0.364
2,500 to 3,500	3,000	122	0.315
3,500 to 4,500	4,000	107	0.262
4,500 to 5,500	5,000	104	0.269
5,500 to 6,500	6,000	98	0.257
6,500 to 7,500	7,000	95	0.261
7,500 to 8,500	8,000	93	0.270

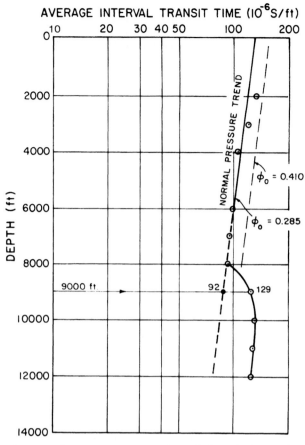

Fig. 6.14—Seismic-derived interval transit time plot for south Texas Frio trend.

practice, a single normal pressure trend line often is applied to sediments of similar lithology but varying geologic age by shifting the normal trend line up or down to fit the observed behavior in the normal pressure region. In the mathematical model developed in Example 6.5, the normal pressure trend line is shifted up or down by decreasing or increasing the value of the surface porosity constant, ϕ_o.

When the interval transit time is significantly greater than predicted by the normal pressure trend line for the given formation, abnormal formation pressure is indicated. The magnitude of the abnormal pressure can be computed by either of the basic approaches illustrated in Fig. 6.10. An empirically developed departure curve such as the one shown in Fig. 6.13 is needed to apply the second basic method. Departure curves developed empirically from interval transit measurements made in shale using a sonic log may be used also.[11] The use of sonic-log interval transit time data for estimating formation pressure is described in detail in Section 6.2.3 (Verification Methods).

Example 6.6. The average interval transit time data shown in Table 6.6 were computed from seismic records at a proposed well location in the south Texas Frio trend. Estimate formation pressure at 9,000 ft using both of the basic approaches discussed in Section 6.2. Extend the mathematical model for the normal pressure trend developed in Example 6.5 to this trend; select an appropriate value of average surface porosity, ϕ_o.

Solution. First, the interval transit time data are plotted vs. depth (Fig. 6.14). The average normal pressure trend line for the Louisiana Upper Miocene trend was determined in Example 6.5 to be

$$t_n = 50 + 339\phi_o e^{-0.000085D} - 180\phi_o^2 e^{-0.00017D},$$

with the surface porosity equal to 0.41. This relationship is plotted in Fig. 6.14. The dashed line compares these data to the south Texas Frio trend data. Since the penetrated formations in the south Texas Frio trend are much older than the formations of the Louisiana Upper Miocene trend, it was necessary to shift the normal pressure trend line upward. This was accomplished by adjusting the value of the surface porosity constant, ϕ_o. Solving the mathematical model of the normal trend line for ϕ_o yields

$$\phi_o = \left\{ \frac{339}{e^{0.000085D}} \right.$$

$$\left. - \sqrt{\left(\frac{339}{e^{0.000085D}}\right)^2 - \left[\frac{720(\ell_n - 50)}{e^{0.00017D}}\right]} \right\}$$

$$\div \frac{360}{e^{0.00017D}}.$$

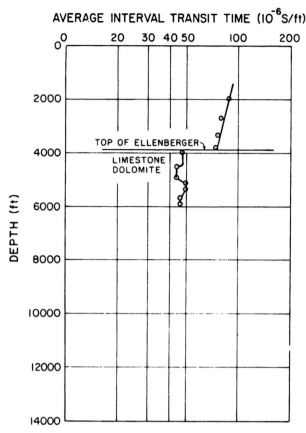

Fig. 6.15—Seismic-derived interval transit time plot for Kendall County, TX, area. [6]

The average depth of the first depth interval shown in Table 6.6 is 2,000 ft and the observed interval transit time is 137. Substitution of these values into the equation above gives

$$\phi_o = \left\{ \frac{339}{e^{0.000085(2,000)}} \right.$$

$$\left. - \sqrt{\left[\frac{339}{e^{0.000085(2,000)}} \right]^2 - \left[\frac{720(137-50)}{e^{0.00017(2,000)}} \right]} \right\}$$

$$\div \frac{360}{e^{0.00017(2,000)}} = 0.364.$$

Results of similar calculations at each depth interval are summarized in Table 6.7. Note that an average value of 0.285 is indicated for the surface porosity constant, ϕ_o. Thus, the normal pressure trend line equation becomes

$$t_n = 50 + 96.6e^{-0.000085D} - 14.6e^{-0.00017D}.$$

This relationship is plotted with a solid line in Fig. 6.14. The first approach that can be used to estimate formation pressure at 9,000 ft is based on the assumption that formations having the same value of interval transit time are under the same vertical effective matrix stress, σ_z. At 9,000 ft, the interval transit time has a value of 129. The

depth of the normally pressured formation having this same value of interval transit time is shown to be 1,300 ft in Fig. 6.14. The vertical overburden stress, σ_{ob}, resulting from geostatic load at a depth of 1,300 ft is defined by Eq. 6.6.

$$(\sigma_{ob})_{1,300} = 0.052\rho_g D_s - \frac{0.052(\rho_g - \rho_{fl})\phi_o}{K}$$

$$\cdot (1 - e^{-KD_s})$$

$$= 0.052(2.60)(8.33)(1,300)$$

$$- \frac{0.052(2.60 - 1.074)(8.33)(0.285)}{0.000085}$$

$$\cdot \left[1 - e^{-0.000085(1,300)} \right]$$

$$= 1,464 - 232 = 1232 \text{ psig.}$$

The formation pore pressure at 1,300 ft is given by $p_{1,300} = 0.465(1,300) = 605$ psig. Thus, the effective matrix stress at both 1,300 and 9,000 ft is

$$\sigma_{9,000} = \sigma_{1,300} = (\sigma_{ob})_{1,300} - p_{1,300}$$

$$= 1,232 - 605 = 627 \text{ psig.}$$

The overburden stress σ_{ob} resulting from geostatic load at 9,000 ft is

$$(\sigma_{ob})_{9,000} = 0.052(2.6)(8.33)(9,000)$$

$$- \frac{0.052(2.6 - 1.074)(8.33)(0.285)}{0.000085}$$

$$\cdot \left[1 - e^{-0.000085(9000)} \right]$$

$$= 10,136 - 1,185 = 8,951 \text{ psi.}$$

This gives, at 9,000 ft, a pore pressure of

$$p_{9,000} = (\sigma_{ob})_{9,000} - \sigma_{9,000} = 8,951 - 627 = 8,324 \text{ psig.}$$

The second method that can be used to estimate formation pressure at 9,000 ft is an empirically determined relationship between interval transit time and formation pressure. (See Fig. 6.13.) The ratio of observed transit time to normal interval transit time at 9,000 ft is

$$\frac{t}{t_n} = \frac{129}{92} = 1.40.$$

From Fig. 6.13, the formation pore pressure gradient is 0.93 psi/ft. Thus, the formation pressure is

$$\rho = 0.93(9,000) = 8,370 \text{ psig.}$$

The previous examples have been concerned with predicting and estimating formation pore pressure in relatively young, shale-dominated formations. Predicting and estimating formation pore pressure is more difficult in older sedimentary basins that generally have a much more complex lithology. Each change in depth characterized by a major change in lithology also manifests a large shift in the normal pressure trend line. Very thick sections of limestone, dolomite, and sandstone (which may have much lower matrix transit time than the shales) are common. Changes in average porosity become less predictable with depth, since forces other than compaction resulting from continuous sedimentation may not be the predominant geologic process. However, the derived seismic interval transit time plot often can be used to determine the depth of known formations, some of which may be known to be abnormally pressured.

An example of an interval transit time plot for a complex lithology is shown in Fig. 6.15. These data were taken in Kendall County (TX) in normally pressured sediments. The large shift to the left at 3,800 ft marks the top of the Ellenberger formation. Dolomite sections within the Ellenberger formation give even lower interval transit time readings than the limestone readings.

6.2.2 Estimation of Formation Pressure While Drilling

As drilling progresses into a transition zone of normal and abnormal formation pressure, variations in rock properties and bit performance often provide many indirect indications of changes in formation pressure. To detect these changes, drilling parameters related to bit performance are monitored continuously and recorded by surface instruments. In addition, many variables associated with the drilling fluid and rock fragments being circulated from the well are monitored carefully and logged using special mud logging equipment and personnel. Ideally, surface instruments used to monitor bit performance plus mud logging equipment are consolidated into a single well-monitoring unit.

Recent developments in subsurface data transmission have enabled continuous subsurface logging of several formation properties while drilling. Such a service can be of great benefit in the estimation of formation pore pressure while drilling.

Occasionally, the wellbore pressure is inadvertently allowed to fall below the pore pressure in a permeable formation. As discussed in Chap. 4, this results in a kick—i.e., an influx of formation fluid into the well. When well-control operations are initiated, the shut-in drillpipe pressure provides a direct indication of the formation pressure. These data are extremely useful in calibrating the more indirect methods of estimating formation pressure.

If the wellbore pressure is inadvertently allowed to fall below the formation pore pressure in low-permeability formations, the influx of formation fluids into the wellbore will not occur rapidly; however, there may be slow seepage of formation fluids into the well, which can be detected in the drilling fluid at the surface. Also, pressure differential in the wellbore may promote spalling of shale fragments from the sides of the wellbore. This also can be seen in the drilling fluid at the surface.

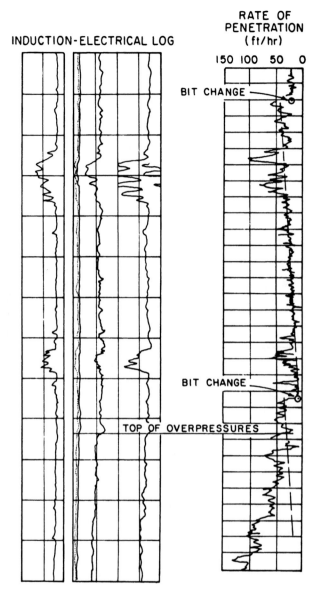

Fig. 6.16—Example comparison of IES and penetration rate logs in the U.S. gulf coast area (after Jorden and Shirley [13]).

Analysis of Drilling Performance Data. Changes in bit behavior can be detected through measurements made at the surface. Commonly, measurements include (1) penetration rate, (2) hook load, (3) rotary speed, and (4) torque. Since the drilling fluid properties and circulating rate affect penetration rate, they also are monitored frequently. In addition, several companies are experimenting with the measurement of longitudinal drillstring vibration.

The bit penetration rate usually changes significantly with formation type. Thus, a penetration-rate log frequently can be used to aid in a lithology correlation with nearby wells with known formation pressures. In addition, the penetration rate in a given type of formation normally tends to decrease with increasing depth. However, when a transition zone into abnormal pressure is encountered, this normal trend is altered. Just above the transition zone to a higher formation pore pressure gradient, a hard, often limey, formation frequently is encountered that yields a lower-than-normal penetration

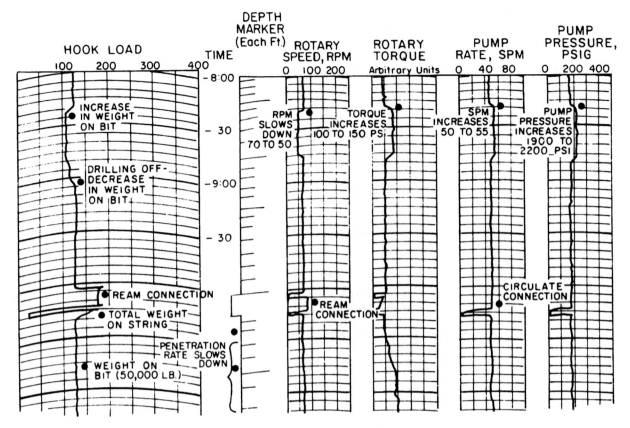

Fig. 6.17—Example elapsed-time recorder chart.
Courtesy of Totco

rate. Many people feel that these formations are extremely low permeability formations that form the pressure seal for the abnormal pressure gradients. These seals may vary in thickness from a few feet to several hundred feet. Just below this abnormal-pressure caprock, the normal penetration rate trend reverses, and an increase in penetration rate with depth may be observed.

Example penetration-rate data [13] in a transition zone to abnormal formation pressure for a well drilled in the U.S. gulf coast area are given in Fig. 6.16. In this area, the lithology is composed primarily of sand and shale formations, with the sands yielding the faster penetration rate. Note the possible correlation between the penetration rate log and an induction-electrical log. Note also the observed reversal of the trend of decreasing penetration rate with depth for shale formations in the transition zone.

The reason for the usual increase in penetration rate in the transition zone of low permeability formations is felt to result from (1) a decrease in the pressure differential across the bottom of the hole and (2) a decrease in the rock strength caused by undercompaction. As discussed in Chap. 5, the term *overbalance* frequently is used for the difference between the bottomhole hydrostatic pressure and the fluid pressure in the pore space of the formation. The effect of overbalance is much more important than the effect of undercompaction. A discussion of the available laboratory and field data on the effect of overbalance on penetration rate was presented in Section 5.7.

Several types of well monitoring services are available that can provide a penetration rate log. The elapsed-time recorder, which records the time required to drill a given depth interval, is a relatively simple and inexpensive mechanical device used for this purpose and is standard equipment on almost all rotary rigs. As shown in Fig. 6.17, the device makes a tick mark on a time chart after each depth interval drilled. However, since the vertical scale is based on time rather than depth, the log is not as convenient to use for lithology correlations as the format shown in Fig. 6.16. Also, it is more difficult to recognize trends in the penetration-rate data with elapsed-time records. Penetration-rate logs often are provided as part of a larger mud-logging and well-monitoring service and involve the use of specialized data units and personnel. When a mud-logging service is desired, penetration rate logs (in almost any format desired) usually are available as part of this service.

Many drilling variables other than formation type and formation pore pressure affect the bit penetration rate. Some additional parameters are: (1) bit type, (2) bit diameter, (3) bit nozzle sizes, (4) bit wear, (5) weight on bit, (6) rotary speed, (7) mud type, (8) mud density, (9) effective mud viscosity, (10) solids content and size distribution in mud, (11) pump pressure, and (12) pump rate. Changes in the variables affecting penetration rate can mask the effect of changing lithology or increasing formation pore pressure. Thus, it is often difficult to detect formation pressure changes using only penetration rate data. It should be emphasized that penetration rate changes are often difficult to interpret and should be used in conjunction with other indicators of formation pressure.

When mill tooth bits are used, the effect of tooth wear

can influence penetration rate during each bit run. When other drilling variables are not changing, the effect of bit dulling can be partially compensated for by establishing the expected dulling trend from past bit performance in normal-pressure formations. Notice that this behavior is exhibited in the example penetration-rate log shown in Fig. 6.16. In some cases, because of tooth wear, the penetration rate still decreases with increasing depth in the transition zone but at a much lower rate than anticipated. Unfortunately, changes in other drilling variables can cause a similar effect and be misinterpreted as a pressure increase. In particular, changes in bit type make changing pore pressure difficult to detect from penetration rate data.

Empirical models of the rotary drilling process have been proposed to mathematically compensate for the effect of changes in the more important variables affecting penetration rate. One of the first empirical models of the rotary drilling process was published by Bingham in 1965. The Bingham drilling model was defined in Chap. 5 by Eq. 5.20. In 1966, Jorden and Shirley[13] proposed using the Bingham model to normalize penetration rate, R, for the effect of changes in weight on bit, W, rotary speed, N, and bit diameter, d_b, through the calculation of a d-exponent defined by

$$d_{exp} = \frac{\log\left(\frac{R}{60N}\right)}{\log\left(\frac{12W}{1,000d_b}\right)}. \qquad \ldots\ldots\ldots\ldots\ldots\ldots (6.9)$$

In this equation, units for R, N, W, and d_b are ft/hr, rpm, k-lbf, and in., respectively. Eq. 6.9 is not a rigorous solution for the d-exponent of Eq. 5.20 because (1) the formation drillability constant, a, was assigned a value of unity and (2) a scaling constant, 10^3, was introduced in the weight-on-bit term. Jorden and Shirley felt that this simpification would be permissible in the U.S. gulf coast area for a single formation type since in this area there are "few significant variations in rock properties other than variations due to increased compaction with depth."[13]

The d-exponent equation can be used to detect the transition from normal to abnormal pressure if the drilling fluid density is held constant. The technique involves plotting values of d obtained in a given type of low-permeability formation as a function of depth. Shale is nearly always the formation type selected. Drilling data obtained in other formation types simply are omitted from the calculation. In normally pressured formation, the d-exponent tends to increase with depth. After abnormally pressured formations are encountered, a departure from the normal pressure trend occurs in which the d-exponent increases less rapidly with depth. In many cases, a complete reversal of the trend occurs and the d-exponent begins decreasing with depth.

Jorden and Shirley also attempted a correlation between the d-exponent and differential pressure. The results of their study are shown in Fig. 6.18. They concluded that the scatter of the data was too wide for quantitative field application.

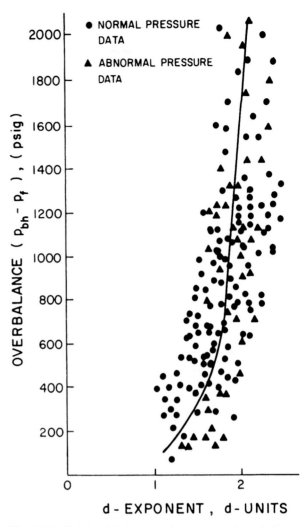

Fig. 6.18—Relationship between d-exponent and overbalance pressure.[13]

In 1971, Rehm and McClendon[14] proposed modifying the d-exponent to correct for the effect of mud-density changes as well as changes in weight on bit, bit diameter, and rotary speed. After an empirical study, Rehm and McClendon computed a modified d-exponent, d_{mod}, using

$$d_{mod} = d_{exp}\frac{\rho_n}{\rho_e}, \qquad \ldots\ldots\ldots\ldots\ldots\ldots\ldots (6.10)$$

where ρ_n is the mud density equivalent to a normal formation pore pressure gradient and ρ_e is the equivalent mud density at the bit while circulating.

Example 6.7. A penetration rate of 23 ft/hr was observed while drilling in shale at a depth of 9,515 ft using a 9.875-in. bit in the U.S. gulf coast area. The weight on the bit was 25,500 lbf and the rotary speed was 113 rev/min. The equivalent circulating density at the bit was 9.5 lbm/gal. Compute the d-exponent and the modified d-exponent.

TABLE 6.8—EXAMPLE MODIFIED d-EXPONENT DATA TAKEN IN U.S. GULF COAST SHALES[15]

Depth (ft)	Modified d-Exponent
8,100	1.52
9,000	1.55
9,600	1.57
10,100	1.49
10,400	1.58
10,700	1.60
10,900	1.61
11,100	1.57
11,300	1.64
11,500	1.48
11,600	1.61
11,800	1.54
12,100	1.58
12,200	1.67
12,300	1.41
12,700	1.27
12,900	1.18
13,000	1.13
13,200	1.22
13,400	1.12
13,500	1.12
13,600	1.07
13,700	1.00
13,800	0.98
13,900	1.00
14,000	0.91
14,200	0.93
14,400	0.86
14,600	0.80
14,800	0.86
14,900	0.80
15,000	0.90
15,200	0.82
15,300	0.87
15,400	0.92
15,500	0.87
15,700	0.80
16,200	0.80
16,800	0.65

Solution. The d-exponent is defined by Eq. 6.9:

$$d_{exp} = \frac{\log\left[\dfrac{23}{60(113)}\right]}{\log\left[\dfrac{(12)(25.5)}{(1,000)(9.875)}\right]} = 1.64 \ d\text{-units.}$$

The modified d-exponent is defined by Eq. 6.10. Recall that the normal pressure gradient in the U.S. gulf coast area is 0.465 psi/ft.

$$\rho_n = \frac{0.465}{0.052} = 8.94 \text{ lbm/gal}$$

and

$$d_{mod} = 1.64\left(\frac{8.94}{9.50}\right) = 1.54 \ d\text{-units.}$$

The modified d-exponent often is used for a quantitative estimate of formation pore pressure gradient as

well as for the qualitative detection of abnormal formation pressure. Numerous empirical correlations have been developed in addition to the equivalent matrix stress concept. Often these correlations are presented in the form of graphical overlays constructed on a transparent plastic sheet that can be placed directly on the d_{mod} plot to read the formation pressure.

Rehm and McClendon[14] recommend using linear scales for both depth and d_{mod} values when constructing a graph to estimate formation pore pressure quantitatively. A straight-line normal pressure trend line having intercept $(d_{mod})_o$ and slope m is assumed such that

$$(d_{mod})_n = (d_{mod})_o + mD. \qquad (6.11)$$

According to the authors, the value of slope m is fairly constant with changes in geologic age. Examples given were plotted with a slope, m, of 0.000038 ft^{-1}. The following empirical relation was presented for the observed departure of the d_{mod} plot and the formation pressure gradient, g_p.

$$g_p = 7.65 \log[(d_{mod})_n - (d_{mod})] + 16.5, \qquad (6.12)$$

where $(d_{mod})_n$ is the value of d_{mod} read from the normal pressure trend line at the depth of interest. In this equation, g_p is given in equivalent mud density units of lbm/gal.

Zamora[15] recommends using a linear scale for depth but a logarithmic scale for d_c values when constructing a graph to estimate formation pore pressure quantitatively. A straight-line normal pressure trend line having intercept $(d_{mod})_o$ and exponent m is assumed such that

$$(d_{mod})_n = (d_{mod})_o e^{mD}. \qquad (6.13)$$

Zamora reports that the slope of the normal pressure trend line "varied only slightly and without apparent regard to location or geological age." The slope of the normal trend was reported to be the slope of a line connecting d_c values of 1.4 and 1.7 that were 5,000 ft apart. This corresponds to an m value of 0.000039 ft^{-1}. Zamora used the following empirical relation for the observed departure on the d_{mod} plot and the formation pressure gradient g_p.

$$g_p = g_n \frac{(d_{mod})_n}{d_{mod}}, \qquad (6.14)$$

where g_n is the normal pressure gradient for the area.

Example 6.8. The modified d-exponent data shown in Table 6.8 were computed from penetration-rate data obtained in shale formations in the gulf coast area. Estimate the formation pressure at 13,000 ft using (1) the empirical correlation of Rehm and McClendon, and (2) the empirical correlation of Zamora.

Solution.

1. The modified d-exponent data given in Table 6.8 are plotted first as in Fig. 6.19 using cartesian coordinates as recommended by Rehm and McClendon. A

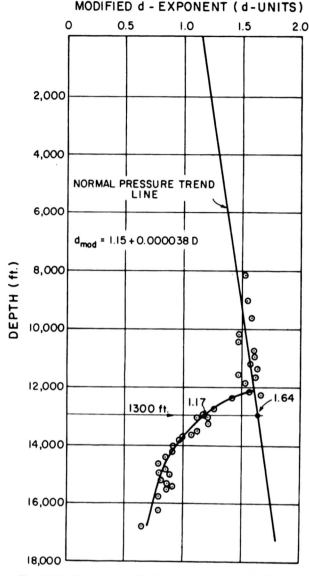

Fig. 6.19—Example modified *d*-exponent plot with Cartesian coordinates.

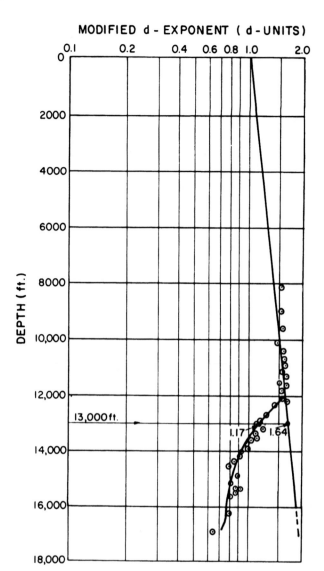

Fig. 6.20—Example modified *d*-exponent plot with semi-logarithmic coordinates.

normal trend line having a slope of 0.000038 was drawn through the data available in the normally pressured region. At a depth of 13,000 ft, values of d_{mod} and $(d_{mod})_n$ are read from Fig. 6.19 as 1.17 and 1.64, respectively. Using these values in Eq. 6.12 yields

$$g_p = 7.65 \log(1.64 - 1.17) + 16.5 = 14 \text{ lbm/gal}$$

and

$$p = 0.052(14)(13,000) = 9,464 \text{ psig.}$$

2. The use of Zamora's empirical correlation requires plotting the modified *d*-exponent data using semilogarithmic coordinates as shown in fig. 6.20. A normal trend line $m = 0.000039$ was drawn through the data available in the normally pressured region. At a depth of 13,000 ft, values of d_{mod} and $(d_{mod})_n$ are read from Fig. 6.20 as 1.17 and 1.64 ft, respectively. Note that at this depth, there is no significant difference resulting from the different plotting procedures used in

Figs. 6.19 and 6.20. Eq. 6.14 gives

$$g_p = g_n \frac{(d_{mod})_n}{d_{mod}} \quad \dots \dots \dots \dots \dots \dots \dots (6.14)$$

$$= 0.465 \left(\frac{1.64}{1.17} \right) = 0.652 \text{ psi/ft}$$

and

$$p = 0.652(13,000) = 8,476 \text{ psig.}$$

Since the d_c parameter considers only the effects of bit weight, bit diameter, rotary speed, and mud density, changes in other drilling variables such as bit type, bit wear, mud type, etc., still may create problems in interpreting the obtained plots. In addition, extreme changes in the variables included in the d_c calculation can create problems. Usually a new trend must be established for the changed conditions. The utility of the *d*-exponent is diminished especially when the mud density is several pounds per gallon greater than the formation pore pressure gradient. Because of the excessive overbalance, the penetration rate no longer responds significantly to changes in formation pressure. Under these conditions, increases in drilling fluid density cause an erroneous shift in the modified d-exponent plot, which yields higher pore pressure readings. This is unfortunate since it tends to confirm erroneously the need for the increase in drilling fluid density.

In 1974, Bourgoyne and Young[16] proposed using a more complex drilling model than the Bingham model, to compensate mathematically for changes in the various drilling parameters. The drilling model adopted by Bourgoyne and Young was presented in Chapter 5 by Eqs. 5.28a through 5.28d and is repeated here in a more concise form for a threshold bit weight of zero.

$$R = \exp\{2.303[a_1 + a_2(10,000 - D)$$

$$+ a_3 D^{0.69}(g_p - \rho_n)$$

$$+ a_4 D(g_p - \rho_c)]\} \times \left[\left(\frac{W}{4d_b}\right)^{a_5}\left(\frac{N}{60}\right)^{a_6}\right.$$

$$\left. \times e^{-a_7 h}\left(\frac{F_j}{1,000}\right)^{a_8}\right], \quad\quad\quad (6.15)$$

where $\exp(x)$ is used to represent the exponential function e^x. The fractional tooth dullness, h, must be computed for each depth interval using a tooth-wear equation as presented in Chap. 5. Also, the jet impact force, F_j, must be computed for current mud density, nozzle sizes, and pump rate. Because of the complexity of the drilling model used and the large number of computations required, the model is best suited for use on a computer.

The penetration rate can be normalized for the effect of bit weight, W, bit diameter, d_b, rotary speed, N, tooth dullness, h, and jet impact force, F_j, by dividing by the second bracketed term in Eq. 6.15.

$$R^* = \frac{R \exp(a_7 h)}{\left(\frac{W}{4d_b}\right)^{a_5}\left(\frac{N}{60}\right)^{a_6}\left(\frac{F_j}{1,000}\right)^{a_8}} . \quad\quad (6.16)$$

The normalized penetration rate, R^*, corresponds to the theoretical penetration rate that would be observed for a new bit (zero tooth dullness) with a bit weight per unit bit diameter, W/d_b, of 4 k-lbf/in., a rotary speed, N, of 60 rpm, and a jet impact force, F_j, of 1,000 lbf.

It was found that the relation between overpressure and penetration rate could be represented approximately by a straight line on a semilogarithmic plot over a reasonable range of overbalance. This was discussed in Chap. 5 and illustrated in Figure 5.35. However, for excessive overbalance, the accuracy of the straight line diminishes. Since overbalance is related more directly to the logarithm of the penetration rate, Bourgoyne and Young defined a drillability parameter, K_p, given by

$$K_p = \log(R^*)$$

$$= \log\left[\frac{R \exp(a_7 h)}{\left(\frac{W}{4d_b}\right)^{a_5}\left(\frac{N}{60}\right)^{a_6}\left(\frac{F_j}{1,000}\right)^{a_8}}\right].$$

$$\quad\quad\quad\quad\quad\quad\quad\quad\quad\quad (6.17)$$

This parameter is somewhat analogous to the d-exponent. To account for changes in mud density and depth, a modified drillability parameter was introduced:

$$K_p' = K_p + a_4 D(\rho_c - \rho_n). \quad\quad\quad\quad (6.18)$$

The modified drillability parameter, K_p', also is analogous to the modified d-exponent.

Example 6.9. A penetration rate of 31.4 ft/hr was observed while drilling in shale at a depth of 12,900 ft using a 9.875-in. bit in the U.S. gulf coast area. The bit weight was 28 k-lbf/in. and the rotary speed was 51 rpm. The computed fractional tooth dullness was 0.42 and the computed jet impact force was 1150 lbf. The equivalent circulating density at the bit was 16.7 lbm/gal. Compute the values of the drillability parameter, K_p, and the modified drillability parameter, K_p', using the following values for a_2 through a_8: $a_2 = 74 \times 10^{-6}$, $a_3 = 100 \times 10^{-6}$, $a_4 = 35 \times 10^{-6}$, $a_5 = 0.80$, $a_6 = 0.40$, $a_7 = 0.41$, and $a_8 = 0.30$.

Solution. The drillability parameter, K_p, is defined by Eq. 6.17.

$$K_p = \log\left\{\frac{31.4 e^{0.41(0.42)}}{\left[\frac{28}{(4)(9.875)}\right]^{0.8}\left[\frac{51}{60}\right]^{0.4}\left[\frac{1150}{1000}\right]^{0.3}}\right\}$$

$$= 1.70 \, K_p \text{ units.}$$

The modified drillability parameter is defined by Eq. 6.18 with the normal pressure gradient, ρ_n, equal to 8.94 for the U.S. gulf coast area.

$$K_p' = 1.70 + (35 \times 10^{-6})(12,900)(16.7 - 8.94)$$

$$= 1.70 + 3.50 = 5.2 \, K_p' \text{ units.}$$

The modified drillability parameter K_p' can be related to the formation pressure gradient using Eq. 6.15. Substituting the definition of K_p' in Eq. 6.15 and solving for the formation pressure gradient, g_p, yields

$$g_p = \rho_n + \frac{K_p' - a_1 - a_2(10,000 - D)}{a_3 D^{0.69} + a_4 D} . \quad\quad (6.19)$$

The coefficients a_1 through a_8 must be chosen according to local drilling conditions. Bourgoyne and Young[16] presented a multiple regression technique for computing the value of these constants from previous drilling data obtained in the area. In addition, the coefficients a_3 through a_8 often can be computed on the basis

TABLE 6.9—AVERAGE VALUES OF REGRESSION COEFFICIENTS OF BOURGOYNE-YOUNG DRILLING MODEL FOR SHALE FORMATIONS IN U.S. GULF COAST AREA

Regression Coefficients						
a_2	a_3	a_4	a_5	a_6	a_7*	a_8
90×10^{-6}	100×10^{-6}	35×10^{-6}	0.9	0.5	0.3	0.4

*Values given are for milled tooth bits only. Use $a_7 = 0$ for insert bits.

TABLE 6.10—EXAMPLE MODIFIED DRILLABILITY PARAMETER OBTAINED IN U.S. GULF COAST SHALES

Depth (ft)	Modified Drillability Parameter
9,515	1.76
9,830	1.82
10,130	1.80
10,250	1.58
10,390	1.80
10,500	1.85
10,575	1.72
10,840	1.82
10,960	1.83
11,060	1.83
11,475	1.92
11,775	2.49
11,940	3.95
12,070	3.99
12,315	4.50
12,900	5.15
12,975	5.22
13,055	5.28
13,250	5.43
13,795	5.27
14,010	5.65
14,455	5.55
14,695	5.69
14,905	5.86

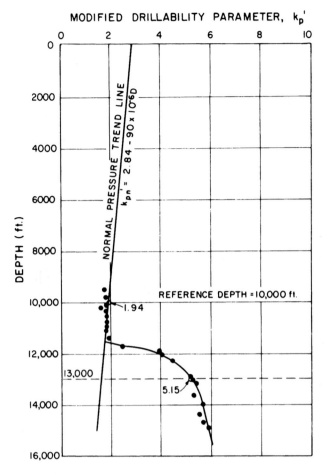

Fig. 6.21—Example modified drillability-parameter plot.

of observed changes in penetration rate caused by a change in only one of the drilling variables. Examples 5.7 and 5.8 (Chap. 5) illustrate the basic technique that can be used. Coefficients a_1 and a_2 usually can be determined graphically from drillability data obtained in normally pressured formations. If no previous data are available to determine coefficients a_2 through a_8, the average values given in Table 6.9 can be used.

Example 6.10. The modified drillability parameter data shown in Table 6.10 were computed from penetration rate data obtained in shale formations in the U.S. gulf coast area. The values of K_p' were computed for 50-ft-depth intervals to dampen fluctuations in the computed results.

Estimate the formation pressure at 13,000 ft using the Bourgoyne-Young drilling model. The slope of the normal trend line, a_2, was determined to be 90×10^{-6}. The average overbalance exponent, a_4, was determined to be 35×10^{-6} by a regression analysis of drilling data collected on previous wells in the area.

Solution. The modified drillability parameter data first are plotted as shown in Fig. 6.21 using cartesian coordinates. The normal trend line was drawn with a slope of

90×10^{-6} K_p units/ft. Coefficient a_1 is read to be 1.94 from the normal trend line at the reference depth of 10,000 ft. At a depth of 13,000 ft, a K_p' value of 5.15 is read from the plot. Eq. 6.19 with a normal pressure gradient of 8.94 lbm/gal for the U.S. gulf coast area yields

$$g_p = 8.94$$

$$+ \frac{5.15 - 1.94 - (90 \times 10^{-6})(10,000 - 13,000)}{(100 \times 10^{-6})(13,000)^{0.69} + (35 \times 10^{-6})(13,000)}$$

$$= 15.6 \text{ lbm/gal},$$

where

$$p = 0.052(15.6)(13,000) = 10,546 \text{ psig}.$$

Drilling performance data other than penetration rate that sometimes give an indication of formation pore-pressure increase include (1) rotary *torque* during drilling, (2) frictional *drag* during vertical drillstring movements, and (3) *hole fill* or accumulations of rock fragments in the lower part of the borehole. Normally, both torque and drag tend to increase slowly with well

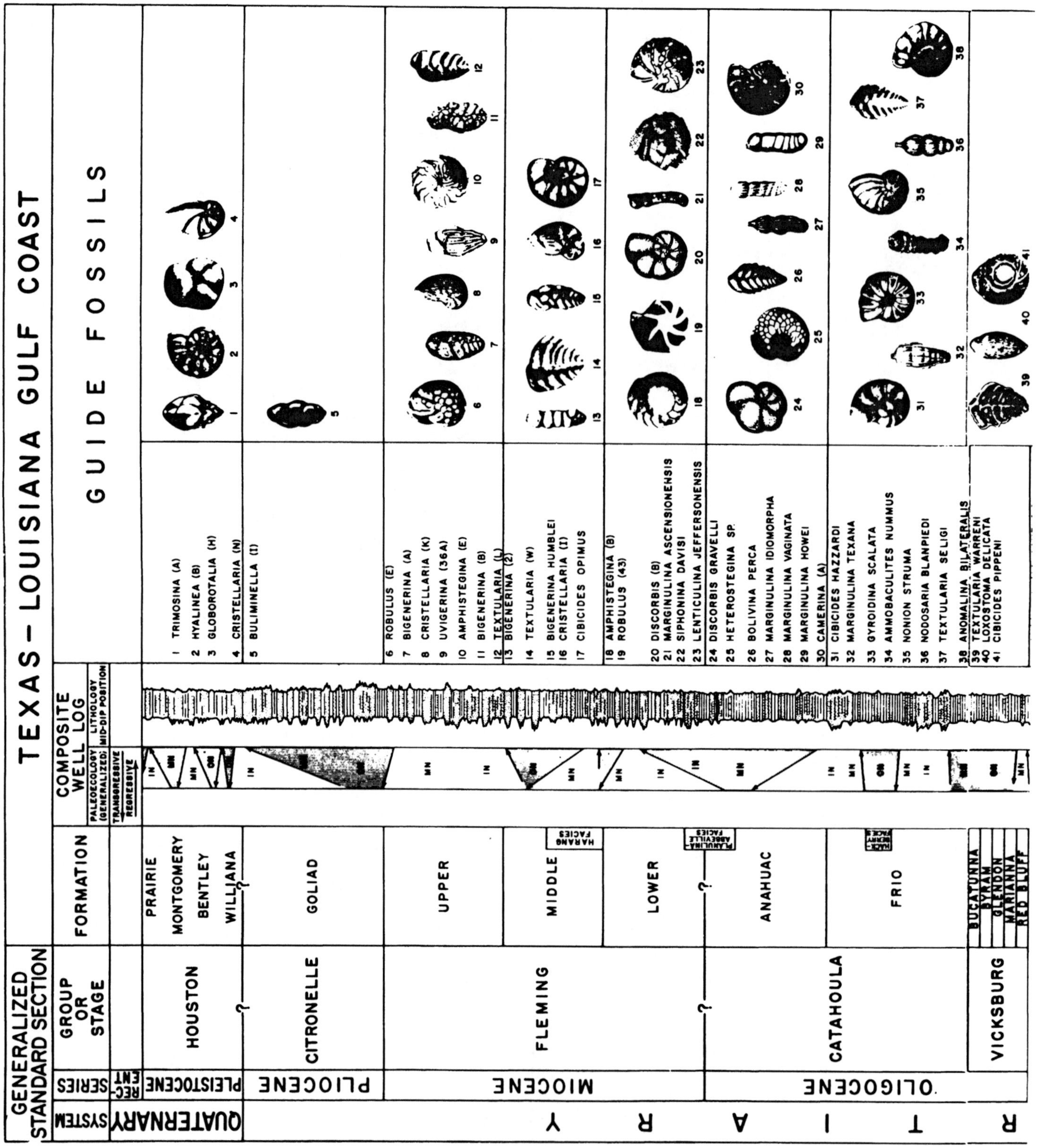

Fig. 6.22—General lithology of Louisiana gulf coast area.
Courtesy of Mobil Oil Corp.

42 BULIMINA JACKSONENSIS
43 GLOBOROTALIA CERROAZULENSIS
44 HANTKENINA ALABAMENSIS
45 TEXTULARIA HOCKLEYENSIS
46 TEXTULARIA DIBOLLENSIS
47 CAMERINA MOODYBRANCHENSIS
48 NONIONELLA COCKFIELDENSIS
49 DISCORBIS YEGUAENSIS
50 EPONIDES YEGUAENSIS
51 CERATOBULIMINA EXIMIA
52 OPERCULINOIDES SABINENSIS
53 TEXTULARIA SMITHVILLENSIS
54 CYCLAMMINA CANERIVERENSIS
55 DISCOCYCLINA ADVENA
56 CYTHERIDEA SABINENSIS
57 GLOBOROTALIA PSEUDOMENARDII
58 DISCORBIS WASHBURNI
59 VAGINULINA LONGIFORMA
60 VAGINULINA MIDWAYANA
61 ROBULUS PSEUDOCOSTATUS
62 GLOBOTRUNCANA ARCA
63 GLOBOTRUNCANA FORNICATA
64 BOLIVINOIDES DECORATA
65 HEDBERGELLA BRITTONENSIS
66 ROTALIPORA CUSHMANI
67 HEDBERGELLA WASHITENSIS
68 ROTALIPORA EVOLUTA
69 NUMMOLOCULINA HEIMI
70 DICTYOCONUS WALNUTENSIS
71 ORBITOLINA TEXANA

Stratigraphic column:

EOCENE — T, E

JACKSON: WHITSETT; McELROY, YAZOO; WELLBORN; CADDELL

CLAIBORNE: MOODYS BRANCH; YEGUA-COOKFIELD, CROCKETT-COOK MTN.; SPARTA; WECHES; QUEEN CITY, CANE RIVER; REKLAW

PALEOCENE

WILCOX: CARRIZO, SABINE-TOWN; ROCK-DALE; SEGUIN; HATCHE-TIGBEE; TUSCA-HOMA; NANA-FALIA; SALT MTN.

MIDWAY: WILLS POINT; NAHEOLA; PORTERS CREEK; KINCAID; CLAYTON

GULF — U. CRETACEOUS

NAVARRO — MAESTR
TAYLOR — CAMPAN
AUSTIN — SAN, TAC, TON
EAGLEFORD — TURON
WOODBINE — CENOMANIAN

COMANCHE — LOWER CRETACEOUS

WASHITA: BUDA; DEL RIO, GRAYSON; GEORGE-TOWN; KIAMICHI; GOODLAND; COMANCHE PEAK; WALNUT — ALBIAN

FREDERICKS-BURG: EDWARDS

TRINITY: PALUXY; GLEN ROSE; MOORINGS-PORT; FERRY LAKE ANHYDRITE; PEAR-SALL; RODESSA; JAMES; PINE ISLAND; SLIGO; HOSSTON — APTIAN, NEO, COM

COAHUILA

depth. However, if the well becomes underbalanced over an interval of impermeable shale, a sudden increase in torque and drag sometimes is observed. After a connection or a trip is made, hole fill also may be observed. A pressure differential into the well can cause large shale fragments to break away from the sides of the borehole and overload the upward carrying capacity of the drilling fluid, resulting in these symptoms. However, drilling problems other than abnormal formation pressure also can cause increased torque, drag, or hole fill.

Analysis of Mud Logging Data. A continuous evaluation of the formation rock fragments and formation fluids in the drilling fluid pumped from the well can provide valuable information about subsurface formations. The information provided is not as timely as the drilling performance data discussed previously because several hours may be required for the drilling fluid and rock cuttings to travel from the bottom of the well to the surface. The approximate depth from which the formation fragments and fluids were drilled must be computed from careful records of drilled depth and cumulative pump strokes. In spite of the time delay or *lag time* required, the additional information that can be obtained is extremely valuable because (1) it can reinforce the indication of an increase in formation pressure gradient by the drilling performance data and (2) it can provide a warning of a possible increase in formation pressure gradient that was not evident from the drilling performance data. None of the available indirect methods of ascertaining changes in formation pressure gradient can be applied with complete confidence. Thus, the tendency is to look at the collective results obtained using as many abnormal pressure indicators as possible.

Graphical presentations of information collected by monitoring the drilling fluid circulated from the well are called *mud logs*. Usually, the mud log displays information about the lithology drilled and the formation fluids present in the drilling fluid. A knowledge of lithology aids in correlating the bottom of the well in progress with formations penetrated in previous wells in the area for which the formation pressures are known. Information about the composition and concentration of formation fluids in the drilling fluid helps in the detection of commercial hydrocarbon accumulations as well as in the detection of abnormal formation pressure.

Cuttings Analysis. The lithology is determined by collecting fresh rock fragments from the shale shaker at regular depth intervals. The fragments then are washed and studied under a microscope to determine the type of minerals present. A portion of the rock fragments are soaked in detergent solutions or kerosene so that further fragmentation occurs, allowing any microfossils present to be separated by screening. Identification of the minerals and microfossils often allows identification of the formation being drilled. In some cases, it may be known from other wells drilled in the area that abnormal formation pressure generally is encountered just below a certain marker formation, which can be identified by the presence of a particular microfossil. The general lithology of the Louisiana gulf coast area along with the key microfossils present is shown in Fig. 6.22.

Variations in size, shape, and volume of shale

fragments in the drilling fluid also can provide indications of abnormal formation pressures. As formation pressure in the transition zone increases while drilling with a constant drilling fluid density, the pressure overbalance across the hole bottom decreases continually. At a reduced overbalance, the shale cuttings sometimes become longer, thinner, more angular, and more numerous. If the formation pressure becomes greater than the drilling fluid pressure while low permeability shale is drilled, large shale fragments begin to spall off the sides of the borehole. Fragments greater than an inch in length often can be observed at the surface. *Spalling shale* appears, longer, thinner, and more splintery than *sloughing shale*, which results from a chemical incompatibility between the borehole wall and the drilling fluid. Spalling shale also has a concoidal fracture pattern that is apparent under a microscope. Examples of spalling shale and sloughing shale are shown in Fig. 6.23. [12]

The mud logger also makes physical and chemical measurements on shale cuttings, which can indicate changes in formation pressure gradient. Commonly measured porosity-dependent physical properties include (1) bulk density and (2) moisture content. Shale cutting resistivity also has been used successfully on an experimental basis, [18] but has not received widespread application. The most common chemical measurement made on shale cuttings is the determination of the cation-exchange capacity of the shale, which is greater when the shale is composed primarily of montmorillonite clays rather than illite, chlorite, or kaolinite. As discussed in Sec. 6.1.3, the diagenesis of montmorillonite to illite is thought to be related to the origin of abnormal formation pressure.

The bulk density of shale cuttings commonly is measured by (1) a *mercury pump*, (2) a *mud balance*, or (3) a *variable-density liquid column*. The procedure used to prepare the sample is similar for all these methods. Approximately one quart of cuttings is taken from the drilling fluid. The cuttings then are placed on a series of screens and washed through the screens with either water or diesel oil, depending on whether a water- or oil-base drilling fluid is being used. Only shale cuttings that pass through a 4-mesh screen and are held on a 20-mesh screen are retained for further processing. The larger cuttings may be spalling or sloughing shale from the borehole walls at an unknown depth. Also, the bulk density of the larger cuttings is thought to be affected to a greater extent by the release of pressure as the cuttings are brought to the surface.

The cuttings caught on a 20-mesh screen are blotted quickly on paper towels and then blown with warm air until the surface liquid sheen reduces to a dark, dull appearance. Care must be exercised not to remove pore water from the shale fragments.

Mercury Pump. A sample of cuttings weighing approximately 25 g usually is used to determine bulk volume in the upper air chamber of a mercury pump (Fig. 6.24). The mercury level in the lower chamber of the mercury pump first is lowered to a reference level by withdrawing the piston to a marked starting point on the piston-position indicator. The piston-position indicator is calibrated in 0.01-cm³ displaced mercury volume increments. An empty sample cup is placed in the upper

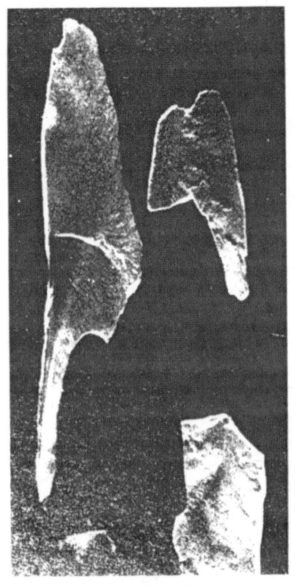

(a) Spalling shale.

(b) Sloughing shale.

Fig. 6.23—Examples of spalling and sloughing shale. [12]

chamber and the chamber is closed. The piston is advanced until the chamber air is pressurized to 24 psig. The piston position indicator is read to the nearest 0.01-cm³ and denoted as V_1. This sequence is repeated with the 25-g sample of shale cuttings in the sample cup and the second reading is denoted as V_2. The difference between the two readings gives the volume of the cuttings. Thus, the bulk density is given by

$$\rho_{sh} = \frac{m_{sh}}{V_1 - V_2}, \qquad \dots\dots\dots\dots\dots\dots(6.20)$$

where m_{sh} is the mass of shale cuttings used in the sample.

Example 6.11. A mercury injection pump gave a scale reading of 45.30 cm³ at 24 psig with an empty sample cup in the air chamber. When a 25.13-g sample of shale

cuttings was placed in the sample cup, a scale reading of 34.24 cm³ was obtained. Compute the average bulk density of the sample.

Solution. The average bulk density of the sample of shale cuttings can be computed with Eq. 6.20:

$$\rho_{sh} = \frac{25.13}{45.3 - 34.24} = 2.27 \text{ g/cm}^3.$$

Mud Balance. The standard mud balance described in Sec. 1, Chap. 2, sometimes is used to measure the density of shale cuttings. Shale cuttings prepared in a manner similar to that for the mercury pump are placed in a clean, dry, mud balance until the density indicated by the balance is equal to the density of water. Thus, the mass of the shale cuttings in the balance is equal to the mass of

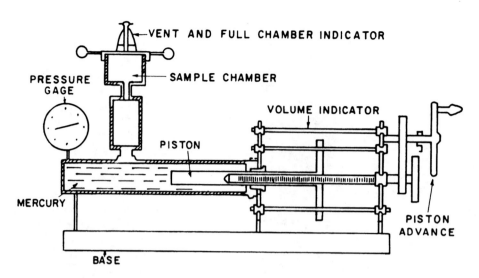

Fig. 6.24—Mercury pump used in determining bulk volume of shale cuttings.

a volume of water equal to the total cup volume, V_t, of the balance:

$$m_{sh} = \rho_{sh} V_{sh} = \rho_w V_t,$$

where ρ_w is the density of water.

Solving this equation for shale volume, V_{sh}, gives

$$V_{sh} = \frac{\rho_w}{\rho_{sh}} V_t. \quad \ldots\ldots\ldots\ldots\ldots\ldots\ldots (6.21)$$

When enough shale cuttings have been added to obtain a balance with the mud cap on and when the rider indicates the density of water, fresh water is added to fill the cup. The mixture is stirred to remove any air. The mud cap then is replaced and the average density, $\bar{\rho}_m$, of the cuttings/water mixture is determined. $\bar{\rho}_m$ can be expressed by

$$\bar{\rho}_m = \rho_{sh} \frac{V_{sh}}{V_t} + \rho_w \frac{(V_t - V_{sh})}{V_t}.$$

Substitution of shale volume, V_{sh} (defined by Eq. 6.21), into the above equation and solving for the shale density, ρ_{sh}, yields

$$\rho_{sh} = \frac{\rho_w{}^2}{2\rho_w - \bar{\rho}_m}. \quad \ldots\ldots\ldots\ldots\ldots\ldots (6.22)$$

Example 6.12. Shale cuttings are added to a clean, dry, mud balance until a balance is achieved with the mud cap in place and the density indicator reads 1.0 g/cm^3. Fresh water is added to the cup and the mixture is stirred until all air bubbles have been removed. The mixture density is determined to be 1.55 g/cm^3. Compute the average density of the shale cuttings.

Solution. Use of Eq. 6.22 gives

$$\rho_{sh} = \frac{(1.0)^2}{2.0 - 1.55} = 2.22 \text{ g/cm}^3.$$

Variable-Density Column. The variable-density column contains a liquid that has an increasing density with depth. Shale fragments dropped into the column fall until they reach a depth at which the density of the shale fragments and the liquid are the same. Since some cuttings may be altered by prolonged contact with the column liquid, the initial rest point is recorded. Usually, five shale fragments are selected from each prepared sample and the average density of the five fragments is reported. When air bubbles are observed clinging to a shale fragment in the column, the results are disregarded.

A variable-density column can be obtained by carefully mixing Bromoform, a dense liquid having a specific gravity of 2.85, with a low-density solvent, such as carbon-tetrachloride or trichloro-ethane, in a graduated cylinder. Bromoform first is poured into a tilted graduated cylinder until it is 60% filled. The solvent is poured slowly on top of the Bromoform while keeping the graduated cylinder tilted to prevent excessive mixing at the liquid interface. Twenty percent of the capacity of the graduated cylinder is filled with solvent, leaving a 20% air space at the top for an air-tight stopper.

Calibration beads with known densities that are spread as evenly as possible over the range of 2.0 to 2.8 are dropped into the column. A clean, 10-mL pipette is inserted slowly to the bottom of the graduated cylinder with the thumb over the end of the pipette. After the pipette is inserted, the thumb is lifted, allowing the pipette to fill with Bromoform. Again with the thumb over the end of the pipette, the top of the pipette is lifted just above the elevation where the calibration density beads are grouped. Slowly, the Bromoform is released as the pipette is lifted into the solvent-rich liquid. As this is done, the calibration beads begin to separate. This sequence is repeated until the calibration beads indicate approximately a linear density variation with depth (Fig. 6.25). The column then is stirred slightly with a stirring rod and allowed to stand for about 1 hour for more uniform mixing across the cross section of the column.

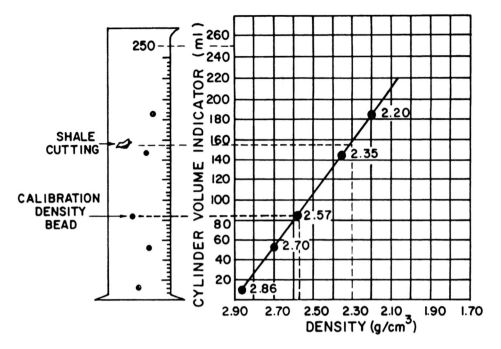

Fig. 6.25—Variable-density column used in determining bulk density of shale cuttings.

The variable-density column should be prepared and used in a fume hood. The halogenated hydrocarbons used in the column are toxic and should not be inhaled. The column should be sealed tightly when not in use.

Example 6.13. Five shale fragments dropped into the variable-density column shown in Fig. 6.25 initially stopped at the following reference marks on the 250-mL graduated cylinder: 150, 155, 160, 145, and 155. Determine the average bulk density of the cuttings.

Solution. By use of the calibration curve constructed in Fig. 6.25 and the calibration density beads, the following shale densities are indicated.

Reading (mL)	Bulk Density (g/cm³)
150	2.32
155	2.30
160	2.28
145	2.34
155	2.30

The average shale density for the five bulk density values shown is 2.31 g/cm³.

Shale density is a porosity-dependent parameter that often is plotted vs. depth to estimate formation pressure. When the bulk density of a cutting composed of pure shale falls significantly below the normal pressure trend line for shale, abnormal pressure is indicated. The magnitude of pore abnormal pressure can be estimated by either of the two basic approaches discussed previously for the generalized example illustrated in Fig. 6.10. An empirically developed departure curve such as the one

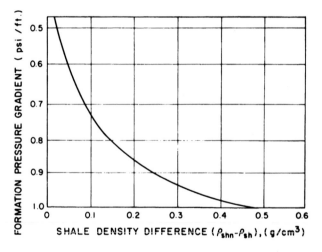

Fig. 6.26—Boatman relationship between formation pore pressure and bulk density of shale cuttings. [17]

shown in Fig. 6.26 is needed to apply the second basic approach.

A mathematical model of the normal compaction trend for the bulk density of shale cuttings can be developed by substituting the exponential porosity expression defined by Eq. 6.4 for porosity in Eq. 6.3a. After rearranging terms, this substitution yields

$$\rho_{shn} = \rho_g - (\rho_g - \rho_{fl})\phi_o e^{-KD}, \quad \ldots\ldots\ldots\ldots (6.23)$$

where ρ_{shn} is the shale density for normally pressured shales. The grain density of pure shale is 2.65. Average pore fluid density, ρ_{fl}, can be found from Table 6.1. Constants ϕ_o and K can be based on shale-cutting bulk density measurements made in the normally pressured formations.

TABLE 6.11—BULK DENSITY DATA FOR SHALE CUTTINGS OBTAINED ON A SOUTH LOUISIANA WELL USING A VARIABLE DENSITY LIQUID COLUMN [19]

Depth (ft)	Bulk Density (g/cm^3)
6,500	2.38
6,600	2.37
6,700	2.33
6,800	2.34
6,900	2.39
7,000	2.39
7,100	2.35
7,500	2.41
7,700	2.37
8,000	2.39
8,200	2.38
8,300	2.34
8,500	2.34
8,900	2.41
9,400	2.41
9,500	2.41
9,700	2.44
10,000	2.44
10,500	2.42
10,600	2.46
10,800	2.44
11,100	2.44
11,200	2.45
11,400	2.45
11,600	2.44
11,900	2.46
11,950	2.42
12,100	2.44
12,300	2.30
12,400	2.21
12,500	2.23
12,600	2.22
13,000	2.29
13,100	2.26
13,200	2.42
13,400	2.25
13,600	2.40
13,800	2.28
14,300	2.38
15,000	2.38
15,700	2.39
16,100	2.42

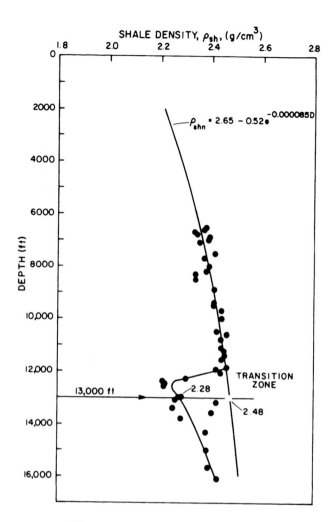

Fig. 6.27—Example shale-density plot.

Example 6.14. The bulk density data shown in Table 6.11 were determined for shale cuttings on a well drilled in south Louisiana using a variable-density liquid column. Determine an equation for the normal pressure trend line. Also, compute the formation pore pressure at a depth of 13,000 ft using the empirical relationship determined by Boatman (Fig. 6.26).

Solution. The shale density first is plotted vs. depth as shown in Fig. 6.27 to establish the depth interval for which the formations are normally pressured. The zone of normal formation pressure appears to extend to a depth of 12,000 ft.

The shale density data in the normal pressure region can be expressed in terms of porosity by Eq. 6.3b with a grain density of 2.65 g/cm^3 and an average pore fluid density of 1.074 g/cm^3. This gives the following equation for shale porosity.

$$\phi_{sh} = \frac{2.65 - \rho_{sh}}{1.576}$$

Application of this equation to the shale density data in the normally pressured formations above 11,000 ft gives the results shown in Table 6.12 and Fig. 6.28. An acceptable straight-line representation of the data was obtained by shifting upward the average porosity trend line for the Louisiana gulf coast obtained in Example 6.2. The upward shift corresponds to the use of a surface porosity constant, ϕ_o, of 0.33.

Substitution of appropriate values for ρ_g, ρ_{fl}, ϕ_o, and K in Eq. 6.23 yields the following expression for the normal pressure trend line.

$$\rho_{shn} = 2.65 - (2.65 - 1.074)(0.33)e^{-0.000085D}$$

$$= 2.65 - 0.52e^{-0.000085D}.$$

The line defined by this equation is plotted as the normal pressure trend line in Fig. 6.27.

At a depth of 13,000 ft, values of 2.28 and 2.48 for ρ_{sh} and ρ_{shn} are obtained from Fig. 6.27. Entering the Boatman correlation given in Fig. 6.26 for $(\rho_{shn} - \rho_{sh}) = 2.48 - 2.28 = 0.2$ gives a formation pressure gradient of 0.86 psi/ft. Thus, the formation pressure at 13,000 ft is estimated to be

$$p = 0.86(13,000) = 11,180 \text{ psig.}$$

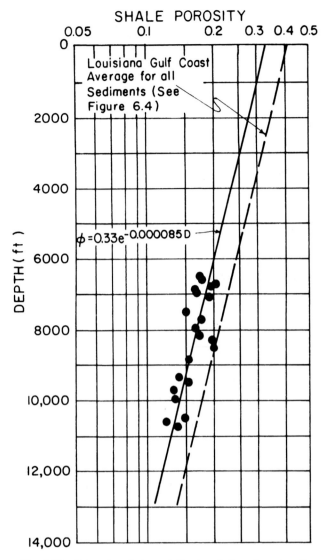

Fig. 6.28—Example shale-porosity trend line in normally pressured formation.

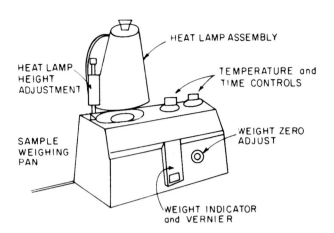

Fig. 6.29—Ohaus moisture-determination balance.

TABLE 6.12—EXAMPLE COMPUTATION OF AVERAGE SHALE POROSITY IN NORMALLY PRESSURED FORMATIONS

Sediment Thickness (ft)	Bulk Density (g/cm^3)	Average Porosity ϕ_s
6,500	2.38	0.171
6,600	2.37	0.178
6,700	2.33	0.203
6,800	2.34	0.197
6,900	2.39	0.165
7,000	2.39	0.165
7,100	2.35	0.190
7,500	2.41	0.152
7,700	2.37	0.178
8,000	2.39	0.165
8,200	2.38	0.171
8,300	2.34	0.197
8,500	2.34	0.197
8,900	2.41	0.152
9,400	2.43	0.140
9,500	2.41	0.152
9,700	2.44	0.133
10,000	2.44	0.133
10,500	2.42	0.146
10,600	2.46	0.121
10,800	2.44	0.133

Moisture Content. The moisture content of shale cuttings can be determined with a moisture-determination balance such as the one shown in Fig. 6.29. Shale cuttings are collected, washed, screened, and dried in the same manner as for a bulk density measurement. A 10-g sample is placed on the balance. The balance is designed to show a moisture content of zero for a 10-g sample size. The drying lamp is placed above the sample and the weight loss resulting from pore-water loss is noted. The sample weight stabilizes after about 5 minutes, indicating that all the water has been lost. The balance is scaled to read the moisture content by weight directly to the nearest 0.1%.

The volume of water loss in milliters is equal to the weight loss in grams. If the effect of the dissolved salts left in the sample is neglected, the shale porosity can be determined as the product of the moisture content and the bulk density in grams per cubic centimeter. This allows the porosity of the cutting to be determined without assuming a grain density. Some of the older shales have a high concentration of heavier minerals, such as pyrite and calcite, mixed within the shale structure, and the assumption of a grain density of 2.65 g/cm^3 cannot be made.

Example 6.15. Exactly 10 g of shale cuttings are placed in a mercury pump and the bulk volume is determined to be 4.20 cm^3. A 10-g sample then is placed in a moisture-determination balance. After 5 minutes of drying, the sample weight stabilizes at 9.28 g, rendering a moisture-content reading of 7.2%. Compute the porosity of the sample.

Solution. The water volume in cubic centimeters is approximately equal to the weight of water loss in grams. Thus, the water volume is 0.72 cm^3. Since the bulk volume was 4.2 cm^3, the porosity is

$$\phi_{sh} = \frac{0.72}{4.2} \times 100\% = 17.1\%$$

Alternatively, since the bulk density is

$$\rho_{sh} = \frac{10}{4.2} = 2.381 \ g/cm^3,$$

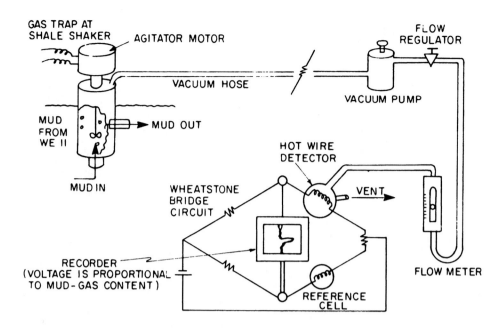

Fig. 6.30—Mud-gas detection system.

the porosity is given by

$$\phi_{sh} = (7.2\%)(2.381) = 17.1\%.$$

Cation-Exchange Capacity. The cation exchange capacity of the shale cuttings can be determined with the titration procedure discussed in Sect. 1, Chap. 2, for clay/water drilling fluids. Also, more detailed instructions are given in API RP13B, which is included in the *Drilling Fluids Laboratory Manual*. The cation-exchange capacity of the shale cuttings reported in milliliters of 0.01 N methylene blue required to titrate 100 g of shale sample is called the *shale factor*.

In normally pressured sediments, the diagenesis of montmorillonite to illite causes a gradual decline in montmorillonite content with depth. In the transition zone, the montmorillonite content as measured by the shale factor usually is observed to decrease at a much faster rate. One hypothesis to explain this relationship is that the release of the more tightly held interlayer water to pore water in the conversion of montmorillonite to illite is a primary cause of the abnormally high pore pressure.

Mud Gas Analysis. Formation gases circulated from the well in the drilling fluid usually are detected by a system such as the one shown in Fig. 6.30. A *gas trap* is placed in the drilling fluid returning from the well. A vacuum hose draws a mixture of air and gas from the gas trap to a gas detector. An agitator usually is built into the gas trap to increase the *gas-trap efficiency*. Gas-trap efficiency is defined as the percentage of gas in the mud that is removed and transmitted to the gas detector. Typical gas-trap efficiency values range from 50 to 85%.

The *hot wire* gas detector shown in Fig. 6.30 employs a cataytic filament that responds to all the combustible gases present. Some of the newer units employ a hydrogen flame detector in place of the hot wire detector. The gas recorder usually is scaled in terms of ar-

bitrary *gas units*, which are defined differently by the various gas-detector manufacturers. In practice, significance is placed only on relative changes in the gas concentrations detected.

An analysis of the composition of the gases removed from the mud is made by means of a gas chromatograph. The technique used by one company is illustrated in Fig. 6.31. A mud sample is placed in a steam-still reflux chamber, where most of the lighter hydrocarbons are separated from the mud as vapor. This method can be used successfully for oil muds as well as water-base muds. A sample of the vapor then is withdrawn from the steam still and injected into a gas chromatograph to determine the concentrations of C_1 through C_5. The concentration of each component in parts per million parts of mud then is plotted at the computed formation depth on the mud log.

Formation gases enter the drilling fluid from (1) the pore fluids of the rock destroyed by the bit, and (2) the seepage of fluids from exposed formations into the borehole. The seepage of fluids into the borehole is an indication that the formation pressure has increased to the point where it exceeds the pressure caused by the drilling fluid during at least a portion of the drilling operations. This can be detected by characteristic zones of high gas concentration being circulated to the surface that correspond to periods where drilling fluid circulation was stopped and upward vertical pipe movements occurred. The pressure caused by the drilling fluid is minimal during such periods and the static drilling fluid allows any seepage to be concentrated in a relatively small volume of drilling fluid. Common examples of such behavior are the detection of (1) *connection gas* peaks, which occur at time intervals corresponding to the time required to drill down one joint of drill pipe and make a connection and (2) *trip gas* peaks, which occur after making a trip for a new bit. *Background gas* is a term used to denote the base-line gas detector readings between peaks.

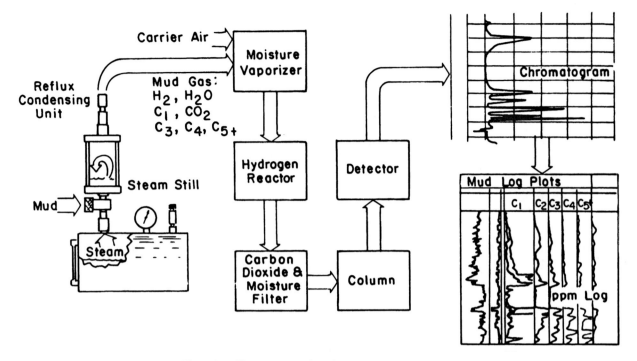

Fig. 6.31—Flow diagram of mud-gas separation and analysis.

Connection gas and trip gas can be suppressed by increasing the drilling fluid density. Gas that enters the mud from the pore fluid of the rock destroyed by the bit is relatively unaffected by increases in drilling fluid density. Methane peaks can occur when drilling aquifers as well as formations containing mostly oil and gas because of the dissolved gas in the formation water. Simultaneous increases in the heavier hydrocarbons in the C_2 through C_5 range are more indicative that a commercial hydrocarbon deposit has been penetrated.

The drilled cuttings can be crushed and the liquids and gases present can be analyzed as a further test for the possible presence of a commercial hydrocarbon deposit. Cuttings also are examined under ultraviolet light for traces of oil. Both refined and crude oils exhibit fluorescence under certain ultraviolet wavelengths. The use of leaching agents is often necessary to bring oil to the surface of the cuttings, where it can be detected by fluorescence.

A sample mud log showing both total combustible gases and the results of the chromatographic analysis plotted as a function of depth is shown in Fig. 6.32. To plot these parameters vs. depth, the mud logger must make allowances for the lag time required for the sample to reach the surface. The depth-lag calculation determines the depth of the bit when the sample observed at the surface originated at the bottom of the hole. This is accomplished most easily by keeping a record of cumulative pump strokes at depth increments of 5 ft and of the strokes required to pump a sample from the bottom of the hole to the surface. Note that in Fig. 6.32, the connection gas (CG) peaks occur approximately 30 ft apart. Note also the larger peak corresponding to the sand drilled at 6,300 ft and the trip gas (TG) peak at 6,500 ft.

Drilling Fluid Analysis. In addition to the analysis procedures performed on the formation fragments and gas samples separated from the drilling fluid, other mud properties that are measured to detect abnormally pressured formations include (1) salinity or resistivity, (2) temperature, and (3) density.

When the formation water salinity is much greater than the salinity of the drilling fluid, a slow influx of formation water from abnormally pressured formations into the wellbore can cause the salinity of the mud returning from the well to increase significantly.

Formation water that enters the drilling fluid from the rock destroyed by the bit causes a much more gradual change in drilling fluid salinity. Periodic mud treatments, such as dilution, cause a decrease in salinity. The salinity increase is determined most accurately by the titration of a mud sample with an $AgNO_3$ solution as discussed in Sec. 1, Chap. 2. Since a change in salinity also causes a change in resistivity, resistivity probes often are placed in the mud stream to monitor continually for salinity changes.

The abnormally high water content of abnormally pressured formations tends to cause these formations to have an abnormally low thermal conductivity and an abnormally high heat capacity. This causes the geothermal gradient to increase in the transition zone. In some cases, the increase with time in the temperature of the mud returning from the well during a bit run reflects the higher geothermal gradient of the transition zone. Unfortunately, many other variables also affect the temperature of the mud returning from the well, frequently causing the temperature data to be difficult to interpret.

Drilling fluid density returning from the well decreases significantly when formation gas is entrained in the drilling fluid. In some cases, the extent by which the drilling fluid density is reduced at the surface by entrained gas is used as a rough indicator of the mud gas content. Of course, the use of gas detector generally gives a more

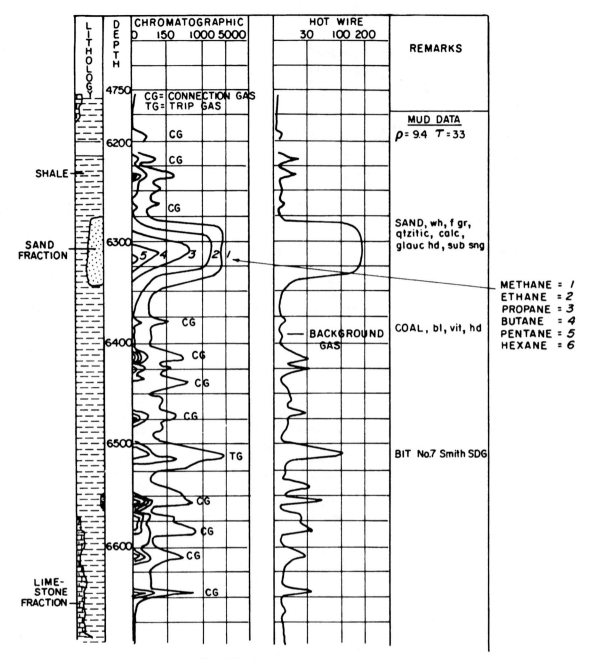

Fig. 6.32—Example mud-gas log.
Courtesy of Core Laboratories Inc.

satisfactory presentation of the mud gas content. As discussed in Chap. 4, Sec. 3, the effect of gas-cut mud on hydrostatic pressure in the well is quite small.

6.2.3 Verification of Formation Pressure Using Well Logs

The decision of when to stop drilling temporarily and cement casing in the well before proceeding with deeper drilling operations is a key decision in both the technical and economic success of a drilling venture. If casing is set too high, an unplanned additional casing string will be required to reach the depth objective, resulting in much higher well costs and a greatly reduced final well size. If casing is not set when it is needed, an underground blowout may occur, which can be very costly to stop and could necessitate plugging and aban-

doning most of the borehole. An accurate knowledge of formation pressure is necessary to select the best casing-setting depth.

The open borehole generally is logged with conventional wireline devices to provide permanent records of the formations penetrated prior to running casing. Empirical methods have been developed for estimating formation pressure from some of the porosity-dependent parameters measured by the well-logging sonde. The pressure estimates made in this manner allow verification of the previous pressure estimates made during well planning and drilling. These pressure estimates are also extremely valuable in planning future wells in the area.

The porosity-dependent formation parameters usually obtained from well logs for the estimation of formation pore pressure are either (1) interval transit time, t, or (2)

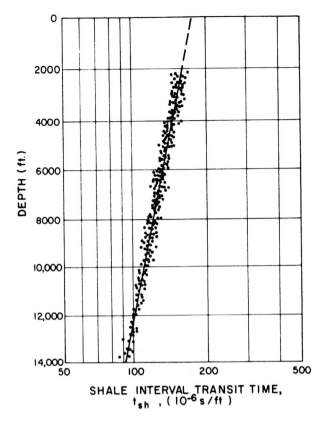

Fig. 6.33—Interval transit time in normally pressured Miocene and Oligocene shales of the Texas and Louisiana gulf coast area.[20]

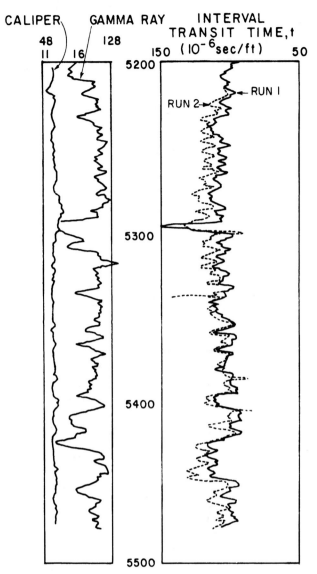

Fig. 6.34—Acoustic log composite showing effect of shale hydration on interval transit time.[21]

conductivity, C_0. Nuclear density logs also have been used, but to a much lesser extent. The acoustic travel time is less affected by other variables and is thought to give the most accurate results.

Pore pressure plots constructed using porosity-dependent formation parameters obtained from logging data include only points obtained in "pure" shales. Criteria that can be applied in selecting the more pure shales from the logging data include the following.

1. Minimal base line values of spontaneous potentials with essentially no fluctuations.

2. Maximum values of gamma ray counts.

3. Maximum conductivity (minimum resistivity) values with a small and constant separation between the shallow and deep radius-of-investigation devices.

4. Maximum values of interval transit time.

5. Use of values obtained in shales having a thickness of 20 ft or more.

It is often difficult to find a sufficient number of shale points in the shallow normally pressured formations to establish the normal pressure trend line with data from a single well. Published average normal trend lines for areas of active drilling obtained from a large number of wells provide a useful guide in interpreting the small amount of normal pressure data available in pure shale on a given well.

Interval Transit Time. The procedure for estimation of formation pressure from log-derived interval transit time data is essentially the same as the procedure discussed in Sec. 6.2.1 for seismic-derived interval transit time data.

The primary difference is that when one uses well log data, only the shale formations are included in the analysis. The lithology cannot be determined accurately enough for this to be done with only seismic data; therefore average interval travel times for all formations present must be used.

A considerable amount of well log data may be necessary to establish a mathematical model of the normal compaction trend for interval transit time in shale. An example plot of interval transit time in shale for the Miocene and Oligocene shales of the Texas and Louisiana gulf coast area is shown in Fig. 6.33. Using the procedure described in Example 6.5, a good fit of these data is obtained using the following relation for the matrix travel time in pure shale, t_{ma}.

$$t_{ma} = 62 + 202\phi. \dotfill (6.24)$$

Substituting this equation for matrix travel time into Eq. 6.7 using a t_{fl} of 207 (105,000 ppm NaCl) and

TABLE 6.13—ABNORMAL PRESSURE AND INTERVAL TRANSIT TIME DEPARTURE IN THE MIOCENE AND OLIGOCENE FORMATIONS OF THE TEXAS AND LOUISIANA GULF COAST[20]

Parish or County and State	Well	Depth (ft)	Pressure (psi)	FPG* (psi/ft)	$(t_{sh})_p - (t_{sh})_n$ (μs/ft)
Terrebonne, LA	1	13,387	11,647	0.87	22
Offshore Lafourche, LA	2	11,000	6,820	0.62	9
Assumption, LA	3	10,820	8,872	0.82	21
Offshore Vermilion, LA	4	11,900	9,996	0.84	27
Offshore Terrebonne, LA	5	13,118	11,281	0.86	27
East Baton Rouge, LA	6	10,980	8,015	0.73	13
St. Martin, LA	7	11,500	6,210	0.54	4
Offshore St. Mary, LA	8	13,350	11,481	0.86	30
Calcasieu, LA	9	11,800	6,608	0.56	7
Offshore St. Mary, LA	10	13,010	10,928	0.84	23
Offshore St. Mary, LA	11	13,825	12,719	0.92	33
Offshore Plaquemines, LA	12	8,874	5,324	0.60	5
Cameron, LA	13	11,115	9,781	0.88	32
Cameron, LA	14	11,435	11,292	0.90	38
Jefferson, TX	15	10,890	9,910	0.91	39
Terrebonne, LA	16	11,050	8,951	0.81	21
Offshore Galveston, TX	17	11,750	11,398	0.97	56
Chambers, TX	18	12,080	9,422	0.78	18

*Formation fluid pressure gradient.

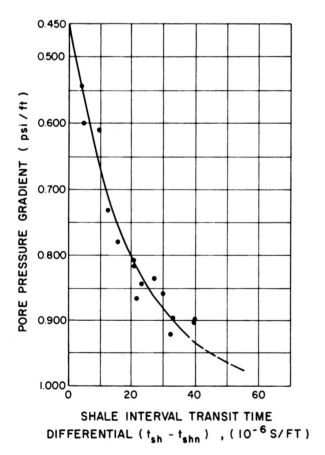

Fig. 6.35—Hottman and Johnson relationship between formation pore pressure and shale interval transit time.[20]

substituting Eq. 6.4 for ϕ yields the following equation for the normal pressure trend line.

$$t_{shn} = 62 + 409\phi_o e^{-KD} - 202\phi_o^2 e^{-2KD}. \quad \ldots (6.25)$$

An excellent fit of the data shown in Fig. 6.33 is obtained using values of 0.33 and 0.0001 for surface porosity, ϕ_o, and compaction constant, K, respectively. A good fit of the data shown also could be obtained with a simple straight-line relationship over the depth interval shown. However, Eq. 6.25 yields much more accurate results at greater depth, while a straight line extrapolation would yield values of interval transit time less than the transit time of solid matrix material.

The geologic age of the shale sediments has been found to affect the normal pressure relationship between interval travel time and depth. Older sediments that have had a longer time for compaction to occur result in a upward shift in the normal pressure trend line. Similarly, younger sediments result in a downward shift in the normal pressure trend line. In practice, this can be handled by changing the surface porosity constant, ϕ_o, in the mathematical model so that the model is brought into agreement with the available data in normally pressured formations in the well of interest.

The shape of the normal trend line observed on a given well also may be affected by the alteration of water-sensitive shales by the drilling fluid. Fig. 6.34 is a comparison of two logging runs made in the same well at different times. Note the significant differences. Because of this problem, many people place more emphasis on the data from just above the transition zone when establishing a normal trend line. There is less time of shale exposure to mud at the deeper depths just above the transition zone.

When the shale interval transit time falls significantly above the normal pressure trend line near the formation of interest, abnormal formation pressure is indicated. The magnitude of the abnormal pressure can be computed by either of the two basic approaches discussed for

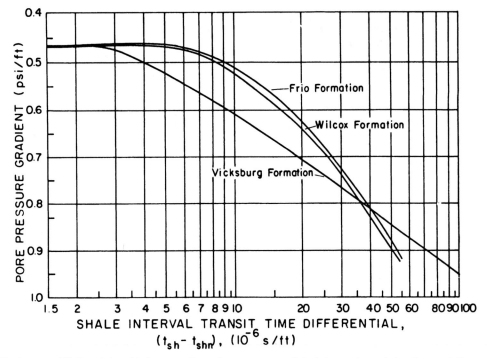

Fig. 6.36—Matthews and Kelly relationship between formation pressure and shale interval transit time for south Texas gulf coast. [23]

the generalized example illustrated in Fig. 6.10, provided an empirically developed departure curve is available for application of the second technique.

Hottman and Johnson presented one of the first empirical relationships between measured formation pressures in permeable sandstones and interval transit time in the adjacent shales. Their basic data are given in Table 6.13 and plotted in Fig. 6.35. This correlation is still widely used today in the Louisiana gulf coast area. Mathews and Kelly [23] published similar correlations (Fig. 6.36) for the Frio, Wilcox, and Vicksburg trends of the Texas gulf coast area. More recent authors have developed similar correlations for the North Sea and South China Sea areas. These correlations are presented in Fig. 6.37.

Example 6.16. The shale interval transit time data shown in Table 6.14 were read from a sonic log made in a well in Jefferston County, TX. Estimate formation pressure at a depth of 12,000 ft using the Hottman and Johnson correlation shown in Fig. 6.35.

Solution. The interval transit time data first are plotted vs. depth as shown in Fig. 6.38. The average normal pressure trend line is given by Eq. 6.25 with $\phi_o = 0.33$ and $K = 0.0001$ ft^{-1}. This relationship was plotted by a dashed line in Fig. 6.37. Since the dashed line falls significantly above the data in the normally pressured formations, it is necessary to shift the average normal pressure trend line downward by adjusting the value of ϕ_o. Solving Eq. 6.25 for ϕ_o using an average K value of 0.0001 yields

$$\phi_o = \frac{\dfrac{409}{e^{0.0001D}} - \sqrt{\left(\dfrac{409}{e^{0.0001D}}\right)^2 - \dfrac{808(t_{shn}-62)}{e^{0.0002D}}}}{\dfrac{404}{e^{0.0002D}}}.$$

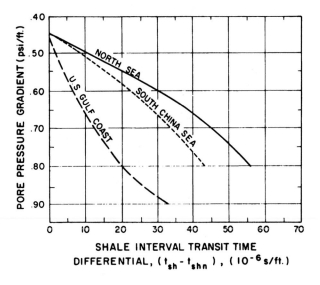

Fig. 6.37—Comparison of relationship between formation pore pressure and shale interval transit time in several territory basins. [21]

The first shale transit time value given in Table 6.14 is 160 μs/ft at a depth of 2,775 ft. Substitution of these values into the above equation gives a surface porosity value ϕ_o of 0.367. Similar calculations at each depth interval of the normally pressured region above 9,000 ft are summarized in Table 6.15. Note that an average value of 0.373 is indicated for ϕ_o. Thus, the normal pressure trend line equation becomes

$$t_{shn} = 62 + 152.6e^{-0.0001D} - 28.1e^{-0.0002D}.$$

This relationship was plotted by a solid line in Fig. 6.38.

TABLE 6.14—SHALE INTERVAL TRANSIT TIME DATA FROM SONIC LOG OF WELL IN JEFFERSON COUNTY, TX[2]

Depth (ft)	Shale Interval Transit Time (10^{-6} s/ft)
2,775	160
3,175	156
3,850	151
4,075	153
4,450	147
5,150	143
5,950	139
6,175	137
6,875	137
7,400	131
7,725	125
7,975	120
8,300	124
8,400	121
8,950	121
8,975	118
9,175	118
9,250	119
9,325	122
9,350	125
9,400	125
9,575	127
9,650	131
9,775	131
9,850	140
9,975	142
10,050	146
10,150	149
10,325	147
10,475	147
11,140	148
11,325	143
11,725	148
12,300	142
13,000	138

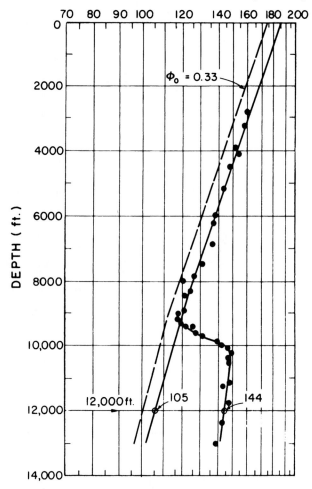

SHALE INTERNAL TRANSIT TIME (10^{-6} s/ft.)

Fig. 6.38—Example of sonic-log interval transit time plot for shale formations in Jefferson County, TX.

TABLE 6.15—EXAMPLE CALCULATION OF SURFACE POROSITY CONSTANT ϕ_o, JEFFERSON COUNTY, TX

Depth (ft)	t_{sh} (10^{-6} s/ft)	ϕ_o
2,775	160	0.367
3,175	156	0.363
3,850	151	0.364
4,075	153	0.382
4,450	147	0.367
5,150	143	0.372
5,950	139	0.381
6,175	137	0.378
6,875	137	0.405
7,400	131	0.390
7,725	125	0.367
7,975	120	0.341
8,300	124	0.379
8,400	121	0.362
8,950	121	0.383
8,975	118	0.362
Average		0.373

At a depth of 12,000 ft, values of 105 and 144 for t_{shn} and t_{sh}, respectively, can be read from the plot shown in Fig. 6.38. Using $t_{sh} - t_{shn} = 144 - 105 = 39$ μs/ft in the Hottman and Johnson correlation given in Fig. 6.35 gives a formation pressure gradient of 0.93 psi/ft. Thus, the formation pressure is $p = 0.93(12,000) = 11,160$ psig.

Conductivity. Well logging devices that measure formation conductivity or resistivity (the reciprocal of conductivity) are used on almost every well drilled. Since the data are almost always readily available, conductivity is the most common porosity-dependent parameter used in the estimation of formation pore pressure from well logs. The term *formation factor*, F_R, generally is used to refer to the ratio of the resistivity of the water-saturated formation, R_0, to the resistivity of the water, R_w. The formation factor also can be expressed in terms of a conductivity ratio:

$$F_R = \frac{R_0}{R_w} = \frac{C_w}{C_0}. \quad\dots\dots\dots\dots\dots\dots\dots(6.26)$$

The relation between formation factor F and porosity ϕ has been defined empirically by

$$\phi = F_R^{-1/m}, \quad \ldots \ldots \ldots \ldots \ldots \ldots \ldots \ldots (6.27)$$

where the exponent m varies between 1.4 and 3.0. An average value of 2.0 generally is used in practice when laboratory data are not available.

Formation conductivity C_o or resistivity R_o varies with lithology, water salinity, and temperature as well as porosity. To avoid changes caused by lithology, only values obtained in essentially pure shales are used. Shales containing some limestone are avoided because of the large effect of the limestone fraction on observed conductivity. The effect of changes in salinity and temperature can be taken into account in the calculation of the formation factor through use of the correct in-situ value of the water conductivity C_w or resistivity R_w for the given temperature and salinity at the depth of interest. Foster and Whalen[22] proposed calculating the water resistivity R_w from measurements of the spontaneous potential (SP) generally made at the same time that formation conductivity or resistivity is measured. A standard well log interpretation technique is available for computing R_w from SP measurements made in a clean (nonshaly) sandstone. The value of R_w in shales must be assumed equal to the value obtained in a nearby sand.

Formation conductivity or resistivity near the borehole also is affected significantly by exposure to the drilling fluid. Even though shale formations are relatively impermeable to the invasion of mud filtrate, changes in the shale properties gradually occur as a result of chemical interaction between the drilling fluid and the borehole wall. Sections of the borehole composed of highly water-sensitive shales give different log readings on logging runs made at different times. This problem can be minimized by using a well logging device with a deep radius of investigation.

A mathematical model of the normal compaction trend for shale formation factor can be obtained by substituting the exponential porosity equation defined by Eq. 6.4 for porosity in Eq. 6.27. After rearrangement of terms, this substitution yields

$$\ln F_R = mKD - m \ln \phi_o. \quad \ldots \ldots \ldots \ldots \ldots (6.28)$$

The constants ϕ_o and K must be chosen on the basis of conductivity data obtained in normally pressured formations in the area of interest.

Example 6.17. The well log shale resistivity values shown in Table 6.16 were obtained from an offshore Louisiana well. Water resistivity values computed from the SP log at all available water sands are given in Table 6.17. Using these data, estimate the formation pressure at 14,000-ft depth using the equivalent matrix stress concept. Assume that mean sediment bulk density varies with depth as shown in Fig. 6.3.

Solution. The water resistivity data first are plotted vs. depth as shown in Fig. 6.39. The formation factor then is computed at each depth listed in Table 6.16 by reading the water resistivity from Fig. 6.39 at the depth of in-

TABLE 6.16—SHALE RESISTIVITY DATA FROM OFFSHORE LOUISIANA[22]

Depth (ft)	Shale Resistivity R_o ($\Omega m^2/m$)
3,110	0.55
3,538	0.55
4,135	0.55
4,544	0.50
4,890	0.50
5,175	0.55
5,363	0.50
5,867	0.50
6,041	0.50
6,167	0.54
6,482	0.55
6,577	0.55
6,955	0.70
7,113	0.70
7,255	0.70
7,696	0.71
8,200	0.76
8,342	0.85
8,767	0.80
9,113	0.85
9,492	0.91
9,665	0.86
9,996	0.80
10,217	0.85
10,485	0.92
10,659	0.91
10,989	0.90
11,162	0.91
11,478	0.90
11,588	1.20
11,776	1.16
11,966	1.10
12,265	1.11
12,470	0.96
12,550	0.90
12,785	1.06
13,069	0.91
13,385	1.10
13,573	1.05
13,778	1.06
13,983	0.96
14,188	0.96
14,487	0.71
14,566	0.80
14,833	0.80
14,960	0.90
15,275	1.06

terest and then dividing the shale resistivity listed in Table 6.16 by the water resistivity read from the graph. For example, at the first depth entry in Table 6.16 of 3,110 ft, a water resistivity of 0.91 is read from Fig. 6.39. This gives a formation factor of

$$F_R = \frac{R_0}{R_w} = \frac{0.55}{0.091} = 6.0.$$

Formation factors obtained in this manner at all depths listed in Table 6.16 have been plotted in Fig. 6.40.

The normally pressured region appears to extend at least to a depth of 10,000 ft. Representative values of the normal pressure trend line are selected as 6.0 at 3,000 ft and 40.0 at 10,000 ft. Use of these two points in Eq. 6.28 yields

$$\ln 6.0 = mK(3,000) - m \ln \phi_o$$

WATER RESISTIVITY, $R\omega$, ($\Omega m^2/m$)

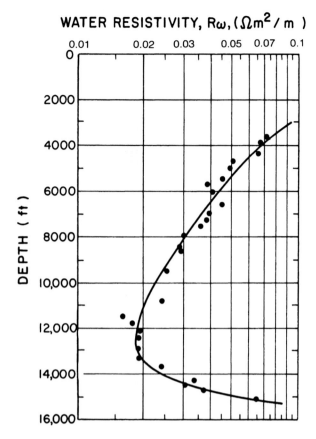

Fig. 6.39—Example formation water resistivity profile.

SHALE FORMATION FACTOR (F_R)

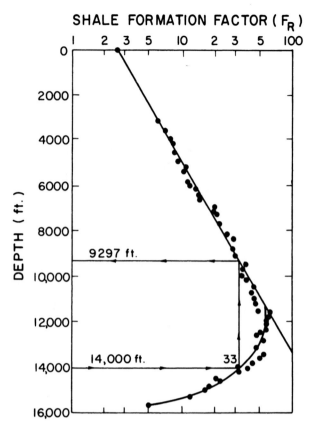

Fig. 6.40—Example plot of shale formation factor vs. depth in offshore Louisiana well.

and

$$\ln 40.0 = mK(10,000) - m \ln \phi_o.$$

Solving these two equations simultaneously gives

$$mK = \frac{\ln(40/6)}{10,000 - 3,000} = 0.000271$$

and

$$m \ln \phi_o = 10,000\, mK - \ln 40 = -0.977.$$

Thus, the normal pressure trend line is given by

$$\ln F_{Rn} = 0.000271D + 0.977.$$

An equivalent expression in a more convenient form is

$$F_{Rn} = 2.656 e^{0.000271D}.$$

The line defined by this equation was plotted on Fig. 6.40 and was found to fit the data accurately in the normally pressured region.

To compute the formation pressure at a depth of 14,000 ft using the equivalent matrix stress concept, the shale formation factor first is read from the plot given in Fig. 6.40 at 14,000 ft. An F_R value of 33 was obtained. The depth of the normally pressured shale formation

having this value of F_R then is determined from the normal pressure trend line.

$$D_n = \frac{\ln F_R - 0.977}{0.000271}$$

$$= \frac{\ln 33 - 0.977}{0.000271} = 9,297 \text{ ft.}$$

The overburden stress, σ_{ob}, due to the geostatic load at a depth of 9,297 ft was obtained using Eq. 6.6 and the average values of ϕ_o and K for all sediments (including nonshales) determined previously in Example 6.2:

$$(\sigma_{ob})_{9,297} = 0.052 \rho_g D_n - \frac{0.052(\rho_g - \rho_{fl})\phi_o}{K}$$

$$\cdot (1 - e^{-KD_n})$$

$$= 0.052(2.6)(8.33)(9297)$$

$$- \frac{0.052(2.6 - 1.074)(8.33)(0.41)}{0.000085}$$

$$\cdot \left[1 - e^{0.000085(9,297)}\right] = 10,470 - 1,741$$

$$= 8,729 \text{ psig.}$$

TABLE 6.17—WATER RESISTIVITY VALUES COMPUTED FROM SPONTANEOUS POTENTIAL LOG ON OFFSHORE LOUISIANA WELL[22]

Depth (ft)	Water Resistivity ($\Omega m^2/m$)
3,611	0.072
3,830	0.068
4,310	0.066
4,625	0.051
4,950	0.049
5,475	0.045
5,630	0.038
6,100	0.041
6,540	0.045
6,910	0.039
7,280	0.038
7,460	0.036
7,900	0.030
8,400	0.028
8,600	0.029
9,460	0.025
10,700	0.024
11,400	0.016
11,800	0.018
12,020	0.019
12,350	0.019
12,880	0.019
13,290	0.019
13,700	0.024
14,300	0.034
14,500	0.030
14,680	0.037
15,090	0.065

The formation pore pressure at 9,297 ft is given by

$$p_{9,297} = 0.465(9,297) = 4,323.$$

Thus, the effective matrix stress at both 9,297 and 14,000 ft is

$$\sigma_{14,000} = \sigma_{9,297} = 8,729 - 4,323 = 4,406 \text{ psig.}$$

Since the overburden stress σ_{ob} at 14,000 ft is

$$(\sigma_{ob})_{14,000} = 0.052(2.6)(8.33)(14,000)$$

$$- \frac{0.052(2.6-1.074)(8.33)(0.41)}{0.000085}$$

$$\cdot \left[1 - e^{-0.000085(14,000)} \right]$$

$$= 15,767 - 2,218 = 13,549 \text{ psig,}$$

the pore pressure is equal to

$$p_{14,000} = 13,549 - 4,406 = 9,143 \text{ psig.}$$

In practice, it is often difficult to obtain reasonable estimates of formation water conductivity over the entire depth range of interest. Formation water conductivity can be estimated from SP logs only in relatively clean and thick sandstone formations, which are rare or nonex-

TABLE 6.18—AVERAGE VALUES OF SLOPE CONSTANT, K_2, FOR PLOT OF log(C_0) VS. DEPTH

Louisiana gulf coast	0.000135
South Texas gulf coast	
Frio trend	0.000139
Wilcox trend	0.000120
Vicksburg trend	0.000132

TABLE 6.19—PRESSURE AND SHALE RESISTIVITY RATIOS, OVERPRESSURED MIOCENE–OLIGOCENE WELLS

Parish or County and State	Well	Depth	Pressure (psi)	FPG* (psi/ft)	Shale Resistivity Ratio** ($\Omega \cdot m$)
St. Martin, LA	A	12,400	10,240	0.83	2.60
Cameron, LA	B	10,070	7,500	0.74	1.70
Cameron, LA	B	10,150	8,000	0.79	1.95
	C	13,100	11,600	0.89	4.20
	D	9,370	5,000	0.53	1.15
Offshore St. Mary, LA	E	12,300	6,350	0.52	1.15
	F	12,500	6,440	0.52	1.30
		14,000	11,500	0.82	2.40
Jefferson Davis, LA	G	10,948	7,970	0.73	1.78
	H	10,800	7,600	0.70	1.92
	H	10,750	7,600	0.71	1.77
Cameron, LA	I	12,900	11,000	0.85	3.30
Iberia, LA	J	13,844	7,200	0.52	1.10
		15,353	12,100	0.79	2.30
Lafayette, LA	K	12,600	9,000	0.71	1.60
		12,900	9,000	0.70	1.70
	L	11,750	8,700	0.74	1.60
	M	14,550	10,800	0.74	1.85
Cameron, LA	N	11,070	9,400	0.85	3.90
Terrebonne, LA	O	11,900	8,100	0.68	1.70
		13,600	10,900	0.80	2.35
Jefferson, TX	P	10,000	8,750	0.88	3.20
St. Martin, LA	Q	10,800	7,680	0.71	1.60
Cameron, LA	R	12,700	11,150	0.88	2.80
		13,500	11,600	0.86	2.50
		13,950	12,500	0.90	2.75

*Formation fluid pressure gradient.
**Ratio of resistivity of normally pressured shale to observed resistivity of overpressured shale.

istent in many abnormally pressured regions. Thus, it may be necessary to ignore the effect of salinity changes with depth in the formation pore water. When this is done, formation conductivity C_0 or resistivity R_0 can be used as the porosity-dependent parameter in the calculation of formation pore pressure.

A mathematical model of the normal compaction trend of shale conductivity can be obtained by substituting the conductivity ratio, C_w/C_0, for F_R in Eq. 6.28 and assuming a constant value of formation water conductivity, C_w. If this is done, Eq. 6.28 becomes

$$\ln C_{0n} = K_1 - K_2 D, \quad \ldots \ldots \ldots \ldots \ldots \ldots (6.29a)$$

where constants K_1 and K_2 are defined by

$$K_1 = \ln C_w + m \ln \phi_o \quad \ldots \ldots \ldots \ldots \ldots \ldots (6.29b)$$

and

$$K_2 = mK. \quad \ldots \ldots \ldots \ldots \ldots \ldots \ldots \ldots (6.29c)$$

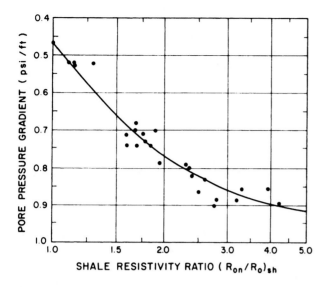

Fig. 6.41—Hottman and Johnson relationship between formation pore pressure and shale resistivity for Miocene and Oligocene formations of the Texas and Louisiana gulf coasts. [20]

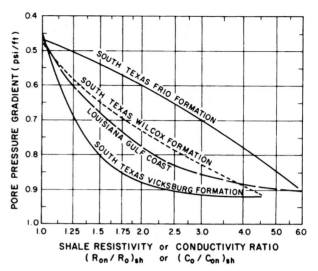

Fig. 6.42—Matthews and Kelly relationship between formation pore pressure and shale resistivity for the south Texas gulf coast. [23]

TABLE 6.20—SHALE RESISTIVITY DATA FROM A WELL DRILLED IN FRIO TREND OF SOUTH TEXAS [23]

Depth (ft)	Shale Conductivity $(10^{-3} \text{ m}/\Omega\text{m}^2)$
2,665	998
3,062	1,020
3,767	1,197
4,273	1,144
4,493	1,225
4,747	1,262
5,100	1,005
5,143	1,206
5,319	1,170
5,639	1,803
5,826	1,311
6,421	1,013
6,498	1,179
6,840	904
6,938	983
7,060	1,005
7,224	877
7,400	904
7,480	714
7,575	794
7,960	812
8,390	698
8,910	693
9,185	595
9,504	560
9,900	703
10,030	1,076
10,320	1,321
10,500	1,252
10,860	1,480
10,990	1,252
11,475	1,831
11,750	1,723
12,235	1,845
12,860	1,404
13,140	1,436
13,460	1,187
13,890	1,351
14,086	1,060
15,000	918

The constants K_1 and K_2 must be chosen on the basis of conductivity data obtained in normally pressured formations in the area of interest. Values of K_2, which were computed from average normal pressure trend lines published by Matthews and Kelly,[23] are given in Table 6.18.

When shale conductivity falls significantly above (or shale resistivity falls significantly below) the normal pressure trend line near the formation of interest, abnormal formation pressure is indicated. The magnitude of the abnormal pressure can be computed with either of the two basic approaches discussed previously for the generalized example illustrated in Fig. 6.10, provided an empirically developed departure curve is available for application of the second technique in the area of interest.

Hottman and Johnson presented one of the first empirical relationships between measured formation pressure in permeable sandstones and formation resistivity in the adjacent shales. Their basic data are listed in Table 6.19 and plotted in Fig. 6.41. Another commonly used empirical relationship published by Mathews and Kelly[23] for the south Texas gulf coast area is shown in Fig. 6.42. Many similar correlations for other areas exist in the private literature of major oil companies.

Example 6.18. The shale conductivity values shown in Table 6.20 were read from a well log recorded in a well drilled in the Frio trend of south Texas. Estimate the formation pressure at a depth of 13,000 ft using the Mathews and Kelly empirical relationship between formation pressure and shale conductivity for this area.

Solution. The shale conductivity data first are plotted vs. depth as shown in Fig. 6.43. The region of normal formation pressure appeared to extend to a depth of 9,500 ft. In determining the normal pressure trend line, emphasis is placed on data obtained in the formations just above the transition zone, which have been altered the least by the drilling fluid and probably have salinities closest to those of the formations of interest. At 9,500 ft, a value of 600×10^{-3} m/$\Omega \cdot$m^2 appeared representative of the normally pressured formations. Use of these values along with an average value for K_2 of 0.000139 (See Table 6.18) in Eq. 6.29a yields

$$K_1 = \ln C_{0n} + K_2 D$$

$$= \ln 600 + 0.000139(9,500)$$

$$= 6.40 + 1.32 = 7.72$$

Thus, the normal pressure trend line is defined by

$$\ln C_{0n} = 7.72 - 0.000139D.$$

After rearrangement into a more convenient form, this equation becomes

$$C_{0n} = 2,250e^{-0.000139D}.$$

The normal pressure trend represented by this equation was drawn on the plot shown in Fig. 6.43 and was found to fit the data accurately just above the transition zone.

At a depth of 13,000 ft, values of 369 and 1,700 are indicated in the plot shown in Fig. 6.43 for C_{0n} and C_0, respectively. Thus, the conductivity ratio is

$$\frac{C_0}{C_{0n}} = \frac{1,700}{369} = 4.6.$$

Use of this value for conductivity ratio in the empirical correlation for the Frio trend of south Texas shown in Fig. 6.42 yields a formation pressure gradient of 0.82 psi/ft. Therefore, the formation pressure is

$$p = 0.82(13,000) = 10,660 \text{ psig.}$$

6.3 Formation Fracture Resistance

When abnormal formation pressure is encountered, the density of the drilling fluid must be increased to maintain the wellbore pressure above the formation pore pressure to prevent the flow of fluids from permeable formations into the well. However, since the wellbore pressure must be maintained below the pressure that will cause fracture in the more shallow, relatively weak, exposed formations just below the casing seat, there is a maximum drilling fluid density that can be tolerated. This means that there is a maximum depth into the abnormally pressured zone to which the well can be drilled safely without cementing another casing string in the well. This is illustrated in Fig. 6.44. Note that for the typical behavior of formation pressure, p_f, and formation fracture pressure, p_{ff}, shown, the mud density, ρ_2, needed to control the formation pressure at D_{max} causes a pressure at the existing casing seat just below the fracture pressure. Thus, a knowledge of the pressure at which

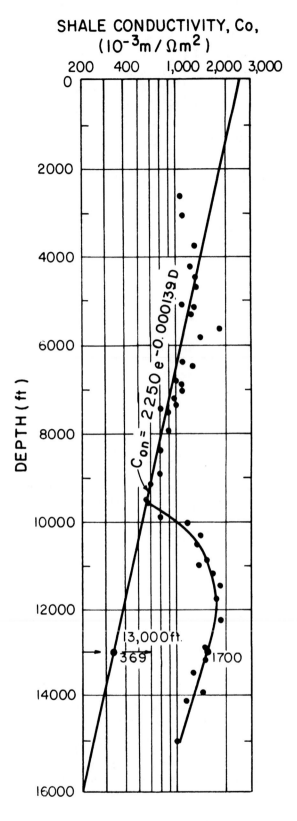

Fig. 6.43—Example shale conductivity plot for Frio trend, south Texas.

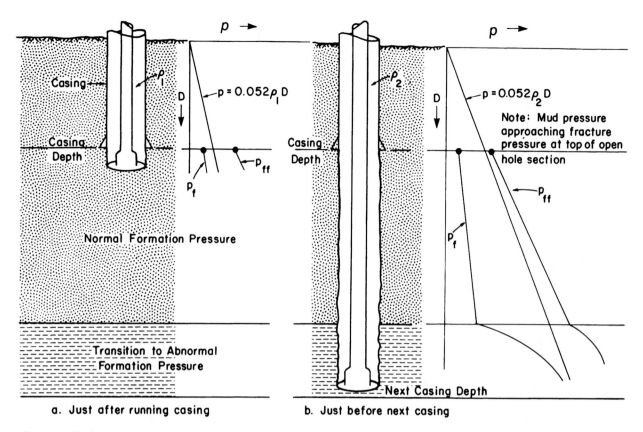

a. Just after running casing   b. Just before next casing

Fig. 6.44—Typical behavior of formation pressure, well pressure, and formation fracture resistance in abnormally pressured well.

formation fracture will occur at all depths in the well is essential for planning and drilling a well into abnormally pressured formations.

To understand underground stresses that resist formation fracture, consider the geologic processes that have occurred. One of the simplest and most common subsurface stress states occurs in relatively young sediments laid down in a deltaic depositional environment (Fig. 6.45). As deposition continues and the vertical matrix stress σ_z increases because of the increased loading at the grain-to-grain contacts, the sediments tend to expand laterally, but essentially are prevented from doing so by the surrounding rock. This tendency causes horizontal matrix stresses that are transmitted laterally through grain-to-grain contact points. If we designate as principal matrix stresses those stresses that are normal to planes with no shear, the general subsurface stress condition can be defined in terms of σ_x, σ_y, and σ_z, as shown in Fig. 6.45.

In a relatively relaxed geologic region, such as a young deltaic sedimentary basin, the horizontal matrix stresses σ_x and σ_y tend to be approximately equal and much smaller than the vertical matrix stress, σ_z. If the sediments are assumed to behave elastically, the horizontal strain, ϵ_x, can be expressed using Hooke's law:

$$\epsilon_x = \frac{\sigma_x}{E} - \mu \frac{\sigma_y}{E} - \mu \frac{\sigma_z}{E},$$

where E is Young's modulus of elasticity and μ is Poisson's ratio. For compressed rock caused by

sedimentation, the horizontal strain, ϵ_x, is essentially zero, and, since the horizontal stresses σ_x and σ_y are approximately equal,

$$\sigma_x = \sigma_y = \sigma_H = \frac{\mu}{1-\mu} \sigma_z, \quad \ldots\ldots\ldots\ldots\ldots (6.30)$$

where σ_H denotes the average horizontal stress. For measured values of μ for consolidated sedimentary rocks,[25] which range from 0.18 to 0.27, the horizontal matrix stress varies from 22% to 37% of the vertical matrix stress. However, if the assumption of elastic rock behavior is not valid, the horizontal matrix stress is higher.

The relative magnitude of the horizontal and vertical matrix stresses can be inferred from naturally occurring fracture patterns in the geologic regions. In geologic regions such as the Louisiana gulf coast, where normal faulting occurs, the horizontal matrix stress tends to be considerably smaller than the vertical matrix stress — usually between 25 and 50% of the vertical matrix stress. On the other hand, in regions that are being shortened, either by folding or thrust faulting, such as California, the horizontal matrix stress tends to be much larger than the vertical matrix stress—between 200 and 300% of the vertical matrix stress. Of course, local structures can cause departures from such regional trends. For example, the stresses near a salt dome in the Louisiana gulf coast area may be altered considerably.

Hydraulic fracturing of rock is a complex phenomenon that is very difficult to describe mathematically. To present the basic principles involved, we consider first a

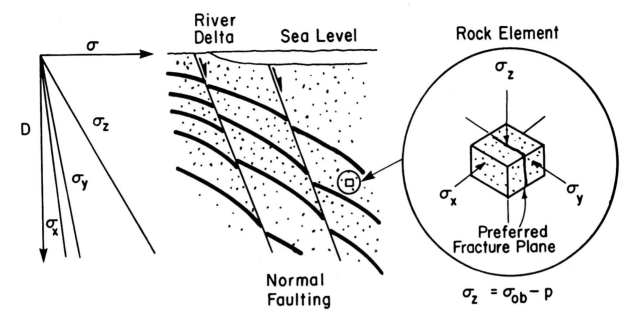

Fig. 6.45—Example underground-stress distribution in relatively young deltaic sediments.

very simplified situation in which a nonpenetrating fracture fluid is introduced into a small cavity located in the center of the rock element (Figs. 6.45 and 6.46) that is assumed to have zero tensile strength. A nonpenetrating fluid is one that will flow into the created fracture but will not flow a significant distance into the pore spaces of the rock. For fracture fluid to enter the cavity, the pressure of the fracture fluid must exceed the pressure of the formation fluid in the pore spaces of the rock. As the pressure of the fracture fluid is increased above the formation pore pressure, the rock matrix begins to be compressed. As shown in Fig. 6.46, the compression is greatest in the direction of the minimum matrix stress. When the fracture fluid pressure exceeds the sum of the minimum matrix stress and the pore pressure, parting of the rock matrix occurs and the fracture propagates. The preferred fracture orientation is perpendicular to the least principal stress.

A cylindrical borehole through the formation significantly alters the horizontal state of stress near the borehole. Hubbert and Willis[24] made an approximate calculation of the stress concentration near the borehole by assuming a smooth and cylindrical borehole with the axis vertical and parallel to one of the three principal stresses. The elastic theory was applied for stresses in an infinite impermeable plate with regional stresses σ_x and σ_y and containing a circular hole with its axis perpendicular to the plate. It was found that horizontal matrix stresses at the borehole wall could be much higher than the undisturbed regional horizontal matrix stresses, but in all cases, the stress concentrations were quite local and rapidly approached the undisturbed regional stresses within a few hole diameters. Thus, once a fracture is propagated a small distance from the wellbore, the *fracture extension pressure* is controlled by the undisturbed minimum regional stress. For the case in which horizontal matrix stresses σ_x and σ_y were equal, the circumferential stress at the borehole wall was twice the regional horizontal stress.

More rigorous mathematical treatments of hydraulic fracturing within a wellbore by both penetrating and nonpenetrating fluids have been developed.[26,27] In addition, equations have been developed for directional wells in which the wellbore axis is not parallel to any of the directions of principal stresses.[26] Unfortunately, these more complex solutions have not been used widely because required information about the principal stresses and formation characteristics is generally not available. Also, the equations for directional wells are extremely lengthy and not conveniently used without a computer.

6.4 Methods for Estimating Fracture Pressure

Prior knowledge of how formation fracture pressure varies with depth can be just as important as prior knowledge of how the formation pore pressure varies with depth when planning and drilling a deep well that will penetrate abnormal formation pressures. Techniques for determining formation fracture pressure, like those for determining pore pressure, include (1) predictive methods, and (2) verification methods. Initial well planning must be based on formation fracture data obtained by a predictive method. After casing is cemented in place, the anticipated fracture resistance of the formations just below the casing seat must be verified by a pressure test before drilling can be continued to the next planned casing depth.

6.4.1 Prediction of Fracture Pressure

Estimates of formation fracture pressure made before setting casing in the well are based on empirical correlations. Since formation fracture pressure is affected greatly by the formation pore pressure, one of the previously described pore pressure prediction methods must be applied before use of a fracture pressure correlation. The more commonly used fracture pressure equations and

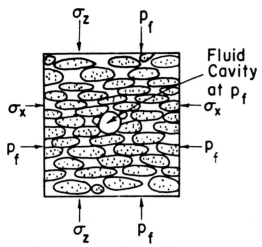

a. Pressure = Pore Pressure

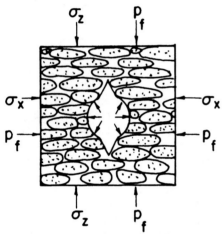

b. Pressure > Pore Pressure

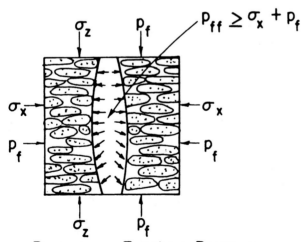

c. Pressure = Fracture Pressure

Fig. 6.46—Fracture initiation opposes least principle stress.

correlations include: (1) the Hubbert and Willis equation, (2) the Mathews and Kelly correlation, (3) the Pennebaker correlation, (4) the Eaton correlation, (5) the Christman equation, and (6) the MacPherson and Berry correlation.

Hubbert and Willis Equation. Hubbert and Willis[24] introduced many fundamental principals that are still used widely today. The minimum wellbore pressure required to extend an existing fracture was given as the pressure needed to overcome the minimum principal stress:

$$p_{ff} = \sigma_{min} + p_f. \qquad (6.31a)$$

Since the earth is so inhomogeneous and anisotropic, with many existing joints and bedding planes, this fracture extension pressure generally is used for well planning and casing design. However, if the minimum principal stress occurs in the horizontal plane and if horizontal stresses σ_x and σ_y are equal, the local stress concentration at the borehole wall, σ_{Hw}, is twice the regional horizontal stress, σ_H. Thus, the pressure required to initiate fracture in a homogeneous, isotropic formation is

$$p_{ff} = \sigma_{Hw} + p_f = 2\sigma_H + p_f. \qquad (6.31b)$$

On the basis of laboratory experiments analyzed using the Mohr failure criteria presented in Chap. 5, Hubbert and Willis concluded that in regions of normal faulting, such as the U.S. gulf coast area, the horizontal matrix stress is the minimum stress. It was also concluded that the minimum matrix stress in the shallow sediments is approximately one-third the vertical matrix stress resulting from weight of the overburden. Thus, the fracture extension pressure for this situation is approximately

$$p_{ff} = \sigma_{min} + p_f = \frac{\sigma_z}{3} + p_f.$$

Since the matrix stress σ_z is given by

$$\sigma_z = \sigma_{ob} - p_f,$$

the fracture extension pressure is expressed by

$$p_{ff} \approx (\sigma_{ob} + 2p_f)/3. \qquad (6.32)$$

Example 6.19. Compute the maximum mud density to which a normally pressured U.S. gulf coast formation at 3,000 ft can be exposed without fracture. Use the Hubbert and Willis equation for fracture extension. Assume an average surface porosity constant ϕ_o of 0.41, a porosity decline constant K of 0.000085, and an average grain density ρ_g of 2.60, as computed in Example 6.2.

Solution. The U.S. gulf coast area has regional normal faults, which indicates that the horizontal matrix stress is much smaller than the vertical matrix stress. Hubbert and Willis presented Eq. 6.32 as an approximate relationship for shallow formations in this type of geologic region.

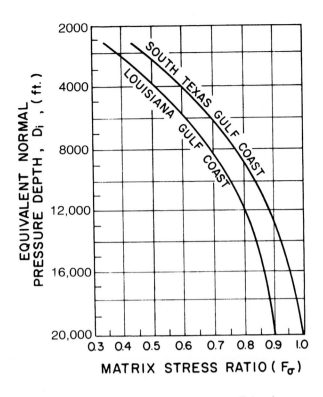

Fig. 6.47—Mathews and Kelly matrix-stress coefficient for normally pressured formations.[23]

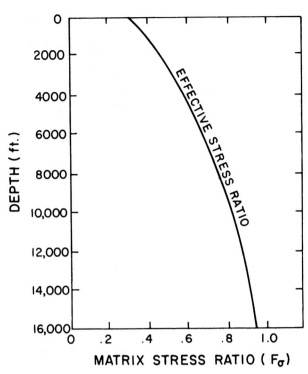

Fig. 6.48—Pennebaker correlation for effective stress ratio.[6]

Also, from Table 6.2, the normal formation pressure gradient is 0.465 psi/ft or 1.074 g/cm³. The vertical overburden stress, σ_{ob}, is computed by Eq. 6.6.

$$\sigma_{ob} = 0.052 \,(2.6)\,(8.33)\,(3,000)$$

$$- \frac{0.052\,(2.6 - 1.074)\,(8.33)\,(0.41)}{0.000085}$$

$$\cdot \left[1 - e^{-0.000085\,(3,000)}\right]$$

$$= 3,378 - 718 = 2,660 \text{ psig.}$$

The formation pore pressure is given by

$$p_f = 0.465\,(3,000) = 1,395 \text{ psig.}$$

Thus, the fracture pressure given by Eq. 6.32 is

$$p_{ff} \approx \left[2,660 + 2\,(1,395)\right]/3 = 1,817 \text{ psig.}$$

This pressure at 3,000 ft would exist in a static well for a drilling fluid density of

$$\rho = \frac{1,817}{(0.052)\,(3,000)} = 11.6 \text{ lbm/gal.}$$

Matthews and Kelly Correlation. Drilling experience showed that formation fracture gradients increased with depth, even in normally pressured formations, and that

Eq. 6.32 was generally not valid for the deeper formations. Matthews and Kelly[23] replaced the assumption that the minimum matrix stress was one-third the overburden stress by

$$\sigma_{min} = F_\sigma \sigma_z, \quad \dots \dots \dots \dots \dots \dots \dots (6.33)$$

where the matrix stress coefficient F_σ was determined empirically from field data taken in normally pressured formations. Fig. 6.47 shows the empirical correlations that were presented for the south Texas gulf coast and Louisiana gulf coast areas. To use these correlation curves for abnormally pressured formations, the depth D_i at which a normally pressured formation would have the same vertical matrix stress as the abnormally pressured formation of interest is used in Fig. 6.47 instead of actual depth when determining the matrix stress coefficient, F_σ. For simplicity, an average overburden stress σ_{ob} of 1.0 psi/ft and an average normal pressure gradient of 0.465 psi/ft are assumed. Thus, the normal vertical matrix stress becomes

$$\sigma_n = \sigma_{ob} - p_{fn} = D_i - 0.465\,D_i$$

$$= 0.535\,D_i.$$

The depth D_i at which a normally pressured formation has the vertical matrix strength present in the abnormally pressured formation of interest is

$$D_i = \frac{\sigma_z}{0.535} = \frac{\sigma_{ob} - p_f}{0.535} = \frac{D - p_f}{0.535}. \quad \dots \dots \dots (6.34)$$

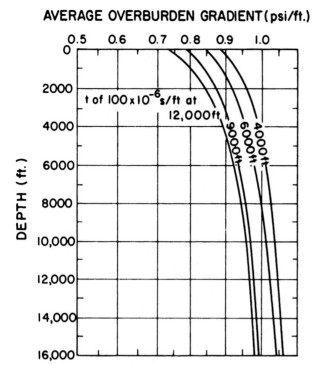

Fig. 6.49—Pennebaker correlation for vertical overburden stress.[6]

Example 6.20. A south Texas gulf coast formation at 10,000 ft was found to have a pore pressure of 8,000 psig. Compute the formation fracture gradient using the Matthews and Kelly correlation.

Solution. The equivalent depth of a normal pressure formation is computed with Eq. 6.34:

$$D_i = \frac{10,000 - 8,000}{0.535} = 3,738 \text{ ft.}$$

Entering Fig. 6.47 at a depth of 3,738 ft yields a value of matrix stress coefficient F_σ of 0.59. Thus, from Eq. 6.33, the minimum matrix stress is

$$\sigma_{min} = F_\sigma \sigma_z = 0.59 \, (10,000 - 8,000)$$

$$= 1,180 \text{ psig.}$$

The fracture pressure is given by Eq. 6.31:

$$p_{ff} = \sigma_{min} + p_f = 1,180 + 8,000 = 9,180 \text{ psig.}$$

Thus, the fracture gradient is

$$\frac{9,180}{10,000} = 0.918 \text{ psi/ft.}$$

Pennebaker Correlation. The Pennebaker correlation[6] is similar to the Matthews and Kelly correlation in that Eq. 6.33 is used to compute the minimum matrix stress. Pennebaker called the coefficient F_σ the effective stress ratio and correlated this ratio with depth, regardless of

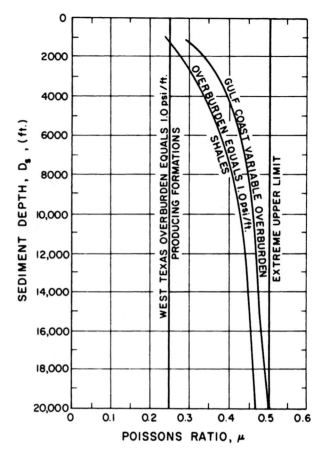

Fig. 6.50—Eaton correlation for Poisson's ratio.[1]

pore pressure gradient. Thus, the actual depth of the formation always is used in the Pennebaker correlation, which is shown in Fig. 6.48. Field data for this correlation come primarily from the south Texas gulf coast area. Pennebaker did not assume a constant value for vertical overburden stress σ_{ob}, and developed the correlation shown in Fig. 6.49 for determining this parameter. The effect of geologic age on overburden stress is taken into account by a family of curves for various depths at which the seismic-derived interval transit time is 100 μs/ft.

Example 6.21. A south Texas gulf coast formation at 10,000 ft was found to have a pore pressure of 8,000 psi. Seismic records indicate an interval transit time of 100 μs/ft at a depth of 6,000 ft. Compute the formation fracture gradient using the Pennebaker correlation.

Solution. Entering Fig. 6.48 at a depth of 10,000 ft yields a value of effective stress ratio F_σ of 0.82. Entering Fig. 6.49 at a depth of 10,000 ft and using the 6,000-ft depth line for 100 μs/ft yields a value of vertical overburden gradient, σ_{ob}, of 1.02 psi/ft. Use of these values in Eq. 6.33 gives

$$\sigma_{min} = F_\sigma \sigma_z = F_\sigma (\sigma_{ob} - p_f)$$

$$= 0.82[(1.02)(10,000) - 8,000]$$

$$= 1,804 \text{ psig.}$$

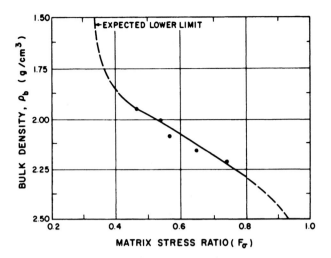

Fig. 6.51—Christman correlation for effective stress ratio.[28]

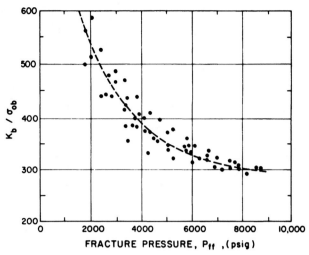

Fig. 6.52—MacPherson and Berry correlation for fracture pressure.

The fracture pressure is given by Eq. 6.31:

$$p_{ff} = \sigma_{min} + p_f = 1,804 + 8,000 = 9,804 \text{ psig.}$$

Thus, the fracture gradient is

$$\frac{9,804}{10,000} = 0.98 \text{ psi/ft.}$$

Eaton Correlation. The Eaton correlation assumes that the relationship between horizontal and vertical matrix stress is described accurately by Eq. 6.30. Values of Poisson's ratio needed to predict observed fracture gradients were computed from field data, resulting in the correlation shown in Fig. 6.50. Data from west Texas and the Texas and Louisiana gulf coast were included in the analysis. The gulf coast data were analyzed assuming (1) a constant vertical overburden stress σ_{ob} of 1.0 psi/ft and (2) a variable overburden stress σ_{ob} obtained by integration of bulk density logs. The resulting correlations are shown in Fig. 6.50.

Example 6.22. An offshore Louisiana well to be drilled in 2,000 ft of water will penetrate a formation at 10,000 ft (subsea) having a pore pressure of 6,500 psig. Compute the fracture gradient of the formation assuming the semisubmersible used to drill the well will have an 80-ft air gap between the drilling fluid flow line and the sea surface. Compute the vertical overburden gradient assuming a seawater density of 8.5 lbm/gal, an average sediment grain density of 2.6 g/cm^3, a surface porosity ϕ_o of 0.45, and a porosity decline constant of 0.000085 ft^{-1}.

Solution. At a depth of 10,000 ft subsea, the depth into the sediments is 8,000 ft. Entering Fig. 6.50 with a depth of 8,000 ft yields a value of Poisson's ratio of

0.44. The vertical overburden gradient is given by Eq. 6.6:

$$\sigma_{ob} = 0.052 \, (8.5) \, (2,000)$$

$$+ 0.052 \, (2.6) \, (8.33) \, (8,000)$$

$$- \frac{0.052 \, (2.6 - 1.074) \, (8.33) \, (0.45)}{0.000085}$$

$$\cdot [1 - e^{-0.000085 \, (8,000)}]$$

$$= 884 + 9,010 - 1,727 = 8,167 \text{ psig.}$$

The horizontal matrix stress, which is the minimum matrix stress, is computed by Eq. 6.30:

$$\sigma_{min} = \frac{\mu}{1-\mu} \sigma_z$$

$$= \frac{0.44}{1-0.44} (8,167 - 6,500) = 1,310 \text{ psig.}$$

The fracture pressure is given by Eq. 6.31:

$$p_{ff} = 1,310 + 6,500 = 7,810 \text{ psig.}$$

Thus, the fracture gradient is

$$\frac{7,810}{10,000 + 80} = 0.775 \text{ psi/ft.}$$

Christman Correlation. Christman,[28] working in the Santa Barbara channel off the California coast, found that stress ratio F_σ could be correlated to the bulk density of the sediments. The Christman correlation is shown in

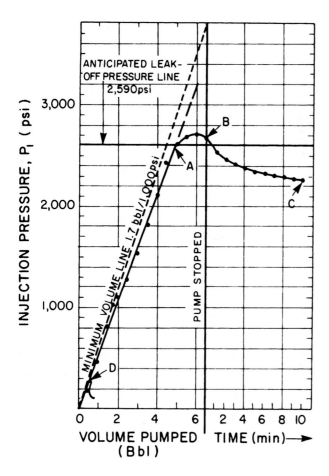

Fig. 6.53—Example leakoff test results taken after drilling the first sand below the casing seat.[30]

Fig. 6.51. The bulk density of the sediments tends to increase with increasing depth, overburden stress, and geologic age. All of these variables appear to affect the formation fracture gradient.

Example 6.23. Apply the Christman correlation obtained in the offshore California area to the offshore Louisiana well described in Example 6.22.

Solution. The average porosity of the sediments at a depth of 10,000 ft is

$$\phi = \phi_o e^{-KD} = 0.45 \, e^{-0.000085 \, (10,000)}$$

$$= 0.192.$$

This corresponds to a bulk density of

$$\rho_b = \rho_{fl}\phi + \rho_g (1-\phi)$$

$$= 1.074 \, (0.192) + 2.6 \, (1-0.192) = 2.31 \text{ g/cm}^3.$$

Entering the Christman correlation shown in Fig. 6.51 with a bulk density of 2.31 gives a value of 0.8 for stress ratio F_σ. Since the vertical overburden stress σ_{ob} is 8,167 psig and the pore pressure p_f is 6,500 psig (see Example 6.22),

$$\sigma_{min} = 0.8 \, (8,167 - 6,500) = 1,334 \text{ psig}$$

---

and

$$p_{ff} = 1,334 + 6,500 = 7,834 \text{ psig}.$$

Thus, the fracture gradient is

$$\frac{7,834}{10,000 + 80} = 0.777 \text{ psi/ft.}$$

MacPherson and Berry Correlation. With a novel approach, MacPherson and Berry[29] developed a correlation between elastic modulus K_b for a compressional wave and formation fracture pressure. Using measurements of interval transit time by means of a sonic log and bulk density by means of a density log, the elastic modulus K_b is computed using the following equation.

$$K_b = 1.345 \times 10^{10} \frac{\rho_b}{t^2}. \quad \ldots\ldots\ldots\ldots\ldots (6.35)$$

An empirical correlation between K_b/σ_{ob} and fracture pressure developed for the offshore Louisiana area is shown in Fig. 6.52.

Example 6.24. The interval transit time in an abnormally pressured sand formation at 8,000 ft was 105 μs/ft. The bulk density log gave a reading of 2.23 g/cm^3. Vertical overburden stress σ_{ob} is 7,400 psig. Compute the fracture pressure using the MacPherson and Berry correlation.

Solution. The elastic modulus can be computed by Eq. 6.35:

$$K_b = 1.345 \times 10^{10} \frac{2.23}{(105)^2} = 2,720,000 \text{ psi.}$$

The K_b/σ_{ob} ratio is

$$\frac{2,720,000}{7400} = 368.$$

Use of this K_b/σ_{ob} ratio in the MacPherson and Berry correlation shown in Fig. 6.52 gives a fracture pressure of about 4,500 psig.

6.4.2 Verification of Fracture Pressure

After each casing string is cemented in place, a pressure test called a *leakoff test* is used to verify that the casing, cement, and formations below the casing seat can withstand the wellbore pressure required to drill safely to the next depth at which casing will be set. In general, a leakoff test is conducted by closing the well at the surface with a blowout preventer and pumping into the closed well at a constant rate until the test pressure is reached or until the well begins to take whole mud, causing a departure from the increasing pressure trend. The pump then is stopped and the pressure is observed for at least 10 minutes to determine the rate of pressure decline. The casing is tested for leaks in this manner before the cement is drilled from the bottom joints. The cement and formations just below the casing seat are tested in this manner after the cement is drilled from the bottom joints of casing and about 10 ft into the formations below the casing seat. Subsequent tests can be conducted periodically after drilling through formations that may have a lower fracture gradient. A common practice in the U.S. gulf coast area is to pressure test the first sand below the casing seat since the fracture gradient is often lower for sandstone than for shale.

The results of a typical leakoff test are shown in Fig. 6.53 for a well that has a short section of open hole exposed. As shown, there is a constant pressure increase for each incremental drilling fluid volume pumped, so that the early test results fall on a relatively straight line. The straight line trend continues until Point A, where the formation grains start to move apart and the formation begins to take whole mud. The pressure at Point A is called the leakoff pressure and is used to compute the formation fracture gradient. Pumping is continued during the leakoff test long enough to ensure that the fracture pressure has been reached. At Point B, the pump is stopped, and the well left shut in to observe the rate of pressure decline. The rate of pressure decline is indicative of the rate at which mud or mud filtrate is being lost.

Also shown in Fig. 6.53 are lines corresponding to the anticipated leakoff pressure and the anticipated slope line for the early test results. These lines are extremely helpful to the person conducting the leakoff test while the test is in progress.

The anticipated surface leakoff pressure is based on the formation fracture pressure predicted by one of the empirical correlations presented in the previous section. The anticipated surface leakoff pressure, p_{lo}, is given by

$$p_{lo} = p_{ff} - 0.052\rho D + \Delta p_f, \quad \dots\dots\dots\dots (6.36)$$

where Δp_f is the frictional pressure loss in the well between the surface pressure gauge and the formation during the leakoff test. This equation also is used to compute the observed fracture pressure, p_{ff}, from the observed leakoff pressure, p_{lo}. Since leakoff tests usually are conducted at a low pump rate, the frictional pressure-loss term is small and is often neglected. Chenevert[30] recommends using the pressure required to break the gel strength and initiate circulation of the well for the frictional pressure loss. This can be done by use of Eqs. 4.77 and 4.78 presented in Chap. 4, Sec. 12.

When using Eqs. 4.77 and 4.78, it is difficult to obtain representative values for gel strength, τ_g. Normally, this parameter is obtained in a rotational viscometer after the mud has been quiescent for 10 minutes. This method has been criticized because it is not performed at downhole temperature and pressure and does not reflect the properties of any contaminated mud that may be in the annulus. To avoid this problem, the gel strength may be computed from observed pump pressure required to initiate circulation of the well after a 10-minute quiescent period (Point D in Fig. 6.53). Circulation is initiated using the same pump rate used in the leakoff test and the additional test is run just after the leakoff test is performed. The

TABLE 6.21—AVERAGE COMPRESSIBILITY VALUES FOR DRILLING FLUID COMPONENTS

Component	Compressibility (psi^{-1})
water	3.0×10^{-6}
oil	5.0×10^{-6}
solids	0.2×10^{-6}

pressure required to initiate circulation is obtained by combining Eqs. 4.77 and 4.78:

$$p = \frac{\tau_g D}{300\, d} + \frac{\tau_g D}{300\,(d_2 - d_1)}.$$

Solving for gel strength τ_g yields

$$\tau_g = \frac{300\, p\, d\,(d_2 - d_1)}{D(d + d_2 - d_1)}. \quad \dots\dots\dots\dots (6.37)$$

The anticipated slope line for the early leakoff test results is determined from the compressibility of the system. The compressibility caused by the expansion of the casing and borehole is small compared with the compressibility of the drilling fluid and can be neglected. The effective compressibility, c_e, of drilling fluid composed of water, oil, and solids having compressibilities c_w, c_o, and c_s, respectively, is given by

$$c_e = c_w f_w + c_o f_o + c_s f_s, \quad \dots\dots\dots\dots (6.38)$$

where f_w, f_o, and f_s denote the volume fractions of water, oil, and solids. Since compressibility is defined by

$$c_e = -\frac{1}{V}\frac{dV}{dp}$$

and since the volume pumped is approximately equal in magnitude and opposite in sign to the change in volume of the drilling fluid already in the well, the slope of the pressure leakoff plot is given by

$$\left(\frac{dp}{dV}\right) = -\frac{1}{c_e V}, \quad \dots\dots\dots\dots (6.39)$$

where V is the initial drilling fluid volume in the well. Approximate compressibility values for water, oil, and solids are given in Table 6.21.

When a leakoff test is conducted, a pump rate should be selected that yields early test results only slightly lower than the anticipated slope line. If too slow a pump rate is used, filtration fluid losses mask the effect of other leaks. Pumping rates between 0.25 bbl/min and 1.50 bbl/min are typical, with the higher rates applicable to tests conducted with large intervals of open hole. A small pump such as a cementing pump provides good flow-rate control over this flow-rate range. Several tests may be required to obtain meaningful results. Results of a properly run leakoff test that indicated a poor cement bond are shown in Fig. 6.54. Such results indicate that the casing shoe should be squeeze-cemented before continuing with the drilling operations.

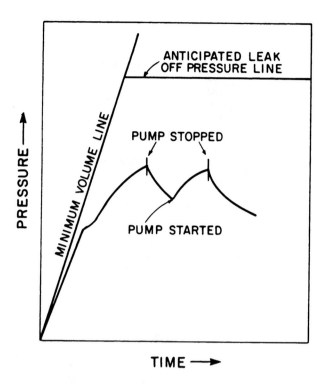

Fig. 6.54—Leakoff test results indicative of a poor cement bond.

Before a leakoff test is initiated, the well should be circulated until the drilling fluid density is uniform throughout the well. This should be verified by removing the kelly and observing a static column of fluid both in the drillstring and in the annulus. Cuttings in the annulus or a slug of heavy mud in the drillpipe can cause density differentials in the well, which will introduce errors in the fracture pressure determined by a leakoff test.

After the conclusion of a leakoff test, a good practice is to monitor the volume of drilling fluid bled from the well when the pressure is released. The volume recovered should be approximately equal to the total volume injected if only filtration fluid losses occurred. The fluid volume recovered thus provides an additional check on the observed pressure behavior.

Many operators prefer not to test a formation to the point of fracture, fearing that such a test will lower the fracture resistance of the formation. However, the fracture resistance of a formation results almost entirely from the stresses created by the compressive pressure of the surrounding rock. The tensile strength of most rocks is so small that it can be neglected. Also, naturally occurring fissures and fractures generally are present. Once the wellbore pressure is released, the fracture will close. Essentially the same fracture pressure will be required again to overcome the compressional stress holding the fracture closed.

Example 6.25. The leakoff test shown in Fig. 6.53 was conducted in 9.625-in. casing having an internal diameter of 8.835 in., which was cemented at 10,000 ft. The test was conducted after drilling to 10,030 ft—the depth of the first sand—with an 8.5-in. bit. Drillpipe having an external diameter of 5.5 in. and an internal diameter of 4.67 in. was placed in the well to a depth of

10,000 ft for the test. A 13.0 lbm/gal water-based drilling fluid containing no oil and having a total volume fraction of solids of 0.20 was used. The gel strength of the mud was 10 lbm/100 sq ft. Verify the anticipated slope line shown in Fig. 6.53 and compute the formation fracture pressure.

Solution. The effective compressibility is computed with Eq. 6.38:

$$c_e = c_w f_w + c_o f_o + c_s f_s$$
$$= (3.0 \times 10^{-6})(0.8) + 0 + (0.2 \times 10^{-6})(0.2)$$
$$= 2.44 \times 10^{-6} \text{ psi}^{-1}.$$

The capacity of the annulus, drillstring, and open hole are

$$A_a = 0.97135 \times 10^{-3} (8.835^2 - 5.5^2)$$
$$= 0.0464 \text{ bbl/ft,}$$
$$A_{dp} = 0.97135 \times 10^{-3} (4.67)^2 = 0.0212 \text{ bbl/ft,}$$

and

$$A_h = 0.97135 \times 10^{-3} (8.5)^2 = 0.0702 \text{ bbl/ft.}$$

The volume of drilling fluid in the well is

$$V = 0.0464 (10,000) + 0.0212 (10,000) + 0.0702 (30)$$
$$= 678 \text{ bbl.}$$

Thus, the anticipated slope line predicted by Eq. 6.39 is

$$\frac{dp}{dV} = \frac{1}{678 \text{ bbl} (2.44 \times 10^{-6}) \text{ psi}^{-1}}$$
$$= 604 \text{ psi/bbl.}$$

The frictional pressure loss is assumed approximately equal to the pressure needed to break circulation. Since flow was down a drillpipe having a 4.67-in. internal diameter, the pressure drop predicted by Eq. 4.77 is

$$\Delta p_f \approx \frac{\tau_g D}{300 d} = \frac{10 (10,000)}{300 (4.67)} = 71 \text{ psi.}$$

The fracture pressure is obtained by use of Eq. 6.36. Using the leakoff pressure of 2,540 psig shown in Fig. 6.53 yields

$$p_{ff} = p_{lo} + 0.052 \rho D - \Delta p_f$$
$$= 2,540 + 0.052 (13)(10,015) - 71 = 9,239 \text{ psig.}$$

Exercises

6.1 Compute the normal formation pressure expected at a depth of 8,000 ft for these areas: west Texas, U.S. gulf coast, California, Rocky Mountains, and Anadarko Basin. *Answer:* 3,464; 3,720; 3,512; 3,488; and 3,464 psig.

6.2 Determine values for surface porosity ϕ_o and porosity decline constant K for the Santa Barbara channel. Use the average bulk density data shown in Fig. 6.55, an average grain density of 2.60, and an average pore fluid density of 1.014 g/cm³. *Answer:* $\phi_o = 0.34$; $K = 0.00019$ ft⁻¹.

6.3 Show that substitution of the exponential porosity expression defined by Eq. 6.4 into Eq. 6.7 yields the normal compaction model given by Eq. 6.8.

6.4 Compute the vertical overburden stress σ_{ob} and the vertical matrix stress σ_z resulting from geostatic load in a normally pressured formation of the U.S. gulf coast area at depths of: 500, 1,000, 2,000, 4,000, and 8,000 ft. Assume a water depth of zero. *Answer:* $(\sigma_{ob})_{500} = 430$ psig; $(\sigma_z)_{500} = 198$ psig; $(\sigma_{ob})_{8,000} = 7,439$ psig; and $(\sigma_z)_{8,000} = 3,716$ psig.

6.5 A tilted gas sand encountered at 4,500 ft is known to have a pore pressure of 2,700 psig. A well is to be drilled near the top of the structure, which is expected to penetrate the sand at 3,500 ft. The gas is known to have a density of 1.0 lbm/gal at reservoir conditions. Compare the mud density required to drill the second well safely with that of the first. *Answer:* $\rho_2 = 14.5+$ lbm/gal; $\rho_1 = 11.5+$ lbm/gal.

6.6 Discuss three situations that can lead to abnormally pressured shallow formations as a result of upward fluid migration.

6.7 Graph the function developed in Example 6.5 between average interval transit time and a depth for normally pressured U.S. gulf coast sediments. Use a depth interval of 0 to 30,000 ft and (a) semilogarithmic graph paper (depth on linear scale) and (b) logarithmic graph paper.

6.8 a. Develop an equation for the normal pressure trend line for the interval transit time data of Table 6.4 assuming a straight-line representation on semilogarithmic graph paper. *Answer:* $t_n = 161e^{-0.000043D}$.

b. Compare results with plots obtained in Exercise 6.7.

c. Compute the porosity of a normally pressured shale at 28,000 ft using the straight-line interval transit time extrapolation. *Answer:* negative ϕ is predicted.

6.9 a. Develop an equation for the normal pressure trend line for the interval transit time data of Table 6.4 assuming a straight-line representation on logarithmic graph paper. *Answer:* $t_n = 1,100D^{-0.257}$.

b. Compare results with plots obtained in Exercise 6.7.

c. Compute the porosity of a normally pressured shale at 28,000 ft using the straight-line extrapolation. *Answer:* 0.09.

6.10 a. Using the straight-line relationship developed in Exercise 6.9, derive an equation for computing pore pressure from interval transit time ratio t/t_n. Use the equivalent matrix stress concept and assume a vertical overburden stress σ_{ob} of 1.0 psi/ft and a normal formation pore pressure gradient of 0.465 psi/ft. *Answer:*

$$p = D\left[1 - 0.535\left(\frac{1}{t/t_n}\right)^{3.89}\right].$$

b. Compare results obtained in Part a with Fig. 6.13.

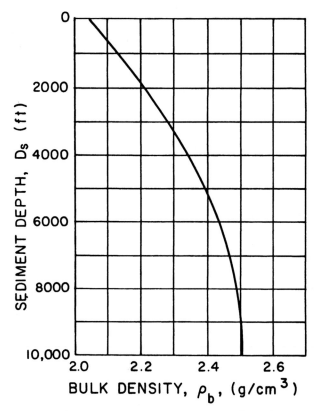

Fig. 6.55—Bulk-density curve from density logs, Santa Barbara (CA) channel.[1]

6.11 Using the data given in Example 6.6 and the Pennebaker correlation presented in Fig. 6.13, compute the formation pressure at 10,000, 11,000, and 12,000 ft. *Answers:* 9,600, 10,600, and 11,600 psig.

6.12 Develop a graphical overlay for reading pore pressures from a plot of interval transit time vs. depth. Assume a normal compaction trend line as plotted previously in Exercise 6.7a. Use the Pennebaker correlation for relating the interval transit time departure to pore pressure.

6.13 a. Compute the pore pressure at 13,000 ft using the data of Fig. 6.19 and the equivalent matrix stress concept. Use an overburden stress of 1.0 psi/ft and a normal pore pressure gradient of 0.456 psi/ft. *Answer:* 12,570 psig.

b. Rehm and McClendon indicated that the use of an equivalent matrix stress concept with d-exponent data resulted in an inaccurately high pore pressure value. Do the data answers in Part a support this statement? *Answer:* Yes.

6.14 The average interval transit time data shown in Table 6.22 were computed from seismic records at a proposed well location in the Pleistocene trend of the offshore Louisiana area. Using the mathematical model for the normal compaction trend developed in Example 6.5, estimate the formation pressure at 1,000-ft depth increments using the equivalent matrix stress concept and the Pennebaker correlation presented in Fig. 6.13.

6.15 Using the data given in Example 6.8, construct a plot of pore pressure vs. depth. Compute the pore pressure using the method of Zamora.

6.16 Repeat Exercise 6.15 using the method of Rehm and McClendon.

TABLE 6.22—AVERAGE INTERVAL TRANSIT TIME DATA COMPUTED FROM SEISMIC RECORDS OBTAINED AT A PROPOSED WELL LOCATION IN THE PLEISTOCENE TREND, OFFSHORE LOUISIANA [12]

Depth (ft)	Average Interval Transit Time (10^{-6} s/ft)
1,500 to 2,500	160
2,500 to 3,000	147
3,000 to 3,750	140
3,750 to 4,250	137
4,250 to 5,500	121
5,500 to 6,500	117
6,500 to 7,500	112
7,500 to 8,500	113
8,500 to 9,500	115
9,500 to 10,500	115
10,500 to 11,500	118
11,500 to 12,500	118

6.17 The penetration rate obtained in shale at 12,000 ft decreased from 20 to 8 ft/hr when the mud density was increased by 1.0 lbm/gal. Estimate the effective value of a_4. *Answer:* 33×10^{-6}.

6.18 At a depth of 10,000 ft in the U.S. gulf coast area, a value of 3.0 was obtained for the modified drillability parameter $K_p{}'$ when drilling a shale formation thought to have a pore pressure gradient of 11.5 lbm/gal. The normal pressure trend line value of $K_p{}'$ was 2.0. If the value of a_4 is known to be 35×10^{-6}, what is the value of a_3? *Answer:* 87×10^{-6}.

6.19 Using the data given in Example 6.10, construct a plot of pore pressure vs. depth with the method of Bourgoyne and Young.

6.20 The data in Table 6.23 were taken in shale on a well drilled in south Louisiana.

a. Using the short-interval drilling data of Table 6.23 between 10,000 and 10,050 ft, estimate values for a_5, a_6, a_7, and a_8. *Answer:* 0.9; 0.5; 1.2; and 0.3.

b. Make a plot of penetration rate vs. depth, using Cartesian coordinates.

c. Make a plot of d-exponent vs. depth using Cartesian coordinates.

d. Make a plot of modified d-exponent vs. depth using Cartesian coordinates.

e. Make a plot of drillability parameter K_p vs. depth using Cartesian coordinates.

f. Make a plot of modified drillability parameter $K_p{}'$ vs. depth using Cartesian coordinates. (Note decrease in K_p between 10,040 and 10,050 ft due to mud weight increase.)

g. Make a plot of pore pressure vs. depth using the method of Rehm and McClendon and the modified d-exponent plot.

h. It is known that the pore pressure at 11,000 ft is 11.5 lbm/gal. Compute a value for a_3 using this known pressure point. *Answer:* 120×10^6.

i. Make a plot of pore pressure vs. depth using pore pressures computed from the modified $K_p{}'$ parameter plot.

j. Do you think the mud density should be increased before the next sand is drilled? *Answer:* Yes.

6.21 A mercury injection pump gave a scale reading of 43.2 cm³ at 24 psig with an empty sample cup in the air chamber. When a 23.4-g sample of shale cuttings was placed in the sample cup, a scale reading of 31.4 cm³ was obtained. Compute the average bulk density of the sample. *Answer:* 1.98 g/cm³.

6.22 Shale cuttings are added to a clean, dry mud balance until a balance is achieved with the density indicator reading 8.3 lbm/gal. Fresh water is added to the cup and the mixture is stirred until all air bubbles are removed. The mixture density is determined to be 13.3 lbm/gal. Compute the average density of the shale cuttings. *Answer:* 2.48 g/cm³.

6.23 The data in Table 6.24 were obtained in a south Louisiana well using a shale density column.

a. Determine the shale density in grams per cubic centimeter at each depth using the calibration curve given in Fig. 6.25.

b. Plot shale density vs. depth as shown in example of Fig. 6.27.

c. Determine the normal pressure trend line using shale porosities computed from shale densities obtained above the apparent transition zone. Assume an average grain density of 2.65 g/cm³ and a pore fluid density of 1.074 g/cm³.

d. Estimate the formation pore pressure gradient at various depths using the concept of equivalent effective

TABLE 6.23—PENETRATION RATE DATA FOR EXERCISE 6.20

Depth (ft)	Penetration Rate (ft/hr)	Bit Weight (1,000 lbm)	Bit Size (in.)	Rotary Speed (rpm)	Tooth Wear (fraction)	Hydraulic Parameter Ratio (fraction)	ECD (lbm/gal)
5,000	26.6	40	9.875	66	0.6	1.150	9.5
6,000	32.5	40	9.875	60	0.2	1.100	9.5
7,000	11.6	30	9.875	42	0.5	1.050	9.5
8,000	28.0	55	9.875	84	0.3	0.950	9.5
9,000	24.8	60	9.875	90	0.0	0.900	10.5
10,000	6.4	50	9.875	60	0.9	0.850	11.0
10,010	2.9	20	9.875	60	0.9	0.850	11.0
10,020	8.2	20	9.875	60	0.0	0.850	11.0
10,030	10.0	20	9.875	90	0.0	0.850	11.0
10,040	8.0	20	9.875	90	0.0	0.400	11.0
10,050	12.1	50	9.875	60	0.0	0.850	12.0
11,000	9.3	50	9.875	60	0.8	0.800	12.0
12,000	19.0	30	6.5	60	0.0	0.750	14.0
13,000	13.1	20	6.5	42	0.5	0.700	14.2

TABLE 6.24—SHALE DENSITY COLUMN DATA FOR EXERCISE 6.23

Density (ft)	Density Column Readings
4,000	172, 176, 178, 174
5,000	165, 168, 163, 164
6,000	156, 158, 154, 155
7,000	150, 145, 147, 148
8,000	140, 144, 143, 142
9,000	138, 140, 139, 137
10,000	133, 135, 137, 134
11,000	130, 133, 129, 132
12,000	130, 132, 134, 128
13,000	165, 166, 163, 167
14,000	166, 167, 165, 164

TABLE 6.25—KENEDY COUNTY (TX) SHALE RESISTIVITY DATA FOR EXERCISE 6.30

Depth (ft)	Shale Resistivity ($\Omega m^2/m$)	Depth (ft)	Shale Resistivity ($\Omega m^2/m$)
2,200	1.0	8,600	1.6
2,400	1.0	8,800	1.6
2,600	1.2	8,900	2.1
2,800	1.2	9,000	2.0
3,000	1.3	9,200	2.5
3,200	1.2	9,400	2.2
3,400	1.1	9,450	3.1
3,600	1.1	9,600	2.5
3,800	1.3	9,800	2.6
4,000	1.3	9,900	2.6
4,200	1.2	10,000	3.2
4,400	1.3	10,200	2.7
4,600	1.4	10,400	1.8
4,800	1.1	10,450	1.5
5,000	1.0	10,500	2.8
5,200	1.4	10,600	1.1
5,400	1.2	10,700	1.3
5,600	1.4	10,800	1.4
5,800	1.5	11,000	1.9
5,900	1.3	11,100	1.2
6,000	1.3	11,200	1.2
6,200	1.5	11,300	1.4
6,400	1.2	11,400	1.5
6,600	1.6	11,700	1.2
6,800	1.3	11,900	0.8
6,900	1.6	12,100	0.8
7,000	1.5	12,300	1.0
7,100	1.4	12,500	1.0
7,200	1.5	12,900	1.0
7,300	1.2	13,200	1.0
7,400	1.4	13,300	1.2
7,450	1.3	13,500	1.1
7,600	1.4	13,600	0.8
7,650	1.2	13,700	0.7
7,800	1.4	14,000	0.8
7,850	1.6	14,100	0.9
7,900	1.3	14,300	0.7
8,000	1.4	14,400	1.0
8,050	1.5	14,700	1.4
8,200	1.3	14,900	1.4
8,250	1.5	15,100	1.3
8,400	1.7	15,400	1.5
8,450	1.8	15,600	1.6

TABLE 6.26—KENEDY COUNTY (TX) MUD DENSITY DATA FOR EXERCISE 6.30

Depth (ft)	Mud Density (lbm/gal)
2,200	8.7
10,000	9.0
11,000	12.6
14,600	18.5
16,000	18.4

overburden stress. Assume the overburden stress is 1.0 psi/ft and the normal pore pressure gradient is 0.465 psi/ft. *Answer:* 11,400 psig at 14,000 ft.

e. Estimate the formation pore pressure gradient at various depths using the Boatman relationship given in Fig. 6.26. *Answer:* 12,300 psig at 14,000 ft..

6.24 Exactly 10 g of shale cuttings are placed in a mercury pump and the bulk volume is determined to be 4.09 cm^3. The 10-g sample then is placed in a moisture determination balance. After 5 minutes of drying, the sample weight stabilizes at 9.15 g. Compute the porosity and the bulk density of the sample. *Answer:* 0.208; 2.44 g/cm^3.

6.25 Using the data of Example 6.16, make a plot of pore pressure vs. depth.

6.26 Using the data of Example 6.17, make a plot of pore pressure vs. depth.

6.27 Using the data of Example 6.18, make a plot of pore pressure vs. depth.

6.28 A south Texas gulf coast formation at 12,000 ft was found to have a pore pressure of 7,500 psi and a bulk density of 2.35 g/cm^3. Compute the fracture gradient using the following:

(a) The Matthews and Kelly correlation. *Answer:* 11,000 psig.

(b) The Eaton correlation (assume variable overburden). *Answer:* 10,700 psig.

(c) The Pennebaker correlation (100 μs/ft at 6,000 ft). *Answer:* 11,700 psig.

(d) The Christman correlation. *Answer:* 10,700 psig.

6.29 The interval transit time for a sand at 14,000 ft was 90 μs/ft. The bulk density log gave a reading of 2.45 g/cm^3. Compute the fracture pressure using the MacPherson and Berry correlation. The overburden stress was calculated from bulk density logs to be 13,500 psig. *Answer:* 8,000 psig.

6.30 a. The shale resistivity data shown in Table 6.25 were obtained on a well drilled in Kenedy County, TX. Using these data and the method of Matthews and Kelly, make a plot of pore pressure and fracture gradient vs. depth.

b. Plot the mud density (Table 6.26) actually used to drill the well on the graph constructed in Part a.

c. A drillstem test at 14,350 ft indicated a formation pore pressure of 12,775 psig. How does this value compare to the pore pressure computed from shale resistivity in Part a?

6.31 A leakoff test will be conducted in 13⅜-in. casing having an internal diameter of 12.515 in. set at 3,000 ft. The test will be conducted after drilling to 3,030 ft—the depth of the first sand—with a 12.25-in. bit. Drillpipe having an external diameter of 5 in. and an internal diameter of 4.276 in. will be inserted to a depth of 3,000 ft to conduct the test. A 10-lbm/gal water-base drilling fluid containing no oil and a total volume fraction of solids of 0.09 is used. The gel strength of the mud is 14 lbm/100 sq ft. The well is located in the south Texas area and is normally pressured. Prepare a leakoff test chart by placing the anticipated leakoff pressure line and slope line on a plot of pressure vs. depth. Use the Matthews and Kelly fracture gradient correlation.

6.32 Compute the gel strength indicated by the pressure test conducted to break circulation in Fig. 6.53 (see Point D). Use the well data given in Example 6.25.

6.33 Compute the formation fracture pressure using the data of Example 6.25 and the gel strength computed in Exercise 6.31.

References

1. Eaton, B.A.: "Fracture Gradient Prediction and its Application in Oilfield Operations," *J. Pet. Tech.* (Oct. 1969) 1353-1360.
2. Powers, M.C.: "Fluid Release Mechanisms in Compacting Marine Mudrocks and their Importance in Oil Exploration," *Bull.* AAPG (1967) **51**, 1245.
3. Burst, J.F.: "Diagenesis of Gulf Coast Clayey Sediments and its Possible Relation in Petroleum Migration," *Bull.*, AAPG (1969), **53**, 80.
4. "Abnormal Subsurface Pressure—A Group Study Report," Houston Geological Soc. (1971) 16.
5. Jones, P.H.: "Hydrology of Neogene Deposits in the Northern Gulf of Mexico Basin," *Proc.* the Louisiana State U. First Symposium on Abnormal Subsurface Pressure, Baton Rouge (1967) 132.
6. Pennebaker, E.S.: "An Engineering Interpretation of Seismic Data," paper SPE 2165 presented at the SPE 43rd Annual Fall Meeting, Houston, Sept. 29–Oct. 2, 1968.
7. Faust, L.Y.: "Seismic Velocity as a Function of Depth and Geologic Time," *Geophysics* (1950) **16**, 192.
8. Kaufman, H.: "Velocity Functions in Seismic Prospecting," *Geophysics* (1953) **18**, 289.
9. West, S.S.: "Dependence of Seismic Wave Velocity Upon Depth and Lithology," *Geophysics* (1950) **15**, 653.
10. Sariento, R.: "Geological Factors Influencing Porosity Estimates from Velocity Logs," *Bull.*, AAPG (1960) **45**, 633.
11. Reynolds, E.B.: "The Application of Seismic Techniques to Drilling Techniques," paper SPE 4643, presented at the SPE 48th Annual Fall Meeting, Las Vegas, Sept. 30–Oct. 3, 1973.
12. McClure, L.J.: "Drill Abnormal Pressure Safely," L.J. McClure, Houston (1977).
13. Jordan, J.R. and Shirley, O.J.: "Application of Drilling Performance Data to Overpressure Detection," *J. Pet. Tech.* (Nov. 1966) 1387-1399.
14. Rehm, W.A. and McClendon, M.T.: "Measurement of Formation Pressure From Drilling Data," paper SPE 3601 presented at the SPE Annual Fall Meeting, New Orleans, Oct. 3–6, 1971.
15. Zamora, M.: "Slide-rule Correlation Aids 'd' Exponent Use," *Oil and Gas J.* (Dec. 18, 1972).
16. Bourgoyne, A.T. and Young, F.S.: "A Multiple Regression Approach to Optimal Drilling and Abnormal Pressure Defection," *Soc. Pet. Eng. J.* (Aug. 1974) 371-384; *Trans.*, AIME (1974) **257**.
17. Boatman, W.A.: "Measuring and Using Shale Density to Aid in Drilling Wells in High-Pressure Areas," *API Drilling and Production Practices Manual*, Dallas (1967) 121.
18. Borel, W.J. and Lewis, R.L.: "Ways to Detect Abnormal Formation Pressures," *Pet. Eng.* (July–Nov. 1969); "Part 3—Surface Shale Resistivity" (Oct. 1969) 82.
19. Rogers, L.: "Shale-Density Log Helps Detect Overpressure," *Oil and Gas J.* (Sept. 12, 1966).
20. Hottman, C.E. and Johnson, R.K.: "Estimation of Formation Pressure From Log-Derived Shale Properties," *J. Pet. Tech.* (June 1965) 717-727; *Trans.*, AIME (1965) 234-254.
21. Reynolds, E.B., Timko, D.J., and Zanier, A.M.: "Potential Hazards of Acoustic Log-Shale Pressure Plots," *J. Pet. Tech.* (Sept. 1973) 1039-1048.
22. Foster, J.B. and Whalen, H.E.: "Estimation of Formation Pressures From Electrical Surveys—Offshore Louisiana," *J. Pet. Tech.* (Feb. 1966) 165-171.
23. Matthews, W.R. and Kelly, J.: "How to Predict Formation Pressure and Fracture Gradient from Electric and Sonic Logs," *Oil and Gas J.* (Feb. 20, 1967).
24. Hubbert, M.K. and Willis, D.G.: "Mechanics of Hydraulic Fracturing," *Trans.*, AIME (1957) **210**, 153-160.
25. Birch, F., Shairer, J.F. and Spicer, H.C.: *Handbook of Physical Constants*, Geologic Soc. of America, Special Paper No. 36.
26. Daneshy, A.A.: "A Study of Inclined Hydraulic Fractures," *Soc. Pet. Eng. J.* (April 1973) 61-68.
27. Bradley, W.B.: "Mathematical Concept—Stress Cloud Can Predict Borehole Failure," *Oil and Gas. J.* (Feb. 19, 1979) 92.
28. Christman, S.: "Offshore Fracture Gradients," *J. Pet. Tech.* (Aug. 1973) 910-914.
29. MacPherson, L.A. and Berry, L.N.: "Prediction of Fracture Gradients," *Log Analyst* (Oct. 1972) 12.
30. Chenevert, M.E.: "How to Run Casing and Open-Hole Pressure Tests," *Oil and Gas J.* (March 6, 1978).

Nomenclature

$a_1 - a_8$ = exponents in penetration rate equation

A_a = capacity (area) of annulus

A_{dp} = capacity (area) of drillstring

A_h = capacity (area) of open hole

c_e = effective compressibility

c_o = compressibility of oil

c_s = compressibility of solids

c_w = compressibility of water

C_0 = conductivity of formation

C_w = conductivity of water

d = diameter

d_b = bit diameter

d_1 = inner diameter of annulus

d_2 = outer diameter of annulus

D = depth

D_i = depth of interest

D_s = depth into sediment

D_w = depth into water

E = Young's modulus

f_o = volume fraction of oil

f_s = volume fraction of solids

f_w = volume fraction of water

F = formation factor

F_j = jet impact force

F_R = formation resistivity factor

F_σ = matrix stress coefficient

g = gravitational constant

g_n = normal pressure gradient

g_p = formation pressure gradient, expressed as equivalent density

h = fractional tooth dullness

K = porosity decline constant

K_b = elastic modulus

K_p = drillability parameter

K'_p = modified drillability parameter

K_1, K_2 = constants

m = saturation exponent

m_{sh} = mass of shale
N = rotary speed
p = pressure
p_f = formation pore pressure
p_{ff} = formation fracture pressure
p_{fn} = normal formation pore pressure
p_{lo} = leakoff pressure
Δp_f = frictional pressure drop
P_H = hydraulic power
q = flow rate
r = radius
R = penetration rate
R^* = normalized penetration rate
R_w = resistivity of water
R_0 = resistivity of water-saturated
      formation
t = time
ι = interval transit time
ι_{fl} = interval transit time of pore fluid
ι_{ma} = interval transit time of matrix
       material
ι_n = normal interval transit time
V = volume
V_{sh} = shale volume
V_t = total volume
W = weight
W_b = weight on bit
x, y, z = spatial coordinates
X = general porosity-dependent parameter
ϵ = strain
μ = Poisson's ratio
ρ = density
ρ_b = bulk density
ρ_g = grain density
ρ_{fl} = pore fluid density

$\bar{\rho}_m$ = average mixture density
ρ_s = density of solid matrix (grains)
ρ_{sw} = seawater density
ρ_w = water density
σ = stress
σ_H = horizontal matrix stress
σ_{ob} = weight of overburden; vertical over-
      burden stress
σ_z = matrix stress
τ_g = gel strength
ϕ = porosity
ϕ_o = porosity constant

Subscripts

e = effective; equivalent
exp = exponent
fl = fluid
n = normal
ma = matrix
mod = modified
o = observed; oil; intercept value
s = solid
sh = shale
w = water

SI Metric Conversion Factors

bbl/ft	× 5.216 119	E−01	=	m^3/m
ft	× 3.048*	E−01	=	m
in.	× 2.54*	E+00	=	cm
lbf	× 4.448 222	E+00	=	N
lbm	× 4.535 924	E−01	=	kg
lbm/gal	× 1.198 264	E+02	=	kg/m^3
psi	× 6.894 757	E+00	=	kPa
psi^{-1}	× 1.450 377	E−04	=	Pa^{-1}
psi/ft	× 2.262 059	E+01	=	kPa/m

*Conversion factor is exact.

Chapter 7
Casing Design

The purpose of this chapter is to present (1) the primary functions of oilwell casing, (2) the various types of casing strings used, and (3) the procedures used in the design of casing strings.

Introduction

Casing serves several important functions in drilling and completing a well. It prevents collapse of the borehole during drilling and hydraulically isolates the wellbore fluids from the subsurface formations and formation fluids. It minimizes damage of both the subsurface environment by the drilling process and the well by a hostile subsurface environment. It provides a high-strength flow conduit for the drilling fluid to the surface and, with the blowout preventers (BOP), permits the safe control of formation pressure. Selective perforation of properly cemented casing also permits isolated communication with a given formation of interest.

As the search for commercial hydrocarbon deposits reaches greater depths, the number and sizes of the casing strings required to drill and to complete a well successfully also increases. Casing has become one of the most expensive parts of a drilling program; studies have shown that the average cost of tubulars is about 18% of the average cost of a completed well.[1] Thus, an important responsibility of the drilling engineer is to design the least expensive casing program that will allow the well to be drilled and operated safely throughout its life. The savings that can be achieved through an optimal design, as well as the risk of failure from an improper design, justify a considerable engineering effort on this phase of the drilling program.

Fig. 7.1 shows typical casing programs for deep wells in several different sedimentary basins. A well that will not encounter abnormal formation pore pressure gradients, lost circulation zones, or salt sections may require only *conductor casing* and *surface casing* to drill to the depth objective for the well. The conductor casing is need-ed to circulate the drilling fluid to the shale shaker without eroding the surface sediments below the rig and rig foundations when drilling is initiated. The conductor casing also protects the subsequent casing strings from corrosion and may be used to support structurally some of the wellhead load. A diverter system can be installed on the conductor casing to divert flow from rig personnel and equipment in case of an unexpected influx of formation fluids during drilling to surface casing depth. The surface casing prevents cave-in of unconsolidated, weaker, near-surface sediments and protects the shallow, fresh-water sands from contamination. Surface casing also supports and protects from corrosion any subsequent casing strings run in the well. In the event of a kick, surface casing generally allows the flow to be contained by closing the BOP's.

The BOP's should not be closed unless the casing to which the BOP's are attached has been placed deep enough into the earth to prevent a pressure-induced formation fracture initiated below the casing seat from reaching the surface. Subsequent flow through such fractures eventually can erode a large crater, up to several hundred feet in diameter, which could completely engulf the rig. Surface-casing-setting depths are usually from 300 to 5,000 ft into the sediments. Because of the possibility of contamination of shallow-water-supply aquifers, surface-casing-setting depths and cementing practices are subject to government regulations.

Deeper wells that penetrate abnormally pressured formations, lost circulation zones, unstable shale sections, or salt sections generally will require one or more strings of *intermediate casing* between the surface-casing depth and the final well depth (Fig. 7.1b). When abnormal formation pore pressures are present in the deeper portions of a well, intermediate casing is needed to protect formations below the surface casing from the pressures created by the required high drilling-fluid density. Similarly, when normal pore pressures are found below sections having abnormal pore pressure, an additional intermediate casing

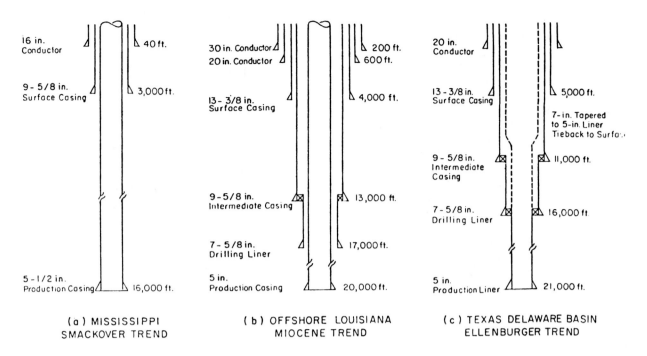

(a) MISSISSIPPI
SMACKOVER TREND

(b) OFFSHORE LOUISIANA
MIOCENE TREND

(c) TEXAS DELAWARE BASIN
ELLENBURGER TREND

Fig. 7.1—Example casing programs.

permits lowering the mud density to drill deeper formations economically. Intermediate casing may also be required after a troublesome lost-circulation zone or an unstable shale or salt section is penetrated, to prevent well problems while drilling below these zones.

Liners are casing strings that do not extend to the surface but are suspended from the bottom of the next larger casing string (Fig. 7.1c). Several hundred feet of overlap between the liner top and the casing seat are provided to promote a good cement seal. The principal advantage of a liner is its lower cost. However, problems sometimes arise from hanger seal and cement leakage. Also, using a liner exposes the casing string above it to additional wear during subsequent drilling. A drilling-liner is similar to intermediate casing in that it serves to isolate troublesome zones that tend to cause well problems during drilling operations.

Production casing is casing set through the productive interval. This casing string provides protection for the environment in the event of a failure of the tubing string during production operations and permits the production tubing to be replaced or repaired later in the life of a well. A *production liner* is a liner set through the productive interval of the well. Production liners generally are connected to the surface wellhead using a *tie-back* casing string when the well is completed. Tie-back casing is connected to the top of the liner with a specially designed connector. Production liners with tie-back casing strings are most advantageous when exploratory drilling below the productive interval is planned. Casing wear resulting from drilling operations is limited to the deeper portion of the well, and the productive interval is not exposed to potential damage by the drilling fluid for an extended period. Use of production liners with tie-back casing strings also results in lower hanging weights in the upper part of the well and thus often permits a more economical design.

7.1 Manufacture of Casing

The three basic processes used in the manufacture of casing include (1) the seamless process, (2) electric-resistance welding, and (3) electric-flash welding. In the seamless process, a billet is first pierced by a mandrel in a rotary piercing mill. The heated billet is introduced into the mill, where it is gripped by two obliquely oriented rolls that rotate and advance the billet into a central piercing plug (Fig. 7.2a). The pierced billet is processed through plug mills, where the wall thickness of the tube is reduced by central plugs with two single-groove rolls (Fig. 7.2b). Reelers similar in design to the piercing mills are then used to burnish the pipe surfaces and to form a more uniform wall thickness (Fig. 7.2c). Finally, sizing mills similar in design to the plug mills produce the final uniform pipe dimensions and roundness (Fig. 7.2d).

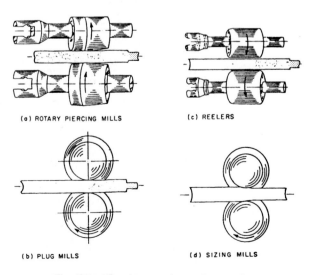

(a) ROTARY PIERCING MILLS

(c) REELERS

(b) PLUG MILLS

(d) SIZING MILLS

Fig. 7.2—Manufacture of seamless casing.

In the electric welding processes, flat sheet stock is cut and formed, and the two edges are welded together, without the addition of extraneous metal, to form the desired tube. The electric-resistance process continuously makes casing from coiled sheet stock that is fed into the machine, formed, and welded by an electric arc. The pipe leaving the machine is then cut to the desired lengths. The electric-flash welding technique processes a sheet by cutting it to the desired dimensions, simultaneously forming the entire length to a tube, and flashing and pressing the two edges together to make the weld. Some welded pipe is passed through dies that deform the steel sufficiently to exceed its elastic limit. This process raises the elastic limit in the direction stressed and reduces it in perpendicular directions.

The nominal size of casing is its OD. The strength of a given size casing is controlled by the yield strength and wall thickness of the steel. Steel used in casing is relatively mild (0.3 carbon) and can be normalized with small amounts of manganese to increase its strength. Strength can also be increased by a quenching and tempering (Q & T) process, which is favored by most manufacturers because of its lower cost.

7.2 Standardization of Casing

The American Petroleum Inst. (API) has developed standards for casing and other tubular goods that have been accepted internationally by the petroleum-producing industry. Casing is defined as tubular pipe with a range of OD's of 4.5 to 20 in. Among the properties included in the API standards[2] for both pipe and couplings are strength, physical dimensions, and quality-control test procedures. In addition to these standards, API provides bulletins on the recommended minimum-performance properties[3] and formulas[4] for the computation of minimum-performance properties. The minimum-performance properties must be used in the design of casing strings to minimize the possibility of casing failure.

API has adopted a casing *grade* designation to define the strength characteristics of the pipe. The grade code consists of a letter followed by a number. The letter designation in the API grade was selected arbitrarily to pro-

TABLE 7.1—GRADES OF CASING RECOGNIZED BY THE API

API Grade	Yield Strength (psi) Minimum	Yield Strength (psi) Maximum	Minimum Ultimate Tensile Strength (psi)	Minimum* Elongation (%)
H-40	40,000	80,000	60,000	29.5
J-55	55,000	80,000	75,000	24.0
K-55	55,000	80,000	95,000	19.5
C-75	75,000	90,000	95,000	19.5
L-80	80,000	95,000	95,000	19.5
N-80	80,000	110,000	100,000	18.5
C-90	90,000	105,000	100,000	18.5
C-95	95,000	110,000	105,000	18.0
P-110	110,000	140,000	125,000	15.0

*Test specimen with area greater than 0.75 sq. in.

vide a unique designation for each grade of casing adopted in the standards. The number designates the minimum yield strength of the steel in thousands of psi. The yield strength is defined by API as the tensile stress required to produce a total elongation per unit length of 0.005 on a standard test specimen. This strain is slightly beyond the elastic limit. Since there are significant variations in the yield strengths measured on manufactured pipe, a minimum yield strength criterion, rather than an average yield stress, was adopted. Based on considerable test data, the minimum yield strength should be computed as 80% of the average yield strength observed. In addition to specifying the minimum acceptable yield strength of each grade of casing, API specifies the maximum yield strength, the minimum ultimate tensile strength, and the minimum elongation per unit length at failure (Table 7.1). It also stipulates that the amount of phosphorus in the steel must not exceed 0.04% and that the amount of sulfur must not exceed 0.06%.

In addition to the API grades, there are many proprietary steel grades that do not conform to all API specifications but are widely used in the petroleum-producing industry. Strength properties of commonly used non-API grades are given in Table 7.2. These steel grades are used

TABLE 7.2—COMMONLY USED NON-API GRADES OF CASING

Non-API Grade	Manufacturer	Yield Strength (psi) Minimum	Yield Strength (psi) Maximum	Minimum Ultimate Tensile Strength (psi)	Minimum* Elongation (%)
S-80	Lone Star Steel	75,000** 55,000†	—	75,000	20.0
Mod. N-80	Mannesmann Tube Co.	80,000	95,000	100,000	24.0
C-90‡	Mannesmann Tube Co.	90,000	105,000	120,000	26.0
SS-95	Lone Star Steel	95,000** 75,000†	—	95,000	18.0
S00-95	Mannesmann Tube Co.	95,000	110,000	110,000	20.0
S-95	Lone Star Steel	95,000** 92,000†	—	110,000	16.0
S00-125	Mannesmann Tube Co.	125,000	150,000	135,000	18.0
S00-140	Mannesmann Tube Co.	140,000	165,000	150,000	17.0
V-150	U.S. Steel	150,000	180,000	160,000	14.0
S00-155	Mannesmann Tube Co.	155,000	180,000	165,000	20.0

*Test specimen with area greater than 0.75 sq in.
**Circumferential.
†Longitudinal.
‡Maximum ultimate tensile strength of 120,000 psi.

for special applications that require very high tensile strength, special collapse resistance, or high-strength steels that are more resistant to hydrogen sulfide.

The API Standards recognize three length ranges for casing. Range 1 (R-1) includes joint lengths in the range of 16 to 25 ft. Range 2 (R-2) is the 25- to 34-ft range, and Range 3 (R-3) is 34 ft and longer. It is also specified that when casing is ordered from the mill in amounts greater than one carload, 95% of the pipe must have lengths greater than 18 ft for R-1, 28 ft for R-2, and 36 ft for R-3. In addition, 95% of the shipment must have a maximum length variation no greater than 6 ft for R-1, 5 ft for R-2, and 6 ft for R-3. Casing is run most often in R-3 lengths to reduce the number of connections in the string. Since casing is made up in single joints, R-3 lengths can be handled easily by most rigs.

To meet API specifications, the OD of casing must be held within a tolerance of ±0.75%. However, casing manufacturers generally will try to prevent the pipe from being undersized to ensure adequate thread run-out when machining a connection. Casing usually is found to be within the API tolerance but slightly oversized. The minimum permissible pipe-wall thickness is 87.5% of the nominal wall thickness. The maximum ID is controlled by the combined tolerances for OD and minimum wall thickness. The minimum ID is controlled by a specified *drift* diameter—the minimum mandrel diameter that must pass unobstructed through the pipe. The length of a casing drift mandrel is 6 in. for casing sizes in the range of 4.5 to 8.625 in. For larger casing sizes, a 12-in. drift mandrel must be used. The drift mandrel is not long enough to ensure a straight pipe, but it will ensure the passage of a bit a size less than the drift diameter.

In some instances, it is desirable to run casing with a drift diameter slightly greater than the API drift diameter for that casing size. In these instances, casing that has passed an oversized drift mandrel can be specially ordered. Some of the more commonly available oversized drift diameters are given in Table 7.3. When non-API drift requirements are specified, they should be made known to the mill, the distributor, and the threading company before the pipe manufacture.

Casing dimensions can be specified by casing size (OD) and nominal wall thickness. However, it is conventional to specify casing dimensions by size and weight per foot. In discussing casing weights, one should differentiate between nominal weight, plain-end weight, and average weight for threads and couplings. The nominal weight per foot is not a true weight per foot but is useful for identification purposes as an approximate average weight per foot. The plain-end weight per foot is the weight per foot of the pipe body, excluding the threaded portion and coupling weight. The average weight per foot is the total weight of an average joint of threaded pipe, with a coupling attached power-tight at one end, divided by the total length of the average joint. In practice, the average weight per foot sometimes is calculated to obtain the best possible estimate of the total weight of a casing string. However, the variation between nominal weight per foot and average weight per foot is generally small, and most design calculations are performed with the nominal weight per foot.

API provides specifications for the following four types of casing connectors.

1. Short round threads and couplings (CSG).
2. Long round threads and couplings (LCSG).
3. Buttress threads and couplings (BCSG).
4. Extremeline threads (XCSG).

Before development of API threads, most manufacturers used a sharp V-shaped thread that proved unsatisfactory with increases in well depth.

Schematics of each of the API connectors are shown in Fig. 7.3. The CSG and LCSG connectors have the same basic thread design. Threads have a rounded shape and are spaced to give eight threads per inch. Because of this, they are sometimes referred to as API 8-Round threads. The threads are cut with a taper of ¾ in./ft on diameter for all pipe sizes. A longer thread run-out and coupling of the LCSG provide a greater strength when needed. These are very commonly used connectors because of their proven reliability, ease of manufacture, and low cost.

As can be seen in Fig. 7.3a, the API Round Thread is cut with a 60° included angle and has rounded peaks and roots. When the coupling is formed, small voids exist at the roots and crests of each thread. Thread compound must be used to fill these voids to obtain a seal. This connection is not designed to be a dependable, high-pressure seal for gases and solid-free, low-viscosity liquids. If the seal is ineffective, internal pressure acts to separate the threaded surfaces further.

Because the threads are cut on a taper, stress rapidly increases as the threads are made up. The proper amount of make-up is best determined by monitoring both the torque and the number of turns. A loose connection can leak and will have reduced strength. An over-tight connection can leak because of galling of the threads or a cracked coupling. It can also have reduced strength and can produce a reduced drift diameter as a result of excessive yielding of the threaded casing end.

Special thread compounds containing powdered metals are used to reduce frictional forces during connection make-up and to provide filler material for assisting in

TABLE 7.3—SPECIAL DRIFT DIAMETERS
(Courtesy of Lone Star Steel)

OD Size (in.)	Weight T&C (lbf/ft)	Wall Thickness (in.)	Drift Diameter (in.) API	Drift Diameter (in.) Special
7	23.00	0.317	6.241	6.250
	32.00	0.453	5.969	6.000
7¾	46.10	0.595	—	6.500
8⅝	32.00	0.352	7.796	7.875
	40.00	0.450	7.600	7.625
8¾	49.70	0.557	—	7.500
9⅝	40.00	0.395	8.679	8.750
	43.50	0.435	8.599	8.625
	47.00	0.472	8.525	8.625
	58.40	0.595	8.279	8.375
9¾	59.20	0.595	—	8.500
9⅞	62.80	0.625	—	8.500
10¾	45.50	0.400	9.794	9.875
	55.50	0.495	9.604	9.625
	65.70	0.595	9.404	9.504
11¾	60.00	0.489	10.616	10.625
	65.00	0.534	10.526	10.625
11⅞	71.80	0.582	—	10.625
13⅜	72.00	0.514	12.191	12.250
	86.00	0.625	11.969	12.000
13½	81.40	0.580	—	12.250
13⅝	88.20	0.625	—	12.250

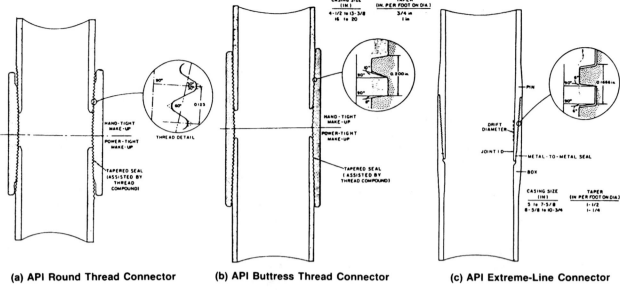

(a) API Round Thread Connector   **(b) API Buttress Thread Connector**   **(c) API Extreme-Line Connector**

Fig. 7.3—API connectors.

plugging any remaining small voids around the roots and crests in the threads. The compound used is critical to prevent galling and to obtain a leak-proof, properly made-up connection. Care must be exercised to ensure that a proper thread compound for the given connector is used.

Threaded connections are often rated according to their *joint efficiency,* which is the tensile strength of the joint divided by the tensile strength of the pipe body. Although the joint efficiency of the API LCSG connector is greater than the CSG connector, neither are 100% efficient. Because of the tapering on the threads, as well as the 60° included angle of the threads, the threaded end of the casing sometimes begins to yield and to collapse (Fig. 7.4). This can produce an unzipping effect and, upon failure, the pin appears to jump out of the coupling. In addition to this jump-out, fracture of the pin or coupling also can occur.

The API BCSG is shown in Fig. 7.3b. The joint efficiency of this connector is 100% in most cases. The basic thread design is similar to that of the API Round Thread in that it is tapered. However, longer coupling and thread run-out are used and the thread shape is squarer, so the unzipping tendency is greatly reduced. Five threads are cut to the inch, and the thread taper is ¾-in./ft for casing sizes up to 7⅝ in. and 1 in./ft for 16-in. or larger casings. As with API Round Threads, the placement of thread compound at the roots of the teeth provides the sealing mechanism. It is not, however, a good choice when a leak-proof connection is needed.

The API XCSG connector is shown in Fig. 7.3c. It differs from the other API connectors in that it is an *integral joint* (i.e., the box is machined on the pipe wall). On an integral-joint connection, the pipe wall must be thicker near the ends of the casing to provide the necessary metal to machine a stronger connector. The OD of an XCSG connector is significantly less than the other API couplings, thus providing an alternative when the largest possible casing size is run in a restricted-clearance situation. Also, only half as many threaded connections exist; therefore, there are fewer potential sites for leakage. However, the minimum ID will be less for the XCSG connector.

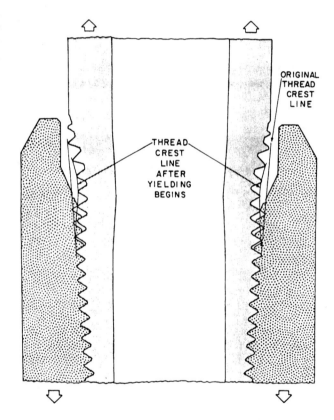

Fig. 7.4—Joint pull-out failure mode for API round thread.

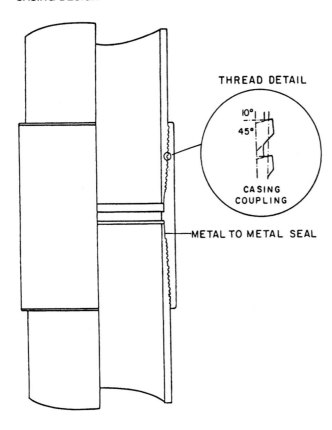

Fig. 7.5—Armco seal-lock connector.

The sealing mechanism used in the XCSG connector is a metal-to-metal seal between the pin and the box (Fig. 7.3c). This connector does not depend only on a thread compound for sealing, although a compound is still needed for lubrication. Because of the required thicker pipe walls near the ends and the closer machining tolerances needed for the metal-to-metal seal, XCSG connectors are much more expensive than the other API connectors.

In addition to the API connections, many proprietary connections are available that offer premium features not available on API connections. Among the special features offered are the following items.

1. Flush joints for maximum clearance.
2. Smooth bores through connectors for reduced turbulence.
3. Threads designed for fast make-up with low tendency to cross-thread.
4. Multiple metal-to-metal seals for improved pressure integrity.
5. Multiple shoulders for improved torque strength.
6. High compressive strengths for special loading situations.
7. Resilient rings for secondary pressure seals and connector corrosion protection.

Several examples of premium non-API connectors are shown in Figs. 7.5 through 7.7, which illustrate the special features listed above.

7.3 API Casing Performance Properties

The most important performance properties of casing include its rated values for *axial tension, burst pressure,* and *collapse pressure.* Axial tension loading results primarily from the weight of the casing string suspended below the joint of interest. *Body yield strength* is the tensional force required to cause the pipe body to exceed its elastic limit. Similarly, *joint strength* is the minimum tensional force required to cause joint failure (Fig. 7.8a). Burst pressure rating is the calculated minimum internal pressure that will cause the casing to rupture in the absence of external pressure and axial loading (Fig. 7.8b). Collapse pressure rating is the minimum external pressure that will cause the casing walls to collapse in the absence of internal pressure and axial loading (Fig. 7.8c). API provides recommended formulas[4] for computing these performance properties.

7.3.1 Tension

Pipe-body strength in tension can be computed by use of the simplified free-body diagram shown in Fig. 7.9. The

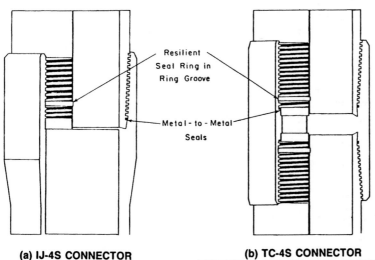

(a) IJ-4S CONNECTOR
(INTEGRAL JOINT CONNECTOR)

(b) TC-4S CONNECTOR
(THREADED AND COUPLED CONNECTOR)

(c) FL-4S CONNECTOR
(FLUSH INTEGRAL JOINT)

Fig. 7.6—Sample Atlas Bradford connectors with resilient seals and smooth bores.

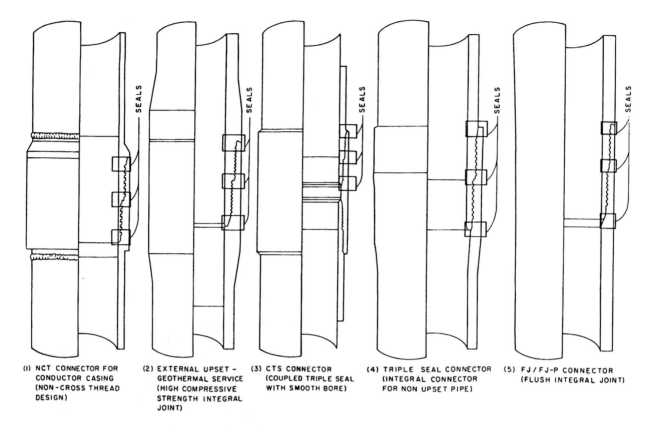

(I) NCT CONNECTOR FOR
    CONDUCTOR CASING
    (NON-CROSS THREAD
    DESIGN)

(2) EXTERNAL UPSET –
    GEOTHERMAL SERVICE
    (HIGH COMPRESSIVE
    STRENGTH INTEGRAL
    JOINT)

(3) CTS CONNECTOR
    (COUPLED TRIPLE SEAL
    WITH SMOOTH BORE)

(4) TRIPLE SEAL CONNECTOR
    (INTEGRAL CONNECTOR
    FOR NON UPSET PIPE)

(5) FJ/FJ-P CONNECTOR
    (FLUSH INTEGRAL JOINT)

Fig. 7.7—Sample Hydril two-step connectors with three metal- to-metal seals.

force F_{ten} tending to pull apart the pipe is resisted by the strength of the pipe walls, which exert a counterforce, F_2. F_2 is given by

$$F_2 = \sigma_{yield} A_s,$$

where σ_{yield} is the minimum yield strength and A_s is the cross-sectional area of steel. Thus, the pipe-body strength is given by

$$F_{ten} = \frac{\pi}{4} \sigma_{yield} (d_n^2 - d^2). \quad \ldots \ldots \ldots \ldots \ldots \ldots (7.1)$$

The pipe-body strength computed with Eq. 7.1 is the minimum force that would be expected to cause permanent deformation of the pipe. The expected minimum force required to pull the pipe in two would be significantly higher than this value. However, the nominal wall thickness rather than the minimum acceptable wall thickness is used in Eq. 7.1. Because the minimum acceptable wall thickness is 87.5% of the nominal wall thickness, the absence of permanent deformation cannot be assured.

Joint-strength formulas based on theoretical considerations and partially on empirical observations have been accepted by API. For API Round Thread connections, formulas for computing the minimum joint fracture force and the minimum joint pull-out force are presented (Fig. 7.10a). The lower values are recommended for use in casing design. Similarly, for Buttress connections, formulas are presented for minimum pipe-thread strength and for minimum coupling-thread strength (Fig. 7.10b). Three formulas are presented for Extreme-line connections, depending on whether the steel area is minimal in the box, pin, or pipe body (Fig. 7.10c).

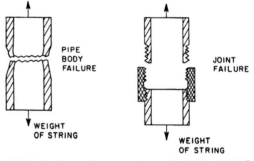

(a) TENSION FAILURE IN PIPE BODY OR JOINT

(b) BURST FAILURE FROM INTERNAL PRESSURE

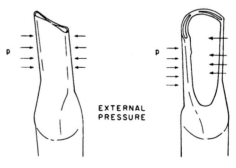

(c) COLLAPSE FAILURE FROM EXTERNAL PRESSURE

Fig. 7.8—Tension, burst, and collapse modes of failure.

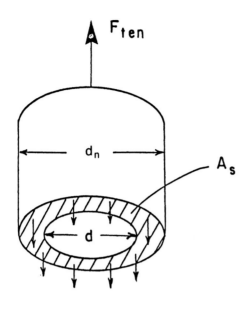

$$F_2 = \sigma_{yield} A_s$$

Fig. 7.9—Tensional force balance on pipe body.

Example 7.1 Compute the body-yield strength for 20-in., K-55 casing with a nominal wall thickness of 0.635 in. and a nominal weight per foot of 133 lbf/ft.

Solution. This pipe has a minimum yield strength of 55,000 psi and an ID of

$$d = 20.00 - 2(0.635) = 18.730 \text{ in.}$$

Thus, the cross-sectional area of steel is

$$A_s = \frac{\pi}{4}(20^2 - 18.73^2) = 38.63 \text{ sq in.}$$

and minimum pipe-body yield is predicted by Eq. 7.1 at an axial load of

$$F_{ten} = 55,000(38.63) = 2,125,000 \text{ lbf.}$$

7.3.2. Burst Pressure

As shown in the simplified free-body diagram of Fig. 7.11, the tendency for the force, F_1, to burst a casing string is resisted by the strength of the pipe walls, which exert a counterforce, F_2. The force, F_1, which results from the internal pressure, p_{br}, acting on the projected area (LdS) is given by

$$F_1 = p_{br} L \frac{d}{2} d\theta.$$

The resisting force, F_2, resulting from the steel strength, σ_s, acting over the steel area (tL) is given by

$$F_2 = \sigma_s t L \frac{d\theta}{2}.$$

Area under last perfect thread

$$A_{jp} = \frac{\pi}{4}[(d_n - 0.1425)^2 - d^2]$$

Tensional force for fracture

$$F_{ten} = 0.95 A_{jp} \sigma_{ult}$$

Tensional force for joint pull-out

$$F_{ten} = 0.95 A_{jp} L_{et} \left(\frac{0.74 d_n^{-0.59} \sigma_{ult}}{0.5 L_{et} + 0.14 d_n} + \frac{\sigma_{yield}}{L_{et} + 0.14 d_n} \right)$$

(a) Round Thread Connector

Area of Steel in Pipe Body

$$A_p = \frac{\pi}{4}(d_n^2 - d^2)$$

Area of Steel in Coupling

$$A_{sc} = \frac{\pi}{4}(d_{c2}^2 - d_{c1}^2)$$

Tensional Force for Pipe Thread Failure

$$F_{ten} = 0.95 A_p \sigma_{ult} \left[1.008 - 0.0396 \left(1.083 - \frac{\sigma_{yield}}{\sigma_{ult}} \right) d_n \right]$$

Tensional Force for Coupling Thread Failure

$$F_{ten} = 0.95 A_{sc} \sigma_{ult}$$

(b) Buttress Thread Connector

Tensional Force for Pipe Failure

$$F_{ten} = \frac{\pi \sigma_{ult}}{4}(d_n^2 - d^2)$$

Tensional Force for Box Failure

$$F_{ten} = \frac{\pi \sigma_{ult}}{4}(d_{j2}^2 - d_b^2)$$

Tensional Force for Pin Failure

$$F_{ten} = \frac{\pi \sigma_{ult}}{4}(d_{pin}^2 - d_{j1}^2)$$

(c) Extreme-Line Connector

Fig. 7.10—API joint-strength formula.[3,4]

Totaling forces for static conditions gives

$$F_1 - 2F_2 = 0.$$

Substituting the appropriate expressions for F_1 and F_2 and solving for the burst pressure rating, p_{br}, yields

$$p_{br} = \frac{2\sigma_s t}{d}.$$

This equation is valid only for thin-wall pipes with d_n/t values greater than those of most casing strings.

Barlow's[5] equation for thick-wall pipe is identical to the above equation for thin-wall pipe if the OD, d_n, is used in place of the ID, d. Barlow's equation results from

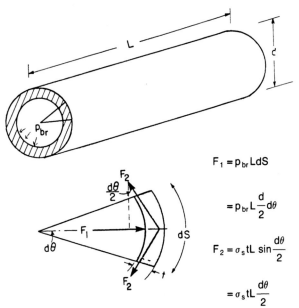

$F_1 = p_{br} L dS$

$= p_{br} L \dfrac{d}{2} d\theta$

$F_2 = \sigma_s t L \sin \dfrac{d\theta}{2}$

$= \sigma_s t L \dfrac{d\theta}{2}$

Fig. 7.11—Free-body diagram for casing burst.

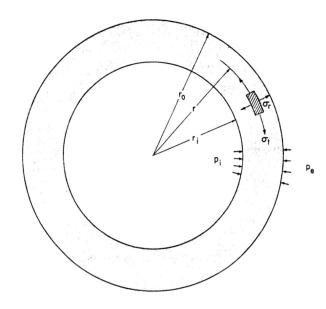

Fig. 7.12—Two-dimensional wall stress.

a nonrigorous solution but is a fairly accurate (slightly conservative) thick-wall formula. The API burst-pressure rating is based on Barlow's equation. Use of 87.5% of the minimum yield strength for steel, σ_s, takes into account the minimum allowable wall thickness and gives

$$p_{br} = 0.875 \frac{2\sigma_{\text{yield}} t}{d_n}. \qquad \ldots\ldots\ldots\ldots\ldots\ldots (7.2)$$

API recommends use of this equation with wall thickness rounded to the nearest 0.001 in. and the results rounded to the nearest 10 psi.

Example 7.2. Compute the burst-pressure rating for 20-in., K-55 casing with a nominal wall thickness of 0.635 in. and a nominal weight per foot of 133 lbf/ft.

Solution. The burst-pressure rating is computed by use of Eq. 7.2.

$$p_{br} = 0.875(2)(55,000)(0.635)/20.00 = 3,056 \text{ psi}.$$

Rounded to the nearest 10 psi, this value becomes 3,060 psi. This burst-pressure rating corresponds to the minimum expected internal pressure at which permanent pipe deformation could take place, if the pipe is subjected to no external pressure or axial loads.

7.3.3 Collapse Pressure

The collapse of steel pipe from external pressure is a much more complex phenomenon than pipe burst from internal pressure. A simplified free-body diagram analysis, such as the one shown in Fig. 7.11, does not lead to useful results. However, a more complex, classical elasticity theory can be used to establish the radial stress and tangential hoop stress in the pipe wall. Consider, for ex-

ample, the casing cross section shown in Fig. 7.12 with any external pressure, p_e, and internal pressure, p_i. Application of the classical elasticity theory for this two-dimensional problem at any radius, r, between the inner radius, r_i, and outer radius, r_o, gives[6]

$$\sigma_r = \frac{-p_i r_i^2 (r_o^2 - r^2) - p_e r_o^2 (r^2 - r_i^2)}{r^2 (r_o^2 - r_i^2)} \quad \ldots (7.3a)$$

and

$$\sigma_t = \frac{p_i r_i^2 (r_o^2 + r^2) - p_e r_o^2 (r_i^2 + r^2)}{r^2 (r_o^2 - r_i^2)}, \quad \ldots\ldots (7.3b)$$

where σ_r and σ_t are the radial and tangential stresses at radius r. For both collapse and burst conditions, stress will be a maximum in the tangential direction. If it is assumed that the pipe is subjected only to an external pressure, p_e, then for $r = r_i$, Eq. 7.3b reduces to

$$\sigma_t = -\frac{2p_e r_o^2}{t(r_o + r_i + r_i - r_i + 2t - 2t)}.$$

Use of the effective compressive yield strength for $-\sigma_t$ and rearranged terms reduces the above equation to the following formulas for collapse pressure rating, p_{cr}.

$$p_{cr} = 2(\sigma_{\text{yield}})_e \left[\frac{d_n/t - 1}{(d_n/t)^2} \right]. \qquad \ldots\ldots\ldots\ldots (7.4a)$$

It can also be shown that Eq. 7.3b reduces to Eq. 7.2 when the pipe is subjected only to internal pressure. The proof of this is left as a student exercise.

The collapse that occurs in approximate agreement with Eq. 7.4a is called *yield-strength* collapse. It has been shown experimentally that yield-strength collapse occurs

TABLE 7.4—EMPIRICAL COEFFICIENTS USED FOR COLLAPSE-PRESSURE DETERMINATION [4]

Grade*	Empirical Coefficients				
	F_1	F_2	F_3	F_4	F_5
H-40	2.950	0.0465	754	2.063	0.0325
-50	2.976	0.0515	1,056	2.003	0.0347
J-K 55 & D	2.991	0.0541	1,206	1.989	0.0360
-60	3.005	0.0566	1,356	1.983	0.0373
-70	3.037	0.0617	1,656	1.984	0.0403
C-75 & E	3.054	0.0642	1,806	1.990	0.0418
L-80 & N-80	3.071	0.0667	1,955	1.998	0.0434
C-90	3.106	0.0718	2,254	2.017	0.0466
C-95	3.124	0.0743	2,404	2.029	0.0482
-100	3.143	0.0768	2,553	2.040	0.0499
P-105	3.162	0.0794	2,702	2.053	0.0515
P-110	3.181	0.0819	2,852	2.066	0.0532
-120	3.219	0.0870	3,151	2.092	0.0565
-125	3.239	0.0895	3,301	2.106	0.0582
-130	3.258	0.0920	3,451	2.119	0.0599
-135	3.278	0.0946	3,601	2.133	0.0615
-140	3.297	0.0971	3,751	2.146	0.0632
-150	3.336	0.1021	4,053	2.174	0.0666
-155	3.356	0.1047	4,204	2.188	0.0683
-160	3.375	0.1072	4,356	2.202	0.0700
-170	3.412	0.1123	4,660	2.231	0.0734
-180	3.449	0.1173	4,966	2.261	0.0769

*Grades indicated without letter designation are not API grades but are grades in use or grades being considered for use and are shown for information purposes.

only for the lower range of d_n/t values applicable for oil-well casing. The upper limit of the yield-strength collapse range is calculated with

$$\frac{d_n}{t} = \frac{\sqrt{(F_1-2)^2 + 8[F_2 + F_3/(\sigma_{\text{yield}})_e]} + (F_1-2)}{2[F_2 + F_3/(\sigma_{\text{yield}})_e]}$$

$$\dotfill (7.4b)$$

where F_1, F_2, and F_3 are given in Table 7.4. Values computed by Eq. 7.4b for zero axial stress are shown in Table 7.5. The effective yield strength, $(\sigma_{\text{yield}})_e$, is equal to the minimum yield strength when the axial stress is zero. Table 7.4 is based on the minimum yield strength.

At high values of d_n/t, collapse can occur at lower pressures than predicted by Eq. 7.4a because of a geometric instability. Application of elastic stability theory [7] leads to the following collapse formula:

$$p_{cr} = \frac{2E}{(1-\mu^2)(d_n/t)(d_n/t-1)^2}.$$

After an adjustment for statistical variations in the properties of manufactured pipe is applied, this equation becomes [4]

$$p_{cr} = \frac{46.95 \times 10^6}{(d_n/t)(d_n/t-1)^2}. \dotfill (7.5a)$$

Collapse that occurs in approximate agreement with Eq. 7.5a is called *elastic* collapse. The applicable range of d_n/t values recommended by API for elastic collapse are given in Table 7.5. The lower limit of the elastic collapse range is calculated by

$$(d_n/t) = \frac{2 + F_2/F_1}{3F_2/F_1}, \dotfill (7.5b)$$

where F_1 and F_2 are given in Table 7.4.

The transition from yield-strength collapse to elastic collapse is not sharp but covers a significant range of d_n/t values. Based on the results of many experimental tests,

TABLE 7.5—RANGE OF d_n/t FOR VARIOUS COLLAPSE-PRESSURE REGIONS WHEN AXIAL STRESS IS ZERO [4]

Grade*	←Yield Strength→ Collapse	←Plastic→ Collapse	←Transition→ Collapse	←Elastic→ Collapse
H-40		16.40	27.01	42.64
-50		15.24	25.63	38.83
J-K-55 & D		14.81	25.01	37.21
-60		14.44	24.42	35.73
-70		13.85	23.38	33.17
C-75 & E		13.60	22.91	32.05
L-80 & N-80		13.38	22.47	31.02
C-90		13.01	21.69	29.18
C-95		12.85	21.33	28.36
-100		12.70	21.00	27.60
P-105		12.57	20.70	26.89
P-110		12.44	20.41	26.22
-120		12.21	19.88	25.01
-125		12.11	19.63	24.46
-130		12.02	19.40	23.94
-135		11.92	19.18	23.44
-140		11.84	18.97	22.98
-150		11.67	18.57	22.11
-155		11.59	18.37	21.70
-160		11.52	18.19	21.32
-170		11.37	17.82	20.60
-180		11.23	17.47	19.93

*Grades indicated without letter designation are not API grades but are grades in use or grades being considered for use and are shown for information purposes.

API has adopted two additional collapse-pressure equations to cover the transition region. A plastic collapse rating for d_n/t values just above the yield-strength collapse region is predicted with

$$p_{cr} = (\sigma_{\text{yield}})_e \left(\frac{F_1}{d_n/t} - F_2 \right) - F_3. \qquad \ldots \ldots \ldots (7.6a)$$

The upper limit of the plastic collapse range is calculated by

$$d_n/t = \frac{(\sigma_{\text{yield}})_e (F_1 - F_4)}{F_3 + (\sigma_{\text{yield}})_e (F_2 - F_5)}, \qquad \ldots \ldots \ldots \ldots (7.6b)$$

where F_1 through F_5 are given in Table 7.4.

A *transition* collapse region between the plastic collapse and elastic collapse regions is defined by use of

$$p_{cr} = (\sigma_{\text{yield}})_e \left(\frac{F_4}{d_n/t} - F_5 \right). \qquad \ldots \ldots \ldots \ldots (7.7)$$

Values of d_n/t computed with Eq. 7.6b for zero axial stress are shown in Table 7.5.

Example 7.3. Compute the collapse-pressure rating for 20-in., K-55 casing with a nominal wall thickness of 0.635 in. and a nominal weight per foot of 133 lbf/ft.

Solution. This pipe has a d_n/t ratio given by

$$d_n/t = 20/0.635 = 31.496.$$

Table 7.5 indicates that this value for d_n/t falls in the range specified for transition collapse. Thus, the collapse-pressure rating can be computed with Eq. 7.7.

$$p_{cr} = 55,000 \left(\frac{1.989}{31.496} - 0.036 \right) = 1,493 \text{ psi.}$$

Rounded to the nearest 10 psi, this value becomes 1,490 psi. This collapse-pressure rating corresponds to the minimum expected external pressure at which the pipe would collapse if the pipe were subjected to no internal pressure or axial loads.

7.3.4 Casing Performance Summary

The values for tensional strength, burst resistance, and collapse resistance given in Table 7.6 were computed in accordance with theoretical and empirical formulas adopted by API. The last entry in this table corresponds to the casing properties determined in Examples 7.1 through 7.3. Such tables generally are found to be extremely useful and convenient for casing design applications.

7.3.5 Effect of Combined Stress

The performance properties given in Table 7.6 apply only for zero axial tension and no pipe bending. Unfortunately, many of the casing performance properties are altered

significantly by axial tension or compression and by bending stresses. Thus, the table values for the performance properties often must be corrected before they are used in a casing design application.

The generally accepted relationship for the effect of axial stress on collapse or burst was presented by Holmquist and Nadia[8] in 1939. Application of classical distortion energy theory to casing gives the following equation.

$$(\sigma_t - \sigma_z)^2 + (\sigma_r - \sigma_t)^2 + (\sigma_z - \sigma_r)^2 = 2\sigma_{\text{yield}}^2, \quad . (7.8)$$

where σ_r, σ_t, and σ_z are the principal radial, tangential, and vertical stresses, respectively. The application of the distortional energy theorem is based on the yield stress value, and the surface that is developed denotes the onset of yield, not a physical failure of the casing. After regrouping, Eq. 7.8 takes the form of either an ellipse or a circle.

$$(\sigma_t - \sigma_r)^2 - (\sigma_z - \sigma_r)(\sigma_t - \sigma_r) + (\sigma_z - \sigma_r)^2 = \sigma_{\text{yield}}^2$$

$$\ldots \ldots \ldots \ldots \ldots \ldots \ldots \ldots \ldots \ldots \ldots (7.9a)$$

or

$$\frac{3(\sigma_t - \sigma_r)^2}{4} + \left(\sigma_z - \frac{\sigma_t + \sigma_r}{2} \right)^2 = \sigma_{\text{yield}}^2. \quad \ldots (7.9b)$$

The ellipse of plasticity was chosen for this book because it is more commonly used in current drilling engineering practice.

Recall that the radial and tangential stresses of Eq. 7.9a were defined previously by Eqs. 7.3a and 7.3b. The maximum stress will occur at the inner pipe wall. Substitution of $r = r_i$ in Eq. 7.3a gives a value of $(-p_i)$ for the radial stress at this point. Use of this value in Eq. 7.9a and rearranged terms yields

$$\left(\frac{\sigma_r + p_i}{\sigma_{\text{yield}}} \right)^2 - \left(\frac{\sigma_z + p_i}{\sigma_{\text{yield}}} \right) \left(\frac{\sigma_t + p_i}{\sigma_{\text{yield}}} \right)$$

$$+ \left[\frac{(\sigma_z + p_i)^2}{\sigma_{\text{yield}}^2} - 1 \right]. \qquad \ldots \ldots \ldots \ldots \ldots \ldots (7.10)$$

Solving this quadratic equation gives

$$\left(\frac{\sigma_t + p_i}{\sigma_{\text{yield}}} \right) = \pm \sqrt{1 - \frac{3}{4} \left(\frac{\sigma_z + p_i}{\sigma_{\text{yield}}} \right)^2} + \frac{1}{2} \left(\frac{\sigma_z + p_i}{\sigma_{\text{yield}}} \right).$$

$$\ldots \ldots \ldots \ldots \ldots \ldots \ldots \ldots \ldots \ldots \ldots (7.11)$$

This is the equation for the ellipse of plasticity shown in Fig. 7.13. With substitution of Eq. 7.3b with $r = r_i$ for σ_t, Eq. 7.11 defines the combinations of internal pressure, external pressure, and axial stress that will result in a *yield strength* mode of failure. It can be shown that for $p_i = 0$ and $\sigma_z = 0$, Eq. 7.11 reduces to Eq. 7.4a. The proof of this is left as a student exercise.

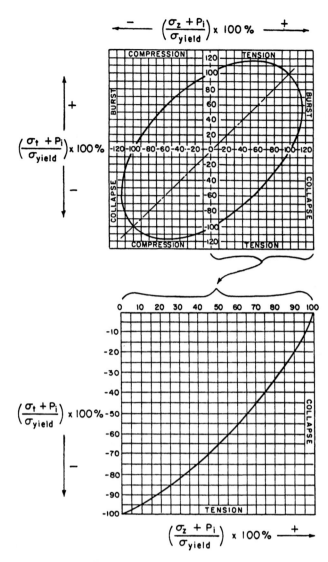

Fig. 7.13—Ellipse of plasticity.

Examination of the ellipse of plasticity (Fig. 7.13) shows that axial tension has a detrimental effect on collapse-pressure rating and a beneficial effect on burst-pressure rating. In contrast, axial compression has a detrimental effect on burst-pressure rating and a beneficial effect on collapse-pressure rating. In casing-design practice, it is customary to apply the ellipse of plasticity only when a detrimental effect would be observed.

Example 7.4. Compute the nominal collapse-pressure rating for 5.5-in., N-80 casing with a nominal wall thickness of 0.476 in. and a nominal weight per foot of 26 lbf/ft. In addition, determine the collapse pressure for in-service conditions in which the pipe is subjected to a 40,000-psi axial tension stress and a 10,000-psi internal pressure. Assume a yield strength mode of failure.

Solution. For a yield strength mode of failure, Eqs. 7.3b and 7.11 can be applied. Use of Eq. 7.3b with $r=r_i$ gives

$$\sigma_t = \frac{p_i(r_o^2+r_i^2)-2p_e r_o^2}{r_o^2-r_i^2}.$$

The ID of the casing is 4.548 in. Evaluation of the terms present in Eq. 7.11 for nominal conditions of zero axial stress and internal pressure gives

$$\left(\frac{\sigma_t+p_i}{\sigma_{\text{yield}}}\right) = \left(\frac{2r_o^2}{r_o^2-r_i^2}\right)\left(\frac{p_i-p_e}{\sigma_{\text{yield}}}\right)$$

$$= \left[\frac{2(5.5)^2}{(5.5^2-4.548^2)}\right]\left(\frac{p_i-p_e}{80,000}\right)$$

$$= \left(\frac{p_i-p_e}{12,649}\right)$$

$$= \frac{-p_e}{12,649};$$

$$\frac{\sigma_z+p_i}{\sigma_{\text{yield}}} = 0.$$

Use of these terms in Eq. 7.11 or Fig. 7.13 yields

$$\frac{-p_e}{12,649} = -1;$$

$$p_e = 12,649.$$

Note that after it is rounded to the nearest 10 psi, this value agrees with that given by Eq. 7.4 and shown in Table 7.6.

For in-service conditions of $\sigma_z=40,000$ psi and $p_i=10,000$ psi,

$$\left(\frac{\sigma_t+p_i}{\sigma_{\text{yield}}}\right) = \left(\frac{10,000-p_e}{12,649}\right);$$

$$\left(\frac{\sigma_z+p_i}{\sigma_{\text{yield}}}\right) = \frac{40,000+10,000}{80,000} = 0.625.$$

Use of these terms in Eq. 7.11 or Fig. 7.13 yields

$$\frac{10,000-p_e}{12,649} = -0.5284;$$

$$p_e = 10,000+0.5284(12,649) = 16,684 \text{ psi.}$$

This analysis indicates that, because of the combined stresses present, the pressure difference (external pressure minus internal pressure) required for collapse failure was reduced to 52.84% of the nominal collapse pressure rating given in Table 7.5.

In casing design practice, the ellipse of plasticity cannot be applied unless the assumption of a yield-strength mode of failure is known to be valid. For a simple stress state in which the internal pressure and axial tension are

TABLE 7.6—MINIMUM PERFORMANCE PROPERTIES OF CASING

1	2	3	4	5	6	7	8	9	10	11	12
					Threaded and Coupled			Extreme Line			
Size Outside Diameter (in.)	Nominal Weight Threads and Coupling (lbm/ft)	Grade	Wall Thickness (in.)	Inside Diameter (in.)	Drift Diameter (in.)	Outside Diameter of Coupling (in.)	Outside Diameter Special Clearance Coupling (in.)	Drift Diameter (in.)	Outside Diameter of Box Powertight (in.)	Collapse Resistance (psi)	Pipe Body Yield Strength (1,000 lbf)
4½	9.50	H-40	0.205	4.090	3.965	5.000	—	—	—	2,760	111
	9.50	J-55	0.205	4.090	3.965	5.000	—	—	—	3,310	152
	10.50	J-55	0.224	4.052	3.927	5.000	4.875	—	—	4,010	165
	11.60	J-55	0.250	4.000	3.875	5.000	4.875	—	—	4,960	184
	9.50	K-55	0.205	4.090	3.965	5.000	—	—	—	3,310	152
	10.50	K-55	0.224	4.052	3.927	5.000	4.875	—	—	4,010	165
	11.60	K-55	0.250	4.000	3.875	5.000	4.875	—	—	4,960	184
	11.60	C-75	0.250	4.000	3.875	5.000	4.875	—	—	6,100	250
	13.50	C-75	0.290	3.920	3.795	5.000	4.875	—	—	8,140	288
	11.60	L-80	0.250	4.000	3.875	5.000	4.875	—	—	6,350	267
	13.50	L-80	0.290	3.920	3.795	5.000	4.875	—	—	8,540	307
	11.60	N-80	0.250	4.000	3.875	5.000	4.875	—	—	6,350	267
	13.50	N-80	0.290	3.920	3.795	5.000	4.875	—	—	8,540	307
	11.60	C-90	0.250	4.000	3.875	5.000	4.875	—	—	6,820	300
	13.50	C-90	0.290	3.920	3.795	5.000	4.875	—	—	9,300	345
	11.60	C-95	0.250	4.000	3.875	5.000	4.875	—	—	7,030	317
	13.50	C-95	0.290	3.920	3.795	5.000	4.875	—	—	9,660	364
	11.60	P-110	0.250	4.000	3.875	5.000	4.875	—	—	7,580	367
	13.50	P-110	0.290	3.920	3.795	5.000	4.875	—	—	10,680	422
	15.10	P-110	0.337	3.826	3.701	5.000	4.875	—	—	14,350	485
5	11.50	J-55	0.220	4.560	4.435	5.563	—	—	—	3,060	182
	13.00	J-55	0.253	4.494	4.369	5.563	5.375	—	—	4,140	208
	15.00	J-55	0.296	4.408	4.283	5.563	5.375	4.151	5.360	5,560	241
	11.50	K-55	0.220	4.560	4.435	5.563	—	—	—	3,060	182
	13.00	K-55	0.253	4.494	4.369	5.563	5.375	—	—	4,140	208
	15.00	K-55	0.296	4.408	4.283	5.563	5.375	4.151	5.360	5,560	241
	15.00	C-75	0.296	4.408	4.283	5.563	5.375	4.151	5.360	6,940	328
	18.00	C-75	0.362	4.276	4.151	5.563	5.375	4.151	5.360	9,960	396
	21.40	C-75	0.437	4.126	4.001	5.563	5.375	—	—	11,970	470
	23.20	C-75	0.478	4.044	3.919	5.563	5.375	—	—	12,970	509
	24.10	C-75	0.500	4.000	3.875	5.563	5.375	—	—	13,500	530
	15.00	L-80	0.296	4.408	4.283	5.563	5.375	4.151	5.360	7,250	350
	18.00	L-80	0.362	4.276	4.151	5.563	5.375	4.151	5.360	10,500	422
	21.40	L-80	0.437	4.126	4.001	5.563	5.375	—	—	12,760	501
	23.20	L-80	0.478	4.044	3.919	5.563	5.375	—	—	13,830	543
	24.10	L-80	0.500	4.000	3.875	5.563	5.375	—	—	14,400	566
	15.00	N-80	0.296	4.408	4.283	5.563	5.375	4.151	5.360	7,250	350
	18.00	N-80	0.362	4.276	4.151	5.563	5.375	4.151	5.360	10,500	422
	21.40	N-80	0.437	4.126	4.001	5.563	5.375	—	—	12,760	501
	23.20	N-80	0.478	4.044	3.919	5.563	5.375	—	—	13,830	543
	24.10	N-80	0.500	4.000	3.875	5.563	5.375	—	—	14,400	566
	15.00	C-90	0.296	4.408	4.283	5.563	5.375	4.151	5.366	7,840	394
	18.00	C-90	0.362	4.276	4.151	5.563	5.375	4.151	5.366	11,530	475
	21.40	C-90	0.437	4.126	4.001	5.563	5.375	—	—	14,360	564
	23.20	C-90	0.478	4.044	3.919	5.563	5.375	—	—	15,560	611
	24.10	C-90	0.500	4.000	3.875	5.563	5.375	—	—	16,200	636
	15.00	C-95	0.296	4.408	4.283	5.563	5.375	4.151	5.360	8,110	416
	18.00	C-95	0.362	4.276	4.151	5.563	5.375	4.151	5.360	12,030	501
	21.40	C-95	0.437	4.126	4.001	5.563	5.375	—	—	15,160	595
	23.20	C-95	0.478	4.044	3.919	5.563	5.375	—	—	16,430	645
	24.10	C-95	0.500	4.000	3.875	5.563	5.375	—	—	17,100	672
	15.00	P-110	0.296	4.408	4.283	5.563	5.375	4.151	5.360	8,850	481
	18.00	P-110	0.362	4.276	4.151	5.563	5.375	4.151	5.360	13,470	580
	21.40	P-110	0.437	4.126	4.001	5.563	5.375	—	—	17,550	689
	23.20	P-110	0.478	4.044	3.919	5.563	5.375	—	—	19,020	747
	24.10	P-110	0.500	4.000	3.875	5.563	5.375	—	—	19,800	778

13	14	15	16	17	18	19	20	21	22	23	24	25	26	27
									*Joint Strength—1,000 lbf					
	**Internal Pressure Resistance, psi								Threaded and Coupled					
			Buttress Thread						Buttress Thread					
Plain End or Extreme Line	Round Thread		Regular Coupling		Special Clearance Coupling		Round Thread		Regular Coupling	Regular Coupling Higher Grade†	Special Clearance Coupling	Special Clearance Coupling Higher Grade†	Extreme Line	
	Short	Long	Same Grade	Higher Grade	Same Grade	Higher Grade	Short	Long					Standard Joint	Optional Joint
3,190	3,190	—	—	—	—	—	77	—	—	—	—	—	—	—
4,380	4,380	—	—	—	—	—	101	—	—	—	—	—	—	—
4,790	4,790	—	4,790	4,790	4,790	4,790	132	—	203	203	203	203	—	—
5,350	5,350	5,350	5,350	5,350	5,350	5,350	154	162	225	225	225	225	—	—
4,380	4,380	—	—	—	—	—	112	—	—	—	—	—	—	—
4,790	4,790	—	4,790	4,790	4,790	4,790	146	—	249	249	249	249	—	—
5,350	5,350	5,350	5,350	5,350	5,350	5,350	170	180	277	277	277	277	—	—
7,290	—	7,290	7,290	—	7,290	—	—	212	288	—	288	—	—	—
8,460	—	8,460	8,460	—	7,490	—	—	257	331	—	320	—	—	—
7,780	—	7,780	7,780	7,780	7,780	7,780	—	212	291	—	291	—	—	—
9,020	—	9,020	9,020	9,020	7,990	9,020	—	257	334	—	320	—	—	—
7,780	—	7,780	7,780	7,780	7,780	7,780	—	223	304	304	304	304	—	—
9,020	—	9,020	9,020	9,020	7,990	9,020	—	270	349	349	337	349	—	—
8,750	—	8,750	8,750	—	8,750	—	—	223	309	—	309	—	—	—
10,150	—	10,150	10,150	—	9,000	—	—	270	355	—	337	—	—	—
9,240	—	9,240	9,240	—	9,240	—	—	234	325	325	325	—	—	—
10,710	—	10,710	10,710	—	9,490	—	—	284	374	374	353	—	—	—
10,690	—	10,690	10,690	10,690	10,690	10,690	—	279	385	385	385	385	—	—
12,410	—	12,410	12,410	12,410	10,990	12,410	—	338	443	443	421	443	—	—
14,420	—	14,420	13,460	14,420	10,990	13,910	—	406	509	509	421	509	—	—
4,240	4,240	—	—	—	—	—	133	—	—	—	—	—	—	—
4,870	4,870	4,870	4,870	4,870	4,870	4,870	169	182	252	252	252	252	—	—
5,700	5,700	5,700	5,700	5,700	5,130	5,700	207	223	293	293	287	293	328	—
4,240	4,240	—	—	—	—	—	147	—	—	—	—	—	—	—
4,870	4,870	4,870	4,870	4,870	4,870	4,870	186	201	309	309	309	309	—	—
5,700	5,700	5,700	5,700	5,700	5,130	5,700	228	246	359	359	359	359	416	—
7,770	—	7,770	7,770	—	6,990	—	—	295	375	—	364	—	416	—
9,500	—	9,500	9,290	—	6,990	—	—	376	452	—	364	—	446	—
11,470	—	10,140	9,290	—	6,990	—	—	466	510	—	364	—	—	—
12,550	—	10,140	9,290	—	7,000	—	—	513	510	—	364	—	—	—
13,130	—	10,140	9,290	—	6,990	—	—	538	510	—	364	—	—	—
8,290	—	8,290	8,290	8,290	7,460	8,290	—	295	379	—	364	—	416	—
10,140	—	10,140	9,910	10,140	7,460	10,140	—	376	457	—	364	—	446	—
12,240	—	10,810	9,910	—	7,460	—	—	466	510	—	364	—	—	—
13,380	—	10,820	9,910	—	7,460	—	—	513	510	—	364	—	—	—
14,000	—	10,810	9,910	—	7,460	—	—	538	510	—	364	—	—	—
8,290	—	8,290	8,290	8,290	7,460	8,290	—	311	396	396	383	396	437	—
10,140	—	10,140	9,910	10,140	7,460	10,140	—	396	477	477	383	477	469	—
12,240	—	10,810	9,910	12,240	7,460	10,250	—	490	537	566	383	479	—	—
13,380	—	10,820	9,910	13,380	7,460	10,260	—	540	537	614	383	479	—	—
14,000	—	10,810	9,910	13,620	7,460	10,250	—	567	537	639	383	479	—	—
9,320	—	9,320	9,320	—	8,400	—	—	311	404	—	383	—	430	—
11,400	—	11,400	11,150	—	8,400	—	—	396	487	—	383	—	469	—
13,770	—	12,170	11,150	—	8,400	—	—	490	537	—	383	—	—	—
15,060	—	12,170	11,150	—	8,400	—	—	540	537	—	383	—	—	—
15,750	—	12,170	11,150	—	8,400	—	—	567	537	—	383	—	—	—
9,840	—	9,840	9,840	—	8,850	—	—	326	424	—	402	—	459	—
12,040	—	12,040	11,770	—	8,850	—	—	416	512	—	402	—	493	—
14,530	—	12,840	11,770	—	8,850	—	—	515	563	—	402	—	—	—
15,890	—	12,850	11,770	—	8,850	—	—	567	563	—	402	—	—	—
16,630	—	12,850	11,770	—	8,850	—	—	595	563	—	402	—	—	—
11,400	—	11,400	11,400	11,400	10,250	11,400	—	388	503	503	479	503	547	—
13,940	—	13,940	13,620	13,940	10,250	13,940	—	495	606	606	479	606	587	—
16,820	—	14,870	13,620	16,820	10,250	13,980	—	613	671	720	479	613	—	—
18,400	—	14,880	13,630	18,400	10,260	13,990	—	675	671	780	479	613	—	—
19,250	—	14,870	13,620	18,580	10,250	13,980	—	708	671	812	479	613	—	—

TABLE 7.6—MINIMUM PERFORMANCE PROPERTIES OF CASING (cont.)

1	2	3	4	5	6	7	8	9	10	11	12
						Threaded and Coupled		Extreme Line			
Size Outside Diameter (in.)	Nominal Weight Threads and Coupling (lbm/ft)	Grade	Wall Thickness (in.)	Inside Diameter (in.)	Drift Diameter (in.)	Outside Diameter of Coupling (in.)	Outside Diameter Special Clearance Coupling (in.)	Drift Diameter (in.)	Outside Diameter of Box Powertight (in.)	Collapse Resistance (psi)	Pipe Body Yield Strength (1,000 lbf)
5½	14.00	H-40	0.244	5.012	4.887	6.050	—	—	—	2,620	161
	14.00	J-55	0.244	5.012	4.887	6.050	—	—	—	3,120	222
	15.50	J-55	0.275	4.950	4.825	6.050	5.875	4.653	5.860	4,040	248
	17.00	J-55	0.304	4.892	4.767	6.050	5.875	4.653	5.860	4,910	273
	14.00	K-55	0.244	5.012	4.887	6.050	—	—	—	3,120	222
	15.50	K-55	0.275	4.950	4.825	6.050	5.875	4.653	5.860	4,040	248
	17.00	K-55	0.304	4.892	4.767	6.050	5.875	4.653	5.860	4,910	273
	17.00	C-75	0.304	4.892	4.767	6.050	5.875	4.653	5.860	6,040	372
	20.00	C-75	0.361	4.778	4.653	6.050	5.875	4.653	5.860	8,410	437
	23.00	C-75	0.415	4.670	4.545	6.050	5.875	4.545	5.860	10,470	497
	17.00	L-80	0.304	4.892	4.767	6.050	5.875	4.653	5.860	6,280	397
	20.00	L-80	0.361	4.778	4.653	6.050	5.875	4.653	5.860	8,830	466
	23.00	L-80	0.415	4.670	4.545	6.050	5.875	4.545	5.860	11,160	530
	17.00	N-80	0.304	4.892	4.767	6.050	5.875	4.653	5.860	6,280	397
	20.00	N-80	0.361	4.778	4.653	6.050	5.875	4.653	5.860	8,830	466
	23.00	N-80	0.415	4.670	4.545	6.050	5.875	4.545	5.860	11,160	530
	17.00	C-90	0.304	4.892	4.767	6.050	5.875	4.653	5.860	6.740	447
	20.00	C-90	0.361	4.778	4.653	6.050	5.875	4.653	5.860	9,630	525
	23.00	C-90	0.415	4.670	4.545	6.050	5.875	4.545	5.860	12,380	597
	26.00	C-90	0.476	4.548	4.423	6.050	5.875	—	—	14,240	676
	35.00	C-90	0.650	4.200	4.075	6.050	5.875	—	—	18,760	891
	17.00	C-95	0.304	4.892	4.767	6.050	5.875	4.653	5.860	6,940	471
	20.00	C-95	0.361	4.778	4.653	6.050	5.875	4.653	5.860	10,010	554
	23.00	C-95	0.415	4.670	4.545	6.050	5.875	4.545	5.860	12,940	630
	17.00	P-110	0.304	4.892	4.767	6.050	5.875	4.653	5.860	7,480	546
	20.00	P-110	0.361	4.778	4.653	6.050	5.875	4.653	5.860	11,100	641
	23.00	P-110	0.415	4.670	4.545	6.050	5.875	4.545	5.860	14,540	729
6⅝	20.00	H-40	0.288	6.049	5.924	7.390	—	—	—	2,520	229
	20.00	J-55	0.288	6.049	5.924	7.390	7.000	—	—	2,970	315
	24.00	J-55	0.352	5.921	5.796	7.390	7.000	5.730	7.000	4,560	382
	20.00	K-55	0.288	6.049	5.924	7.390	7.000	—	—	2,970	315
	24.00	K-55	0.352	5.921	5.796	7.390	7.000	5.730	7.000	4,560	382
	24.00	C-75	0.352	5.921	5.796	7.390	7.000	5.730	7.000	5,550	520
	28.00	C-75	0.417	5.791	5.666	7.390	7.000	5.666	7.000	7,790	610
	32.00	C-75	0.475	5.675	5.550	7.390	7.000	5.550	7.000	9,800	688
	24.00	L-80	0.352	5.921	5.796	7.390	7.000	5.730	7.000	5,760	555
	28.00	L-80	0.417	5.791	5.666	7.390	7.000	5.666	7.000	8,170	651
	32.00	L-80	0.475	5.675	5.550	7.390	7.000	5.550	7.000	10,320	734
	24.00	N-80	0.352	5.921	5.796	7.390	7.000	5.730	7.000	5,760	555
	28.00	N-80	0.417	5.791	5.666	7.390	7.000	5.666	7.000	8,170	651
	32.00	N-80	0.475	5.675	5.550	7.390	7.000	5.550	7.000	10,320	734
	24.00	C-90	0.352	5.921	5.796	7.390	7.000	5.730	7.000	6,140	624
	28.00	C-90	0.417	5.791	5.666	7.390	7.000	5.666	7.000	8,880	732
	32.00	C-90	0.475	5.675	5.550	7.390	7.000	5.550	7.000	11,330	826
	24.00	C-95	0.352	5.921	5.796	7.390	7.000	5.730	7.000	6,310	659
	28.00	C-95	0.417	5.791	5.666	7.390	7.000	5.666	7.000	9,220	773
	32.00	C-95	0.475	5.675	5.550	7.390	7.000	5.550	7.000	11,810	872
	24.00	P-110	0.352	5.921	5.796	7.390	7.000	5.730	7.000	6,730	763
	28.00	P-110	0.417	5.791	5.666	7.390	7.000	5.666	7.000	10,160	895
	32.00	P-110	0.475	5.675	5.550	7.390	7.000	5.550	7.000	13,220	1009

13	14	15	16	17	18	19	20	21	22	23	24	25	26	27
							*Joint Strength—1,000 lbf							
**Internal Pressure Resistance, psi							Threaded and Coupled							
			Buttress Thread						Buttress Thread					
Plain End or Extreme Line	Round Thread		Regular Coupling		Special Clearance Coupling		Round Thread		Regular Coupling	Regular Coupling Higher Grade†	Special Clearance Coupling	Special Clearance Coupling Higher Grade†	Extreme Line	
	Short	Long	Same Grade	Higher Grade	Same Grade	Higher Grade	Short	Long					Standard Joint	Optional Joint
3,110	3,110	—	—	—	—	—	130	—	—	—	—	—	—	—
4,270	4,270	—	—	—	—	—	172	—	—	—	—	—	—	—
4,810	4,810	4,810	4,810	4,810	4,730	4,810	202	217	300	300	300	300	339	339
5,320	5,320	5,320	5,320	5,320	4,730	5,320	229	247	329	329	318	329	372	372
4,270	4,270	—	—	—	—	—	189	—	—	—	—	—	—	—
4,810	4,810	4,810	4,810	4,810	4,730	4,810	222	239	366	366	366	366	429	429
5,320	5,320	5,320	5,320	5,320	4,730	5,320	252	272	402	402	402	402	471	471
7,250	—	7,250	7,250	—	6,450	—	—	327	423	—	403	—	471	471
8,610	—	8,610	8,430	—	6,450	—	—	403	497	—	403	—	497	479
9,900	—	9,260	8,430	—	6,450	—	—	473	550	—	403	—	549	479
7,740	—	7,740	7,740	7,740	6,880	7,740	—	338	428	—	403	—	471	471
9,190	—	9,190	8,990	9,190	6,880	9,190	—	416	503	—	403	—	497	479
10,560	—	9,880	8,990	10,560	6,880	9,460	—	489	550	—	403	—	549	479
7,740	—	7,740	7,740	7,740	6,880	7,740	—	348	446	446	424	446	496	496
9,190	—	9,190	8,990	9,190	6,880	9,190	—	428	524	524	424	524	523	504
10,560	—	9,880	8,990	10,560	6,880	9,460	—	502	579	596	424	530	577	504
8,710	—	8,710	8,710	—	7,740	—	—	356	456	—	424	—	496	496
10,340	—	10,340	10,120	—	7,740	—	—	438	536	—	424	—	523	504
11,880	—	11,110	10,120	—	7,740	—	—	514	580	—	424	—	577	504
13,630	—	11,110	10,120	—	7,740	—	—	598	580	—	424	—	—	—
18,610	—	11,110	10,120	—	7,740	—	—	614	580	—	424	—	—	—
9,190	—	9,910	9,190	—	8,170	—	—	374	480	—	445	—	521	521
10,910	—	10,910	10,680	—	8,170	—	—	460	563	—	445	—	549	530
12,540	—	11,730	10,680	—	8,170	—	—	540	608	—	445	—	606	530
10,640	—	10,640	10,640	10,640	9,460	10,640	—	445	568	568	530	568	620	620
12,640	—	12,640	12,360	12,640	9,460	11,880	—	548	667	667	530	667	654	630
14,520	—	13,160	12,360	14,520	9,460	11,880	—	643	724	759	530	668	772	630
3,040	3,040	—	—	—	—	—	184	—	—	—	—	—	—	—
4,180	4,180	4,180	4,180	4,180	4,060	4,180	245	266	374	374	374	374	—	—
5,110	5,110	5,110	5,110	5,110	4,060	5,110	314	340	453	453	390	453	477	477
4,180	4,180	4,180	4,180	4,180	4,060	4,180	267	290	453	453	453	453	—	—
5,110	5,110	5,110	5,110	5,110	4,060	5,110	342	372	548	548	494	520	605	605
6,970	—	6,970	6,970	—	5,540	—	—	453	583	—	494	—	605	605
8,260	—	8,260	8,260	—	5,540	—	—	552	683	—	494	—	648	644
9,410	—	9,410	9,200	—	5,540	—	—	638	771	—	494	—	717	644
7,440	—	7,440	7,440	—	5,910	—	—	473	592	—	494	—	605	605
8,810	—	8,810	8,810	—	5,910	—	—	576	693	—	494	—	648	644
10,040	—	10,040	9,820	—	5,910	—	—	666	783	—	494	—	717	644
7,440	—	7,440	7,440	7,440	5,910	7,440	—	481	615	615	520	615	637	637
8,810	—	8,810	8,810	8,810	5,910	8,120	—	586	721	721	520	650	682	678
10,040	—	10,040	9,820	10,040	5,910	8,120	—	677	814	814	520	650	755	678
8,370	—	8,370	8,370	—	6,650	—	—	520	633	—	520	—	637	637
9,910	—	9,910	9,910	—	6,650	—	—	633	742	—	520	—	682	678
11,290	—	11,290	11,050	—	6,650	—	—	732	837	—	520	—	755	678
8,830	—	8,830	8,830	—	7,020	—	—	546	665	—	546	—	668	668
10,460	—	10,460	10,460	—	7,020	—	—	665	780	—	546	—	716	712
11,920	—	11,830	11,660	—	7,020	—	—	769	880	—	546	—	793	712
10,230	—	10,230	10,230	10,230	8,120	8,310	—	641	786	786	65	786	796	796
12,120	—	11,830	12,120	12,120	8,120	8,310	—	781	922	922	650	832	852	848
13,800	—	11,830	13,500	13,800	8,120	8,310	—	904	1,040	1,040	650	832	944	848

TABLE 7.6—MINIMUM PERFORMANCE PROPERTIES OF CASING (cont.)

1	2	3	4	5	6	7	8	9	10	11	12
					Threaded and Coupled			Extreme Line			
Size Outside Diameter (in.)	Nominal Weight Threads and Coupling (lbm/ft)	Grade	Wall Thickness (in.)	Inside Diameter (in.)	Drift Diameter (in.)	Outside Diameter of Coupling (in.)	Outside Diameter Special Clearance Coupling (in.)	Drift Diameter (in.)	Outside Diameter of Box Powertight (in.)	Collapse Resistance (psi)	Pipe Body Yield Strength (1,000 lbf)
7	17.00	H-40	0.231	6.538	6.413	7.656	—	—	—	1,420	196
	20.00	H-40	0.272	6.456	6.331	7.656	—	—	—	1,970	230
	20.00	J-55	0.272	6.456	6.331	7.656	—	—	—	2,270	316
	23.00	J-55	0.317	6.366	6.241	7.656	7.375	6.151	7.390	3,270	366
	26.00	J-55	0.362	6.276	6.151	7.656	7.375	6.151	7.390	4,320	415
	20.00	K-55	0.272	6.456	6.331	7.656	—	—	—	2,270	316
	23.00	K-55	0.317	6.366	6.241	7.656	7.375	6.151	7.390	3,270	366
	26.00	K-55	0.362	6.276	6.151	7.656	7.375	6.151	7.390	4,320	415
	23.00	C-75	0.317	6.366	6.241	7.656	7.375	6.151	7.390	3,750	499
	26.00	C-75	0.362	6.276	6.151	7.656	7.375	6.151	7.390	5,220	566
	29.00	C-75	0.408	6.184	6.059	7.656	7.375	6.059	7.390	6,730	634
	32.00	C-75	0.453	6.094	5.969	7.656	7.375	5.969	7.390	8,200	699
	35.00	C-75	0.498	6.004	5.879	7.656	7.375	5.879	7.530	9,670	763
	38.00	C-75	0.540	5.920	5.795	7.656	7.375	5.795	7.530	10,680	822
	23.00	L-80	0.317	6.366	6.241	7.656	7.375	6.151	7.390	3,830	532
	26.00	L-80	0.362	6.276	6.151	7.656	7.375	6.151	7.390	5,410	604
	29.00	L-80	0.408	6.184	6.059	7.656	7.375	6.059	7.390	7,020	676
	32.00	L-80	0.453	6.094	5.969	7.656	7.375	5.969	7.390	8,610	745
	35.00	L-80	0.498	6.004	5.879	7.656	7.375	5.879	7.530	10,180	814
	38.00	L-80	0.540	5.920	5.795	7.656	7.375	5.795	7.530	11,390	877
	23.00	N-80	0.317	6.366	6.241	7.656	7.375	6.151	7.390	3,830	532
	26.00	N-80	0.362	6.276	6.151	7.656	7.375	6.151	7.390	5,410	604
	29.00	N-80	0.408	6.184	6.059	7.656	7.375	6.059	7.390	7,020	676
	32.00	N-80	0.453	6.094	5.969	7.656	7.375	5.969	7.390	8,610	745
	35.00	N-80	0.498	6.004	5.879	7.656	7.375	5.879	7.530	10,180	814
	38.00	N-80	0.540	5.920	5.795	7.656	7.375	5.795	7.530	11,390	877
	23.00	C-90	0.317	6.366	6.241	7.656	7.375	6.151	7.390	4,030	599
	26.00	C-90	0.362	6.276	6.151	7.656	7.375	6.151	7.390	5,740	679
	29.00	C-90	0.408	6.184	6.059	7.656	7.375	6.059	7.390	7,580	760
	32.00	C-90	0.453	6.094	5.969	7.656	7.375	5.969	7.390	9,380	839
	35.00	C-90	0.498	6.004	5.879	7.656	7.375	5.879	7.530	11,170	915
	38.00	C-90	0.540	5.920	5.795	7.656	7.375	5.795	7.530	12,820	986
	23.00	C-95	0.317	6.366	6.241	7.656	7.375	6.151	7.390	4,140	632
	26.00	C-95	0.362	6.276	6.151	7.656	7.375	6.151	7.390	5,880	717
	29.00	C-95	0.408	6.184	6.059	7.656	7.375	6.059	7.390	7,830	803
	32.00	C-95	0.453	6.094	5.969	7.656	7.375	5.969	7.390	9,750	885
	35.00	C-95	0.498	6.004	5.879	7.656	7.375	5.879	7.530	11,650	966
	38.00	C-95	0.540	5.920	5.795	7.656	7.375	5.795	7.530	13,440	1,041
	26.00	P-110	0.362	6.276	6.151	7.656	7.375	6.151	7.390	6,230	830
	29.00	P-110	0.408	6.184	6.059	7.656	7.375	6.059	7.390	8,530	929
	32.00	P-110	0.453	6.094	5.969	7.656	7.375	5.969	7.390	10,780	1,025
	35.00	P-110	0.498	6.004	5.879	7.656	7.375	5.879	7.530	13,020	1,119
	38.00	P-110	0.540	5.920	5.795	7.656	7.375	5.795	7.530	15,140	1,205
7⅝	24.00	H-40	0.300	7.025	6.900	8.500	—	—	—	2,030	276
	26.40	J-55	0.328	6.969	6.844	8.500	8.125	6.750	8.010	2,890	414
	26.40	K-55	0.328	6.969	6.844	8.500	8.125	6.750	8.010	2,890	414
	26.40	C-75	0.328	6.969	6.844	8.500	8.125	6.750	8.010	3,280	564
	29.70	C-75	0.375	6.875	6.750	8.500	8.125	6.750	8.010	4,650	641
	33.70	C-75	0.430	6.765	6.640	8.500	8.125	6.640	8.010	6,300	729
	39.00	C-75	0.500	6.625	6.500	8.500	8.125	6.500	8.010	8,400	839
	42.80	C-75	0.562	6.501	6.376	8.500	8.125	—	—	10,240	935
	45.30	C-75	0.595	6.435	6.310	8.500	8.125	—	—	10,790	986
	47.10	C-75	0.625	6.375	6.250	8.500	8.125	—	—	11,290	1,031

13	14	15	16	17	18	19	20	21	22	23	24	25	26	27
										*Joint Strength—1,000 lbf				
	Internal Pressure Resistance, psi								Threaded and Coupled					
			Buttress Thread						Buttress Thread					
Plain End or Extreme Line	Round Thread		Regular Coupling		Special Clearance Coupling		Round Thread		Regular Coupling	Regular Coupling Higher Grade†	Special Clearance Coupling	Special Clearance Coupling Higher Grade†	Extreme Line	
	Short	Long	Same Grade	Higher Grade	Same Grade	Higher Grade	Short	Long	Regular Coupling	Higher Grade†	Special Clearance Coupling	Higher Grade†	Standard Joint	Optional Joint
2,310	2,310	—	—	—	—	—	122	—	—	—	—	—	—	—
2,720	2,720	—	—	—	—	—	176	—	—	—	—	—	—	—
3,740	3,740	—	—	—	—	—	234	—	—	—	—	—	—	—
4,360	4,360	4,360	4,360	4,360	3,950	4,360	284	313	432	432	421	432	499	499
4,980	4,980	4,980	4,980	4,980	3,950	4,980	334	367	490	490	421	490	506	506
3,740	3,740	—	—	—	—	—	254	—	—	—	—	—	—	—
4,360	4,360	4,360	4,360	4,360	3,950	4,360	309	341	522	522	522	522	632	632
4,980	4,980	4,980	4,980	4,980	3,950	4,980	364	401	592	592	533	561	641	641
5,940	—	5,940	5,940	—	5,380	—	—	416	557	—	533	—	632	632
6,790	—	6,790	6,790	—	5,380	—	—	489	631	—	533	—	641	641
7,650	—	7,650	7,650	—	5,380	—	—	562	707	—	533	—	685	674
8,490	—	8,490	7,930	—	5,380	—	—	633	779	—	533	—	761	674
9,340	—	8,660	7,930	—	5,380	—	—	703	833	—	533	—	850	761
10,120	—	8,660	7,930	—	5,380	—	—	767	833	—	533	—	917	761
6,340	—	6,340	6,340	6,340	5,740	6,340	—	435	565	—	533	—	632	632
7,240	—	7,240	7,240	7,240	5,740	7,240	—	511	641	—	533	—	641	641
8,160	—	8,160	8,160	8,160	5,740	7,890	—	587	718	—	533	—	685	674
9,060	—	9,060	8,460	9,060	5,740	7,890	—	661	791	—	533	—	761	674
9,960	—	9,240	8,460	9,960	5,740	7,890	—	734	833	—	533	—	850	761
10,800	—	9,240	8,460	10,800	5,740	7,890	—	801	833	—	533	—	917	761
6,340	—	6,340	6,340	6,340	5,740	6,340	—	442	588	588	561	588	666	666
7,240	—	7,240	7,240	7,240	5,740	7,240	—	519	667	667	561	667	675	675
8,160	—	8,160	8,160	8,160	5,740	7,890	—	597	746	746	561	702	721	709
9,060	—	9,060	8,460	9,060	5,740	7,890	—	672	823	823	561	702	801	709
9,960	—	9,240	8,460	9,960	5,740	7,890	—	746	876	898	561	702	895	801
10,800	—	9,240	8,460	10,800	5,740	7,890	—	814	876	968	561	702	965	801
7,130	—	7,130	7,130	—	6,450	—	—	447	605	—	561	—	666	666
8,150	—	8,150	8,150	—	6,450	—	—	563	687	—	561	—	675	675
9,180	—	9,180	9,180	—	6,450	—	—	648	768	—	561	—	721	709
10,190	—	9,520	9,520	—	6,450	—	—	729	847	—	561	—	801	709
11,210	—	9,520	9,520	—	6,450	—	—	809	876	—	561	—	895	801
12,150	—	9,520	9,520	—	6,450	—	—	883	876	—	561	—	965	801
7,530	—	7,530	7,530	—	6,810	—	—	505	636	—	589	—	699	699
8,600	—	8,600	8,600	—	6,810	—	—	593	722	—	589	—	709	709
9,690	—	9,520	9,690	—	6,810	—	—	683	808	—	589	—	757	744
10,760	—	9,520	10,050	—	6,810	—	—	768	891	—	589	—	841	744
11,830	—	9,520	10,050	—	6,810	—	—	853	920	—	589	—	940	841
12,820	—	9,520	10,050	—	6,810	—	—	931	920	—	589	—	1,013	841
9,960	—	9,520	9,960	9,960	7,480	7,480	—	693	853	853	702	853	844	844
11,220	—	9,520	11,220	11,220	7,480	7,480	—	797	955	955	702	898	902	886
12,460	—	9,520	11,640	11,790	7,480	7,480	—	897	1,053	1,053	702	898	1,002	886
13,700	—	9,520	11,640	11,790	7,480	7,480	—	996	1,096	1,150	702	898	1,118	1,002
14,850	—	9,520	11,640	11,790	7,480	7,480	—	1,087	1,096	1,239	702	898	1,207	1,002
2,750	2,750	—	—	—	—	—	212	—	—	—	—	—	—	—
4,140	4,140	4,140	4,140	4,140	4,140	4,140	315	346	483	483	483	483	553	553
4,140	4,140	4,140	4,140	4,140	4,140	4,140	342	377	581	581	581	581	700	700
5,650	—	5,650	5,650	—	5,650	—	—	461	624	—	624	—	700	700
6,450	—	6,450	6,450	—	6,140	—	—	542	709	—	709	—	700	700
7,400	—	7,400	7,400	—	6,140	—	—	635	806	—	735	—	766	744
8,610	—	8,610	8,610	—	6,140	—	—	751	929	—	735	—	851	744
9,670	—	9,670	9,190	—	6,140	—	—	852	1,035	—	735	—	—	—
10,240	—	9,840	9,180	—	6,140	—	—	905	1,090	—	764	—	—	—
10,760	—	9,840	9,190	—	6,140	—	—	953	1,140	—	735	—	—	—

TABLE 7.6—MINIMUM PERFORMANCE PROPERTIES OF CASING (cont.)

1	2	3	4	5	6	7	8	9	10	11	12
					Threaded and Coupled			Extreme Line			
Size Outside Diameter (in.)	Nominal Weight Threads and Coupling (lbm/ft)	Grade	Wall Thickness (in.)	Inside Diameter (in.)	Drift Diameter (in.)	Outside Diameter of Coupling (in.)	Outside Diameter Special Clearance Coupling (in.)	Drift Diameter (in.)	Outside Diameter of Box Powertight (in.)	Collapse Resistance (psi)	Pipe Body Yield Strength (1,000 lbf)
7⅝	26.40	L-80	0.328	6.969	6.844	8.500	8.125	6.750	8.010	3,400	602
	29.70	L-80	0.375	6.875	6.750	8.500	8.125	6.750	8.010	4,790	683
	33.70	L-80	0.430	6.765	6.640	8.500	8.125	6.640	8.010	6,560	778
	39.00	L-80	0.500	6.625	6.500	8.500	8.125	6.500	8.010	8,820	895
	42.80	L-80	0.562	6.501	6.376	8.500	8.125	—	—	10,810	998
	45.30	L-80	0.595	6.435	6.310	8.500	8.125	—	—	11,510	1,051
	47.10	L-80	0.625	6.375	6.250	8.500	8.125	—	—	12,040	1,100
	26.40	N-80	0.328	6.969	6.844	8.500	8.125	6.750	8.010	3,400	602
	29.70	N-80	0.375	6.875	6.750	8.500	8.125	6.750	8.010	4,790	683
	33.70	N-80	0.430	6.765	6.640	8.500	8.125	6.640	8.010	6,560	778
	39.00	N-80	0.500	6.625	6.500	8.500	8.125	6.500	8.010	8,820	895
	42.80	N-80	0.562	6.501	6.376	8.500	8.125	—	—	10,810	998
	45.30	N-80	0.595	6.435	6.310	8.500	8.125	—	—	11,510	1,051
	47.10	N-80	0.625	6.375	6.250	8.500	8.125	—	—	12,040	1,100
	26.40	C-90	0.328	6.969	6.844	8.500	8.125	6.750	8.010	3,610	677
	29.70	C-90	0.375	6.875	6.750	8.500	8.125	6.750	8.010	5,040	769
	33.70	C-90	0.430	6.765	6.640	8.500	8.125	6.640	8.010	7,050	875
	39.00	C-90	0.500	6.625	6.500	8.500	8.125	6.500	8.010	9,620	1,007
	42.80	C-90	0.562	6.501	6.376	8.500	8.125	—	—	11,890	1,122
	45.30	C-90	0.595	6.435	6.310	8.500	8.125	—	—	12,950	1,183
	47.10	C-90	0.625	6.375	6.250	8.500	8.125	—	—	13,540	1,237
	26.40	C-95	0.328	6.969	6.844	8.500	8.125	6.750	8.010	3,710	714
	29.70	C-95	0.375	6.875	6.750	8.500	8.125	6.750	8.010	5,140	811
	33.70	C-95	0.430	6.765	6.640	8.500	8.125	6.640	8.010	7,280	923
	39.00	C-95	0.500	6.625	6.500	8.500	8.125	6.500	8.010	10,000	1,063
	42.80	C-95	0.562	6.501	6.376	8.500	8.125	—	—	12,410	1,185
	45.30	C-95	0.595	6.435	6.310	8.500	8.125	—	—	13,660	1,248
	47.10	C-95	0.625	6.375	6.250	8.500	8.125	—	—	14,300	1,306
	29.70	P-110	0.375	6.875	6.750	8.500	8.125	6.750	8.010	5,350	940
	33.70	P-110	0.430	6.765	6.640	8.500	8.125	6.640	8.010	7,870	1,069
	39.00	P-110	0.500	6.625	6.500	8.500	8.125	6.500	8.010	11,080	1,231
	42.80	P-110	0.562	6.501	6.376	8.500	8.125	—	—	13,920	1,372
	45.30	P-110	0.595	6.435	6.310	8.500	8.125	—	—	15,430	1,446
	47.10	P-110	0.625	6.375	6.250	8.500	8.125	—	—	16,550	1,512
8⅝	28.00	H-40	0.304	8.017	7.892	9.625	—	—	—	1,610	318
	32.00	H-40	0.352	7.921	7.796	9.625	—	—	—	2,200	366
	24.00	J-55	0.264	8.097	7.972	9.625	—	—	—	1,370	381
	32.00	J-55	0.352	7.921	7.796	9.625	9.125	7.700	9.120	2,530	503
	36.00	J-55	0.400	7.825	7.700	9.625	9.125	7.700	9.120	3,450	568
	24.00	K-55	0.264	8.097	7.972	9.625	—	—	—	1,370	381
	32.00	K-55	0.352	7.921	7.796	9.625	9.125	7.700	9.120	2,530	503
	36.00	K-55	0.400	7.825	7.700	9.625	9.125	7.700	9.120	3,450	568
	36.00	C-75	0.400	7.825	7.700	9.625	9.125	7.700	9.120	4,000	775
	40.00	C-75	0.450	7.725	7.600	9.625	9.125	7.600	9.120	5,330	867
	44.00	C-75	0.500	7.625	7.500	9.625	9.125	7.500	9.120	6,660	957
	49.00	C-75	0.557	7.511	7.386	9.625	9.125	7.386	9.120	8,180	1,059
	36.00	L-80	0.400	7.825	7.700	9.625	9.125	7.700	9.120	4,100	827
	40.00	L-80	0.450	7.725	7.600	9.625	9.125	7.600	9.120	5,520	925
	44.00	L-80	0.500	7.625	7.500	9.625	9.125	7.500	9.120	6,950	1,021
	49.00	L-80	0.557	7.511	7.386	9.625	9.125	7.386	9.120	8,580	1,129
	36.00	N-80	0.400	7.825	7.700	9.625	9.125	7.700	9.120	4,100	827
	40.00	N-80	0.450	7.725	7.600	9.625	9.125	7.600	9.120	5,520	925
	44.00	N-80	0.500	7.625	7.500	9.625	9.125	7.500	9.120	6,950	1,021
	49.00	N-80	0.557	7.511	7.386	9.625	9.125	7.386	9.120	8,580	1,129
	36.00	C-90	0.400	7.825	7.700	9.625	9.125	7.700	9.120	4,250	930
	40.00	C-90	0.450	7.725	7.600	9.625	9.125	7.600	9.120	5,870	1,040
	44.00	C-90	0.500	7.625	7.500	9.625	9.125	7.500	9.120	7,490	1,149
	49.00	C-90	0.557	7.511	7.386	9.625	9.125	7.386	9.120	9,340	1,271

13	14	15	16	17	18	19	20	21	22	23	24	25	26	27
									*Joint Strength—1,000 lbf					
		**Internal Pressure Resistance, psi							Threaded and Coupled					
			Buttress Thread						Buttress Thread					
Plain End or Extreme Line	Round Thread		Regular Coupling		Special Clearance Coupling		Round Thread		Regular Coupling	Regular Coupling Higher Grade†	Special Clearance Coupling	Special Clearance Coupling Higher Grade†	Extreme Line	
	Short	Long	Same Grade	Higher Grade	Same Grade	Higher Grade	Short	Long					Standard Joint	Optional Joint
6,020	—	6,020	6,020	6,020	6,020	6,020	—	482	635	—	635	—	700	700
6,890	—	6,890	6,890	6,890	6,550	6,890	—	566	721	—	721	—	700	700
7,900	—	7,900	7,900	7,900	6,550	7,900	—	664	820	—	735	—	766	744
9,180	—	9,180	9,180	9,180	6,550	9,000	—	786	945	—	735	—	851	744
10,320	—	10,320	9,790	—	6,550	—	—	892	1,053	—	735	—	—	—
10,920	—	10,500	9,790	—	6,550	—	—	947	1,109	—	764	—	—	—
11,480	—	10,490	9,790	—	6,550	—	—	997	1,160	—	735	—	—	—
6,020	—	6,020	6,020	6,020	6,020	6,020	—	490	659	659	659	659	737	737
6,890	—	6,890	6,890	6,890	6,550	6,890	—	575	749	749	749	749	737	737
7,900	—	7,900	7,900	7,900	6,550	7,900	—	674	852	852	773	852	806	784
9,180	—	9,180	9,180	9,180	6,550	9,000	—	798	981	981	773	967	896	784
10,320	—	10,320	9,790	10,320	6,550	9,000	—	905	1,093	1,093	773	967	—	—
10,920	—	10,500	9,790	10,920	6,550	8,030	—	962	1,152	1,152	804	1,005	—	—
11,480	—	10,490	9,790	11,480	6,550	9,000	—	1,013	1,205	1,204	773	967	—	—
6,780	—	6,780	6,780	—	6,780	—	—	532	681	—	681	—	737	737
7,750	—	7,750	7,750	—	7,370	—	—	625	773	—	773	—	737	737
8,880	—	8,880	8,880	—	7,370	—	—	733	880	—	804	—	806	784
10,330	—	10,330	10,330	—	7,370	—	—	867	1,013	—	804	—	896	784
11,610	—	11,610	11,020	—	7,370	—	—	984	1,129	—	804	—	—	—
12,290	—	11,800	11,020	—	7,370	—	—	1,045	1,189	—	804	—	—	—
12,910	—	11,800	11,020	—	7,370	—	—	1,100	1,239	—	804	—	—	—
7,150	—	7,150	7,150	—	7,150	—	—	560	716	—	716	—	774	774
8,180	—	8,180	8,180	—	7,780	—	—	659	813	—	812	—	774	774
9,380	—	9,380	9,380	—	7,780	—	—	772	925	—	812	—	846	823
10,900	—	10,900	10,900	—	7,780	—	—	914	1,065	—	812	—	941	823
12,250	—	11,800	11,620	—	7,780	—	—	1,037	1,187	—	812	—	—	—
12,970	—	11,800	11,630	—	7,780	—	—	1,101	1,251	—	854	—	—	—
13,630	—	11,800	11,620	—	7,780	—	—	1,159	1,300	—	812	—	—	—
9,470	—	9,470	9,470	9,470	8,030	8,030	—	769	960	960	960	960	922	922
10,860	—	10,860	10,860	10,860	8,030	8,030	—	901	1,093	1,093	967	1,093	1,008	979
12,620	—	11,800	12,620	12,620	8,030	8,030	—	1,066	1,258	1,258	967	1,237	1,120	979
14,190	—	11,800	12,680	12,680	8,030	8,030	—	1,210	1,402	1,402	967	1,237	—	—
15,020	—	11,800	12,680	12,680	8,030	8,030	—	1,285	1,477	1,477	1,005	1,287	—	—
15,780	—	11,800	12,680	12,680	8,030	8,030	—	1,353	1,545	1,545	967	1,237	—	—
2,470	2,470	—	—	—	—	—	233	—	—	—	—	—	—	—
2,860	2,860	—	—	—	—	—	279	—	—	—	—	—	—	—
2,950	2,950	—	—	—	—	—	244	—	—	—	—	—	—	—
3,930	3,930	3,930	3,930	3,930	3,930	3,930	372	417	579	579	579	579	686	686
4,460	4,460	4,460	4,460	4,460	4,060	4,460	434	486	654	654	654	654	688	688
2,950	2,950	—	—	—	—	—	263	—	—	—	—	—	—	—
3,930	3,930	3,930	3,930	3,930	3,930	3,930	402	452	690	690	690	690	869	869
4,460	4,460	4,460	4,460	4,460	4,060	4,460	468	526	780	780	780	780	871	871
6,090	—	6,090	6,090	—	5,530	—	—	648	847	—	839	—	871	871
6,850	—	6,850	6,850	—	5,530	—	—	742	947	—	839	—	942	886
7,610	—	7,610	7,610	—	5,530	—	—	834	1,046	—	839	—	1,007	886
8,480	—	8,480	8,480	—	5,530	—	—	939	1,157	—	839	—	1,007	886
6,490	—	6,490	6,490	6,490	5,900	6,490	—	678	864	—	839	—	871	871
7,300	—	7,300	7,300	7,300	5,900	7,300	—	776	966	—	839	—	942	886
8,120	—	8,120	8,120	8,120	5,900	8,110	—	874	1,066	—	839	—	1,007	886
9,040	—	9,040	9,040	9,040	5,900	8,110	—	983	1,180	—	839	—	1,007	886
6,490	—	6,490	6,490	6,490	5,900	6,340	—	688	895	895	883	895	917	917
7,300	—	7,300	7,300	7,300	5,900	6,340	—	788	1,001	1,001	883	1,001	992	932
8,120	—	8,120	8,120	8,120	5,900	6,340	—	887	1,105	1,105	883	1,103	1,060	932
9,040	—	9,040	9,040	9,040	5,900	6,340	—	997	1,222	1,222	883	1,103	1,060	932
7,300	—	7,300	7,300	—	6,340	—	—	749	928	—	883	—	917	917
8,220	—	8,220	8,220	—	6,340	—	—	858	1,038	—	883	—	992	992
9,130	—	9,130	9,130	—	6,340	—	—	965	1,146	—	883	—	1,060	932
10,170	—	10,170	10,170	—	6,340	—	—	1,085	1,268	—	883	—	1,060	932

TABLE 7.6—MINIMUM PERFORMANCE PROPERTIES OF CASING (cont.)

1	2	3	4	5	6	7	8	9	10	11	12
						Threaded and Coupled		Extreme Line			
Size Outside Diameter (in.)	Nominal Weight Threads and Coupling (lbm/ft)	Grade	Wall Thickness (in.)	Inside Diameter (in.)	Drift Diameter (in.)	Outside Diameter of Coupling (in.)	Outside Diameter Special Clearance Coupling (in.)	Drift Diameter (in.)	Outside Diameter of Box Powertight (in.)	Collapse Resistance (psi)	Pipe Body Yield Strength (1,000 lbf)
8⅝	36.00	C-95	0.400	7.825	7.700	9.625	9.125	7.700	9.120	4,350	982
	40.00	C-95	0.450	7.725	7.600	9.625	9.125	7.600	9.120	6,020	1,098
	44.00	C-95	0.500	7.625	7.500	9.625	9.125	7.500	9.120	7,740	1,212
	49.00	C-95	0.557	7.511	7.386	9.625	9.125	7.386	9.120	9,710	1,341
	40.00	P-110	0.450	7.725	7.600	9.625	9.125	7.600	9.120	6,390	1,271
	44.00	P-110	0.500	7.625	7.500	9.625	9.125	7.500	9.120	8,420	1,404
	49.00	P-110	0.557	7.511	7.386	9.625	9.125	7.386	9.120	10,740	1,553
9⅝	32.30	H-40	0.312	9.001	8.845	10.625	—	—	—	1,370	365
	36.00	H-40	0.352	8.921	8.765	10.625	—	—	—	1,720	410
	36.00	J-55	0.352	8.921	8.765	10.625	10.125	—	—	2,020	564
	40.00	J-55	0.395	8.835	8.679	10.625	10.125	8.599	10.100	2,570	630
	36.00	K-55	0.352	8.921	8.765	10.625	10.125	—	—	2,020	564
	40.00	K-55	0.395	8.835	8.679	10.625	10.125	8.599	10.100	2,570	630
	40.00	C-75	0.395	8.835	8.679	10.625	10.125	8.599	10.100	2,990	859
	43.50	C-75	0.435	8.755	8.599	10.625	10.125	8.599	10.100	3,730	942
	47.00	C-75	0.472	8.681	8.525	10.625	10.125	8.525	10.100	4,610	1,018
	53.50	C-75	0.545	8.535	8.379	10.625	10.125	8.379	10.100	6,350	1,166
	40.00	L-80	0.395	8.835	8.679	10.625	10.125	8.599	10.100	3,090	916
	43.50	L-80	0.435	8.755	8.599	10.625	10.125	8.599	10.100	3,810	1,005
	47.00	L-80	0.472	8.681	8.525	10.625	10.125	8.525	10.100	4,760	1,086
	53.50	L-80	0.545	8.535	8.379	10.625	10.125	8.379	10.100	6,620	1,244
	40.00	N-80	0.395	8.835	8.679	10.625	10.125	8.599	10.100	3,090	916
	43.50	N-80	0.435	8.755	8.599	10.625	10.125	8.599	10.100	3,810	1,005
	47.00	N-80	0.472	8.681	8.525	10.625	10.125	8.525	10.100	4,760	1,086
	53.50	N-80	0.545	8.535	8.379	10.625	10.125	8.379	10.100	6,620	1,244
	40.00	C-90	0.395	8.835	8.679	10.625	10.125	8.599	10.100	3,250	1,031
	43.50	C-90	0.435	8.755	8.599	10.625	10.125	8.599	10.100	4,010	1,130
	47.00	C-90	0.472	8.681	8.525	10.625	10.125	8.525	10.100	5,000	1,221
	53.50	C-90	0.545	8.535	8.379	10.625	10.125	8.379	10.100	7,120	1,399
	40.00	C-95	0.395	8.835	8.679	10.625	10.125	8.599	10.100	3,320	1,088
	43.50	C-95	0.435	8.755	8.599	10.625	10.125	8.599	10.100	4,120	1,193
	47.00	C-95	0.472	8.681	8.525	10.625	10.125	8.525	10.100	5,090	1,289
	53.50	C-95	0.545	8.535	8.379	10.625	10.125	8.379	10.100	7,340	1,477
	43.50	P-110	0.435	8.755	8.599	10.625	10.125	8.599	10.100	4,420	1,381
	47.00	P-110	0.472	8.681	8.525	10.625	10.125	8.525	10.100	5,300	1,493
	53.50	P-110	0.545	8.535	8.379	10.625	10.125	8.379	10.100	7,950	1,710
10¾	32.75	H-40	0.279	10.192	10.036	11.750	—	—	—	840	367
	40.50	H-40	0.350	10.050	9.894	11.750	—	—	—	1,390	457
	40.50	J-55	0.350	10.050	9.894	11.750	11.250	—	—	1,580	629
	45.50	J-55	0.400	9.950	9.794	11.750	11.250	9.794	11.460	2,090	715
	51.00	J-55	0.450	9.850	9.694	11.750	11.250	9.694	11.460	2,700	801
	40.50	K-55	0.350	10.050	9.894	11.750	11.250	—	—	1,580	629
	45.50	K-55	0.400	9.950	9.794	11.750	11.250	9.794	11.460	2,090	715
	51.00	K-55	0.450	9.850	9.694	11.750	11.250	9.694	11.460	2,700	801
	51.00	C-75	0.450	9.850	9.694	11.750	11.250	9.694	11.460	3,110	1,092
	55.50	C-75	0.495	9.760	9.604	11.750	11.250	9.604	11.460	3,920	1,196
	51.00	L-80	0.450	9.850	9.694	11.750	11.250	9.694	11.460	3,220	1,165
	55.50	L-80	0.495	9.760	9.604	11.750	11.250	9.604	11.460	4,020	1,276
	51.00	N-80	0.450	9.850	9.694	11.750	11.250	9.694	11.460	3,220	1,165
	55.50	N-80	0.495	9.760	9.604	11.750	11.250	9.604	11.460	4,020	1,276

13	14	15	16	17	18	19	20	21	22	23	24	25	26	27
									*Joint Strength—1,000 lbf					
	**Internal Pressure Resistance, psi								Threaded and Coupled					
			Buttress Thread						Buttress Thread					
Plain End or Extreme Line	Round Thread		Regular Coupling		Special Clearance Coupling		Round Thread		Regular Coupling	Regular Coupling Higher Grade†	Special Clearance Coupling	Special Clearance Coupling Higher Grade†	Extreme Line	
	Short	Long	Same Grade	Higher Grade	Same Grade	Higher Grade	Short	Long					Standard Joint	Optional Joint
7,710	—	7,710	7,710	—	6,340	—	—	789	976	—	927	—	963	936
8,670	—	8,670	8,670	—	6,340	—	—	904	1,092	—	927	—	1,042	979
9,640	—	9,640	9,640	—	6,340	—	—	1,017	1,206	—	927	—	1,113	979
10,740	—	10,380	10,740	—	6,340	—	—	1,144	1,334	—	927	—	1,113	979
10,040	—	10,040	10,040	10,040	6,340	6,340	—	1,055	1,288	1,288	1,103	1,288	1,240	1,165
11,160	—	10,380	11,160	11,160	6,340	6,340	—	1,186	1,423	1,423	1,103	1,412	1,326	1,165
12,430	—	10,380	11,230	11,230	6,340	6,340	—	1,335	1,574	1,574	1,103	1,412	1,326	1,165
2,270	2,270	—	—	—	—	—	254	—	—	—	—	—	—	—
2,560	2,560	—	—	—	—	—	294	—	—	—	—	—	—	—
3,520	3,520	3,520	3,520	3,530	3,520	3,520	394	453	639	639	639	639	—	—
3,950	3,950	3,950	3,950	3,950	3,660	3,950	452	520	714	714	714	714	770	770
3,520	3,520	3,520	3,520	3,520	3,520	3,520	423	489	755	755	755	755	—	—
3,950	3,950	3,950	3,950	3,950	3,660	3,950	486	561	843	843	843	843	975	975
5,390	—	5,390	5,390	—	4,990	—	—	694	926	—	926	—	975	975
5,930	—	5,930	5,930	—	4,990	—	—	776	1,016	—	934	—	975	975
6,440	—	6,440	6,440	—	4,990	—	—	852	1,098	—	934	—	1,032	1,032
7,430	—	7,430	7,430	—	4,990	—	—	999	1,257	—	934	—	1,173	1,053
5,750	—	5,750	5,750	—	5,140	—	—	727	947	—	934	—	975	975
6,330	—	6,330	6,330	—	5,140	—	—	813	1,038	—	934	—	975	975
6,870	—	6,870	6,870	—	5,140	—	—	893	1,122	—	934	—	1,032	1,032
7,930	—	7,930	7,930	—	5,140	—	—	1,047	1,286	—	934	—	1,173	1,053
5,750	—	5,750	5,750	5,750	5,140	5,140	—	737	979	979	979	979	1,027	1,027
6,330	—	6,330	6,330	6,330	5,140	5,140	—	825	1,074	1,074	983	1,074	1,027	1,027
6,870	—	6,870	6,870	6,870	5,140	5,140	—	905	1,161	1,161	983	1,161	1,086	1,086
7,930	—	7,930	7,930	7,930	5,140	5,140	—	1,062	1,329	1,329	983	1,229	1,235	1,109
6,460	—	6,460	6,460	—	5,140	—	—	804	1,021	—	983	—	1,027	1,027
7,120	—	7,120	7,120	—	5,140	—	—	899	1,119	—	983	—	1,027	1,027
7,720	—	7,720	7,720	—	5,140	—	—	987	1,210	—	983	—	1,086	1,086
8,920	—	8,460	8,920	—	5,140	—	—	1,157	1,386	—	983	—	1,235	1,109
6,820	—	6,820	6,820	—	5,140	—	—	847	1,074	—	1,032	—	1,078	1,078
7,510	—	7,510	7,510	—	5,140	—	—	948	1,178	—	1,032	—	1,078	1,078
8,150	—	8,150	8,150	—	5,140	—	—	1,040	1,273	—	1,032	—	1,141	1,141
9,410	—	8,460	8,460	—	5,140	—	—	1,220	1,458	—	1,032	—	1,297	1,164
8,700	—	8,700	8,700	8,700	5,140	5,140	—	1,106	1,388	1,388	1,229	1,388	1,283	1,283
9,440	—	9,440	9,160	9,160	5,140	5,140	—	1,213	1,500	1,500	1,229	1,500	1,358	1,358
10,900	—	9,670	9,160	9,160	5,140	5,140	—	1,422	1,718	1,718	1,229	1,573	1,544	1,386
1,820	1,820	—	—	—	—	—	205	—	—	—	—	—	—	—
2,280	2,280	—	—	—	—	—	314	—	—	—	—	—	—	—
3,130	3,130	—	3,130	3,130	3,130	3,130	420	—	700	700	700	700	—	—
3,580	3,580	—	3,580	3,580	3,290	3,580	493	—	796	796	796	796	975	—
4,030	4,030	—	4,030	4,030	3,290	4,030	565	—	891	891	822	891	1,092	—
3,130	3,130	—	3,130	3,130	3,130	3,130	450	—	819	819	819	819	—	—
3,580	3,580	—	3,580	3,580	3,290	3,580	528	—	931	931	931	931	1,236	—
4,030	4,030	—	4,030	4,030	3,290	4,030	606	—	1,043	1,043	1,041	1,043	1,383	—
5,490	5,490	—	5,490	—	4,150	—	756	—	1,160	—	1,041	—	1,383	—
6,040	6,040	—	6,040	—	4,150	—	843	—	1,271	—	1,041	—	1,515	—
5,860	5,860	—	5,860	—	4,150	—	794	—	1,190	—	1,041	—	1,383	—
6,450	6,450	—	6,450	—	4,150	—	884	—	1,303	—	1,041	—	1,515	—
5,860	5,860	—	5,860	5,860	4,150	4,150	804	—	1,228	1,228	1,096	1,228	1,456	—
6,450	6,450	—	6,450	6,450	4,150	4,150	895	—	1,345	1,345	1,096	1,345	1,595	—

TABLE 7.6—MINIMUM PERFORMANCE PROPERTIES OF CASING (cont.)

1	2	3	4	5	6	7	8	9	10	11	12
					Threaded and Coupled			Extreme Line			
Size Outside Diameter (in.)	Nominal Weight Threads and Coupling (lbm/ft)	Grade	Wall Thickness (in.)	Inside Diameter (in.)	Drift Diameter (in.)	Outside Diameter of Coupling (in.)	Outside Diameter Special Clearance Coupling (in.)	Drift Diameter (in.)	Outside Diameter of Box Powertight (in.)	Collapse Resistance (psi)	Pipe Body Yield Strength (1,000 lbf)
10¾	51.00	C-90	0.450	9.850	9.694	11.750	11,250	9.694	11,460	3,400	1,310
	55.50	C-90	0.495	9.760	9.604	11.750	11.250	9.604	11,460	4,160	1,435
	51.00	C-95	0.450	9.850	9.694	11.750	11.250	9.694	11.460	3,480	1,383
	55.50	C-95	0.495	9.760	9.604	11.750	11.250	9.604	11.460	4,290	1,515
	51.00	P-110	0.450	9.850	9.694	11.750	11.250	9.694	11.460	3,660	1,602
	55.50	P-110	0.495	9.760	9.604	11.750	11.250	9.604	11.460	4,610	1,754
	60.70	P-110	0.545	9.660	9.504	11.750	11.250	9.504	11.460	5,880	1,922
	65.70	P-110	0.595	9.560	9.404	11.750	11.250	—	—	7,500	2,088
11¾	42.00	H-40	0.333	11.084	10.928	12.750	—	—	—	1,040	478
	47.00	J-55	0.375	11.000	10.844	12.750	—	—	—	1,510	737
	54.00	J-55	0.435	10.880	10.724	12.750	—	—	—	2,070	850
	60.00	J-55	0.489	10.772	10.616	12.750	—	—	—	2,660	952
	47.00	K-55	0.375	11.000	10.844	12.750	—	—	—	1,510	737
	54.00	K-55	0.435	10.880	10.724	12.750	—	—	—	2,070	850
	60.00	K-55	0.489	10.772	10.616	12.750	—	—	—	2,660	952
	60.00	C-75	0.489	10.772	10.616	12.750	—	—	—	3,070	1,298
	60.00	L-80	0.489	10.772	10.616	12.750	—	—	—	3,180	1,384
	60.00	N-80	0.489	10.772	10.616	12.750	—	—	—	3,180	1,384
	60.00	C-90	0.489	10.772	10.616	12.750	—	—	—	3,360	1,557
	60.00	C-95	0.489	10.772	10.616	12.750	—	—	—	3,440	1,644
	60.00	P-110	0.489	10.772	10.616	12.750	—	—	—	3,610	1,903
13⅜	48.00	H-40	0.330	12.715	12.559	14.375	—	—	—	740	541
	54.50	J-55	0.380	12.615	12.459	14.375	—	—	—	1,130	853
	61.00	J-55	0.430	12.515	12.359	14.375	—	—	—	1,540	962
	68.00	J-55	0.480	12.415	12.259	14.375	—	—	—	1,950	1,069
	54.50	K-55	0.380	12.615	12.459	14.375	—	—	—	1,130	853
	61.00	K-55	0.430	12.515	12.359	14.375	—	—	—	1,540	962
	68.00	K-55	0.480	12.415	12.259	14.375	—	—	—	1,950	1,069
	68.00	C-75	0.480	12.415	12.259	14.375	—	—	—	2,220	1,458
	72.00	C-75	0.514	12.347	12.191	14.375	—	—	—	2,600	1,558
	68.00	L-80	0.480	12.415	12.259	14.375	—	—	—	2,260	1,556
	72.00	L-80	0.514	12.347	12.191	14.375	—	—	—	2,670	1,661
	68.00	N-80	0.480	12.415	12.259	14.375	—	—	—	2,260	1,556
	72.00	N-80	0.514	12.347	12.191	14.375	—	—	—	2,670	1,661
	68.00	G-90	0.480	12.415	12.259	14.375	—	—	—	2,320	1,750
	72.00	G-90	0.514	12.347	12.191	14.375	—	—	—	2,780	1,869
	68.00	C-95	0.480	12.415	12.259	14.375	—	—	—	2,330	1,847
	72.00	C-95	0.514	12.347	12.191	14.375	—	—	—	2,820	1,973
	68.00	P-110	0.480	12.415	12.259	14.375	—	—	—	2,330	2,139
	72.00	P-110	0.514	12.347	12.191	14.375	—	—	—	2,890	2,284

13	14	15	16	17	18	19	20	21	22	23	24	25	26	27
									*Joint Strength—1,000 lbf					
**Internal Pressure Resistance, psi							Threaded and Coupled							
			Buttress Thread						Buttress Thread					
Plain End or Extreme Line	Round Thread		Regular Coupling		Special Clearance Coupling		Round Thread		Regular Coupling	Regular Coupling Higher Grade†	Special Clearance Coupling	Special Clearance Coupling Higher Grade†	Extreme Line	
	Short	Long	Same Grade	Higher Grade	Same Grade	Higher Grade	Short	Long					Standard Joint	Optional Joint
6,590	6,590	—	6,590	—	4,150	—	692	—	1,287	—	1,112	—	1,465	—
7,250	6,880	—	7,250	—	4,150	—	771	—	1,409	—	1,112	—	1,595	—
6,960	6,880	—	6,960	—	4,150	—	927	—	1,354	—	1,151	—	1,529	—
7,660	6,880	—	7,450	—	4,150	—	1,032	—	1,483	—	1,151	—	1,675	—
8,060	7,860	—	7,450	7,450	4,150	4,150	1,080	—	1,594	1,594	1,370	1,594	1,820	—
8,860	7,860	—	7,450	7,450	4,150	4,150	1,203	—	1,745	1,745	1,370	1,745	1,993	—
9,760	7,860	—	7,450	7,450	4,150	4,150	1,338	—	1,912	1,912	1,370	1,754	2,000	—
10,650	7,860	—	7,450	7,450	4,150	4,150	1,472	—	2,077	2,077	1,370	1,754	—	—
1,980	1,980	—	—	—	—	—	307	—	—	—	—	—	—	—
3,070	3,070	—	3,070	3,070	—	—	477	—	807	807	—	—	—	—
3,560	3,560	—	3,560	3,560	—	—	568	—	931	931	—	—	—	—
4,010	4,010	—	4,010	4,010	—	—	649	—	1,042	1,042	—	—	—	—
3,070	3,070	—	3,070	3,070	—	—	509	—	935	935	—	—	—	—
3,560	3,560	—	3,560	3,560	—	—	606	—	1,079	1,079	—	—	—	—
4,010	4,010	—	4,010	4,010	—	—	693	—	1,208	1,208	—	—	—	—
5,460	5,460	—	5,460	—	—	—	869	—	1,361	—	—	—	—	—
5,830	5,820	—	5,830	—	—	—	913	—	1,399	—	—	—	—	—
5,830	5,820	—	5,830	—	—	—	924	—	1,440	1,440	—	—	—	—
6,550	5,820	—	6,300	—	—	—	1,011	—	1,517	—	—	—	—	—
6,920	5,820	—	6,300	—	—	—	1,066	—	1,596	—	—	—	—	—
8,010	5,820	—	6,300	6,300	—	—	1,242	—	1,877	1,877	—	—	—	—
1,730	1,730	—	—	—	—	—	322	—	—	—	—	—	—	—
2,730	2,730	—	2,730	2,730	—	—	514	—	909	909	—	—	—	—
3,090	3,090	—	3,090	3,090	—	—	595	—	1,025	1,025	—	—	—	—
3,450	3,450	—	3,450	3,450	—	—	675	—	1,140	1,140	—	—	—	—
2,730	2,730	—	2,730	2,730	—	—	547	—	1,038	1,038	—	—	—	—
3,090	3,090	—	3,090	3,090	—	—	633	—	1,169	1,169	—	—	—	—
3,450	3,450	—	3,450	3,450	—	—	718	—	1,300	1,300	—	—	—	—
4,710	4,550	—	4,710	—	—	—	905	—	1,496	—	—	—	—	—
5,040	4,550	—	4,930	—	—	—	978	—	1,598	—	—	—	—	—
5,020	4,550	—	4,930	—	—	—	952	—	1,545	—	—	—	—	—
5,380	4,550	—	4,930	—	—	—	1,029	—	1,650	—	—	—	—	—
5,020	4,550	—	4,930	4,930	—	—	963	—	1,585	1,585	—	—	—	—
5,380	4,550	—	4,930	4,930	—	—	1,040	—	1,693	1,693	—	—	—	—
5,650	4,550	—	4,930	—	—	—	1,057	—	1,683	—	—	—	—	—
6,050	4,550	—	4,930	—	—	—	1,142	—	1,797	—	—	—	—	—
5,970	4,550	—	4,930	—	—	—	1,114	—	1,772	—	—	—	—	—
6,390	4,550	—	4,930	—	—	—	1,204	—	1,893	—	—	—	—	—
6,910	4,550	—	4,930	4,930	—	—	1,297	—	2,079	2,079	—	—	—	—
7,400	4,550	—	4,930	4,930	—	—	1,402	—	2,221	2,221	—	—	—	—

TABLE 7.6—MINIMUM PERFORMANCE PROPERTIES OF CASING (cont).

1	2	3	4	5	6	7	8	9	10	11	12
					Threaded and Coupled			Extreme Line			
Size Outside Diameter (in.)	Nominal Weight Threads and Coupling (lbm/ft)	Grade	Wall Thickness (in.)	Inside Diameter (in.)	Drift Diameter (in.)	Outside Diameter of Coupling (in.)	Outside Diameter Special Clearance Coupling (in.)	Drift Diameter (in.)	Outside Diameter of Box Powertight (in.)	Collapse Resistance (psi)	Pipe Body Yield Strength (1,000 lbf)
16	65.00	H-40	0.375	15.250	15.062	17.000	—	—	—	630	736
	75.00	J-55	0.438	15.124	14.936	17.000	—	—	—	1,020	1,178
	84.00	J-55	0.495	15.010	14.822	17.000	—	—	—	1,410	1,326
	75.00	K-55	0.438	15.124	14.936	17.000	—	—	—	1,020	1,178
	84.00	K-55	0.495	15.010	14.822	17.000	—	—	—	1,410	1,326
18⅝	87.50	H-40	0.435	17.775	17.567	20.000	—	—	—	‡630	994
	87.50	J-55	0.435	17.755	17.567	20.000	—	—	—	‡630	1,367
	87.50	K-55	0.435	17.755	17.567	20.000	—	—	—	‡630	1,367
20	94.00	H-40	0.438	19.124	18.936	21.000	—	—	—	‡520	1,077
	94.00	J-55	0.438	19.124	18.936	21.000	—	—	—	‡520	1,480
	106.50	J-55	0.500	19.000	18.812	21.000	—	—	—	‡770	1,685
	133.00	J-55	0.635	18.730	18.542	21.000	—	—	—	1,500	2,125
	94.00	K-55	0.438	19.124	18.936	21.000	—	—	—	‡520	1,480
	106.50	K-55	0.500	19.000	18.812	21.000	—	—	—	‡770	1,685
	133.00	K-55	0.635	18.730	18.542	21.000	—	—	—	1,490	2,125

zero, the mode of failure can be predicted by use of Table 7.5. However, for combined stress, the table should not be applied.

API recommends the following procedure for determining collapse pressure in the presence of a significant axial stress, σ_z. The effective yield strength, $(\sigma_{yield})_e$, is first computed by

$$\frac{(\sigma_{yield})_e}{\sigma_{yield}} = \sqrt{1 - \frac{3}{4}\left(\frac{\sigma_z}{\sigma_{yield}}\right)^2} - \frac{1}{2}\left(\frac{\sigma_z}{\sigma_{yield}}\right).$$

$$\dots\dots\dots\dots\dots\dots\dots(7.12)$$

Note that this equation can be obtained from the ellipse of plasticity with an internal pressure of zero. The effective yield strength is then used in Eq. 7.4b, 7.5b, and 7.6b to determine the mode of failure. Eq. 7.4a, 7.5a, 7.6a, or 7.7 is then used to determine the effective collapse pressure. For an elastic mode of failure, collapse pressure is independent of effective yield strength, and a corrected collapse pressure does not have to be computed. The nominal collapse pressure shown in Table 7.6 can be used.

The API-recommended equations ignore the effect of internal pressure on the correction for collapse-pressure rating. The collapse-pressure rating is the minimum pressure difference across the pipe wall required for failure. Thus, the minimum pressure difference required for failure is assumed to be independent of internal pressure.

Example 7.5. Compute the corrected collapse-pressure rating for the casing of Example 7.3 for in-service conditions where the axial tension will be 1,000,000 lbf. Also,

compute the minimum external pressure required for failure if the internal pressure will be 1,000 psig.

Solution. Calculate collapse resistance for 20-in., 133-lbf/ft, K-55 casing for in-service conditions of 1,000,000-lbf axial tension and 1,000-psig internal pressure. Table 7.6 lists the following values for this pipe: body tension rating: 2,125,000 lbf; nonstressed collapse rating: 1,490 psi; burst rating: 3,060 psi; wall thickness: 0.635 in.; ID: 18.730 in.; $D/t = 20/0.635 = 31.496$; steel area = 38.631 in.²

$\sigma_z = 1,000,000/38.632 = 25,886$ psi.

$$\frac{\sigma_z + p_i}{\sigma_{yield}} = \frac{25,886 + 1,000}{55,000} = 0.48883.$$

Inserting 0.48883 into Eq. 7.12 yields

$$\frac{(\sigma_{yield})_e}{\sigma_{yield}} = \sqrt{1 - 0.75 \times 0.48883^2} - 0.5 \times 0.48883$$
$$= \sqrt{0.82078} - 0.24442 = 0.66155$$
$$= 55.000 \times 0.66155 = 36,385 \text{ psi.}$$

At equivalent yield strength of 36,385 psi and $D/t = 38.6$, collapse will be in transition mode. Then, from Eq. 7.7, collapse pressure is

$$p_{cr} = (\sigma_{yield})_e [F_4/(D/t) - F_5],$$

where F is determined at 36,385 psi rather than at the original 55,000 psi. F at different values of yield strength can be obtained from the following equations shown in API *Bull. 5C3*, fourth edition.

For convenience in writing, set $(\sigma_{yield})_e = (Y)$

$$F_1 = 2.8762 + 0.10679 \times 10^{-5}(Y)$$
$$+ 0.21301 \times 10^{-10}(Y)^2 - 0.53132 \times 10^{-16}(Y)^3,$$

13	14	15	16	17	18	19	20	21	22	23	24	25	26	27
							*Joint Strength—1,000 lbf							
**Internal Pressure Resistance, psi							Threaded and Coupled							
			Buttress Thread						Buttress Thread					
Plain End or Extreme Line	Round Thread		Regular Coupling		Special Clearance Coupling		Round Thread		Regular Coupling	Regular Coupling Higher Grade†	Special Clearance Coupling	Special Clearance Coupling Higher Grade†	Extreme Line	
	Short	Long	Same Grade	Higher Grade	Same Grade	Higher Grade	Short	Long					Standard Joint	Optional Joint
1,640	1,640	—	—	—	—	—	439	—	—	—	—	—	—	—
2,630	2,630	—	2,630	2,630	—	—	710	—	1,200	1,200	—	—	—	—
2,980	2,980	—	2,980	2,980	—	—	817	—	1,351	1,351	—	—	—	—
2,630	2,630	—	2,630	2,630	—	—	752	—	1,331	1,331	—	—	—	—
2,980	2,980	—	2,980	2,980	—	—	865	—	1,499	1,499	—	—	—	—
1,630	1,630	—	—	—	—	—	559	—	—	—	—	—	—	—
2,250	2,250	—	2,250	2,250	—	—	754	—	1,329	1,329	—	—	—	—
2,250	2,250	—	2,250	2,250	—	—	794	—	1,427	1,427	—	—	—	—
1,530	1,530	1,530	—	—	—	—	581	—	—	—	—	—	—	—
2,110	2,110	2,110	2,110	2,110	—	—	784	907	1,402	1,402	—	—	—	—
2,410	2,400	2,400	2,320	2,320	—	—	913	1,057	1,596	1,596	—	—	—	—
3,060	2,400	2,400	2,320	2,320	—	—	1,192	1,380	2,012	2,012	—	—	—	—
2,110	2,110	2,110	2,110	2,110	—	—	824	955	1,479	1,479	—	—	—	—
2,410	2,400	2,400	2,320	2,320	—	—	960	1,113	1,683	1,683	—	—	—	—
3,060	2,400	2,400	2,320	2,320	—	—	1,253	1,453	2,123	2,123	—	—	—	—

*Some joint strengths listed in Col. 20 through 27 are greater than the corresponding pipe body yield strength listed in Col. 12.
**Internal pressure resistance is the lowest of the internal yield pressure of the pipe, the internal yield pressure of the coupling, or the internal pressure leak resistance at the E_1 or E_7 plane.
†For P-110 casing the next higher grade is 150YS, a non-API steel grade having a minimum yield strength of 150,000 psi.
‡Collapse resistance values calculated by elastic formula.
Courtesy of API, Bulletin 5C2, March 1982.

$$F_2 = 0.026233 + 0.50609 \times 10^{-6}(Y),$$

$$F_3 = -465.93 + 0.030867(Y) - 0.10483 \times 10^{-7}(Y)^2 + 0.36989 \times 10^{-13}(Y)^3,$$

$$F_4 = \frac{46.95 \times 10^6 \left[\dfrac{3(F_2/F_1)}{2+(F_2/F_1)} \right]^3}{(Y) \left[\dfrac{3(F_2/F_1)}{2+(F_2/F_1)} - (F_2/F_1) \right] \left[1 - \dfrac{3(F_2/F_1)}{2+(F_2/F_1)} \right]^2},$$

and $F_5 = F_4(F_2/F_1)$.

For $(\sigma_{yield})_e = 36,386$ psi, $F_1 = 2.941$, $F_2 = 0.0446$, $F_3 = 645.1$, $F_4 = 2.101$, and $F_5 = 0.0319$.

$$p_{cr} = 36,385 \left(\frac{2.101}{31.496} - 0.0319 \right) = 1,267 \text{ psi.}$$

p_{cr} is then the corrected pressure differential (external p minus internal p) for the in-service conditions. Collapse pressure rating is $1,267 + 1,000 = 2,267$ psi, or 2,260 psi.

7.3.6 Effect of Bending

In directional wells, the effect of the wellbore curvature and vertical deviation angle on axial stress in the casing and couplings must be considered in the casing design. When casing is forced to bend, the axial tension on the convex side of the bend can be increased greatly. On the other hand, in relatively straight sections of hole with a significant vertical deviation angle, the axial stress caused by the weight of the pipe is reduced. Axial stress is also affected significantly by increased friction between the casing and the borehole wall. In current design practice, the detrimental effect of casing bending is considered, but the favorable effect of the vertical deviation angle is neglected. Wall friction, which is favorable for downward pipe movement and unfavorable for upward pipe movement, generally is compensated for by addition of a minimum acceptable *overpull force* to the free-hanging axial tension.

The curvature of a directional well generally is expressed in terms of the change in angle of the borehole axis per unit length. The *dogleg-severity* angle, α, is the change in angle, in degrees, per 100 ft of borehole length. The relation between dogleg severity and increased axial tensile stress caused by bending is illustrated in Fig. 7.14. Note that the maximum increase in axial stress, $\Delta\sigma_z$, on the convex side of the pipe is given by

$$(\Delta\sigma_z)_{max} = 218\alpha d_n. \quad \ldots\ldots\ldots\ldots\ldots\ldots (7.13)$$

This equation is valid for *pure bending,* where the bending moment is constant along the pipe length and the pipe takes the form of a circular arc with radius of curvature r_c (Fig. 7.14).

It is often convenient to express the increased axial stress caused by bending in terms of an equivalent axial force, F_{ab}, where

$$F_{ab} = (\Delta\sigma_z)_{max} A_s = 218\alpha d_n A_s, \quad \ldots\ldots\ldots (7.14a)$$

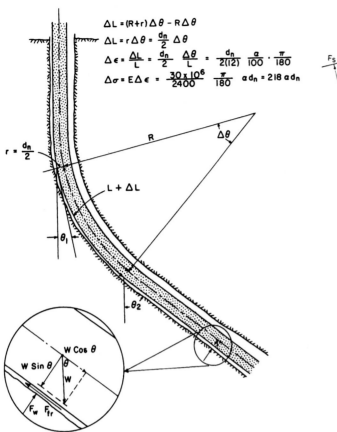

Fig. 7.14—Incremental stress caused by bending of casing in a directional well.

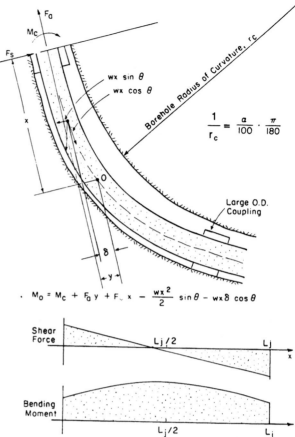

Fig. 7.15—Bending moments in casing with large OD couplings.

The area of steel, A_s, can be expressed conveniently as the weight per foot of pipe divided by the density of steel. For common field units, Eq. 7.14a becomes

$$F_{ab} = 64\alpha d_n w, \qquad\qquad\qquad (7.14b)$$

where F_{ab}, α, d_n, and w have units of lbf, degrees/100 ft, in., and lbf/ft, respectively. Use of a nominal weight per foot for w generally will give acceptable accuracy. The use of Eq. 7.14b has been recommended by several authors,[1,9] is used widely in current design practice, and is felt to be valid when the pipe wall is uniformly in close contact with the borehole wall (i.e., when the size of the upset in OD at the casing connectors is small compared with the borehole irregularities).

When the casing is in contact with the borehole wall only at the connectors, the radius of curvature of the pipe is not constant (Fig. 7.15). In this case, the maximum axial stress can be significantly greater than that predicted by Eq. 7.13. An analysis of the shape of the moment and shear diagrams for this situation indicates that the shear will be nil and the bending moment a maximum near the center of the joint.

Lubinski[10] has shown that the classical beam deflection theory can be applied to this case to determine the maximum axial stress. Recall that when a beam is bent within the elastic range of its material,

$$\frac{d^2 y}{dx^2} = \pm \frac{M}{EI}, \qquad\qquad\qquad (7.15)$$

where x and y are the spatial coordinates defined in Fig. 7.15, M is the bending moment, E is Young's modulus, and I is the moment of inertia of the beam. For circular pipe, I is given by

$$I = \frac{\pi}{64}(d_n{}^4 - d^4). \qquad\qquad\qquad (7.16)$$

The equation for M at any given distance, x, that is less than the joint length is given by

$$M = M_c + F_a y + F_{sc} x - \frac{wx^2}{2} \sin\theta - wx\delta \cos\theta, \qquad\qquad\qquad (7.17a)$$

where M_c is the bending moment at Point 0, F_a is the force of axial tension, and F_{sc} is the side force exerted by the borehole wall on the coupling. The last two terms of this equation are small and, for simplicity, will be neglected. Thus, the weight of the pipe joint under consideration will be neglected, and the axial tension will be assumed constant throughout the joint.

Because of symmetry, the radius of curvature of the pipe at the coupling is equal to the radius of curvature of the borehole. Thus, at the coupling ($x=0$), the pipe is in pure bending, and

$$\frac{1}{r_c} = \frac{M_c}{EI} = \frac{(\Delta\sigma_z)_{max}}{E(d_n/2)}. \qquad\qquad\qquad (7.18)$$

Solving Eq. 7.18 for M_c, substituting into Eq. 7.17a, and neglecting the last two terms yields

$$M = \frac{2(\Delta\sigma_z)_{max}EI}{Ed_n} + F_a y + F_{sc}x. \quad \ldots\ldots\ldots (7.17b)$$

Substituting Eq. 7.17b into 7.15 gives

$$\frac{d^2 y}{dx^2} - \frac{F_a}{EI}y = \frac{2(\Delta\sigma_z)_{max}}{Ed_n} + \frac{F_{sc}x}{EI}, \quad \ldots\ldots\ldots (7.19a)$$

for which the solution is

$$y = \frac{1}{K^2}\left[\frac{2(\sigma_z)_{max}}{Ed_n}\right](\cosh Kx - 1)$$

$$+ \frac{F_{sc}}{KEI}(\sinh Kx - Kx), \quad \ldots\ldots\ldots\ldots (7.19b)$$

where

$$K = \sqrt{\frac{F_a}{EI}}.$$

If there is no pipe-to-wall contact, symmetry suggests that the shear at the midpoint of the joint is nil, and

$$\left(\frac{d^3 y}{dx^3}\right)_{x=L_{j/2}} = 0. \quad \ldots\ldots\ldots\ldots\ldots (7.20a)$$

Similarly, symmetry requires that the pipe be parallel to the borehole wall at the midpoint of the joint. Thus, the slope of the pipe form, dy/dx, is equal to the slope of the borehole at this point. Because the borehole has a radius of curvature r_c, and the abscissa, x, was chosen parallel to the borehole at $x=0$, then

$$\left(\frac{dy}{dx}\right)_{x=L_j/2} = \frac{L_j}{2}\times\frac{1}{r_c}. \quad \ldots\ldots\ldots\ldots (7.20b)$$

Applying the boundary conditions of Eq. 7.20a to Eq. 7.19 gives

$$\frac{F_s}{KEI} = \frac{2(\Delta\sigma_z)_{max}}{Ed_n}\tanh\left(\frac{KL_j}{2}\right). \quad \ldots\ldots\ldots (7.21)$$

$$\frac{1}{r_c} = \frac{2(\Delta\sigma_z)_{max}}{Ed_n}\frac{\tanh\left(\dfrac{KL_j}{2}\right)}{\dfrac{KL_j}{2}}.$$

Upon solving for maximum stress, expressing the radius of curvature in terms of dogleg severity, and converting to common field units, we derive

$$(\sigma_z)_{max} = 218\alpha d_n\frac{6KL_j}{\tanh(6KL_j)}, \quad \ldots\ldots\ldots (7.22a)$$

where K and L_j have units of inches^{-1} and feet, respectively. If the increased axial stress caused by bending is expressed as an equivalent axial force, F_{ab}, then

$$F_{ab} = 64\alpha d_n w\frac{6KL_j}{\tanh(6KL_j)}, \quad \ldots\ldots\ldots\ldots (7.22b)$$

where F_{ab} would be the force required to create the same maximum stress level in a straight section of pipe.

In the previous discussion, the effect of bending on casing failure was handled by consideration of the maximum stress present under the combined loading situation experienced. In this analysis, a possibility of failure is indicated when the maximum stress level exceeds the yield strength of the steel. An alternative approach sometimes used is to express the axial strength of the material in terms of combined tension and bending. The approach is used most commonly in rating the tensional joint strength of a coupling subjected to bending. API formulas[4] have been developed for the joint strength of round-thread casing subjected to bending. When the axial tension strength divided by the cross-sectional area of the pipe wall under the last perfect thread is greater than the minimum yield strength, the joint strength is given by

$$F_{cr} = 0.95A_{jp}\left\{\sigma_{ult} - \left[\frac{140.5\alpha d_n}{(\sigma_{ult} - \sigma_{yield})^{0.8}}\right]^5\right\},$$

$$\ldots\ldots\ldots\ldots\ldots\ldots\ldots\ldots (7.23a)$$

where

$$F_{cr}/A_{jp} \geq \sigma_{yield}$$

and

$$A_{jp} = \pi/4[(d_n - 0.1425)^2 - (d_n - 2t)^2].$$

When the axial tension strength divided by the cross-sectional area of the pipe wall under the last perfect thread is less than the minimum yield strength, then

$$F_{cr} = 0.95A_{jp}\left(\frac{\sigma_{ult} - \sigma_{yield}}{0.644} + \sigma_{yield} - 218.15\alpha d_n\right).$$

$$\ldots\ldots\ldots\ldots\ldots\ldots\ldots\ldots (7.23b)$$

These empirical correlations were developed from experimental tests conducted with 5.5-in., 17-lbf/ft, K-55 casing with short, round-thread couplings.

Example 7.6. Determine the maximum axial stress for a 36-ft joint of 7.625-in., 39-lbf/ft, N-80 casing with API long, round-thread couplings, if the casing is subjected to a 400,000-lbf axial-tension load in a portion of a directional wellbore having a dogleg severity of 4°/100 ft. Compute the maximum axial stress assuming (1) uniform contact between the casing and the borehole wall, and (2) contact between the casing and the borehall wall only at the couplings. Also compute the joint strength of the API round-thread couplings.

Solution. Nominal pipe-body yield strength for this casing is 895,000 lbf, nominal joint strength is 798,000 lbf, and the ID is 6.625 in. (Table 7.6). The cross-sectional area of steel in the pipe body is

$$\pi/4(7.625^2 - 6.625^2) = 11.192 \text{ sq in.}$$

The axial stress without bending is

$$400,000/11.192 = 35,740 \text{ psi.}$$

The additional stress level on the convex side of the pipe caused by bending can be computed with Eq. 7.13 for the assumption of uniform contact between the casing and the borehole wall.

Use of Eq. 7.13 gives a maximum bending stress of

$$(\Delta\sigma_z)_{max} = 218(4.0)(7.625) = 6,649 \text{ psi,}$$

and a total stress of

$$35,740 \text{ psi} + 6,649 \text{ psi} = 42,389 \text{ psi.}$$

If it is assumed that contact between the casing and the borehole wall occurs only at the couplings, then Eqs. 7.16 and 7.22a must be used.

$$I = \pi/64(7.625^4 - 6.625^4) = 71.37 \text{ in.}^4$$

$$K = \sqrt{\frac{400,000}{30(10^6)(71.37)}} = 0.01367 \text{ in.}^{-1}$$

$$(\Delta\sigma_z)_{max} = \frac{6,649(6)(0.01367)(36)}{\tanh[(6)(0.01367)(36)]} = 19,732 \text{ psi.}$$

Thus, for this assumption, the calculated maximum axial stress is

$$35,740 \text{ psi} + 19,732 \text{ psi} = 55,472 \text{ psi.}$$

For either assumption, the maximum axial stress is well below the 80,000-psi minimum yield strength.

The joint strength for a dogleg severity of 4°/100 ft can be computed with Eq. 7.23. Recall that the minimum ultimate strength for N-80 casing is specified in Table 7.1 to be 100,000 psi. First, applying Eq. 7.23a gives

$$\frac{F_{cr}}{A_{jp}} = 0.95 \left\{ 100,000 \right.$$

$$\left. - \left[\frac{140.5(4)(7.625)}{(100,000 - 80,000)^{0.8}} \right]^5 \right\} = 94,991.$$

Because $F_{cr}/A_{jp} > 80,000$ psi, this result is valid and Eq. 7.23b does not apply. The steel area under the last perfect thread is

$$A_{jp} = \frac{\pi}{4}[(7.625 - 0.1425)^2 - 6.625^2] = 9.501 \text{ sq in.}$$

and the calculated joint strength is

$$F_{cr} = (9.501)(94,991) = 902,500 \text{ lbf.}$$

Because this value is above the nominal table value of 798,000 lbf, the nominal table value must be based on joint pull-out strength. Thus, for these conditions, joint strength is controlled by the minimum pull-out force.

7.3.7 Effect of Hydrogen Sulfide

Hydrogen sulfide in the presence of water can have a major effect upon casing strength. When hydrogen atoms are formed on a metal surface by a corrosion reaction, some of these atoms combine to form gaseous molecular hydrogen. When hydrogen sulfide is present, the rate at which the hydrogen atoms (H) combine to form hydrogen gas (H_2) is reduced. As a result, atomic hydrogen (H) may enter the metal at a significant rate before recombining. The presence of this molecular hydrogen within the steel reduces its ductility and causes it to break in a brittle manner rather than yield. This phenomenon is known as hydrogen embrittlement. The resulting failure is called sulfide cracking. Water must be present for the corrosion reaction to occur, which generates hydrogen atoms. Dry hydrogen sulfide does not cause embrittlement.

Hydrogen embrittlement is especially significant in high-strength steels at low temperature. Common carbon steels with yield strengths below 90,000 psi generally will not fail by sulfide cracking for temperatures above 100°F. This corresponds to a Rockwell hardness number (RHN) of 22. Steel that is alloyed with other materials (such as nickel) can fail by sulfide cracking at a lower RHN; certain heat treatments can raise the RHN at which sulfide cracking can occur. In casing design practice, steel grades with minimum yield strengths above 75,000 psi generally are avoided when possible for applications where exposure to hydrogen sulfide is anticipated. Increased casing strength is achieved by selecting casing with a greater wall thickness rather than by selecting a higher grade of steel. When this is not possible, special high-strength hydrogen-sulfide-resistant casings, such as L-80 and C-90, have been used successfully.

There is evidence that, as temperature increases, casings with a higher minimum yield strength than 90,000 psi can be used safely in wells that contain hydrogen sulfide in the produced fluids. In deep, abnormally pressured wells, a practical casing design is difficult to obtain without the use of some high-strength steel. Kane and Greer[11] have presented the results of experimental laboratory and field tests of several steel grades that were exposed to hydrogen sulfide in varying concentrations, at various temperatures, and at various stress levels. Shown in Figs. 7.16 and 7.17 are the maximum safe stress levels observed (expressed as percent of minimum yield strength) for various steel grades, hydrogen sulfide concentrations, and exposure temperatures.

Failures resulting from hydrogen embrittlement often do not occur immediately after exposure to hydrogen sulfide. A time period during which no damage is evident is followed by a sudden failure. During the time period

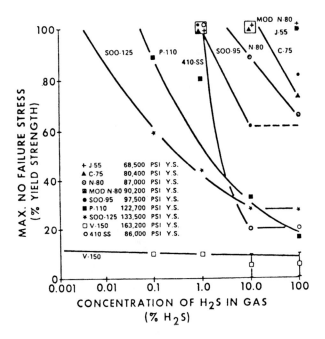

Fig. 7.16—Maximum safe stress level for various grades of casing and H₂S concentrations at 75°F.

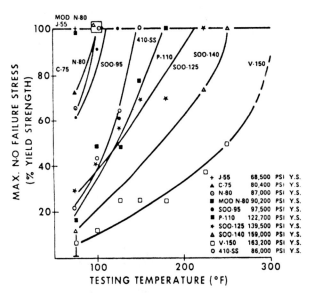

Fig. 7.17—Maximum safe stress level for various grades of casing and temperatures at 100% H₂S concentrations.

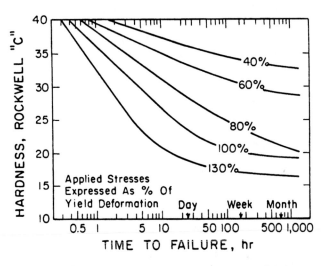

Fig. 7.18—Effects of stress level on time to H₂S-induced failures for various Rockwell hardness numbers.

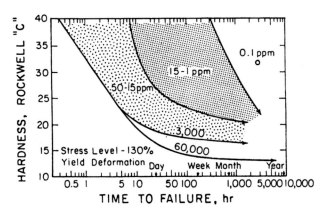

Fig. 7.19—Effects of H₂S concentration on time to failure for various Rockwell hardness numbers.

before failure, called the *incubation period*, hydrogen is diffusing to points of high stress. Fig. 7.18 shows test results of the time to failure for different RHN's and different applied stresses. Fig. 7.19 shows the effect of hydrogen sulfide concentration. [12]

7.3.8 Effect of Field Handling

Performance properties that a given joint of casing will exhibit in the field can be affected adversely by several field operations. For example, burst strength is affected significantly by the procedure and equipment used to make up the pipe. Tests have shown that burst strength can be reduced by as much as 70% by combinations of tong marks that penetrate 17% of the wall thickness and 4% out-of-roundness caused by excessive torque.

Mechanical deformities can also occur while the casing is transported to location or while it is run into the hole. Any mechanical deformity in the pipe normally results in considerable reduction in its collapse resistance. This is especially true for casing with high d_n/t ratios. A thin-wall tube that is deformed by 1% out-of-round will have its collapse resistance lowered by 25%. Thus, the slightest crushing by tongs, slips, or downhole conditions diminishes the collapse resistance by a significant amount. Some of the special hydrogen-sulfide-resistant casings, such as C-90, can be stress-hardened by careless handling. If this occurs, the resistance to hydrogen embrittlement can be lost.

The API Recommended Practice for Care and Use of Casing (RP5C1) [13] lists common causes of problems experienced with casing and tubing. Over half are related to poor shipping, handling, and pipe-running practices. Once casing-design principles are mastered, the student should become familiar with good pipe-handling procedures as recommended in RP5C1 before attempting to implement a casing design plan.

7.4 Casing Design Criteria

The design of a casing program begins with specification of the surface and bottomhole well locations and the size of the production casing that will be used if hydrocarbons are found in commercial quantities. The number and sizes of tubing strings and the type of subsurface artificial lift equipment that may eventually be placed in the well determine the minimum ID of the production casing. These specifications usually are determined for the drilling engineer by other members of the engineering staff. In some cases, consideration must also be given to the possibility of exploratory drilling below an anticipated productive interval. The drilling engineer then must design a program of bit sizes, casing sizes, grades, and setting depths that will allow the well to be drilled and completed safely in the desired producing configuration.

To obtain the most economical design, casing strings often consist of multiple sections of different steel grade, wall thickness, and coupling types. Such a casing string is called a *combination string*. Additional cost savings sometimes can be achieved by the use of liner-tieback combination strings instead of full strings running from the surface to the bottom of the hole. When this is done, reduced tension loads experienced in running the casing in stages often make it possible to use lighter weights or lower grades of casing. Of course, the potential savings must be weighed against the additional risks and costs of a successful, leak-free tieback operation as well as the additional casing wear that results from a longer exposure of the upper casing to rotation and translation of the drillstring.

7.4.1 Selection of Casing Setting Depths

The selection of the number of casing strings and their respective setting depths generally is based on a consideration of the pore-pressure gradients and fracture gradients of the formations to be penetrated. The example shown in Fig. 7.20 illustrates the relationship between casing-setting depth and these gradients. The pore-pressure gradient and fracture gradient data are obtained by the methods presented in Chap. 6, are expressed as an equivalent density, and are plotted vs. depth. A line representing the planned-mud-density program also is plotted. The mud densities are chosen to provide an acceptable *trip margin* above the anticipated formation pore pressures to allow for reductions in effective mud weight caused by upward pipe movement during tripping operations. A commonly used trip margin is 0.5 lbm/gal or one that will provide 200 to 500 psi of excess bottomhole pressure (BHP) over the formation pore pressure.

To reach the depth objective, the effective drilling fluid density shown at Point a is chosen to prevent the flow of formation fluid into the well (i.e., to prevent a kick). However, to carry this drilling fluid density without exceeding the fracture gradient of the weakest formation exposed within the borehole, the protective intermediate casing must extend at least to the depth at Point b, where the fracture gradient is equal to the mud density needed to drill to Point a. Similarly, to drill to Point b and to set intermediate casing, the drilling fluid density shown at Point c will be needed and will require surface casing to be set at least to the depth at Point d. When possible, a kick margin is subtracted from the true fracture-gradient line to obtain a design fracture-gradient line. If no kick margin is provided, it is impossible to take a kick at the casing-setting depth without causing hydrofracture and a possible underground blowout.

Other factors—such as the protection of freshwater aquifers, the presence of vugular lost-circulation zones, depleted low-pressure zones that tend to cause stuck pipe, salt beds that tend to flow plastically and to close the borehole, and government regulations—also can affect casing-depth requirements. Also, experience in an area may show that it is easier to get a good casing-seat cement job in some formation types or that fracture gradients are generally higher in some formation types. When such conditions are present, a design must be found that simultaneously will meet these special requirements and the pore-pressure and fracture-gradient requirements outlined above.

The conductor casing-setting depth is based on the amount required to prevent washout of the shallow borehole when drilling to the depth of the surface casing and to support the weight of the surface casing. The conductor casing must be able to sustain pressures expected during diverter operations without washing around the outside of the conductor. The conductor casing often is driven into the ground, and the length is governed by the resistance of the soil. The casing-driving operation is stopped when the number of blows per foot exceeds some some specified upper limit.

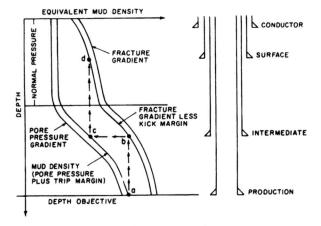

Fig. 7.20—Sample relationship among casing-setting depth, formation pore-pressure gradient, and fracture gradient.

Example 7.7. A well is being planned for a location in Jefferson Parish, LA. The intended well completion requires the use of 7-in. production casing set at 15,000 ft. Determine the number of casing strings needed to reach this depth objective safely, and select the casing setting depth of each string. Pore pressure, fracture gradient, and lithology data from logs of nearby wells are given in Fig. 7.21. Allow a 0.5-lbm/gal trip margin, and a 0.5-lbm/gal kick margin when making the casing-seat selections. The minimum length of surface casing required to protect the freshwater aquifers is 2,000 ft. Approximately 180 ft of

TABLE 7.7—COMMONLY USED BIT SIZES FOR RUNNING API CASING

Casing Size (OD in.)	Coupling Size (OD in.)	Common Bit Sizes Used (in.)
4½	5.0	6, 6⅛, 6¼
5	5.563	6½, 6¾
5½	6.050	7⅞, 8⅜
6	6.625	7⅞, 8⅜, 8½
6⅝	7.390	8½, 8⅝, 8¾
7	7.656	8⅝, 8¾, 9½
7⅝	8.500	9⅞, 10⅝, 11
8⅝	9.625	11, 12¼
9⅝	10.625	12¼, 14¾
10¾	11.750	15
13⅜	14.375	17½
16.0	17.0	20
20.0	21.0	24, 26

conductor casing generally is required to prevent washout on the outside of the conductor. It is general practice in this area to cement the casing in shale rather than in sandstone.

Solution. The planned-mud-density program first is plotted to maintain a 0.5-lbm/gal trip margin at every depth. The design fracture line then is plotted to permit a 0.5-lbm/gal kick margin at every depth. These two lines are shown in Fig. 7.21 by dashed lines. To drill to a depth of 15,000 ft, a 17.6-lbm/gal mud will be required (Point a). This, in turn, requires intermediate casing to be set at 11,400 ft (Point b) to prevent fracture of the formations above 11,400 ft. Similarly, to drill safely to a depth of 11,400 ft to set intermediate casing, a mud density of 13.6 lbm/gal is required (Point c). This, in turn, requires surface casing to be set at 4,000 ft (Point d). Because the formation at 4,000 ft is normally pressured, the usual conductor-casing depth of 180 ft is appropriate.

Only 2,000 ft of surface casing is needed to protect the freshwater aquifers. However, if this minimum casing length is used, intermediate casing would have to be set higher in the transition zone. An additional liner also would have to be set before the total depth objective is reached to maintain a 0.5-lbm/gal kick margin. Because shale is the predominant formation type, only minor variations in casing-setting depth are required to maintain the casing seat in shale.

7.4.2 Selection of Casing Sizes

The size of the casing strings is controlled by the necessary ID of the production string and the number of intermediate casing strings required to reach the depth objective. To enable the production casing to be placed in the well, the bit size used to drill the last interval of the well must be slightly larger than the OD of the casing connectors. The selected bit size should provide sufficient clearance beyond the OD of the coupling to allow for mud cake on the borehole wall and for casing appliances, such as centralizers and scratchers. The bit used to drill the lower portion of the well also must fit inside the casing string above. This, in turn, determines the minimum size

of the second-deepest casing string. With similar considerations, the bit size and casing size of successively more shallow well segments are selected.

Table 7.7 provides commonly used bit sizes for drilling a hole in which various API casing strings generally can be placed safely without getting the casing stuck. Shown in Table 7.8 are casing ID's and *drift diameters* for various standard casing sizes and wall thicknesses. The pipe manufacturer assures that a bit smaller than the drift diameter will pass through every joint of casing bought. In most instances, bits larger than the drift diameter but smaller than the ID will also pass.

Only the most commonly used bit sizes are shown in Tables 7.7 and 7.8. Selection of casing sizes that permit the use of commonly used bits is advantageous because the bit manufacturers make readily available a much larger variety of bit types and features in these common sizes. However, additional bit sizes are available that can be used in special circumstances.

Example 7.8. Using the data given in Example 7.7, select casing sizes (OD) for each casing string.

Solution. A 7-in. production casing string is desired. An 8.625-in. bit is needed to drill the bottom section of the borehole (see Table 7.7). An 8.625-in. bit will pass through most of the available 9.625-in. casings (see Table 7.8). However, a final check will have to be made after the required maximum weight per foot is determined. According to the data presented in Table 7.7, a 12.25-in. bit is needed to drill to the depth of the intermediate casing. As shown in Table 7.8, a 12.25-in. bit will pass through 13.375-in. casing. A 17.5-in. bit is needed to drill

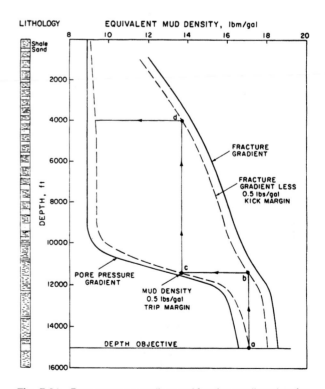

Fig. 7.21—Pore-pressure gradient and fraction gradient data for Jefferson Parish, LA.

TABLE 7.8—COMMONLY USED BIT SIZES THAT WILL PASS THROUGH API CASING

Casing Size (O.D., in.)	Weight Per Foot (lbm/ft)	Internal Diameter (in.)	Drift Diameter (in.)	Commonly Used Bit Sizes (in.)
4½	9.5	4.09	3.965	3⅞
	10.5	4.052	3.927	
	11.6	4.000	3.875	
	13.5	3.920	3.795	3¾
5	11.5	4.560	4.435	4¼
	13.0	4.494	4.369	
	15.0	4.408	4.283	
	18.0	4.276	4.151	3⅞
5½	13.0	5.044	4.919	4¾
	14.0	5.012	4.887	
	15.5	4.950	4.825	
	17.0	4.892	4.764	
	20.0	4.778	4.653	4⅝
	23.0	4.670	4.545	4¼
6⅝	17.0	6.135	6.010	6
	20.0	6.049	5.924	5⅝
	24.0	5.921	5.796	
	28.0	5.791	5.666	
	32.0	5.675	5.550	4¾
7	17.00	6.538	6.413	6¼
	20.00	6.456	6.331	
	23.00	6.366	6.241	
	26.00	6.276	6.151	6⅛
	29.00	6.184	6.059	6
	32.00	6.094	5.969	
	35.00	6.006	5.879	
	38.00	5.920	5.795	5⅝
7⅝	20.00	7.125	7.000	6¾
	24.00	7.025	6.900	
	26.40	6.969	6.844	
	29.70	6.875	6.750	
	33.70	6.765	6.640	6½
	39.00	6.625	6.500	
8⅝	24.00	8.097	7.972	7⅞
	28.00	8.017	7.892	
	32.00	7.921	7.796	6¾
	36.00	7.825	7.700	
	40.00	7.725	7.600	
	44.00	7.625	7.500	
	49.00	7.511	7.386	
9⅝	29.30	9.063	8.907	8¾, 8½
	32.30	9.001	8.845	
	36.00	8.921	8.765	
	40.00	8.835	8.679	8⅝, 8½
	43.50	8.755	8.599	
	47.00	8.681	8.525	8½
	53.50	8.535	8.379	7⅞
10¾	32.75	10.192	10.036	9⅞
	40.50	10.050	9.894	
	45.50	9.950	9.794	9⅝
	51.00	9.850	9.694	
	55.00	9.760	9.604	
	60.70	9.660	9.504	8¾, 8½
	65.37	9.560	9.404	8¾, 8½
11¾	38.00	11.154	10.994	11
	42.00	11.084	10.928	10⅝
	47.00	11.000	10.844	
	54.00	10.880	10.724	
	60.00	10.772	10.616	
13⅜	48.00	12.715	12.559	12¼
	54.50	12.615	12.459	
	61.00	12.515	12.359	
	68.00	12.415	12.259	
	72.00	12.347	12.191	11
16	55.00	15.375	15.188	15
	65.00	15.250	15.062	
	75.00	15.125	14.939	14¾
	84.00	15.010	14.822	
	109.00	14.688	14.500	
18⅝	87.50	17.755	17.567	17½
20	94.00	19.124	18.936	17½

to the depth of the surface casing (see Table 7.7). Finally, as shown in Table 7.8, a 17.5-in. bit will pass through 18.625-in. conductor casing, which will be driven into the ground.

7.4.3 Selection of Weight, Grade, and Couplings

Once the length and OD of each casing string is established, the weight, grade, and couplings used in each string can be determined. In general, each casing string is designed to withstand the most severe loading conditions anticipated during casing placement and the life of the well. The loading conditions that are always considered are burst, collapse, and tension. When appropriate, other loading conditions (such as bending or buckling) must also be considered. Because the loading conditions in a well tend to vary with depth, it is often possible to obtain a less expensive casing design with several different weights, grades, and couplings in a single casing string.

It is often impossible to predict the various loading conditions that a casing string will be subjected to during the life of a well. Thus, the casing design usually is based on an assumed loading condition. The assumed design load must be severe enough that there is a very low probability of a more severe situation actually occurring and causing casing failure. When appropriate, the effects of casing wear and corrosion should be included in the design criteria. These effects tend to reduce the casing thickness and greatly increase the stresses where they occur.

The design loads assumed by the various well operators differ significantly and are too numerous for an exhaustive listing in this text. Instead, example design criteria that are felt to be representative of current drilling engineering practice will be presented. Once the concepts presented in this text are mastered, the student should be able to apply easily any of the other criteria used by his particular company.

To achieve a minimum-cost casing design, the most economical casing and coupling that will meet the design loading conditions must be used for all depths. Because casing prices change frequently, a detailed list of prices in a text of this type is not practical. In general, minimum cost is achieved when casing with the minimum possible weight per foot in the minimum grade that will meet the design load criteria is selected. For this illustration, only API casing and couplings will be considered in the example applications. It will be assumed that the cost per foot of the casing increases with increasing burst strength and that the cost per connector increases with increasing joint strength.

Casing strings required to drill safely to the depth objective serve different functions than the production casing does. Similarly, drilling conditions applicable for surface casing are different from those for intermediate casing or drilling liners. Thus, each type of casing string will have different design-load criteria. Design criteria can also vary with the well environment (e.g., the wells drilled into permafrost on the north slope of Alaska) and with the well application (e.g., geothermal steam wells or steam injection wells). General design criteria will be presented for surface casing, intermediate casing, intermediate casing with a liner, and production casing. Also, additional criteria for thermal wells and arctic wells will be discussed.

7.4.3.1 Surface Casing.

Example design-loading conditions for surface casing are illustrated in Fig. 7.22 for burst, collapse, and tension considerations. The high-internal-pressure loading condition used for the burst design is based on a well-control condition assumed to occur while circulating out a large kick. The high-external-pressure loading condition used for the collapse design is based on a severe lost-circulation problem. The high-axial-tension loading condition is based on an assumption of stuck casing while the casing is run into the hole before cementing operations.

The burst design should ensure that formation-fracture pressure at the casing seat will be exceeded before the burst pressure is reached. Thus, this design uses formation fracture as a safety pressure-release mechanism to ensure that casing rupture will not occur at the surface and endanger the lives of the drilling personnel. The design pressure at the casing seat is equal to the fracture pressure plus a safety margin to allow for an injection pressure that is slightly greater than the fracture pressure. The pressure within the casing is calculated assuming that all of the drilling fluid in the casing is lost to the fractured formation, leaving only formation gas in the casing. The external pressure, or backup pressure outside the casing that helps resist burst, is assumed to be equal to the normal formation pore pressure for the area. The beneficial effect of cement or higher-density mud outside the casing is ignored because of the possibility of both a locally poor cement bond and mud degradation that occurs in time. A safety factor also is used to provide an additional safety margin for possible casing damage during transportation and field-handling of the pipe.

The collapse design is based either on the most severe lost-circulation problem that is felt to be possible or on the most severe collapse loading anticipated when the casing is run. For both cases, the maximum possible external pressure that tends to cause casing collapse results from the drilling fluid that is in the hole when the casing is placed and cemented. The beneficial effect of the cement and of possible mud degradation is ignored, but the detrimental effect of axial tension on collapse-pressure rating is considered. The beneficial effect of pressure inside the casing can also be taken into account by the consideration of a maximum possible depression of the mud level inside the casing. A safety factor generally is applied to the design-loading condition to provide an additional safety margin.

If a severe lost-circulation zone is encountered near the bottom of the next interval of hole and no other permeable formations are present above the lost-circulation zone, the fluid level in the well can fall until the BHP is equal to the pore pressure of the lost-circulation zone. Equating the hydrostatic mud pressure to the pore pressure of the lost-circulation zone gives

$$0.052\rho_{max}(D_{lc}-D_m)=0.052g_pD_{lc}, \quad \ldots\ldots(7.24a)$$

where D_{lc} is the depth (true vertical) of the lost-circulation zone, g_p is the pore-pressure gradient of the lost-circulation zone, ρ_{max} is the maximum mud density anticipated in drilling to D_{lc}, and D_m is the depth to which the mud level will fall. Solving this expression for D_m yields

$$D_m = \frac{(\rho_{max} - g_p)}{\rho_{max}}D_{lc}. \quad \ldots\ldots\ldots\ldots(7.24b)$$

There is usually considerable uncertainty in the selection of the minimum anticipated pore-pressure gradient and the maximum depth of the lost-circulation zone for use in Eq. 7.24b. In the absence of any previously produced and depleted formations, the normal pore-pressure gradient for the area can be used as a conservative estimate for the minimum anticipated pore-pressure gradient. Similarly, if the lithology is not well-known, the depth of the next full-length casing string can be used as a conservative estimate of D_{lc}.

The minimum fluid level in the casing while it is placed in the well depends on field practices. The casing usually is filled with mud after each joint of casing is made up and run in the hole, and an internal casing pressure that is approximately equal to the external casing pressure is maintained. However, in some cases the casing is *floated in* or run at least partially empty to reduce the maximum hook load before reaching bottom. If this practice is anticipated, the maximum depth of the mud level in the casing must be compared to the depth computed with Eq. 7.24b, and the greater value must be used in the collapse-design calculations.

The most difficult part of the collapse design is the correction of the collapse-pressure rating for the effect of axial tension. The difficulty lies in establishing the axial tension present at the time the maximum collapse load is imposed. If the maximum collapse load is encountered when the casing is run, the axial tension is readily calculated from a knowledge of the casing weight per foot and the mud hydrostatic pressure with the principles previously presented in Sec. 4.5.1. However, if the maximum collapse load is encountered after the cement has hardened and the casing has been landed in the wellhead, the determination of axial stress is much more difficult. Some

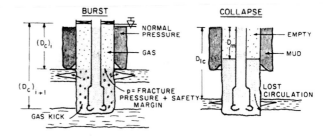

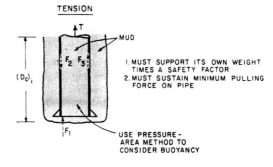

Fig. 7.22—Drilling casing design loads for burst, collapse, and tension.

evidence suggests that, when the cement begins to form a crystalline structure, the hydrostatic pressure exerted by the cement is reduced to that of its water phase. Also, in some cases a microannulus is thought to exist between the casing and the cement sheath that may permit some elongation or contraction of the casing within the sheath in response to changing buoyancy forces. To avoid consideration of these complications, it is recommended that axial tension be computed as the hanging weight for the hydrostatic pressures present when the maximum collapse load is encountered plus any additional tension put in the pipe during and after casing landing. This assumption will result in a maximum tension and a corrected minimum collapse-pressure rating.

Tension design requires consideration of axial stress present when the casing is run, during cementing operations, when the casing is landed in the slips, and during subsequent drilling and production operations throughout the life of the well. In most cases, the design load is based on conditions that could occur when the casing is run. It is assumed that the casing becomes stuck near the bottom and that a minimum acceptable amount of pull, in excess of the hanging weight in mud, is required to work the casing free. A minimum safety-factor criterion is applied so that the design load will be dictated by the maximum load resulting from the use of either the safety factor or the overpull force, whichever is greater. The minimum overpull force tends to control the design in the upper portion of the casing string, and the minimum safety factor tends to control the lower part of the casing string. Once the casing design is completed, maximum axial stresses anticipated during cementing, casing landing, and subsequent drilling operations should also be checked to ensure that the design load is never exceeded.

In the design of a combination string of nonuniform wall thickness, the effect of buoyancy is most accurately included by use of the pressure-area method previously presented in Sec. 4.5. The drilling fluid in use at the time the casing is run is used to compute the hydrostatic pressure at each junction between sections of different wall thicknesses.

In directional wells, the additional axial stress in the pipe body and connectors caused by bending should be added to the axial stress that results from casing weight and fluid hydrostatic pressure. The directional plan must be used to determine the portions of the casing string that will be subjected to bending while the pipe is run. The lower portion of the casing string will have to travel past all the curved sections of the wellbore, but the upper section of the casing string may not be subjected to any bending.

When the selection of casing grade and weight in a combination string is controlled by collapse, a simultaneous design for collapse and tension is best. The greatest depth at which the next most economical casing can be used depends on its corrected collapse-pressure rating, which in turn depends on the axial tension at that depth. Thus, the corrected collapse-pressure rating cannot be computed without the axial tension being computed first. An iterative procedure, in which the depth of the bottom of the next most economical casing section first is selected on the basis of uncorrected table value of collapse resistance, can be used. The axial tension at this point is then computed, and the collapse resistance is corrected. This allows the depth of the bottom on the next casing section to be updated for a second iteration. Several iterations may be required before the solution converges.

7.4.3.2 Intermediate Casing. Intermediate casing is similar to surface casing in that its function is to permit the final depth objective of the well to be reached safely. When possible, the general procedure outlined for surface casing also is used for intermediate casing strings. However, in some cases, the burst-design requirements dictated by the design-loading condition shown in Fig. 7.22 are extremely expensive to meet, especially when the resulting high working pressure is in excess of the working pressure of the surface BOP stacks and choke manifolds for the available rigs. In this case, the operator may accept a slightly larger risk of loosing the well and select a less severe design load. The design load remains based on an underground blowout situation assumed to occur while a gas kick is circulated out. However, the acceptable mud loss from the casing is limited to the maximum amount that will cause the working pressure of the surface BOP stack and choke manifold to be reached. If the existing surface equipment is to be retained, it is pointless to design the casing to have a higher working pressure than the surface equipment.

When the surface burst-pressure load is based on the working pressure of the surface equipment, p_{max}, internal pressure at intermediate depths should be determined, as shown in Fig. 7.23. It is assumed that the upper portion of the casing is filled with mud and the lower portion of the casing is filled with gas. The depth of the mud/gas interface, D_m, is determined with the following relationship.

$$p_{max} + 0.052\rho_m D_m + 0.052\rho_g (D_{lc} - D_m) = p_i,$$

$$\dotfill (7.25a)$$

where p_i is the injection pressure opposite the lost-circulation zone, ρ_m and ρ_g are the densities of the mud and gas, and D_{lc} is the depth of the lost-circulation zone. Solving this equation for D_m gives

$$D_m = \frac{p_i - p_{max}}{0.052(\rho_m - \rho_g)} - \frac{\rho_g D_{lc}}{(\rho_m - \rho_g)}. \quad \dotfill (7.25b)$$

The gas density is estimated with Eq. 4.5 and an assumed average molecular weight. The density of the drilling mud is determined to be the maximum density

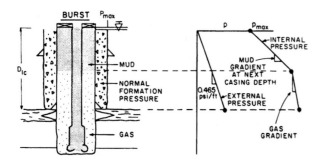

Fig. 7.23—Modified burst design load for intermediate casing.

anticipated while drilling to the depth of the next full-length casing string. This permits the calculation of the maximum intermediate pressures between the surface and the casing seat. The depth of the lost-circulation zone is determined from the fracture gradient vs. depth plot to be the depth of the weakest exposed formation. The injection pressure is equal to the fracture pressure plus an assumed safety margin to account for a possible pressure drop within the hydraulic fracture.

7.4.3.3 Intermediate Casing With a Liner.
The burst-design-load criteria for intermediate casing on which a drilling liner will be supported later must be based on the fracture gradient below the liner. The burst design considers the intermediate casing and liner as a unit. All other design criteria for the intermediate casing are identical to those previously presented.

7.4.3.4 Production Casing.
Example burst- and collapse-design loading conditions for production casing are illustrated in Fig. 7.24. The example burst-design loading condition assumes that a producing well has an initial shut-in BHP equal to the formation pore pressure and a gaseous produced fluid in the well. The production casing must be designed so that it will not fail if the tubing fails. A tubing leak is assumed to be possible at any depth. It generally is also assumed that the density of the completion fluid in the casing above the packer is equal to the density of the mud left outside the casing. If a tubing leak occurs near the surface, the effect of the hydrostatic pressure of the completion fluid in the casing would negate the effect of the external mud pressure on the casing. Mud degradation outside the casing is neglected because the formation pore pressure of any exposed formation would nearly equal the mud hydrostatic pressure.

The collapse-design load shown in Fig. 7.24 is based on conditions late in the life of the reservoir, when reservoir pressure has been depleted to a very low (negligible) abandonment pressure. A leak in the tubing or packer could cause the loss of the completion fluid, so the low internal pressure is not restricted to just the portion of the casing below the packer. Thus, for design purposes, the entire casing is considered empty. As before, the fluid density outside the casing is assumed to be that of the mud in the well when the casing was run, and the beneficial effect of the cement is ignored.

In the absence of any unusual conditions, the tension design load criteria for production casing are the same as for surface and intermediate casing. When unusual conditions are present, maximum stresses associated with these conditions must be checked to determine whether they exceed the design load in any portion of the string.

Example 7.9 Design the surface casing for the proposed well described in Examples 7.7 and 7.8. To achieve a minimum cost design, consider the use of a combination string. However, do not include any section shorter than 500 ft to reduce the logistical problem of shipping and unloading the casing in the proper order in which it is run in the hole. Assume that only the casing shown in Table 7.6 is available.

For burst considerations, use an injection pressure that is equivalent to a mud density 0.3 lbm/gal greater than the fracture gradient and a safety factor of 1.1. Also assume that any gas kick is composed of methane, which has a molecular weight of 16. For simplicity, assume ideal gas behavior. The normal formation pore pressure for the area is 0.465 psi/ft. Formation temperature in degrees Rankine is equal to $(520+0.012D)$.

For collapse considerations, assume that a normal-pressure, lost-circulation zone could be encountered as deep as the next casing seat, that no permeable zones are present above the lost-circulation zone, and use a safety factor of 1.1. Also assume that the casing was landed "as cemented" and that the axial tension results only from the hanging weight of the casing under prevailing borehole conditions.

For tension considerations, use a minimum overpull force of 100,000 lbf or a safety factor of 1.6, whichever is greater.

Solution. The surface casing selected in Examples 7.7 and 7.8 has an OD of 13.375 in. and is to be set at 4,000 ft. The first step in the selection of the casing grade, wall thickness, and connectors is to eliminate the casing that will not meet the burst-design load. The fracture gradient at 4,000 ft is read from Fig. 7.21 to be equivalent to 14.1-lbm/gal mud. For an injection-pressure gradient that is 0.3 lbm/gal higher than the fracture pressure,

$$p_i = 0.052(14.1+0.3)(4,000) = 2,995 \text{ psig.}$$

The gas gradient for methane is given by Eq. 4.5 as

$$0.052\rho_g = 0.052 \frac{(2,995+15)(16)}{80.3(1)(520+0.012 \cdot 4,000)}$$

$$= 0.055 \text{ psi/ft.}$$

BURST

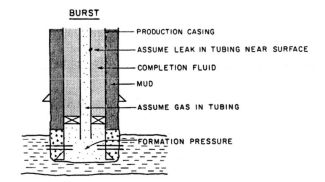

COLLAPSE

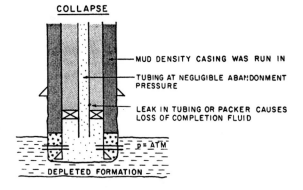

Fig. 7.24—Production casing design loads for burst and collapse.

Thus, the surface casing pressure for the design loading conditions is

$$2{,}995 - 0.055(4{,}000) = 2{,}775 \text{ psig.}$$

The external pressure is zero at the surface. For a normal formation pore pressure of 0.465 psi/ft, the external pressure at the casing seat is

$$(0.465)(4{,}000) = 1{,}860 \text{ psig.}$$

The pressure differential that tends to burst the casing is 2,775 psi at the surface and 1,135 psi (2,995 − 1,860) at the casing seat. Multiplying these pressures by a safety factor of 1.1 yields a burst-design load of 3,053 psi at the surface and 1,249 psi at the casing seat. A graphical representation of the burst-design load is shown in Fig. 7.25. A comparison of the burst-strength requirements to the burst-pressure ratings of 13.375-in. casing in Table 7.6 illustrates that H-40 casing and J-55, 54.5-lbm/ft casing do not meet the design requirements at the top of the string. The H-40 casing, which has a burst rating of 1,730 psi, could be used below.

$$(3{,}053 - 1{,}730)(4{,}000)/(3{,}053 - 1{,}249) = 2{,}933 \text{ ft.}$$

The J-55 casing, which has a burst rating of 2,730 psi, could be used below.

$$(3{,}053 - 2{,}730)(4{,}000)/(3{,}053 - 1{,}249) = 716 \text{ ft.}$$

All the other casings listed have burst-pressure ratings in excess of the design requirements.

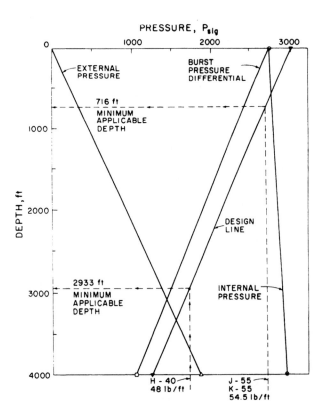

Fig. 7.25—Graphical representation of example burst design load.

The external pressure of the collapse-design load is based on the mud density in the hole when the casing is run. According to Fig. 7.21, the planned mud density is 9.3 lbm/gal and the external pressure at 4,000 ft is

$$(0.052)(9.3)(4{,}000) = 1{,}934 \text{ psig.}$$

The internal pressure for the collapse-design load is controlled by the maximum loss in fluid level that could occur if a severe lost-circulation problem is encountered. The maximum depth of the mud level is calculated with Eq. 7.24b. If it is assumed that a normal-pressure, lost-circulation zone unexpectedly is encountered near the depth of the next casing seat (11,400 ft) while the planned 13.7-lbm/gal mud (Fig. 7.21) is used, and if no permeable zones are exposed above this depth, then

$$D_m = (13.7 - 0.465/0.052)(11{,}400)/13.7 = 3{,}959 \text{ ft.}$$

For these conditions, then, the mud level could fall to within 41 ft of the casing bottom. The internal pressure is assumed to be zero to a depth of 3,959 ft, and

$$(0.052)(13.7)(41) = 29 \text{ psig}$$

at the bottom of the casing.

Note that when casing is run, this design would permit the internal fluid level to fall safely to a depth of 3,959 ft. Thus, if desired, the casing could be partially "floated in" without exceeding the collapse design.

The pressure differential that tends to collapse the casing is zero at the surface.

$$(0.052)(9.3)(3{,}959) = 1{,}915 \text{ psi}$$

at 3,959 ft, and

$$(1{,}934 - 29) = 1{,}905 \text{ psi}$$

at the casing seat of 4,000 ft. Multiplying these pressures by a safety factor of 1.1 yields a collapse-design load of zero at the surface, 2,107 psi at 3,959 ft, and 2,096 psi at 4,000 ft. A graphical representation of the collapse-design load is shown in Fig. 7.26. To meet the collapse-design requirement at the bottom of the casing string, 68-lbm/ft, C-75 casing with a collapse rating of 2,220 psi will be required (see Table 7.6).

The collapse-design load requires the use of stronger casing than the burst-design load at the bottom of the string. In this case, the final casing selection is made most easily beginning with the bottom of the casing string. The bottom section of casing will be composed of C-75, 68-lbm/ft casing. Because the collapse-pressure load decreases toward the top of the casing string, it will be possible to change to a less expensive casing at an intermediate depth between the surface and the casing seat. The next most economical casing will be J-55 or K-55, 68-lbm/ft casing.

To determine the minimum possible length of C-75, 68-lbm/ft casing, consider the free-body diagram shown in Fig. 7.27a. Point 1 is located at the bottom of the casing string, and Point 2 is located at the top of the C-75, 68-lbm/ft section. A force balance on this section gives

$$F_a = 68L_1 - p_1 A_s + p_2 \Delta A_{s2},$$

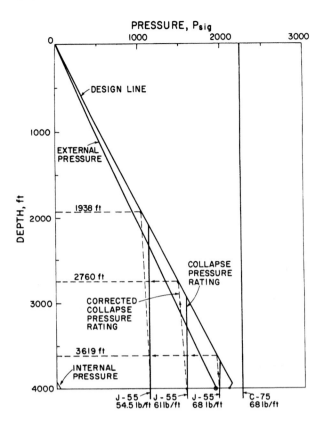

Fig. 7.26—Graphical representation of sample collapse design load.

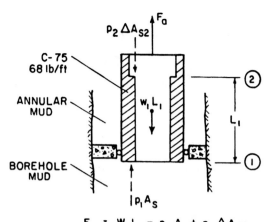

$$F_a = W_1 L_1 - p_1 A_S + p_2 \Delta A_{S2}$$

(a) AXIAL TENSION AT BOTTOM OF SECTION 2

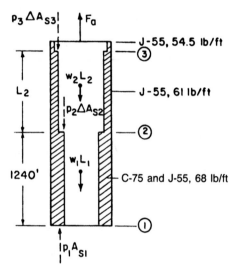

$$F_a = W_1 L_1 + W_2 L_2 - P_1 A_{s1} + P_2 \Delta A_{s2} + P_3 \Delta A_{s3}$$

(b) AXIAL TENSION AT BOTTOM OF SECTION 3

Fig. 7.27—Free-body diagram for axial tension present for collapse design load.

where F_a is the axial tension, L_1 is the length of Sec. 1, p_1 and p_2 are the hydrostatic pressures at the bottom and top of Sec. 1, A_s is the cross-sectional area of steel at the bottom of Sec. 1, and ΔA_s is the difference in steel cross-sectional areas between Secs. 1 and 2. The borehole hydrostatic pressure at 4,000 ft for the collapse-design load was determined previously to be 29 psig. The ID of the casing is given as 12.415 in. in Table 7.6, and the cross-sectional area of steel is

$$\pi/4(13.375^2 - 12.415^2) = 19.445 \text{ sq in.}$$

Because Sec. 2 has the same weight per foot and steel area as Sec. 1, ΔA_s is zero, and the axial tension at the top of Sec. 1 is

$$F_a = 68L_1 - (29)(19.445) = 68L_1 - 564.$$

The minimum length, L_1, must be chosen so that the corrected collapse-pressure rating at the bottom of Sec. 2 will be equal to the collapse-design load.

After inspection of the collapse-load line shown in Fig. 7.26, L_1 is given by

$$L_1 = 4,000 - 3,959 p_{cr}/2,107,$$

where p_{cr} is the corrected collapse-pressure rating.

Unfortunately, the corrected collapse-pressure rating is a function of F_a, and a trial-and-error solution procedure

generally is used to find the root of this equation. An uncorrected collapse-pressure rating can be used for p_{cr} for the first iteration. The F_a computed with the equation above is then used to correct the collapse-pressure rating. The process is continued until the collapse-pressure rating does not change significantly.

The uncorrected collapse-pressure rating for J-55 or K-55, 68-lbm/ft casing is 1,950 psi, and the pipe-body yield strength is 1,069,000 lbf (Table 7.6). Use of 1,950 psi for p_{cr} for the first iteration gives

$$L_1 = 4,000 - 1.879(1,950) = 336 \text{ ft}$$

and

$$F_a = 68(336) - 564 = 22,284 \text{ lbf.}$$

TABLE 7.9—CORRECTION OF COLLAPSE PRESSURE RATING FOR SEC. 1A OF EXAMPLE 7.9

P_{cr}	L_i	F_a	D_i	σ_z/σ_{yield}	$(\sigma_{yield})_e/\sigma_{yield}$
1929	375	24936	3625	0.0233	0.9881
1926	381	25344	3619	0.0237	0.9879
1926					

TABLE 7.10—CORRECTION OF COLLAPSE PRESSURE RATING FOR SEC. 1B OF EXAMPLE 7.9

P_{cr}	L_i	F_a	D_i	σ_z/σ_{yield}	$(\sigma_{yield})_e/\sigma_{yield}$
1540	1106	74,644	2894	0.07759	0.9589
1477	1225	82,736	2775	0.08600	0.9542
1469	1240	83,756	2760	0.08706	0.9536
1469					

TABLE 7.11—CORRECTION OF COLLAPSE PRESSURE RATING FOR SEC. 2 OF EXAMPLE 7.9

P_{cr}	L_i	F_a	D_i	σ_z/σ_{yield}	$(\sigma_{yield})_e/\sigma_{yield}$
1130	637	122,613	2123	0.1437	0.9204
1040	806	132,922	1954	0.1558	0.9130
1031	822	133,898	1938	0.1570	0.9122
1031					

The corrected collapse pressure for a second iteration can be estimated now with the method previously presented in Example 7.5. Since the fluid level inside the casing is below the point of interest, p_i is zero and

$$\frac{\sigma_z}{\sigma_{yield}} = \frac{22,284}{1,069,000} + 0 = 0.0208,$$

$$\frac{(\sigma_{yield})_e}{\sigma_{yield}} = \sqrt{1 - \frac{3}{4}(0.0208)^2} - \frac{1}{2}(0.0208) = 0.9894.$$

The lower limit of the transition mode of failure is given by Eq. 7.6b, with factors F_1 to F_5 taken from Table 7.4.

$$(d_n/t) = \frac{(\sigma_{yield})_e(F_1 - F_4)}{F_3 + (\sigma_{yield})_e(F_2 - F_5)}$$

$$= \frac{(0.9894)(55,000)(2.991 - 1.989)}{1,206 + (0.9894)(55,000)(0.0541 - 0.0360)}$$

$$= 24.89.$$

The upper limit of the transition mode of failure is 37.21 (Table 7.5). Because the wall thickness for 68-lbm/ft, 13.375-in. casing is 0.480 (Table 7.6), the actual d_n/t is

$$\frac{13.375}{0.480} = 27.86,$$

which is within the range for the transition mode of failure. Use of Eq. 7.6a for the collapse-pressure rating gives

$$p_{cr} = (\sigma_{yield})_e \left(\frac{F_4}{d_n/t} - F_5\right)$$

$$= (0.9894)(55,000)\left(\frac{1.989}{27.86} - 0.0360\right)$$

$$= 0.9894 \ (1,950)$$

$$= 1,929 \text{ psi.}$$

Note that the correction factor 0.9894 is applied to the nominal table value of the collapse-pressure rating as long as the mode of failure remains unchanged.

Continuing these calculations through an additional iteration yields the values in Table 7.9. However, because use of J-55 or K-55, 68-lbm/ft pipe would result in Sec. 1 having a length less than the specified minimum length of 500 ft, this type casing will not be chosen, and the next most economical casing (J-55 or K-55, 61 lbm/ft) will be considered for Sec. 2.

The ID for 61-lbm/ft pipe is 12.515 in., the pipe-body yield strength is 962,000 lbf, and the collapse-pressure rating is 1,540 psi (Table 7.6). The change in the steel cross-sectional area at Point 2 becomes

$$\Delta A_s = \pi/4(12.515^2 - 12.415^2) = 1.958 \text{ sq in.}$$

and the axial tension becomes

$$F_a = 68L_i - 564 + 1.958p_2,$$

where p_2 is the internal pressure at distance L_i from bottom. (See Table 10 for calculations for further iterations.)

The next most economical pipe available is J-55 or K-55, 54.5-lbm/ft casing, which has an ID of 12.615 in., a pipe-body strength of 853,000 lbf, and a collapse-pressure rating of 1,130 psi (Table 7.6). The free-body diagram for the bottom portion of Sec. 3 is shown in Fig. 7.27b. The axial tension is given by

$$F_a = w_1L_1 + w_2L_2 - p_iA_{s1} + p_2\Delta A_{s2} + p_3\Delta A_{s3}$$

$$= 68(1,240) + 61L_2 - 564 + 0 + 0$$

$$= 83,756 + 61L_2,$$

and the length of Sec. 2 is given by

$$L_2 = 4,000 - 1.879p_{cr} - 1,240.$$

See Table 7.11 for calculations for further iterations for Sec. 2.

The next most economical pipe available is H-40, 48-lbm/ft casing, which was determined previously not to meet the burst requirements for depths more shallow than 2,933 ft. Furthermore, the J-55 or K-55 54.5-lbm/ft casing will not meet the burst requirements for depths more shallow than 716 ft. Thus, it will be necessary to use 61-lbm/ft casing in the upper 716 ft of the casing string.

The third step in the casing design will be to check the tension design requirements for the preliminary design found to satisfy the burst- and collapse-strength requirements. The design-loading condition for tension was specified to be while the casing is run, when the wellbore contains a 9.3-lbm/gal mud.

A free-body diagram for the tension design calculations is shown in Fig. 7.28. The axial tension at the top of each section is given by

$$(F_a)_n = \Sigma w_i L_i - p_1 A_{s1} + \Sigma p_i \Delta A_{si},$$

where

$$
\begin{aligned}
w_1 L_1 &= (68)(1,240) = 84,320 \text{ lbf}, \\
w_2 L_2 &= (61)(822) = 50,142 \text{ lbf}, \\
w_3 L_3 &= (54.5)(1,222) = 66,599 \text{ lbf}, \\
w_4 L_4 &= (61)(716) = 43,676 \text{ lbf}, \\
p_1 A_{s1} &= 0.052(9.3)(4,000)(19.445) = 37,614 \text{ lbf}, \\
p_2 \Delta A_{s2} &= 0.052(9.3)(2,760)(1.958) = 2,613 \text{ lbf}, \\
p_3 \Delta A_{s3} &= 0.052(9.3)(1,938)(1.974) = 1,850 \text{ lbf, and} \\
p_4 \Delta A_{s4} &= -0.052(9.3)(716)(1.974) = -684 \text{ lbf}.
\end{aligned}
$$

The tension diagram shown in Fig. 7.29 was constructed by computation of the axial tension at each section boundary. The tension design line was obtained by multiplying the tensions of 1.6, or by adding a 100,000-lbf pulling force assuming the casing was stuck in the borehole near bottom, and selecting the larger of the two results. Comparison of the joint strengths and pipe-body strengths given in Table 7.6 to the tension load requirements shown in Fig. 7.29 indicates that the casing selected from burst and collapse considerations will also meet the tension design load, even with the most economical connectors available.

After the design is complete, the bit clearance is calculated and compared to the bit diameter and oversize tolerance. The final design is summarized in Table 7.12. An extra joint of the minimum-ID casing is specified at the top of the casing string for use as a minimum-ID gauge.

Note that the bit clearance calculated from the drift diameter for Sec. 1 is less than the bit manufacturers' tolerance for oversize bits. While this is certainly cause for caution, experience has shown that a bit can pass casing-wall imperfections better than a drift mandrel because of its shorter length. However, the operator generally should check the casing-drift diameter or order casing that has passed an oversized drift mandrel when such close tolerances are used. Another, more conservative option would be the use of a 12-in. bit for drilling below the surface casing. However, since a 12.25-in. bit is more commonly used, a greater variety of bit features would probably be available for this size.

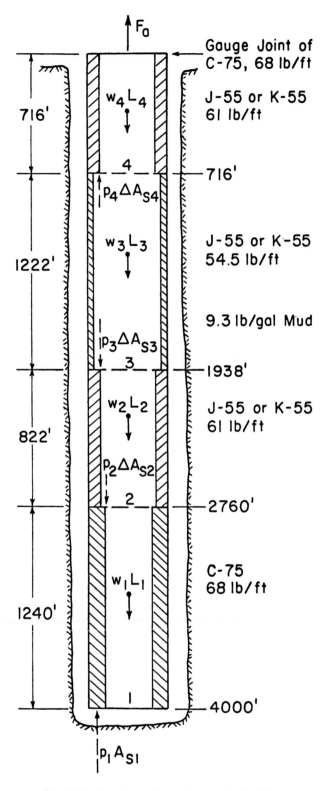

Fig. 7.28—Free-body diagram for tension load line.

7.5 Special Design Considerations

In the previous section, casing design considerations were based on selected burst-, collapse-, and axial-tension-loading conditions. While these loading conditions are important in the design of all casing strings, other loading conditions also can be important and should be recognized by the student. These additional loading conditions can be caused by shock loading, changing internal pressure,

TABLE 7.12—SUMMARY OF EXAMPLE CASING DESIGN

Section	Depth Interval (ft)	Length (ft)	Grade	Weight (lbf/ft)	Coupling	Drift Diameter (in.)	Bit* Clearance (in.)	Maximum O.P. (in.)	Coupling** Clearance (in.)
1	2,760–4,000	1,240	C-75	68	LCSG	12.259	0.009	14.375	3.125
2	1,938–2,760	822	J-55	61	CSG	12.359	0.109	14.375	3.125
3	716–1,938	1,222	J-55	54.5	CSG	12.459	0.209	14.375	3.125
4	34–716	682	J-55	61	CSG	12.359	0.109	14.375	3.125
Top Joint	0–34	34	C-75	68	LCSG	12.259	0.009	14.375	3.125

*Bit size is 12.25 in.
 Bit tolerance is +1/32 or 0.03125 in.
**Hole size is 17.5 in. (no washout).

changing external pressure, thermal effects, subsidence, and casing landing practices. In some cases, additional design-load criteria may be appropriate.

7.5.1. Shock Loading

Significant shock loading can develop if a casing string is suddenly stopped. Axial stresses result from sudden velocity changes in a manner analogous to water-hammer in a pipe caused by a sudden valve closure. Elastic theory leads to the following equation for axial shock loads resulting from instantaneously stopping the casing:

$$\Delta \sigma_z = \Delta v \sqrt{E \rho_s}, \quad \dots\dots\dots\dots\dots\dots (7.26a)$$

where $\Delta \sigma_z$ is the change in axial stress caused by the shock load, Δv is the change in pipe velocity, E is Young's modulus, and ρ_s is the density of steel. After average values for Young's modulus and steel density are substituted, this equation becomes

$$\Delta \sigma_z = 1{,}780 \Delta v, \quad \dots\dots\dots\dots\dots\dots (7.26b)$$

where $\Delta \sigma_z$ is in psi and Δv is in ft/sec. Note that shock loading normally is not severe for modest changes in pipe velocity.

7.5.2 Changing Internal Pressure

In the previous section, design-loading conditions were based on the maximum anticipated internal pressure occurring during well-control operations and during the producing life of the well. Casing was selected to withstand this internal pressure without bursting. However, changes in internal pressure also can cause significant changes in axial stress. These changes in axial stress can occur both during and after the casing has been cemented and landed.

During cementing operations, the casing is exposed to a high internal pressure because of the hydrostatic pressure of the cement slurry and the pump pressure imposed to displace the slurry. This not only creates hoop stresses in the casing wall, which tend to burst the casing, but also creates axial stresses, which tend to pull the casing apart (Fig. 7.30). While the burst tendency generally is recognized and maintained within the burst limits by field personnel, the axial loads sometimes are neglected. This can have disastrous consequences, especially if the cement is beginning to harden toward the end of the displacement and if the pump pressure is increased in an attempt to complete the cement placement. The surface pressure inside the casing causes an axial load given by

$$\Delta F_a = p_i \pi d^2 / 4, \quad \dots\dots\dots\dots\dots\dots (7.27)$$

where d is the ID of the casing. Caution must be exercised during cementing operations to ensure that neither the burst rating nor the tension rating of the casing is exceeded. Cement that sets up inside the casing can be drilled out far more easily than parted casing can be repaired.

As shown in Fig. 7.31, an increase in internal pressure causes an increase in tangential stress; thus the casing tends to contract. Similarly, a reduction in internal pressure tends to cause the casing to elongate. However, once the casing is cemented and landed in the wellhead, the casing may not be free to contract or to elongate in response to changing internal pressure. According to Hook's law, this can cause changes in the axial stress that

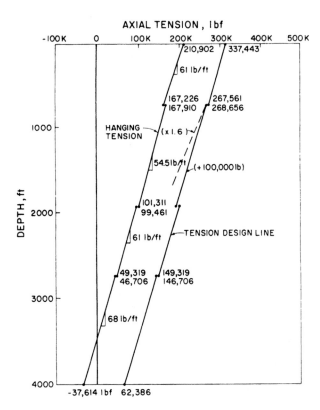

Fig. 7.29—Graphical representation of sample tension design load.

are directly proportional to the suppressed strain to develop. Hook's law is applicable if (1) the casing was landed with sufficient tension to prevent helical buckling from occurring in a portion of the free casing above the cement top, and (2) the maximum axial stress was less than the yield strength of the steel.

The strain that would occur if the casing were free to move is given by

$$\Delta\epsilon_z = \frac{\mu}{E}\Delta(\sigma_r + \sigma_t), \quad\ldots\ldots\ldots\ldots\ldots (7.28)$$

where μ is Poisson's ratio and the other variables are as defined previously. The sum of the radial and tangential stresses is given by

$$(\sigma_r + \sigma_t) = 2\frac{r_i^2 p_i - r_o^2 p_e}{r_o^2 - r_i^2}$$

$$= 2\frac{A_i p_i - A_e p_e}{A_e - A_i}$$

$$= \frac{2}{A_s}(A_i p_i - A_e p_e), \quad\ldots\ldots\ldots\ldots (7.29)$$

and the change in radial and tangential stress caused by a change in internal pressure is given by

$$\Delta(\sigma_r + \sigma_t) = 2\frac{A_i}{A_s}\Delta p_i.$$

This would cause an axial strain given by

$$\Delta\epsilon_z = -2\frac{\mu}{E}\frac{A_i}{A_s}\Delta p_i,$$

where the negative sign denotes a decrease in length for a given increase in internal pressure. If this entire strain is prevented, Hook's law is applicable to the total strain, and an axial (tensional) stress, given by

$$\Delta\sigma_z = 2\mu\frac{A_i}{A_s}\Delta p_i,$$

would develop. Substituting an average value of 0.3 for Poisson's ratio for steel and converting from axial stress to axial force gives

$$\Delta F_a = +0.471 d^2 \Delta p_i, \quad\ldots\ldots\ldots\ldots (7.30)$$

where the positive sign denotes an increase in tension for a given increase in internal pressure.

Eq. 7.30 was derived for a uniform change in external pressure over a given length interval. A uniform change in pressure in a well usually is caused by a change in surface pressure. When the pressure change is not uniform,

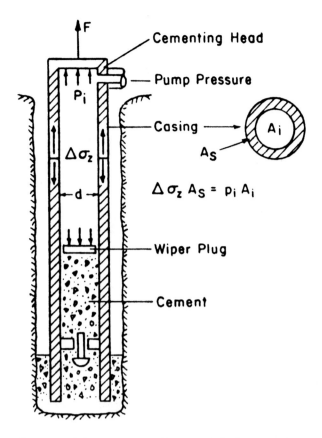

Fig. 7.30—Axial stress caused by cement displacement stress.

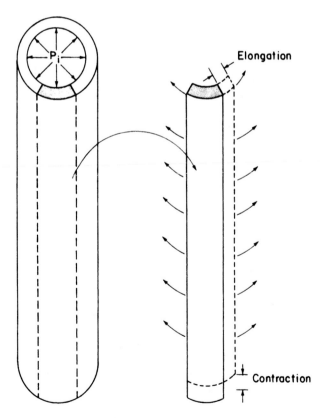

INTERNAL PRESSURE TENDS TO CAUSE INCREASE IN CASING DIAMETER AND DECREASE IN CASING LENGTH

Fig. 7.31—Effect of internal pressure on axial stress.

Eq. 7.30 can still be applied, as long as the average pressure change of the exposed length interval is known. In general, the average pressure change is given by

$$\overline{\Delta p} = \frac{\int_0^L \Delta p \, dx}{L} . \qquad \qquad (7.31)$$

With this relation, the average pressure change for a change in mud density in a vertical well is

$$\overline{\Delta p} = \frac{0.052 \Delta \rho \int_0^L x \, dx}{L} = \frac{0.052 \Delta \rho L}{2} \qquad (7.32)$$

Thus, the average pressure change caused by a change in mud density in a vertical well occurs at the midpoint of the depth interval. The total tendency to shorten would be as though the pressure applied at $L/2$ were applied over the entire length, L.

In the event that the casing has not been landed in sufficient tension to prevent helical buckling, the behavior of the casing would not be governed by Hook's law alone. The helical bending of the casing within the confines of the borehole wall would permit some strain to take place and would reduce the change in stress level caused by a change in internal pressure. According to Hook's law, only the portion of the strain not accommodated by buckling would be converted to stress. Lubinski[14] has shown that, for a pipe of uniform cross-sectional area, the change in the effective length, ΔL_{bu}, of the pipe caused by a buckling force, F_{bu}, is given by

$$\Delta L_{bu} = -\frac{\Delta r^2 F_{bu}{}^2}{8EIw} , \qquad \qquad (7.33)$$

where Δr is the radial clearance between the tube and the confining borehole and F_{bu} is the buckling force at the top of the cement.

Goins[15] introduced the following equation for the buckling force.

$$F_{bu} = F_s - F_a , \qquad \qquad (7.34)$$

where F_s is called the stability force and is defined by

$$F_s = A_i p_i - A_o p_o . \qquad \qquad (7.35)$$

The length change, ΔL_{bu}, is the total caused by buckling above the point at which F_{bu} is calculated. The negative sign in Eq. 7.33 denotes a decrease in length for an increase in internal pressure, a decrease in external pressure, or a decrease in axial tension. If F_{bu} is negative, then buckling will not occur and Eq. 7.33 no longer has meaning.

As will be discussed in Sec. 7.5.6, it is generally desirable to land intermediate casing strings in sufficient tension to prevent buckling. Dellinger and McLean[16] have presented evidence that casing wear that results from drilling operations is much more severe in sections of casing

that have a tendency to buckle, especially when the borehole diameter has been increased greatly because of washout. Borehole enlargement resulting from washout also makes it difficult to employ Eq. 7.33 for casing because the radial clearance must be known.

Example 7.10. Casing having an ID of 12.459 in. and a tension strength of 853,000 lbf is suspended from the surface while being cemented. If the effective weight of the casing being supported is 300,000 lbf and a safety factor of 1.3 is desired, how much surface pump pressure can be applied safely to displace the cement?

Solution. The additional axial force permitted is

$$853,000/1.3 - 300,000 = 356,154 \text{ lbf}.$$

The use of Eq. 7.27 yields

$$356,154 = p_i \pi (12.459)^2 /4;$$

$$p_i = 2,921 \text{ psig}.$$

7.5.3 Changing External Pressure

Design-loading conditions for external pressure were based on the mud density left outside the casing during cementing operations. Casing that could withstand this external pressure without collapsing was selected. Other situations sometimes are encountered when the external pressure can be higher than that caused by the mud. This occurs most commonly when casing is set through sections of formations (such as salt) that can flow plastically, and when casing is set through permafrost, which can alternately thaw and freeze, depending on whether the well is producing or is shut in. Also, changing external pressure can result in significant changes in axial stress.

When casing is set through a salt formation that can flow plastically, it should be assumed that the salt eventually will creep plastically until it transmits the full vertical overburden stress to the casing. Thus, the average density of the sediments above the salt bed should be used in place of the mud density in the collapse design for the portion of the casing penetrating the salt.

When casing is set through permafrost, alternate thawing and freezing of the pore water can build excessive pressures during the volume expansion of the water as it turns to ice. Complex computer models have been used to determine the maximum pressures during freezing, which take into account local permafrost conditions. As in the case of plastic salt flow, an upper limit of the external confining pressure is the overburden stress at the depth of interest.

In permafrost sections, concern must be given to the possible freezing of mud and completion fluids in addition to the freezing of pore water outside the outer casing string. The usual practice in this environment is to attempt to displace all water-base fluids from the casing strings and to replace them with an oil-base fluid. However, it is extremely difficult to remove all of the water-base material by a simple displacement process, and it generally is anticipated that pockets of water-base material may remain in the permafrost region of the well. An additional

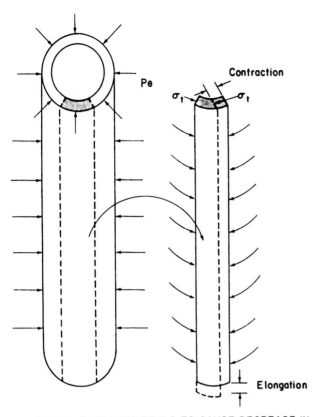

EXTERNAL PRESSURE TENDS TO CAUSE DECREASE IN
CASING DIAMETER AND INCREASE IN CASING LENGTH

Fig. 7.32—Effect of external pressure on axial stress.

precaution is to design successive casing strings so that the burst pressures of successively larger strings are less than the collapse pressure of the inner strings. Even though burst of an outer casing is undesirable, it is more desirable than collapse of the innermost string.

Axial stresses also can result from changing external pressure after the well is completed. A common example of changing external pressure is caused by degradation of the mud left outside the casing.

As shown in Fig. 7.32, an increase in external pressure causes a decrease in tangential tensional stress (i.e., an increase in tangential compressive stress). This may cause the diameter of the casing to decrease and the length of the casing to increase. Similarly, a reduction in external pressure may cause the casing to shorten. If the casing is cemented and landed in the wellhead under sufficient tension to prevent buckling, it may not be free to contract or elongate in response to changing external pressure. As discussed previously for changes in internal pressure, this can cause axial stresses that are directly proportional to the suppressed strain to develop. Eq. 7.29 gives the change in radial and tangential stress caused by a change in external pressure as

$$\Delta(\sigma_r + \sigma_t) = -2 \frac{A_e}{A_s} \Delta p_e.$$

Substituting this expression into Eq. 7.28 yields

$$\epsilon_z = +2 \frac{\mu}{E} \frac{A_e}{A_s} \Delta p_e,$$

where the positive sign denotes an increase in length for a given increase in external pressure. If the entire strain is prevented, Hook's law is applicable to the total strain, and an axial stress (compression) of

$$\Delta\sigma_z = -2\mu \frac{A_e}{A_s} \Delta p_e$$

would develop. Substitution of 0.3 for Poisson's ratio for steel and conversion from axial stress to axial force gives

$$\Delta F_a = -0.471 d_n^2 \Delta p_e, \dots\dots\dots\dots\dots\dots (7.36)$$

where the negative sign denotes a decrease in tension for an increase in average external pressure. If the pressure change is not constant over the entire casing length exposed to a pressure increase, then the average pressure change can be computed with Eq. 7.31. For a change in external mud density resulting from mud degradation, Eq. 7.32 can be used to compute the average pressure change.

In the event that the casing has not been landed in sufficient tension to prevent helical buckling, the relationships given in Eqs. 7.33 through 7.35 would be used in addition to Hook's law to estimate the effective axial-stress/casing-stretch relationship.

Example 7.11. A casing string having an OD of 10.75 in. is cemented in a vertical well containing 14-lbm/gal mud. The mud is left outside the casing above the cement top at 8,000 ft. If the casing was landed in sufficient tension to prevent buckling, compute the maximum change in axial force that could result from degradation of the 14-lbm/gal mud over a long period of time. The pore pressure of the formations in this area is equivalent to a 9-lbm/gal density.

Solution. Assuming that the external pressure on the 8,000-ft interval of casing decreased with time from that of a 14-lbm/gal mud to a 9-lbm/gal pore fluid, the average pressure change given by Eq. 7.32 is

$$0.052(9-14)(8,000/2) = -1,040 \text{ psi.}$$

The change in axial stress caused by this average pressure change is given by Eq. 7.36 as

$$\Delta F_a = -0.471(10.75)^2(-1,040) = +56,607 \text{ lbf.}$$

The positive sign indicates that the axial force would increase by 56,607 lbf because of the loss in external pressure.

7.5.4 Thermal Effects

The example design-loading conditions previously presented did not consider axial stress caused by changes in temperature after the casing is cemented and landed in the wellhead. Temperature changes encountered during the life of the well usually are small and can be neglected. However, when the temperature variations are not small, the resulting axial stress must be considered in the casing-design and casing-landing procedures. Examples of wells that will encounter large temperature variations include (1) steam-injection wells used in thermal recovery processes, (2) geothermal wells used in extracting steam from volcanic areas of the earth, (3) arctic wells completed in permafrost, (4) deep gas wells, (5) offshore wells with significant riser lengths, and (6) wells completed in abnormally hot areas. In arctic regions, the *thaw ball*, or volume of melted permafrost around the well, caused by radial heat flow from the warm oil being produced grows with time. As a result, compaction of the formations and surface subsidence occur. Both of these induce local compression and tensile stresses in the various casing strings.

The axial strain for a temperature change, ΔT, is determined from the thermal coefficient of expansion, α, using

$$\epsilon_z = \alpha_T \Delta T. \quad \ldots\ldots\ldots\ldots\ldots\ldots\ldots\ldots (7.37)$$

The average thermal coefficient of expansion for steel is $6.67 \times 10^{-6}/°F$. Thus, if the casing is cemented and landed in sufficient tension to prevent buckling and if the axial stress is less than the yield stress, then the change in axial stress is given by

$$\sigma_z = -E\alpha_T \Delta T = -200\Delta T. \quad \ldots\ldots\ldots\ldots (7.38)$$

Converting this stress to an axial force yields

$$F_a = -200A_s\Delta T = -58.8w\Delta T. \quad \ldots\ldots\ldots\ldots (7.39)$$

In the event that the casing has not been landed in sufficient tension to prevent helical buckling, then Eqs. 7.33 through 7.35 would be used in addition to Hook's law to estimate the effective axial-stress/casing-stretch relationship.

It may not always be practical to design casing to withstand extreme temperature variations without exceeding the yield strength of the material. Because temperature loading is a limited-strain process, the casing can be designed in such a way that yielding can limit the stress. When this is done, however, close attention must be given to joint selection so that the joints are considerably stronger than the pipe body, thereby effectively limiting the inelastic pipe stretch to the pipe body and preventing joint failure. This type of design is called a *strain-limit design* and is a relatively new concept.[17]

7.5.5 Subsidence Effects

Compressive axial loading of casing generally is not severe and usually can be neglected in the casing design. However, significant compressive loading of casing sometimes can result from subsidence of a formation. Formation subsidence can occur in volumetric reservoirs because of the production of pore fluids and the depletion of formation pressure. Subsidence can also be caused by a thaw/freeze cycle in a well completed through permafrost. An approximate equation sometimes used to estimate the axial strain caused by a pressure drop, Δp, within the producing formation is given by

$$\epsilon_z = \frac{\Delta p(1-2\mu_f)(1+\mu_f)}{E_f(1-\mu_f)}, \quad \ldots\ldots\ldots\ldots (7.40)$$

where Young's modulus, E_f, and Poisson's ratio, μ_f, are used for the formation. Assuming that this strain is also imposed on the casing completed in the subsiding formation and applying Hook's law gives

$$\sigma_z = \frac{E_f\Delta p(1-2\mu_f)(1+\mu_f)}{E_f(1-\mu_f)}. \quad \ldots\ldots\ldots\ldots (7.41)$$

The axial stress resulting from subsidence tends to be greatest in soft soil with a low value for Young's modulus.

In permafrost sections, designing the casing to withstand the large subsidence loads without exceeding the yield strength of the steel may not be practical. In this case, the strain-limit design concept can be applied.

7.5.6 Casing Landing

The loading conditions previously presented for the example casing-design calculations did not consider additional axial stress placed in the casing when it is landed. Casing landing practices vary significantly throughout the industry. In some cases, considerable additional axial stress will be placed in the casing when it is landed in the wellhead. Obviously, when this practice is followed, the axial stress must be considered in the casing design.

In an early study, an API committee identified the following four common methods for landing casing.

1. Landing the casing with the same tension that was present when cement displacement was completed.

2. Landing the casing in tension at the freeze point, which is generally considered to be at the top of the cement.

3. Landing the casing with the neutral point of axial stress ($\sigma_z = 0$) at the freeze point.

4. Landing the casing in compression at the freeze point.

All these general procedures are still used within the industry. In addition, operators differ as to how much tension or compression they place at the freeze point in the second and fourth procedures. The first two procedures are most commonly used. The API study committee[18] recommended that casing be landed by the first procedure in all wells in which the mud density does not exceed 12.5 lbm/gal, where standard design factors are used, and where the wellhead equipment and outer casing strings are of sufficient strength to withstand the landing loads. The second procedure was recommended for wells in which excessive mud weights are anticipated, with the amount of tension at the freeze point being selected to prevent any tendency of the casing to buckle above the freeze point. However, because API has recently withdrawn Bulletin D-7, it currently does not have a recommended casing landing practice.

Dellinger and McLean[16] performed a study that linked casing wear during drilling operations with helical buckling of casing just above the freeze point and, in some cases, below the freeze point. Excessive borehole washout was determined to be present when casing wear was experienced below the top of the cement and when the cement over the entire cemented interval was determined not to support the casing firmly. They recommended landing drilling casing whenever possible, so that no tendency to buckle at any point above the freeze point would exist, and making every economical effort to prevent excessive washout. However, they pointed out that placing

TABLE 7.13—INTERMEDIATE CASING FOR EXAMPLE 7.12

Section	Depth Interval (ft)	Length (ft)	Grade	Weight (lbm/ft)	Internal Diameter (in.)	Internal Area, A_i (sq in.)	Internal Area, A_o (sq in.)	Steel Area, A_s (sq in.)
1	6,200–10,000	3,800	C-95	40.0	8.835	61.306	72.760	11.454
2	1,800–6,200	4,400	C-95	43.5	8.755	60.201	72.760	12.559
3	0–1,800	1,800	C-95	47.0	8.681	59.187	72.760	13.573

enough tension in the casing at the freeze point to prevent buckling was often impossible because the casing becomes stuck before it can be landed and the casing suspension equipment will not permit the needed loads to be applied, e.g., when an ocean bottom suspension system is used. When it is impossible to maintain the freeze point in tension, they recommended circulating cement to a more shallow depth or holding internal pressure on the casing while the cement is hardening. They felt that the use of more cement was best because the use of internal pressure may increase the maximum axial tension to which the casing is subjected, thus requiring the use of more expensive casing. In addition, a microannulus may be formed between the casing and the cement sheath when the internal pressure is released.

Goins[15] has presented the following graphical procedure for determining the portion of the casing string that has a tendency to buckle.

1. Determine the axial force in the casing at the bottom and top of each section and make a plot of axial force vs. depth.

2. Determine the stability force from Eq. 7.35 at the bottom and top of each section, and make a plot of stability force vs. depth.

3. Locate the intersection of the load line and the stability-force line to determine the location of the neutral point of buckling. This is the point where the axial stress is equal to the average of the radial and tangential stresses.

The buckling tendency occurs below the neutral point.

Example 7.12. A 9.625-in. intermediate casing string composed of the three sections shown in Table 7.13 is set at 10,000 ft in a vertical well having an average borehole diameter of 13.0 in. The casing was run in 10-lbm/gal mud and cemented with 2,000 ft of 15.7-lbm/gal cement. The casing was landed "as cemented," i.e., with the same axial tension in the top of the string as when the cement-wiper plug reached bottom. Subsequently, the well was deepened to a depth of 15,000 ft, and the borehole mud density was increased from 10 to 16 lbm/gal. Also, because of the deeper well depth, the circulating mud temperature raises the average temperature of the casing by 30°F over the temperature initially present after cementing. Perform a stability analysis to determine the portion of the casing that may have a tendency to buckle (1) after cement placement and (2) during drilling operations at 15,000 ft. Assume that the surface casing pressure was held at 593 psig from the time the cement wiper plug reached bottom to when the cement hardened.

Solution. A graphical stability analysis will be performed as recommended by Goins.[15] The vertical forces acting

on the casing when placement of the cement slurry was completed is shown in Fig. 7.33. The hydrostatic forces F_1 through F_4 are given by

$$F_1 = -p_1(A_o)_1$$

$$= -[0.052(10)(8,000) + 0.052(15.7)(2,000)](72.760)$$

$$= -(5,792.8)(72.760) = -421,484 \text{ lbf},$$

$$F_2 = +p_2(A_i)_1$$

$$= +[593 + 0.052(10)(10,000)](59.187)$$

$$= +(5,793)(59.187) = +342,870 \text{ lbf},$$

$$F_3 = -p_3[(A_i)_2 - (A_i)_1]$$

$$= -[593 + 0.052(10)(6,200)](60.201 - 59.187)$$

$$= -(3,817)(1.014) = -3,870 \text{ lbf},$$

and

$$F_4 = -p_4[(A_i)_3 - (A_i)_2]$$

$$= -[593 + 0.052(10)(1,800)](61.306 - 60.201)$$

$$= -(1,529)(1.105) = -1,690 \text{ lbf}.$$

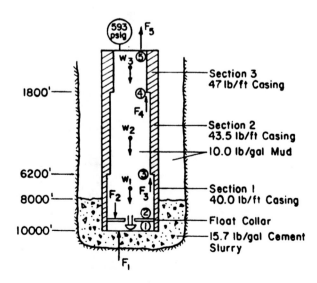

Fig. 7.33—Forces acting on casing after placement of cement slurry.

The casing weights W_1 through W_3 are given by

$$W_1 = 3,800(40) = 152,000 \text{ lbf},$$

$$W_2 = 4,400(43.5) = 191,400 \text{ lbf},$$

and

$$W_3 = 1,800(47) = 84,600 \text{ lbf}.$$

The graph of axial force vs. depth is constructed by starting at the bottom of the casing string and solving the force balance for successive sections upward. The force below the casing float collar is $-421,484$ lbf; the force above the float collar is

$$-421,484 + 342,870 = -78,614 \text{ lbf},$$

and the axial force at the top of Sec. 1 is

$$-78,614 + 152,000 = 73,386 \text{ lbf}.$$

The axial force at the bottom of Sec. 2 is

$$73,386 - 3,870 = 69,516 \text{ lbf},$$

and the axial force at the top of Sec. 2 is

$$69,516 + 191,400 = 260,916 \text{ lbf}.$$

Similarly, the axial force at the bottom of Sec. 3 is

$$260,916 - 1,690 = 259,226 \text{ lbf},$$

and the axial force at the top of Sec. 3 is

$$259,226 + 84,600 = 343,826 \text{ lbf}.$$

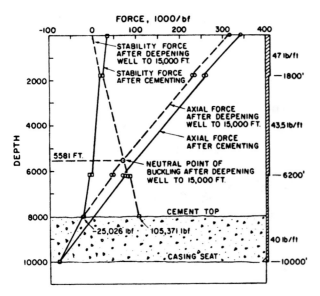

Fig. 7.34—Plot of axial force and stability force vs. depth for Example 7.12.

With the force endpoints of each section calculated above, a plot of axial force vs. depth is made as shown in Fig. 7.34.

The next step in the analysis is to determine the stability force as a function of depth. The stability force given by Eq. 7.35 at the bottom of the float collar is

$$p_1(A_i)_1 - p_1(A_o)_1 = -p_1 A_s = -5,792.8(13.573)$$

$$= -78,626 \text{ lbf},$$

and, since $p_1 = p_2$, the stability force above the float collar is also $-78,626$ lbf.

The stability force at the top of the cement is

$$[593 + 0.052(10)(8,000)](59.187)$$

$$-[0.052(10)(8,000)](72.760)$$

$$= 281,316 - 302,682 = -21,366 \text{ lbf},$$

and the stability force at the top of Sec. 1 is

$$(3,817)(59.187) - [0.052(10)(6,200)](72.760)$$

$$= 225,917 - 234,578 = -8,861 \text{ lbf}.$$

Similarly, the stability force at the bottom of Sec. 2 is

$$(3,817)(60.201) - (3,224)(72.760)$$

$$= 229,787 - 234,578 = -4,791 \text{ lbf},$$

and the stability force at the top of Sec. 2 is

$$(1,529)(60.201) - [0.052(10)(1,800)](72.760)$$

$$= 92,047 - 68,103 = +23,944 \text{ lbf}.$$

Finally, the stability force at the bottom of Sec. 3 is

$$(1,529)(61.306) - 68,103 = 25,634 \text{ lbf},$$

and the stability force at the top of Sec. 3 is

$$(593)(61.306) = 36,354 \text{ lbf}.$$

When the stability forces in Fig. 7.34 are plotted, the intersection of the axial force and stability force occurs essentially at the float collar at the bottom of the casing string. Thus, there is no significant tendency to buckle, and the casing will be held straight by the internal pressure until the cement hardens. Note that the amount of surface internal pressure needed for this example corresponds to the difference in hydrostatic pressure between the cement and mud. Note also that the above calculations assume that the cement does not lose its ability to transmit hydrostatic pressure to the bottom of the float collar after it begins to harden.

The casing that had the same axial force at the surface as after cementing was landed. However, after the borehole was deepened to 15,000 ft, the mud density increased from 10 to 16 lbm/gal, the average temperature increased by 30°F, and the axial stress was changed. To estimate the change in axial stress, the tendency of the casing to change its length will first be calculated, assuming the lower end of the casing is free to move. The change in axial stress required to return the lower end of the casing to its original position then will be computed.

The increase in internal pressure would change the axial stress in the casing above the cement in two ways. First, the change in pressure would cause changes in the forces F_3 and F_4, given by

$$\Delta F_3 = -0.052(6,200)(16-10)(1.014) = -1,961 \text{ lbf}$$

and

$$\Delta F_4 = -0.052(1,800)(16-10)(1.105) = -621 \text{ lbf}.$$

The change in F_3 might cause Sec. 2 to shorten by

$$\Delta L_1 = \frac{\Delta F_3 L}{A_s E} = \frac{-1,961(4,400)}{(12.559)(30 \times 10^6)} = -0.0229 \text{ ft}.$$

Similarly, the sum of the change in F_3 and F_4 might cause Sec. 3 to shorten by

$$\Delta L_2 = -\frac{(1,961+621)(1,800)}{(13.573)(30 \times 10^6)} = -0.0114 \text{ ft}.$$

Second, the increase in average pressure above 8,000 ft would increase the radial and tangential stress, and thus cause the casing to shorten. The average change in internal pressure is given by Eq. 7.31 as

$$0.052(16-10)(8,000/2) = 1,248 \text{ psi}.$$

The length change associated with this change in internal pressure is

$$\Delta L_3 = -2L \frac{\mu}{E} \frac{A_i}{A_s} \Delta p_i,$$

where the average A_i/A_s ratio above the cement is

$$\left(\frac{8,000-6,200}{8,000} \right) \left(\frac{61.306}{11.454} \right)$$

$$+ \left(\frac{4,400}{8,000} \right) \left(\frac{60.201}{12.559} \right) + \left(\frac{1,800}{8,000} \right) \left(\frac{59.187}{13.573} \right)$$

$$= 0.225(5.352) + 0.55(4.793) + 0.225(4.36) = 4.821.$$

Solving for the length change gives

$$\Delta L_3 = \frac{2(8,000)(0.3)(4.821)(1,248)}{30 \times 10^6} = -0.9627 \text{ ft}.$$

The length change caused by the average temperature increase is given by Eq. 7.37 as

$$\Delta L_4 = \alpha_T L \Delta T = 6.667 \times 10^{-6}(8,000)(30) = +1.600 \text{ ft}.$$

The net tendency for the casing to change in length is obtained by totalling the various length changes. Thus, the net ΔL is

$$\Delta L = \Sigma \Delta L_i = -0.0229 - 0.0114 - 0.9627 + 1.6 = 0.603 \text{ ft}.$$

If buckling can occur, a portion of this elongation will be permitted by bending within the confines of the borehole wall, and the remainder will be converted to a decrease in axial stress according to Hook's law. This gives

$$\Delta F_a = \frac{-EA_s}{L} \Delta L_e = \frac{-EA_s}{L}(0.603 - \Delta L_{bu}),$$

where ΔL_{bu} is the length change permitted by buckling and ΔL_e is the effective length change that was prevented and converted to a decrease in axial tension. The average steel area is given by

$$\frac{1,800}{8,000}(11.454) + \frac{4,400}{8,000}(12.559) + \frac{1,800}{8,000}(13.573)$$

$$= 12.539 \text{ sq in}.$$

In Eq. 7.33, the change in length permitted by buckling is

$$\Delta L_{bu} = \frac{\Delta r^2 F_{bu}^2}{8 E I w}.$$

Assuming that buckling occurs only in the bottom section, then the radial clearance is

$$\Delta r = \left(\frac{13-9.625}{2} \right) = 1.6875 \text{ sq in}.$$

and the moment of inertia from Eq. 7.16 is

$$I = \frac{\pi}{64}(9.625^4 - 8.835^4) = 122.2 \text{ in.}^4$$

Substitution of these values into the expression for ΔL_{bu} yields

$$\Delta L_{bu} = \frac{1.6875^2 F_{bu}^2}{8(30 \times 10^6)(122.2)(40)} = 2.427 \times 10^{-12} F_{bu}^2,$$

and the change in axial force can be expressed by

$$\Delta F_a = -\frac{(30 \times 10^6)(12.539)}{8,000}$$

$$\times (0.603 - 2.427 \times 10^{-12} F_{bu}^2).$$

The axial force at the top of the cement is

$$F_a = (F_a)_o + \Delta F_a$$

$$= [-78,614 + 2,000(40)] - 28,354$$

$$+ 1.142 \times 10^{-7} F_{bu}{}^2$$

$$= -26,968 + 1.142 \times 10^{-7} F_{bu}{}^2,$$

and the buckling force at the top of the cement given by Eq. 7.34 is

$$F_{bu} = F_s - F_a = F_s + 26,968 - 1.142 \times 10^{-7} F_{bu}{}^2.$$

Since the stability force at the top of the cement is

$$[0.052(16)(8,000)](61.306)$$

$$- [0.052(10)(8,000)](72.760)$$

$$= 408,053 - 302,682 = 105,371 \text{ lbf,}$$

then

$$F_{bu} = 105,371 + 26,968 - 1.142 \times 10^{-7} F_{bu}{}^2$$

$$= 132,339 - 1.142 \times 10^{-7} F_{bu}{}^2,$$

and upon solving for F_{bu},

$$F_{bu} = 130,397 \text{ lbf.}$$

Thus, the axial force at the top of the cement is

$$F_a = F_s - F_{bu} = 105,371 - 130,397$$

$$= -25,026 \text{ lbf,}$$

and the axial force at the top of Sec. 1 is

$$-25,026 + (8,000 - 6,200)40 = 46,974 \text{ lbf.}$$

The axial force at the bottom of Sec. 2 is

$$46,974 - [0.052(16)(6,200)](1.014) = 41,743 \text{ lbf,}$$

and at the top of Sec. 2 is

$$41,743 + 191,400 = 233,143 \text{ lbf.}$$

Finally, the axial force at the bottom of Sec. 3 is

$$233,143 - [0.052(16)(1,800)](1.105) = 231,488 \text{ lbf,}$$

and at the top of Sec. 3 is

$$231,488 + 84,600 = 316,088 \text{ lbf.}$$

A plot of axial force vs. depth while drilling at 15,000 ft is shown as a dashed line in Fig. 7.34.

The stability force at the top of the cement was determined to be 105,371 lbf. The stability force at the top of Sec. 1 is

$$[0.052(16)(6,200)](61.306)$$

$$- [0.052(10)(6,200)](72.760) = 81,662 \text{ lbf.}$$

The stability force at the bottom of Sec. 2 is

$$[0.052(16)(6,200)](60.201)$$

$$- [0.052(10)(6,200)](72.760) = 76,273 \text{ lbf,}$$

and at the top of Sec. 2 is

$$[0.052(16)(1,800)](60.201)$$

$$- [0.052(10)(1,800)](72.760) = 22,052 \text{ lbf.}$$

Similarly, the stability force at the bottom of Sec. 3 is

$$[0.052(16)(1,800)](59.187)$$

$$- [0.052(10)(1,800)](72.760) = 20,535 \text{ lbf,}$$

and at the top of Sec. 3 is

$$(0)(59.187) - (0)(72.760) = 0 \text{ lbf.}$$

A plot of stability force vs. depth for these conditions was made with dashed lines in Fig. 7.34. Note that the intersection of the stability force and axial force lines occurs at a depth of 5,581 ft. Thus, the casing will have a tendency to buckle helically from this depth to the top of the cement at 8,000 ft. The buckled casing can be expected to increase the rate of casing wear in this interval as a result of drilling and tripping operations. If it were possible to increase the axial tension by

$$105,371 - (-25,026) = 130,397 \text{ lbf}$$

when the casing was landed, the neutral point of buckling could have been lowered to the top of the cement, and the tendency to buckle could have been eliminated.

Exercises

7.1 Discuss the functions of the following casing strings.
   a. Conductor casing,
   b. Surface casing,
   c. Intermediate casing, and
   d. Production casing.

7.2 Discuss the advantages of using a liner rather than a full-length casing string.

7.3 Name the three basic processes used in the manufacture of casing.

7.4 What is the diameter range of API casing?

7.5 Give the three length ranges of casing specified by API.

TABLE 7.14—PORE-PRESSURE-GRADIENT AND FRACTURE-GRADIENT DATA FOR EXERCISE 7.21

Depth (ft)	Pore-Pressure Gradient (lbm/gal)	Fracture-Gradient (lbm/gal)
1,000	9.0	12.0
2,000	9.0	12.9
3,000	9.0	13.6
4,000	9.0	14.2
5,000	9.0	14.8
6,000	9.0	15.2
7,000	9.0	15.6
8,000	9.0	15.9
9,000	12.0	16.8
10,000	14.0	17.4
11,000	15.0	17.8
12,000	16.0	18.1

7.6 Define the following terms.

a. Nominal weight,
b. Plain-end weight,
c. Average weight, and
d. Drift diameter.

7.7 Make a sketch of the following casing connectors and label the fluid seal area. Also indicate whether the seal is based primarily on the use of a thread compound or on a metal-to-metal seal area.

a. Short round threads and couplings,
b. Long round threads and couplings,
c. Buttress threads and couplings,
d. Extreme-line threads,
e. Hydril two-step, and
f. Atlas Bradford TC-4S connector.

7.8 Compute the body yield strength for 10.75-in., J-55 casing having a nominal wall thickness of 0.35 in. and a nominal weight per foot of 40.5 lbm/ft.

7.9 Compute the burst-pressure rating for 10.75-in., J-55 casing having a nominal wall thickness of 0.40 in., and a nominal weight per foot of 45.5 lbm/ft.

7.10 Compute the collapse-pressure ratings for the following casing.

a. 11.75-in., C-95, 60 lbm/ft,
b. 10.75-in., P-110, 51 lbm/ft,
c. 10.75-in., J-55, 40.5 lbm/ft, and
d. 4.50-in., J-55, 11.6 lbm/ft.

7.11 Determine a corrected collapse-pressure rating for the casings listed in Exercise 7.10 for in-service conditions where the internal pressure is 10% of the burst rating and the axial tension is 60% of the pipe body yield strength.

7.12 Determine the maximum axial stress for a 40-ft joint of 10.75-in., 40.5-lbm/ft, J-55 casing having API short round threads if the casing is subjected to a 300,000-lbf axial-tension load in a portion of a directional wellbore having a dogleg severity of 3°/100 ft. Compute the maximum axial stress:

a. assuming uniform contact between the casing and the borehole wall; and
b. assuming contact between the borehole wall and the casing only at the couplings.

7.13 Estimate the maximum stress that can be used safely for P-110 casing that will be exposed to hydrogen sulfide at a temperature of 150°F.

7.14 Discuss the effect that crushing (out-of-roundness) caused by poor field handling will have on the performance properties of casing.

7.15 Determine the bit size needed for drilling to the depth of the surface casing of a well if the casing program calls for a surface string, an intermediate string, and a 5.50-in. production string.

7.16 A combination production casing string is to consist of 15.5 lbm/ft and 17 lbm/ft, J-55 casing to a total depth of 6,300 ft and is to be run in 12-lbm/gal mud. Determine the depth at which the casing weight per foot should change because of collapse-pressure considerations with a design factor of 1.125, assuming that the fluid level could fall as low as 6,300 ft in subsequent drilling operations.

7.17 For what weights of 5.5-in., N-80 casing would failure in tension occur by axial yielding rather than by joint failure? Consider long round threads and couplings, buttress threads and couplings, and extremeline connectors.

7.18 Complete the design of the intermediate casing of Example 7.7. Use the same design factors and other safety margins used in Example 7.9.

7.19 Complete the design of the production casing of Example 7.7. Use the same design factors and other safety margins used in Example 7.9.

7.20 A string of 9.625-in. production casing is to be run in 11.0-lbm/gal mud. Complete the design of this string using the same design factors and other safety margins used in Example 7.9.

7.21 Your company wants to complete a well at 12,000 ft using a 6.625-in. production casing. The pore-pressure- and fracture-gradient data are given in Table 7.14. Design a complete casing program for this well. Use the same design factors and other safety margins used in Example 7.9.

7.22 Casing with an ID of 10.05 in. and a strength in tension of 450,000 lbf in the joints and 629,000 lbf in the pipe body is being suspended from the surface while being cemented. If the effective weight of the casing being supported is 240,000 lbf and a safety factor of 1.4 is desired, how much surface pump pressure can be applied safely to displace the cement? The casing burst is rated at 3,130 psi (for zero axial stress).

7.23 Your company has a policy of slacking off 30% of the hook load when landing the casing after cementing. Also, the internal casing pressure is released to the atmosphere immediately after bumping the cement wiper plug on bottom. Repeat the buckling analysis for Example 7.12 for these operating conditions.

References

1. Greenip, J.F. Jr.: "Designing and Running Pipe," *Oil and Gas J.* (Oct. 9, 16, 30, and Nov. 13 and 27, 1978).
2. "Specifications for Restricted Yield Strength Casing and Tubing," *Spec. 5AC*, 11th edition, API, Dallas (May 1985).
3. "Bulletin on Performance Properties of Casing, Tubing, and Drill Pipe," *Bull. 5C2*, 18th edition, API, Dallas (March 1982).
4. "Bulletin on Formulas and Calculations for Casing, Tubing, Drill Pipe, and Line Pipe Properties," *Bull. 5C3*, fourth edition, API, Dallas (Feb. 1985).
5. Goodman, J.: *Mechanics Applied to Engineering*, eighth edition, Longmans Green, London (1914) 421–23.
6. Timoshenko, S.P. and Goodier, J.N.: *Theory of Elasticity*, third edition, McGraw-Hill Book Co., New York City (1961).
7. Timoshenko, S.P. and Gere, J.M.: *Theory of Elastic Stability*, second edition, McGraw-Hill Book Co., New York (1961).
8. Holmquist, J.L. and Nadia, A.: "A Theoretical and Experimental Approach to the Problem of Collapse of Deep-Well Casing," *Drill. and Prod. Prac.*, API, Dallas (1939) 392.

9. Bowers, C.N.: "Design of Casing Strings," paper SPE 514-G presented at the 1955 SPE Annual Meeting, New Orleans, LA, Oct. 2–5.

10. Lubinski, A.: "Maximum Permissible Doglegs in Rotary Borehead," *Trans.*, AIME (1961) 175.

11. Kane, R.D. and Greer, J.B.: "Sulfide Stress Cracking of High-Strength Steels in Laboratory and Oilfield Environment," *Trans.*, AIME (1977) 1483.

12. Patton, C.C.: "Corrosion Fatigue Causes Bulk of Drill-String Failures," *Caring for Casing and Drill Pipe*, published by the Oil and Gas Journal.

13. "Recommended Practices for Care and Use of Casing and Tubing," *RP 5C1*, API, Dallas.

14. Lubinski, A., Althouse, W.S., and Logan, J.L.: "Helical Buckling of Tubing Sealed in Packers," *Trans.*, AIME (1962) 35.

15. Goins, W.C.: "Better Understanding Prevents Tubular Buckling Problems," *World Oil*, Part 1 (Jan. 1980) 101, Part 2 (Feb. 1980) 35.

16. Dellinger, T.B. and McLean, J.C.: "Preventing Instability in Partially Cemented Intermediate Casing Strings," paper SPE 4606 presented at the 1973 SPE Annual Meeting, Sept. 30–Oct. 3, 1973.

17. Wooley, G.R.: "Strain Limit Design of 13.375-in., 72 lbm/ft, N-80 Buttress Casing," *J. Pet. Tech.*.

18. "Casing Landing Recommendations," *Bull. D7*, API, Dallas (1955).

Nomenclature

A = area
A_i = inner pipe area enclosed by ID
A_{jp} = steel area under last perfect thread
A_o = outer pipe area enclosed by OD
A_p = steel area in pipe body
A_s = steel cross-sectional area
A_{sc} = steel area in coupling
d = ID of pipe
d_b = ID at critical section of joint box
d_{c1} = diameter at root of coupling thread at end of pipe in power-tight position
d_{c2} = OD of coupling
d_n = nominal pipe diameter
d_{j1} = nominal joint ID of made-up connection
d_{j2} = nominal joint OD of made-up connection
d_{pin} = OD at critical section of joint box
d_1 = smaller diameter of annulus
d_2 = larger diameter of annulus
D = depth
D_c = depth of casing
D_{lc} = depth of lost-circulation zone
D_m = depth of mud surface
E = Young's modulus of elasticity
E_f = Young's modulus of elasticity for the formation
F = force, also ratio
F_a = axial force
F_{ab} = equivalent axial force caused by bending
F_{fr} = frictional force
F_s = stability force
F_{sc} = side force at coupling
F_{ten} = tensional force
F_w = wall force
g_p = pore pressure gradient expressed as equivalent mud density
h = thickness
I = moment of inertia

K = square root of 1 over EI
L = length
L_j = joint length
L_t = length of engaged threads
M = bending moment
M_c = bending moment at coupling
p = pressure
p_{br} = burst-pressure rating
p_{cr} = collapse-pressure rating
p_e = external pressure
p_i = internal pressure
r = radius
Δr = radial clearance of annulus, i.e., $(d_2 - d_1)/2$
r_c = radius of curvature of borehole axis
r_i = inner radius
r_o = outer radius
t = thickness
T = temperature
w = weight per foot
W = weight
x,y = spatial coordinates
α = dogleg severity, degrees/100 ft
α_T = temperature coefficient of expansion
Δ = change
ϵ = strain
ϵ_r = radial strain
ϵ_t = tangential strain
ϵ_z = axial strain
θ = angle
μ = Poisson's ratio
μ_f = Poisson's ratio for the formation
ρ = density
ρ_g = gas density
ρ_m = mud density
ρ_s = steel density
σ = stress
σ_r = radial stress
σ_s = nominal steel strength
σ_t = tangential stress
σ_{ult} = ultimate strength
σ_{yield} = yield strength
σ_z = axial stress

Subscripts

e = effective
max = maximum
1,2,3 = sections 1, 2, 3

SI Metric Conversion Factors

°F	(°F−32)/1.8		=	°C
ft	× 3.048*	E−01	=	m
in.	× 2.54*	E+00	=	cm
lbf	× 4.448 222	E+00	=	N
lbf/ft	× 1.355 818	E−03	=	kJ
lbm/gal	× 1.198 264	E+02	=	kg/m³
psi	× 6.894 757	E+00	=	kPa
psi/ft	× 2.262 059	E+01	=	kPa/m

*Conversion factor is exact.

Chapter 8
Directional Drilling and Deviation Control

8.1 Definitions and Reasons for Directional Drilling

Directional drilling is the process of directing the wellbore along some trajectory to a predetermined target. Deviation control is the process of keeping the wellbore contained within some prescribed limits relative to inclination angle, horizontal excursion from the vertical, or both. This chapter discusses the principles and mechanisms associated with directional drilling and deviation control.

The preceding chapters deal with the one-dimensional process of penetrating the earth with the bit to some vertical depth. However, drilling is a three-dimensional process. The bit not only penetrates vertically but is either purposely or unintentionally deflected into the X-Y planes (see Fig. 8.1) The X plane is defined as the direction plane and the Y plane is the inclination plane. The angles associated with the departures in the X and Y planes are called "direction" and "inclination" angles, respectively.

Fig. 8.2 presents a typical example of the trajectory-control situation. Here a structure is located almost entirely under a lake. Well 1, drilled on a part of the structure that is not under the lake, could be treated simply as a deviation-control well drilled on the shore. To develop the rest of the field, however, will necessitate drilling directional wells. The only way vertical wells could be drilled would be from a floating drilling vessel or platform, with the wells being completed on the lake bed (sublake completions), or from a floating or fixed production platform; and the economics of those approaches would be far less attractive than drilling directional wells from some convenient land-based site where a standard land rig can be used. In some situations, there is no alternative to drilling a directional well. For example, the lake may be the only source for drinking water in the area,

and thus there may be environmental restrictions that prohibit the use of power vessels and equipment such as offshore drilling rigs and production facilities.

The early drilling of directional wells was clearly motivated by economics. The oil fields offshore California were the spawning ground for directional drilling practices and equipment, and for a special group of people called "directional drillers." Later discoveries of oil and gas in the Gulf of Mexico and in other countries promoted the expanded application of directional drilling. Offshore field development has accounted for the majority of directional drilling activities. Fig. 8.3 shows a typical offshore platform development. In a number of cases, fields have been discovered beneath population centers, and the only way to develop the fields economically has been to use a drilling pad and to drill directionally (see Fig. 8.4). Natural obstructions such as mountains or other severe topographical features frequently prohibit building a surface location and drilling a near-vertical well (Fig. 8.5). Sidetracking out of an existing wellbore is another application of directional drilling. This sidetracking may be done to bypass an obstruction (a "fish") in the original wellbore (see Fig. 8.6) or to explore for additional producing horizons in adjacent sectors of the field (see Fig. 8.7).

Strong economic and environmental pressures have increased the use of directional drilling. In some areas it is no longer possible to develop a field by making roads to each surface location and drilling a near-vertical well. Instead, as in offshore installations, drilling pads must be built from which a number of wells can be drilled. Not only is directional drilling increasing, but trajectory programs are becoming more complicated and directional drilling is being applied in situations and areas where directional drilling has not been common. In hot-rock developments, for example, directional wells are being

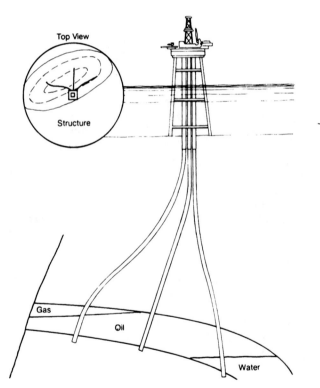

Fig. 8.1—Inclination and direction planes as a wellbore proceeds in the depth plane.

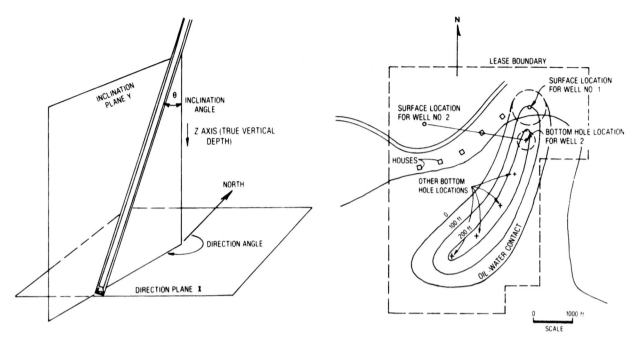

Fig. 8.2—Plan view of a typical oil and gas structure under a lake showing how directional wells could be used to develop it.

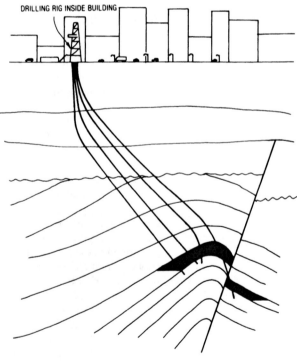

Fig. 8.3—Typical offshore development platform with directional wells.

Fig. 8.4—Developing a field under a city using directionally drilled wells.

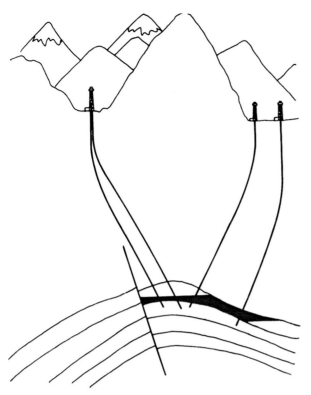

Fig. 8.5—Drilling of directional wells where the reservoir is beneath a major surface obstruction.

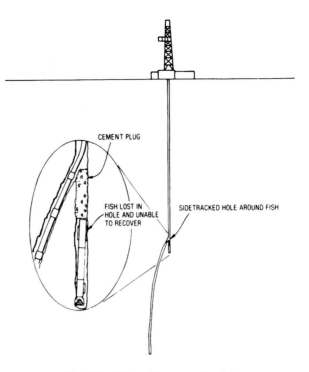

Fig. 8.6—Sidetracking around a fish.

drilled in hard granites and other igneous and metamorphic rocks. Geothermal projects have been developed with directional wells. Wells with extended horizontal reaches of 14,000 ft are being drilled, with goals of going even farther. As the costs of field development increase—in deeper waters, remote locations, hostile environments, and deeper producing zones—the use of directional drilling will also increase.

8.2 Planning the Directional Well Trajectory

The first step in planning any directional well is to design the wellbore path, or trajectory, to intersect a given target. The initial design should propose the various types of paths that can be drilled economically. The second, or refined, plan should include the effects of geology on the bottomhole assemblies (BHA's) that will be used and other factors that could influence the final wellbore trajectory. This section explains how to plan the initial trajectory for most common directional wells.

Fig. 8.8 depicts three types of trajectories that could be drilled to hit the target. Path A is a build-and-hold trajectory: the wellbore penetrates the target at an angle equal to the maximum buildup angle. Path B is a "modified-S" and C is an "S" trajectory. With the S-shape trajectory the wellbore penetrates the target vertically, and with the modified-S trajectory the wellbore penetrates the target at some inclination angle less than the maximum inclination angle in the hold section. For Path D, a "continuous-build trajectory," the inclination keeps increasing right up to or through the target. The build-and-hold path requires the lowest inclination angle to hit the target; the modified-S requires more inclination; and the S-shape requires still more. The continuous-build path requires the highest inclination of all the trajectory types to hit the target.

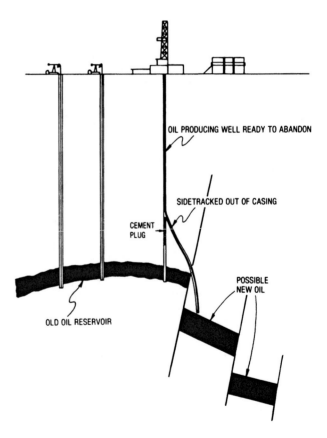

Fig. 8.7—Using an old well to explore for new oil by sidetracking out of the casing and drilling directionally.

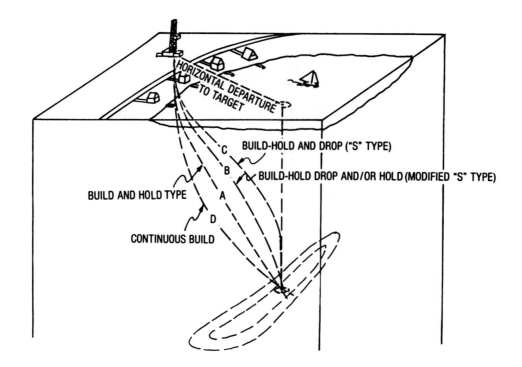

Fig. 8.8—Major types of wellbore trajectories.

8.2.1 Build-and-Hold Trajectory

Fig. 8.9 depicts a simple build-and-hold wellbore trajectory intersecting a target at a true vertical depth (TVD) of D_3 and at a horizontal departure of X_3 (Point B). The kickoff point is at a TVD of depth D_1, where the rate of inclination angle buildup is q in degrees per unit length.

The radius of curvature, r_1, is found thus:

$$r_1 = \frac{180}{\pi} \times \frac{1}{q}. \qquad (8.1)$$

To find the maximum inclination angle, θ, consider in Fig. 8.9 that

$$90° = \theta + (90 - \Omega) + \tau,$$

or

$$\theta = \Omega - \tau. \qquad (8.2)$$

The angle τ can be found by considering Triangle OAB, where

$$\tan \tau = \frac{BA}{AO} = \frac{r_1 - X_3}{D_3 - D_1}, \qquad (8.3a)$$

and

$$\tau = \text{arc tan} \frac{r_1 - X_3}{D_3 - D_1}. \qquad (8.3b)$$

Angle Ω can be found by considering Triangle OBC,

where

$$\sin \Omega = \frac{r_1}{OB} \qquad (8.4)$$

and

$$L_{OB} = \sqrt{(r_1 - X_3)^2 + (D_3 - D_1)^2}.$$

Substituting OB into Eq. 8.4 gives

$$\sin \Omega = \frac{r_1}{\sqrt{(r_1 - X_3)^2 + (D_3 - D_1)^2}}. \qquad (8.5)$$

The maximum inclination angle, θ, for the build-and-hold case, is not limited to $X_3 < r_1$. It is also valid for $X_3 \geqq r_1$.

$$\theta = \text{arc sin} \left[\frac{r_1}{\sqrt{(r_1 - X_3)^2 + (D_3 - D_1)^2}} \right]$$

$$- \text{arc tan} \left(\frac{r_1 - X_3}{D_3 - D_1} \right). \qquad (8.6)$$

The length of the arc, Section DC, is

$$L_{DC} = \frac{\pi}{180} \times r_1 \times \theta,$$

or

$$L_{DC} = \frac{\theta}{q}. \qquad (8.7)$$

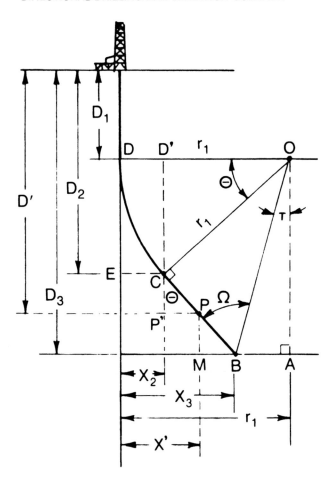

Fig. 8.9—Geometry of build-and-hold-type well path for $X_3 < r_1$.

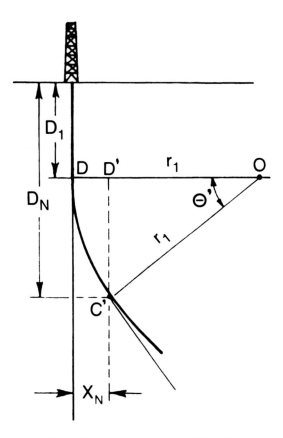

Fig. 8.10—Geometry for the build section.

The length of the trajectory path, CB, at a constant inclination angle can be determined from Triangle BCO as

$$\tan \Omega = \frac{CO}{L_{CB}} = \frac{r_1}{L_{CB}}$$

and

$$L_{CB} = \frac{r_1}{\tan \Omega}.$$

The total measured depth, D_M, for a TVD of D_3 is

$$D_M = D_1 + \frac{\theta}{q} + \frac{r_1}{\tan \Omega}, \quad \dots \dots \dots \dots (8.8)$$

where D_M equals the vertical section to kickoff plus build section plus constant inclination section (Fig. 8.9).

The horizontal departure EC (X_2) at the end of the build can be determined by considering Triangle D'OC, where

$$X_2 = r_1 - r_1 \cos \theta = r_1(1 - \cos \theta). \quad \dots \dots \dots (8.9)$$

To find the measured depth and horizontal departure along any part of the build before reaching maximum angle θ, consider the intermediate inclination angle θ', the inclination angle at C' (Fig. 8.10), which will yield a new horizontal departure, X_N. The distance d_N can be determined considering Triangle D'OC', where

$$D_N = D_1 + r_1 \sin \theta', \quad \dots \dots \dots \dots \dots (8.10)$$

and the horizontal displacement, X_N, is

$$X_N = r_1 - r_1 \cos \theta' = r_1 (1 - \cos \theta). \quad \dots \dots (8.11)$$

The TVD at the end of the build section is D_2, which can be derived from Triangle D'OC (Fig. 8.9):

$$D_2 = D_1 + r_1 \sin \theta. \quad \dots \dots \dots \dots \dots (8.12)$$

The new measured depth for any part of the buildup is

$$D_{MN} = D_1 + \frac{\theta'}{q}. \quad \dots \dots \dots \dots \dots \dots (8.13)$$

The new measured depth at a TVD of D' can be determined from Triangle PP'C:

$$D_{MP} = D_1 + \frac{\theta}{q} + CP, \quad \dots \dots \dots \dots \dots (8.14)$$

where

$$CP = \frac{CP'}{\cos \theta}$$

and

$$CP'=D'-D_2=(D'-D_1-r_1 \sin \theta).$$

Therefore,

$$CP=\frac{(D'-D_1-r_1 \sin \theta)}{\cos \theta}. \quad \ldots \ldots \ldots \ldots (8.15)$$

Substituting Eq. 8.15 into Eq. 8.14,

$$D_{MP}=D_1+\frac{\theta}{q}+\frac{D'-D_1-r_1 \sin \theta}{\cos \theta}. \quad \ldots \ldots (8.16)$$

Eq. 8.16 also can be used instead of Eq. 8.14 to calculate the measured depth by making $D'=D_3$.

The horizontal departure at Point P is

$$X'=X_2+P'P, \quad \ldots \ldots \ldots \ldots \ldots \ldots \ldots \ldots (8.17)$$

where $P'P=CP' \tan \theta$.

Combining Eq. 8.17, Eq. 8.9, and CP' yields

$$X'=r_1(1-\cos \theta)+(D'-D_1-r_1 \sin \theta)\tan \theta. \quad .(8.18)$$

The preceding derivation is valid only when $X_3 < r_1$.

Another way of expressing the maximum inclination angle, θ, in terms of r_1, D_1, D_3, and X_3 for $X_3 < r_1$ is

$$\theta=\text{arc} \tan\left(\frac{D_3-D_1}{r_1-X_3}\right)-\text{arc} \cos\left\{\left(\frac{r_1}{D_3-D_1}\right)\right.$$

$$\left.\times \sin\left[\text{arc} \tan\left(\frac{D_3-D_1}{r_1-X_3}\right)\right]\right\}. \quad \ldots \ldots \ldots (8.19)$$

Example 8.1. It is desired to drill under the lake to the location designated for Well 2. For this well, a build-and-hold trajectory will be used. Horizontal departure to the target is 2,655 ft at a TVD of 9,650 ft. The recommended rate of build is 2.0°/100 ft. The kickoff depth is 1,600 ft. Determine (1) the radius of curvature, R_1; (2) the maximum inclination angle, θ; (3) the measured depth to the end of the build; (4) the total measured depth; (5) the horizontal departure to the end of the build; (6) the measured depth at a TVD of 1,915 ft; (7) the horizontal displacement at a TVD of 1,915 ft; (8) the measured depth at a TVD of 7,614 ft; and (9) the horizontal departure at a TVD of 7,614 ft.

Solution. From Eq. 8.1

$$r_1=\frac{180}{\pi} \frac{1}{2°/100 \text{ ft}}=2,865 \text{ ft}.$$

Since $X_3 < r_1$, $\theta=\Omega-\tau$. From Eq. 8.3a,

$$\tan \tau=\frac{r_1-X_3}{D_3-D_1}=\frac{210 \text{ ft}}{8,050 \text{ ft}}=0.0261,$$

$$\tau=\text{arc} \tan 0.0261=1.5°.$$

From Eq. 8.5,

$$\sin \Omega=\frac{r_1}{\sqrt{(r_1-X_3)^2+(D_3-D_1)^2}}$$

$$=\frac{2,865 \text{ ft}}{8,053 \text{ ft}}=0.3558,$$

$$\Omega=\text{arc} \sin 0.356=20.84°.$$

The maximum inclination angle is

$$\theta=20.84°-1.5°=19.34°.$$

Using Eq. 8.19,

$$\theta=\text{arc} \tan\left(\frac{9,650 \text{ ft}-1,600 \text{ ft}}{2,865 \text{ ft}-2,655 \text{ ft}}\right)$$

$$-\text{arc} \cos\left(\frac{2,865 \text{ ft}}{9,650 \text{ ft}-1,600 \text{ ft}}\right)$$

$$\times \sin\left[\text{arc} \tan\left(\frac{9,650 \text{ ft}-1,600 \text{ ft}}{2,865 \text{ ft}-2,655 \text{ ft}}\right)\right]=19.34°.$$

The measured depth to the end of the build at an inclination of 19.34° is

$$D_M=1,600 \text{ ft}+\frac{19.34°}{2°}\times 100 \text{ ft}=2,565 \text{ ft},$$

and the total measured depth to the target TVD of 9,650 ft, using Eq. 8.14, is

$$D_{tar}=2,565 \text{ ft}+\frac{R_1}{\tan \Omega}=2,565 \text{ ft}+\frac{2,865 \text{ ft}}{\tan(20.84°)}$$

$$=10,091 \text{ ft}.$$

The horizontal departure to the end of the build, from Eq. 8.9, is

$$X_2=r_1(1-\cos \theta)=2,865 \text{ ft} [1-\cos(19.34)]=161 \text{ ft}.$$

At a TVD of 1,915 ft, the measured depth at a rate of build of 2°/100 ft can be determined by first calculating the inclination at 1,915 ft using Eq. 8.10:

$$1,915 \text{ ft} = 1,600 \text{ ft} + 2,865 \text{ ft} \sin \theta$$

$$\theta = \text{arc } \sin \left(\frac{315 \text{ ft}}{2,865 \text{ ft}} \right) = 6.31°.$$

The arc length of the build to 6.31° can be calculated using Eq. 8.7:

$$L_{DC} = \frac{6.31°}{2.0°} \times 100 \text{ ft} = 315.5 \text{ ft}.$$

The measured depth for a TVD of 1,915 ft is

$$D_M = 315.5 \text{ ft} + 1,600 \text{ ft} = 1,915.5 \text{ ft},$$

which is only 0.5 ft more than the TVD.

The horizontal departure at a TVD of 1,915 ft is found from Eq. 8.11:

$$X_{1,915} = 2,865 \text{ ft} (1.0 - \cos 6.31) = 17.36 \text{ ft}.$$

The measured depth at a TVD of 7,614 ft is

$$D_M = 1,600 \text{ ft} + \frac{19.34°}{2°} \times 100 \text{ ft}$$

$$+ \frac{7,614 \text{ ft} - 1,600 \text{ ft} - 2,865 \text{ ft} \sin(19.34°)}{\cos(19.34°)}$$

$$= 7,934 \text{ ft}.$$

The horizontal departure at a TVD of 7,614 ft is calculated with Eq. 8.18:

$$X'_{7,614} = 2,865 \text{ ft} (1 - \cos 19.34)$$

$$+ (7,614 \text{ ft} - 1,600 \text{ ft} - 2,865 \text{ ft} \sin 19.34°)$$

$$\times \tan 19.34 = 1,935.5 \text{ ft}.$$

The preceding derivation and example calculation is for the case where $r_1 > X_3$ for a simple build-and-hold trajectory. For the case where $r_1 < X_3$, the maximum angle, θ, can be calculated by

$$\theta = 180 - \text{arc } \tan \left(\frac{D_3 - D_1}{X_3 - r_1} \right) - \text{arc } \cos \left\{ \left(\frac{r_1}{D_3 - D_1} \right) \right.$$

$$\left. \times \sin \left[\text{arc } \tan \left(\frac{D_3 - D_1}{X_3 - r_1} \right) \right] \right\}. \quad \dots \dots \dots (8.20)$$

8.2.2 Build-Hold-and-Drop ("S") Trajectory

The second type of trajectory is the build, hold, and drop—or S-shape curve—which is depicted by Fig. 8.11 for the cases were $r_1 < X_3$ and $r_1 + r_2 > X_4$, and in Fig. 8.12 for the cases where $r_1 < X_3$ and $r_1 + r_2 < X_4$. In both cases, the maximum inclination is reduced to zero at D_4 with drop radius r_2, which is derived in the same manner as the build radius, r_1. The following equations are used to calculate the maximum inclination angles for $r_1 + r_2 > X_4$ and for $r_1 + r_2 < X_4$.

$$\theta = \text{arc } \tan \left(\frac{D_4 - D_1}{r_1 + r_2 - X_4} \right) - \text{arc } \cos \left\{ \left(\frac{r_1 + r_2}{D_4 - D_1} \right) \right.$$

$$\left. \times \sin \left[\text{arc } \tan \left(\frac{D_4 - D_1}{r_1 + r_2 - X_4} \right) \right] \right\}. \quad \dots \dots \dots (8.21)$$

$$\theta = 180° - \text{arc } \tan \left[\frac{D_4 - D_1}{X_4 - (r_1 + r_2)} \right]$$

$$- \text{arc } \cos \left(\left(\frac{r_1 + r_2}{D_4 - D_1} \right) \right.$$

$$\left. \times \sin \left\{ \text{arc } \tan \left[\frac{D_4 - D_1}{X_4 - (r_1 + r_2)} \right] \right\} \right) . \quad \dots \dots \dots (8.22)$$

8.2.3 Build, Hold, Partial Drop, and Hold (Modified "S") Trajectory

The build, hold, partial drop, and hold (Fig. 8.13) is the modified S type of wellbore path. Consider that the arc length

$$L_{CA} = \frac{\theta'}{q}.$$

From the Right Triangle CO′B, the following relationships can be written.

$$L_{CB} = r_2 \sin \theta' \quad \dots \dots \dots \dots \dots \dots \dots \dots \dots (8.23a)$$

and

$$s_{BA} = r_2 - r_2 \cos \theta' = r_2 (1 - \cos \theta'). \quad \dots \dots \dots (8.23b)$$

Eqs. 8.21 and 8.22 can be rewritten by substituting $D_5 + r_2 \sin \theta'$ for D_4 and $X_5 + r_2 (1 - \cos \theta')$ for X_4.

For any of the S-shape curves, the measured depths and horizontal departures can be calculated in the same way they are calculated for the build-and-hold trajectory by deriving the appropriate relationships for the various geometries.

8.2.4 Multiple Targets

When a directional well is being planned, the depth and horizontal departure of the target are given, as well as its dimensions. Targets may be rectangular, square, or circular. If the target is a circle, a radius is designated.

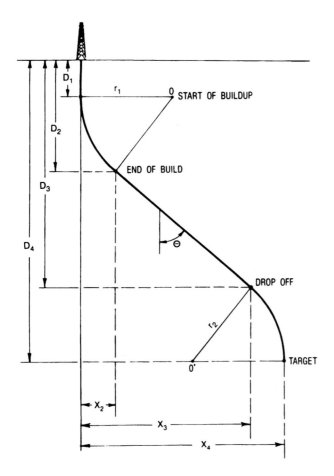

Fig. 8.11—Build-hold-and-drop for the case where $r_1 < X_3$ and $r_1 + r_2 < X_4$.

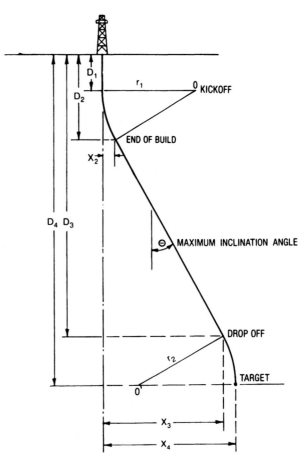

Fig. 8.12—Build-hold-and-drop for the case where $r_1 < X_3$ and $r_1 + r_2 > X_4$.

Sometimes there are multiple targets, as shown by Figs. 8.14a and 8.14b. If they are favorably positioned, multiple targets can be economically penetrated with one of the aforementioned types of trajectories (Fig. 8.14a). Sometimes, however, they are unfavorably aligned (Fig. 8.14b) and expensive trajectory alterations are required. The trajectory in Fig. 8.14b could be difficult and expensive to drill even though the vertical section appears the same as that in Fig. 8.14a. The direction change to hit Target 3 would in most situations be extremely difficult to execute.

8.2.5 Direction Quadrant and Compass Schemes

In the previous discussions all the trajectory planning has been reduced to a two-dimensional problem, considering only depth and horizontal departure. All directional wells also have an X component that is associated with direction. For example, Well 2 in Fig. 8.2 has a target direction of 100° east of north by a normal compass reading. In directional drilling, a 90° quadrant scheme is used to cite directions and the degrees are always read from north to east or west, and from south to east or west. For example, the direction angle in Fig. 8.15a by compass (always read clockwise from due north) is 18°, and by the quadrant scheme it is N18E. The well in the second quadrant (Fig. 8.15b) at 157° is read S23E. In quadrant three (Fig. 8.15c), the well is S20W, for a measured angle of 200°. In quadrant four (Fig. 8.15d), the compass angle of 305° is read N55W.

Example 8.2. What are the directions, in the alternative format, of each of the following wells?

Well A	N15E
Well B	225°
Well C	N0E

Solution. Well A is in the first quadrant and is 15°; Well B is in the third quadrant and should be read as S45W; and Well C represents 0° or north.

8.2.6 Planning the *X-Y* Trajectory

The first step in planning a well is to determine the two-dimensional *Y-Z* trajectory (Fig. 8.1). The next step is to account for the *X* component of the trajectory that departs from the vertical plane section between the surface location and the bottomhole target. Fig. 8.16 is a plan view, looking down on the straight line projected path from Well 2's surface location to the bull's-eye of a target with a 100-ft radius. The dashed line indicates a possible path the bit could follow because of certain influences exerted by the bit, the BHA configuration, the geology, general hole conditions, and other factors that are covered later in this chapter.

The target area provides a zone of tolerance for the wellbore trajectory to pass through. The size and dimensions of the target are usually based on factors pertaining to the drainage of a reservoir, geological criteria, and lease boundary constraints.

When a well is kicked off, the practice is to orient the trajectory to some specific direction angle called "lead." This lead usually is to the left of the target departure line and ranges from 5 to 25°. The value used is generally based on local experience or some rule of thumb. More recent research on direction variation (or, to use an older term, "bit walk") indicates that the lead can be selected on the basis of analysis of offset wells and of factors that might cause bit walk.

As the drilling progresses after the lead is set, the trajectory varies in the X and Y planes as the bit penetrates in the Z plane. Figs. 8.17 and 8.18 are vertical and horizontal (elevation and plan) views of a typical trajectory path. Past the lead angle, the trajectory shows a clockwise, or right-hand, tendency or bit walk. A counter-clockwise curvature is called left-hand tendency or bit walk.

The initial trajectory design did not account for the excursion of the bit away from the vertical plane that goes through the surface location and the target's bull's-eye. There are many ways to calculate the three-dimensional path of the wellbore.[1-3] The most common method used in the field is "angle averaging," which can be performed on a hand calculator with trigonometric functions.

Consider the vertical section as depicted by Fig. 8.17. The distance from the surface to the kickoff point is D_1. At A_1 the well is kicked off and drilled to A_2. The inclination angle at the kickoff is zero. Fig. 8.18 shows the

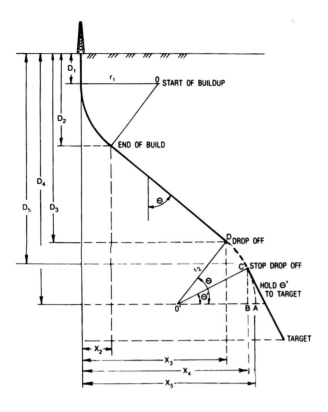

Fig. 8.13—Build-hold-and-drop and hold (modified-S) where $r_1 < X_3$ and $r_1 + r_2 < X_4$.

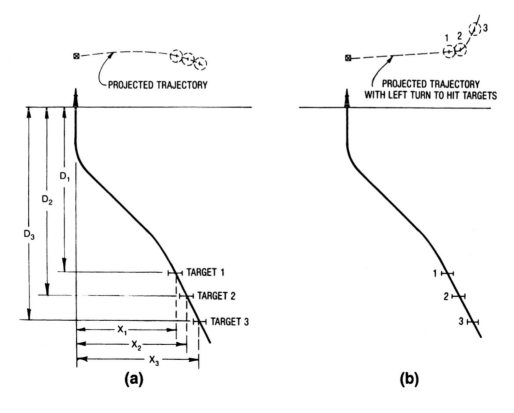

(a)   **(b)**

Fig. 8.14—Directional well used to intersect multiple targets.

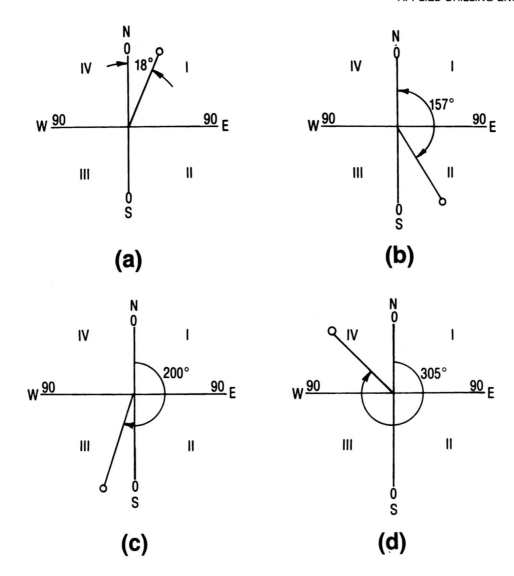

Fig. 8.15—Directional quadrants and compass measurements.

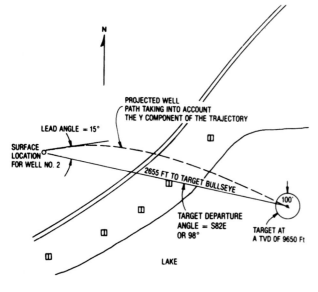

Fig. 8.16—Plan view.

top, or plan, view of the trajectory; Point A_1 on the vertical section corresponds to the starting point, A_1, on the plan view. Using the angle-averaging method, the following equations can be derived for the north/south (L) and east/west (M) coordinates.

$$L = \Delta D_M \sin\left(\frac{\alpha_A + \alpha_{A-1}}{2}\right)\cos\left(\frac{\epsilon_A + \epsilon_{A-1}}{2}\right)$$
.........................(8.24)

and

$$M = \Delta D_M \sin\left(\frac{\alpha_A + \alpha_{A-1}}{2}\right)\sin\left(\frac{\epsilon_A + \epsilon_{A-1}}{2}\right).$$
.........................(8.25)

The TVD can be calculated by

$$D = \Delta D_M \cos\left(\frac{\alpha_A + \alpha_{A-1}}{2}\right), \qquad \text{.............(8.26)}$$

where ΔD_M is the measured depth increment.

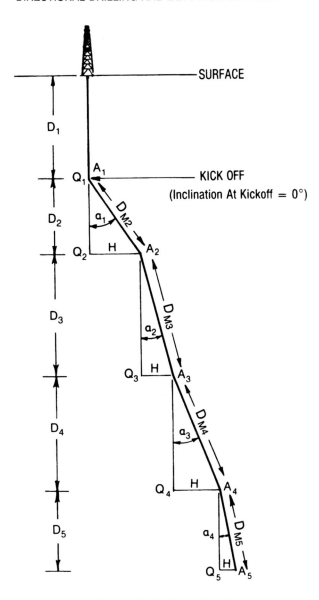

Fig. 8.17—Vertical calculation.

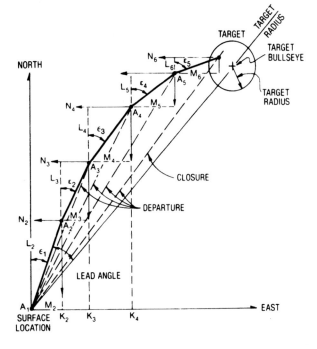

Fig. 8.18—Horizontal calculation.

Example 8.3. Calculate the trajectory for the well from 8,000 to 8,400 ft, where the kickoff is at 8,000 ft and the rate of build is 1°/100 ft, using a lead of 10° and a right-hand walk rate of 1°/100 ft. Direction to the bull's-eye is N30E. Assume that the first 200 ft is to set the lead, where the direction is held constant to 8,200 ft and then turns right at a rate of 1°/100 ft.

Solution. The north and east coordinates are calculated using Eqs. 8.24 and 8.25, and the TVD from 8,000 to 8,100 ft is calculated from Eq. 8.26.

$$L_2 = 100 \text{ ft } \sin\left(\frac{1°+0}{2}\right)\cos(20)* = 0.82 \text{ ft},$$

$$M_2 = 100 \text{ ft } \sin\left(\frac{1°+0}{2}\right)\sin(20) = 0.30 \text{ ft},$$

*For the first point the direction should not be averaged.

and

$$D_2 = 100 \text{ ft } \cos\left(\frac{1+0}{2}\right) = 99.996 \text{ ft},$$

$$D = 8,000 \text{ ft} + 99.996 \text{ ft} = 8,099.996 \text{ ft}.$$

From 8,100 to 8,200 ft:

$$L_3 = 100 \text{ ft } \sin\left(\frac{1+2}{2}\right)\cos\left(\frac{20+20}{2}\right) = 2.46 \text{ ft}.$$

Total north $= 0.82 + 2.46 = 3.28$ ft.

$$M_3 = 100 \text{ ft}\sin\left(\frac{1+2}{2}\right)\sin\left(\frac{20+20}{2}\right) = 0.90 \text{ ft}.$$

Total east $= 0.30 + 0.90 = 1.20$ ft.

$$D_3 = 100 \text{ ft}\cos\left(\frac{1+2}{2}\right) = 99.966 \text{ ft}.$$

$$D = 8,099.996 + 99.966 = 8,199.962 \text{ ft}.$$

From 8,200 to 8,300 ft, the direction changes by 1°/100 ft from N20E to N21E.

$$L_4 = 100 \text{ ft } \sin\left(\frac{2+3}{2}\right)\cos\left(\frac{20+21}{2}\right) = 4.09 \text{ ft}.$$

Total north $= 3.28 + 4.09 = 7.37$ ft.

$$M_4 = 100 \text{ ft} \sin\left(\frac{2+3}{2}\right) \sin\left(\frac{20+21}{2}\right) = 1.53 \text{ ft}.$$

Total east $= 1.20 + 1.53 = 2.73$ ft.

$$D_4 = 100 \text{ ft} \cos\left(\frac{2+3}{2}\right) = 99.90 \text{ ft}.$$

$$D = 8,199.962 + 99.90 = 8,299.862 \text{ ft}.$$

From 8,300 to 8,400 ft, the direction further changes to N22E.

$$L_5 = 100 \text{ ft} \sin\left(\frac{3+4}{2}\right) \cos\left(\frac{21+22}{2}\right) = 5.68 \text{ ft}.$$

Total north $= 7.37 + 5.68 = 13.05$ ft.

$$M_5 = 100 \text{ ft} \sin\left(\frac{3+4}{2}\right) \sin\left(\frac{21+22}{2}\right) = 2.24 \text{ ft}.$$

Total east $= 2.73 + 2.24 = 4.97$ ft.

$$D_5 = 100 \text{ ft} \cos\left(\frac{3+4}{2}\right) = 99.81 \text{ ft},$$

$$D = 8,299.862 + 99.81 = 8,399.672 \text{ ft}.$$

The total departure at each depth can be calculated from each triangle—$A_1A_2K_2$, $A_2A_3K_3$, $A_3A_4K_4$, and $A_4A_5K_5$—and the departure angle can be determined from the target of each triangle:

$$\text{total departure} = \sqrt{(\text{total north})^2 + (\text{total east})^2},$$

$$\text{departure angle} = \arctan\left(\frac{\text{total east}}{\text{total north}}\right).$$

Table 8.1 is a tabulation of the foregoing calculations.

Using the rate of build of 1°/100 ft, calculate the TVD going from 0 to 4° inclination. Calculate first the radius of curvature, r_1, using Eq. 8.1:

$$r_1 = \frac{180°}{\pi} \times \frac{100 \text{ ft}}{1} = 5,730 \text{ ft}.$$

Then find the TVD, D, using Eq. 8.26:

$$D = 8,000 \text{ ft} + 5,730 \text{ ft} \sin 4.0° = 8,399.70 \text{ ft}.$$

Example 8.3 showed us how the trajectory variation in the direction plane is determined. The departure angle in Table 8.1 shows the effect of the 1°/100-ft rate of direction change, where the angle is increasing from the north to the east in a clockwise manner. These same calculations can be used in any of the quadrants as long as the

TABLE 8.1—DATA FOR EXAMPLE 8.3

D_M (ft)	TVD (ft)	N North (ft)	N East (ft)	Departure (ft)	Departure Angle* (degrees)
8,000	8,000.00	0.00	0.00	0.00	—
8,100	8,099.99	0.82	0.30	0.87	20.1
8,200	8,199.96	3.28	1.20	3.49	20.1
8,300	8,299.86	7.37	2.73	7.86	20.33
8,400	8,399.67	13.05	4.97	13.97	20.85

*Note that the statement of the problem requires the departure angle to be 20° to 8,200 ft. Roundoff error in the very small early-departure distances can cause the calculated departure angle to be different.

proper sign convention is observed. In planning a trajectory that is near 0° (first quadrant) and 360° (fourth quadrant), special care must be taken. For example, the average of 359° and 1° would be 180°, whereas 358° and 360° would be 359°. Logically, the way to handle this problem is to continue with the clockwise angle notation, so that 359° and 1° are 359° and 361°, which average 360°.

The steps to plan a trajectory are as follows.

1. From geological or other considerations, establish the target depth, number of targets, target radius, and horizontal departure to the target.

2. Select a kickoff point that seems appropriate and choose a type of trajectory such as a build and hold, an S-shape or a modified S-curve, or a continuous build. Make a two-dimensional plan.

3. Calculate the maximum inclination point and other trajectory information.

4. Determine the lead angle and estimate the rate of direction change.

5. Calculate the three-dimensional well path to hit the target by using the initial two-dimensional well plan as a guide. This reduces the number of trial-and-error calculations.

With this procedure a trajectory can be devised for most directional wells. The practical considerations in designing a directional well are covered later in this chapter.

8.3 Calculating the Trajectory of a Well

The normal method for determining the well path is to ascertain the coordinates by using some type of surveying instrument to measure the inclination and direction at various depths (stations) and then to calculate the trajectory. Sec. 8.5 discusses the various instruments in detail. All that must be known in this section is that values of inclination and direction can be obtained at preselected depths.

Fig. 8.19 depicts part of a trajectory path where surveys are taken at Stations A_2, A_3, and A_4. At each station, inclination and direction angles are measured as well as the course length between stations. Each direction angle obtained by a magnetic type of survey must be corrected to true north, and each gyroscope must be corrected for drift. This is explained in Sec. 8.5. Assume in this section that all the direction readings are corrected for the declination void of magnetic interference, and that the drift conversion is performed for the gyroscopic surveys.

There are 18 or more calculation techniques for determining the trajectory of the wellbore. The main difference in all the techniques is that one group uses straight line approximations and the other assumes the wellbore is more of a curve and is approximated with curved segments. It is beyond the scope of this chapter to derive every method. References at the end of the chapter cite sources for some of the more common ones.

8.3.1 Tangential Method

The simplest method used for years has been the *tangential method*. The original derivation or presentation to industry is unknown. The mathematics uses the inclination and direction at a survey station A_2 (Fig. 8.19) and assumes the projected angles remain constant over the preceding course length D_{M2} to A_2. The Angles are A_1 are not considered.

It can be shown that the latitude north/south coordinate, L, can be calculated using Eq. 8.27 for each course length D_M.

$$L_i = D_{Mi} \ \sin(\alpha_i) \cdot \cos(\epsilon_i). \quad \ldots\ldots\ldots\ldots\ldots (8.27)$$

Likewise, the east/west coordinate M is determined by Eq. 8.28:

$$M_i = D_{Mi} \ \sin(\alpha_i) \cdot \sin(\epsilon_i). \quad \ldots\ldots\ldots\ldots\ldots (8.28)$$

The TVD segment is calculated by Eq. 8.29:

$$D_i = D_{Mi} \ \cos \alpha_i. \quad \ldots\ldots\ldots\ldots\ldots\ldots (8.29)$$

To calculate the total north/south and east/west coordinates and the TVD,

$$L_n = \sum_{i=1}^{n} L_i, \quad \ldots\ldots\ldots\ldots\ldots\ldots (8.30)$$

$$M_n = \sum_{i=1}^{n} M_i, \quad \ldots\ldots\ldots\ldots\ldots\ldots (8.31)$$

and

$$D_n = \sum_{i=1}^{n} D_i. \quad \ldots\ldots\ldots\ldots\ldots\ldots (8.32)$$

8.3.2 Average Angle or Angle Averaging Method

It was recognized that the tangential method caused a sizable error by not considering the previous inclination and direction. The average angle and angle averaging method considers the average of the angles α_1, ϵ_1, and α_2, ϵ_2 over a course length increment D_2 to calculate L_2, M_2, and D_2. Eqs. 8.24 through 8.26 are the angle averaging and average angle relationships:

$$L_i = D_{Mi} \ \sin\left(\frac{\alpha_i + \alpha_{i-1}}{2}\right) \cos\left(\frac{\epsilon_i + \epsilon_{i-1}}{2}\right),$$

$$M_i = D_{Mi} \ \sin\left(\frac{\alpha_i + \alpha_{i-1}}{2}\right) \sin\left(\frac{\epsilon_i + \epsilon_{i-1}}{2}\right),$$

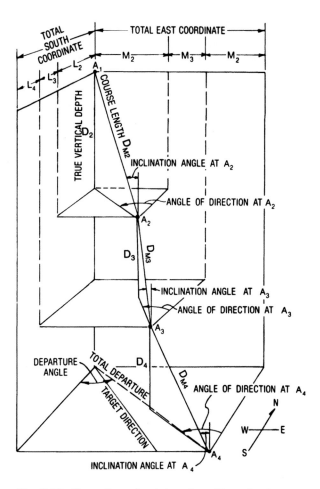

Fig. 8.19—Three-dimensional view of a wellbore showing components that comprise the X, Y, and Z parts of the trajectory.

and

$$D_i = D_{Mi} \ \cos\left(\frac{\alpha_i + \alpha_{i-1}}{2}\right),$$

and

$$L_n = \sum_{i=1}^{n} L_i,$$

$$M_n = \sum_{i=1}^{n} M_i, \text{ and}$$

$$D_n = \sum_{i=1}^{n} D_i.$$

On the basis of Eqs. 8.24 through 8.26, the trajectory calculation can easily be set in tabular form (Fig. 8.20) or set up as a program on a programmable hand calculator. Table 8.2 shows a sequence of steps used in the angle-averaging technique to determine the trajectory coordinates from measured values of inclination and direction.

SURVEY DATA			VERTICAL DEPTH CALCULATIONS							COURSE COORDINATES		TOTAL COORDINATES		TOTAL DEPARTURE	DEPARTURE ANGLE
MEASURED DEPTH	INCL ANGLE	DIRECTION	COURSE LENGTH	AVERAGE INCLIN $\frac{\alpha_x + \alpha_{(x-1)}}{2}$	(D)COS(E)	T.V.D. Σ(d)	(D)SIN(E)	SURVEY AZIMUTH	AVERAGE AZIMUTH $\frac{\epsilon_x + \epsilon_{(x-1)}}{2}$	(H)COS(K) NORTH -SOUTH	(H)SIN(K) EAST -WEST	Σ(L) NORTH -SOUTH	Σ(M) EAST -WEST	$\sqrt{(N)^2 + (O)^2}$	ARCTAN $\frac{(O)}{(N)}$
(A)	α_x	(C)	(D)	(E)	(d)	(G)	(H)	ϵ	(K)	(L)	(M)	(N)	(O)	(P)	(Q)
* 1 7100	0	0	7100	0	7100 00	7100 00	0	0	0	0	0	0 0	0	0	0
** 2 7200	10.1	S68W	100	5.05	99 61	7199 61	8 80	248	248	-3 30	-8 16	-3 30	-8 16	8 80	S68W
3 7300	13.4	S65W	100	11.75	97 90	7297 51	20 36	245	246.5	-8 12	-18 67	-11 42	-26 83	29 16	S67W
4 7400	16.3	S57W	100	14.85	96 66	7394 17	25 63	237	241	-12 43	-22 42	-23 85	-49 25	54 72	S64W
5 7500	19.6	S61W	100	17.95	95 13	7489 3	30 82	241	239	-15 87	-26 42	-39 72	-75 67	85 46	S62W
6															
7															
8															
9															
10															
11															
12															
13															
14															
15															
16															
17															

* AT POINT X1 (FOR KICK OFF POINT) ENTER VALUE OF ZERO FOR INCLINATION IN COLUMNS (B),(C),(E)
COLUMNS (H) THROUGH (S) WILL ALSO BE ZERO.

** AT POINT X2 (FIRST SURVEY STATION) ENTER AVERAGE VALUE FOR INCLINATION (E).
USE ACTUAL SURVEY DIRECTION IN COLUMNS (J) AND (K).
DO NOT USE AVERAGE AZIMUTH IN COLUMN (K) FOR CALCULATIONS AT POINT X2.

Fig. 8.20—Angle averaging method.

Example 8.4. Determine the trajectory coordinates for the corrected survey points given in Table 8.3.

Solution. Using the step-by-step procedure and the form depicted by Fig. 8.20, the solution is given as the filled-in values presented by Fig. 8.20.

8.3.3 Minimum Curvature Method

The minimum curvature method[4] uses the angles at A_1 and A_2 and assumes a curved wellbore over the course length D_2 and not a straight line as shown by Fig. 8.21. Fig. 8.22 shows the curved course length and the two surveying stations A_1 and A_2. This method includes the overall angle change of the drillpipe β between A_1 and A_2. The overall angle, which is discussed and derived in Sec. 8.4, can be written for the minimum curvature method as

$$\cos \beta = \cos(\alpha_2 - \alpha_1)$$
$$- \{\sin(\alpha_1) \sin(\alpha_2)[1 - \cos(\epsilon_2 - \epsilon_1)]\}. \quad ...(8.33)$$

As depicted in Fig. 8.21, the straight line segments $A_1B + BA_2$ adjoin the curve segments $A_1Q + QA_2$ at Points A_1 and A_2. It follows that

$$A_1Q = OA_1 \cdot \beta/2,$$

$$QA_2 = OA_2 \cdot \beta/2,$$

$$A_1B = OA_1 \cdot \tan(\beta/2),$$

$$BA_2 = OA_2 \cdot \tan(\beta/2),$$

and that

$$A_1B/A_1Q = \tan(\beta/2)/(\beta/2) = 2/\beta \tan(\beta/2)$$

and

$$BA_2/QA_2 = \tan(\beta/2)/(\beta/2) = 2/\beta \tan(\beta/2).$$

A factor of the straight line section vs. the curved section ratios is defined as F, where

$$F = 2/\beta_i \tan(\beta_i/2). \quad ...(8.34)$$

If β is less than 0.25 radians, it is reasonable to set $F = 1.0$. Once F is known, the remaining north/south and

TABLE 8.2—TRAJECTORY COORDINATE DETERMINATION

Worksheet Letter	Value to be Obtained	Source or Equation for Obtaining Value
A	*Measured depth:* The actual length of the wellbore from its surface location to any specified station.	survey
α	*Inclination angle:* The angle of the wellbore from the vertical.	survey
C	*Direction:* The direction of the course.	survey
D	*Course length:* The difference in measured depth from one station to another.	$A_x - A_{(x-1)}$
E	*Average inclination:* The arithmetic average of the inclination angles at the upper and lower ends of each course.	$\dfrac{\alpha_x + \alpha_{(x-1)}}{2}$
d	*Course vertical depth:* The difference in vertical depth of the course from one station to another.	$(D)\cos(E)$
G	*True vertical depth:* The summation of the course vertical depths of an inclined wellbore.	$\Sigma(d)$
H	*Course departure:* The distance between two points that are projected onto a horizontal plane.	$(D)\sin(E)$
ϵ	*Survey azimuth:* The direction of a course measured in a clockwise direction from 0° to 360°; 0° is north.	survey, in degrees (ϵ)
K	*Average azimuth:* The arithmetic average of the azimuths at the ends of the course.	$\dfrac{\epsilon_x + \epsilon_{(x-1)}}{2}$
L	*North/south course coordinate:* The course component displacement from one station to another; negative value = south.	$(H)\cos(K)$
M	*East/west course coordinate:* The course component displacement from one station to another; negative value = west.	$(H)\sin(K)$
N	*North/south total coordinate:* The summation of the course displacements in the north/south direction (south is negative).	$\Sigma(L)$
O	*East/west total coordinate:* The summation of the course displacements in the east/west direction (west is negative).	$\Sigma(M)$
P	*Total departure:* Shortest distance from vertical wellbore to each station point.	$\sqrt{(N)^2 + (O)^2}$
Q	*Departure direction:* The direction of vertical projection onto horizontal plane from station to surface location. Must take value calculated and put into proper quadrant. See sign convention below.	$\arctan\dfrac{(O)}{(N)}$

east/west coordinates and TVD can be calculated using the following equations.

$$M_i = (D_i/2)[\sin(\alpha_{i-1}) \cdot \sin(\epsilon_{i-1}) + \sin(\alpha_i)$$

$$*\sin(\epsilon_i)] \cdot F_i. \qquad \ldots\ldots\ldots\ldots\ldots\ldots (8.35)$$

$$L_i = D_i/2[\sin(\alpha_{i-1}) \cdot \cos(\epsilon_{i-1}) + \sin(\alpha_i)$$

$$*\cos(\epsilon_i)] \cdot F_i. \qquad \ldots\ldots\ldots\ldots\ldots\ldots (8.36)$$

$$D_i = D_i/2[\cos(\alpha_{i-1}) + \cos(\alpha_i)] \cdot F_i. \qquad \ldots\ldots\ldots (8.37)$$

The total departures and TVD are calculated using Eqs. 8.30 through 8.32.

Other calculation methods that have been commonly used are the balanced tangential method,[2] the radius of curvature method,[3] the mercury method,[1] acceleration method, trapezoidal method, and vector averaging method. It is interesting to note that the balanced tangential, trapezoidal, vector averaging, and acceleration methods, even though derived differently, yield the identical mathematical formulas for the north/south and east/west coordinates and TVD.

As to which method yields the most accurate results, Table 8.4 compares six of the different methods using data taken from a test hole. Note that the tangential method shows considerable error for M, L, and *D*. This is why the tangential method is no longer used. The differences among the average angle, the minimum curvature, and the balanced tangential methods are so small that any of the methods could be used for calculating the trajectory.

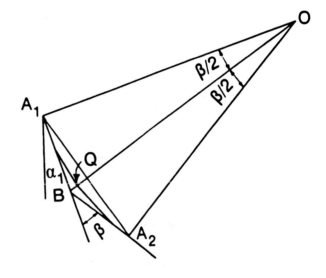

Fig. 8.21—Representation of minimum curvature ratio factor, *F*.

TABLE 8.3—DATA FOR EXAMPLE 8.4

D_M (ft)	Inclination Angle (degrees)	Direction Angle
7,100	0	0
7,200	10.1	S68W
7,300	13.4	S65W
7,400	16.3	S57W
7,500	19.6	S61W

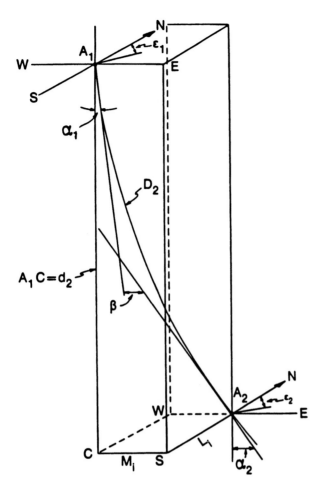

Fig. 8.22—A curve representing a wellbore between Survey Stations A_1 and A_2.

It will be shown in Sec. 8.5 that the systematic errors overpower the variation in the survey calculation differences when comparing the multistation surveying methods.

With the advent of the programmable hand calculator, the minimum curvature method has become the most common.

8.4 Planning the Kickoff and Trajectory Change

In kicking off a well, setting the lead angle, or making a controlled trajectory change, some method must be used to force the bit in the desired direction. The first tool used to deflect the bit was a simple whipstock (Fig. 8.23a). Later, mud motors equipped with bent subs or bent housings were introduced (Fig. 8.23b), as well as jetting bits (Fig. 8.23c). These deflection tools are discussed in detail in Sec. 8.6. This section explains how to control a change in direction and inclination angle to achieve a change in trajectory.

All deflection methods depend on manipulating the drillpipe (rotation and downward motion) to cause a departure of the bit in either the direction plane or the inclination plane, or both. In Fig. 8.24, showing a drillstring with a deflection sub, an arrow indicates the direction the sub will cause the drillstem to face—the toolface direction. The magnitude of the deflection is controlled by the depar-

ture from the centerline of the deflecting tool. This includes the angle of the whipstock toe, bent sub, and the nozzle offset in a jetting bit.

8.4.1. Orientation of the Bit

For years the common ways to design a bit orientation were to refer to a "Ouija Board" (see Fig. 8.25) and to use tables of values relating the deflection-sub offset to a particular orientation. Later, Millheim et al.[5] demonstrated how to derive the equations necessary to design and analyze a kickoff and trajectory change.

Fig. 8.26 depicts the three-dimensional deflection of the course of a wellbore. At a measured depth (a), an inclination α and direction ϵ are measured. With a deflecting device, the course is to be deflected to a new inclination angle, α_N, and a new direction, ϵ_N. This will cause an overall angle change, β, which is directly related to what is called "dogleg severity." If the wellbore in Fig. 8.26 were not deflected, it would continue on its present course to Point b. Deflected, the wellbore will follow the new desired course in Plane A to Point c or d.

8.4.2 Derivation of the Direction Change, $\Delta\varepsilon$

The three-dimensional geometrical representation of the trajectory is presented in Fig. 8.27, where the wellbore is deflected at Point O. A vertical plane is projected through Points MOCEM' (see Fig. 8.28), normal to the horizontal plane O'A'B'E, and tending in a southeast direction. The plane cutting through DAB is parallel to the plane cutting MOCEM'. Lines AA', BB', and ODO' are vertical and parallel. If Line MOC is rotated 360°, it will transcribe a circle of radius r. Fig. 8.29 represents the circle cutting through Plane ACE with a rotation angle γ and normal to Planes OCE, OCA, etc.

The change in direction, $\Delta\epsilon$, shown by Fig. 8.27 is the angle of Triangle A'O'B', where

$$\tan \Delta\epsilon = \frac{A'B'}{O'B'} = \frac{AB}{O'E+EB'}. \quad \dots\dots\dots\dots (8.38)$$

Fig. 8.29 shows the rotation of Line EA (radius r) from Point A to Point B and transcribes an angle γ. It follows that

$$AB = r \sin \gamma.$$

TABLE 8.4—COMPARISON OF ACCURACY OF VARIOUS CALCULATION METHODS (after Craig and Randall[1])

Direction: Due north
Survey interval: 100 ft
Rate of build: 3°/100 ft
Total inclination: 60° at 2,000 ft

Calculation Method	Total Vertical Depth. Actual (ft)	Difference From Actual (ft)	North Displacement. Actual (ft)	Difference From Actual (ft)
Tangential	1,628.61	−25.38	998.02	+43.09
Balanced tangential	1,653.61	−0.38	954.72	−0.21
Angle-averaging	1,654.18	+0.19	955.04	+0.11
Radius of curvature	1,653.99	0.0	954.93	0.0
Minimum curvature	1,653.99	0.0	954.93	0.0
Mercury*	1,153.62	−0.37	954.89	0.04

*Fifteen foot survey tool

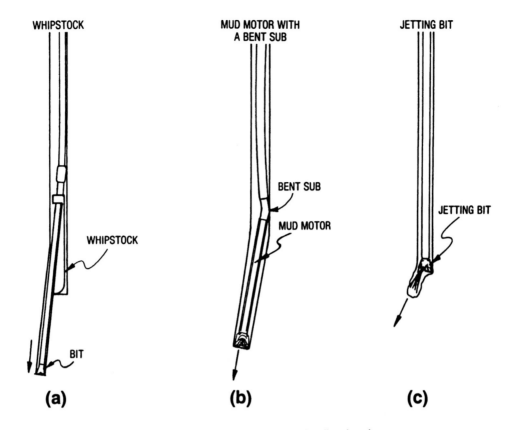

Fig. 8.23—Techniques for making a positive direction change.

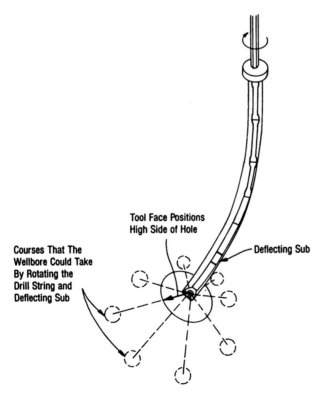

Fig. 8.24—System for deflecting the wellbore trajectory.

The next step is to determine relationships in terms of OE and EB. Triangle EBB′ (Fig. 8.28) relates EB′ to the angle α as

$$EB' = EB \cos \alpha, \quad \ldots\ldots\ldots\ldots\ldots\ldots (8.39)$$

and from Triangle EAB (Fig. 8.29), EB is related to angle γ:

$$EB = r \cos \gamma.$$

Substituting EB in Eq. 8.39 forms Eq. 8.40:

$$EB' = r \cos \gamma \cos \alpha. \quad \ldots\ldots\ldots\ldots\ldots (8.40)$$

To determine O′E, consider Triangles OEC and OO′E, where

$$r = \ell \tan \beta \quad (\text{Fig. 8.28})$$

and

$$O'E = \ell \sin \alpha.$$

Substituting these last two relationships for terms AB, O′E, and EB′ into Eq. 8.38 and eliminating r yields

$$\tan \Delta\epsilon = \frac{\tan \beta \sin \gamma}{\sin \alpha + \tan \beta \cos \alpha \cos \gamma} \quad \ldots\ldots\ldots (8.41)$$

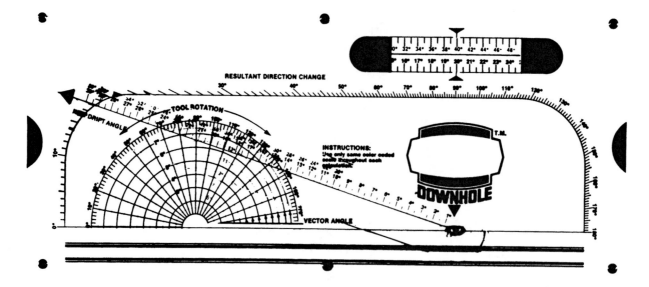

Fig. 8.25—Directional drilling Ouija Board.

and

$$\Delta\epsilon = \text{arc tan} \frac{\tan \beta \sin \gamma}{\sin \alpha + \tan \beta \cos \alpha \cos \gamma}. \quad \ldots \ldots (8.42)$$

The overall angle change, β, is directly related to dogleg severity, δ, by the following relationship.

$$\delta = \frac{\beta}{L_C}(i), \quad \ldots \ldots \ldots \ldots \ldots \ldots \ldots \ldots (8.43)$$

where L_C is the course length between the measured surveys and i is the index of angle change. For example, if i is 100, δ could be degrees/100 meters if course length is reported in meters, or δ could be degrees/100 ft if course length is in feet.

Example 8.5. Determine the new direction for a whipstock set at 705 m with a tool-face setting of 45° right of high side for a course length of 10 m. The inclination

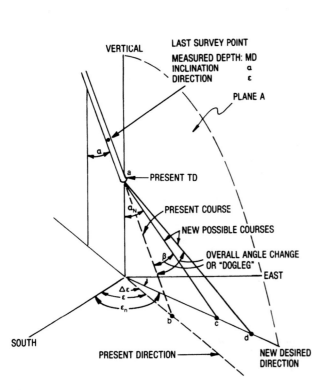

Fig. 8.26—Three-dimensional deflection of the course of a borehole (after Millheim *et al.*[5]).

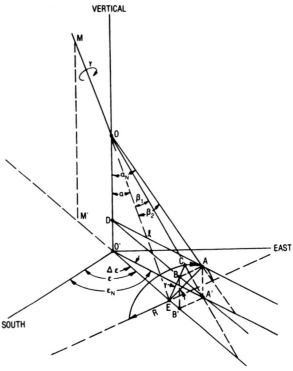

Fig. 8.27—Three-dimensional trajectory change model (after Millheim *et al.*[5]).

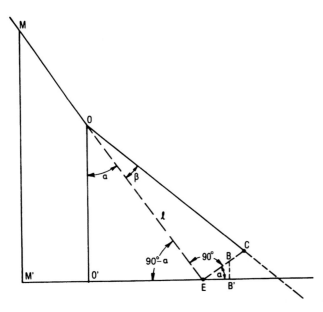

Fig. 8.28—Vertical plane through MOCEM' (after Millheim *et al.*[5]).

is 7° and the direction is N15W of 705 m. The curve of the whipstock will cause a total angle change of 3°/30 m over a course length.

Solution. The overall angle change, β, can be determined from Eq. 8.43:

$$\beta = \frac{\delta L_C}{i} = \frac{3°10\ m}{30\ m} = 1°,$$

and the direction change can be calculated using Eq. 8.42:

$$\Delta\epsilon = \arc \tan \frac{\tan(1)\ \sin(45)}{\sin(7) + \tan(1)\ \cos(7)\ \cos(45)}$$

$$= \arc \tan 0.09199 = 5.3°.$$

If β is right of high side of the hole, the new direction, ϵ_N, is $\epsilon + \Delta\epsilon$; and if the tool-face setting is left of high, $\epsilon_N = \epsilon - \Delta\epsilon$. Therefore, in this example, the new direction is

$$\epsilon_N = 345° + 5.3° = 350.3° \text{ or N9.7W.}$$

8.4.3 Derivation of the New Inclination Angle, α_N

The new inclination angle, α_N, can be derived by considering Triangle AOD in Plane OAA'O' (Fig. 8.27):

$$\cos \alpha_N = \frac{OD}{OA} = \frac{OO' - O'D}{OA} = \frac{OO' - AA'}{OA}$$

$$= \frac{OO' - BB'}{OA}. \quad\quad\quad (8.44)$$

Using Triangles OO'E and EB'E (Fig. 8.28), the inclination angle, α, can be obtained from

$$OO' = \ell \cos \alpha$$

and

$$BB' = EB \sin \alpha.$$

Substituting for EB in the above equation yields

$$BB' = r_1 \cos \gamma \sin \alpha.$$

Triangles AOE and COE are equal, and AO equals OC. From Triangle AOE,

$$OA = OC = \frac{\ell}{\cos \beta}.$$

Substituting for OA, OO', and BB' in Eq. 8.44 yields

$$\cos \alpha_N = \cos \alpha \cos \beta - \sin \alpha \sin \beta \cos \gamma \quad\quad (8.45)$$

and

$$\alpha_N = \arc \cos(\cos \alpha \cos \beta - \sin \alpha \sin \beta \cos \gamma), \quad (8.46)$$

where

$$\Delta\alpha = \alpha_N - \alpha. \quad\quad\quad\quad\quad (8.47)$$

8.4.4 Derivation of the Tool-Face Angle, γ

The tool-face angle, γ, can be calculated by rearranging Eq. 8.45 if the initial and final inclination angles and the overall angles are known or set as desired values.

$$\gamma = \arc \cos\left(\frac{\cos \alpha \cos \beta - \cos \alpha_N}{\sin \alpha \sin \beta}\right). \quad\quad (8.48)$$

The tool-face angle also can be calculated if the final inclination, direction change, and overall angle change are known or selected as desired values.

$$\gamma = \arc \sin\left(\frac{\sin \alpha_N \sin \Delta\epsilon}{\sin \beta}\right). \quad\quad\quad (8.49)$$

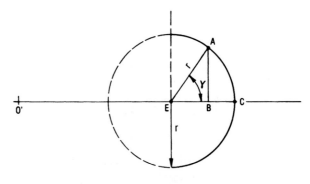

Fig. 8.29—Tool-face plane (after Millheim *et al.*[5]).

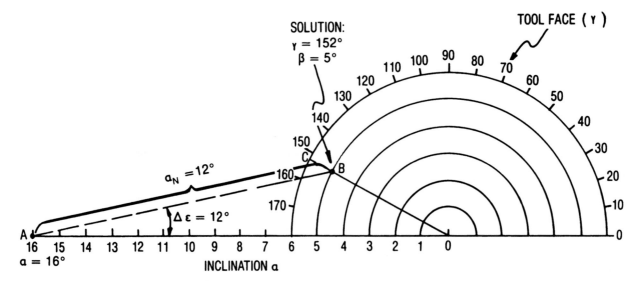

Fig. 8.30—Graphical Ouija analysis.

Eqs. 8.42, 8.46, 8.48, and 8.49 can be used to determine the direction change ($\Delta\epsilon$), new inclination (α_N), overall angle change (β), and tool-face angle (δ).

8.4.5 Derivation of the Ouija Board Nomograph

A graphic technique that is the basis of the Ouija Board type of nomograph (Fig. 8.30) can be used to determine the same parameters as from the derived equations for α_N, $\Delta\epsilon$, β, and γ for the condition where the value of the overall angle, β, is small—i.e., less than 6°. By specifying the small angle condition, the following identities can be written.

$$\cos \beta \approx 1,$$

$$\sin \beta \approx \beta \text{ (radians)},$$

and

$$\tan \beta \approx \beta \text{ (radians)}.$$

Substituting the identities into Eq. 8.42 and Eq. 8.45 yields the following relationships.

$$\Delta\epsilon = \text{arc tan} \frac{\beta \sin \gamma}{\sin \alpha + \beta \cos \alpha \cos \gamma}, \quad \dots\dots\dots (8.50)$$

$$\beta = \frac{\cos \alpha - \cos \alpha_N}{\sin \alpha \cos \gamma}, \quad \dots\dots\dots\dots\dots (8.51)$$

and

$$\gamma = \text{arc sin} \left[\frac{1 - \cos \alpha \cos \alpha_N}{\beta \sin \alpha} \tan(\Delta\epsilon) \right]. \quad \dots\dots (8.52)$$

In the nomograph (Fig. 8.30), the abscissa is the original inclination angle (except when $\Delta\epsilon=0$ and $\beta=0$, in which case $\alpha=\alpha_N$). The angle between the abscissa and

Line AB is $\Delta\epsilon$; the point at which Line 0°B is projected through the outer semicircle at C is read as the tool-face angle, γ, left or right of high side. The number of degrees traversed by Line 0°B is the overall angle change, β. Line AB is the new inclination angle, α_N.

The values in Fig. 8.30 are $\alpha(0°A)=16°$, $\alpha_N(AB) =12°$, $\gamma(C)=152°$, and $\beta(0°B)=5°$.

A nomograph like Fig. 8.30 can be created by drawing a scale of degrees (at increments of, say, ¼ in. per degree) on the abscissa to represent the number of degrees in the original inclination angle, in this case 16°. The degrees decrease to zero from left to right. Placing the needle of the compass at 0°, draw enough 1° concentric semicircles to equal the degrees in β (in this case, 6°). With a protractor, scale the outer semicircle from 0° (right) to 180° (left) in 10° increments.

Example 8.6. Determine where to set the tool face (the tool-face angle, γ) for a jetting bit to go from a direction of 10° to 30° and from an inclination of 3° to 5°. Also calculate the dogleg severity, δ, assuming that the trajectory change takes 60 ft.

Solution. Scale 3° on the abscissa. At the 3° mark (Point A), measure $\Delta\epsilon$, which is 20°; from Point A, project the α_N line 5° units of length. The end of the line is Point B. Draw a line from 0° to Point B (this line is equal to β). Continue Line 0°B to C to find the tool-face setting: 45°. Fig. 8.31 depicts the graphical solution.

$$\delta = \frac{\beta}{L_c} i = \frac{2.4°}{60 \text{ ft}} 100 \text{ ft} = 4.00°/100 \text{ ft}.$$

Example 8.7. Using the data of Example 8.6, calculate the tool-face setting and the dogleg severity.

Solution. Since Eq. 8.41 is in terms of β, which is an unknown, and Eqs. 8.48 and 8.49 are in terms of γ, it is necessary to rearrange the equations in terms of α, α_N,

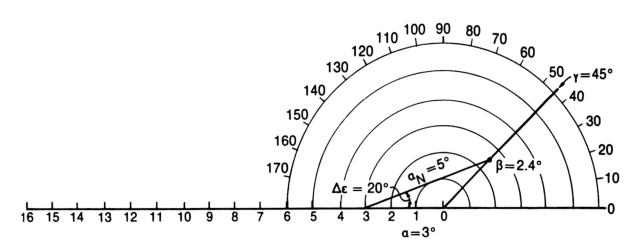

Fig. 8.31—Solution to Example 8.6.

and $\Delta\epsilon$ only. By combining Eq. 8.41 with Eqs. 8.48 and 8.49 in terms of $\sin\gamma$ and $\cos\gamma$ and doing some manipulations, the total angle change, β, can be written in terms of only $\Delta\epsilon$, and α and α_N.

$$\beta = \text{arc } \cos(\cos\Delta\epsilon\,\sin\alpha_N\,\sin\alpha + \cos\alpha\,\cos\alpha_N),$$

$$\ldots\ldots\ldots\ldots\ldots\ldots\ldots\ldots\ldots\ldots\ldots(8.53)$$

or, in another form,

$$\cos\beta = \cos\Delta\epsilon\,\sin\alpha_N\,\sin\alpha + \cos\Delta\alpha - \sin\alpha\,\sin\alpha_N.$$

$$\ldots\ldots\ldots\ldots\ldots\ldots\ldots\ldots\ldots\ldots\ldots(8.54)$$

For Example 8.7, $\Delta\epsilon = 20°$, $\alpha = 3°$, $\alpha_N = 5°$, and $\Delta\alpha = 2°$.

$$\beta = \text{arc } \cos[\cos(20)\,\sin(5)\,\sin(3) + \cos(3)\,\cos(5)] = 2.4°.$$

8.4.6 Overall Angle Change and Dogleg Severity

Eq. 8.55 derived by Lubinski[6] is used to construct Fig. 8.32, a nomograph for determining the total angle change, β, and the dogleg severity, δ.

$$\beta = 2 \text{ arc } \sin\sqrt{\sin^2\left(\frac{\Delta\alpha}{2}\right) + \sin^2\left(\frac{\Delta\epsilon}{2}\right)\sin^2\left(\frac{\alpha+\alpha_N}{2}\right)}.$$

$$\ldots\ldots\ldots\ldots\ldots\ldots\ldots\ldots\ldots\ldots\ldots(8.55)$$

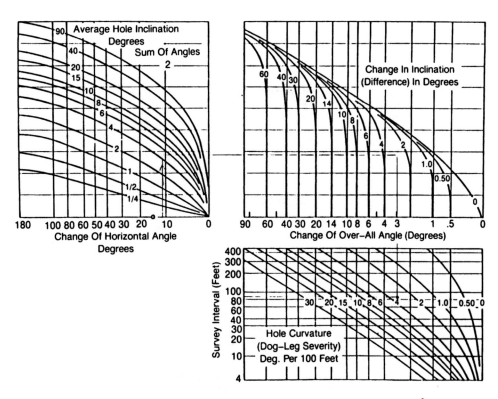

Fig. 8.32—Chart for determining dogleg severity (after Lubinski[6]).

OC – Old Borehole
OB – Parallel To New Borehole
<BOC – Overall Angle

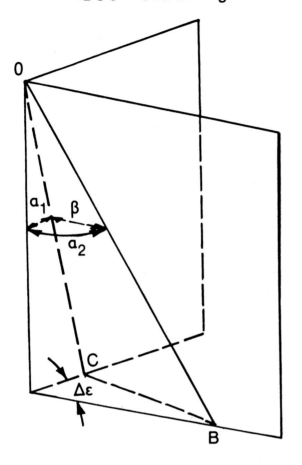

Fig. 8.33—Basis of chart construction is a trigonometric relationship illustrated by two intersecting planes (after Lubinski[6]).

The overall angle, β, calculated by Eq. 8.51, is the same as calculated by Eq. 8.50. However, Lubinski originally derived Eq. 8.51, which is based on measuring two consecutive sets of inclination and direction and is not concerned with the tool-face setting. Fig. 8.33 depicts the geometric basis for the derivation of the total angle change, β.

Example 8.8. Determine the dogleg severity following a jetting run where the inclination was changed from $4.3°$ to $7.1°$ and the direction from N89E to S80E over a drilled interval of 85 ft.

Solution. Using Eq. 8.55, the total angle change is

$$\beta =$$

$$2 \text{ arc } \sin\sqrt{\sin^2\left(\frac{2.8}{2}\right) + \sin^2\left(\frac{11}{2}\right)\sin^2\left(\frac{7.1+4.3}{2}\right)}$$

$$= 3.0°,$$

Construct a line representing the initial inclination, α

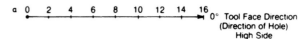

Step Two

Measure the total direction change $\Delta\epsilon$ and new angle change α_r and correct α with the α_r vector.

Step Three

Measure the connecting vector which is the total angle change β and the direction of the vector which is the tool face angle.

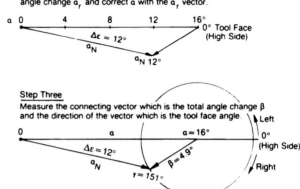

Fig. 8.34—Example solution using the Ragland diagram.

and using Eq. 8.53, the total angle change is

$$\beta = \text{arc } \cos[\cos(11)\sin(7.1)\sin(4.3)$$

$$+\cos(4.3)\cos(7.1)] = 2.99°.$$

Using the nomograph given in Fig. 8.32, the change of horizontal angle is $100° - 89° = 11°$, and the sum of the inclination divided by 2.0 is $5.7°$. The change in inclination is $2.8°$, yielding a total angle change, β, of $3.0°$ and a dogleg severity of $3.5°/100$ ft. Using Eq. 8.40, the dogleg severity calculates

$$\delta = \frac{2.99°(100)}{85 \text{ ft}} = \frac{3.5°}{100 \text{ ft}}.$$

8.4.7 The Ragland Diagram

Another way of solving the trajectory change problem is by using what is called the Ragland diagram.[7] The method is essentially similar to the Ouija Board nomograph in that a graphical process is used to solve for the unknown parameters. Fig. 8.34 shows the steps using the Ragland diagram to solve for the same problem as presented by Fig. 8.30 using the Ouija Board solution.

The first step is to construct a vector that represents the length of the initial inclination, α. This vector always represents the direction of the hole, ϵ, and is the high side ($0°$ tool face) of the hole. For this particular case, α_N and $\Delta\epsilon$ are given as well as the initial inclination. The unknowns are the total angle change and tool-face setting. To find these unknowns, the new inclination vector α_N is drawn from the $0°$ origin at a direction change of $\Delta\epsilon = 12°$. The total angle change β is determined by connecting the α and α_N vectors and measuring the length of the β vector. In this example it is approximately $4.9°$. The last step is to draw a circle around the origin point

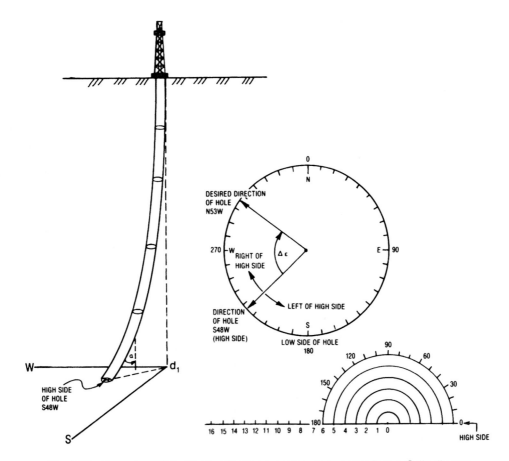

Fig. 8.35—Example of high side of wellbore showing how it corresponds to a Ouija diagram.

where $\alpha = 16°$, and measure with a protractor the tool-face angle, γ. In the example, $\gamma = 151°$ and is right of the high side. For example, this case would represent a direction change from N40E to N52E where N40E is the original direction of the hole. However, if the direction change that is desired is N28E, the α_N vector would be drawn above the α vector, making the tool-face setting on the left of the high side.

Like the Ouija solution, the Ragland vector analysis uses any three knowns to solve for two unknowns. In Figs. 8.30 and 8.33, α, α_{N1}, and $\Delta\epsilon$ are knowns. β and γ are unknowns.

8.4.8 Planning a Trajectory Change and Compensating for Reverse Torque of a Mud Motor

A kickoff or trajectory change is designed according to which of these plans is desired: (1) maximum build or drop without any direction change, (2) maximum direction change to the right or left of high side with no inclination change, (3) build with direction change to the right or left of high side, and (4) drop with direction change to the right or left of high side.

The design also depends on whether a jetting bit is used, or a mud motor with a bent sub or bent housing. Except in special circumstances, openhole whipstocks are seldom used to kick off or change a trajectory. They are mainly used now to kick off or sidetrack out of casing. This is discussed in Sec. 8.5.

Fig. 8.35 depicts a wellbore at a depth D_1 where it is desired to build angle from α to α_N and to change the direction from S48W to N53W. The high side of the hole

at depth D_1 is S48W. (The high side is defined as the direction of the wellbore and opposite the gravity side or low side of the wellbore.) Changing the direction of the wellbore will require some type of deflection technique whereby resetting the face of the deflecting tool will cause the wellbore to rotate to the right (clockwise) to a new direction of N53W. If a jetting bit is used to change direction, the tool-face angle can be calculated directly without any other corrections. However, if a bent sub or housng is used in combination with a mud motor, a correction must be made to compensate for reverse torque. When a mud motor in a borehole is activated and starts to drill, the tool face will always rotate to the left by some amount, depending on a number of factors that are discussed in Sec. 8.5. The left rotation, or reverse torque, must be compensated for by reorienting the tool face enough to offset the reverse torque effects.

The rotation of the drillstring because of the bit face torque can be estimated using Eq. 8.56:

$$\theta_M = \frac{ML_{\text{motor}}}{G_{\text{motor}}J_{\text{motor}}} + \frac{ML_{\text{BHA}}}{G_{\text{BHA}}J_{\text{BHA}}}$$

$$+ \frac{ML_{\text{drillstring}}}{G_{\text{drillstring}}J_{\text{drillstring}}} \quad \dots \dots \dots \dots \dots (8.56)$$

and

$$J = \frac{\pi}{32}(\text{OD}^4 - \text{ID}^4)(\text{in.}^4),$$

where M is the torque generated at the bit and L is the length of any drillstring section (L_{motor} is length of motor and L_{BHA} is the length of the bottomhole assembly). G is the shear modulus of each drillstring element, and J is the polar moment of inertia.

Furthermore, the BHA and the drillstring could be subdivided according to different size drill collars, stabilizers, and drillpipe, and

$$\theta_M = M \sum \frac{L_i}{G_i J_i}. \qquad \ldots\ldots\ldots\ldots\ldots\ldots (8.57)$$

Example 8.9. Calculate the total angle change of 3,650 ft of 4½-in. [3.826-in. ID] Grade E 16.60-lbm/ft drillpipe and 300 ft of 7-in. drill collars [2¹³⁄₁₆-in. ID] for a bit-generated torque of 1,000 ft-lbf. Assume that the motor has the same properties as the 7-in. drill collars. Use the shear modulus of steel ($G = 11.5 \times 10^6$ psi) for the BHA and drillstring. Calculate the total angle change if 7,300 ft of drillpipe was used.

Solution.

$$\theta = \frac{M}{11.5 \times 10^6 \text{ psi}} \left(\frac{43,800 \text{ in.}}{19.2} + \frac{3,600 \text{ in.}}{229.6} \right)$$

$$= 1.997 \times 10^{-4} M = 1.997 \times 10^{-4} \frac{1}{\text{in. lbm}}$$

$$\times (1,000 \text{ ft-lbf}) \left(\frac{12 \text{ in.}}{\text{ft}} \right) = 2.396,$$

$$\theta_{1,000} = \frac{2.396}{2\pi} 360° = 137.3°,$$

$$\theta_{(long\ drillpipe)} = 3.96 \times 10^{-4} \frac{1}{\text{in. lbm}} (1,000 \text{ ft-lbf})$$

$$\times \left(\frac{12 \text{ in.}}{\text{ft}} \right),$$

$$\theta = \frac{4.772}{2\pi} (360) = 273.4°.$$

Example 8.9 shows that the longer the drillstring the more the bit face torque causes the drillstring to rotate. (In the case of drilling with a motor, the reactive torque is counterclockwise or to the left.)

Eq. 8.57 does not include the friction caused by the bent sub or any other part of the drillstring, especially at higher inclinations where drillstring friction can be considerable. However, at shallower depths ($< 3,000$ ft) where the wellbore is at a low inclination (1 to 5°), Eq. 8.57 can give a first approximation for the amount of reverse torque

compensation that is needed. This calculation also requires a knowledge of bit face torque, which is discussed in Sec. 8.5.

The most foolproof way to compensate for the reverse torque problem is to run, right above the bent sub, a surface-indicating measuring tool that transmits the tool-face position frequently. Without relying on the calculational technique or guesswork, one can rotate the drillpipe dynamically to compensate for the reverse torque. If the surface recording system is not used, one must run a conventional single shot to orient the tool face making sure to correct for the reverse torque. For deeper kickoffs where it is difficult to estimate the amount of compensation for the reverse torque, for expensive drilling operations, and when in doubt, some type of surface-indicating tool-face monitor should be used.

Example 8.10. Plan a kickoff for the wellbore in Fig. 8.35 where the depth at kickoff is 2,560 ft, the direction at kickoff is S48W, the course length is 150 ft, and the inclination is 2°, and determine the dogleg severity. It is desired to set the direction at N53W to increase the inclination to 6°. Assume that the mud motor at this depth and inclination has 20° of reverse torque.

Solution. Using the graphic technique exemplified in Fig. 8.35, arrive at the following values: $\Delta\epsilon = 79°$, $\alpha = 2°$, $\alpha_N = 6°$, $\gamma = 97°$, and $\beta = 5.8°$. This is the calculation for the dogleg severity:

$$\delta = \frac{5.8}{150} \times 100 = 3.87°/100 \text{ ft.}$$

Since it is desired to turn the well clockwise (right of high side) and the reverse torque will be 20° to the left, the actual tool-face setting must be 97° + 20°, or 117° right of high side (N15W or 345°). If the correction for the reverse torque is not included, the dynamic tool face setting would be at 97° right of high side and the desired new direction and inclination would be incorrect. Fig. 8.36 shows the graphical solution and the correction for the reverse torque of the motor.

Example 8.10 shows one way to calculate the kickoff using a mud motor and bent sub or housing where the primary factors are the direction and inclination changes. The residual dogleg severity is a result of making the trajectory change. Changing the trajectory without regard for the dogleg severity can cause serious problems in running the BHA through the dogleg; it also can cause casing wear and later production problems. Better, by far, is to determine a maximum tolerable dogleg and to design the trajectory change accordingly. The easiest way to do this is to control the course length. If in Example 8.10 the course length is 50 ft, the dogleg severity is 11.61°/100 ft, whereas if the course length were 200 ft, the dogleg severity would be only 2.9°/100 ft. Once β is fixed, only a limited change in direction and inclination can be achieved for a given course length. If no direction change is desired, the tool face can be oriented to the high or low side of the hole and maximum inclination change can be made.

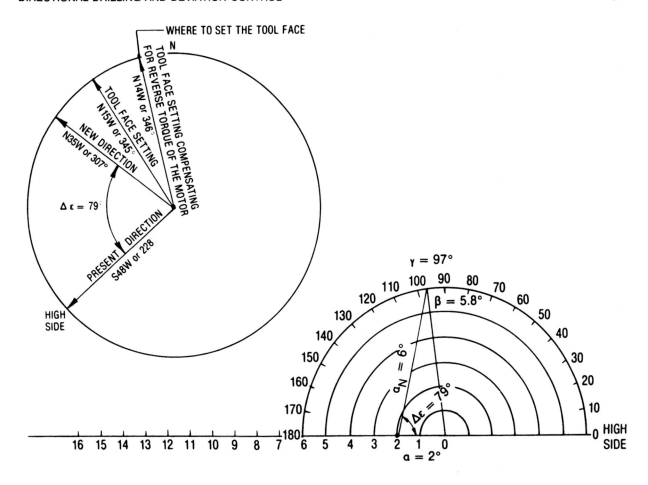

Fig. 8.36—Solution for Example 8.10.

The maximum direction change for a fixed value of β is illustrated by Fig. 8.37. Between 0 and 1.5° inclination, the maximum change is 360°; at 1.5° the maximum change is 90°. Below $\alpha = 1.5°$ the maximum direction change is highly dependent on the inclination. At $\alpha = 2°$ the maximum change is 50° and at $\alpha = 10°$ the maximum change is 10°. Note that as the inclination increases, the tool-face setting for a maximum change reduces from 180° ($\alpha = 1.5°$) to approximately 95° ($\alpha = 10°$). Fig. 8.38 presents the same information for $\beta = 1.5°$ for inclinations between 10 and 45°. At $\alpha = 45°$ the maximum direction change is about 3°. Even though the maximum $\Delta\epsilon$ for $\alpha = 10$ to 45° occurs at one specific tool-face setting, the curves are so flat between $\gamma = 80°$ and $\gamma = 120°$ that an error in the tool-face setting between 80 and 120° would make very little difference in the resultant direction change.

Both figures show the importance of making the direction change at low inclinations. A design practice is to try to set the direction at an inclination between 1 and 6° and then build inclination only, holding the direction of the hole fairly constant. Too much curvature in direction will cause the BHA to respond in a direction opposite to the curvature, thus negating some of the benefits of the original trajectory change.

At higher inclinations where direction must be changed, the course length must be extended to minimize the dogleg severity.

In planning a trajectory change when a jetting bit is to be used, the type of jetting bit and nozzle size (which

should be large) must be selected. A mud motor with a bent sub or bent housing requires the selection of a bent sub with some amount of deflection. The total angle change β is not the bent sub angle, although the bent sub angle can approach β under certain circumstances. Fig. 8.39a shows a combination of bent sub and mud motor without the constraints of the borehole. Fig. 8.39b is the same bent sub assembly in the borehole. The effect of the bent sub depends on the geology, motor stiffness, bent sub angle, inclination, the BHA configuration above the bent sub, and the axial weight applied to the motor.

Manufacturers have published tables that present bent-sub angles in the rate of angle change in degrees per 100 ft. The effective bent sub angle can be derived from trajectory field data if all the necessary parameters are recorded. Example 8.11 presents how this can be done.

Example 8.11. Determine the effective bent sub response for a 1½° bent sub for a motor run where at 6,357 ft α is 1° and ϵ is S85E and at 6,382 ft α_N is 1° and ϵ_N is S20E; the tool face is 160° right of high side.

Solution. Using Eq. 8.53, β can be calculated as

$$\beta = \text{arc } \cos[\cos(65) \sin(1) \sin(1) + \cos(1) \cos(1)]$$

$$= 1.07°.$$

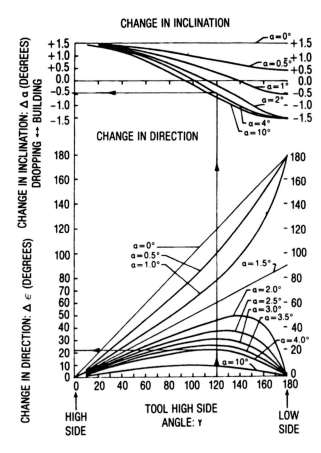

Fig. 8.37—Overall angle change (β = 1.5°); hole inclination from 0.0° to 10.0° (after Millheim *et al.*[5]).

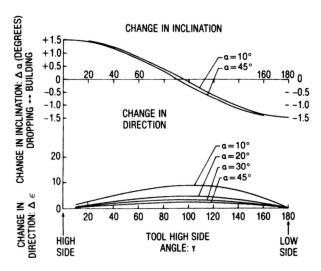

Fig. 8.38—Overall angle change (β = 1.5°); hole inclination from 0.0° to 45.0° (after Millheim *et al.*[5]).

The dogleg severity over the interval drilled is

$$\delta = 1.07° \times \frac{100}{25 \text{ ft}} = 4.28°/100 \text{ ft}.$$

If the mud motor length is 25 ft from the bit face to the bent sub, the maximum angle change that could be reached if the formation is soft enough to allow the bit to drill sideways without restriction is

$$\delta = 1.5° \times \frac{100}{25 \text{ ft}} = 6.0°/100 \text{ ft}.$$

The lower rate of build of 4.3°/100 ft implies that the formation resisted the maximum rate of build by a factor of 4.3°/6.0° = 0.72.

Another way of affecting the rate of angle change for this example is to move the bent sub farther from or closer to the bit. If the motor length is 30 ft instead of 25 ft, the unrestricted angle change is 5°/100 ft instead of 6°/100 ft. A positive displacement motor can also have a bent housing close to the bit. For a 1.5° bent housing 8 ft from the bit, the unrestricted angle change is 18¾°/100 ft.

Jetting bits have a fixed big jet direction which serves much the same purpose as the bent sub—to cause deflection. The effectiveness of the jetted pocket depends on

the length of the jetting run, geology, hydraulics, and BHA configuration. A typical jetting run can be analyzed with the same techniques used for designing a trajectory change.

Example 8.12. A well has been jetted at three different depths: 1,745 to 1,752 ft, 1,850 to 1,862 ft, and 1,925 to 1,931 ft. At 1,722 ft the surveys were S32W at an inclination of 2.25°. The nozzle was oriented at S90E. The survey at 1,799 ft was S30E with an inclination of 2.75°. The second jetting run had a survey at 1,814 ft with the direction at S20E and the inclination at 3.0°. The nozzle orientation was N80E. At 1,877 ft the direction was S36E and the inclination was 3.25°. The last jetting station was at the last survey at 1,877 ft. Orientation of the nozzle was N70E. The direction and inclination at 1,940 ft was S66E and 4.75°. Normal drilling occurred between jetting runs. Determine the direction change $\Delta\epsilon, \beta$, and δ for each run.

Solution. If the beginning and ending inclinations and the tool-face setting are known, the total angle change, dogleg severity, and the change of direction can be determined using the graphical approach (see Figs. 8.40a, b, and c). This is done by drawing a line from $\beta = 0°$ to $\gamma = 122°$ and drawing a line from $\alpha = 2.25°$ that is 2.75° units long where it intercepts the other line. $\Delta\epsilon$ is the angle (77°) between the two line segments at $\alpha = 2.25°$. The total angle change ($\beta = 3.1°$) is the distance along the line from the origin ($\beta = 0$ and $\alpha = 0$) to the point of intersection with the other line ($\alpha_N = 2.75°$). The dogleg can be calculated by knowing β and the course length between surveys at 1,722 and 1,799 ft ($L_C = 77$ ft).

$$\delta = 3.1° \frac{100 \text{ ft}}{71 \text{ ft}} = 4°/100 \text{ ft}.$$

The measured change in direction from S32W to S30E is 62°. Using a similar approach, the other jetting runs can be analyzed. From 1,850 to 1,862 ft $\Delta\epsilon = 14°, \beta = 1.0,$

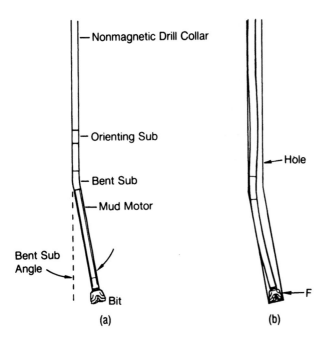

Fig. 8.39—Bent sub unconstrained and constrained in a wellbore.

and $\delta = 1.58°/100$ ft. The measured $\Delta\epsilon$ is 16°. From 1,925 to 1,931 ft, $\Delta\epsilon = 63°$, $\beta = 2.7°$, and $\delta = 4.28°/100$ ft. The measured $\Delta\epsilon$ is 33°.

8.5 Directional Drilling Measurements

The trajectory of a wellbore is determined by the measurement of the inclination and direction at various depths and by one of the calculations presented in Sec. 8.3. A tool-face measurement is required to orient a whipstock, the large nozzle on a jetting bit, an eccentric stabilizer, a bent sub, or bent housing.

Inclination and direction can be measured with a magnetic single- or multishot and a gyroscopic single- or multishot. All these tools are self-contained and are powered with batteries or from the surface. Magnetic tools are run on a slick line (steel wireline), on a sand line (braided steel cable), or in the drill collars while the hole is tripped, or they can be dropped (go-deviled) from the surface. Some gyroscope tools are run on conductor cable, permitting the reading of measurements at the surface and the supplying of power down the conductor cable. The battery-powered gyroscope tools are run on a wireline.

Another way to measure direction, inclination, and tool face is with an arrangement of magnetometers and accelerometers. Power can be supplied by batteries, a conductor cable, or a generator powered by the circulation of the drilling fluid. If the measurement tool is located in the BHA near the bit and the measurements are taken during drilling, it is called a measurement-while-drilling (MWD) tool.

This section presents the various measurement tools, the principles of operation, the factors that affect the measurements, and the necessary corrections.

8.5.1 Magnetic Single-Shot Instruments

The magnetic single-shot instrument records the inclination, direction, and tool face on either sensitized paper or photographic film. Fig. 8.41 shows a typical angle compass unit for 0 to 20° and 0 to 70° inclinations. In this arrangement, the direction and inclination indicator floats in a fluid, thereby minimizing the friction between the center post and the float. Fig. 8.42 shows some of the types of compass units for the various single-shot devices.

The angle inclination (AC) units range from 0 to 3° to 0 to 70° with a scribe line for tool-face orientation. The sample readings for the 3, 20, 70, and 80° units are 2.25° inclination, S80E; 6.5° inclination, S65E; 49° inclination, S73E; and 74° inclination, S13E. The scribe lines indicate that the tool faces of the 20°, the 70°, and the 70° magnetic method of orientation (MMO) compasses are 30° right of high side, 119° right of high side, and 40° right of high side, respectively. A 17 to 125° unit shows the inclination as 78° and the direction as 133° or

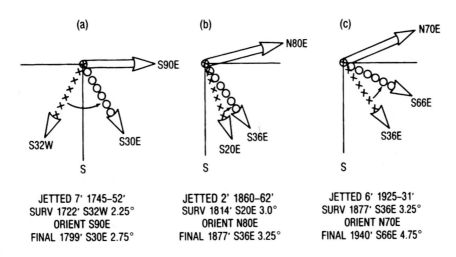

JETTED 7' 1745–52'
SURV 1722' S32W 2.25°
ORIENT S90E
FINAL 1799' S30E 2.75°

JETTED 2' 1860–62'
SURV 1814' S20E 3.0°
ORIENT N80E
FINAL 1877' S36E 3.25°

JETTED 6' 1925–31'
SURV 1877' S36E 3.25°
ORIENT N70E
FINAL 1940' S66E 4.75°

NOTE: DRILLED BETWEEN JETTING RUNS

Fig. 8.40—Example of three jetting stops while trying to kick off and set the wellbore lead (after Millheim et al.[5]).

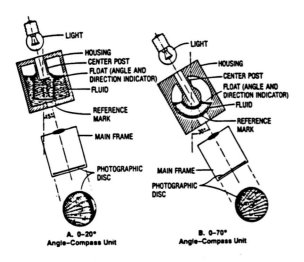

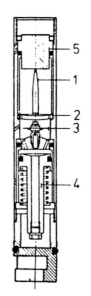

1. Pendulum
2. Circular glass
3. Compass
4. Pressure equalization
5. Cover glass

Fig. 8.41—Schematic diagrams of magnetic single-shot angle compass unit (courtesy Kuster Co.).

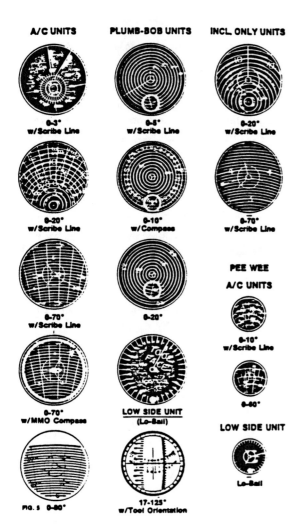

Fig. 8.42—Single-shot film disks (courtesy of Kuster Co.).

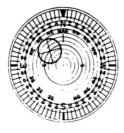

Indicated inclination 5° 0'
Direction of inclination N 45° 0'
or azimuth 45°

Fig. 8.43—Pendulum suspended inclinometer and compass unit for a 0 to 17° single-shot unit (after Eikelberg et al.[8]).

S47E. Other units show plumb bob units to measure only inclination (with and without scribe lines) and units to detect the low side of the borehole.

Fig. 8.43 shows a different type of compass unit that measures inclination by a Cardan-suspended pendulum that moves over a compass rose. The high-range units have a dual Cardan-suspended arrangement in which the compass moves on the main Cardan suspension and the inclinometer moves on an internal Cardan suspension (see Fig. 8.44).

Fig. 8.45 is a diagram of a typical single-shot unit with the bottom landing and orientation assembly. The unit is triggered either by a timer set for a period of time (up to 90 minutes) or by an inertial timer that does not activate until the unit comes to a complete stop. When the unit stops, the light comes on and a picture is taken.

Single-shot instruments are used to monitor the progress of a directional- or deviational-control well and to help orient the tool face for a trajectory change. The usual procedure is to load the film into the instrument, to activate the timer, to make up the tool, and to drop it down the drillpipe. When the timer is activated, a surface stopwatch is started, unless the motion timer is used. The surface stopwatch will indicate when the instrument has taken the picture. The tool is then retrieved with either a wireline overshot or the drillpipe. Fig. 8.46 shows a typical single-shot operation.

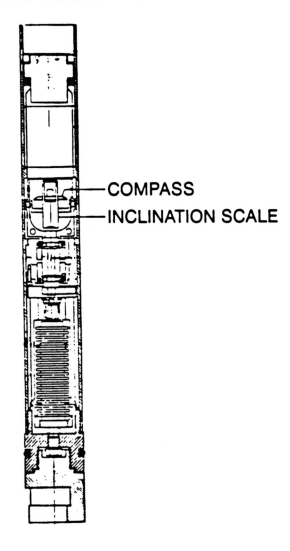

—COMPASS

—INCLINATION SCALE

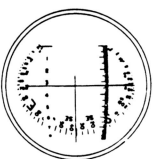

Fig. 8.44—Cardan suspended compass and inclinometer for a single-shot 5 to 90° unit (after Eikelberg *et al.*[8]).

Running a single-shot tool and interpreting data is simple. Valuable rig time, however, is consumed when a survey is run; depending on the depth of the well, the time used can range from a few minutes for shallow depths to more than an hour for a deep hole. Also, if the inclination becomes excessive, the tools must be pumped down. Another problem is the temperature in the area where the survey is taken. If the temperature is too high, the film will be completely exposed, yielding a black picture. To solve that problem, a special protective case is used to retard the temperature buildup in the film unit. The case fits over the single-shot tool and works like a vacuum flask, having a vacuum gap between the case and the tool. However, the tool still must be retrieved quickly because the case only slows the temperature rise.

Orientation of the tool face with a single-shot tool requires the use of either a mule-shoe mandrel and bottomhole-orienting sub or a nonmagnetic orienting collar. Fig. 8.47 shows a mule-shoe orienting arrangement. The mule-shoe orienting sleeve is positioned in the mule-shoe orienting sub to line up with the bent sub or bent housing knee, the large nozzle on a bit used for jetting, the undergauge blade of an eccentric stabilizer, or the whipstock wedge. The single-shot tool has a mule-shoe mandrel on the bottom that is shaped to go in the orienting sleeve only in the direction of the tool face.

If an MMO tool is used, the instrument is spaced so that a shadow graph compass in the single-shot instrument is opposite the magnets placed in the nonmagnetic collar. The location of the magnets is identified by a scribe line on the outside of the collar. When an orienting tool is made up, the positions of the orienting tool and of the scribe line on the nonmagnetic collar are observed and the make-up difference noted. An MMO single-shot picture will show a shadow graph, which indicates the direction of the scribe line on the drill collar; this graph is superimposed on a regular single-shot picture. When the makeup correction is considered, the true tool-face direction is obtained. This method of orientation is rarely used now.

8.5.2 Magnetic Multishot Instruments

The magnetic multishot instrument is capable of taking numerous survey records in one running. It either is dropped down the drillpipe or is run on a wireline in open hole.

Fig. 8.48 is a multishot instrument landed in such a way that the compass unit is spaced adjacent to the nonmagnetic collar. Fig. 8.49 is a depiction of a complete multishot instrument rigged for bottom landing, showing the component parts. Fig. 8.50 shows both sides of the watch and camera sections. The watch is spring wound and uses the power of the mainspring to operate a timer cam. The cam rotates, causing an electrical connection between the batteries and the motor and camera section. The motor drives a series of gears that finally drives the Geneva gear and the wheel assembly, which advances the film and turns on the light long enough for a picture to be taken.

A multishot tool is usually dropped down the drillpipe and landed in the nonmagnetic drill collar(s). During the trip out, a survey is taken approximately every 90 ft, the length of a stand. More closely spaced stations could be obtained, however, by stopping the pipe movement at a

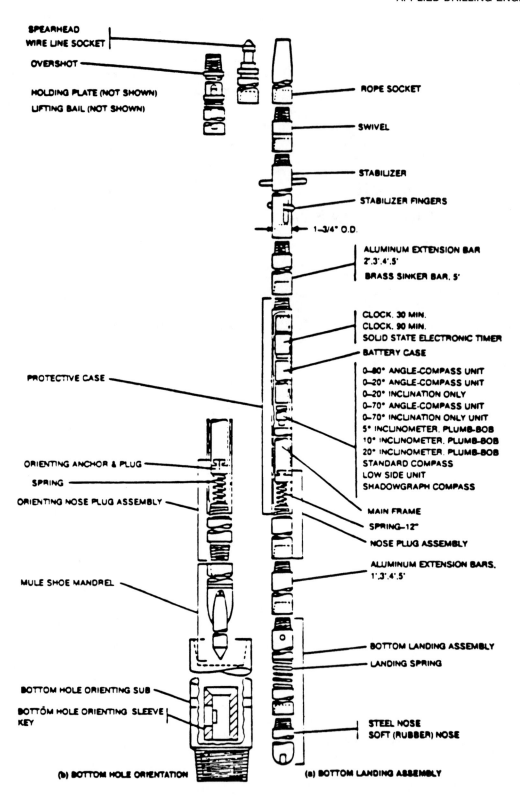

SPEARHEAD
WIRE LINE SOCKET

OVERSHOT

HOLDING PLATE (NOT SHOWN)
LIFTING BAIL (NOT SHOWN)

ROPE SOCKET

SWIVEL

STABILIZER

STABILIZER FINGERS

1—3/4" O.D.

ALUMINUM EXTENSION BAR
2',3',4',5'

BRASS SINKER BAR, 5'

CLOCK, 30 MIN.
CLOCK, 90 MIN.
SOLID STATE ELECTRONIC TIMER
BATTERY CASE

0—80° ANGLE-COMPASS UNIT
0—20° ANGLE-COMPASS UNIT
0—20° INCLINATION ONLY
0—70° ANGLE-COMPASS UNIT
0—70° INCLINATION ONLY UNIT
5° INCLINOMETER, PLUMB-BOB
10° INCLINOMETER, PLUMB-BOB
20° INCLINOMETER, PLUMB-BOB
STANDARD COMPASS
LOW SIDE UNIT
SHADOWGRAPH COMPASS

PROTECTIVE CASE

ORIENTING ANCHOR & PLUG
SPRING
ORIENTING NOSE PLUG ASSEMBLY

MAIN FRAME
SPRING—12"
NOSE PLUG ASSEMBLY

ALUMINUM EXTENSION BARS,
1',3',4',5'

MULE SHOE MANDREL

BOTTOM LANDING ASSEMBLY
LANDING SPRING

BOTTOM HOLE ORIENTING SUB

BOTTOM HOLE ORIENTING SLEEVE
KEY

STEEL NOSE
SOFT (RUBBER) NOSE

(b) BOTTOM HOLE ORIENTATION

(a) BOTTOM LANDING ASSEMBLY

Fig. 8.45—Typical magnetic single-shot tool with landing sub (courtesy of Kuster Co.).

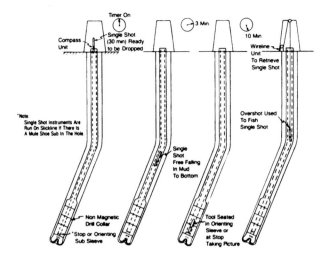

Fig. 8.46—Typical single-shot operation.

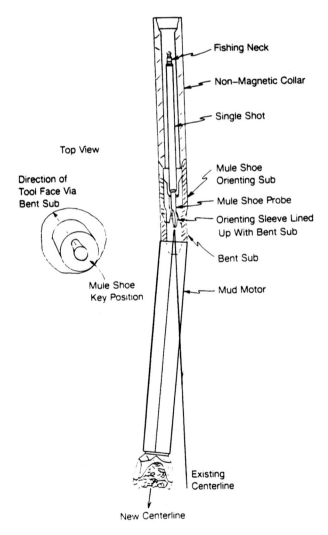

Fig. 8.47—Arrangement of the mule shoe for orienting a mud motor.

desired point and waiting for a picture. When the instrument is assembled at the surface and the timer is turned on, a stopwatch is also started. The watch is synchronized with the timer so that the operator knows exactly when a picture is being taken and how many frames were exposed. Fig. 8.51 illustrates how the stopwatch is used to take a picture. The upper dial records minutes and indicates the number of pictures taken; the lower dial indicates hours and records sets of 60 pictures.

Once a survey is completed, the tool is broken down, and the film is developed and read. Special readers that count the frames and project the picture onto a screen where it can be read easily are used (see Figs. 8.52A and B). Direction and inclination readings are identical to those of the single-shot units, except that there are no tool-face readings.

Multishot surveys typically are run at the end of a particular section of hole before casing is run. Because the surveys are usually closer together than the single-shot surveys and are run with the same instrument, multishot surveys are considered more representative of the trajectory of the borehole than a series of single-shot surveys. The accuracy of the multishot and single-shot surveys is affected by the inclination, the general trajectory direction, the position on the earth, and the magnetism of the wellbore and drillstring. All these factors will be covered later in this section.

8.5.3 Steering Tools

When a mud motor with a bent sub or housing is used, running a steering tool is sometimes wiser and more economical. Fig. 8.53 is a schematic of a typical operation with a steering tool. An instrument probe is lowered by a wireline unit and is seated in the mule-shoe orienting sleeve. The wireline can be passed through a circulating head mounted on the drillpipe. If this is done every 90 ft of drilled hole, the steering tool should be retrieved into the top stand. The top stand is set back and another stand is added. Then the stand with the steering tool is connected, and the steering tool instrument is run to the orienting sub and reseated. A way to overcome the need to pull the instrument every 90 ft is to insert a side-entry sub on the last drillpipe after tripping into the hole (Fig. 8.54). The stuffing box arrangement that prevents fluid leakage is built into the side of the sub. The steering tool is run conventionally and is seated in the orienting sleeve. The wireline is secured in the side-entry sub and let out the side, as illustrated in Fig. 8.55. As drilling continues, new joints of drillpipe can be added conventionally with the kelly. The wireline is clipped to the side of the drillpipe as more and more drillpipe is added. With care, this technique can be used to drill hundreds of feet without the need for tripping the instrument to the surface. Generally, drilling is continued until the side-entry sub reaches a point in the borehole where the inclination is so high that there is a risk of damaging the wireline. When this depth is reached, the drillpipe can be short tripped and the steering tool retrieved. The side-entry sub can be reset at the top of the string and the operation restarted.

The steering tool uses electronic means to measure direction and inclination. Direction is measured by fluxgate magnetometers that measure the earth's magnetic field in the x, y, and z planes. From this measurement, the vector components can be summed up to determine

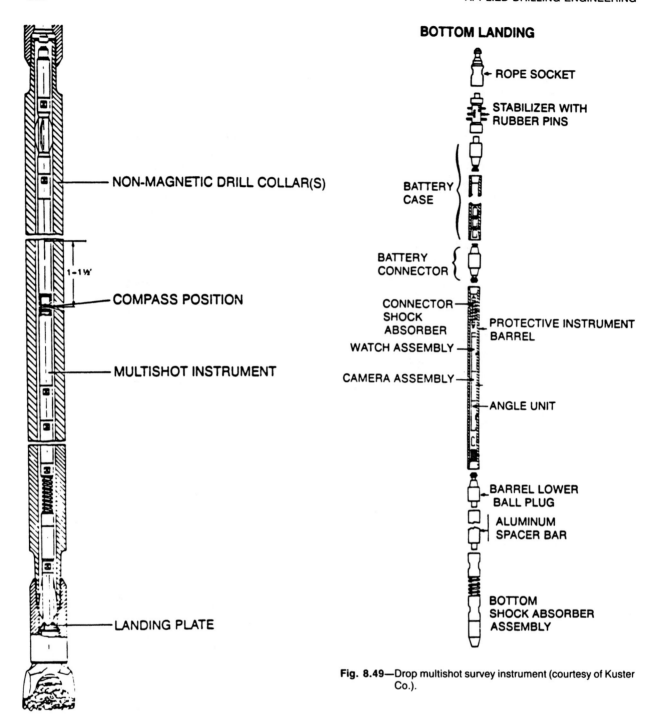

BOTTOM LANDING

Fig. 8.48—Typical arrangement for landing a multishot instrument (after Eikelberg *et al.*[8]).

Fig. 8.49—Drop multishot survey instrument (courtesy of Kuster Co.).

the wellbore direction. Inclination is measured by accelerometers that measure the gravity component along two axes. Fig. 8.56 shows the arrangement. The angle of the tool face below 3 to 7° can be determined by a computer with the magnetometer data. Above 3 to 7°, the angle of the tool face is referenced to the hole direction and is related to the gravity readings of the accelerometers.

Figs. 8.57A and B picture a typical steering-tool surface recorder and tool-face indicator mounted near the driller. Most steering tools constantly sense inclination, direction, and tool-face angle. Therefore, the steering tool

gives the directional operator more information with which to adjust the tool face and to achieve better control of the mud motor. The steering tool takes the guesswork out of correcting the tool-face angle for the expected amount of reverse torque. Having a constant, continuous tool-face reading, the operator can make minor adjustments and even use the readings to slack off weight on the bit (WOB). As weight is applied, the reverse torque increases, and the tool face rotates back to the right as the bit drills off.

A steering tool is one of the most economical means of making a trajectory change when a mud motor and bent sub or bent housing are used for drilling, especially when rig costs are high and the trajectory change is below 3,000 to 4,000 ft. Table 8.5 sets out a typical steering-tool trajectory change.

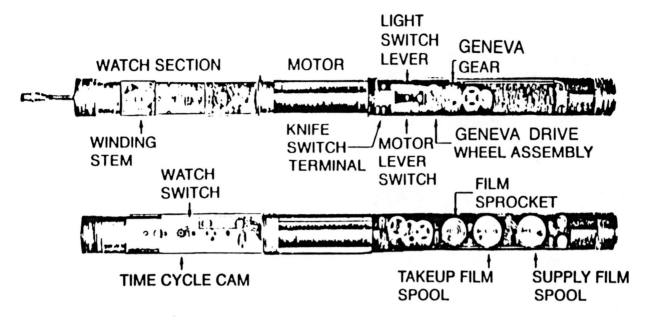

Fig. 8.50—Views of the watch and camera unit of a typical multishot tool (courtesy of Kuster Co.).

8.5.4 Tools for Measuring Trajectory During Drilling
As early as the 1960's, companies were experimenting with ways to log formations during drilling, but, technologically, it was difficult to build tools that could withstand the harsh downhole environment and transmit reliable data. A spinoff of the effort to overcome the problem was a recognition that inclination, direction, and toolface angle could be measured during drilling and the data could be transmitted to the surface.

Various transmission methods were used—such as electromagnetic, acoustic, pressure pulse, pressure-pulse modulation, or cable and drillpipe. Of all the transmission methods, the pressure-pulse and pressure-pulse-modulation methods have evolved into commercial systems often used by the directional drilling community.

One of the earliest commercial systems offered to the industry was the Teledrift, which was designed as a sub that could be placed near the bit to record and to transmit the inclination of the wellbore to the surface. Fig. 8.58 depicts the teledrift tool at various transmission positions. An initial setting of the inclination range must be made before the tool is run into the hole. The range is for 2.5°,

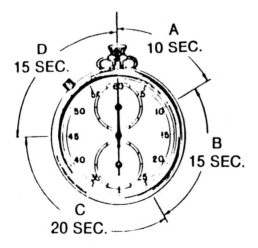

SURFACE WATCH

SYNCHRONIZE WITH INSTRUMENT WATCH
BY STARTING AT THE INSTANT CAMERA
LIGHTS GO ON.

Time Intervals:

A. 10 Seconds – Lights Are On, Exposing Film.

B. 15 Seconds – Delay Before Moving. This Is An Allowance For Instrument Watch Lag During Survey.

C. 20 Seconds – Instrument Is Idle Allowing Movement Of Drill String Without Affecting Picture. Most Moves Require Sufficient Time For Taking One or More Shots While Moving.

D. 15 Seconds – Minimum Time For Plumb Bob and Compass To Settle For Good Picture, Plus Allowance For Instrument Gain During Survey

Fig. 8.51—Use of the surface watch while running a magnetic multishot operation (after Eikelberg et al.[8]).

Fig. 8.52A— Typical multishot film reader (courtesy of NL Sperry Sun Co.).

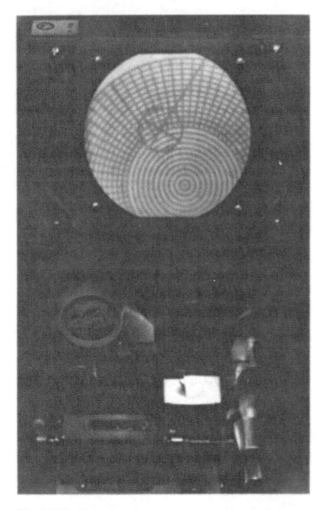

Fig. 8.52B—Projection of one survey frame for determining inclination and direction (courtesy NL Sperry Sun Co.).

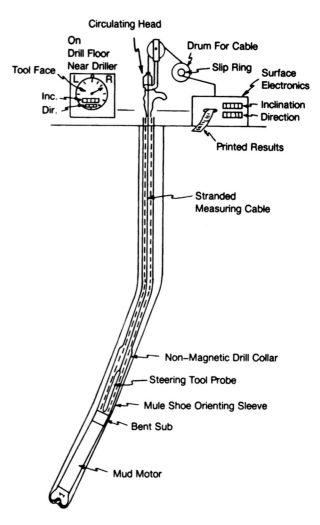

Fig. 8.53—Typical operation of a steering tool for orienting the bit.

in 0.5° increments between 0 and 15°. For example, an initial setting of 0 to 2.5° can be made. If the inclination exceeds 2.5°, the teledrift will report only 2.5°.

During drilling, the fluid velocity keeps the signal piston pressed down to its lowest position outside the pulse rings against the tension of the shaft spring. When a reading is needed, drilling is stopped, the bit is lifted off bottom, and circulation is terminated. The shaft spring forces the signal piston to the position that is proportional to the inclination or to the highest position if the inclination is greater than the maximum range setting. A pendulum controls the setting of the signal piston. As the inclination increases, the pendulum goes farther down the stop rings until it reaches the maximum inclination. The spring tension is released accordingly, allowing the signal piston to advance upward. When pumping begins, the signal piston is forced by each pulse ring, sending from one to seven pressure pulses up the drillstring to the surface, where the pulses are detected by a recorder that picks up the number of pulses and prints the data on a strip chart. Signal strength can vary from 60 to 150 psi and depends on the depth of the well and the condition of the mud. A problem arises if there is air or gas in the mud; either will reduce the signal transmission to a point at which the pulses are difficult to detect. Another problem is that materials in the drilling fluid may plug the tool. Properly

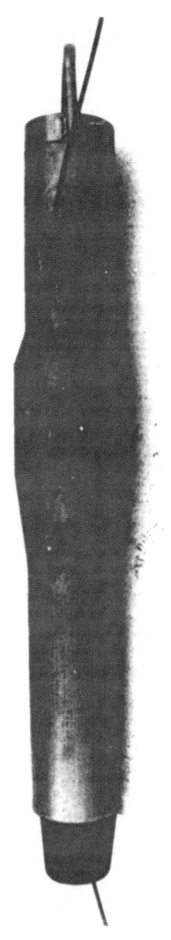

Fig. 8.54—Side-entry sub.

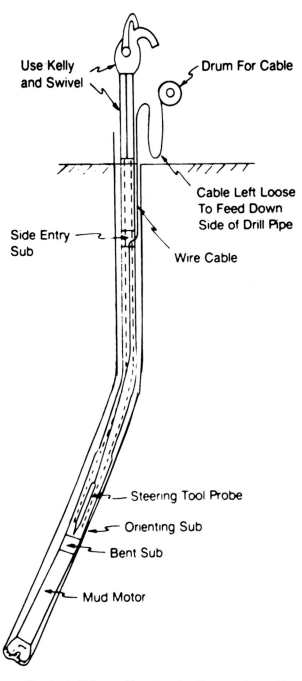

Fig. 8.55—Using a side-entry sub with a steering tool.

maintained and operated, however, the tool has application for deviation-control wells. As many inclination readings as desired can be obtained at any time when drilling and pumping are stopped. Thus decisions can be made before a well deviates to the point that drastic measures must be taken. Even though this tool is still available and is run periodically, its use has diminished with the advent of the various mud-pulse MWD tools that are now reliable and economical for a wide range of drilling situations.

The two most common MWD systems are the pressure-pulse and the modulated-pressure-pulse transmission systems. The pressure-pulse system can be subdivided further into positive- and negative-pressure-pulse systems.

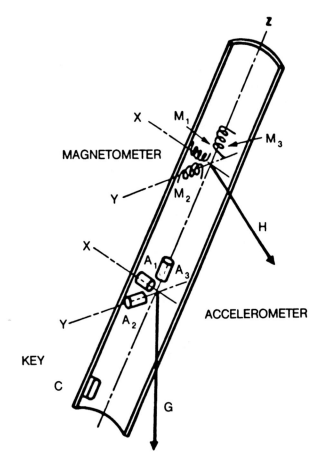

Fig. 8.56—Arrangement of sensors in a steering tool (after Eikelberg et al.[8]).

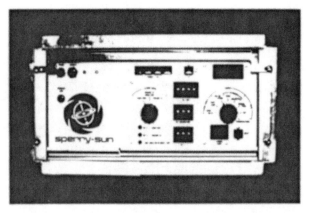

Fig. 8.57A—Steering-tool surface panel (courtesy of NL Sperry Sun Co.).

Fig. 8.57B—Tool-face indicator located on the drill floor (courtesy NL Sperry Sun Co.).

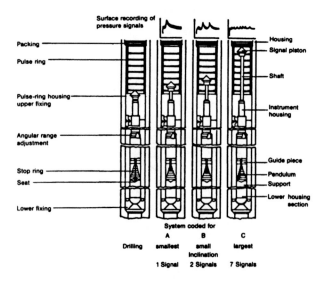

Fig. 8.58—Operation of a teledrift tool (after Eikelberg et al.[8]).

Fig. 8.59 depicts a typical MWD system with the downhole sensor unit, the sensor-to-signal unit, the pulser section, and the power section. At the surface, the signals are received by a pressure transducer and transmitted to a computer that processes and converts the data to inclination, direction, and tool-face angle. This information is transmitted to a terminal, which prints it, and to a rigfloor display similar to the steering-tool surface unit, which displays inclination, direction, and tool-face angle.

Most sensor packages used in an MWD tool consist of three inclinometers (accelerometers) and three flux-gate magnetometers. At low angles, inclination can be read with one of the gravity inclinometers. At higher angles—approaching 90°—another axis (hence another inclinometer) is needed to obtain correct values. Direction measurements are obtained from the three flux-gate magnetometers. The accelerometer readings are needed to correct the direction measurements for the inclination and the position of the magnetometers with the low side of the hole. Tool-face angle is derived from the relationship of hole direction to the low side of the hole, which is measured by the inclinometers.

Once the readings are measured, they are encoded through a downhole electronics package into (1) a series of binary signals that are transmitted by a series of pressure pulses or (2) a modulated signal that is phase shifted to indicate a logic 1 or 0. Fig. 8.60 shows a negative pulser, a positive pulser, and a mud siren that generates a continuous wave.

TABLE 8.5—EXAMPLE OF STEERING TOOL DATA FOR A KICK OFF OF A GULF OF MEXICO DIRECTIONAL WELL

Bit Depth		Magnetic Seat	High-Side Seat	Inclination (degrees)	Azimuth (degrees)	Dip	Magnitude (degrees)	Picture Inclination/Direction	Readout
Start (ft)	End (ft)								
820		105.4	333	7.4	122.1	62.2	50.3		
	910	133.5	356	13.0	122.1	59.6	50.3	11°5' S58E	3°L
910		137.3	359	12.8	122.4	59.7	50.6		0°
	950	83.8	309	16.3	115.0	69.6	49.4	16° S65E	50°L
950		62.9	289	16.7	113.5	59.7	49.4		50°L
	1,031	130.0	353	20.0	111.4	59.7	49.3	21°50' S68E	
1,031		135.0	355	22.7	111.5	59.3	49.4		5°L
	1,156	136.9	356	29.5	111.7	59.2	49.1	29° S67E	4°L
1,156		126.7	345.5	29.7	112.4	58.8	48.9		15°L
	1,212	119.0	337	32.7	111.7	58.8	48.8	32° S68E	30°L
1,212		112.0	330.1	37.3	111.7	58.7	48.9		30°L
	1,306	101.8	322.2	38.5	111.6	58.5	48.7	38° S67E	30°L
1,306		104.9	320.2	38.7	111.7	58.7	48.7		40°L
	1,398	129.3	342.5	44.0	117.3	58.6	48.9	44° S67E	15°L

The negative pulser works by an actuator that opens and closes a small valve that discharges a small amount of the drilling fluid to the annulus. The fluid causes a brief, small pressure decrease in the drillpipe (100 to 300 psi), causing a negative pressure pulse. The duration of the pressure pulse is related to how quickly the valve opens and shuts. Because both valve wear and power consumption must be considered, complex schemes are used to encode the sensor data and to transmit them with the fewest pulses in the shortest time. To transmit a set of data—including time for a turn-on sequence and for a parity check of inclination, direction, and tool-face angle—3 to 5 minutes typically are needed. Table 8.6 is an example of one reading of a negative pulse system for making a PDM motor trajectory change.

The positive pulser with a valve actuator works by restricting the flow of drilling fluid down the drillstring and creating a positive pressure pulse. The positive pressure pulse can be greater than the negative pulse and is easier to detect. The time required to transmit a set of data by the positive pulse system is about the same as that for a negative pulse system—3 to 5 minutes.

The mud siren is based on a mud-driven turbine that turns a generator that powers a motor whose speed varies between 200 and 300 cycles/sec. The motor drives a turbine rotor that, in conjunction with the stator, generates a carrier wave, which is modulated by the turbine rotor's speeding up or slowing down. The phase shift is detected at the surface and is interpreted as a logic 0 or 1.

All commercial MWD systems are powered either by batteries or by a mud-driven turbine. The lithium batteries limit the operating time, depending on the downhole temperature, to less than 300 hours. Because most bit runs last less than 100 hours, the battery pack can be replaced during a bit change. Battery-powered MWD systems have some advantages over a turbine-powered MWD system in that they permit almost full flow of the drilling fluid to the bit without a significant pressure loss. The turbine system is sensitive to the flow rate and to the type of fluid going through the turbine. Lost-circulation material or other debris that normally could be passed through the drill collars and bit is not tolerated as easily by the stator/rotor of a turbine.

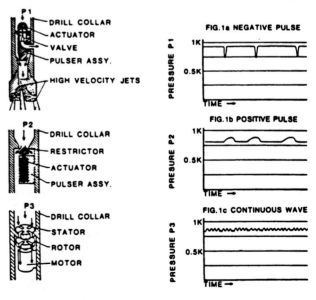

Fig. 8.59—Typical MWD system (after Gearhart et al.[9]).

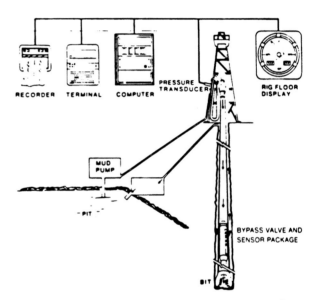

Fig. 8.60—Basic types of mud pulsers (after Gearhart et al.[9]).

TABLE 8.6—TYPICAL OUTPUT OF A NEGATIVE PULSE SYSTEM

April 10, 1984 15:38 Amoco Production Anschutz Ranch   P.01
                Com #81   20:34:20 12-31-83

Drift Angle 1115   Tool Face 128 R   Direction 370:OE
MDIP = 67.7586   TMF (X,Y,Z) = 1.1097   TGF (A,B) = 0.0239

T	3,946	X	2,032	EA	0.019	EX	−0.306	V+	24.62	OTF	0.0		INC	1.37
A	2,602	Y	2,040	EB	−0.015	EY	−0.292	V−	−24.60	MDC	15.0	E	RDT	127.98
B	2,526	Z	4,308	EC	−1.001	EZ	1.226	TEMP	74.4				AZI	110.05
C	2,560					TMF (X,Y)		0.4229	DTMP	49.4				

MWDD, MWD, 0183.03.MSC

Field AFE-West Lobe Development   Well W-20-04
Survey Method RC   Survey Number 130
Elec Number 149
Mag Number 302
12-31-83 20:33:25

Parameter	DGT 1	DGT 2	DGT 3	Counts
T	7	11	5	3,946
A	5	1	5	2,682
B	4	14	15	2,536
C	5	0	0	2,560
X	3	15	3	2,022
Y	3	15	13	2,040
Z	8	6	10	4,308

Depth	10397.00	
Angle	1:15	
Direction	S 70:0E	
Tool Face	128P	
	E 30 A 15	
DAC		
DDO	S 70: O E	
DDC		
CL	4.100	
TUD	10349.45	
RCN/S	152.08 S	
RCE/W	243.97 E	
SECTION	287.32	
DLS/CL	0.12	
DLS/100 FT	0.30	
PDD	S 60: O E	
TEMP	74.42	
TMF	1.1097	
TMF (X,Y)	.4229	
DIP ANGLE	67.7586	
ANG CORR	0: 0	
DIR CORR	0: 0	
MDC	15: 0 E	
OTF	0:00	

Courtesy of Gearhart Industries Inc.

An advantage of the turbine-powered generator, on the other hand, is that it can supply more power to the downhole electronics and valve actuator and is more tolerant of high bottomhole temperatures.

Other developments related to MWD systems are still in the prototype stage: (1) electromagnetic systems, (2) electric cable, and (3) specially designed drill collars and drillpipe to be used in conjunction with electric cable. Advantages of the hard-wired systems are that they can transmit data very rapidly and can communicate with the downhole electronics. The disadvantage is that it is necessary to handle the signal-transmitting electric cables that are suspended in the drillpipe and the slip-ring arrangement on the kelly, which must be used to transmit the data from the rotating kelly to the surface instrumentation.

A new generation of MWD tools that can be run in hole and retrieved with a slick-line unit and overshot is being developed. Also, other downhole sensors are being used to determine drilling parameters, such as torque and WOB; gamma ray and resistivity logging sensors are used for formation evaluation. Fig. 8.61A and B are typical outputs of an MWD system with drilling and logging sensors.

8.5.5 Magnetic Reference and Interference

Surveying instruments that are used to measure the wellbore direction on the basis of the earth's magnetic field must be corrected for the difference between true north and magnetic north. Fig. 8.62 depicts the earth's magnetic field, showing magnetic north and true north. The

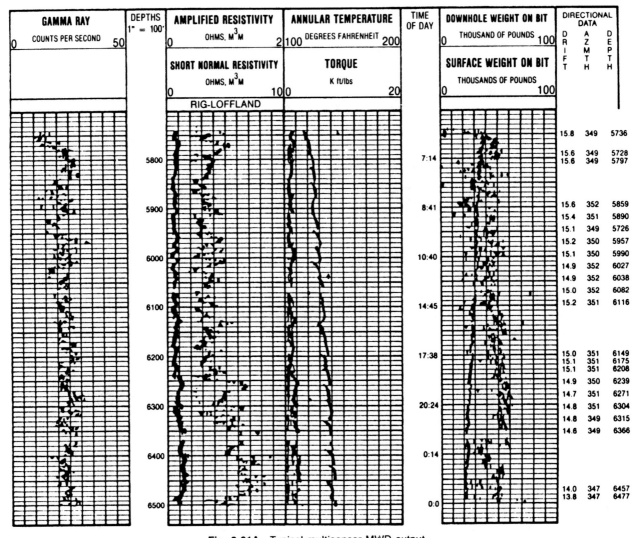

Fig. 8.61A—Typical multisensor MWD output.

compass reacts to the horizontal component of the magnetic field; the reaction decreases when the compass is moved northward. Declination is the angle between magnetic north and true north (see Fig. 8.63). The angle changes with time and depends on the position and surface features of the earth. Fig. 8.64 shows the declination angles for the U.S. The isogonal lines (lines of equal declination) indicate how much correction should be made, depending on where the survey is made (see Table 8.7).

Example 8.13. You are drilling a well near Corpus Christi, TX, and the directions are reading all in the SE quadrant. What is the correction for the direction readings?

Solution. The declination angle near Corpus Christi is 7.75E. Because it is an east declination, 7.75E must be subtracted from the direction readings.

Besides making a correction for true north, one must take special care when running a magnetic survey to prevent the effects of magnetic interference. Such interfer-

ence can be caused by a proximity to steel collars and by adjacent casing, hot spots in nonmagnetic collars, magnetic storms, and formations with diagenetic minerals.

Nonmagnetic drill collars are used to separate the compass from magnetic fields of magnetic steel above and below the compass and to prevent the distortion of the earth's magnetic field. The collars are of four basic compositions: (1) K Monel 500™, an alloy containing 30% copper and 65% nickel, (2) chrome/nickel steels (approximately 18% chrome and more than 13% nickel), (3) austenitic steels based on chromium and manganese (over 18% manganese, and (4) copper beryllium bronzes.

Currently, austenitic steels are used to make most nonmagnetic drill collars. The disadvantage of the austenitic steel is its susceptibility to stress corrosion in a salt-mud environment. The K Monel and copper beryllium steels are too expensive for most drilling operations; both, however, are considerably more resistant to mud corrosion than austenitic steels are. The chrome/nickel steel tends to gall, causing premature damage to the threads, especially for larger collars that require high makeup torques.

Fig. 8.65 shows the compass located in a nonmagnetic collar between the bit and the steel collars. The nonmagnetic collar does not distort the earth's magnetic field lines and isolates the interference field lines from the sections

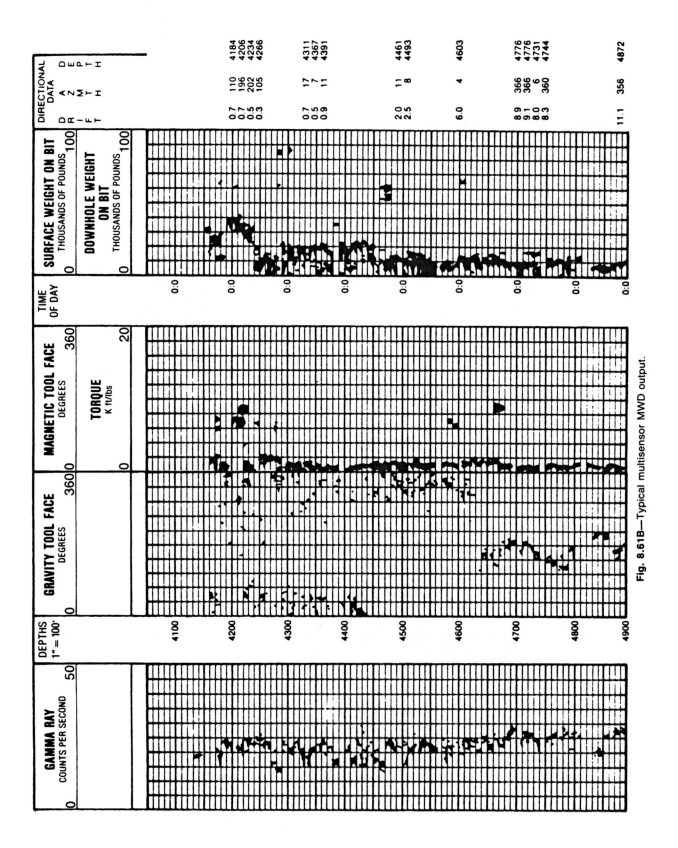

Fig. 8.61B—Typical multisensor MWD output.

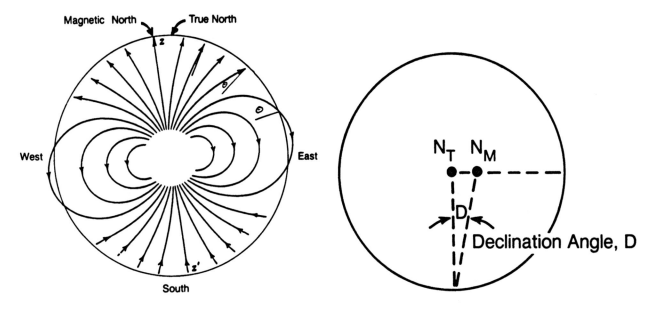

Fig. 8.62—Earth's magnetic field.

Fig. 8.63—Declination.

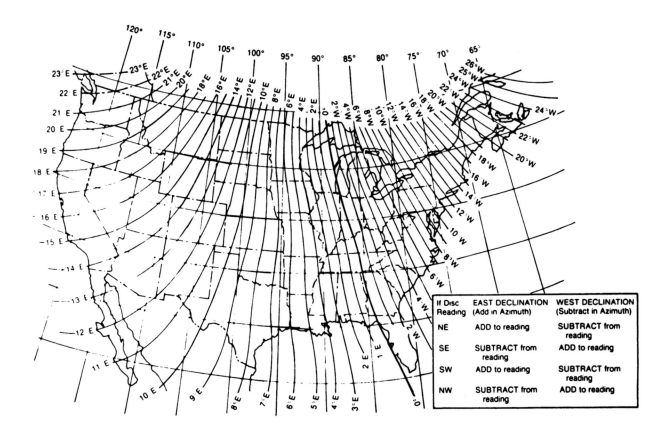

Fig. 8.64—Isogonic chart for the U.S.

TABLE 8.7—CORRECTION FOR DECLINATION

Direction Reading	East Declination	West Declination
NE	Add to reading	Subtract from reading
SE	Subtract from reading	Add to reading
SW	Add to reading	Subtract from reading
NW	Subtract from reading	Add to reading

above and below the compass unit. The number of required nonmagnetic collars depends on the location of the wellbore on the earth and the inclination and direction of the wellbore. Fig. 8.66 is a compilation of empirical data that are fairly reliable in selecting the number of nonmagnetic drill collars.

First, a zone is picked where the wellbore is located. Then the expected inclination and direction are used to locate the curve, either A, B, or C.

Example 8.14. Select the number of nonmagnetic drill collars needed to drill a well to 55° inclination at a direction of N40E on the north slope of Alaska.

Solution. The north slope of Alaska is in Zone III. From the empirical data charts for Zone III at 55° inclination and N40E, the point falls just below Curve B, indicating the need for two nonmagnetic collars with the compass unit 8 to 10 ft below the center.

The effect of the magnetic interference is illustrated further by Fig. 8.67, which shows typical directional errors in the Gulf of Mexico area when 14-, 25-, 31-, and two 31-ft nonmagnetic collars are used for drilling at various inclinations and directions. A well drilled in the Gulf of Mexico with one 31-ft nonmagnetic collar at an inclination of 30° and a direction of S75W has a directional error of approximately 1.3°.

8.5.6 Gyroscopic Measurement

A gyroscopic compass is used when magnetic surveying instruments cannot be used because of the magnetic interference of nearby casing or when a borehole with casing already set is being surveyed.

There are various kinds of gyroscopic instruments: single- and multishot gyroscopes, the surface-recording gyroscope, the rate or north-seeking gyroscope, and the Ferranti tool (a highly precise, inertial guidance tool similar to that used on modern commercial aircraft). Of the gyroscopic instruments used for surveying cased boreholes, the multishot is the most common.

Fig. 8.68 depicts a Cardan-suspended horizontal gyroscope. A high-frequency AC current drives a squirrel-cage rotor at a speed of 20,000 to 40,000 rpm; as long as the rotor runs at its reference speed and there are no external forces, the gyroscope stays fixed.

Fig. 8.69 shows a complete gyroscope assembly. The upper part of the tool holds the batteries, camera assembly, and multishot clock. The lower part of the tool contains the inclinometer, the Cardan-suspended gyroscope motor, electronic components for the gyroscope, and the shock absorber.

Even though the gyroscope is not influenced by magnetic interference, its very design introduces unique problems associated with obtaining accurate survey information. If the gyroscope could be supported exactly at its center of gravity, it would be free of influences by external forces. However, such accuracy is practically impossible to achieve. Consequently, a slightly off-center gyroscope will tend to show a force, F, caused by gravity, in the direction indicated in Fig. 8.68. The gyroscope compensates for the gravitational and frictional forces caused by the bearings by rotating about its vertical axis in a direction commensurate with the right or left side of the downward force on the horizontal gimbal axis. (Fig. 8.68 shows a counterclockwise movement for the force on the right side.) The amount of this rotation determines the accuracy of the gyroscope. The tilt of the horizontal gimbal is corrected by a sensor that detects any departure of the gyroscope from the horizontal axis and sends a signal to a servo motor. This corrects the gyroscope by rotating the vertical axis until the horizontal axis is properly adjusted. The gyroscope is not as rugged as the magnetic instrument and must be handled more carefully. Unlike the magnetic tools, the gyroscope must be run on a wireline. When it is run, the survey stations usually are made going into the hole with a few check shots coming out; this is done mainly to make accounting for drift easier.

Drift is caused by the rotation of the earth. Fig. 8.70 shows the amount of drift per hour, which is dependent on the latitude. At either pole the drift is at a maximum, and at the equator there is no apparent drift. When a gyroscopic survey is run, the effects of drift must always be determined.

The first step in running a gyroscopic survey is to orient the gyroscope. Fig. 8.71 shows the direction face of a gyroscope. The solid triangle is always pointed at the stake or reference point (see Fig. 8.72). Because the bearing or direction of the reference point is known, it is relatively easy to determine the direction of the "zero" spin axis on the inner scale.

The reference direction is determined by making a sighting from some point over the wellbore, either to a stake fixed some distance from the wellbore or to a constant

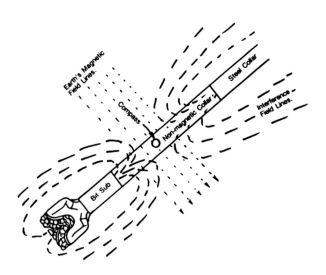

Fig. 8.65—Source of magnetic interference.

The Earth's horizontal magnetic intensity varies geographically, and the length
of nonmagnetic drill collars used in a bottom hole assembly should fit the
requirements of the particular area. This map is used to determine which set
of empirical data should be used for a given area.

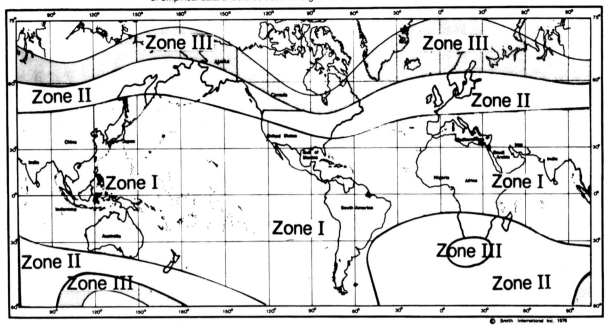

Empirical Data Charts

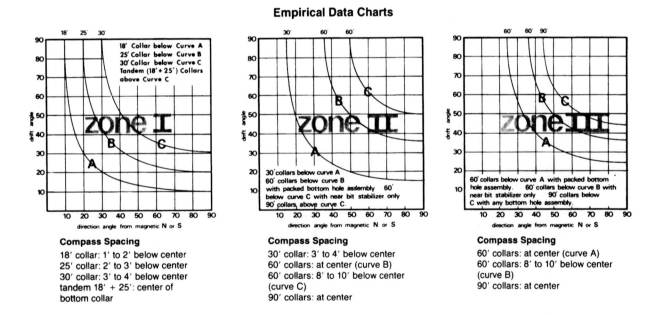

Compass Spacing

18' collar: 1' to 2' below center
25' collar: 2' to 3' below center
30' collar: 3' to 4' below center
tandem 18' + 25': center of
bottom collar

Compass Spacing

30' collar: 3' to 4' below center
60' collars: at center (curve B)
60' collars: 8' to 10' below center
(curve C)
90' collars: at center

Compass Spacing

60' collars: at center (curve A)
60' collars: 8' to 10' below center
(curve B)
90' collars: at center

Fig. 8.66—Zone-selection map and charts to determine how many nonmagnetic drill collars are required (courtesy Smith Intl. Inc.).

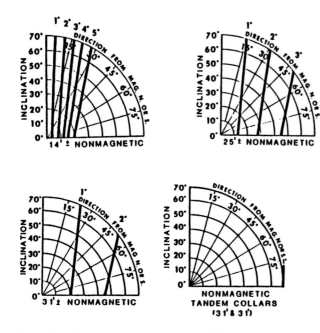

Fig. 8.67—Typical direction errors resulting from magnetic interference—Gulf Coast (courtesy NL Sperry-Sun).

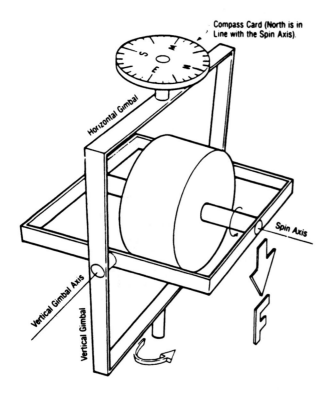

Fig. 8.68—Typical gyroscope.

reference point away from the wellbore, such as a building, another drilling rig, or some other object. Fig. 8.72 shows a typical example of a reference sighting.

When the gyroscope is referenced initially, one of two procedures, depending on the maximum inclination that is expected, is followed. If the inclination is to be less than 10°, the case index (see Fig. 8.71) is lined up with the reference marker—i.e., in the stake direction (D_s). Refer to the example presented by Fig. 8.72, in which the stake is N20E or 20°. The gyroscope spin axis—i.e., index setting (i_s)—is moved until 20° is opposite the case index, thereby aligning the spin axis with north. Below 10° inclination, the 3D instrument correction is negligible. (This will be covered later and is called the intercardinal correction.) Above 10°, the 3D correction must be considered. Again, the case index must be aligned with the reference object, but the index setting is determined by Eq. 8.58.

$$i_s = D_s - \delta_{dr}, \dots\dots\dots\dots\dots\dots\dots\dots (8.58)$$

where δ_{dr} is the assumed hole direction.

For example, if the D_s is S17W or 197° and the δ_{dr} is N20E or 20°, the index setting is determined as follows:

$$197° - 20° = 177°.$$

When the case index is aligned with 177°, the gyroscope north will be aligned with the δ_{dr} of 20°, and errors resulting from gyroscope tilt will be minimized. Every survey reading should be adjusted for the initial offset of 20°, as well as for drift correction and the intercardinal correction.

After the initial gyroscope orientation, or "gyro orientation" (GO), the tool case is oriented to what is called a "case orientation" (CO). From these initial checks, an initial drift is estimated. The tool is run in the wellbore, making stops for survey pictures. At 10-minute intervals, drift is checked by keeping the tool still for 3 minutes or more. Fig. 8.73 shows a typical drift data sheet for a gyroscope survey. Note that most of the check stops were made going into the wellbore; only two were made coming out.

Once the data are obtained, the drift correction must be made by the construction of a drift correction plot. Fig. 8.74 is a plot for the drift data on Fig. 8.73. The vertical axis, measured in degrees, is the scale for the correction values that will be applied to directional data to correct for drift during the survey. The horizontal scale is the surveying time in minutes.

The range of the vertical axis is determined by taking the correction factor, F_C, which is determined from Eq. 8.59,

$$F_C = D_s - i_s, \dots\dots\dots\dots\dots\dots\dots\dots (8.59)$$

and by scaling above and below this factor in 1° increments. For the example presented in Fig. 8.74, F_C is 19.8 (the case was actually indexed at 177.2° rather than the calculated 177.0°), and the scale ranges from +19.0 to +26.0. The horizontal scale should cover the entire survey from the gyroscope start to end; for this survey, the duration was 101.00 minutes.

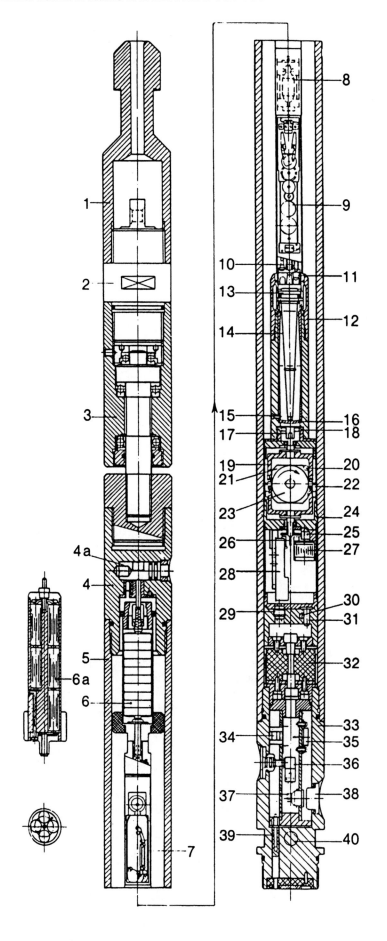

1. Fishing device
2. Rope connection
3. Rope socket
4. Connecting piece
5. Camera housing
6. Battery
6a. Dry-cell batteries
7. Multi-shot switch clock
8. Multi-shot motor
9. Camera
10. Lens
11. Light bulb
12. Pendulum housing
13. Cover glass
14. Pendulum
15. Filler plug
16. Circular glass
17. Compass card
18. Marking ring
19. Cardan suspension for gyroscope
20. Electrolyte level switch
21. Cardan suspension
22. Cardan contact brush
23. Gyroscope motor
24. Gyroscope housing
25. Gear unit
26. Cardan contact brush
27. Servomotor
28. Converter
29. Plug connection
30. Nut
31. Gyroscope adjusting pin
32. Rubber shock absorber
33. Gear casing
34. Safety fuse
35. Voltage regulator
36. Gyroscope ON-OFF switch
37. Multi-range voltmeter
38. Cover plate
39. Gyroscope battery connection
40. Hole for adjusting telescope

Fig. 8.69—Components of a gyroscope compass instrument (after Eikelberg *et al.*[8]).

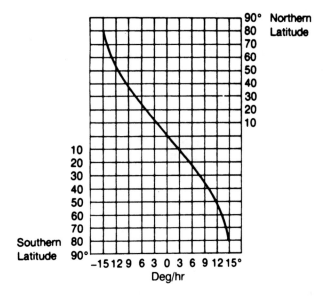

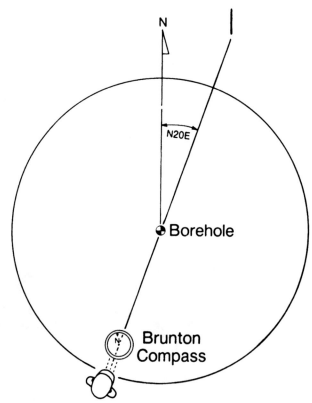

Fig. 8.70—Diagram of apparent drift for every point on Earth, depending on geographical latitude (after Eikelberg et al.[8]).

Fig. 8.72—Sighting of reference object.

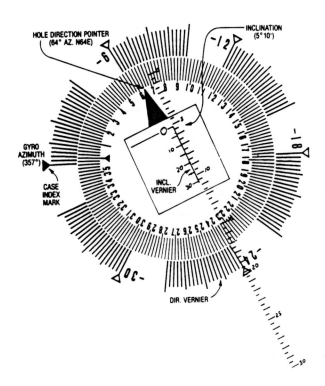

Fig. 8.71—Gyroscope face.

The sequence of events for the survey begins with turning on the gyroscope at 2 minutes : 20 seconds. At 7 minutes : 20 seconds, the CO is performed. Following that are eight drift-check stops interspersed with the survey stops, which are taken every 100 ft up to 1,815 ft and then coming out of the hole at 1,200 ft, 200 ft, and at the surface. At the end of the survey, a final case orientation is performed before the gyroscope is turned off. (Although it was not done on this survey, rotational shots sometimes are taken at the end of the survey to determine the true wellbore center and to correct any misalignment of the survey tool with the wellbore. The rotational shots are obtained by the lowering of the surveying tool into the wellbore until the stabilizers center the tool in the casing. Six survey records are taken at 60° increments. Once the film is processed, the true center of the survey tool can be determined, and the misalignment errors in inclination and direction can be corrected. Problem 8.47 presents this type of data.)

The first drift check—between 7 minutes : 20 seconds and 9 minutes : 40 seconds—lasts 2 minutes : 20 seconds; this is entered as a block on the time scale. The same procedure is followed for each drift check. Next, a centerline is determined by the marking of the halfway point between each two blocks of drift time. The rate of drift (rate/hour) for each drift check is plotted, starting with the CO point at 7 minutes : 20 seconds, by drawing the correct slope from this point to the point where it intersects the next centerline (15 minutes : 0 seconds). The next slope (+3.6°/hr) is plotted from the new point to the next centerline, and so on.

DRIFT DATA SHEET · L.R.

CUSTOMER __SMITH OIL CO.__ RUN NO __1__

JOB NO __SU 73-0802-1__ GYRO NO. __242-LR__ TYPE __3" L.R.__ DATE __8-2-73__

DATE RUN __8-2-73__ READ BY __SWG__ DATE __8-2-73__ CHECKED BY __E.J.B.__ DATE __8-4-73__

SD TRUE NORTH AZIM __197__ IS INDEX NORTH AZIM __177__ C.F. • S.D. - I.S. CORRECTION __+20__ ·/•

TIME				AZIMUTH				
GYRO START	02-00	END 101-00	START	177.0	END	173.1	CORR	+3.9 ·⊙
FILM ORIENT	05-20	99-00		177.2		172.8	CORR	+4.4 ·⊙
CASE ORIENT	07-20	87-20		177.2		172.8	CORR	+4.4 ·⊙

REF PT CHECK TIME __1__ INCL __DIR__ BOTTOM TIME __59-20__ DEPTH __1815'__

DRIFT CHECK NO	DEPTH	TIME	DELTA TIME	GYRO AZIM	INCL	HOLE DIRECT	REAL HOLE DIR.	• −	DELTA HOLE DIR.	− +	RATE/HR. HOLE DIR.
1	0	07-20 / 09-40	140	177.2 / 177.1	0°	— / —	— / —	−	0.1	+	2.57
2	500	20-20 / 23-40	200	042.0 / 041.8	9°-15'	344.0 / 343.8	344.20 / 344.00	−	0.2	+	3.60
3	1000	31-20 / 34-20	180	061.9 / 062.4	31°-20'	344.0 / 344.5	346.24 / 346.67	+	0.43	−	8.60
4	1400	41-40 / 44-40	180	067.8 / 067.6	43°-25'	351.0 / 350.8	353.44 / 353.29	−	0.15	+	3.00
5	1800	52-20 / 55-40	200	097.7 / 097.0	41°-25'	027.0 / 026.3	20.91 / 20.34	−	0.57	+	10.26
6	1200	65-00 / 68-20	200	064.8 / 064.6	37°-20'	346.5 / 346.3	349.19 / 349.03	−	0.16	+	2.88
7	200	77-40 / 83-40	360	104.0 / 102.5	0°-15'	020.0 / 018.5	020.00 / 018.5	−	1.5	+	15.00
8	0	93-40 / 96-20	160	190.7 / 190.7	0°	316.0 / 316.0	316.0 / 316.0	−	⊖	+	⊖

CASE ORIENT AT 87:20

Fig. 8.73—Typical drift data on surveying data sheet.

The plotted slopes yield a calculated drift curve. The straight line connecting the initial CO and the ending CO is the average drift rate for the survey interval. In this case (Fig. 8.74), it is +4.4° over the 80 minutes between the initial and final CO's.

A straight line is drawn between the initial CO at 7 minutes : 20 seconds and the final CO at 87 minutes : 20 seconds. The difference, A, between the two lines is subtracted from the calculated drift curve at each drift correction point. Note that if the average survey line is above the calculated average survey drift rate, the delta increment would be added to the calculated survey points. Connecting each of the new drift-correction points forms the adjusted correction curve, which will be used to correct the observed gyroscopic data for drift.

As mentioned before, whenever the inclination is greater than 10°, a correction must be made to account for

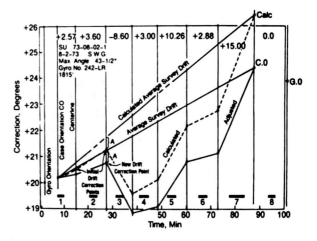

Fig. 8.74—Drift correction plot.

TABLE 8.8—DATA REDUCTION SHEET

Corrected Time	Meas Depth	Obs Incl Deg	Obs Incl Min	Obs Hole Az	Gyro Az	Incl Az	True Incl Deg	True Incl Min	T C Corr	Real or Corr Az	Drift Corr	True Hole Azimuth	True Quad Direction			Remarks
13-00	100	0	05	039						039.	19.9	58.9	N	58.9	E	
15-00	200	0	12	040.5						40.5	20.0	60.5	N	60.5	E	
17-20	300	0	05	356.5						356.5	20.0	16.5	N	16.5	E	
19-20	400	2	55	000.5						000.5	20.1	20.6	N	20.6	E	
20-20	500	9	15	344.						344.20	20.1	4.3	N	4.3	E	
25-20	600	14	45	345.						345.47	20.3	5.8	N	5.8	E	
27-20	700	20	20	345.5						346.37	20.4	6.8	N	6.8	E	
28-20	800	24	32	351.5						352.26	20.2	12.5	N	12.5	E	
30-20	900	27	50	350.0						351.14	19.9	11.04	N	11.04	E	
31-20	1000	31	20	344.						346.24	19.8	6.04	N	6.04	E	
36-40	1100	32	55	355.						355.80	18.9	14.7	N	14.7	E	
38-40	1200	37	17	349.5						351.61	18.7	10.3	N	10.3	E	
40-40	1300	40	25	342.						346.11	18.7	4.8	N	4.8	E	
41-40	1400	43	25	351.						353.44	18.8	12.2	N	12.2	E	
46-40	1500	40	25	007.						005.34	18.9	24.2	N	24.2	E	
48-40	1600	39	13	013.						010.14	19.0	29.1	N	29.1	E	
51-00	1700	38	02	027.5						22.29	19.3	41.6	N	41.6	E	
52-20	1800	41	25	027.						020.91	19.5	40.4	N	40.4	E	
59-40	1815	41	30	025.5						019.66	20.6	40.3	N	40.3	E	
61-20	1700	38	02	024.5						21.0	20.7	41.7	N	41.7	E	
65-00	1200	37	20	346.5						349.19	20.8	10.0	N	10.0	E	
73-00	700	20	25	343.5						345.49	21.0	6.5	N	6.5	E	
77-40	200	0	15	20						38.1	22.1	60.2	N	60.2	E	

OPERATOR _____

JOB NO _____

Read By _____

DATA REDUCTION SHEET
INRUN/OUTRUN

DATE _____

SHEET _____ OF _____

Checked By _____

Courtesy of NL Sperry-Sun; L.R. Surwel

the tilt of the surveying case, while the horizontal gimbal remains level. This is called an intercardinal correction, A_{corr}, and can be calculated using Eq. 8.60:

$$A_{corr} = \text{arc tan } (\tan \ dr_o)$$

$$\times \cos(in_o). \quad \dots\dots\dots\dots\dots\dots \quad (8.60)$$

Table 8.8 shows that the observed hole azimuth is corrected to the gyroscope azimuth by the intercardinal correction for each survey. If the true center correction is significant enough, it would be applied to each survey, yielding a real or corrected azimuth. The drift correction, either added to or subtracted from the real azimuth, gives the final true hole azimuth. In the example presented by Table 8.8, the true center correction was not applied.

Another type of gyroscope that eliminates the drift correction and the initial orientation of the gyroscope assembly is the rate or north-seeking gyroscope. The major difference between the level or horizontal gyroscope and the rate gyroscope is that the rate gyroscope measures the rate of earth rotation and detects the horizontal component of the spin vector of the earth in relation to the hole axis (see Fig. 8.75).

A motor that drives the spin rotor at 12,000 rpm is coupled to the rotor by a universal joint (see Fig. 8.76). Once

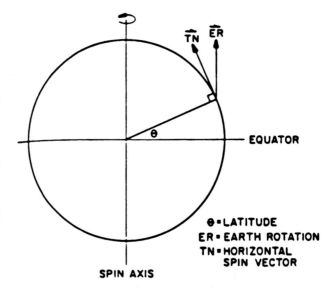

Ө = LATITUDE
ER = EARTH ROTATION
TN = HORIZONTAL SPIN VECTOR

Fig. 8.75—Earth forces sensed by gyro/accelerometer (after Uttrecht and deWardt[10]).

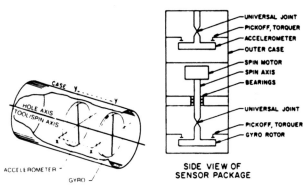

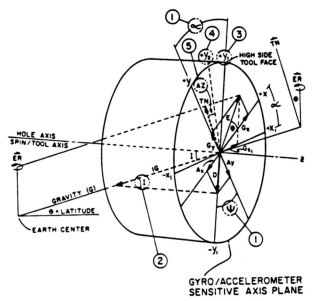

Fig. 8.76—A dual-axis, dynamically tuned rate gyro with dual-axis accelerometer (after Uttrecht and deWardt[10]).

Fig. 8.77A—Vector analysis of combined gyro and accelerometer readings (after Uttrecht and deWardt[10]).

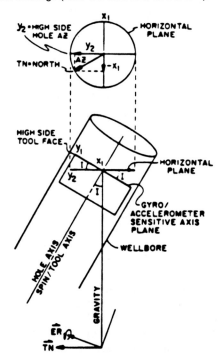

Fig. 8.77B—Mathematically rotating the sensing plane to horizontal creates a compass (after Uttrecht and deWardt[10]).

12,000 rpm is reached, the rotor is free to precess at right angles to the tool axis (x-y planes). The tool always tries to maintain a null position between the rotor and the case with torques and pickoffs. The measured torque necessary to maintain the null position is proportional to the rate of angular motion of the case at right angles to the x-y spin plane. When the tool is still, the earth's spin causes it to detect the movement of the case, to return the rotor automatically to a null position, and to measure the rate of the earth's rotation.

Figs. 8.77A and 8.77B show the relationship between the earth and the tool axis, which is aligned with the hole axis. Uttecht[10] showed that the orientation of the tool's axis to the gravity vector is given by Eq. 8.61.

$$\Psi = \arc\tan\frac{A_x}{A_y}, \quad \dots\dots\dots\dots\dots\dots\dots (8.61)$$

where Ψ is the angle between the y axis and D, which is the direction of the low side of the hole. A_x and A_y are sensed by the accelerometer. The wellbore inclination can be derived since gravity, g, is a known vector:

$$\arc\sin[(A_x{}^2 + A_y{}^2)^{1/2}/g] = \arc\sin(D/g). \quad \dots\dots (8.62)$$

Knowing the gravity vector in the x-y plane, the x-y plane can be rotated by Ψ around the z axis until the y axis is oriented to the high side of the hole and the x axis is the horizontal plane.

The gyroscope detects the earth's spin vector, ER, which is Vector E:

$$\overline{g_y} + \overline{g_x} = \vec{E}. \quad \dots\dots\dots\dots\dots\dots\dots\dots (8.63)$$

The angle ϕ is defined as the angle between E and the x-y plane and by the rotation of Ψ from x and y to x_1 and y_1; E stays constant, and $\phi + \Psi$ is the angle between x_1 and E. g_{x_1} in the horizontal plane can now be determined.

Because the horizontal component of the earth rate, T_e, is known for a given latitude, θ:

$$T_e = ER \cos(\theta). \quad \dots\dots\dots\dots\dots\dots\dots\dots (8.64)$$

The earth-rate component in the horizontal axis is

$$g_{y_2} = (T_e{}^2 - g_{x_1}{}^2)^{1/2}, \quad \dots\dots\dots\dots\dots\dots (8.65)$$

and the hole azimuth A_z is as follows:

$$A_z = \arc\tan\left(\frac{x_1}{y_2}\right). \quad \dots\dots\dots\dots\dots (8.66)$$

Fig. 8.78 shows the rate gyroscope system, consisting of a power source, printer, CPU, and sonde, which is run on electric wireline. Because drift stops are not required, surveying time is shorter than that required for running a normal gyroscope. As with all gyroscopes, the nearer the poles and the higher the inclination, the less accurate the tool. For a typical rate gyroscope, this starts becoming significant at inclinations exceeding 70° and at latitudes of 70°.

Gyroscopic Directional Survey System

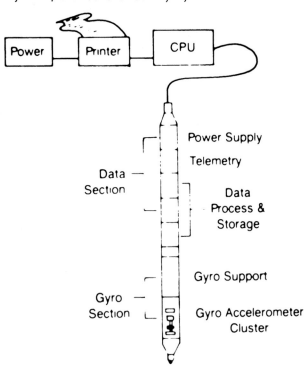

Fig. 8.78—The downhole survey probe and surface equipment (after Uttrecht and deWardt[10]).

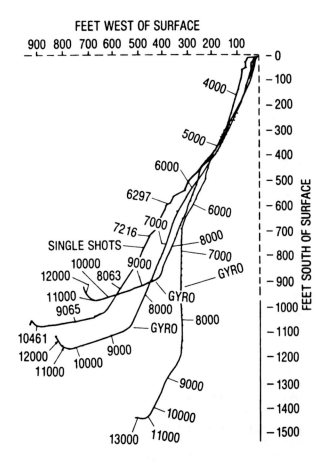

Fig. 8.79—Plan view of wellbore based on four different surveys.

Another type of surface-recording gyroscope uses a gyroscope-stabilized single-axis platform with an orthogonal set of accelerometers to measure direction and inclination. This gyroscope requires short survey station stops of 2 to 4 seconds and conventional 10-minute drift correction checks. A microprocessor performs all the calculations and displays all the results almost instantaneously. The tool is run on electric wireline and is powered at the surface.

8.5.7 Surveying Accuracy and the Position of the Borehole

Fig. 8.79 shows the plan view of a borehole that has been surveyed four times: once with magnetic single-shots while the well was being drilled and later with three separate gyroscope surveys. The position of the wellbore becomes particularly relevant if a well blows out and a relief well must be drilled to intercept the wellbore of the blown-out well at a desired point. Knowing the exact position of the wellbore is also extremely important in other applications: (1) in drilling near a cluster of wellbores from a multiwell directional pad or platform where there is a risk of intersecting other wellbores; (2) in intersecting the target exactly for closely spaced infill drilling where sweep efficiencies are critical; and (3) in drilling a sidetrack to ensure that the new wellbore does not reenter the old wellbore.

Both magnetic and gyroscopic survey instruments have inherent inaccuracies. Magnetic compasses are subject to magnetic interference by the surrounding drillstring and are affected by the position of the survey on the earth. The conventional gyroscope has a drift error because of the spin of the earth and the position of the survey on the earth. Along with the major measuring problems, there may be errors caused by magnetic storms (which can change the north reading), declination variation, hot spots on the nonmagnetic collars, or inaccurate readings. Also, inaccuracy may be caused by compass or inclinometer alignment or by excessive bearing friction that leads to compass and inclinometer drag.

Gyroscope errors can occur from improper orientation at the surface, misalignment of the gimbal assembly, excessive bearing drag, high inclinations (gyroscopes cannot be run at inclinations exceeding 70°), or excessive drifts resulting from lengthy survey times.

All the inaccuracies can be shown to be systematic[10] and can be related to five major categories: compass reference, compass instrument, inclination, misalignment, and depth errors.

Fig. 8.80 is an example of a gyroscopic well survey that shows erratic and excessive drifts. Fig. 8.81 is an example of poor magnetic survey data taken with two different magnetic multishot tools. Which one is correct? Fig. 8.82 is an example of surveys taken off depth.

Fig. 8.83 shows data on a well surveyed first with single-shots, then with an MWD tool, and later with a magnetic multishot at the intermediate casing depth. The multishot surveys indicated that the trajectory would miss the target, while the MWD and single-shot surveys indicated the trajectory was on course. When the BHA and the location of the compass unit of the multishot was checked, the compass was found to be directly opposite a steel clamp-on stabilizer that caused a consistent systematic error in all the directional readings.

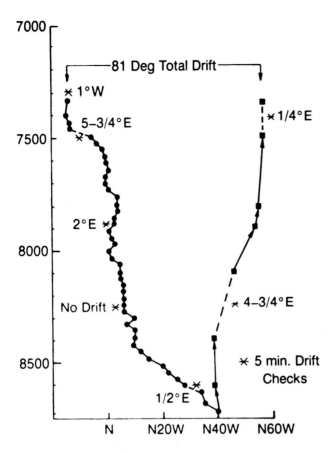

Fig. 8.80—Example of gyro survey with excessive and erratic drift.

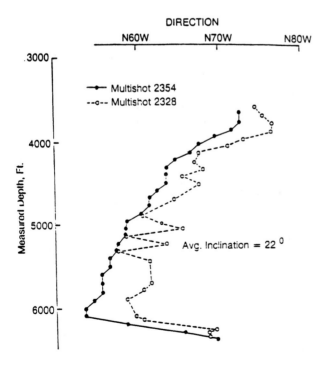

Fig. 8.81—Example of poor-quality survey.

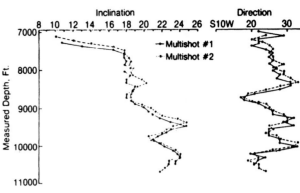

Fig. 8.82—Example of off-depth survey for a Gulf Coast well.

An example of magnetic interference on the compass can be related to an equation derived by Wolfe and deWardt.[11]

$$\Delta C = \sin \alpha \sin \epsilon \left(\frac{\Delta M_E}{M_F} \right), \quad \ldots \ldots \ldots \ldots \ldots (8.67)$$

where ΔC is the actual compass deflection, α is the inclination, ϵ is the azimuth, and M_F is the magnetic field, which varies between 40 μT at the equator and 0 μT at the north and south poles. ΔM_E is the strength of the magnetic error field. Eq. 8.67 implies that the greater the inclination and the more east or west the direction, the higher the change in ΔC. Anything that increases the magnetic error field, such as hot spots or improper positioning of the compass in the nonmagnetic collars, increases ΔC. Moving south or north from the equator also increases ΔC. Maximum deflections occur at greater inclinations east or west in the northern and southern latitudes, such as in the North Sea and Alaska and toward the South Pole.

Example 8.15. Calculate the azimuth error for a north slope well in Alaska where the wellbore is at 60° inclination and N70E. Assume a magnetic strength of 10.2 μT for the north component and 2.0 μT because of the collars and drillstring.

Solution. The azimuth error is calculated with Eq. 8.67:

$$\Delta C = \sin(60°)\sin(70°) \left(\frac{2}{10.2} \right) = 0.1596.$$

Azimuth error = 0.1596 rad.

57.295°/rad = 9.14°.

The maximum survey error can be estimated with Fig. 8.84, which is based on typical measuring for good and poor gyroscope surveys and good and poor magnetic surveys (east/west). Table 8.9 shows the typical values used to construct Fig. 8.84.

TABLE 8.9—TYPICAL VALUES FOR MEASURING ERRORS

	Relative Depth ϵ 10^{-3}	Misalignment ΔI_m (degrees)	True Inclination ΔI_{to} (degrees)	Reference Error ΔC_1 (degrees)	Drillstring Magnification ΔC_2 (degrees)	Gyro Compass ΔC_3 (degrees)
Good Gyro	0.5	0.03	0.2	0.1	—	0.5
Poor Gyro	2.0	0.2	0.5	1.0	—	2.5
Good Magnification	1.0	0.1	0.5	1.5	0.25	—
Poor Magnification	2.0	0.3	1.0	1.5	5.0 ± 5.0	—
Weighting	1	1	sin I	sin I	sin I sin A	$(\cos I)^{-1}$

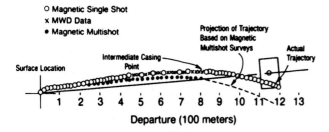

○ Magnetic Single Shot
× MWD Data
● Magnetic Multishot

Fig. 8.83—Example of surveying a hole with a magnetic multishot before running intermediate casing where the multishot does not agree with single-shot and MWD surveys.

Example 8.16. Determine the maximum expected survey error for the well described below. Assume that a good and a poor magnetic survey are available.

Well's inclination:
  0 to 5,000 ft   vertical
  5,000 to 6,500 ft   0 to 30° (constant build)
  6,500 to 10,000 ft   30° average

Solution. According to Fig. 8.84, the first interval from 0 to 5,000 ft shows 0.0018 ft/ft MD; 5,000 to 6,500 ft is 0.0075 ft/ft MD, and 0.014 ft/ft MD for 30° inclination. The following is the maximum survey error:

5,000 ft×0.0018 ft/ft MD= 9 ft
1,500 ft×0.0075 ft/ft MD=11 ft
3,500 ft×0.0140 ft/ft MD=49 ft
                                  ⎯⎯⎯
                                  69 ft

With this method, the error for the poor survey is 193 ft.

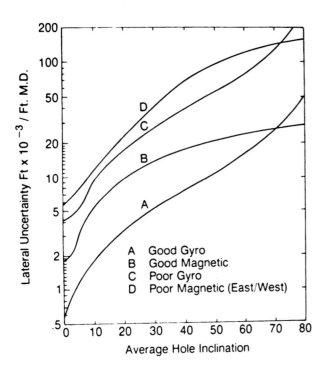

Fig. 8.84—Chart to determine lateral uncertainty for typical good and poor magnetic and gyro surveys (after Wolff and deWardt[11]).

As stated previously, surveying errors appear to be systematic, not random. Data from each survey station form an ellipsoid reflecting the amount of depth, inclination, and directional error. Fig. 8.85 shows the ellipsoids at three different depths for the MWD survey on the basis of the ellipsoids. It was determined that the real wellbore was south of the wellbore derived from the conventionally calculated surveys. Later, it was found that the actual well did come within a few feet of the blowout wellbore at this depth. In this case, the single-shot surveys predicted the most accurate trajectory.

The mathematics needed for the calculation of the three axes of the ellipsoid are presented by Wolfe and deWardt.[11]

To minimize surveying errors, all tools should be test-stand calibrated over the range of inclinations and directions expected. All multishots should be run back to the highest point possible for comparison, and all inclination and directional data should be plotted over the overlapped sections to assure repeatability.

8.6 Deflection Tools

Sec. 8.4 explained how to make a controlled trajectory change. Whether a whipstock, mud motor, or jetting bit is used, the principles for determining the total angle change, dogleg severity, new inclination angle, direction, and tool-face setting are all the same. Once a trajectory is reached, there are various ways of implementing it. One can use a positive displacement motor (PDM) with a bent

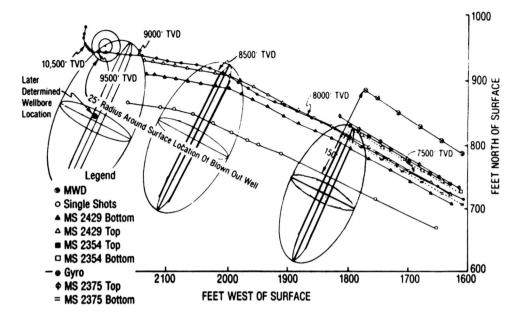

Fig. 8.85—Plan view of various surveys run in a relief well designed to intersect a blowout.

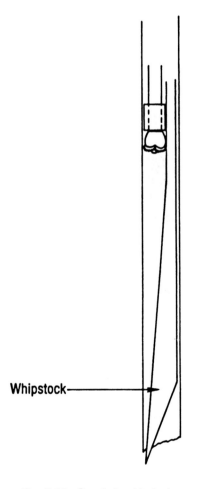

Fig. 8.86—Openhole whipstock.

sub or bent housing and regular tricone bits or diamond or polycrystalline diamond bits. Instead of a PDM, a mud-powered turbine can be used with a bent sub or an eccentric stabilizer. A whipstock or a jetting bit can also be used.

This section describes various tools used in changing trajectories and the principal factors affecting their use.

8.6.1 Openhole Whipstocks

The whipstock was the first widely used deflection tool for changing the wellbore trajectory. Fig. 8.86 shows a typical openhole whipstock, and Fig. 8.87 is a diagram of the principle of operation. A whipstock is selected according to the wedge needed to effect the desired deflection. A bit that is small enough to fit in the hole with the whipstock is then chosen; at the start of the running mode, the bit is locked to the top of the whipstock. When the whipstock is positioned at the kick-off depth, whether it is the total depth of the wellbore or the top of a cement plug, it is carefully lowered to bottom, and the center line of the toe is oriented in the desired direction by a conventional nonmagnetic collar with a mule-shoe sub and by a single-shot survey. With the whipstock assembly oriented, enough weight is applied to the toe of the wedge so that it will not move when rotation begins.

Additional weight is applied to shear the pin that holds the drill collars to the wedge; then rotation can begin. Forcing the bit to cut sideways as well as forward, the wedge deflects the bit in an arc set by the curvature of the whipstock. When the bit reaches the end of the wedge, it ordinarily continues in the arc set by the wedge. Drilling continues until the top of the whipstock assembly reaches the stop (Fig. 8.87). Fig. 8.88 (a through d) depicts the operation.

The entire whipstock assembly is pulled, and a pilot bit and hole opener are run to the kick-off point. The wellbore is enlarged to the original hole size, and the assembly is pulled again. The drilling BHA finally is run, and normal drilling is resumed (Fig. 8.88 e and f).

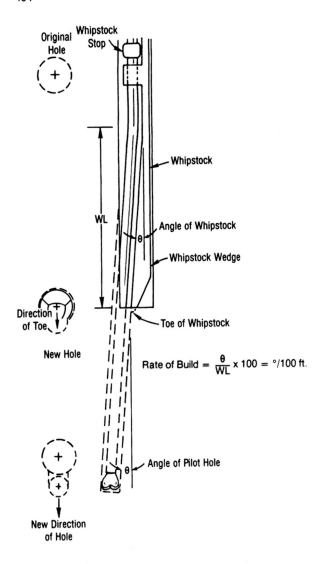

Rate of Build = $\dfrac{\theta}{WL} \times 100 = °/100$ ft.

Fig. 8.87—Diagram of retrievable whipstock operation.

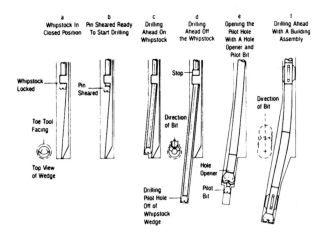

Fig. 8.88—Drilling with a retrievable whipstock.

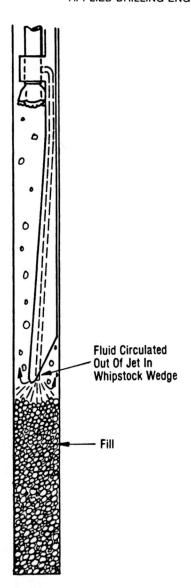

Fig. 8.89—A jetting whipstock.

The foregoing description is of an ideal kick-off with an openhole whipstock. Many factors, however, can cause the whipstock operation to deviate from the norm. If the bottom of the hole is covered with fill and the whipstock is set in the fill, a number of complications can occur. The unstable bottom can cause the toe of the whipstock to rotate when drilling starts. The fill tends to wash away, causing the bit to slide down the side of the wellbore and the entire whipstock assembly to rotate. Even if a kick-off is achieved with fill in the hole, entering the deviated borehole with the drilling assembly is usually impossible because the fill washes away and lowers the bottom of the hole. Fig. 8.89 is a diagram of a jetting whipstock that can be used to jet out the fill so that the whipstock can proceed to the bottom of the hole and, then, can seat properly.

Even when the toe is firmly seated on the bottom of the hole, the driller must be careful not to unseat the toe, to change the orientation, or to rotate the bit off the wedge. The critical stage occurs when the bit leaves the end of the whipstock wedge. If the rock is too soft and the circulation too high, the bit can lose the curvature the whipstock has started and continue drilling nearly straight.

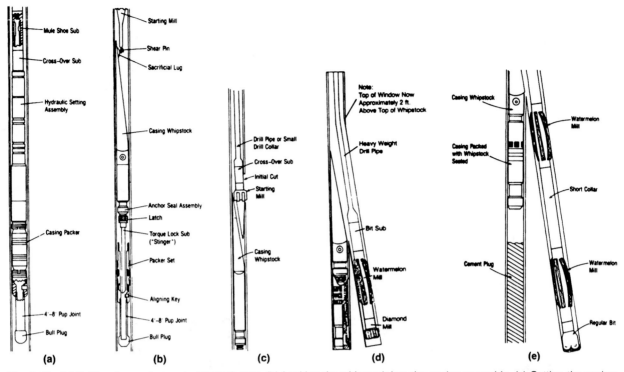

Fig. 8.90—(a) Setting the packer and whipstock seat. (b) Locking the whipstock into the packer assembly. (c) Cutting the casing with the starting mill. (d) Cutting a window in the casing with a side packing mill. (e) Drilling ahead with a tricone bit through a window in the casing.

Openhole whipstocks are almost never used because changing the trajectory is very complicated and because too much experience is required to run the tools properly. In certain circumstances, however, they are still useful, such as in very hard rocks and in wellbores where temperatures are too high for mud motors.

8.6.2 Casing Whipstocks

Another type of whipstock is the casing whipstock. Unlike the openhole whipstock, the casing whipstock is used routinely to sidetrack out of cased wellbores. Cagle *et al.*[12] described the most common technique for the use of a casing whipstock. A permanent packer is run to the desired kick-off point either on wireline (for a blind kick-off) or on drillpipe carrying a mule-shoe sub for orientation. Once the packer is set, a retrievable starting mill is run on the whipstock. The whipstock assembly is locked into the packer (see Figs. 8.90a and b). Weight is applied, shearing the starting mill off the whipstock (Fig. 8.90c). The starting mill is used to start a cut in the casing and then is pulled out of the hole. A sidetracking mill or diamond bit replaces the starting mill (Fig. 8.90d). The whipstock forces the sidetracking bit through the side of the casing, making a window about 8 to 12 ft long. Once outside the casing, the same bit drills a pilot hole. Then that bit is pulled and replaced with a taper mill and a BHA of string and watermelon mills to make the casing window large enough to accommodate a conventional BHA. After the window is dressed, the assembly is pulled and the taper mill is replaced with a conventional tricone or drag bit (Fig. 8.90e). The conventional bit and watermelon mill assembly are used to drill ahead. When the hole is tripped out, the watermelon mills can be used to ream

the window to ensure that it is large enough to accommodate similar BHA's with conventional stabilizers.

The most common method of sidetracking out of casing, especially when considerable drilling is to follow, is to mill a length of casing with a section mill and then to divert the trajectory with a mud motor and bent sub or bent housing.

When considering the use of a section mill, one should first check for cement behind the portion of casing to be milled. If there is no cement or the cement bonding is poor, milling problems are inevitable. The rotating mill causes the unsupported casing to vibrate, which slows progress, and the mill may torque up and jam. Either the section must be cemented remedially, or another section with better cement bonding must be selected.

Assuming that the casing is well cemented, the section chosen should start immediately below a casing coupling, thereby minimizing the number of couplings that must be milled in a normal 30- to 60-ft milling operation.

Fig. 8.91 shows a section mill with the arms retracted. Fig. 8.92 a through c shows the operation, in which a section mill is run into the hole, the arms are extended, and the casing is milled. At the bottom of the milled section, a cement plug is set to isolate the new open hole from the casing below it. After the appropriate section is milled, a mud motor with a deflecting device is used to leave the old wellbore and start a new one (Fig. 8.92 d and e).

Generally, the more drilling in the new wellbore, the longer the section milled. The disadvantage of the casing/whipstock method is that the casing window is too short. Numerous trips and long hours of rotation can wear or damage the casing, sometimes making it difficult to trip out the BHA through the casing window.

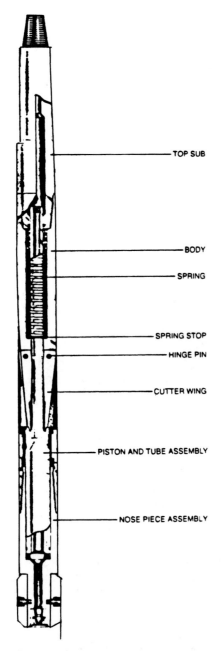

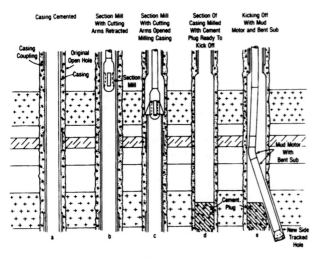

Fig. 8.91—Typical section mill with arms retracted (courtesy of A-Z Intl. Tool Co.).

Fig. 8.92—Using a section mill to prepare for a kick-off.

8.6.3 Jetting Bits

Another effective means of changing the trajectory of a borehole is jetting. A bit with one large nozzle (see Fig. 8.93) is oriented to the desired tool-face setting. The mule-shoe sub is oriented in the same line as the jetting nozzle.

In jetting, the hydraulic energy of the drilling fluid erodes a pocket out of the bottom of the borehole. The drilling assembly is advanced without rotation into the jetted pocket for a distance of 3 to 6 ft. Rotation is started and conventional drilling proceeds until a depth of 20 to 25 ft is reached; at that point, a survey is taken to evaluate the last jetting interval. If more trajectory change is required, the jetting assembly is oriented again, and the jetting sequence is repeated. This procedure is continued until the desired trajectory change is achieved. Fig. 8.94 shows a typical jetting operation.

Geology is the most important influence on where jetting can be used; next in importance is the amount of hydraulic energy available for jetting. Sandstones and oolitic limestones that are weakly cemented are the best candidates for jetting. Unconsolidated sandstones and some other types of very soft rocks can be jetted with some degree of success. Very soft rocks erode too much, making it difficult to jet in the desired direction; when rotation begins, the stabilizers cut away the curved, jetted section and return to a nearly vertical well path. Sometimes this problem can be overcome by the use of smaller drill collars in the jetting assembly than those normally used in a hole of the same size. Another solution is to reduce the circulation rate to a level at which a regular pocket can be eroded.

Even though shales may be soft, they are not good candidates for jetting. Most medium-strength rock is too well cemented to jet with conventional drilling rig pumps, so it limits the depth to which jetting can be applied. Higher pressures and more hydraulic energy can extend the depth to which jetting is practical.

The principal advantage of jetting is that the same BHA can be used to change the trajectory and to drill ahead. If the geology is conducive, jetting is more economical than running a mud motor. An important secondary advantage of jetting is that slight trajectory alterations can be made after the original trajectory has been established.

Typically, jetting operations take place in wells that have alternating sandstones and shales. A two-cone jetting bit and a single-stabilizer building assembly are used for the operation. The kick-off depth is selected, and the large nozzle is oriented in the general direction desired. To set and to maintain a specific direction at very low inclinations (less than 1°) are virtually impossible. The first jetting operation is primarily to build the inclination to 1 to 2°. A drilling break usually indicates a sandstone; in sandstone at shallow depths, jetting an interval of 3 to 6 ft in 3 to 10 minutes is possible. The harder the rock is or the more shale there is in the rock, the slower the jetting will be. If the jetting procedure is not one that is familiar in that particular drilling area, normal drilling is resumed until another drilling break signifies a possible formation for jetting. After jetting begins a curve, normal drilling is continued until a survey can be run to evaluate the success of the previous jetting. When the inclination exceeds 1 to 2°, another jetting interval in which the jetting nozzle can be oriented to achieve the desired

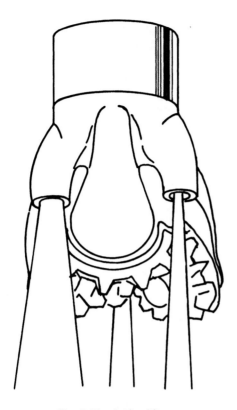

Fig. 8.93—Jetting bit.

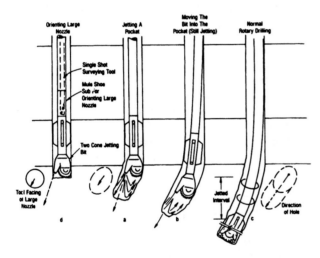

Fig. 8.94—Jetting a trajectory change.

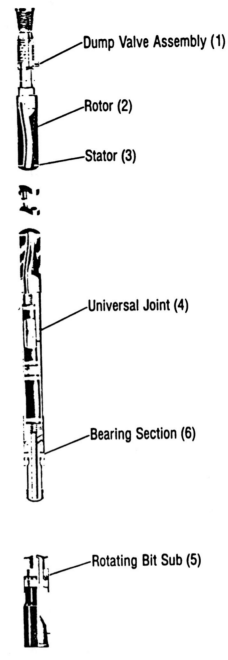

Dump Valve Assembly (1)

Rotor (2)

Stator (3)

Universal Joint (4)

Bearing Section (6)

Rotating Bit Sub (5)

Fig. 8.95—A typical positive-displacement mud motor (PDM) (courtesy Dyna-Drill).

direction and inclination is found. Sometimes this can require as many as four attempts. Hence, a number of jetting intervals should appear over a few hundred feet.

A major drawback to jetting is that the formation must be favorable at a shallow depth or in the desired kick-off interval; otherwise, the technique is no better than the use of a mud motor with a deflecting device. Another problem is that if jetting is continued too long without conventional drilling being resumed, large doglegs can be created. However, if only short intervals are jetted and surveys cover at least 30 ft, the dogleg problem is controllable. If excessive curvature is detected within 30 ft, the borehole can be reamed with the drilling assembly to try to remove the curvature.

8.6.4 Positive Displacement Mud Motors

The most important advancement in trajectory control is the use of the PDM and the turbine with a bent sub, bent housing, or eccentric stabilizer for making a controlled trajectory change.

Without a bent sub or bent housing, both types of motors can be used for normal directional and straight-hole drilling.

The PDM was developed in 1966, and 2 years later the PDM began to be used in the U.S., primarily as a directional tool. Since then the PDM has been used worldwide as both a directional and a straight-hole drilling tool.

The PDM is based on the Moineau principle. Fig. 8.95 is a cross section of a typical half-lobe profile PDM. The

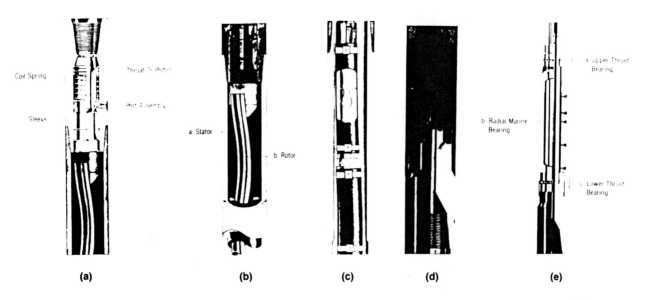

Fig. 8.96—(a) Dump-valve assembly. (b) Multistage motor. (c) Universal joint. (d) Rotating-bit sub. (e) Bearing and drive-shaft assembly (courtesy Dyna-Drill).

dump valve (1) is used to bypass the fluid while it flows in and out of the hole (Fig. 8.96a). When circulation begins, fluid forces the piston down, thereby closing the ports and directing the fluid through the stator. Because of the eccentricity of the rotor (2) in the stator (3) (Fig. 8.96b), the circulated fluid imparts a torque to the rotor, causing the rotor to turn and to pass the fluid from chamber to chamber. Rotation from the stator is transmitted to the bit by a universal joint (Fig. 8.96c) (4) to a rotating sub (5) to which the bit is connected (Fig. 8.96d). Thrust and radial bearings (6) (Fig. 8.96e) are used to withstand axial and normal loads on the bit and rotating sub. An upper-thrust bearing guards against hydraulic loads when the bit is off bottom and when there is circulation (Fig. 8.96E).

The operating life of a PDM is limited primarily by wear of the stator, thrust bearings, and drive components—such as the universal joint load coupling. It is important to maintain operating histories of key components, to conduct thorough inspections after each run, and to replace parts regularly before they fail downhole. Operators most often rent PDM's and thus are dependent on the service tool companies for strict quality-assurance procedures.

The stator is a vulnerable portion of the motor because it is subjected to continuous rubbing and deformation by the rotor. The stator rubber must have the resiliency to provide an effective hydraulic seal around the rotor while permitting the rotor to turn freely. It is essential that the stator consist of a correctly formulated elastomer compound that is bonded securely to the motor housing. Stators are occasionally subjected to chemical attack by aromatic hydrocarbons in the diesel phase of oil mud systems. Diesel fuels are typically "winterized" by the addition of aromatic compounds to lower the temperature at which the fuel gels. The aniline point of a diesel fuel—the temperature at which aniline becomes soluble in the diesel—is an inversely related indicator of aromatic content. Fuels with aniline points less than 155°F are potentially detrimental to PDM stators.

Excessive pressure drops across each motor stage accelerate stator wear. This problem is reduced in multilobed motors because the rotational speed and pressure drop per stage is less. However, the higher operating torques of multilobed motors tend to make the universal joint and related drive train components the weak link in the system.

Motor bearings can fail because of fluid erosion of mud-lubricated (nonsealed) systems, excessive loading of either the off-bottom or on-bottom thrust bearings, and normal attrition. When trajectory changes are made, the motor run is usually short enough that bearing life is not exceeded. Bearing life can be the limiting factor during longer trajectory changes or straight-hole drilling. Early PDM designs permitted only low pressure drops across the bearings; thus bit pressure drop was limited to similarly low values (about 250 psi). Higher pressure drops caused erosion of the restrictor used to control mud flow through the bearings. Newer nonsealed designs permit up to 1,000 psi pressure drop while sealed bearings operate at pressure differentials up to 1,500 psi. Unusual operating practices—such as considerable washing and reaming or running at abnormally low WOB—can hasten wear of off-bottom bearings. Abnormally high WOB accelerates on-bottom bearing wear. Normal attrition is the usual wear mode. Advances in bearing materials technology are making PDM bearing wear a less significant factor than in the past.*

The most common PDM is called a half-lobe motor, which means that the rotor has one lobe or tooth ($n_r = 1$) and the stator has two lobes or teeth ($n_{st} = 2$). A key aspect of PDM design is that the stator always has one more lobe than the rotor, thus forming a series of progressive fluid cavities as the rotor turns:

$$n_{st} = n_r + 1. \dots\dots\dots\dots\dots\dots\dots\dots (8.68)$$

The rotor has a diameter d_r and an eccentricity e_r, as shown in Fig. 8.97. Fig. 8.98 shows the pitch and lead of a half-lobe PDM. The rotor pitch, P_r, is equivalent

---

*The following information has been provided by Baker Service Tools Co.

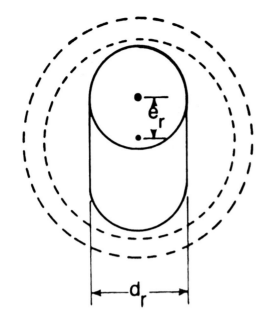

Fig. 8.97—Cross section of a half-lobe PDM showing the diameter and eccentricity of the rotor.

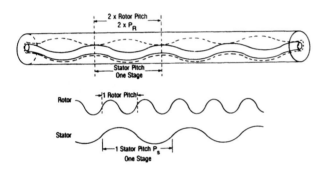

Fig. 8.98—A half-lobe PDM showing the stator and rotor pitches.

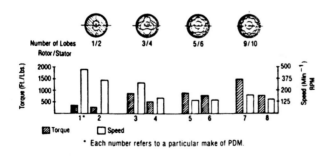

Fig. 8.99—Characteristics of various multilobe PDM profiles (after Jürgens[13]).

to the wavelength of the rotor. The rotor lead, L_r, is the axial distance that a tooth advances during one full rotor revolution. For any PDM the rotor pitch and stator pitch are equal while the rotor and stator leads are proportional to the number of teeth:

$$P_r = P_{st}, \dots \dots \dots \dots \dots \dots \dots (8.69a)$$

$$L_r = n_r P_r, \dots \dots \dots \dots \dots \dots (8.69b)$$

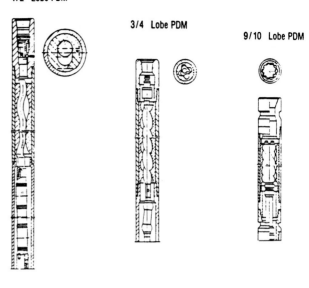

Fig. 8.100—Multilobe PDM designs (after Eickelberg et al.[8]).

and

$$L_s = n_{st} P_{st}. \dots \dots \dots \dots \dots \dots (8.69c)$$

For example, in a half-lobe PDM, the pitch and lead of the rotor are the same while the stator lead is twice the pitch.

In addition to half-lobe PDM's there are multilobe designs with ¾, ⅚, and 9/10 profiles, as shown in Figs. 8.99 and 8.100. Motor torque increases as the number of lobes increases, with a proportionate decrease in bit speed. The bit speed of some multilobe PDM's is low enough to permit lengthy straight-hole runs with journal-bearing roller-cone bits.

Example 8.17. What is the stator pitch (P_{st}), rotor lead (L_r), and stator lead (L_{st}) of a ¾-lobe PDM with a 7-in. rotor pitch?

Solution. The stator pitch is equal to the rotor pitch.

$$P_{st} = P_r = 7 \text{ in.}$$

The rotor lead is equal to the pitch times the number of rotor teeth.

$$L_r = 7 \text{ in.} \times 3 = 21 \text{ in.}$$

The stator lead is equal to the pitch times the number of stator teeth.

$$L_{st} = 7 \text{ in.} \times 4 = 28 \text{ in.}$$

The starting point for PDM-design calculations is to determine the specific displacement, s, per revolution of the rotor. This is equal to the cross-sectional area of the fluid times the distance the fluid advances.

$$s = n_r \times n_{st} \times P_r \times A. \dots \dots \dots \dots (8.70)$$

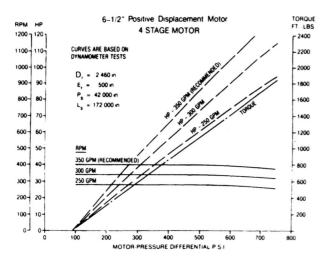

Fig. 8.101A—Data for four-stage 6½-in. positive-displacement motor (courtesy of Dyna-Drill).

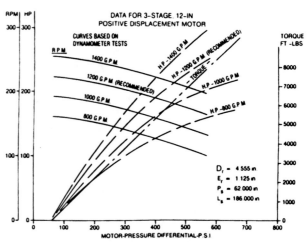

Fig. 8.101B—Data for three-stage 12-in. positive-displacement motor (courtesy of Dyna-Drill).

The fluid cross-sectional area is approximated by

$$A = \frac{\pi d_o^2}{4} \frac{2n_{st}-1}{(n_{st}+1)^2}, \quad \dots \dots \dots \dots \dots (8.71a)$$

where d_e = stator gear OD.

For a half-lobe PDM,

$$A = 2e_r d_r. \quad \dots \dots \dots \dots \dots \dots \dots \dots (8.71b)$$

Bit speed is simply the flow rate divided by the specific displacement.

$$N_b = \frac{231q}{s}, \quad \dots \dots \dots \dots \dots \dots \dots \dots (8.72a)$$

where q is in gallons per minute and s is in cubic inches per revolution.

The bit speed of a half-lobe PDM is equal to

$$\frac{57.754q}{e_r d_r P_r}. \quad \dots \dots \dots \dots \dots \dots \dots \dots (8.72b)$$

Motor torque is obtained by relating mechanical horsepower output to hydraulic horsepower input and substituting Eqs. 8.70 and 8.72a to yield Eq. 8.73b:

$$\frac{M \times \text{rpm}}{5252} = \frac{q \times \Delta p}{1714} \times \eta, \quad \dots \dots \dots \dots (8.73a)$$

$$M = \frac{3.064q\Delta pE}{\text{rpm}} = 0.0133 n_r n_{st} p_r A\Delta p \eta, \quad \dots (8.73b)$$

where M is measured in foot-pounds mass and Δp is measured in pounds per square inch. Motor efficiency rarely exceeds 80% ($\eta = 0.80$) for half-lobe PDM's and 70% for multilobe PDM's.

Example 8.18. Find the torque of a half-lobe 8-in.-OD PDM with 1.75-in. rotor eccentricity, 2.5-in. rotor diameter, and 24-in. rotor pitch; the total pressure drop is 465 psi. What is the bit speed at a flow rate of 600 gal/min?

Solution. Eq. 8.73c is developed by a combination of Eqs. 8.72b and 8.73b.

$$M_{\frac{1}{2}} = \frac{3.064q\Delta p\eta}{57.754q/e_r d_r p_r} = 0.0531 \, e_r d_r p_r \eta \Delta p; \quad (8.73c)$$

$$M_{\frac{1}{2}} = 0.0531 \, (1.75)(2.5)(24)(0.80)(465) = 2{,}074 \text{ ft-lbf.}$$

The bit speed is given by Eq. 8.72b.

$$N_{b\frac{1}{2}} = \frac{57.754q}{e_r d_r p_r} = \frac{57.754 \, (600)}{(1.75)(2.5)(24)} = 330 \text{ rpm.}$$

Notice from Eq. 8.73c that PDM torque is directly proportional to Δp and is independent of rotary speed. Also, torque decreases as eccentricity decreases; if eccentricity is zero, then motor torque is zero.

The number of motor stages is

$$n_s = \frac{L}{P_{st}} - (n_{st}-1), \quad \dots \dots \dots \dots (8.74)$$

where L is the length of motor section only.

Ideally, the bit speed of a PDM should be linear with pump rate, as implied by Eq. 8.72a. The stator is made of an elastomer, and as the pressure drop across the motor increases (i.e., as the torque increases), the elastomer deforms, allowing a small portion of the fluid to bypass, thus reducing the bit speed. This is a nonlinear effect, as shown clearly by Figs. 8.101a and 8.101b.

8.6.5 Turbines

Turbines were first used in the Soviet Union in 1934. The use of turbines increased from 65% in 1953 to 86.5% of all drilling in 1959. Currently, turbines are used in the Soviet Union for 50 to 60% of all drilling.

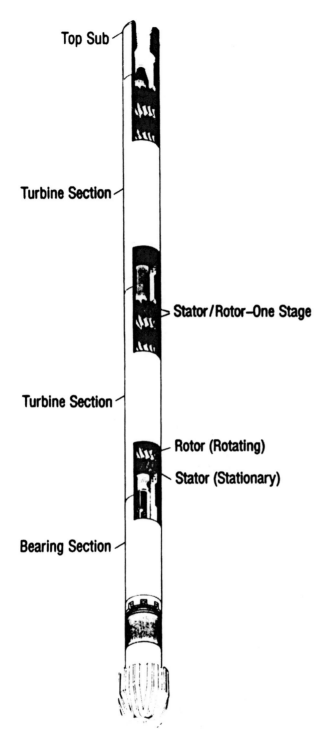

Fig. 8.102—Typical turbine design (courtesy of Baker Tool Co.).

Unlike the PDM, the turbine's power output is optimal over only a limited range of operating conditions. Fig. 8.103 shows a common curve of torque, speed, and power for a typical directional-drilling turbine where the pump rate is 500 gal/min. For the given pump rate, which is the input power, the output power varies, reaching an optimum at 820 rpm. Fig. 8.104 shows a similar curve for a straight-hole turbine. The torque and power curves exhibit a much narrower range of operation.

At lower speeds and higher torques, the efficiency of turbine drilling is reduced significantly. Two- and three-cone rock bits require high axial loads and lower speeds to drill and, therefore, are impractical for use with turbines. Diamond bits and the new polycrystalline diamond cutter (PDC) bits are better suited for the turbine. Diamond bits have not been used with turbines as much as roller-cone bits because it is difficult to match certain diamond bit designs with particular types of formations. Even engineers in the Soviet Union, who usually drill with turbines, use principally roller-cone bits. This approach has forced them to build mud motors with slower speeds.

The type of thrust bearings used also affects the performance of a turbine significantly. Rubber bearings were designed so that the axial thrust load balances the downward velocity of the fluid against the drive section of the turbine. If a properly designed bit is selected for a given formation and allowance is made for the appropriate axial WOB to balance the thrust and to optimize the output power, a successful turbine run can be achieved. This assumes that the operator can keep the turbine drilling at the correct torque and speed. Clearly, without some means of monitoring downhole performance (i.e., torque and speed), it is much harder to drill with a turbine than with a PDM. (Because torque is proportional to the differential operating pressure for a PDM, the standpipe pressure can be used to indicate operating torque; and, because the bit speed is proportional to the pump rate, the bit speed can be monitored by keeping track of the pump strokes.) Rubber bearings, which must be balanced to help prolong motor life, are not as durable as balanced roller bearings. The testing of newer bearing materials promises to increase bearing life and to extend the hours for a turbine run.

In 1959, the first successful drilling with turbines outside the Soviet Union took place in southern France. Turbines were introduced to the U.S. in 1960, but less than 1% of footage drilled in the U.S. has been with turbines. They are used more extensively in parts of Europe and the North Sea, although not as much as they are in the Soviet Union.

Fig. 8.102 shows a typical turbine design. The fluid enters the top sub and travels past the stators and rotors (one stator and one rotor compose one stage). The lower part of the turbine is the main thrust-bearing section.

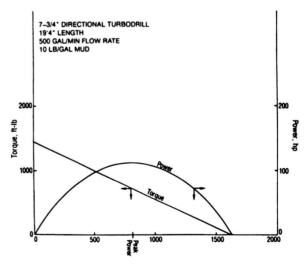

Fig. 8.103—Typical torque/power curves for a turbine.

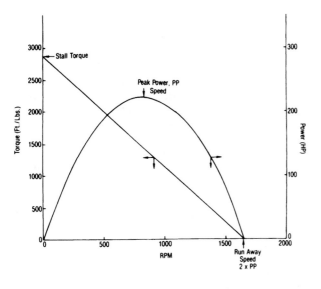

Fig. 8.104—Typical torque/power curves for a turbine at a pump rate of 500 gal/min, 10-lbm/gal mud.

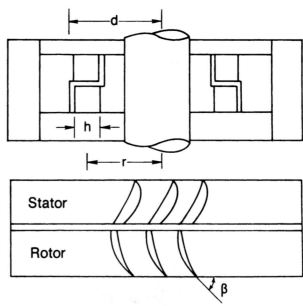

Fig. 8.105—Side view of a single-stage turbine. (after Jürgens[13]).

The stall torque of any turbine can be determined with Eq. 8.75[13]:

$$M = 1.38386 \times 10^{-5} \eta_H \eta_M \tan \beta n_s W_m q^2 / h, \quad \dots (8.75)$$

where

η_H = hydraulic efficiency,
η_M = mechanical efficiency,
β = exit blade angle, degress,
n_s = number of stages,
W_m = mud weight, lbm/gal,
q = circulation rate, gal/min, and
h = height of vane, in.

Fig. 8.105 shows the side view of a single turbine stage.
The runaway bit speed of a turbine can be calculated with Eq. 8.76:

$$N_b = 5.85 \eta_V \tan \beta q / r^2 h, \quad \dots (8.76)$$

where η_V is the volumetric efficiency and r is the median blade radius.

From Fig. 8.104 it is apparent that the function that relates turbine torque to turbine speed is of the form

$$M_i = M_{ts} - B \times N_b, \quad \dots (8.77)$$

where

M_i = instantaneous torque,
M_{ts} = turbine stall torque, and
B = constant.

And if torque is equal to a constant times bit speed and if Eqs. 8.75 and 8.76 are combined, where K_1 is the slope and a constant, then

$$M_i = M_{ts} - K_1 N_b. \quad \dots (8.78)$$

Expanding Eq. 8.78 yields Eq. 8.79:

$$M_i = M_{ts} - \left(\frac{r^2 h}{231 \eta_V \tan \beta} \right)$$

$$\times \left(\frac{4.46 \times 10^{-4} \eta_H \eta_M \tan \beta n_s W_m q}{h} \right) N_b,$$

$$\dots (8.79)$$

where

$$K_2 = \frac{r^2 h}{231 \eta_V \tan \beta},$$

$$K_3 = \frac{4.46 \times 10^{-4} \eta_H \eta_M \tan \beta n_s W_m q}{h},$$

and

$$K_2 K_3 = K_1.$$

It can be seen from Eq. 8.79 that, for a given number of stages, mud weight, and pump rate, the torque is linear with bit speed, as depicted in Figs. 8.103 and 8.104. Eq. 8.75 implies that the addition of stages can increase the torque. This assumes all other variables are held constant. Eq. 8.79 states that as the bit speed increases, the overall turbine torque decreases. However, increasing the flow rate and/or the mud weight will increase the overall torque stall point. The necessity of higher pump rates required to drive the turbine precluded the use of turbines on most land rigs and on some offshore drilling rigs until the

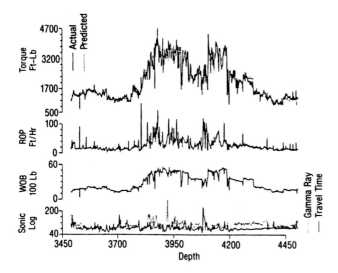

Fig. 8.106—Observed and predicted torque for an overthrust well with a 12¼-in. IADC Series 5-1-7 bit.

mid-1970's; finally, drilling contractors upgraded their pump sizes to accommodate the deeper drilling. Now, in both the U.S. and Europe, more drilling rigs can run turbines than ever before.

Once the torque/speed relationship of a turbine is known, the mechanical power output can be calculated by Eq. 8.71, and the hydraulic power can be determined from Eq. 8.70 if the pressure drop across the motor (Δp) is known. The ratio of the two is the overall efficiency.

Example Problem 8.19. Determine the mechanical efficiency of the turbine presented by Fig. 8.103. The following information is also known:

Number of stages = 100,
Radius of blade = 3.0 in.,
$\eta_V = 0.80$, and
$\eta_H = 0.45$.

Solution. From the torque curve presented in Fig. 8.103, the slope is 1.67 ft-lbf/rpm. Because the slope is K_1 and $K_1 = K_2 K_3$, the mechanical efficiency can be calculated by Eq. 8.76.

$$\eta_M = \frac{(K_1)(5.85)\eta_V}{1.38386 \times 10^{-5} n_s W_m q E_H r^2}$$

$$= \frac{(1.67)(5.85)(0.80)}{1.38386 \times 10^{-5}(100)(10)(500)(0.45)(3.0)^2},$$

$$= 0.279 = 27.9\% \text{ efficiency at 500 gal/min.}$$

8.6.6 Using the PDM for Directional and Straight-Hole Drilling

Drilling with the PDM is much easier than with a turbine because the surface standpipe pressure reflects the PDM

torque. As the motor torque increases, the standpipe pressure increases; as the motor torque decreases, the standpipe pressure also decreases. Therefore, the driller should use the standpipe pressure or a downhole torque indicator as a primary output indicator to advance the bit. The tool face of a wireline orienting tool or an MWD tool is influenced by the bit torque and, therefore, can indicate the bit torque. This will be covered later in this section.

The relationship between the motor torque and the torque used to drill a given formation with a tricone bit is developed by Warren.[14]

$$M = \left[C_3 + C_4 \sqrt{\frac{q_p}{N_b \cdot d_b}} \right] (d_b W_b) f_{(tooth\ wear)}, \quad (8.80)$$

where

C_3 = bit constant (dimensionless),
C_4 = bit constant (dimensionless),
q_p = penetration rate (ft/hr),
N_b = bit speed (rpm),
d_b = bit diameter (in.),
W_b = weight on bit (1,000 lbf), and
$f_{(tooth\ wear)}$ = function to relate tooth wear to footage drilled.

Warren[14] cites a derived torque relationship for a 12¼-in. bit drilling a section between 3,484 and 4,510 ft. Fig. 8.106 shows the data for such a bit run. From a regression analysis of the data the constants C_3, C_4, and $f_{(bit\ wear)}$ are obtained:

$$M = \left(3.79 + 19.17 \sqrt{\frac{R}{N_b \cdot d_b}} \right) (d_b W_b)$$

$$\cdot \left(\frac{1}{1 + 0.0021 L_t} \right), \quad \dots\dots\dots\dots\dots (8.81)$$

where L_t is the total footage drilled with this series 5-1-7 bit. Warren also cites other experimental work that verifies Eq. 8.80 (see Fig. 8.107a and b).

The torque the PDM experiences is (1) the torque required to overcome the off-bottom torque so that the rotor can rotate against the stator and against the friction of the bearings and (2) the torque required to drill a given formation with a specific bit, bit diameter, bit speed, and WOB. Rearranging Eq. 8.68, combining it with Eq. 8.80, and adding a term for off-bottom rotational torque yields

$$\Delta p = \frac{M}{K_3} = \frac{1}{K_3} \left[\left(C_3 + C_4 \sqrt{\frac{q_p}{N_b \cdot d_b}} \right) \right.$$

$$\left. \times (d_b W_b) f_{(toothwear)} \right] + p_m, \quad \dots\dots\dots (8.82)$$

where

$$K_3 = \frac{1}{0.636 e_r d_r P_s},$$

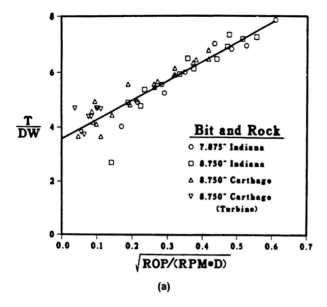

(a)

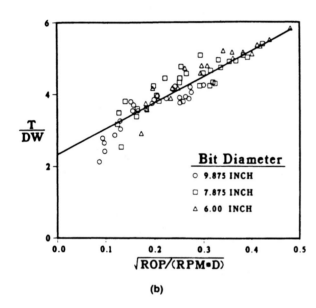

(b)

Fig. 8.107—Results of bit tests to verify torque relationship for three-cone rotary bits.

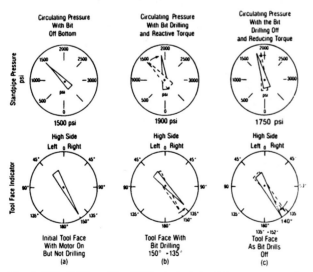

Fig. 8.108—Drilling with a PDM—typical standpipe pressure and tool-face indicator responses.

and p_m is the off-bottom pressure drop caused by the friction of the stator and rotor and bearings. The standpipe pressure can reflect the pressure variation of the motor torque if all the pressure losses are considered:

$$p_{sp} = p_{dl} + \Delta p_{(PDM)} + p_{bl} + p_{al}, \quad \ldots \ldots \ldots \ldots (8.83)$$

where p_{sp} is the pressure of the standpipe, p_{dl} is the pressure of the drillstring loss, $\Delta p_{(PDM)}$ is the pressure difference of a PDM, p_{bl} is the pressure of a bit loss, and p_{al} is the pressure of the annulus loss. If all the pressures are constant for a given circulation rate, p_{sp} will directly reflect $\Delta p_{(PDM)}$.

Fig. 8.108 shows what a driller sees while drilling with a PDM when there is no noticeable sideforce on the bit. The p_{sp} will go up nearly linearly as WOB is added. This is true for roller cones or for polycrystalline diamond bits with jets but not for natural diamond bits. (The reason for this nonlinear increase in pressure for natural diamond bits will be covered later when bit pump-off is discussed.) As the penetration rate increases and rock strength decreases, the torque and the p_{sp} go up. Conversely, the torque and p_{sp} decrease in harder formations. The p_{sp} increase for variations in penetration rate response is less than an equivalent increase in WOB, because Δp is proportional to the square root of penetration rate.

While the PDM driller feeds out line through the brake, WOB is increased; torque also increases and causes the p_{sp} to increase. If the maximum p_{sp} is not exceeded and the driller stops adding WOB, the torque and p_{sp} will decrease as the bit drills off. If WOB is increased, causing maximum p_{sp}, the bit speed and penetration rate will decrease. The PDM will stall if too much weight is applied. If the motor is left in the stall position too long, seals can break, allowing the circulating fluid to bypass and necessitating the tripping out of the PDM.

In the foregoing discussion, it is assumed that all the torque is developed by the penetration of the formation by the bit and the overcoming of the internal friction in the PDM. In directional drilling with a PDM, a bent sub, bent housing, or a PDM as part of a directional assembly, torque is developed as the bit cuts sideways out of the vertical plane. This additional torque is related to the total torque used by a PDM as

$$M_m = M_b + M_{mf} + M_s, \quad \ldots \ldots \ldots \ldots \ldots \ldots (8.84)$$

where M_m is motor torque, M_b is bit torque, M_{mf} is motor-friction torque, and M_s is sidecutting torque.

When a directional driller monitors the tool-face angle on a surface display with a PDM with a bent sub, he observes the following occurrences (see Fig. 8.108). Circulation begins with the bit off bottom. As the motor reacts to the clockwise rotation of the rotor against the stator and the bearings, the tool face begins to rotate counterclockwise. As the bit engages the side of the borehole, a right readjustment occurs; and finally, as the bit face engages, a counterclockwise rotation of the tool face takes place. As the bit drills off, the left rotation decreases. If the data are obtained frequently enough, the driller can use this information to advance the bit more accurately than he can with the coarser-reading pressure gauge. If

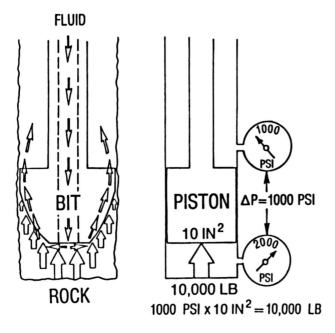

Fig. 8.109—Pressure differential caused by circulation of the drilling fluid.

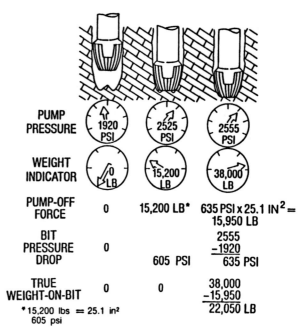

Fig. 8.110—Pumpoff pressure and weight (after Winters and Warren[16]).

the data are obtained only infrequently, however, as with most MWD tools, the use of the tool-face readings for drilling is difficult.

The higher the inclination angle, the greater the bit's sidecutting effect; sometimes the tool-face orientation varies more than 90°. The tool face oscillates so severely that a visual running average is required to direct the orientation of the bit. This oscillation usually occurs when the inclination is greater than 50°, the formation is medium soft, and aggressive bits are used.

8.6.7 Drilling With a PDM

When a drag-type bit is used (diamond, polycrystalline diamond, or a combination of the two), the pump-off of the bit must be considered in the design of the motor bit system and in the determination of the amount of effective WOB to be applied at the surface to obtain a desired WOB at the bit. Fig. 8.109 illustrates how the circulation of the drilling fluid causes a 1,000-psi pressure difference across the bit. This equates to a pump-off force of 10,000 lbf across a 10-sq-in. piston.

Winters and Warren[15,16] verified the pump-off force in the field and with controlled laboratory tests. Their work showed that the pump-off force can be obtained in the field by observing the hook load and standpipe pressures for what is called a "pump-off test." Eq. 8.85 relates the pump-off force F_h to the difference between the hook load with the bit off bottom, F_{ob}, and the hook load with the bit just drilled off, F_{do}.

$$F_h = F_{do} - F_{ob}. \qquad (8.85)$$

The difference between the standpipe pressure with the bit off bottom, p_{ob}, and that with the bit just drilled off, p_{do}, is the pressure drop across the bit, Δp_b, and is related by Eq. 8.86:

$$\Delta p_b = p_{do} - p_{ob}. \qquad (8.86)$$

Fig. 8.110 shows how the pump-off pressure and weight are obtained and how the effective pump-off area of the bit can be determined.

With the bit off bottom, the pump pressure is 1,920 psi and the Δ hook load is 0.0 lbf. The pump-off force is 0.0 lbf and the bit pressure drop is 0.0 psi. There is no WOB. The pump-off force and pressure can be found in one of two ways: the bit can be advanced until it drills or the brake can be locked so that the bit can drill off. The point at which the bit will be just drilled off is the point of pump-off pressure and weight. Fig. 8.111A plots a carefully controlled drill-off test. However, the plot of WOB vs. time does not define the pump-off clearly. When pump pressure vs. WOB is replotted, the true pump-off point is discernable (see Fig. 8.111B). As the drillstring is lowered slowly, the Δ hook load and the standpipe pressure are recorded until the pump-off is noted. Fig. 8.112 shows how this is done. At 0.0 lbf Δ hook load, the Δ standpipe pressure is 0.0 psi. As the bit is lowered, the pressure starts to increase and continues to increase almost linearly until a pressure difference of 595 psi is reached at a Δ hook load of 9,000 lbf. Further addition of Δ hook load causes the bit to drill ahead, as indicated by the change in slope of the data. The pump-off force of the new bit is 9,000 lbf, and its pump-off area is determined to be 9,000 lbf/595 psi = 15.1 sq in. Fig. 8.112 indicates that as the bit dulled, its pump-off area increased to 25.1 sq in.

As the example in Fig. 8.110 shows, the determined pump-off point is 2,525 psi at a Δ hook load of 15,200 lbf. The bit pressure drop is

$$\Delta p_b = 2{,}525 \text{ psi} - 1{,}920 \text{ psi} = 605 \text{ psi}.$$

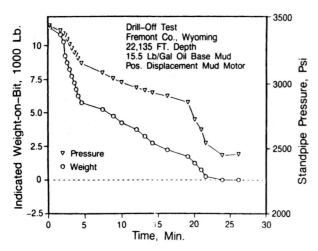

Fig. 8.111A—Drill-off test with an 8½-in. standard crossflow bit.

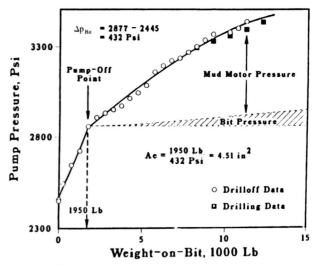

Fig. 8.111B—Point of pump-off (after Winters and Warren[16]).

The true WOB is still 0.0 lbf at this point. The pump-off area, A_{po}, of the bit is calculated with

$$A_{po} = \frac{F_h}{\Delta p_b}. \quad\quad\quad\quad\quad\quad (8.87)$$

For the values presented by Fig. 8.110, the pump-off area is

$$A_{po} = \frac{15,200 \text{ lbf}}{605 \text{ psi}} = 25.1 \text{ sq in.}$$

If the pump-off area and the off-bottom pump pressure are known, the true WOB for any pump rate can be determined.

Fig. 8.110 shows a drilling example in which the surface-indicated WOB is 38,000 lbf and the standpipe pressure is 2,555 psi. The bit pressure drop is

$$\Delta p_b = 2,555 \text{ psi} - 1,920 \text{ psi} = 635 \text{ psi};$$

the pump-off force is

$$F_b = 635 \text{ psi} \times 25.1 \text{ sq in.} = 15,939 \text{ lbf};$$

and the true WOB is

$$W_b = 38,000 - 15,939 \text{ lbf} = 22,061 \text{ lbf}.$$

A typical example of drilling with a sidetracking diamond bit and a PDM with a bent housing is shown by Fig. 8.113. (Note that the downward arrows indicate each time the driller slacked off weight.)

At time T_1 the bit was off bottom with 0.0 lbf WOB. The off-bottom pressure was 1,540 psi. At T_2 the string was lowered slowly, causing the standpipe pressure to increase until T_3, when the pump-off point was noted. Additional WOB (T_4) caused the motor to drill, reaching a Δp of 100 psi for a WOB of 13,400 lbf. As the bit drilled off, the standpipe pressure and WOB decreased until the pump-off point was reached. Then the slope changed (T_5). WOB was again applied (T_6), causing the standpipe pressure to increase to 2,270 psi and giving a motor Δp of 210 psi (T_7). The bit drilled off at a fairly constant rate until the pump-off point was again reached at T_8 when, again, the rate of decline of the standpipe pressure and WOB increased. The driller slacked off weight

three times, but there was no apparent torque response because of hole drag. Then at T_9, the excessive weight reached the bit, and the motor stalled. The standpipe stall pressure was 2,880 psi, and the Δp motor stall pressure was 820 psi.

This typical field example shows the principal operating features of the motor and diamond-bit system. The pump-off force is 13,700 lbf at a standpipe pressure of 2,060 psi. The bit pressure drop is 2,060 psi minus the off-bottom pressure of 1,540 psi, which is 520 psi. The pump-off area is 26 sq in. The actual motor Δp is any pressure above that at the pump-off point, in this case 2,060 psi. The actual WOB can be calculated by multiplication of the bit Δp and the pump-off area.

The motor Δp relates to the amount of torque necessary for a given type of bit to drill a particular type of formation. The torque relationships for drag bits are similar to those for three-cone bits. Fig. 8.114A is a typical plot of diamond bit torque and penetration rate vs. WOB for a given pump rate and rotary speed.

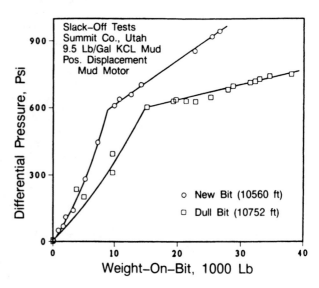

Fig. 8.112—Slack-off test data with 12¼-in. flat profile radial flow bit.

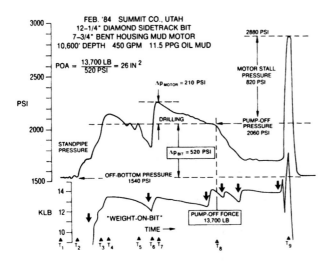

Fig. 8.113—Drilling with a sidetracking diamond bit and bent housing mud motor.

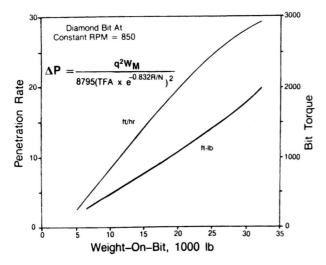

Fig. 8.114A—Diamond-bit penetration rate and torque analysis at 850 rpm.

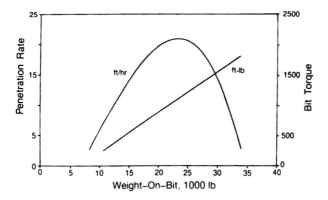

Fig. 8.114B—130-stage turbine with diamond bit.

8.6.8 Effect of Standpipe Pressure Change on Drillpipe Elongation

When a PDM is used, the change in the internal drillpipe pressure causes an elongation of the drillstring. This effect can be related by

$$\Delta L = \left[\frac{-\Delta W_b}{E \times A_s} + \frac{\Delta p_i \times A_i}{E \times A_s} - \frac{2\mu}{E} \frac{\Delta p_i}{(F_d{}^2 - 1)} \right] L,$$

.... (8.88)

where ΔL is the change in drillpipe length, ΔW_b is the change in WOB, Δp_i is the change in internal drillpipe pressure, A_s is the cross-sectional area of steel in drillpipe, A_i is the internal drillpipe area, F_d is the ratio of drillpipe, E is the Young's modulus for steel (29×10^6 psi), μ is the Poisson's ratio for steel (0.3), and L is the length of drillpipe.

For a 5-in. drillpipe, Eq. 8.88 reduces to

$$\Delta L = \left(\frac{-\Delta W_b}{15.82 \times 10^7} + \frac{\Delta p_i}{2.76 \times 10^7} \right) L. \quad \ldots \ldots (8.89)$$

Example 8.20. If the driller slacks off 0.95 ft of 15,000 ft of 5-in. drillpipe, what is the indicated WOB? This slack-off causes an increase of 155 psi, which causes an increase in WOB. How much must be drilled off to maintain the indicated WOB?

Solution. With no change in motor pressure ($\Delta p = 0$), the slack-off weight is

$$W_b = \frac{15.82 \times 10^7}{15,000} (0.95) = 10,000 \text{ lbf.}$$

An increase of 10,000 lbf WOB causes a Δp of 155 psi. This increases the WOB, so (155 psi/2.76×10^7 psi) $\times 15,000$ ft $= 0.08$ ft must be drilled off to maintain 10,000 lbf WOB.

8.6.9 Planning a Trajectory Change With a PDM

The following are steps for planning and executing a trajectory change when a PDM is used.

1. Design the hydraulics in such a way that the pressure drop across the bit does not exceed the manufacturer's limits and supplies enough pressure and circulation rate to power the motor throughout the trajectory change. Select a PDM with enough power to rotate a bit of the size and type necessary both to drill a given series of formations and to cause the trajectory change.

2. Once the motor, bit, and hydraulics are designed, select the appropriate bent sub (depending on the desired trajectory change).

3. Trip the bit, motor, bent sub, mule-shoe sub, and the remainder of the BHA into the hole.

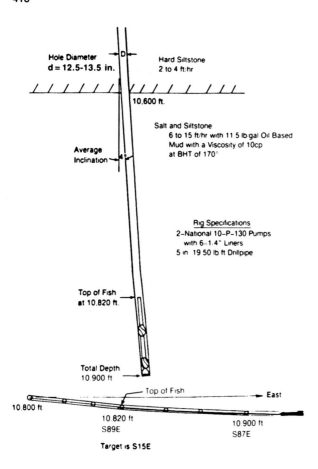

Fig. 8.115—Wellbore situation for Example 8.21.

limit. In such cases, the bit run must be optimized to last the life of the motor to maximize the interval drilled.

Example 8.21. A bit and two stabilizers are stuck in the hole and cannot be retrieved economically. It is decided to sidetrack around the fish and to continue with the drilling. Fig. 8.115 shows the wellbore situation. The siltstone section above the salt is extremely hard to drill with the existing 11.6-lbf/gal mud. Penetration rates vary in this interval from 2 to 4 ft/hr. The salt-and-siltstone section below 10,600 ft is easier to drill, with penetration rates varying from 10 ft/hr for the 100% siltstone to 30 ft/hr for 100% salt. The mud is a low-fluid-loss oil/water emulsion. The average wellbore diameter varies between 12.5 and 13.5 in. Surveys at 10,820 ft and 10,900 ft report the inclination at 4.0° and 4.5°, and the directions S84E and S87E, respectively. The top of the fish is at 10,820 ft. Design an optimum sidetrack to get around the fish without drilling any of the harder siltstone above the salt. To be safe at the top of the fish, the new wellbore should be two diameters laterally displaced from the old wellbore. The drillpipe is 5.0-in., 19.50-lbm/ft Grade E.

Solution. First, place a hard cement plug from the top of the fish to at least 200 ft above the salt. This ensures that at the top of the salt where the kick-off needs to start, there is good, consistently hard cement. A regular 12¼-in. Series 1-1-1 bit or a good rerun bit can be used to drill the cement to the top of the salt. Next, select the proper size and type of motor, the bit, and the bent-sub angle, and design the hydraulics program to run the motor over the expected range of pressures.

There are a number of possible motor types and sizes that could be used for the sidetracking operation (see Tables 8.10 and 8.11).

Because this sidetrack will be done over a minimum section of hole (approximately 220 ft to miss the top of the fish; Fig. 8.115), the shortest motor offers the best chance of changing the angle off the plug. This assumes that a bent sub, not a bent housing, will be used for the deflection. Of the eight possible choices, the 7¾-in. Type D motor, which is 21 ft long, is the shortest. This motor develops 50.8 to 73.3 hp with a maximum torque of 1,160 ft-lbf. The maximum pressure differential is 360 psi. Pump rates of 325 to 450 gal/min drive the bit at 230 to 332 rpm.

4. On the basis of the calculations presented in Sec. 8.4 and allowing for the reactive torque of the motor, orient the tool face before starting the motor by orienting the pipe at the surface while moving the drillpipe up and down to reduce the static friction of the drillstring and the bent-sub or bent-housing knee. Or the bit face can be adjusted and the pipe worked to transmit the torque to the bit.

5. Start the motor by circulating the mud and bringing up the circulation rate to the desired level.

6. Advance the bit until a reactive torque is indicated by the standpipe pressure and/or by the tool-face indicator; this implies a bit/formation interaction. (If Step 4 is performed correctly, there should be very little readjustment of the tool face at the surface. If that step is omitted, however, the driller must keep readjusting the tool face until the final trajectory change is obtained. Such changing can cause severe, unplanned doglegs.)

7. Generally, plan to make a direction change when the inclination exceeds 2°. To control the dogleg severity, change the direction over a drilled section and use the motor to hold the direction as constant as possible while building inclination over a course length that covers the controlling section of the next BHA. If a bent housing is used, the general strategy is to change the trajectory over a course length that does not exceed the dogleg criteria and then to replace the PDM bent housing with a PDM bent-sub arrangement.

8. If a trajectory change is required at a higher inclination, use a longer tool run (multiple tool runs may be required) to keep the dogleg severity to a predetermined

TABLE 8.10—TYPICAL MOTOR SIZES USED FOR SIDETRACKING

Type	OD (in.)	Stator/Rotor	Length (ft)
A	8	⅚	23
B	9½	⅚	24
B	8	½	26.5
B	9½	½	33
C	8	½	23.8
C	9½	½	24.9
D	7¾	½	21.0
D	9⅜	½	26.5

TABLE 8.11—TYPICAL OPERATING PARAMETERS FOR VARIOUS MOTORS

Tool Size OD (in.)	Recommended Hole Size (in.)	Pump Rate (gal/min) min.	Pump Rate (gal/min) max.	Bit Speed Range	Maximum Differential Pressure	Maximum Torque (ft/lbf)	Horsepower Range	Thread Connection Bit Sub, Box Down	Length (ft)	Weight (lbm)
					TYPE A					
4¾	6 to 7⅞	80	155	90-180	580	920	17 to 32	3½ (in.)-Reg.	17.4	750
6¾	7⅞ to 9⅞	185	370	85-185	580	2,065	36 to 73	4½ (in.)-Reg.	19.8	1,760
8	9½ to 12¼	300	600	75-150	465	3,400	49 to 97	6⅝ (in.)-Reg.	23.0	2,430
9½	12¼ to 17½	425	845	80-160	465	4,490	69 to 137	6⅝ (in.)-Reg.	24.6	3,970
11¼	17½ to 26	525	1,050	65-130	465	6,850	85 to 170	7⅝ (in.)-Reg.	26.6	5,950
					TYPE B					
3¾	4¼ to 5⅞	80	190	325-800	580	320	20 to 52	2⅞ (in.)-Reg.	20.8	470
4¾	6 to 7⅞	100	240	245-600	580	585	27 to 67	3½ (in.)-Reg.	21.5	840
6¼	7⅞ to 9⅞	170	345	200-510	580	1,015	39 to 98	4½ (in.)-Reg.	24.0	1,760
6¾	8⅜ to 9⅞	200	475	205-485	580	1,500	58 to 138	4½ (in.)-Reg.	26.6	2,160
8	9½ to 12¼	245	635	165-380	465	2,085	66 to 151	6⅝ (in.)-Reg.	26.5	2,800
9½	12¼ to 17½	395	635	230-380	810	3,730	163 to 270	6⅝ (in.)-Reg.	33.0	5,200
11¼	17½ to 26	525	1,055	120-250	465	5,385	123 to 256	7⅝ (in.)-Reg.	30.0	7,300
					TYPE C					
3¾	4¼ to 5⅞	60	145	340-855	580	245	16 to 40	2⅞ (in.)-Reg.	16.8	400
4¾	6 to 7⅞	80	185	270-680	580	415	21 to 53	3½ (in.)-Reg.	17.5	680
6¼	7⅞ to 9⅞	170	345	200-510	580	1,015	39 to 98	4½ (in.)-Reg.	25.0	1,760
6¾	8⅜ to 10⅝	160	395	140-480	465	995	27 to 91	4½ (in.)-Reg.	21.9	1,585
8	9½ to 12¼	200	475	160-400	465	1,475	45 to 112	6⅝ (in.)-Reg.	23.8	2,430
9½	12¼ to 17½	240	610	135-340	465	2,280	59 to 148	6⅝ (in.)-Reg.	24.9	3,970
11¼	17½ to 26	290	690	115-290	465	2,990	65 to 165	7⅝ (in.)-Reg.	26.0	5,950
					TYPE D					
3⅞	4⅝ to 6	100	150	380-580	625	412	29.8 to 45.5	2⅞ (in.)-Reg.	22.5	530
5	6½ to 7⅞	180	250	350-482	360	480	32.0 to 44.1	3½ (in.)-Reg.	19.9	911
6½	8⅜ to 9⅞	250	350	292-431	360	801	44.5 to 65.7	4½ (in.)-Reg.	19.9	1,582
7¾	9⅞ to 12¼	325	450	230-332	360	1,160	50.8 to 73.3	6⅝ (in.)-Reg.	21	2,350
9⅝	12¼ to 17½	500	800	200-420	360	1,775	67.6 to 142.1	7⅝ (in.)-Reg.	26.5	4,350
12	17½ to 26	800	1,200	125-188	360	5,666	134.8 to 202.8	7⅝ (in.)-Reg.	33.2	8,100

Next, determine the horsepower necessary to drive the motor and to provide enough pressure across the bit to satisfy the pressure-drop requirements for the bearings and for drilling with a Series 5-1-7 bit.

Table 8.10 shows that the maximum recommended pump rate of the 7¾-in. PDM is 450 gal/min. A 10-P-130 pump operating at 112 strokes/min will pump 448 gal/min (this assumes 100% volumetric efficiency and 85% mechanical efficiency), which is well within the operating range of the pump.

To determine how many drill collars might be needed, consider how much torque will be necessary to drill the salt at a maximum penetration rate of 15 ft/hr. With Eq. 8.81, the maximum WOB for the maximum PDM torque is

$$1{,}160=\left[3.79+19.7\sqrt{\frac{15(\text{ft/hr})}{332(\text{rpm})(12.25\text{ in.})}}\right]$$

$$\times 12.25W_b,$$

where $W_b=19{,}000$ lbf.

The drill collars are 7¾ in. OD by 3 in. ID. The weight is 116 lbm/ft in air and 0.825×116 lbm/ft in 11.5 lbm/gal mud or 95.7 lbf/ft. Because 19,000 lbf are needed, and because a 20% safety factor is needed to keep the drillpipe in tension, the number of collars (n_C) needed is

$$n_C=\left[\frac{19{,}000\text{ lbm}}{(95.7\text{ lbm/ft})(30\text{ ft/collar})}\right]\frac{1}{0.80}$$

$$=8.3\text{ collars}.$$

For convenience, nine collars (three stands) could be used. This leaves 10,330 ft of 5-in., 19.5-lbm/ft XH drillpipe with an ID of 4.276 in. The pressure losses in the drillstring and up the annulus minus the bit and PDM pressure at a pump rate of 448 gal/min are as follows.

Pressure loss* through surface equipment (Case 4: 45 ft of 4-in.-ID standpipe, 55 ft of 3-in.-ID hose, 6 ft of 3-in.-ID swivel, 40 ft of 4-in.-ID kelly)	24 psi
Pressure loss through drillpipe	710 psi
Pressure loss through drill collars (3-in. ID)	89 psi
Pressure loss up the drillpipe annulus	161 psi
Pressure loss up the drill collar annulus	56 psi

*Pressure-loss calculations are based on a power-law model; a yield value of 9 lbf/100 sq ft is assumed.

The total pressure drop for the 10,600-ft string is calculated as

$$\Delta p_t = 24 \text{ psi} + 710 \text{ psi} + 89 \text{ psi} + 161 \text{ psi} + 56 \text{ psi}$$

$$= 1{,}040 \text{ psi circulating pressure open ended.}$$

To design the optimum PDM motor run correctly, the maximum bit pressure drop should be determined on the basis of the hydraulic thrust and bit-weight balance. The maximum WOB for this application is 20,000 lbf. Fig. 8.116 shows that the on-bottom bearing load at a WOB of 20,000 lbf is 4,600 lbf for a bit Δp of 750 psi and 8,000 lbf for a bit Δp of 500 psi, the maximum recommended bit pressure differential. If a balanced bearing load is desired at a Δp of 500 psi, the maximum WOB should not exceed 12,000 lbf. Because this will be a short motor run, the on-bottom bearing load can be increased as much as 5,000 lbf (to 17,000 lbf WOB). The actual pressure drop of the bit must be corrected for mud weight before the use of Table 8.12.

$$500 \text{ psi} = \frac{\text{pressure loss} \times \text{actual mud weight}}{10 \text{ lbm/gal}},$$

and

$$\text{pressure loss} = 500 \text{ psi} \frac{10 \text{ lbm/gal}}{11.5 \text{ lbm/gal}} = 435 \text{ psi.}$$

The nozzles should be sized for a bit Δp of 435 psi and a mud weight of 11.5 lbm/gal. Table 8.12 shows that $2^{16}\!/\!_{32}$- and $1^{18}\!/\!_{32}$-in. nozzles are required to give the approximate Δp.

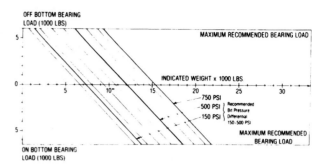

Fig. 8.116—Hydraulic thrust and bit weight balance (courtesy of Dyna-Drill).

The total standpipe pressure, including the PDM Δp, would be as follows (at 450 gal/min):

Drillstring	= 1,040 psi
Δp_b	= 500 psi
Maximum Δp for motor torque	= 360 psi
Total standpipe pressure	= 1,900 psi

The next step is to design a trajectory for the sidetrack that will miss the top of the fish by at least two bit diameters, which means that the side of the new wellbore will be about 24 in. from the side of the old wellbore. The minimum average change to offset the fish by two bit diameters is 0.78°. Because the desired target for the well is S15E, the plan should call for a right turn away from the old wellbore. A simple direction change with no inclination change would be risky; therefore, a drop and right turn should be planned to ensure that the new borehole will not re-enter the old borehole. It must be

TABLE 8.12—PRESSURE LOSS THROUGH THE JET NOZZLES (PSI)

Flow Rate (gal./min.)	15 15 15 (0.5177 sq in.)	15 15 16 (0.5415 sq in.)	15 16 16 (0.5653 sq in.)	16 16 16 (0.5890 sq in.)	16 16 18 (0.5412 sq in.)	16 18 18 (0.6934 sq in.)	18 18 18 (0.7455 sq in.)	18 18 18 (0.8038 sq in.)	18 18 20 (0.8621 sq in.)	18 20 20 (0.9204 sq in.)	20 20 20 (0.9048 sq in.)	20 20 22 (0.0492 sq in.)	20 22 22
410	578	528	485	446	377	322	279	240	208	183	160	141
420	606	554	508	468	395	338	292	251	219	192	168	148
430	635	581	533	491	414	354	306	264	229	201	176	155
440	665	608	558	514	434	371	321	276	240	210	184	162
450	696	636	584	537	454	388	336	289	251	220	192	169
460	727	665	610	562	474	405	351	302	262	230	201	177
470	759	694	637	586	495	423	366	315	274	240	210	185
480	792	724	664	612	516	441	382	328	286	250	219	193
490	825	754	692	637	538	460	398	342	298	261	228	201
500	859	785	721	664	560	479	414	356	310	272	237	209
510	894	817	750	690	583	498	431	371	322	283	247	218
520	929	849	779	718	606	518	448	385	335	294	257	226
530	965	882	810	746	629	538	465	400	348	305	267	235
540	1,002	916	840	774	653	559	483	416	361	317	277	244
550	1,039	950	872	803	678	580	501	431	375	329	287	253
560	1,078	985	904	832	702	601	520	447	389	341	298	262
570	1,116	1,021	936	862	728	622	538	463	403	353	309	272
580	1,156	1,057	970	893	754	644	557	480	417	366	319	281
590	1,196	1,093	1,003	924	780	667	577	496	431	378	331	291
600	1,237	1,131	1,038	956	806	690	597	513	446	391	342	301
610	1,279	1,169	1,073	988	834	713	617	530	461	405	353	311
620	1,321	1,207	1,108	1,020	861	736	637	548	476	418	365	322
630	1,364	1,247	1,144	1,053	889	760	658	566	492	432	377	332

*Nozzle Size (32/100 in.) Nozzle Area (sq in.)

(Courtesy of Dyna Drill)

TABLE 8.13—DEFLECTION ANGLE RESULTING FROM BENT-SUB ANGLE AND HOLE SIZE

BENT-SUB ASSEMBLY

Bent-Sub Angle (degrees)	3⅞ in.		5 in.		6½ in.		7¾ in.		9⅝ in.	
	Hole Size (in.)	Deflection Angle (deg/100 ft)	Hole Size (in.)	Deflection Angle (deg/100 ft)	Hole Size (in.)	Deflection Angle (deg/100 ft)	Hole Size (in.)	Deflection Angle (deg/100 ft)	Hole Size (in.)	Deflection Angle (deg/100 ft)
1	4¼	4°00′	6	3°30′	8¾	2°30′	9⅞	2°30′	13½	2°00′
1½		4°30′		4°45′		3°30′		3°45′		3°00′
2		5°30′		5°30′		4°30′		5°00′		4°30′
1	4¾	3°00′	6¾	3°00′	9⅞	1°45′	10⅝	2°00′	15	1°45′
1½		3°30′		4°15′		3°00′		3°30′		2°30′
2		4°00′		5°00′		3°45′		4°15′		3°45′
2½		5°00′		5°45′		5°00′		5°30′		5°00′
1	5⅞	2°00′	7⅞	2°30′	10⅝	1°15′	12¼	1°45′	17½	1°15′
1½		2°30′		3°30′		2°00′		2°30′		2°15′
2		3°00′		4°30′		3°00′		3°30′		3°00′
2½		3°30′		5°30′		4°00′		5°00′		4°30′

remembered when sidetracking in harder formations that the bent-sub assembly will not always respond as predicted. Therefore, a safe design would be based on a 2°/100-ft right turn and drop away from the old wellbore.

Based on the 2°/100 ft dogleg severity over 220 ft, the total angle change is

$$\beta = \frac{2(220)}{(100)} = 4.4°.$$

The current inclination, α, is 4°. Assuming a 2° inclination drop, what should be the tool-face setting to maintain a total angle change of 4.4° and achieve a maximum direction change $\Delta\epsilon$? First, calculate the tool face setting with Eq. 8.48:

$$\gamma = \text{arc cos} \left[\frac{\cos(4) \cos(4.4) - \cos(2)}{\sin(4) \sin(4.4)} \right] = 153°.$$

With a tool-face setting of 153° right of the high side, the maximum direction change can be calculated with Eq. 8.42.

$$\Delta\epsilon = \text{arc tan} \left[\frac{\tan(4.4) \sin(153°)}{\sin(4) + \tan(4.4) \cos(4) \cos(153)} \right]$$

$$= 87.8°$$

The reactive torque for the bit and the PDM can be calculated from Eq. 8.57 with a maximum motor torque of 1,160 ft-lbf.

$$\theta = \frac{13,920 \text{ in. lbf}}{11.5 \times 10^6 \text{ psi}} \left(\frac{125,160 \text{ in.}}{-32.8 \text{ in.}^4} + \frac{3,240 \text{ in.}}{346 \text{ in.}^4} \right)$$

$$= 4.61.$$

$$\theta = \frac{4.61(36)}{2\pi} = 264°.$$

Note that the PDM is omitted because its short length makes it a negligible term.

The reactive torque from the PDM and the Series 5-1-7 bit is significant and requires that a steering tool be run so that the tool face will always be at the proper setting. Because the reactive torque calculation allows for no friction, it would be wise to plan initially to set the tool face at N48E, to engage the bit with near-maximum WOB (20,000 lbf), and to observe the reactive torque response. The full reactive torque probably will not be observed because of friction, and the tool face will need to be adjusted less than the calculated 264° to obtain the tool-face setting of 153°. A few tries with the steering tool should be ample to set the tool face and to start the sidetrack. The calculated tool-direction change is more than enough to head the wellbore toward the target. Because the drilling is slow, the direction can be watched and corrected to a lesser tool-face setting (90 to 120°), which will result in a smaller $\Delta\epsilon$.

The last part of the design is to select a bent sub that will give an angle change of approximately 2.0°/100 ft. From Table 8.13, a 1° bent sub on a 7¾-in. PDM in a 12¼-in. hole should yield an angle change of nearly 1.75°/100 ft. A 1.5° bent sub would give an angle change of 2.5°/100 ft. Because the formations are harder and the response is less than those included in Table 8.13, the safest plan would be to run the 1.5° bent sub.

8.6.10 The Use of a Turbine for Directional and Straight-Hole Drilling

Drilling economically with a turbine is more complicated than with a PDM. The bit must be chosen for a given turbine, and the drilling rig must be able to deliver the required flow rates at pressures that will operate the turbine at maximum efficiency. In addition, the operator must be capable of operating the turbine at speed and torque ranges that will achieve maximum horsepower most of the time.

Because, unlike the PDM, neither the bit speed nor the torque is related to the standpipe pressure, the only way to be certain a turbine is performing properly is to measure downhole bit speed and/or torque. A surface-reading tachometer is needed to control a turbine accurately, unless the operator is very familiar with the formations in the drilling area, can pick the appropriate bit and motor

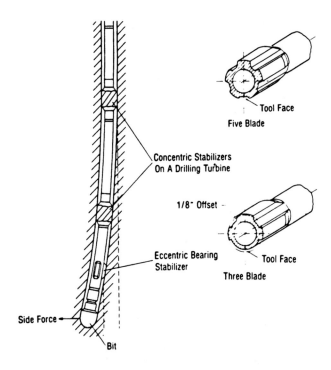

Fig. 8.117—Eccentric stabilizer (after Feenstra and Kamp[17]).

size, and knows the capabilities of the turbine thoroughly. Even with such experience, the operator must watch the turbine constantly to keep it from stalling and to drill with the right WOB to ensure optimum drilling conditions. This is another reason for the slow acceptance of turbines outside the Soviet Union and Europe.

To emphasize the problem, consider Eq. 8.78 rewritten as

$$N_b = \frac{1}{K}(M_{ts} - M_i).$$

The torque relationship for an 8.5-in. polycrystalline diamond bit is similar to any drag bit. Neglecting the torque to overcome friction yields the following relationship:

$$M_i = \left(8.537 + 203 \times \frac{\sqrt{F_d}}{n}\right) d_b W_b, \quad \ldots \ldots (8.90)$$

If the pump stroke, bit size, and mud weight are maintained at constant levels, the addition of WOB will increase the turbine torque, which in turn reduces both the bit speed and the penetration rate. The reduction in penetration rate causes the torque to decrease and the bit speed to increase. If the bit speed is not controlled so that it is kept at the peak of the power curve, optimal performance of the turbine and the bit is not obtained. Unlike using the hook-load indicator in drilling with a rotary system and a tricone bit or using the standpipe pressure gauge in drilling with a PDM, drilling with a turbine requires a bit-speed indicator.

Properly engineered, the turbine can be economically competitive in many drilling situations, especially with the advanced designs of the new polycrystalline diamond bits. For kicking off and making trajectory changes, turbines have not been as widely used as PDM's. In areas of soft formations, turbines have been used with tricone bits to make successful (though not necessarily economical) trajectory changes. For most trajectory changes, however, operating a turbine with a tricone bit is too difficult because of the previously mentioned bit speed and bit side-force problems. Where formations dictate a less aggressive bit, such as a diamond or other type of drag bit, the turbine is more likely to be successful. A properly designed turbine and bit combination with a tachometer can be as successful in making a trajectory change as a PDM.

One way of making controlled trajectory changes is to use an eccentric stabilizer on the turbine near the bit (see Fig. 8.117). The undergauge blade is oriented as a bent sub or bent housing and the drillstring does not rotate. After the desired trajectory change is achieved, drillstring rotation is begun with the eccentric stabilizer, a part of the controlling BHA. This system is frequently used with a building BHA. (See Sec. 8.7 for a description of a building assembly.)

The turbine is commonly used as part of the BHA for normal directional drilling. Unlike most rotary directional drilling systems, the reactive torque of the turbine gives a strong left or counterclockwise component to the bit direction. Rotating the drillstring counteracts this reactive torque to varying degrees, dependent on the bit type and BHA, and it reduces some of the left or counterclockwise tendency.

Example 8.22. Determine whether it is economical to run a diamond bit with a turbine in a situation where 8½-in. Series 6-2-7 bits average 3 ft/hr at a depth of 17,520 ft. The bits usually drill 120 to 140 ft and are graded 4 to 6 and ⅛-in. in gauge. Use the same rig as in Example 8.21. The oil-mud density is 16.2 lbm/gal, the plastic viscosity is 22 cp, and the yield value is 15 lbf/100 sq ft. A Series 6-2-7 bit costs $4,400. The diamond bit costs $19,500 assuming 50% salvage. A 7-in. turbodrill costs $500 per rotating hour. Round-trip time is 16 hours. Rig cost is $800/hr. Fig. 8.118A is a nomograph for a 7-in. turbine with 50, 100, 150, and 200 stages. The turbine develops maximum horsepower at about 850 rpm. Fig. 8.114A shows the theoretical torque, penetration rate, and pressure-drop characteristics of the new diamond bit under downhole conditions at a constant rotary speed of 850 rpm. The bit has a pump-off area of 10 sq in. Assume that the penetration rate and torque of the diamond bit when dull is about half that of the new bit. Select the appropriate flow rate and number of stages for the turbine, and size the total flow area (A_{tf}) of the diamond bit to provide at least 2.0 to 2.5 hhp/sq in.

Solution. Refer to the nomograph in Fig. 8.118A to find that the turbine requires 440 gal/min to develop its maximum power at 850 rpm. The nomograph and Fig. 8.114A provide the information found in Table 8.14.

Roughly calculate the number of drill collars required, assuming that 25,000 lbf will be applied initially and that it will take 50% additional weight to maintain bit torque as the bit dulls and as its pump-off effect increases.

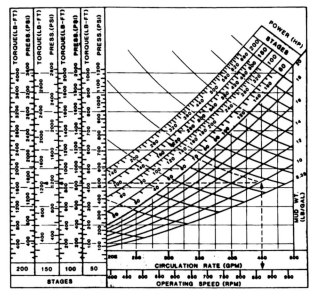

Fig. 8.118A—Seven-inch turbine drilling motor operating characteristics (courtesy Baker Service Tools).

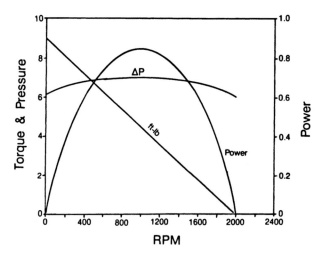

Fig. 8.118B—One-stage turbodrill characteristics.

Calculate the collars required for 37,500 lbf, using a 20% safety factor for collars having a buoyed weight of 81.3 lbf/ft.

$$n_C = \frac{37{,}500 \text{ lbf}}{(87.3 \text{ lbf/ft})(30 \text{ ft/collar})(0.80)}$$

$$= 17.9 \text{ collars} = 6 \text{ stands.}$$

Using a penetration rate of 18 ft/hr, determine the diamond bit A_{tf} required for 2.0 hhp/sq in.

A minimum of 113 hhp is required, which at 440 gal/min equates to 442 psi Δp at the bit. Use of the equation from Fig. 8.114A yields

$$A_{tf} = \left[\sqrt{\frac{(440)^2 16.2}{8{,}795(442)}} \right] \div 0.983 = 0.91 \text{ sq in.}$$

The A_{tf} should be no larger than 0.91 sq in.

Can the turbine be operated at 440 gal/min without exceeding the pressure limit of the drilling rig? It depends on the size of the liners that are run in the pump and on the pressure capacity of all other components in the circulating system. For the 10-P-130 pump, a good choice is to install 6-in. liners. The pump is rated to 140 strokes/min, and only 125 strokes/min will be required.

440 gal/min ÷ (3.7 gal/stroke × 0.95 pump efficiency)

= 125 strokes/min.

The pump is rated to 3,900 psi maximum with 6-in. liners, so 3,705 psi will be the maximum planned standpipe pressure.

3,900 psi × 0.95 pump safety factor = 3,705 psi.

Now calculate the component pressure losses in the circulating system for a power law fluid:

Surface equipment, psi	24
Drillpipe bore, psi	1,638
Drill collar bore, psi	219
Drill bit, psi	442
Drill collar annulus, psi	310
Drillpipe annulus, psi	456
Total, psi	3,089

**TABLE 8.14—TURBINE OPERATING RANGE
SELECTED FROM FIGS. 8.114A and 8.118A**

No. Stages	Turbine Δp (psi)	Turbine Torque (ft/lbf)	Turbine (hp)	Bit Weight (lbm)	Rate of Penetration (ft/hr)
50	800	680	110	13.5	12.2
100	1340	1360	220	24.5	24.1
150	1890	2040	330	32.5	29.3

TABLE 8.15—DRILLSTRING PRESSURES LOSSES AT VARIOUS FLOW RATES FOR PROBLEM 8.22

Components	Pressures (psi)				
	at 100 gal/min	at 200 gal/min	at 300 gal/min	at 400 gal/min	at 500 gal/min
Surface equipment	—	4	11	19	31
Drillpipe bore	77	460	878	1,401	2,021
Drill collar bore	20	60	117	188	272
Drill bit	23	92	207	369	576
Drill collar annulus	24	48	91	144	207
Drillpipe annulus	78	133	248	391	561
Total standpipe	222	797	1,552	2,512	3,668
Available pressure, psi	3,483	2,908	2,153	1,193	37
Available power, hp	203	339	377	278	11

Available pressure for the turbine: $3,705-3,089=616$ psi.

The pressure losses in the system are too great for even the 50-stage turbine, which consumes 800 psi at 440 gal/min with 16.2-lbm/gal mud. This does not mean that the problem is solved and that a turbine can be run. The problem cannot be solved until it is analyzed properly as follows.

The correct procedure is to analyze the entire circulating system.

1. Compute the component pressure losses at several flow rates, such as 100, 200, 300, 400, and 500 gal/min. Add the component losses to determine the total standpipe pressure for each flow rate.

2. Plot standpipe pressure vs. flow rate. Mark a horizontal line at the maximum recommended standpipe pressure.

3. Determine the remaining available standpipe pressure at each flow rate (the difference between maximum and calculated standpipe pressure). On a separate graph, plot the available pressure vs. flow rate.

4. Compute the available hydraulic power at each flow rate, and plot on the same graph available power equals available psi times gallons per minute divided by 1,714.

5. Locate the maximum available power on the curve, and note the flow rate at which it occurs. This is the flow rate at which the turbine should be operated. The corresponding available pressure is the pressure drop for which the turbine should be sized. It may not necessarily agree with the flow rate and pressure at which a particular turbine was designed to operate, in which case the turbine is mismatched to the system and a different turbine must be considered.

The foregoing procedure is applied to this problem (see Table 8.15) and is shown graphically in Figs. 8.119A and B.

The optimal flow rate for this system is approximately 300 gal/min, as shown in Fig. 8.119B. That is the flow rate at which the most hydraulic horsepower is delivered to a downhole motor for conversion to mechanical horsepower at the bit. Without changing something in the system, such as switching to larger drillpipe, no other flow rate will affect mechanical horsepower more strongly at the bit. The nomograph (Fig. 8.118A) shows that at 300 gal/min, turbine speed is only 575 rpm and the resultant power is reduced to 115 hp with 150 stages and 153 hp with 200 stages. Fig. 8.119B shows that at 300 gal/min, 377 hhp is available from the system. Assuming roughly a 55% power conversion factor, a turbine that develops about 207 hp ($377\times0.55=207$) at 300 gal/min with 16.2-lbm/gal mud is best suited to this system. It becomes clear that even if there were no standpipe-pressure limitation, even the 200-stage turbine in Fig. 8.118A would be less than optimal with its 153 hp. The problem now becomes one of selecting a turbine that is matched to the system.

Fig. 8.118B shows the characteristics for one stage of a 7-in. turbodrill designed to operate in the range of 275 to 325 gal/min. This turbine can deliver more power to the bit in this particular situation than the turbine represented previously in Fig. 8.118A. Fig. 8.118B is based on 300 gal/min with water. Notice that peak power occurs at 1,000 rpm and that $\Delta p_1 =7$ psi (for water). Therefore Δp_1 equals 14 psi for 16.2-lbm/gal mud ($16.2\times 7/8.33=14$).

Resize the pump liners, if possible, and adjust the A_{tf} of the diamond bit to give 2.5 hhp/sq. in. at 300 gal/min flow rate. For the 10-P-130 pump, a good choice is 5½-in. liners. The pump is rated to 140 strokes/min, but only 102 will be required.

300 gal/min $\div$ (3.1 gal/stroke $\times$ 0.95 pump efficiency)

= 102 strokes/min.

The pump is rated to 4,645 psi with 5½-in. liners; however, although 95% of 4,645 psi is 4,413 psi, both the contractor and operator agree in this case not to exceed 3,950 psi because of the limitations of the standpipe, kelly hose, and swivel.

TABLE 8.16—SYSTEM PRESSURE DROP AT VARIOUS FLOW RATES WITH A DIAMOND BIT TFA OF 0.42 SQ IN.

Flow Rate (gal/min)	Bit Δp (psi)	Standpipe Pressure (psi)	Available Standpipe Pressure (psi)	Available Hydraulic Horsepower (hp)
100	90	289	3,661	214
200	361	1,066	2,884	337
300	811	2,156	1,794	314
400	1,443	3,586	364	85
500	2,254	5,346	—	—

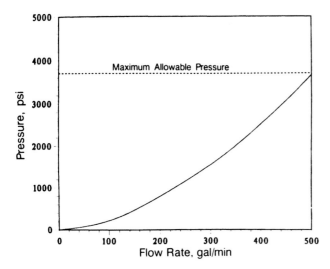

Fig. 8.119A—Standpipe pressure vs. flow rate.

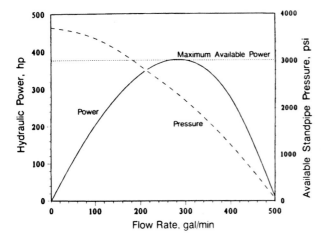

Fig. 8.119B—Available pressure and power vs. flow rate.

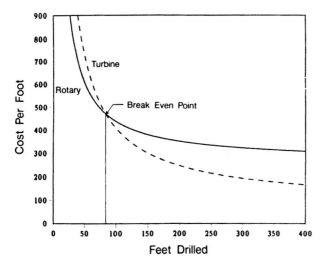

Fig. 8.119C—Analysis of cost per foot—rotary vs. turbine drilling.

The A_{tf} of the diamond bit must be reduced to provide 2.5 hhp/sq in. at the bit. A pressure drop of 810 psi is required at 300 gal/min to give 2.5 hhp/sq in.:

$$2.5 \text{ hhp/sq in.} \times 56.7 \text{ sq in.} = 142 \text{ hhp},$$

$$142 \text{ hhp} \times 1{,}714/300 = 811 \text{ psi}.$$

A minimum A_{tf} of 0.46 sq in. is required. Now recalculate the system pressures using the 0.46-sq-in. A_{tf}. See Table 8.16 for system pressure drops at various flow rates. With the A_{tf} reduced, the optimal flow rate of the system has shifted to about 260 gal/min, but it is decided to operate at 300 gal/min to operate the turbine at its optimal design rate and to provide adequate bit cleaning and annular cuttings transport. Bit cleaning is important because the turbine is designed to divert at least 5% of the mud flow through the bearings; therefore, no more than 285 gal/min will be flowing through the bit, thus giving a bit Δp of 730 psi rather than the 811 psi shown in the previous table. Bit hydraulic horsepower is 2.1 hhp/sq in., which is marginal, and the available turbine pressure will be adjusted by 81 psi because of the diversion of fluid. Available turbine pressure (p_{at}) is calculated as

$$p_{at} = 1{,}794 + 81 = 1{,}875.$$

Calculate the number of turbine stages:

$$\Delta p_E = 1{,}875 \text{ psi},$$

$$\Delta p_1 = 14 \text{ psi/stage},$$

$$n_s = \Delta p_E/\Delta p_1 = 1{,}875/14 = 134 \text{ stages}.$$

The 7-in. turbine represented in Fig. 8.118B is readily available as a 130-stage tool.

Determine the power output of a 130-stage turbine. Note in Fig. 8.118B that one turbine stage develops a maximum of 0.87 hp at 300 gal/min with water. Therefore,

$$0.87 \times (16.2 \text{ lbm/gal})/(8.33 \text{ lbm/gal})$$

$$= 1.7 \text{ hp/stage with 16.2-lbm/gal mud.}$$

$$1.7 \text{ hp/stage} \times 130 \text{ stages} = 223 \; H_m \text{ at the bit.}$$

The 130-stage turbine will deliver 221 hp at the bit, compared with 153 hp with the 200-stage turbine referred to in Fig. 1.118A. This does not mean that the 200-stage turbine is not a good tool; however, it is highly mismatched to the system in this example.

It must be determined now whether the 130-stage turbine and natural diamond bit will be economical.

Fig. 8.114B shows the penetration rate and torque curves for a new diamond bit on the 130-stage turbine. This figure indicates that at about 7,000 lbf, the bit is pumped off and at 35,000 lbf the turbine is stalled. Peak penetration rates of 20 to 21 ft/hr are achieved in the 20,000- to 26,000-lbf range. Remember that the rate of penetration of the bit will decrease by about half when the bit is dulled, so assume an average penetration rate of 15 ft/hr over the entire run. Also, remember that bit torque will decrease by 50% as the bit dulls; therefore, WOB must be increased throughout the run to maintain turbine operation at peak power. Also, diamond bit pump-off force is likely to increase about 50% as the bit dulls, thus requiring still more weight to maintain torque. The result is that initial WOB should be about 23,000 lbf and the final WOB should be about 36,000 lbf, which means that the original calculation calling for six stands of drill collars still applies.

The rotary drilling cost per foot, using the Series 6-2-7 bit, is

$$C_{rot} = \frac{(\text{trip time} + \text{rotating time})(\text{rig cost}) + \text{bit cost}}{\text{feet drilled}}$$

$$= \frac{(16 \text{ hr} + 40 \text{ hr})(\$800/\text{hr}) + \$4,400}{120}$$

$$= 410 \text{ \$/ft,}$$

and for 140 ft, the cost per foot is \$390.

The cost per foot for the turbine is

$$C_{turbine} =$$

$$\frac{(\text{trip time} + \text{rotating time})(\text{rig cost}) + \text{bit cost}}{+ (\text{rotating time})(\text{turbine cost})}{\text{feet drilled}}.$$

Assuming the penetration rate averages 15 ft/hr, the costs per foot for 120 and 140 ft are

$$C_{120} = \frac{(16 + 8)(\$800/\text{hr}) + \$19,500 + (8 \text{ hrs})(\$500/\text{hr})}{120}$$

$$= \$356/\text{ft,}$$

and

$$C_{140} = \$317/\text{ft.}$$

The break-even point for the turbine/diamond-bit combination is at a drilled interval of 84 ft. The cost per foot and interval drilled can be calculated with the equation for turbine cost per foot. Fig. 8.119C shows the break-even point at 5.6 hours for a drilled interval of 84 ft. If the average penetration rate decreases to less than 15 to 12 ft/hr, it will take 7.9 hours and a drilled interval of 95 ft for the turbine/diamond-bit combination to break even.

8.7 Principles of the BHA

The BHA is the portion of the drillstring that affects the trajectory of the bit and, consequently, of the wellbore. Its construction could be simple, having only a drill bit, collars, and drillpipe, or it may be complicated, having a drill bit, stabilizers, magnetic collar, telemetry unit, shock sub, collars, reamers, jars, crossover subs, heavyweight drillpipe, and regular drillpipe. Fig. 8.120 depicts two BHA's.

In the earlier days of drilling, the slick assembly (bit with drill collars) was the most common. Later, Lubinski and Woods[18] showed that the pendulum assembly

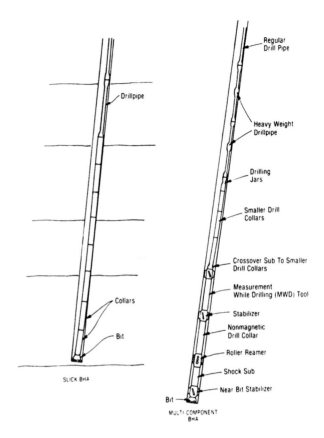

Fig. 8.120—Bottomhole assemblies.

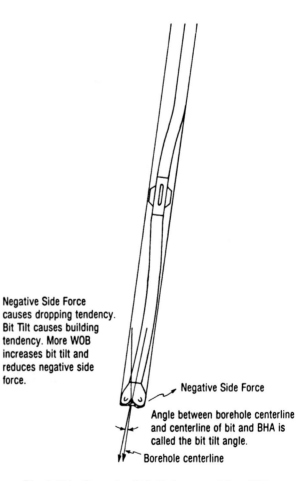

Negative Side Force causes dropping tendency. Bit Tilt causes building tendency. More WOB increases bit tilt and reduces negative side force.

Negative Side Force

Angle between borehole centerline and centerline of bit and BHA is called the bit tilt angle.

Borehole centerline

Fig. 8.121—Example of bit tilt for a pendulum BHA.

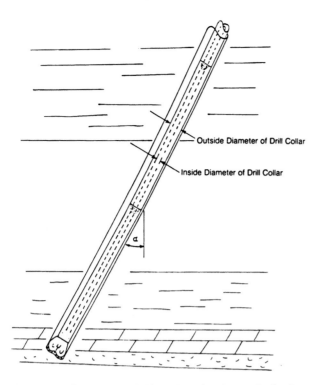

Fig. 8.122—Typical slick BHA drilling ahead at an inclination alpha.

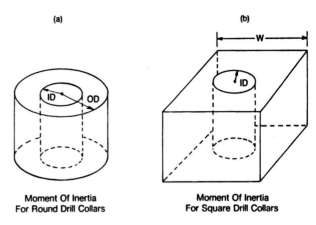

Fig. 8.123—Moment of inertia representation.

could be used for deviation control. Multistabilizer BHA's became popular because of directional drilling and, later, were shown effective in some attempts for deviation control.

All BHA's cause a side force at the bit that makes the bit build or drop or makes it hold an angle and turn to the left or right. Furthermore, the stabilizers and parts of the BHA that contact the wellbore exert side forces on the formation or casing. Sometimes these forces are so great that the contacting equipment wears a hole in the casing, mechanically wears or cuts the formation, and wears the pipe or stabilizer blades touching the wellbore. The forces and displacements for a given WOB and the rotary speed for any BHA can be determined accurately if the physical properties of each BHA component are known, and if the shape, size, and trajectory of the wellbore can be described.

Bit tilt is another factor in BHA mechanics that influences bit direction and inclination, especially in drilling softer formations. The curvature of the BHA centerline is transmitted to the bit, causing some tilt and movement in the direction of the centerline (see Fig. 8.121). The softer the rock, the more the bit tilt controls the trajectory of the bit. On the other hand, the harder the rock, the more the bit side force predominates. A bent housing works on the bit-tilt principle; with a bent sub, however, either a bit tilt or a side-force mechanism or both may exist.

This section will present the principles that govern BHA design and performance. The properties that govern the elastic behavior of a simple BHA will provide the basis for the more complex analysis of the single- and multistabilizer BHA's. Most of the BHA (bit and drill collars) analysis will apply to a 2D static system. The effects of drillstring rotation on the inclination and direction bit forces will be covered briefly at the end of this section.

8.7.1 Statistics of the Tubular Column

Fig. 8.122 is a BHA consisting of drill collars and a bit in an inclined borehole. The types of metals that compose the collars dictate the weight of the collars and their elastic behavior.

From the shape and dimensions of the collars, the axial moment of inertia, I, and the polar moment of inertia, J, can be determined. Most drilling components used in a BHA can be represented as a thick-walled cylinder or as a square column with a cylindrical hole in the center (see Fig. 8.123).

The axial moment of inertia for a thick-walled cylinder is expressed as follows.

$$I = \frac{(d_e^4 - d_i^4)\pi}{64}. \qquad (8.91)$$

The polar moment of inertia for the cylinder,

$$J = \frac{1}{32}\pi(d_e^4 - d_i^4) \quad \text{or} \quad J = 2I. \qquad (8.92)$$

Example 8.23. Calculate the axial and polar moments of inertia for a 6-in. round collar with a 2³⁄₁₆-in. ID (I_6 and J_6, respectively) and for an 11-in. collar with a 3-in. ID (I_{11} and J_{11}, respectively).

Solution.

$$I_6 = (6.0^4 - 2.1875^4)\pi/64 = 62.5 \text{ in.}^4$$

$$J_6 = 2(62.5) = 125.0 \text{ in.}^4$$

$$I_{11} = (11.0^4 - 3.0^4)\pi/64 = 715 \text{ in.}^4$$

$$J_{11} = 2(715 \ m^4) = 1,430 \text{ in.}^4.$$

In Example 8.24, the moment of inertia is increased an order of magnitude by increasing the OD from 6 to 11 in. If the ID's in both examples are neglected, the difference would be small (i.e., $I_6 = 63.6$ in.[4] and

TABLE 8.17—TYPICAL PROPERTIES OF SOME COMMON ALLOYS AND METALS

Metal	Melting Point (°F)	Density (lbm/cu ft)	Modulus of Elasticity (10^6 psi)	Tensile Strength (10^3 psi)	Yield Strength (10^3 psi)	Electrical Resistance Microhms (cm^3)	Thermal Conductivity (BTU/hr/sq ft/°F/ft)	Brinnel Hardness
Iron Base Alloys								
Steel, Low Carbon	2760	491	29.0	60	40	10	30	100 to 300
Cast Iron	2150	449	13.5 to 21	18 to 60	8 to 40	66	—	50
Ni Resist-Type 1	2250	456	—	—	—	140	—	30
Cr-Mo Steels	2500	491	27.4 to 29.9	63 to 84	40 to 56	10	15 to 19	70
12 Cr Steel	2720	484	29.2	89	47	9	13.0	70
Stainless 304	2590	501	27.4	85	35	72	9.4	160
Stainless 316	2550	501	28.1	90	42	74	9.4	165
Stainless 317	2550	501	—	85	40	74	—	165
Worthite Alloy 20	2650	501	—	—	—	75	—	160
25 Cr-12 Ni Steel	2650	501	—	80	57	78	8.0	160
25 Cr-12 Ni Steel	2650	501	28.2	88	33	78	8.0	165
Incoly 800	2525	503	28.5	82	43	93	8.0	184
Nickel Base Alloys								
Monel 400	2460	551	26.0	81	32	48	12.6	44
Monel k-500	2460	529	26.0	160	111	48	10.1	50
Nickel 200	2640	556	30.0	67	22	7	32.5	20
Inconel 600	2600	526	31.0	85	36	103	8.6	40
Hastelloy B	2460	577	31.0	131	56	135	6.5	23
Hastelloy C	2380	558	29.8	121	58	133	7.3	23
Super Alloys								
Nimonic 80	2590	515	27.0	155	87	124	7.0	185
Inconel X-750	2600	515	31.0	178	122	122	6.92	176
Refractaloy 26	2450	513	30.6	154	91	92	—	250
Haynes Alloy 31	2500	538	28.0	172	87	98	8.6	340
Other Metals								
Aluminum	1220	170	10.6	28	25	3	131	20
Titanium	3135	281	16.0	85	63	61	11.5	150
Tungsten	6200	1205	51.5	200	—	60 ± 18	95	230

$I_{11} = 718.7$ in.^{4}). If the ID in the 6-in. example is increased to 3 in., the moment is decreased only to 59.6 sq in. or 4.6%.

Young's modulus, E, relates the amount of strain of a material to a given amount of stress. This assumes that the material strains linearly when stressed and is in the Hooke's-law region. Most of the time the drillstring and BHA are in the elastic region. Sometimes, however, the pipe can be pulled beyond its elastic limit, resulting in plastic deformation and possible failure. Table 8.17 presents the values of some common alloys and metals. Note that the modulus of aluminum is nearly one-third that of mild steel and that the modulus of tungsten is nearly double. The modulus slightly decreases with an increase in temperature.

The product of the moment of inertia and the modulus of elasticity is called the stiffness of a material, EI. Fig. 8.124 is a plot of the stiffness of various drillstring and BHA components as a function of OD and ID.

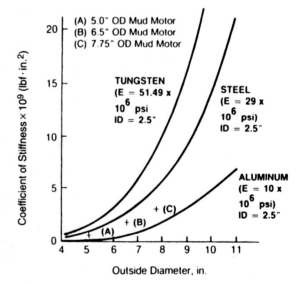

Fig. 8.124—Drill-collar stiffness (after Millheim[19]).

Example 8.24. Determine the stiffness of a tungsten collar having an OD of 6¼ in. and an ID of 2³⁄₁₆ in.

Solution.

$$EI = (51.5 \times 10^6 \text{ lbf/sq in.})$$

$$\times \frac{(\pi)[(6.25 \text{ in.})^4 - (2.1875 \text{ in.})^4]}{64}$$

$$= 3.80 \times 10^9 \text{ sq in.-lbf.}$$

The air weight of any BHA component can be determined if the density, cross-sectional area, and length are known (see Table 8.11). For most round drill collars, this can be determined if the ID, OD, and length are known. However, even the round drill collars can weigh less than the calculated air weight if, for example, they have machined grooves or if the OD at both ends has been decreased so that the collar can be picked up with elevators. Determining air weights of other BHA components—such as stabilizers, reamers, motors, shock subs, telemetry collars, jars, thick-walled drillpipe, and other downhole tools—is more complex because the cross-sectional geometries vary with length. Another reason the calculated air weight can differ from the actual air weight is that the wear on the outside of the component may not be uniform, so the cross-sectional shape of the component may be more elliptical than circular.

When any BHA component is placed in a fluid-filled hole, the air weight is reduced by the buoyancy of the component. The buoyancy correction factor, B_c, can be determined from Eq. 8.93:

$$B_c = \frac{(\rho - W_m)}{\rho}, \quad \dots \dots \dots \dots \dots \dots \dots (8.93)$$

where ρ is the density of the metal of the BHA component and W_m is the weight of the mud in consistent units.

Example 8.25. Determine the weight of 45 steel collars whose OD is 10 in., ID is $3\frac{1}{16}$ in., and ρ_s is 490 lbm/cu ft. Each collar is 31 ft long, and the mud weight is 16 lbm/gal.

Solution.

Weight of String in Air $= W_s$

$= (490 \text{ lbm/cu ft}) \left(\frac{\pi}{4}\right) \left(\frac{1 \text{ sq ft}}{144 \text{ sq in.}}\right)$

$\times (10^2 - 3.06^2) \text{ sq in. } (45) \, (31 \text{ ft})$

$= 337,909 \text{ lbm.}$

Weight of String in Mud $= W_s B_c$

$= (337,909 \text{ lbm}) \left[490 \text{ lbm/cu ft} - \frac{16 \text{ lbm/gal}}{8.33 \text{ lbm/gal}} \right.$

$\left. \times (62.4 \text{ lbm/cu ft}) \right] \div (490 \text{ lbm/cu ft}) = 255,255 \text{ lbm.}$

8.7.2 Slick BHA

Fig. 8.125 depicts a slick BHA (a) and a typical pendulum assembly (b). Each BHA has a negative component of side force, F_B, caused by gravity. At the bit, this component can be determined from

$$F_B = -0.5 W_c L_T B_c \sin \alpha, \quad \dots \dots \dots \dots (8.94)$$

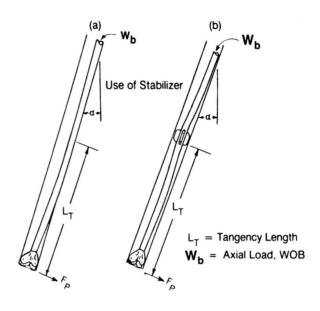

Fig. 8.125—Tangencies for slick and pendulum BHA's.

where W_c is the weight of the collar in lbm/linear ft in air, L_T is the length of the BHA between the bit and the first point of tangency (Fig. 8.125a and b) in feet, and α is the inclination angle.

Example 8.26. Determine the negative side force, in a 9-lbm/gal mud, for a slick BHA whose air weight is 98.6 lbm/ft. The wellbore is at an inclination of 4° and length to the point of the tangency (L_T) is 25 ft.

Solution.

$F_B = -(98.6 \text{ lbf/ft})(25 \text{ ft}) \{[489 \text{ lbf/cu ft}$

$- \frac{9 \text{ lbf/gal}}{8.33 \text{ lbf/gal}} (62.4 \text{ lbf/cu ft})]/489 \text{ lbf/cu ft}\}$

$\times \frac{\sin(4°)}{2} = -74.1 \text{ lbf.}$

If axial weight is applied to the bit, a positive force component, called the bending force, must be considered. Fig. 8.126 shows the zero axial load (a) and a BHA with the pipe bending as an axial load is applied (b). To determine the positive component of any BHA, one must assess the bending moments occurring over the active portion of the BHA. (Active portion refers to all parts of the BHA below the main tangency point.)

Eq. 8.95, presented by Jiazhi[20] and based on Timoshenko's[21] method of "Three Moment Equations," shows both the negative and the positive components, which are functions of WOB or applied axial load.

$F_B = -0.5 W_c B_c L_T \sin \alpha$

$+ (P_B - 0.5 W_c B_c L_T \cos \alpha) \ell/L_T. \quad \dots \dots \dots (8.95)$

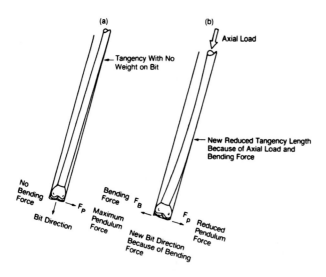

Fig. 8.126—Slick BHA without axial load and with axial load.

(Note that the sign convention used by Jiazhi is changed to be consistent with the present practice in BHA analysis of using a positive sign to mean a building side force and a negative sign to indicate a dropping side force.)

In Eq. 8.95 F_B is the bit side force (lbf), p_B is the axial or compressive load on the bit (lbf), and ℓ is the clearance radius of the drill collars (in.). Because L_T is unknown in both the negative and positive sides of the equations, it must be determined before F_B can be calculated.

The clearance of the drill collar is related by

$$\ell = 0.5(d_b - d_{dc}), \qquad \dots \dots \dots \dots \dots \dots \dots (8.96)$$

where d_b is the diameter of the bit and d_{dc} is the diameter of the drill collars.

For Jiazhi's solution, one must initially guess the tangency length, L_T. If the guess agrees with the length calculated by Eq. 8.97, Eq. 8.95 can be used to calculate the side force for a slick BHA with a constant inclination and drill collars of the same diameter:

$$L_T{}^4 = \frac{24EI\ell}{W_c B_c \sin \alpha \, X}, \qquad \dots \dots \dots \dots (8.97)$$

where X is a transcendental function related by Eq. 8.98.

$$X = \frac{3(\tan u - u)}{u^3}, \qquad \dots \dots \dots \dots \dots \dots (8.98)$$

where u is in radians and is given by

$$u = \frac{L_T}{2} \left(\frac{p_c}{EI}\right)^{0.5}. \qquad \dots \dots \dots \dots \dots \dots (8.99)$$

The compressive load on the collars, p_c, can be determined by

$$p_c = p_B - 0.5 W_c B_c L_T \, \cos \alpha. \qquad \dots \dots \dots \dots (8.100)$$

Example 8.27. Determine the resultant side forces for a slick BHA for 0-, 10,000-, 30,000-, 50,000-, 70,000-, and 80,000-lbf WOB. At what WOB will the BHA start building for a formation force of 0 and 525 lbf? Plot the negative and positive side-force components and the resultant side force, considering the 0- and 525-lbf formation forces. The drill bit diameter is 8.75 in.; the steel collars have 7.0-in. OD's and 2³⁄₁₆-in. ID's. Mud weight is 9.2 lbm/gal. Inclination is 3.2°.

Solution. An initial estimate of 40 ft is made for the case of 30,000-lbf WOB. The weight of the drill collars in 9.2-lbm/gal mud is

$$W_c = \frac{\pi}{4(144)} \frac{\text{sq ft}}{\text{sq in.}} (7.0^2 - 2.188^2) \text{ sq in.}$$

$$\times (489 \text{ lbm/cu ft}) B_c,$$

$$W_c = 118 \text{ lbm/ft } B_c,$$

$$B_c = \left[489 \text{ lbm/cu ft} - \frac{9.2 \text{ lbm/gal}}{8.33 \text{ lbm/gal}} \right.$$

$$\left. \times (62.4 \text{ lbm/cu ft}) \right] / (489 \text{ lbm/cu ft}) - 0.859,$$

$$W_c B_c = 118(0.859) = 101.4 \text{ lbm/ft},$$

$$\ell = 0.5(8.75 - 7.0)\text{in.} \left(\frac{\text{in.}}{12 \text{ ft}}\right) = 0.0729 \text{ ft},$$

$$P_c = 30,000 \text{ lbf} - 0.5(101.4 \text{ lbf/ft})$$

$$\times (40 \text{ ft})(\cos 3.2) = 27,977 \text{ lbf},$$

$$u = \frac{40}{2} \left[\frac{27,977 \text{ lbf}}{(4.18 \times 10^9 \text{ lbf/sq ft})(5.63 \times 10^{-3} \text{ ft}^4)} \right]^{0.5}$$

$$= 0.69,$$

$$X = \frac{3[\tan(0.69) - 0.69]}{(69)^3} = 1.24,$$

and

$$L_T =$$

$$\left[\frac{24(4.18 \times 10^9 \text{ lbf/sq ft})(5.63 \times 10^{-3} \text{ ft}^4)(0.0729 \text{ ft})}{(101.4 \text{ lbf/ft})(1.24)(\sin 3.2)} \right]^{0.25}$$

$$= 49.2 \text{ ft}.$$

Because the calculated L_T does not agree with the initial estimate of L, a second estimate of the average of the initial estimate and the calculated value should be used.

$$L_T = \frac{49.2 + 40}{2} = 44.6 \text{ ft}.$$

TABLE 8.18—RESULTING SIDE FORCES
AND TANGENCY LENGTH FOR VARIOUS WOB

Weight on Bit (lbf)	0 Formation Force F_B (lbf)	525 lbf Formation Force F_B (lbf)	L_T (ft)
0	−147	378	51.9
10,000	−133	392	50.8
30,000	−94	431	48.0
50,000	−51	474	45.3
70,000	−6	519	42.9
80,000	18	543	41.7

Successive iterations yield the value of L_T=48.0 ft.
The side force at the bit is found as follows:

$$F_B = -0.5(101.4 \text{ lbf/cu ft})(48 \text{ ft})(\sin 3.2)$$
$$+[30,000 \text{ lbf}-0.5(101.4 \text{ lbf/cu ft})(48 \text{ ft})(\cos 3.2)]$$
$$\times(0.0729 \text{ ft})/48 \text{ ft},$$
$$= -93.8 \text{ lbf}.$$

The negative sign indicates a net dropping tendency with 0 lbf of formation forces and 431.2 lbf of build tendency with a 525-lbf formation force.

Similar calculations yield the results found in Table 8.18 for the other WOB's.

Fig. 8.127 is a plot of the results of Example 8.28 for 0-lbf formation force and 525-lbf formation force. The bending or positive side force and negative side force components also are plotted vs. WOB. It appears that, for this particular BHA with 0-lbf formation force, nearly 80,000-lbf WOB would be required to start a slight build, and for 525-lbf formation force there would be a moderate building tendency, even at the very low bit weights.

The tendency of the bit in the foregoing example to build, to hold, or to drop angle is based on a positive, zero, or negative side force. Essentially, this would be the case for hard formations (i.e., drilling rates of 1 to 10 ft/hr). When the formation is soft to medium-hard, the side-force tendency is not the only component that will influence the inclination and direction of the bit. Because of the curvature of the BHA near the bit, the bit is canted or tilted in some resultant direction and inclination, somewhat like the bent housing and bent sub. The magnitude of the tilt is directly influenced by the strength of the formation. Just as a deflection tool will not obtain the maximum curvature for which it was designed in harder formations, so it is with a BHA and a given bit tilt. In very soft formations (drilling rates exceeding 100 ft/hr), the side force again can be the predominant mechanism and will, in many cases, mitigate the effects of BHA bit tilt. This is especially true when the larger, stiffer collars are run.

When the formations are soft to medium-hard (drilling rates from 10 to 100 ft/hr), effects of the bit tilt can be significant. To determine the bit tilt, one must know the curvature of both the wellbore and the BHA near the bit. Analytical BHA solutions, such as Jiazhi's, are difficult and cumbersome to use with the varying wellbore incli-

nations and directions and to describe the curvature of the BHA. To calculate bit tilt, one of the finite-element BHA algorithms or similar solutions are better suited.[20,21] To be truly accurate, the dynamic effects of the BHA must be considered. However, a strong understanding of the basic BHA mechanics with Jiazhi's technique will give an insight to the ideal 2D behavior of most common BHA's.

In the special case of the slick BHA, no stabilizers are used. In the previous example calculation, 7-in. collars are used in an 8.75-in. wellbore, which generates side forces ranging from a low of −147 lbf at 0 lbf WOB to a high of 18 lbf at 80,000 lbf WOB. If the formation forces are 525 lbf, the slick BHA is going to build angle, no matter what weight is used on the bit, until an equilibrium inclination angle is reached at which the formation force is offset by the negative side force.

Example 8.28. For the previous example, estimate the inclination at which the slick BHA will cease building angle (assume no WOB).

Solution. Assuming the calculated tangency length will not change substantially,

$$525=0.5(101.4 \text{ lbf/ft})(48 \text{ ft})\sin \alpha$$

and

$$\alpha=12.46° \text{ inclination.}$$

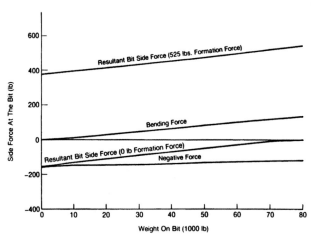

Fig. 8.127—Results of BHA calculation for slick BHA.

If it is desired to maintain a low inclination angle with a strong formation force of 525 lbf in the 8.75-in.-OD hole, the 7-in.-OD collars are as large as they can be for safe use. An alternative is to drill a larger wellbore so that collars with larger diameters can be used.

Example 8.29. Estimate the collar OD, assuming a $2^{13}\!/_{16}$-in.-ID collar and a $9^7\!/_8$-in.-diameter wellbore, 9.2-lbm/gal mud, and a maximum tolerable inclination of 3.2° to offset the 525-lbf formation force.

Solution.

$$525 \text{ lbf} = \frac{\pi}{4(144)} \left(\frac{\text{sq ft}}{\text{sq in.}}\right)(d_e{}^2 - 2.8125^2)\text{sq in.}$$

$$\times (48 \text{ ft})(489 \text{ lbm/cu ft})(0.859)(\sin 3.2°),$$

$$d_e = 9.67 \text{ in.} \approx 9.5 \text{ in.-OD collars.}$$

The 9.5-in. collars would be too large for a $9^7\!/_8$-in. wellbore but could be used in a 12¼-in. wellbore. If the hole size cannot be enlarged, the only other possibility is to increase the tangency length of the BHA. With the use of a stabilizer to move the tangency point farther up the wellbore, more negative force can be obtained. A pendulum assembly is one that has a stabilizer to control the tangency length.

8.7.3 Single-Stabilizer BHA

The same type of analysis performed for a slick BHA can be applied to a single-stabilizer BHA (see Fig. 8.128). Again, one must estimate a tangency length that agrees with the calculated length $L_T{}^2$ in Eq. 8.101.

$$L_T{}^4 = \frac{24EI_2(\ell_2 - \ell_1)}{q_2 x_2} - \frac{4m_1 L_T{}^2 W_2}{q_2 x_2}. \quad \ldots \ldots (8.101)$$

The unknown bending moment, m, is calculated from the relationship

$$2m_1\left(V_1 + \frac{L_2 I_1}{L_1 I_2} V_2\right) = -\frac{q_1 L_1{}^2}{4}x_1 - \frac{q_2 L_2{}^3 I_1}{4L_1 I_2}x_2$$

$$+ \frac{6EI_1\ell_1}{L_1{}^2} + \frac{6EI_1(\ell_1 - \ell_2)}{L_1 L_2}, \quad \ldots \ldots \ldots (8.102)$$

where $q_1 = W_{c1} B_c \sin \phi$, $q_2 = W_{c2} B_c \sin \phi$, W_{c1} is the weight of the drill collars from the bit to the stabilizers, and W_{c2} is the weight of the collars from the stabilizers to the point of tangency.

The coefficients W_i and V_i can be calculated from Eqs. 8.103 and 8.104.

$$W_i = \frac{3}{u_i}\left[\frac{1}{\sin(2u_i)} - \frac{1}{2u_2}\right] \quad \ldots \ldots \ldots \ldots (8.103)$$

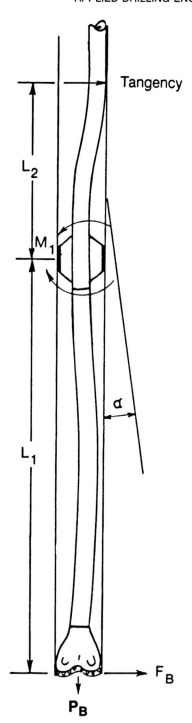

Fig. 8.128—Typical single-stabilizer BHA.

and

$$V_i = \frac{3}{2u_i} \left[\frac{1}{2u_i} - \frac{1}{\tan(2u_i)} \right], \quad \dots \dots \dots (8.104)$$

where $i = 1$ or 2. Coefficients X_i and u_i are determined from Eqs. 8.98 and 8.99, considering the collars from the bit to the stabilizer and from the stabilizer to the point of tangency.

$$X_i = \frac{3[\tan(u_i) - u_i]}{u_i^3},$$

where

$$u_i = \frac{L_i}{2} \left(\frac{p_{ci}}{EI_i} \right)^{0.5}.$$

Eq. 8.96 is used to calculate the clearances for each section of collar where d_s is the diameter of the stabilizer and d_2 is the diameter of the collars that achieve tangency.

$$\ell_1 = 0.5(d_b - d_s)$$

and

$$\ell_2 = 0.5(d_b - d_2).$$

The compressive load for the first section is calculated with Eq. 8.103, and

$$p_{c1} = p_B - \left[\left(\frac{L_1}{2} W_{c1} B_c \right) \cos \phi \right],$$

while the compressive load of the second section is given by Eq. 8.105.

$$p_{c2} = p_B - \{ [(W_{c1} B_c L_1) + (0.5 W_{c2} B_c L_2)] \cos \phi \}.$$

$$\dots \dots \dots \dots \dots \dots \dots (8.105)$$

If the estimated value of L_2 equals L_T, Eq. 8.106 can be used to calculate the side force at the bit.

$$F_B = -0.5 B_c W_{c1} L_1 \sin \phi + \frac{p_{c1} \ell_1}{L_1} - \frac{m}{L_1}. \quad \dots (8.106)$$

If the estimated value of L_2 does not agree with the value of L_T, the new assigned value should be an average of L_2 and L_T and the same calculation procedure should be repeated until $L_2 = L_T$.

Example 8.30. Consider the case of a single-stabilizer building BHA on which the distance between the bit and the first stabilizer is 5.0 ft and the wellbore diameter is 12¼ in. All the drill collars have an OD of 8.0 in. and an ID of 2¹³⁄₁₆ in. The diameter of the stabilizer blade is 12.21875 in.; the inclination angle is 10°, and the mud weight is 10.5 lbm/gal.

Calculate the bit side forces (F_B) for WOB's of 10,000, 20,000, 30,000, 40,000, and 50,000 lbf.

Solution. For 30,000-lbf WOB, the initial estimate for L_2 is 40 ft. The stepwise calculation is shown below. Note that for $W_{c1} B_c$, the weight of the collars in mud,

$$W_{c1} = W_{c2} = 149.6 \text{ lbf and } B_c = 0.839,$$

$$W_{c1} B_c = (149.6 \text{ lbf})(0.839) = 125.5 \text{ lbf/ft},$$

$$E = 4.176 \times 10^9 \text{ lbf/sq ft},$$

$$I = 9.55 \times 10^{-3} \text{ ft}^4,$$

$$\ell_1 = 0.5 \left(\frac{\text{ft}}{12 \text{ in.}} \right) (12.25 - 12.21875) \text{in.} = 0.00130 \text{ ft},$$

$$\ell_2 = 0.5 \left(\frac{\text{ft}}{12 \text{ in.}} \right) (12.25 - 8.0) \text{in.} = 0.17708 \text{ ft},$$

$$p_{c1} = 30,000 - 0.5(125.5 \text{ lbf/ft})(5 \text{ ft})(\cos 10°)$$

$$= 29,691 \text{ lbf},$$

$$p_{c2} = 30,000 - [(125.5 \text{ lbf/ft})(5 \text{ ft})$$

$$+ 0.5(125.5 \text{ lbf/ft})(40 \text{ ft})](\cos 10°) = 26,900 \text{ lbf}.$$

$$u_1 = \left(\frac{5 \text{ ft}}{2} \right)$$

$$\times \left[\frac{29,691 \text{ lbf}}{(4.176 \times 10^9 \text{ lbf/sq ft})(9.55 \times 10^{-3} \text{ ft}^4)} \right]^{0.5}$$

$$= 6.82 \times 10^{-2},$$

$$u_2 = \left(\frac{40 \text{ ft}}{2} \right)$$

$$\times \left[\frac{26,900 \text{ lbf}}{(4.176 \times 10^9 \text{ lbf/sq ft})(9.55 \times 10^{-3} \text{ ft}^4)} \right]^{0.5}$$

$$= 5.20 \times 10^{-1},$$

$$x_1 = 3 \frac{\tan(0.0682) - 0.0682}{(0.0682)^3} = 1.00185,$$

$$x_2 = 3 \frac{\tan(0.520) - 0.520}{(0.520)^3} = 1.12117,$$

$$W_2 = \frac{3}{(0.520)} \frac{1}{\sin(2 \times 0.520)} - \frac{1}{2(0.520)} = 1.142,$$

$$V_1 = \frac{3}{2(0.0682)} \frac{1}{2(0.0682)} - \frac{1}{\tan(2 \times 0.0682)}$$

$$= 1.00124,$$

$$V_2 = \frac{3}{2(0.520)} \frac{1}{2(0.520)} - \frac{1}{\tan(2 \times 0.520)} = 1.08025,$$

and

$$q_1 = 125.5 \text{ lbf/ft } (\sin 10°) = 21.8 \text{ lbf/ft}.$$

Therefore,

$$2m \left[1.00124 + \frac{(40)(1.08025)}{5} \right]$$

$$= - \frac{(21.8 \text{ lbf/ft})(5 \text{ ft})^2}{4}(1.00185)$$

$$- \frac{(21.8 \text{ lbf/sq ft})(40 \text{ ft})^3}{4(5 \text{ ft})}(1.12117)$$

$$+ \frac{6(3.987 \times 10^7 \text{ lbf/sq ft})(0.00130 \text{ ft})}{(5 \text{ ft})^2}$$

$$+ \frac{6(3.987 \times 10^7 \text{ lbf/sq ft})(0.00130 - 0.17708)}{(5 \text{ ft})(40 \text{ ft})}.$$

Therefore,

$$m = -14{,}318 \text{ ft-lbf}.$$

$$L_T{}^4 = \frac{24(3.987 \times 10^7 \text{ lbf/sq ft})(0.17708 - 0.00130)\text{ft}}{(21.8 \text{ lbf/ft})(1.12117)}$$

$$- \frac{4(-14{,}318 \text{ ft/lbf})(L_T{}^2)(1.142)}{(21.8 \text{ lbf/ft})(1.12117)}.$$

Therefore,

$$L_T = 57.8 \text{ ft}.$$

Because L_T does not equal L_2, 48 ft should be used for L_2 in the next try.

Successive tries will show that $L_T = L_2 = 65.5$ ft and that

$$F_B = - \{ [0.5(125.5 \text{ lbf/sq ft})(5 \text{ ft})(\sin 10°)] \}$$

$$+ \frac{(29{,}691 \text{ lbf})(0.00130 \text{ ft})}{(5 \text{ ft})} - \frac{(-15{,}552)}{5 \text{ ft}}$$

$$= 3{,}058 \text{ lbf}.$$

Table 8.19 summarizes the solutions according to bit weight.

This solution indicates that the additional WOB of 20,000 lbf increases the bit side force from 3,058 to 3,086 lbf and reduces the tangency length from 65.5 to 64.2 ft.

Fig. 8.129 shows bit side force as a function of distance of the stabilizer from the bit for WOB's of 10,000 to 60,000 lbf for the BHA cited in Example 8.31. Fig. 8.130 is a similar plot for 6½-in. collars. (A finite-element BHA code[20] was used to generate these plots. The tech-

TABLE 8.19—SUMMARY OF BIT SIDE FORCES FOR VARIOUS WOB'S

Weight on bit (lbf)	Bit Side Force F_B (lbf)	Tangency Length L_T (ft)	Bending Moment M (ft-lbf)
10,000	3,030	66.9	15,410
20,000	3,044	66.2	15,466
30,000	3,058	65.5	15,522
40,000	3,072	64.9	15,579
50,000	3,086	64.2	15,638

nique used in Example 8.31, however, can be used to calculate the same data up to the point where tangency between the bit and the stabilizer occurs.)

The stabilizer near the bit causes a building or positive side force. As the stabilizer is moved away from the bit, between 30 and 35 ft, a 0-lbf bit side force is achieved. This assembly is called a neutral BHA. If the stabilizer is positioned beyond 30 ft, the bit side force becomes negative and decreases to a maximum negative value. The single-stabilizer negative assembly is called a pendulum or dropping assembly. The maximum pendulum or negative bit side force is reached at the point where the drill collars achieve tangency between the bit and the stabilizer. The BHA solution by Jiazhi cannot predict this tangency. Other BHA algorithms that can calculate the maximum pendulum force when tangency occurs indicate that the maximum negative side force occurs between 75 and 85 ft for bit loads of 10,000 to 60,000 lbf. For the smaller-diameter wellbores and drill collars, the maximum negative side force occurs between 55 and 80 ft for bit loads of 10,000 to 60,000 lbf.

8.7.4 Two-Stabilizer BHA's

The two-stabilizer BHA's can also be solved with Jiazhi's[20] technique. Fig. 8.131 depicts a typical two-stabilizer building BHA; L_1 and L_2 are known lengths between the bit and the first stabilizer and between the first and second stabilizers. The distance, L_3, between the second stabilizer and the point of tangency is unknown, and, as in the case of the slick and single-stabilizer solutions, L_3 must be estimated initially.

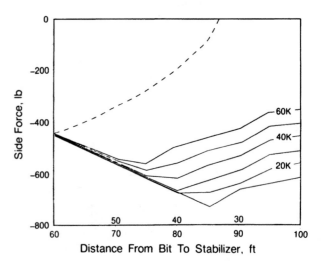

Fig. 8.129—Side force vs. pendulum collar length; 12¼-in. hole, 8-in. collars, 10½-lbm/gal mud; 10° inclination.

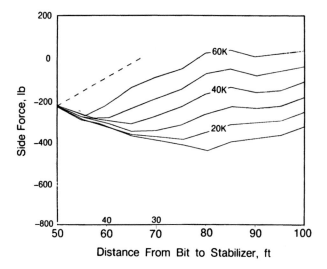

Fig. 8.130—Side force vs. pendulum collar length; 12¼-in. hole, 6½-in. collars, 10½-lbm/gal mud, 10° inclination.

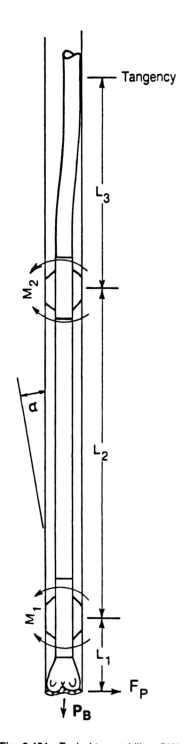

Fig. 8.131—Typical two-stabilizer BHA.

Three different collar diameters and material types can be used in the solution. Once a correct tangency length is obtained, the two moments, M_1 and M_2, can be determined, and the bit side force can be calculated.

As in Eq. 8.96, the clearance between the stabilizers and the wellbore and the last collar and the wellbore are given by

$$\ell_1 = 0.5(d_b - DS_1)/12,$$

$$\ell_2 = 0.5(d_b - DS_2)/12,$$

and

$$\ell_3 = 0.5(d_b - DC_3)/12.$$

Eq. 8.107 can be used to calculate the bit side force.

$$F_B = -0.5W_c B_c L_1 \sin \phi + P_{c1}\ell_1/L_1 - m_1/L_1.$$

$$\dots\dots\dots\dots\dots\dots\dots\dots\dots (8.107)$$

Everything is known in Eq. 8.108 except m_2. To determine m_2, a value of L_3 must be estimated. If the estimated L_3 agrees with L_T in Eq. 8.108, the equivalent value of m_2 can be used in Eq. 8.108, and m_1 can be used in the bit-side-force calculations.

$$L_T{}^4 = \frac{24EI_3(\ell_3 - \ell_2)}{q_3 X_3} - \frac{4m_2 L_3{}^2 W_3}{q_3 X_3}, \quad \dots\dots (8.108)$$

and

$$q_i = W_c B_c \sin \phi,$$

where $i = 1, 2,$ or 3.

The relationships for the first and second moments are given by Eqs. 8.109 and 8.110.

$$2m_1\left(V_1 + \frac{L_2 I_1}{L_1 I_2}V_2\right) + m_2\frac{L_2 I_1}{L_1 I_2}W_2$$

$$= -\frac{q_1 L_1^2}{4}X_1 - \frac{q_2 L_2^3 I_1}{4L_1 I_2}X_2 + \frac{6EI_1 \ell_1}{L_1^2}$$

$$+ \frac{6EI_1(\ell_1 - \ell_2)}{L_1^2}. \quad\dots\dots\dots\dots\dots (8.109)$$

$$m_1 W_2 + 2m_2\left(V_2 + \frac{L_3 I_2}{L_2 I_3}V_3\right)$$

$$= -\frac{q_2 L_2^2}{4}X_2 - \frac{q_3 L_3^3 I_2}{4L_2 I_3}$$

$$- \frac{6EI_2(\ell_1 - \ell_2)}{L_2^2} - \frac{6EI_2(\ell_3 - \ell_2)}{L_2 L_3}. \quad\dots\dots (8.110)$$

X_i, W_i, and V_i can be calculated from Eqs. 8.98, 8.103, and 8.104, where $i = 1$, 2, or 3.

$$X_i = \frac{3[\tan(u_i) - (u_i)]}{u_i^3},$$

$$W_i = \frac{3}{u_i}\left[\frac{1}{\sin(2u_i)} - \frac{1}{2u_i}\right],$$

and

$$V_i = \frac{3}{2u_i}\left[\frac{1}{2u_i} - \frac{1}{\tan(2u_i)}\right],$$

where

$$u_i = \frac{L_i}{2}[(p_{ci}/EI_i)^{0.5}];$$

and

$$p_1 = p_B - [(0.5W_{c1}B_c L_1)\cos \phi], \quad\dots\dots (8.111)$$

$$p_2 = p_B - \{[(W_{c1}B_c L_1) + 0.5W_{c2}L_2]\cos \phi\},$$

$$\dots\dots\dots\dots\dots (8.112)$$

and

$$p_3 = p_B - \{[(W_{c1}B_c L_1) + (W_{c2}B_c L_2)$$

$$+ (0.5W_{c3}L_3)]\cos \phi\}. \quad\dots\dots (8.113)$$

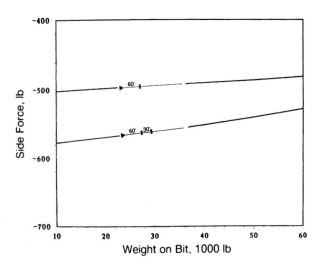

Fig. 8.132—Effect of adding a stabilizer; 12¼-in. hole, 8-in. collars, 9-lbm/gal mud, 10° inclination.

Fig. 8.132 shows the solution for a single-stabilizer, 60-ft pendulum assembly at 10° inclination vs. the two-stabilizer pendulum assembly. The second stabilizer increases the negative side force by reducing the effect of the positive bending force.

The slick, the single-stabilizer, and the two-stabilizer BHA's can be analyzed with the foregoing algorithms. Also, the scheme proposed by Jiazhi[20] can be expanded to handle multistabilizer BHA's, including those with three, four, and five stabilizers.

For the slick BHA there is no solution, except at the tangency length L_T, when $u_i < 1.57$ and $P_{c1} > 0$. This means that the solution technique is valid only as long as the lower part of the BHA is in compression. The same applies for the single stabilizer. When $u_i > 1.57$ and $p_{c1} < 0$, there is no solution; and when $u_2 > 1.57$ and $p_2 < 0$, there is no solution except at the final tangency length, L_T, where $u_2 < 1.57$ and $p_2 > 0$. The two-stabilizer BHA has no solution when $u_1 > 1.57$, $p_{c1} < 0$, $u_2 > 1.57$ and $p_{c2} < 0$. When $u_3 > 1.57$ and $p_3 < 0$, there is no solution except at the final tangency length, where $u_3 < 1.57$ and $p_3 > 10$.

This BHA analysis technique does not allow for hole curvature and cannot handle cases in which tangency occurs between the bit and the first stabilizer or between the two stabilizers. Furthermore, the wellbore must be a constant gauge. Adding too much WOB can result in no solution because the collars usually reach tangency between the bit and the first stabilizer or between the two stabilizers. This technique is only 2D and static and does not give a directional side-force component. If the neutral point is below the tangency length, no solution is obtained. In this analytical technique, blade lengths are ignored, and it is assumed that point-contact stabilizers are used.

Even with all those restrictions, the technique can provide basic insights into the mechanics of a number of BHA configurations and can help explain why BHA's behave in a certain manner for different hole sizes, inclinations, collar diameters, and applied WOB. Certain programmable hand calculators can be used to solve the slick, single-stabilizer, and multistabilizer problems.

8.7.5 Multistabilizer BHA Analysis

A solution technique for the slick, one-, and two-stabilizer BHA's has been presented. Other techniques, developed by Walker[22] and by Millheim and Apostal,[23] solve the three-dimensional (3D) BHA case. Both techniques yield inclination and direction side-force components. They also handle wellbore curvature, variable gauge holds, and combination BHA components. Unlike the analytical solution, these more generalized solutions can handle situations in which tangency occurs between the bit and stabilizer or between the stabilizers, as well as the cases in which increases in WOB force the creation of additional points of tangency.

Fig. 8.133A shows the two-stabilizer pendulum cases in which tangency occurs between the bit and the stabilizer because of the pendulum length and inclination angle. Fig. 8.133B shows a two-stabilizer, 90-ft building BHA in which the tangency occurs between the two stabilizers; Fig. 8.133C shows the effect of increasing the WOB.

Hole curvature also can influence the response of a BHA significantly. Fig. 8.134 shows how a curvature affects bit side force for a build rate of 1°/100 ft to 12° inclination. A case of constant inclination of 12° is shown also. All BHA's try to reach equilibrium for a given set of conditions—i.e., geology, penetration rate, WOB, speed, BHA configuration, inclination, and hole condition. As long as the conditions remain essentially constant, the average curvature is constant (see Fig. 8.135).

Whenever a BHA is run in a section of hole that has not been created by a BHA of that configuration, the curvature of the hole can cause various consequences: the new BHA may not be able to reach the bottom of the wellbore; the bit may stop rotation; or the BHA responds in a manner counter to that for which it was designed. Curvature can also accelerate a BHA response, especially with building BHA's.

If a formation is soft, the hole curvature caused by a mud motor with a bent sub or a bent housing usually causes the BHA to drill ahead with increased torque and to ream out the hole. The harder the formation, the greater the effects of hole curvature. When a trajectory has been changed with a mud motor in harder formations, the BHA that is used later usually reverses the original curvatures of the wellbore; this is called "bounce back."

Hole curvature can significantly affect more than just the bit side forces. A good example is a pendulum assembly that is run in a hole to reduce angle. Even when the pendulum side force is adequate to reduce angle, the curvature effects can cause the BHA to build angle beyond the point of equilibrium. The curvature of the wellbore can contribute to a bit-tilt effect that is stronger than the side-force effect.

$$G=(F_B C_1 + C_2)100, \quad\ldots\ldots\ldots\ldots\ldots\ldots(8.114)$$

where $C_1 = $ deg/lbf-ft, $C_2 = $ deg/ft, and G is the resultant bit curvature in degrees/100 ft.

Example 8.31. Determine the resultant bit angle for a bit side force of −162 lbf, where the side-force side-cutting response is 0.0001°/lbf-ft and the bit tilt effect caused by the BHA bit tilt and hole curvature is 0.025°/ft.

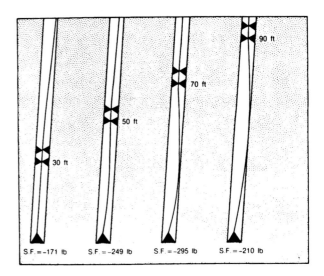

Fig. 8.133A—Tangency between bit and stabilizer resulting from pendulum length and inclination; 9⅞-in. hole, 6¾-in. collars, 8° inclination, 30,000-lbf WOB.

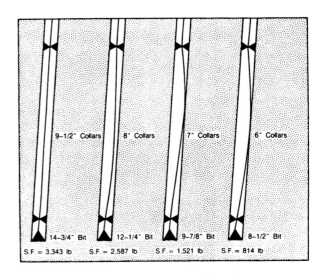

Fig. 8.133B—Tangency between bit and stabilizer; 90-ft building assembly, 10° inclination, 30,000-lbf WOB.

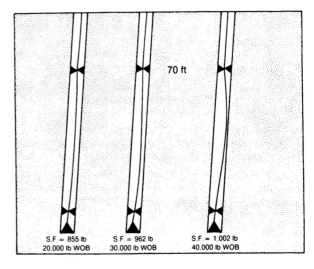

Fig. 8.133C—Tangency resulting from increasing WOB; 8½-in. hole, 6-in. collars, 10° inclination, 70-ft tangency length.

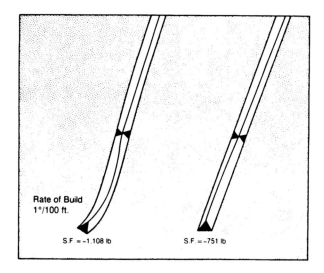

Fig. 8.134—Curvature and its effect on bit side force; 12¼-in. hole, 9-in. collars, 30,000-lbf WOB, 14° constant inclination.

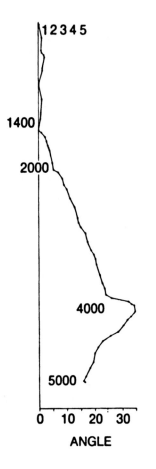

Fig. 8.135—Data showing constant inclination.

Solution.

$$G = [(-162 \text{ lbf})(0.0001°/\text{lbf ft}) + 0.025°/\text{ft}]100$$

$$= 0.88°/100 \text{ ft}.$$

8.7.6 BHA's for Building Inclination Angle

Fig. 8.136 presents various commonly used BHA's for building inclination angle. Fig. 8.137 shows the side-force response of the building BHA's for 8-in. drill collars in a 12¼-in. wellbore as a function of inclination angle from 5 to 60° for a WOB of 30,000 lbf. These cases were solved with a finite-element algorithm presented by Millheim. [19]

The most building side force is generated by the 90-ft building BHA, except at lower inclinations, where the single-stabilizer building assembly can generate more side force. Rates of build ranging from 2 to 5°/100 ft can be achieved with these building assemblies. Addition of WOB, depending on collar size, increases the rate of build. This is caused not so much by the bit side force as by the increase in bit tilt. The smaller the collar size relative to the diameter of the hole, the greater the influence of bit tilt.

In the earlier days of directional drilling in the Gulf of Mexico, smaller drill collars were used for most BHA's (6 to 6½-in. drill collars in 9⅞- to 12½-in. wellbores). Such building assemblies were very sensitive to WOB, sometimes responding to changes of less than 5,000 lbf.

In harder rocks, the 90-ft BHA is not as responsive as in softer rocks and is less affected by bit tilt. Rates of build of 1 to 2°/100 ft are fairly common.

The single-stabilizer building assembly can achieve a response approaching that of the 90-ft BHA, especially when the smaller drill collars are used. With larger collars—8 to 11 in.—the response is usually less than that for the 90-ft BHA. The single-stabilizer building BHA is more responsive than the intermediate—55- to 75-ft—building assemblies that generate rates of build from 1 to 3°/100 ft. Response also depends on the geology, inclination, wellbore diameter, and collar diameter. As the inclination increases, the general response of all building assemblies increases (see Fig. 8.137).

The rate of build with the three-stabilizer, 30- to 50-ft building assemblies varies from slight to moderate; in some situations those assemblies could even be considered holding assemblies. The two-stabilizer, 30- to 50-ft BHA's with 8-in. collars function as dropping BHA's for the complete range of inclinations. These assemblies are almost unaffected by bit tilt. They are generally used to regain inclination in the hold section. Similarly, the BHA with an undergauge stabilizer in the middle (No. 3 in Fig. 8.136) is used as a slight- to medium-building assembly, depending on how much under gauge the mid-stabilizer is and how responsive to weight the BHA is.

In modern directional drilling, especially in soft formations, the practice is to use the fewest drill collars and stabilizers possible to accomplish a given objective. Thick-walled or heavy drillpipe replaces regular drill collars, eliminating the need for stabilizers that hold the drill collars off the wellbore wall.

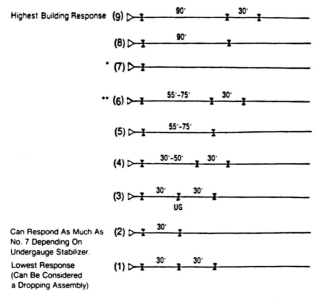

* At lower inclinations this BHA is the most responsive.

** Fig 8.137 shows that the level of building tendency changes with inclination where BHA's 6 and 7 generate more side force at higher angles.

Fig. 8.136—BHA's for building inclination angle.

A typical building BHA has a bit or near-bit stabilizer placed 3 to 5 ft from the bit face to the leading edge of the stabilizer blade. Beyond the last stabilizer are three to six drill collars and enough heavy drillpipe to satisfy the WOB requirements.

8.7.7 BHA's for Holding Inclination Angle

Holding BHA's do not maintain inclination angle; rather, they minimize angle build or drop. All the BHA's in Figs. 8.138 and 8.139 have either a slight building or a slight dropping tendency. The four-stabilizer holding BHA (No. 7) shows the least change with side force as the inclination increases (see Fig. 8.139). Using more than five stabilizers for deviation control has no added effect on the neutrality of the BHA. At lower inclinations, however, the five-stabilizer BHA is most effective. At higher

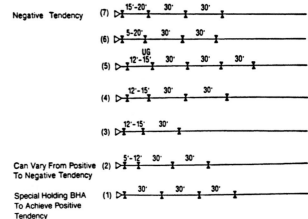

Fig. 8.138—BHA's for holding inclination angle.

inclinations, the fifth stabilizer can add too much torque for the rotary system, so the three- or four-stabilizer BHA's usually are used. With 8-in. collars, they have a negative or dropping tendency. Fig. 8.139 shows the bit-side-force tendencies as a function of inclination for the BHA's depicted in Fig. 8.138. The holding BHA undergauge second stabilizer is used when a slight positive side force or building tendency is required: for example, where the geology or hole conditions are such that the normal BHA's drop too quickly. Also, an undergauge second stabilizer causes a slight bit tilt.

A holding BHA is actually a BHA designed to build or to drop inclination slightly, opposing the formation characteristics in such a way as to prevent a rapid change in inclination angle. Minimal bit tilt, as well as stiffness of the BHA near the bit, also helps maintain inclination angle. Also characteristic of the holding BHA is the small variation in bit side force as a function of WOB change. (See Fig. 8.140.)

Assembly No. 1 in Fig. 8.138 can have either a building or a dropping tendency, which is dependent on a variety of conditions. If the near-bit stabilizer becomes undergauge or the formation around the bit and the stabilizer is eroding, this BHA can respond similarly to

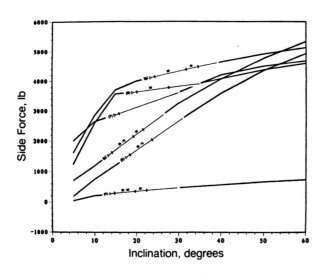

Fig. 8.137—Side-force response as a function of inclination angle, for building BHA's referred to in Fig. 8.136.

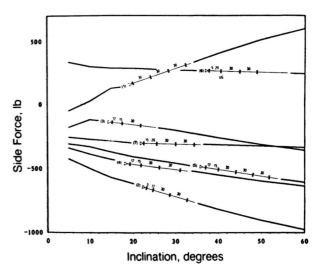

Fig. 8.139—Bit side force tendency as a function of inclination angle, for holding BHA's referred to in Fig. 8.138.

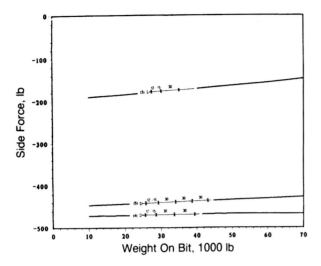

Fig. 8.140—Bit side force as a function of WOB, for holding BHA's.

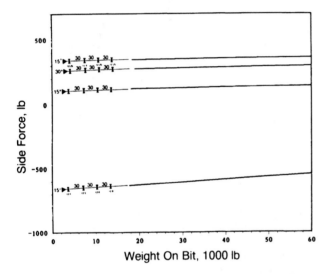

Fig. 8.141—Effect of geology on performance of Holding BHA No. 2 (12½-in. hole, 9-lbm/gal mud, 8-in. collar).

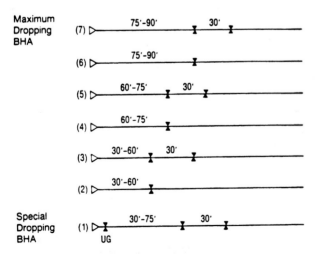

Fig. 8.142—BHA's for dropping inclination.

a 30-ft pendulum BHA. If those conditions exist at higher inclinations, this BHA will drop inclination at a moderate rate. Fig. 8.141 indicates the tendencies of this BHA under various conditions.

8.7.8 BHA's for Dropping Inclination Angle

Fig. 8.142 presents common dropping assemblies. The 75- to 90-ft BHA with two stabilizers (No. 7) achieves the greatest dropping response, except at the higher inclinations, where it approaches BHA No's. 4 and 5. As the inclination increases, more and more of the collars make contact with the wellbore between the bit and the first stabilizer, causing a reduction in the negative side force at the bit. Table 8.20 shows the tangency points for various hole sizes, collar sizes, and inclinations. For example, it shows that, for a 90-ft pendulum BHA with 8-in. collars at 10° in a 12¼-in. wellbore, the tangency occurs at 38.6 ft from the bit.

As previously discussed, the second stabilizer, which is 30 ft from the first stabilizer, increases the negative side force. The dropping assembly with the undergauge near-bit stabilizer is used when a drop is initiated at higher inclination angles. Except for the type with an undergauge stabilizer (like BHA No. 1), pendulum BHA's are rarely used for directional drilling. They are used more for deviation control and are discussed in the section on that subject.

TABLE 8.20—PENDULUM-ASSEMBLY TANGENCY POINTS

Inclination (Degrees)	Tangency Point, ft—WOB, 1,000 lbf		
	7⅞-in. Hole. 6¼ × 2¼ Collar	8¾-in. Hole. 6¾ × 2¼ Collar	12¼-in. Hole. 8 × 2¼ Collar
30-ft Pendulum Assembly			
10	30— 0 to 50	30— 0 to 50	30—0 to 50
20	—	—	—
40	—	—	—
60	—	—	—
80	—	—	—
45-ft Pendulum Assembly			
10	45— 0 to 50	45— 0 to 50	45— 0 to 50
20	45— 0 to 50	45— 0 to 50	45— 0 to 50
40	18— 0 to 50	45— 0 to 30	45— 0 to 50
	—	18—30 to 50	—
60	18— 0 to 50	18— 0 to 50	45— 0 to 50
80	18— 0 to 50	18— 0 to 50	18— 0 to 50
60-ft Pendulum Assembly			
10	60— 0	60— 0 to 30	60— 0 to 50
	30—10	30—50	—
	20—30 to 50	—	—
20	20— 0 to 50	30— 0 to 10	60— 0 to 50
	—	20—20 to 50	—
40	20— 0 to 50	20— 0 to 50	60— 0 to 30
	—	—	24—50
60	20— 0 to 50	20— 0 to 50	24— 0 to 50
80	20— 0 to 50	20— 0 to 50	24— 0 to 50
90-ft Pendulum Assembly			
10	30— 0 to 50	40— 0 to 10	38.6— 0 to 50
	—	30—20 to 50	—
20	30— 0 to 50	30— 0 to 50	38.6— 0 to 50
40	20— 0 to 50	30— 0 to 50	38.6— 0 to 50
60	20— 0 to 50	30— 0	38.6— 0 to 50
		20—10 to 50	
80	20— 0 to 50	20— 0 to 50	38.6— 0
	—	—	25.7—10 to 50

TABLE 8.21—COMPARISON OF LOW RPM RESULTS BETWEEN A STATIC BHA ALGORITHM WITH AN EQUIVALENT TORQUE AND A FULL DYNAMIC SOLUTION

Rotary Speed	Solution	Side Force, Direction (lbf)	Side Force, Inclination (lbf)
50	Dynamic	−809	1,934
50	quasi Dynamic (Static)	−819	1,934
100	Dynamic	−880	1,825
100	quasi Dynamic (Static)	−895	1,847
125	Dynamic	−955	1,497
125	quasi Dynamic (Static)	−919	1,815

*Solution no longer approximates full 3D dynamic solution.

8.7.9 Rotation of the Drillstring

The BHA analysis discussed earlier assumes that the drillstring is static and 2D. This is adequate for estimating the inclination tendency of the BHA and the bit. To find the direction or the "bit-walk" component of the bit trajectory, however, the rotation of the drillstring and the 3D forces and displacements of the BHA must be considered, as must the effects of bit-face torque and the rotating friction of the stabilizer(s). Dynamic analysis by Millheim and Apostal[23] showed also that the 3D static analysis, which includes inclination and direction forces, can be in error if the drillstring is rotated faster than 100 to 125 rpm. If the rotation is at or below 100 rpm, a quasidynamic solution that includes torque and neglects inertial effects can be used to approximate a 3D dynamic solution. Table 8.21 gives results of a static quasidynamic solution and a full dynamic solution for a slick BHA at 50, 100, and 125 rpm. Note the close agreement of the 50- and 100-rpm cases and the difference between the two 125-rpm solutions.

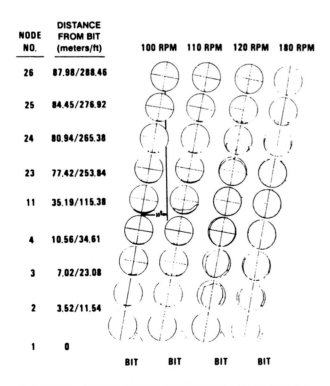

NODE NO.	DISTANCE FROM BIT (meters/ft)	100 RPM	110 RPM	120 RPM	180 RPM
26	87.98/288.46				
25	84.45/276.92				
24	80.94/265.38				
23	77.42/253.84				
11	35.19/115.38				
4	10.56/34.61				
3	7.02/23.08				
2	3.52/11.54				
1	0	BIT	BIT	BIT	BIT

Fig. 8.143—Effect of rotation on orbital path; 9⅞-in. hole, 7-in. collars (after Millheim and Apostal[23]).

The rotation of the drillstring with an axial load on the bit causes a number of occurrences: (1) the bit generates a bit-face torque and a side-cutting friction or torque; (2) the stabilizers generate a torque or side-cutting friction; and (3) the inertial energy of the drill collars causes a certain orbital path that changes as the speed increases or decreases (see Fig. 8.143). The inertial energy and the varying orbit followed by each component of the BHA can cause the inclination side force to change with changes in speed.

Consider a slick BHA with a tricone bit, drill collars, and drillstring. The total torque that the rotary motor must supply to rotate this system is

$$M_{rd} = M_{bf} + M_{bsc} + M_{dc} + M_{ds}, \quad \ldots\ldots\ldots\ldots (8.115)$$

where M_{rd} is the torque of a rotary drive, M_{bf} is the torque of a bit face, M_{bsc} is the torque of a bit side cutting, M_{dc} is the torque of a drill collar, and M_{ds} is the torque of a drillstring. The torques that affect the bit trajectory most are the bit torque and the bit side cutting torque. The torque needed to rotate the drill collars and drillstring is negligible. The speed and inclination, however, contribute to the resultant direction and inclination vector, as well as the final tilt of the bit.

The bit-face-torque relationship presented in Sec. 8.6 for tricone bits by Warren[14] can also be used for dynamic BHA analysis. For polycrystalline diamond compact and diamond bits, another torque function similar to Eq. 8.90 represents the bit-face torque.

Bit-face torque, like the PDM- and turbine-torque response discussed in Sec. 8.6, causes a reactive left, or counterclockwise, bit tendency. Bit side cutting can have either a left or a right tendency, dependent on whether the inclination side forces are building (left) or dropping (right). This also holds true for stabilizer side cutting. For example, a slick BHA with a dropping tendency and with a bit-face torque less than the resultant bit-side torque will turn right. If this same assembly starts building, it could turn left (see Figs. 8.144A and B).

If a single stabilizer is added as a building assembly, the bit has a left tendency, even though the stabilizer itself can have a strong right tendency (see Fig. 8.145). The resultant direction tendency can be predicted, in this case, only with the aid of a BHA algorithm that can resolve the magnitudes of the various side forces and torques. A single-stabilizer dropping assembly has a strong right tendency because both the bit and the stabilizer have side-force negative components yielding right-direction force tendencies (see Fig. 8.146).

Determining the side forces for a multistabilizer building BHA is not as easy as for slick and single-stabilizer BHA's. Fig. 8.147 shows the inclination and direction side forces at the bit for a 60-ft building BHA as a function of stabilizer side friction and bit side friction for constant WOB and speed. Notice that, as the stabilizer friction increases, the building assembly goes from a left-direction tendency to a right-direction tendency, dependent on the bit side friction. Bit-face torque is not included. If it had been, it would have negated some of the right tendency, dependent on the rock type, bit type, WOB, speed, and bit diameter. The greater the bit-side cutting torque, the greater the tendency of the bit to resist the effects of the increased stabilizer friction. This is logical because the

a. <u>Bit</u> (Dropping Tendency)

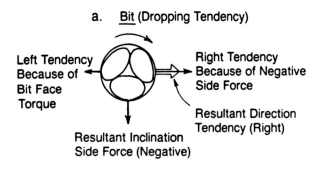

Left Tendency Because of Bit Face Torque

Right Tendency Because of Negative Side Force

Resultant Direction Tendency (Right)

Resultant Inclination Side Force (Negative)

b. <u>Bit</u> (Building Tendency)

Resultant Inclination Side Force (Positive)

Left Tendency Because of Positive Side Force

Left Tendency Because of Bit Face Torque

Resultant Direction Tendency (Left)

Fig. 8.144—Rotation of slick BHA causing direction (bit turn) tendency.

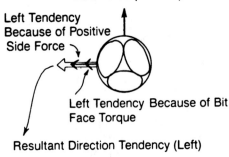

Fig. 8.145—Rotation of single-stabilizer pendulum BHA causing direction (bit turn) tendency.

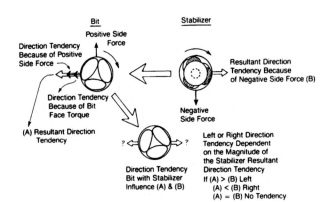

Fig. 8.146—Rotation of single-stabilizer building BHA causing direction (bit turn) tendency.

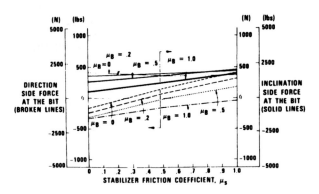

Fig. 8.147—Inclination and direction side force at the bit as a function of stabilizer and bit side friction for constant WOB and rpm. Building assembly. (After Millheim and Apostal[23].)

bit side force is positive, making the direction component left. As the stabilizer influence increases, however, the overall rotational forces favor a right-direction tendency, except for a bit friction, $\mu_B = 1.0$. This is why adding two near-bit stabilizers (increasing μ_s) can cause more of a right-direction tendency for building BHA's.

Multistabilizer holding BHA's are influenced more by the bit side friction and force than by the stabilizer friction. Fig. 8.148 shows the bit forces for a holding BHA. Because the resultant inclination force is negative, the bit direction force is positive or right. As the bit side friction increases, the direction tendency also increases (opposite that of the building BHA). Bit torque can reduce or even alter the overall tendency, dependent on its magnitude.

Example 8.32. Determine the direction tendency for a building BHA that has the following configuration, side forces, and bit-face torque. There is no formation tendency.

Bit side force:	+1,200 lbf
Bit face torque:	1,000 ft-lbf (equal to 200 lbf of bit side force)
First stabilizer:	−2,000 lbf (5 ft from bit face)
Second stabilizer:	−1,500 lbf (15 ft from bit face)
Third stabilizer:	−1,000 lbf (100 ft from bit face)

Solution. The +1,250 lbf of bit side force has a left tendency, as does the bit-face torque. However, all the stabilizers have strong negative or right-turn tendencies. The closer the stabilizer is to the bit, the more influence it has on the turning tendency. Because there are two near-bit stabilizers with nearly 2,100 lbf more of negative side force, the probability that this assembly will turn right is very high.

Following the rules and analytical procedures presented in this section will show how direction and inclination tendencies of the bit can be estimated for the basic building, dropping, and holding BHA's. Accurate predictions of a BHA's inclination and direction tendencies require

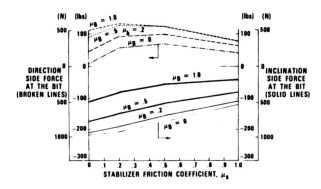

Fig. 8.148—Inclination and direction side force at the bit as a function of stabilizer and bit side friction for constant WOB and rpm. Holding assembly. (After Millheim and Apostal[23].)

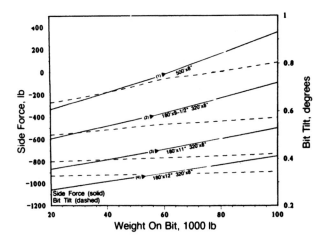

Fig. 8.149—Side force and bit tilt for a slick BHA; 17½-in. hole, 10-lbm/gal mud, 5° inclination.

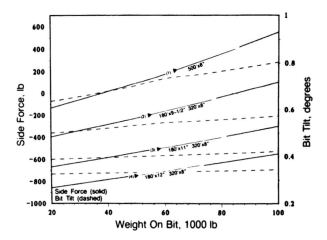

Fig. 8.150—Side force and bit tilt for a geological force of 200 lbf; 17½-in. hole, 10-lbm/gal mud, 5° inclination.

the use of an algorithm that can account for all the variables influencing the BHA forces and displacements, such as bit-force torque, hole curvature, BHA configuration, bit type, penetration rate, hole shape, and other factors. Even the mud properties and hydraulics can affect the final trajectory tendencies of a BHA.

The total interplay of the bit, BHA, geology, and other factors make directional drilling a subsystem of the overall drilling system. An understanding of each component of the drilling system presented in the previous chapters and of the basics presented in this chapter will bring about an understanding of the overall system that controls the bit's trajectory and how it is used in directional-drilling engineering.

8.8 Deviation Control

Directional drilling was used first for deviation control, which is concerned specifically with limiting the inclination or horizontal departure of the wellbore within some predescribed limits. At first, such problems as premature sucker-rod failure and damage caused when sucker rods rubbed the tubing were thought to result from the greater wellbore inclinations. Later experience proved that the cause was not the magnitude of inclination but the severity of the doglegs.

The principal use of deviation control is to limit the inclination angle for such reasons as keeping the wellbore from crossing lease lines or remaining within specific drainage boundaries. The practice of hitting a target is considered directional drilling and not deviation control, even though the inclinations and departures might be small.

This section will present the typical deviation-control practices for contending with the following drilling situations: (1) controlling the large-diameter hole (from 12¼ to 26 in.), (2) drilling complex geologies, and (3) general deviation control.

8.8.1 Deviation Control for Drilling Large-Diameter Wellbores

In most exploration and development wells that are deeper than 10,000 ft, portions of the surface and intermediate hole are from 12¼ to 26 in. in diameter. Deeper wells and wells that require multiple casing strings usually require even larger intermediate and surface holes. When the trajectory for such wells is planned, the aim is to hold the dogleg severity to less than 1°/100 ft to prevent casing wear and failure. Other factors that determine inclination are lease boundaries and reservoir drainage constraints.

Deviation control problems associated with large holes usually result from the larger size of the bit and the conditions necessary to make the bit drill. For an optimal WOB, a 17½-in. bit might require from 2,000 to 5,000 lbf/in. If the BHA is not designed properly for the optimal WOB, it is possible that the bit will build inclination angle.

Fig. 8.149 presents the side forces and bit tilts for four slick BHA's used for drilling a 17½-in. hole. Collars with OD's of 8, 9½, 11, and 12 in. are used. The WOB varies from 20,000 to 100,000 lbf. Both bit side force and bit tilt increase with increased WOB. Fig. 8.150 presents the same data with geological forces of +200 lbf. If the geological forces are substantial, the 11- and 12-in. collar diameters provide significantly more bit side force and less bit tilt than the 8- and 9½-in. collars. However, even

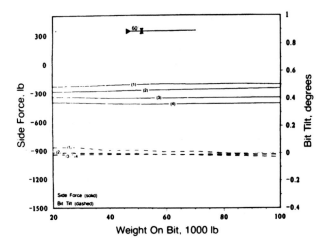

Fig. 8.151A—Single-stabilizer BHA; 17½-in. hole, 10-lbm/gal mud, 5° inclination.

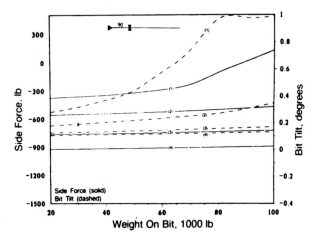

Fig. 8.151B—Single-stabilizer BHA; 17½-in. hole, 10-lbm/gal mud, 5° inclination.

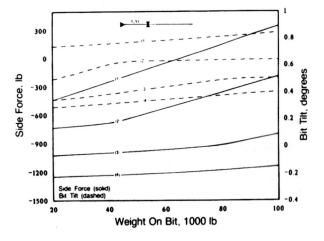

Fig. 8.151C—Single-stabilizer BHA; 17½-in. hole, 10-lbm/gal mud, 5° inclination.

the 12-in. collars will build angle if the geological force is strong enough. The inclination will continue to increase until some equilibrium angle is reached.

A common alternative to the slick BHA is the pendulum. In the larger-diameter wellbore where the big collars are used, tangency lengths can be as great as 120 ft when the inclination is low. Figs. 8.151A through 8.151C show the side forces and bit tilts for the same collars and conditions presented by Figs. 8.149 and 8.150, except that a stabilizer is placed at 60, 90, and 120 ft. Figs. 8.152A through 8.152C show the same cases, except that a two-stabilizer pendulum is used. The second stabilizer is 30 ft from the first stabilizer.

Deviation control with a slick or a pendulum BHA in a large-diameter hole relies on the negative side force to offset the bit tilt and geological effects that tend to build angle. The pendulum and slick BHA's work best when the geological forces are small, the formations are fairly homogeneous, and the formation strength is low. When the geological forces are significant, when the lithology is varied, and when the formation strengths are medium to hard, the multistabilizer BHA usually effects better results.

Fig. 8.153 presents the typical packed BHA's used for deviation-control drilling in large holes. Figs. 8.154A through 8.154C show the side forces and bit tilts for the same BHA's with 8-, 9½-, 11-, and 12-in.-OD drill collars.

If the geological forces are strong, all the packed BHA's will build inclination at some constant rate until the geological force ceases. Once a building curvature is started with a packed BHA, however, there is a tendency for the build to continue until a major change occurs to break the curvature. This can be a change in the BHA configuration, in the geology, or in operating practices.

The use of multistabilizer BHA's has positive and negative aspects for deviation-control drilling. Stabilizers—especially large-diameter stabilizers—add torque and, in some instances, reduce the amount of weight that can be applied at the bit. Chemical and hydraulic actions can weaken certain formations, and stabilizers can accelerate the erosion of a wellbore by mechanically cutting, wearing, and striking the wellbore wall. Trip times are increased because of the handling of the stabilizers; with larger-diameter holes, this is especially true because of the greater makeup torques of the larger tool joints. Tables 8.22 and 8.23 present some of the makeup torques for various drill collars. If the makeup torques are not achieved, tool joints can be overtorqued during drilling; this can crack the boxes and can make breaking the tool joint at the surface difficult. In unstable wellbores, there is a risk of excessive drag in tripping out of the hole with multistabilizer BHA's. Moreover, fishing is more difficult.

The virtues of multistabilizer BHA's are many. Bit performance is usually better because of the small bit tilts and better dynamics. If the bit becomes worn in drilling an abrasive formation, the near-bit stabilizer will ream the hole to the original bit diameter. When these stabilizer blades become worn and undergauge, the rotary torque increases and the penetration rate decreases, signaling that something is wrong and that the bit should be pulled. The chances are reduced, therefore, of drilling undergauge holes and pinching new bits in tripping back into the hole.

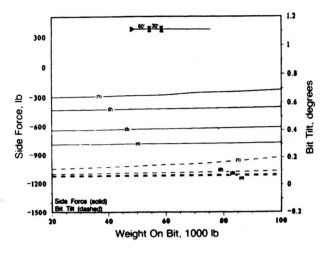

Fig. 8.152A—Two-stabilizer BHA; 17½-in. hole, 10-lbm/gal mud, 5° inclination.

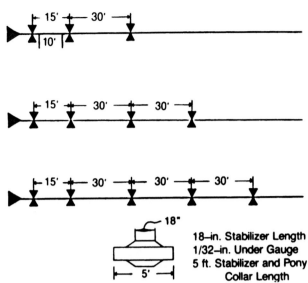

Fig. 8.153—Packed BHA's used for deviation control.

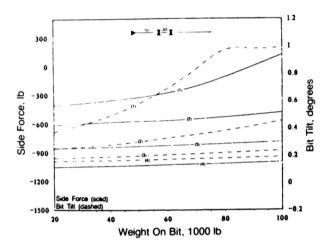

Fig. 8.152B—Two-stabilizer BHA; 17½-in. hole, 10-lbm/gal mud, 5° inclination.

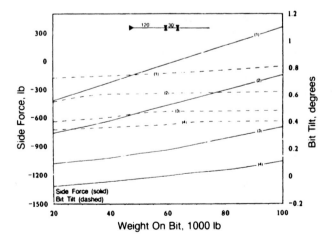

Fig. 8.152C—Two-stabilizer BHA; 17½-in. hole, 10-lbm/gal mud, 5° inclination.

Multistabilizer BHA's permit the use of extreme WOB without incurring appreciable changes in the bit side force or bit tilt. This does not mean that the packed BHA will not build angle; it will at a constant curvature. Variation in lithology does not affect the packed BHA nearly as much as it does a slick or pendulum BHA. Also, the potential for a severe dogleg is much less with a packed BHA. Usually it is easier to place casing into a section of hole newly drilled with a packed BHA than a pendulum or slick BHA.

The selection of a BHA to drill larger wellbores depends on such factors as geological forces and stratigraphy, depth of hole, inclination and dogleg severity constraints, desired penetration rates, trip times, risk of fishing, and drilling-rig capabilities. Of all those factors, geological effects have the greatest bearing on deviation control and on whether wellbores are to be large or small.

8.8.2 Geological Forces and Deviation Control

Geological forces that cause deviation control problems are associated with medium-soft to medium-hard formations with significant interbedding and dip angles ranging from 5 to 90°. Such conditions are usually present where there is folding and faulting.

Fig. 8.155 depicts a typical anticlinal structure where most of the wellbores trend toward the crest of the structure. For dip angles less than about 45°, the tendency of the bit is to build angles perpendicular to the strike. The opposite tendency to drill downdip occurs when the beds dip more than about 60°. When the beds are 80 to 90°, the bit tends to follow the dip of the formation, but at a reduced penetration rate. At formation dip angles between 45 and 75°, the bit tends to follow the strike of the formation.

The magnitude of the dips, the formation hardness, and the amount of interbedding dictate the overall formation strength. Another complication is that dips and strike (dip direction) vary with depth. The variation in dip is a function of the position of the well on the structure and the attack angle of the bit with the dipping formations. Fig. 8.156A is an example of a typical series of geological

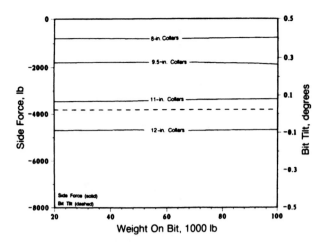

Fig. 8.154A—Three-stabilizer BHA; 17½-in. hole, 10-lbm/gal mud, 5° inclination.

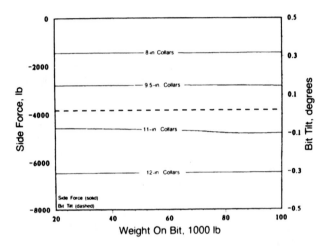

Fig. 8.154B—Four-stabilizer BHA; 17½-in. hole, 10-lbm/gal mud, 5° inclination.

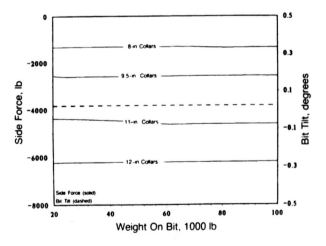

Fig. 8.154C—Five-stabilizer BHA; 17½-in. hole, 10-lbm/gal mud, 5° inclination.

events that caused various dips and dip directions (Fig. 8.156B). The initial 4,000 ft is medium-soft Tertiary sediments deposited over an erosional surface, having only a slight regional dip of 2 to 5° with a westerly dip direction. Below the unconformity is an anticlinal structure that has been thrust faulted, causing dips as high as 50° on the eastern side. The Cretaceous sediments are highly interbedded, while the Jurassic and Triassic sediments are fairly uniform shales. Below the shales, the formations dip severely and are also very hard.

During the drilling of the main structure, there is very little bit-building tendency through the fairly flat Tertiary formations. Below the unconformity, the dips are greater and vary in direction and angle. If a slick, pendulum, or packed BHA is used to drill through the interface of the unconformity where the dip direction changes, the borehole, more than likely, will change direction and trend perpendicular to the strike and updip. If the inclination of the wellbore is between 1 and 5°, there is a strong probability that the wellbore direction will not change during drilling through the interface. Above 5°, the change in direction may not be too drastic. Once in the interbedded dipping beds, the slick and pendulum BHA's will continue to build angle until equilibrium is achieved. This equilibrium depends on a balance between the resultant bit side force and the formation force, as discussed in Sec. 8.7.4. A packed BHA will reach an equilibrium curvature and continue on this curvature until some major change occurs.

Drilling into shales that are fairly uniform and of medium strength, the slick, pendulum, and packed BHA's will all tend to drop angle even though the dips of the shale formations are similar to those of the overlying sediments. The underlying sandstones and limestones are much harder, and they dip more than the overlying sediments. Any change in direction or inclination will be slow and slight, regardless of the BHA used. For example, if an inclination occurs when the harder formations are being penetrated, it will probably remain unchanged whether a pendulum, slick, or packed BHA is used. Starting a drop of inclination angle usually requires a mud motor.

8.8.3 Theoretical Investigations Pertaining to Deviation Control

Lubinski and Woods[24] defined the anisotropy index, h, as a way to account for the geological forces that cause a bit to deviate when the rock properties change abruptly. Their theory, however, could not explain downdip deviation in steeply dipping formations. Other investigators tried to postulate the deviation of a bit in a variety of ways. The miniature whipstock theory proposed by Rollins[25] and further researched by Murphey and Cheatham[26] explained that brittle material subjected to compressional forces will fail normal to the bedding planes. In dipping formations, this failure leads to miniature whipstocks that cause the bit to deviate (see Fig. 8.157). The miniature-whipstock theory does not explain downdip deviation in steeply dipping beds.

Fig. 8.158 illustrates a theory presented by Knapp,[27] who suggests that drilling in a soft formation and intersecting a dipping, hard formation will deviate downdip. This theory is not consistent with field results. In contrast to Knapp's theory, Sultanov and Shandalov's[28] theory states that a bit drilling in a soft formation and

TABLE 8.22—MAKE-UP TORQUE FOR DRILL COLLARS (ft-lbf)

CONNECTION			BORE OF DRILL COLLAR SIZE (in.)									+ WEAK MEMBER
SIZE (in.)	TYPE	OD (in.)	1	1-1/4	1-1/2	1-3/4	2	2-1/4	2-1/2	2-13/16	3	
API	NC 23	3	2,500	2,500	2,500							BOX
		3-1/8	3,300	3,300	2,600							PIN
		3-1/4	4,000	3,400	2,600							PIN
2-7/8	PAC (See Note 4)	3		3,800	3,800	2,900						PIN
		3-1/8		4,900	4,200	2,900						PIN
		3-1/4		5,200	4,200	2,900						PIN
2-3/8 API 2-7/8	API IF or NC 26 or SLIM HOLE	3-1/2		4,600	4,600	3,700						PIN
		3-3/4		5,500	4,700	3,700						PIN
2-7/8 3-1/2 2-7/8	EXTRA HOLE or DBL STREAMLINE or MOD. OPEN	3-3/4		4,100	4,100	4,100						BOX
		3-7/8		5,300	5,300	5,300						BOX
		4-1/8		8,000	8,000	7,400						PIN
2-7/8 API 3-1/2	API IF or NC 31 or SLIM HOLE	3-7/8		4,600	4,600	4,600						BOX
		4-1/8		7,300	7,300	7,300						BOX
		4-1/4		8,800	8,800	8,100						PIN
		4-1/2		10,000	9,300	8,100						PIN
API	NC 35	4-1/2				8,900	8,900	8,900	7,400			PIN
		4-3/4				12,100	10,800	9,200	7,400			PIN
		5				12,100	10,800	9,200	7,400			PIN
3-1/2 4 3-1/2	EXTRA HOLE or SLIM HOLE or MOD. OPEN	4-1/4				5,100	5,100	5,100	5,100			BOX
		4-1/2				8,400	8,400	8,400	8,200			PIN
		4-3/4				11,900	11,700	10,000	8,200			PIN
		5				13,200	11,700	10,000	8,200			PIN
		5-1/4				13,200	11,700	10,000	8,200			PIN
3-1/2 API 4-1/2	API IF or NC 38 or SLIM HOLE	4-3/4				9,900	9,900	9,900	9,900	8,300		PIN
		5				13,800	13,800	12,800	10,900	8,300		PIN
		5-1/4				16,000	14,600	12,800	10,900	8,300		PIN
		5-1/4				16,000	14,600	12,800	10,900	8,300		PIN
3-1/2	H-90 (See Note 3)	4-3/4				8,700	8,700	8,700	8,700	8,700		BOX
		5				12,700	12,700	12,700	12,700	10,400		PIN
		5-1/4				16,900	16,700	15,000	13,100	10,400		PIN
		5-1/2				16,500	16,700	15,000	13,100	10,400		PIN
4 API 4 4-1/2	FULL HOLE or NC 40 or MOD. OPEN or DBL STREAMLINE	5				10,800	10,800	10,800	10,800	10,800		BOX
		5-1/4				15,100	15,100	15,100	14,800	12,100		PIN
		5-1/2				19,700	18,600	16,900	14,800	12,100		PIN
		5-3/4				20,400	18,600	16,900	14,800	12,100		PIN
		6				20,400	18,600	16,900	14,800	12,100		PIN
4	H-90 (See Note 3)	5-1/4					12,500	12,500	12,500	12,500		BOX
		5-1/2					17,300	17,300	17,300	16,500		PIN
		5-3/4					23,300	21,500	19,400	16,500		PIN
		6					23,500	21,500	19,400	16,500		PIN
		6-1/4					23,500	21,500	19,400	16,500		PIN
4-1/2	API REGULAR	5-1/2					15,400	15,400	15,400	15,400		BOX
		5-3/4					20,300	20,300	19,400	16,200		PIN
		6					23,400	21,600	19,400	16,200		PIN
		6-1/4					23,400	21,600	19,400	16,200		PIN
API	NC 44	5-3/4	FOR LARGER OD's NOT SHOWN, ADD 10% AS STATED IN NOTE 2.				20,600	20,600	20,600	18,000		PIN
		6					25,000	23,300	21,200	18,000		PIN
		6-1/4					25,000	23,300	21,200	18,000		PIN
		6-1/2					25,000	23,300	21,200	18,000		PIN
4-1/2	API FULL HOLE	5-1/2					12,900	12,900	12,900	12,900	12,900	BOX
		5-3/4					17,900	17,900	17,900	17,700	17,700	PIN
		6					23,300	23,300	22,800	19,800	17,700	PIN
		6-1/4					27,000	25,000	22,800	19,800	17,700	PIN
		6-1/2					27,000	25,000	22,800	19,800	17,700	PIN
4-1/2 API 4 4-1/2 5 4-1/2	EXTRA HOLE or NC 46 or API IF or SEMI IF or DBL STREAMLINE or MOD. OPEN	5-3/4					17,600	17,600	17,600	17,600		BOX
		6					23,200	23,200	22,200	20,200		PIN
		6-1/4					28,000	25,500	22,200	20,200		PIN
		6-1/2					28,000	25,500	22,200	20,200		PIN
		6-3/4					28,000	25,500	22,200	20,200		PIN
4-1/2	H-90 (See Note 3)	5-3/4					17,600	17,600	17,600	17,600		BOX
		6					23,400	23,400	23,000	21,000		PIN
		6-1/4					28,500	26,000	23,000	21,000		PIN
		6-1/2					28,500	26,000	23,000	21,000		PIN
		6-3/4					28,500	26,000	23,000	21,000		PIN
5	H-90 (See Note 3)	6-1/4					25,000	25,000	25,000	25,000		BOX
		6-1/2					31,500	31,500	29,500	27,000		PIN
		6-3/4					35,000	33,000	29,500	27,000		PIN
		7					35,000	33,000	29,500	27,000		PIN
5-1/2	H-90 (See Note 3)	6-3/4					34,000	34,000	34,000	34,000		PIN
		7					41,500	40,000	36,500	34,000		PIN
		7-1/4					42,500	40,000	36,500	34,000		PIN
		7-1/2					42,500	40,000	36,500	34,000		PIN
5-1/2	API REGULAR	6-3/4					31,500	31,500	31,500	31,500		BOX
		7					39,000	39,000	36,000	33,500		PIN
		7-1/4					42,000	39,500	36,000	33,500		PIN
		7-1/2					42,000	39,500	36,000	33,500		PIN

(1) Basis of calculations for recommended make-up torque assumed the use of a thread compound containing 40-60% by weight of finely powdered metallic zinc or 60% by weight of finely powdered metallic lead applied thoroughly to all threads and shoulders and using the modified jack screw formula as shown in the IADC Tool Pusher's Manual Sec B1, p. 7 and API RP 7G, second edition April 1971 Appendix A Sec A9 and a unit stress of 62,500 PSI in the box or pin whichever is weaker.

(3) H-90 connection make-up torque based on 56,250 PSI stress and other factors as stated in note (1).

(2) Normal torque range — tabulated minimum value to 10% greater. Largest diameter shown for each connection is the maximum recommended for that connection. If connections are used on drill collars larger than the maximum shown, increase the torque values shown by 10% for a minimum value. In addition to the increased minimum torque value it is also recommended that a fishing neck be machined to the maximum diameter shown.

(4) 2-7/8 PAC make up torque based on 87,500 PSI stress and other factors as stated in note (1).

TABLE 8.23—MAKE-UP TORQUE FOR DRILL COLLARS (ft-lbf)

| CONNECTION | | | BORE OF DRILL COLLAR SIZE (in.) | | | | | | | + WEAK |
SIZE (in.)	TYPE	OD (in.)	2-1/4	2-1/2	2-13/16	3	3-1/4	3-1/2	3-3/4	MEMBER
4-1/2 API / API / 5 / 5 / 5-1/2 / 5	API IF or NC 50 or EXTRA HOLE or MOD. OPEN or DBL STREAMLINE or SEMI-IF	6-1/4	22,800	22,800	22,800	22,800	22,800			BOX
		6-1/2	29,500	29,500	29,500	29,500	26,500			PIN
		6-3/4	36,000	35,500	32,000	30,000	26,500			PIN
		7	38,000	35,500	32,000	30,000	26,500			PIN
		7-1/4	38,000	35,500	32,000	30,000	26,500			PIN
5-1/2	API FULL HOLE	7		32,500	32,500	32,500	32,500			BOX
		7-1/4		40,500	40,500	40,500	40,500			BOX
		7-1/4		49,000	47,000	45,000	41,500			PIN
		7-3/4		51,000	47,000	45,000	41,500			PIN
API	NC 56	7-1/4		40,000	40,000	40,000	40,000			BOX
		7-1/2		48,500	48,000	45,000	42,000			PIN
		7-3/4		51,000	48,000	45,000	42,000			PIN
		8		51,000	48,000	45,000	42,000			PIN
6-5/8	API REGULAR	7-1/2		46,000	46,000	46,000	46,000			BOX
		7-3/4		55,000	53,000	50,000	47,000			PIN
		8		57,000	53,000	50,000	47,000			PIN
		8-1/4		57,000	53,000	50,000	47,000			PIN
6-5/8	H-90 (See Note 3)	7-1/2		46,000	46,000	46,000	46,000			BOX
		7-3/4		55,000	55,000	51,000	49,500			PIN
		8		59,500	56,000	53,000	49,500			PIN
		8-1/4		59,500	56,000	53,000	49,500			PIN
API	NC 61	8		54,000	54,000	54,000	54,000			BOX
		8-1/4		64,000	64,000	64,000	61,000			PIN
		8-1/2		72,000	68,000	65,000	61,000			PIN
		8-3/4		72,000	68,000	65,000	61,000			PIN
		9		72,000	68,000	65,000	61,000			PIN
5-1/2	API IF	8		56,000	56,000	56,000	56,000			BOX
		8-1/4		66,000	66,000	66,000	63,000	59,000		PIN
		8-1/2		74,000	70,000	67,000	63,000	59,000		PIN
		8-3/4		74,000	70,000	67,000	63,000	59,000		PIN
		9		74,000	70,000	67,000	63,000	59,000		PIN
		9-1/4		74,000	70,000	67,000	63,000	59,000		PIN
6-5/8	API FULL HOLE	8-1/2			67,000	67,000	67,000	67,000	66,500	PIN
		8-3/4			78,000	78,000	72,000	72,000	66,500	PIN
		9			83,000	80,000	76,000	72,000	66,500	PIN
		9-1/4			83,000	80,000	76,000	72,000	66,500	PIN
		9-1/2			83,000	80,000	76,000	72,000	66,500	PIN
API	NC 70	9			75,000	75,000	75,000	75,000	75,000	BOX
		9-1/4			88,000	88,000	88,000	88,000	88,000	BOX
		9-1/2			101,000	101,000	100,000	95,000	90,000	PIN
		9-3/4			107,000	105,000	100,000	95,000	90,000	PIN
		10			107,000	105,000	100,000	95,000	90,000	PIN
		10-1/4			107,000	105,000	100,000	95,000	90,000	PIN
API	NC 77	10				107,000	107,000	107,000	107,000	BOX
		10-1/4				122,000	122,000	122,000	122,000	BOX
		10-1/2				138,000	138,000	133,000	128,000	PIN
		10-3/4				143,000	138,000	133,000	128,000	PIN
		11				143,000	138,000	133,000	128,000	PIN
CONNECTIONS WITH FULL FACE										
7	H-90 (See Note 3)	8			53,000	53,000	53,000	53,000		BOX
		8-1/4			63,000	63,000	63,000	60,500		PIN
		8-1/2			71,500	68,500	65,000	60,500		PIN
7-5/8	API REGULAR	8-1/2			60,000	60,000	60,000	60,000		BOX
		8-3/4			71,000	71,000	71,000	71,000		BOX
		9			83,000	83,000	79,000	74,000		PIN
		9-1/4			88,000	83,000	79,000	74,000		PIN
		9-1/2			88,000	83,000	79,000	74,000		PIN
7-5/8	H-90 (See Note 3)	9			72,000	72,000	72,000	72,000		BOX
		9-1/4			85,500	85,500	85,500	85,500		BOX
		9-1/2			98,000	98,000	98,000	95,500		PIN
8-5/8	API REGULAR	10			108,000	108,000	108,000	108,000		BOX
		10-1/4			123,000	123,000	123,000	123,000		BOX
		10-1/2			135,000	134,000	129,000	123,000		PIN
8-5/8	H-90 (See Note 3)	10-1/4			112,500	112,500	112,500	112,500		BOX
		10-1/2			128,500	128,500	128,500	128,500		BOX
CONNECTIONS WITH LOW TORQUE FACE										
7	H-90 (See Note 3)	8-3/4			67,500	67,500	66,500	62,000		PIN
		9			74,000	71,000	66,500	62,000		PIN
7-5/8	API REGULAR	9-1/4			72,000	72,000	72,000	72,000		BOX
		9-1/2			85,000	85,000	82,000	77,000		PIN
		9-3/4			91,000	87,000	82,000	77,000		PIN
		10			91,000	87,000	82,000	77,000		PIN
7-5/8	H-90 (See Note 3)	9-3/4			91,000	91,000	91,000	91,000		BOX
		10			105,000	105,000	103,500	98,000		PIN
		10-1/4			112,500	108,000	103,500	98,000		PIN
		10-1/2			112,500	108,000	103,500	98,000		PIN
8-5/8	API REGULAR	10-3/4			112,000	112,000	112,000	112,000		BOX
		11			129,000	129,000	129,000	129,000		BOX
8-5/8	H-90 (See Note 3)	10-3/4			92,500	92,500	92,500	92,500		BOX
		11			110,000	110,000	110,000	110,000		BOX
		11-1/4			128,000	128,000	128,000	128,000		BOX

FOR LARGER OD's NOT SHOWN, ADD 10% AS STATED IN NOTE 2.

* Largest diameter shown is the maximum recommended for these full face connections. If larger diameters are used, machine connections with low torque faces and use the torque values shown under low torque face tables. If low torque faces are not used see note (2) for increased torque values. 84

+ Notation in this column indicates cross section area (3/4" from base on pin or 3/8" from shoulder on box) is smaller on the member indicated.

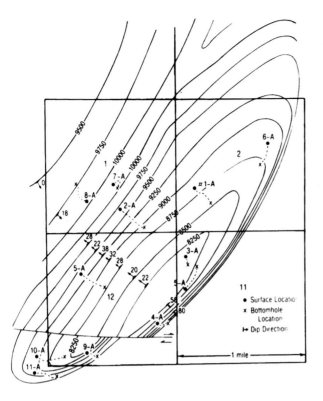

Fig. 8.155—Typical anticlinal structure that can cause deviation control problems.

A theory that does explain both updip and downdip deviation was initially proposed by McLamore[30] and later expanded by Bradley.[31] It is called the ''preferred chip formation'' theory. Fig. 8.159 summarizes McLamore's idea of the interaction of a bit's tooth with a brittle formation. At time T_1, a vertical force is imparted to a single wedge with a wedge angle of Ω into a formation with a dip, ϕ. At time T_2, a tensile crack is formed, and a pulverized or plastic zone is immediately below the tooth. At time T_3 a chip breaks loose on the updip side, and at time T_4 a crater is left with more volume removed from the updip side than from the downdip side.

The force to create a chip on either side of the tooth was derived by McLamore.

$$F_i = \frac{2\tau(\rho_i)D_w \sin \Omega \cos \phi}{\cos^2\left(\dfrac{\pi/2+\Omega+\phi}{2}\right)}, \quad \ldots \ldots \ldots \ldots (8.116)$$

where i is 1 for the updip side of the wedge and 2 for the downdip side, D_w is the depth of the wedge penetration, Ω is the included wedge half angle, ϕ is the bedding dip angle relative to the bottom of the hole, ρ_1 equals ϕ plus Ω, α_{dh} is the bedding dip angle relative to the bottom of the hole, ρ_2 equals ϕ minus Ω, $\tau(\rho_i)$ is the shear strength of the anisotropic rock along the plane of failure, and ϕ is the angle of internal friction of the anisotropic rock.

The deviation force can be calculated by Eq. 8.117, assuming that the chip will form on the side of the wedge having the lesser amount of fracture force.

$$F_{DEV} = \frac{\tau(\rho_i)D_w \cos \Omega \cos \phi}{\cos^2\left(\dfrac{\pi/2+\Omega+\phi}{2}\right)} \cdot \quad \ldots \ldots \ldots (8.117)$$

intersecting a hard formation will deviate updip. Again, the tendency to deviate downdip in steeply dipping beds is not explained. Another theory, presented by Smith and Cheatham,[29] considers the plastic fracture of a formation under a single wedge. The conclusions of their work imply downdip deviation in all cases. This, too, is inconsistent with field observations.

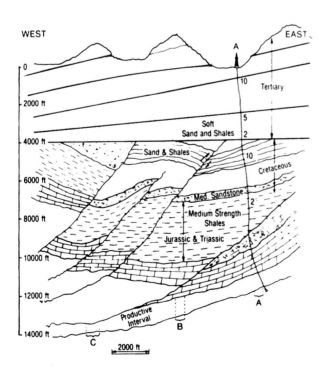

Fig. 8.156A—Typical cross section of complex geology that causes deviation control problems.

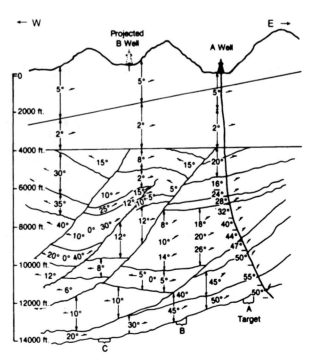

Fig. 8.156B—Projected dips based on Well A and geophysical information.

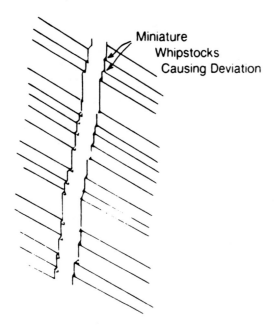

Fig. 8.157—Miniature-whipstock theory (after Rollins[25]).

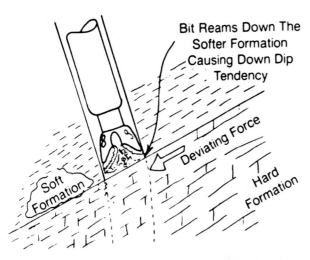

Fig. 8.158—Soft-to-hard downdip theory (after Knapp[27]).

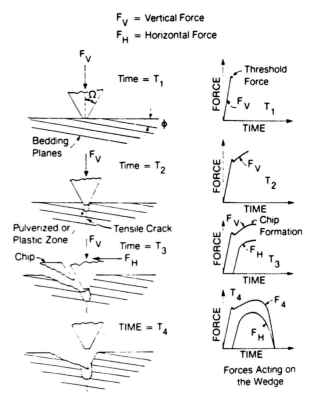

Fig. 8.159—Preferred chip formation model (after McLamore[30]).

Both parts of Fig. 8.160 were calculated by McLamore with Eq. 8.115. Fig. 8.160a shows that for wedge angles (2Ω) of 35 to 45° and for bed dips of 30°, the tendency is for a positive deviation force (updip); for bed dips of more than 30°, there is a negative deviation force (downdip). Fig. 8.160b shows that, for the large wedge angles, the tendency shifts more to the downdip with the transition dip angle being between 30 and 40°.

McLamore's theory implies that the tooth-wedge angle affects the deviation force. The narrower the wedge, the greater the tendency to drill updip.

8.8.4 General Deviation Control

Sec. 8.8.1 presented deviation control for large-diameter wellbores using BHA's with 8- to 12-in.-diameter drill collars. Drilling intermediate-sized holes—7⅞ to 12½-in.—requires 6- to 10-in. collars. Larger collars can be used if the fishability is not a concern. Usually, 1½ in. is subtracted from the wellbore diameter to determine

the largest collar that can be used if a standard overshot might be required for a fishing job.

In many drilling situations, the smaller drill collars are not stiff enough to offset moderate to severe formation forces. Even with low WOB's to ''fan the formation,'' it is difficult to keep from building an inclination angle when the formation forces are greater than the maximum 0-lbf WOB case where the negative side force is at its maximum level. Because of the lesser collar stiffness, shorter pendulum BHA's should be used. A 9⅞-in. hole with 8-in. collars would have a maximum pendulum length of about 80 ft, while a 7⅞-in. wellbore with 6⅛-in. collars would have a maximum pendulum length of about 45 ft.

In most situations, using low WOB's to control deviation is not economical, except when a turbine or positive-displacement motor is used. The correct strategy is to determine how much the formation forces will cause the wellbore to deviate over the depth of a well while optimum WOB's and rotary speeds are used. Fig. 8.155 shows surface locations positioned in such a way that the natural tendency of the bit will be to drill updip toward the bottomhole location, which is now a pseudotarget. The positioning of the surface target requires a knowledge of the dips and strikes of all the formations from surface to total depth. Many times a surface location is chosen in expectation of encountering a certain dip direction, but the well is found to be going in a different direction. This usually happens because the near-surface structure is different from that of the deeper formations. Such misjudgments generally occur when the geology of a new structure is undefined, when there is a great deal of faulting (as in the case of piercement domes), and when the drilling is on the flanks of complex structures, such as those depicted in Fig. 8.156.

The best way to determine the horizontal departure of a wellbore from the bottomhole target to the surface location is to determine the optimum WOB's and hole-size requirements and to design the BHA's that will resist the formation forces best. This horizontal distance can be used to determine how far to set the surface location back. This distance can range from a few hundred to more than a thousand feet, depending on the total depth.

Knowing the geological forces and how well a given BHA will respond to them, one can adjust the surface location to take advantage of those factors and optimally drill a well. A packed BHA, for example, will provide a smoother trajectory than a pendulum or a slick BHA. Square drill collars can increase the stiffness of a BHA for a given hole size and can reduce the bending force effects, allowing for more applied WOB than would be possible for round drill collars. As discussed previously, it is not the magnitude of the inclination that causes casing failures, worn pipe, key seats, and production problems with sucker rods and tubing; it is dogleg severity. Accordingly, the same practices used in drilling directional wells to minimize dogleg severity should be used in deviation control.

Example 8.33. Your company wants to drill a second well to Bottomhole Location B, as shown in Fig. 8.156A. You need to determine where to position the surface location on the basis of information from Well A and the geological cross section (Fig. 8.156A). You are also required to design the BHA's to hit the bottomhole target. The production string that is required is 7-in., 29-lbm/ft casing. The offset information on Well A is found in Table 8.24. (Note: A valley location costs $100,000; a mountain location costs $750,000.)

Solution. The calculated trajectory of Well A is presented in Table 8.25.

The total departure of the well is 2,534 ft, which misses the target by 600+ ft. From the surface to the measured depth of 4,558 ft, the rate of angle build is low—i.e.,

less than 0.1°/100 ft. At 4,558 ft, the rate of angle build increases to 0.6°/100 ft and continues at rates between 0.4 and 0.7°/100 ft to a measured depth of 9,805 ft, where the rate decreases to 0.2 to 0.3°/100 ft. From 11,050 to 11,800 ft, the rate of angle build varies between 0.5 and 0.6°/100 ft. The plot of the trajectory data shows that most of the departure occurs in the Jurassic and Triassic shales and limestone.

Fig. 8.156B shows the interpreted dips based on the Well A dipmeter and geophysical data. Given the 2,540 ft of departure, the surface location of Well B should be 5,000 ft south of Well A. The formation dips that are projected for Well B are much less severe than those for Well A; therefore, if the same BHA's are used, the BHA's would not deviate the trajectory the 2,500 ft north of the surface location. In fact, the wellbore should be nearly vertical down to 6,000 ft. From 6,000 to 10,000 ft, it is possible that the wellbore would deviate much less than that of Well A. Below 10,000 ft, it is possible that the trajectory of Well B would kick to the west away from the target if the wellbore is not east of the 0° dips. Therefore, the trajectory control plan should provide for at least 1,400 ft of north departure when true vertical depth (TVD) of 10,000 ft is reached. This is 400 ft less than Well A. To do this, it might be necessary to design a building BHA that will provide just enough angle build to obtain the necessary departure.

There are three possible approaches for hitting the designated target. The first approach, using a continuous-build trajectory, assumes there will be negligible departure down to the top of the Cretaceous (approximately 4,000 ft). With the continuous build of 0.5/100 ft starting at 5,239 ft, the desired departure of 2,540 ft would be reached at the top of the target (12,450 ft TVD) with a final inclination of 39.0° and a measured depth of 13,039 ft. The second approach, using a build-and-hold trajectory, could also be used to hit the target. For example, starting at 4,039 ft and maintaining the same build rate as the continuous-build (0.5°/100 ft), the inclination would be increased to 22.5°, where it would be held constant to total depth. At the top of the target at 12,450 ft, the

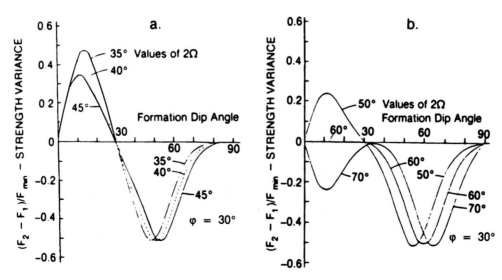

Fig. 8.160—Chip direction preference as a function of wedge angle and formation dip angle (after McLamore[30]).

TABLE 8.24—OFFSET INFORMATION ON WELL A FOR EXAMPLE 8.33

Hole Sizes and Casing Sizes

Hole Size (in.)	Casing Size (in.)	Depth
17½ in.	13⅜ in.	Surface to 4,020 ft
12¼ in.	9⅝ in.	4,020 to 9,290 ft
8½ in.	7 in.	9,290 to 11,180 ft

Drilling Results

Depth (ft)		Geology	Average Penetration Rate Over Interval (ft/hr)	Inclination	Formation Dip Average Over Interval
0 to	600	Tertiary	30	1°	5°
600 to	1,800	Tertiary	20	1½°	5°
1,800 to	2,801	Tertiary	14	1¾°	2°
2,801 to	3,405	Tertiary	12	2°	2°
3,405 to	4,020	Top of Cretaceous	10	2°	2°
4,020 to	4,558	Cretaceous	35	5°	20°
4,558 to	5,240	Cretaceous	33	8°	20°
5,240 to	5,800	Cretaceous	37	5°	16°
5,800 to	6,350	Cretaceous	8	8.5°	24°
6,350 to	7,010	Cretaceous	24	12.5°	28°
7,010 to	7,500	Jurassic and Triassic	20	16.0°	32°
7,500 to	8,100	Jurassic and Triassic	17	19.5°	40°
8,100 to	8,640	Jurassic and Triassic	15	23.0°	44°
8,640 to	9,290	Jurassic and Triassic	12	27.0°	47°
9,290 to	9,805	Jurassic and Triassic	9	28.0°	50°
9,805 to	10,500	Jurassic and Triassic	8.5	30.0°	55°
10,500 to	11,050		7.0	33.0°	55°
11,050 to	11,470		15.0	35.0°	50°
11,470 to	11,800		12.0	37.0°	50°

Bottomhole Assemblies

Interval	Quantity	Equipment
0 to 4,020 ft		17½-in. bit
	9	9.5- × 3.0-in. Drill Collars
	21	8.0- × 2¹³⁄₁₆-in. Drill Collars
	12	6.5- × 2³⁄₁₆-in. Drill Collars
		5.0-in. Grade E Drillpipe
4,020 to 9,290 ft		12¼-in. bit
	30	8.0- × 2¹³⁄₁₆-in. Drill Collars
	12	6.5- × 2³⁄₁₆-in. Drill Collars
		5.0-in. Grade E Drillpipe
9,290 to 11,800 ft		8.5-in bit
	36	6.5- × 2³⁄₁₆-in. Drill Collars
		5.0-in. Grade E Drillpipe

Drilling Parameters

Measured Depth (ft)	Weight on Bit (1000 lb)	Rotary Speed	Pump Pressure (psig)	Pump Rate (gal/min)	Mud Weight (lb/gal)	Viscosity (cp)
600	25	80	3,434	850	9.2	12
1,800	55	80	3,458	850	9.2	12
2,800	70	75	3,433	850	9.2	12
3,405	70	70	3,532	850	9.2	12
4,020	70	70	3,456	850	9.2	12
4,558	55	70	3,431	600	9.2	12
5,240	55	70	3,390	600	9.4	15
5,800	55	70	3,443	600	9.4	15
6,350	55	70	3,491	600	9.4	15
7,040	55	70	3,393	600	9.4	15
7,500	60	65	3,438	600	9.4	15
8,100	60	65	3,489	600	9.6	15
8,640	60	65	3,420	600	9.6	18
9,290	60	65	3,550	600	9.6	18
9,805	35	75	3,401	600	9.6	18
10,500	40	70	3,427	350	9.5	15
11,500	40	70	3,451	350	9.5	15
11,470	40	70	3,471	350	9.5	15
11,800	35	70	3,481	350	9.5	15

departure is 2,540 ft, and the measured depth is 12,896 ft. A third possibility is to drill a build-and-hold trajectory, assuming the kickoff is toward the base of the Jurassic and Triassic section at 8,419 ft. The dips predicted at that depth would cause very little departure. At that point, using a build rate of 2°/100 ft, building to an inclination of 40.5° at a TVD of 10,280 ft, and holding the inclination at 40.5° will achieve the required horizontal departure at 12,450 ft. The measured depth at a TVD of 12,450 ft is 13,298 ft.

Let us examine the three approaches and determine which one would be optimal.

Case 1—The Continuous Build. The continuous build would start at 5,239 ft and continue to total depth. If the BHA can be designed to maintain a constant build of 0.5°/100 ft to maximize the WOB, an overall optimization could be achieved. A two-stabilizer building BHA with a spacing between 45 and 55 ft should achieve the desired results. If the second stabilizer is an adjustable slip-on type, the constant rate of build probably could be maintained. The final inclination of 39.0° at total depth is acceptable. The total amount of extra casing necessary for this plan is 589 ft.

Case 2—Build-and-Hold Starting at 4,039 ft. The build would start at 4,039 ft at 0.5°/100 ft and would continue to a final inclination of 22.5° at 8,424 ft TVD and 8,539 ft measured depth. At 12,450 ft TVD the measured depth is 12,896 ft, which requires 446 ft of extra casing. However, the final inclination is 16.5° less than that for the continuous-build case. The BHA's used for the continuous-build approach would be suitable for this plan. Maximum WOB could be maintained over the build section. To hold angle, a three-stabilizer BHA could be used, with a spacing of 12 to 15 ft between the lead and second stabilizers and with a 30-ft spacing between the second and third stabilizers.

The problem with this program lies in the hold section. It is doubtful that a constant 22.5° inclination can be maintained against the formation forces below the Jurassic and Triassic section. Because the dips go to the south, the bit probably would have a strong tendency to turn and drop. This would require going back to some type of building assembly and possibly making a motor run to correct the direction.

Case 3—Build and Hold Starting at 8,419 ft. It might be possible to start at 8,419 ft and to maintain a rate of build at 2°/100 ft in the 12¼-in. hold if a strongly building BHA is used. A two-stabilizer BHA with 60- to 70-ft spacing might work, as would a single-stabilizer building BHA. Before the bottom of the Jurassic-Triassic interval is reached and the harder limestone is penetrated, the required inclination of 40.5° should have been achieved (10,280 ft TVD). This angle should be held constant to a total measured depth of 13,298 ft. In this case, 848 ft of extra casing would be required, mostly in the 8½-in. hole (with 8-in. casing). This program also calls for a long hold section, but mostly in the hard limestone. It would be much easier to maintain the 40.5° with the holding BHA used in Case 2. The main drawbacks to this approach are that the higher inclinations would cause more drag (which might lead to more time-consuming trips and

TABLE 8.25—CALCULATED TRAJECTORY OF WELL A

Measured Depth (ft)	Depth (TVD) (ft)	North-South	East-West (ft)	Vertical Section (ft)	Departure (ft)	Inclination (degree)	Dog-Leg °/100 ft
600	599.98	0.00	5.24	5.24	5.2	1.00	0.2
1,800	1,799.69	0.00	31.41	31.41	31.4	1.50	0.0
2,801	2,800.29	0.00	59.80	59.80	59.8	1.75	0.0
3,405	3,403.97	0.00	79.56	79.56	79.6	2.00	0.0
4,020	4,018.59	0.00	101.03	101.03	101.0	2.00	0.0
4,558	4,555.59	0.00	133.97	133.87	133.9	5.00	0.6
5,340	5,332.56	0.00	222.39	222.39	222.4	8.00	0.4
5,800	5,789.60	0.00	274.47	274.47	274.5	5.00	0.7
6,350	6,335.79	0.00	339.11	339.11	339.1	8.50	0.6
7,010	6,984.73	0.00	459.39	459.39	459.4	12.50	0.6
7,500	7,459.66	0.00	580.00	580.00	580.0	16.00	0.7
8,100	8,031.09	0.00	762.92	762.92	762.9	19.50	0.6
8,640	8,534.37	0.00	958.64	958.64	958.6	23.00	0.6
9,290	9,123.47	0.00	1,233.34	1,233.34	1,233.3	27.00	0.6
9,805	9,580.28	0.00	1,471.14	1,471.14	1,471.1	28.00	0.2
10,500	10,188.14	0.00	1,808.08	1,808.08	1,808.1	30.00	0.3
11,050	10,657.09	0.00	2,095,45	2,095.45	2.095.5	33.00	0.5
11,470	11,005.29	0.00	2,330.31	2,330.31	2,330.3	35.00	0.5
11,800	11,272.26	0.00	2,524.28	2,524.28	2,524.3	37.00	0.6

damage from key seats and stuck pipe) and that footage would need to be drilled in the harder limestone formations.

Conclusion. The analyses of these cases indicate that the continuous-build program is the most attractive approach. Simple two-stabilizer building assemblies can be used with optimal WOB. Because the BHA could always be building angle, the effects of the formation forces could be controlled easily by the adjustment of the second stabilizer either closer to or farther from the lead stabilizer.

Drilling through the abrasive sandstone at the base of the Cretaceous would require 3-point roller reamers instead of stabilizers. These reamers would be necessary in the hard limestones also. To drill through the shale, regular spiral or straight-bladed stabilizers could be used.

The risk in the continuous-build program is that the build might close in the harder limestones and motors would be needed to complete the hole. There is also a danger that the mud program would not stabilize the shales and the hole would enlarge, causing hole instability, key-seating, and other problems that could keep the building assembly from responding in the desired manner. However, good planning and careful selection of the bit, mud system, etc., should minimize those risks.

The foregoing approaches are only a few of the possibilities that can be deduced from the principles presented in Chap. 8. There is no absolute way of drilling a directional or deviation-control well. However, there are better, optimal ways to drill any well. By applying sound drilling engineering analyses, as presented in this textbook, and by treating drilling as a system and using good deductive logic, the drilling engineer can greatly enhance the success of drilling operations.

Exercises*

8.1. Derive Eq. 8.19 for $r_1 > X_3$.

8.2. Plan a build-and-hold directional well whose surface location on a 640-acre section is 780 ft from the north line (FNL) and 1,000 ft from the east line (FEL), and whose bottomhole location is 1,250 ft from the south line (FSL) and 2,000 ft FEL. TVD to the target is 10,500 ft. Rate of build should be 1.5°/100 ft. Kick-off depth is 2,450 ft. Your plan should include the following: (1) maximum inclination angle reached, (2) measured depth to the end of the build and to the target, (3) horizontal departure to the end of the build and at TVD's of 5,450, 6,000, and 8,600 ft, and (4) measured depth at TVD's of 5,450, 6,000, and 8,600 ft. *Answer.* (1) Maximum inclination is 25.3°; (2) measured depth at end of build is 4,137 ft, and measured depth at TVD is 11,235 ft; (3) horizontal departures are 3,662; 1020; 1,273; and 2,502 ft, respectively; (4) measured depths are 5,649; 6,258, and 9,134 ft, respectively.

8.3. Derive Eq. 8.20 for $r_1 < X_3$ and the relationships necessary to calculate (1) measured depth at any TVD and (2) horizontal departure at any TVD or measured depth.

8.4. Plan a build-and-hold trajectory where the kick-off depth is at 2,000 ft, and the target bull's-eye is 5,500 ft from the surface location at a TVD of 8,100 ft. Use a rate of build of 2°/100 ft. Your plan should include the following: (1) maximum inclination angle reached, (2) measured depth to the end of the build and to the target, (3) horizontal departure to the end of the build and at TVD's of 3,100, 5,100, and 7,100 ft, and (4) measured depths at TVD's of 3,100, 5,100, and 7,100 ft.

8.5. Derive the build, hold, and drop relationships (Eq. 8.21) for $r_1 + r_2 > X_4$ and $r_1 < X_3$. Also derive the equations necessary to calculate the measured depth and horizontal departure at any TVD.

8.6. Plan a directional program using a build, hold, and drop trajectory path for which the kick-off depth is 3,100 ft and the maximum required departure is 2,100 ft. Rate of build should be 2°/100 ft, and the rate of drop should be 2°/100 ft. (1) If the TVD is 10,200 ft where the hole returns to vertical, what is the maximum inclination reached? (2) How much casing will you have to purchase for a well with a TVD bottomhole target of 10,200 ft? (3) How long is the hold section? (4) If the drilling plan calls for an intermediate string to be run to a TVD of 8,210 ft, how much casing is needed, and what is the horizontal departure at the casing shoe?

TABLE 8.26—COSTS FOR FIELD DEVELOPMENT

	Case 1	Case 2	Case 3
Cost of Subsea Completion Per Well	$75 Million	$7 Million	$60 Million
Cost of Offshore Platform	$500 Million	$15 Million	$280 Million
Cost of a Drillship, Jackup or Semi-submersible	$80,000/Day (Semi-submersible or drillship)	$35,000/Day (Jackup)	$120,000/Day (Semi-submersible)
Cost of Drilling Rig and Full Support On Fixed Platform	$40,000/Day	$30,000/Day	$80,000/Day
Special Considerations	None	Can drill only 4 months of year because of environmental considerations	None

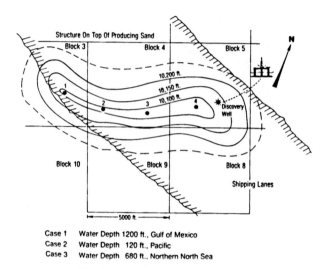

Case 1   Water Depth 1200 ft., Gulf of Mexico
Case 2   Water Depth 120 ft., Pacific
Case 3   Water Depth 680 ft., Northern North Sea

Fig. 8.161—Plot of offshore discovery.

8.7. Derive the build, hold, and drop relationships (Eq. 8.22) $r_1 + r_2 < X_4$ and $r_1 < X_3$. Also derive the equations necessary to calculate the measured depth and horizontal departure.

8.8. Plan a directional well using a build, hold, and drop trajectory for which the horizontal departure to the target is 9,010 ft and the TVD to the top of the target is 14,100 ft. The rate of build and drop is 2°/100 ft. (1) What should be the kick-off depth if the maximum possible angle for hole drop is 52°? (2) If the intermediate casing is to be run at the end of the build, how much casing is needed, and what are the horizontal departure and the TVD at the casing shoe? (3) Where should the drop begin (TVD and measured depth)? (4) What is the horizontal departure at the beginning and at the end of the drop? (5) How much casing is required from surface to a TVD of 14,100 ft?

8.9. Write all the equations for a build, hold, drop, and hold (modified "S") using Eqs. 8.21 and 8.22. Derive all the relationships to calculate horizontal departures, measured depths, and inclination angle.

8.10. Plan a modified "S" trajectory where the target must intersected at a constant inclination of 20°. Kick-off depth is 1,500 ft. Rate of build and drop is 2°/100 ft. The desired horizontal departure is 3,100 ft from the surface location at a TVD of 9,075 ft. (1) What is the maximum inclination reached? (2) How much casing is

needed to go from surface to the end of the build-and-hold section? (3) How much casing is needed from surface to a TVD of 9,075 ft? (4) If the producing interval is logged between 9,355 and 9,420 ft (MD), what are the TVD and the horizontal departure at the top and bottom of the pay?

8.11. Repeat Problem 8.10 with a rate of build and drop of 1°/100 ft.

8.12. An offshore discovery has been made under an active shipping lane (see Fig. 8.161). Further seismic work has defined the anticlinal structure as being almost completely under the shipping lane. (You cannot locate any drilling structure within the shipping lane.) Reservoir analysis dictates drilling the structure with the four bottomhole locations indicated on the plot. Your company owns Blocks 3, 4, and 5. You have permission to drill anywhere outside the shipping lane from Blocks 3, 5, 8, and 10. It is your assignment to determine the best economics to develop this field. (Note: This problem will be referred to in later problems.)

The first part of the study is to determine the economics of different approaches to drilling Wells 1, 2, 3, and 4—i.e., whether to use a semisubmersible and drill some or all of the wells for subsea completions or to set a platform. Table 8.26 gives the various conditions for Cases 1, 2, and 3.

Fig. 8.162 summarizes the results from the discovery well. Determine whether you would set a platform to drill all the wells, set a platform to drill some of the wells and drill the others with a movable drilling vessel and complete them subsea, or drill all the wells with a movable drilling vessel and complete them subsea. You must do the following:

A. Determine the surface location for each well.
B. Calculate the trajectory for each well.
C. Calculate the casing costs on the basis of setting 20-in. surface casing to 3,500 ft (TVD), intermediate 13⅜-in. casing to cover the build section and 9⅝-in. casing to cover the overpressured interval (8,500 ft TVD), and 7-in. casing to total depth. Below is the cost of the casing.

Size (in.)	Cost ($/ft)
20	105.24
13⅜	74.58
9⅝	44.73
7	28.82

D. Using the information in Table 8.27, determine the time needed to drill each well for each case. (Maximum inclination is 70° for Case 1, 80° for Case 2, and 60° for Case 3.)

8.13. Replan Problem 8.4 using a target direction of N25E and 18° left of N25E as the amount of lead necessary to offset the expected bit walk. The direction of N7E should be held constant until an inclination of 10° is reached. At what constant rate of right-hand walk will the lead of 18° hit within a 50-ft radius of the target bull's-eye? How much extra casing will be needed because of the bit walk? *Answer.* For Problem 8.4, at 8,085.27 ft, measured depth is 10,423.16 ft; for Problem 8.13, at 8,085.27 ft, measured depth is 10,491 ft.

8.14. Given the trajectory data from an offset well, determine the amount of lead necessary to drill a well to intersect two targets with the following target requirements. How much casing must you buy?

Target 1: Direction of bull's-eye: S75E
Horizontal departure: 3,570 ft
TVD to top of target: 10,400 ft
Radius of target: 150 ft

Target 2: Direction of bull's-eye: S70E
Horizontal departure: 4,250 ft
TVD to top of second target: 10,900 ft
Radius of target: 100 ft

Hint: Plot the offset data in Table 8.28 to determine the total amount of bit walk. Divide the total bit walk by two to estimate the amount of lead to account for the right-hand walk.

8.15. Derive the angle-averaging Eqs. 8.24 through 8.26.

8.16. Using Table 8.29, determine the extra amount of casing you must buy because of bit walk for the trajectories you determined for Problem 12. Use a target radius of 150 ft. How much lead must you have to hit each target?

8.17. Derive the tangential method Eqs. 8.27 through 8.29.

8.18. Determine the trajectory of the well using the tangential method. Did the wellbore intersect the target? Include TVD, north/south and east/west coordinates, the total coordinates, total departure, and the departure angle. The target is a circle with a radius of 600 ft. The top of the target is 7,800 ft TVD at a direction of N13E. See Table 8.30 for survey data.

8.19. Using the angle-averaging method, determine the trajectory of the well on the basis of the surveying data presented in Problem 8.18. How much difference is there in TVD and horizontal departure between the angle-averaging method and the tangential method at total depth? *Answer.* The difference in TVD is +19.03 ft; the difference in horizontal departure is −39 ft.

8.20. Set up a step-by-step procedure using the minimum-curvature method to calculate the TVD, the north/south and east/west coordinates, the total coordinates, the total departures, and the departure angle.

8.21. Determine the trajectory of the well, using the survey data presented in Problem 8.18 and the format derived in Problem 8.20 for the minimum-curvature method. How do the results compare with those of the tangential and angle-averaging methods?

TABLE 8.27—DATA FOR PROBLEM 8.12D

Case 1 Time to Drill = [0.016(D_M)-91](Maximum Inclination Factor)
Case 2 Time to Drill = [0.013(D_M)-95](Maximum Inclination Factor)
Case 3 Time to Drill = [0.015(D_M)-42](Maximum Inclination Factor)

	Maximum Inclination Factor		
Inclination	Case 1	Case 2	Case 3
0°-10°	0.9	0.9	0.9
10°-24°	1.0	1.0	1.0
24°-48°	1.3	1.2	1.5
48°-60°	1.5	1.7	2.0
60°-80°	2.0	2.5	—

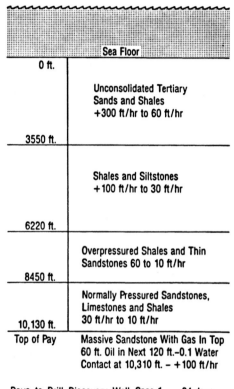

Days to Drill Discovery Well Case 1 84 days
Case 2 37 days
Case 3 110 days

The Maximum Inclination for the Discovery Well in Each Case is 22°

Fig. 8.162—Discovery well information.

8.22. Derive the balanced tangential method from the minimum-curvature method (hint: assume $L_t = 1.0$). Calculate the survey data from Problem 8.18 and determine the trajectory of the well. Is the balanced tangential method comparable with the other methods?

8.23. Using the survey data from two wells drilled from an offshore platform, determine how far apart the wellbores are at their closest point. The survey data are found in Tables 8.31 and 8.32. The surface location of Well 2 is 6 ft due east of Well 1.

motor from 2,140 ft (3° inclination and N5W) to 2,260 ft. The bent sub causes an angle change of 3.5°/100 ft. Tool-face setting is 20° right of high side. *Answer.* N6.7E.

8.25. Determine the new inclination angle for Problem 8.24.

8.26. Determine the tool-face setting to sidetrack out of a wellbore at 8,520 ft for a motor run of 220 ft. The

TABLE 8.28—OFFSET DATA FOR PROBLEM 8.14

Measured Depth (ft)	Inclination (degrees)	Direction	TVD (ft)	Vertical Section	East-West Coordinate	North-South Coordinate	Dogleg Severity (°/100 ft)
646	3	S60W	646	15.6	−14.6	−8.5	0.0
955	1	S61W	955	25.6	−24.0	−13.8	0.65
1,206	1	S71W	1,206	29.4	−28.0	−15.5	0.06
1,468	0.5	S60W	1,468	32.5	−31.2	−17.0	0.20
1,963	0.25	S16E	1,963	35.6	−32.4	−20.0	0.10
2,429	0.25	N45E	2,429	34.8	−30.4	−20.5	0.09
2,861	0.25	N22W	2,861	33.1	−30.0	−18.6	0.06
2,959	2	S20W	2,958	34.3	−31.9	−18.7	2.23
3,057	2.25	S20W	3,056	37.8	−33.2	−22.1	0.23
3,155	4.75	S15W	3,154	43.4	−35.0	−27.8	2.57
3,253	6.75	S20W	3,252	52.6	−37.9	−37.2	2.10
3,322	7.25	S17W	3,320	60.5	−40.6	−45.1	0.89
3,420	8.5	S25W	3,417	73.4	−45.4	−57.7	1.69
3,518	10.5	S25W	3,514	89.2	−52.3	−72.3	2.04
3,616	12.5	S23W	3,610	108.2	−60.2	−90.2	2.08
3,714	14	S22W	3,705	129.8	−68.8	−110.9	1.54
3,812	15.5	S19W	3,800	153.7	−77.5	−134.3	1.72
3,910	17.5	S20W	3,894	180.1	−86.8	−160.5	2.06
4,008	20	S19W	3,987	210.0	−97.2	−190.2	2.57
4,157	20	S21W	4,127	258.6	−114.8	−238.1	0.46
4,281	20	S21W	4,243	299.2	−130.0	−277.7	0.0
4,404	19.75	S22W	4,359	339.4	−145.3	−316.6	0.34
4,528	19.25	S23W	4,476	379.3	−161.1	−354.9	0.48
4,649	19.5	S24W	4,590	418.2	−177.1	−391.7	0.33
4,771	19.75	S25W	4,705	458.1	−194.1	−429.0	0.33
4,891	20.75	S27W	4,818	498.8	−212.3	−466.3	1.01
5,018	20.5	S28W	4,937	542.8	−233.0	−506.0	0.33
5,193	20.75	S30W	5,100	603.7	−262.9	−559.9	0.42
5,368	22.25	S31W	5,263	667.4	−295.4	−615.1	0.88
5,461	22	S32W	5,349	702.2	−313.7	−645.0	0.48
5,492	21.5	S31W	5,378	713.6	−319.7	−654.8	1.98
5,592	20.5	S33W	5,471	749.3	−338.7	−685.2	1.23
5,653	21.5	S33W	5,528	771.1	−350.6	−703.5	1.63
5,718	21.25	S33W	5,589	794.7	−363.5	−723.4	0.34
5,781	20	S34W	5,648	816.8	−375.8	−741.9	2.06
5,874	18	S37W	5,736	847.1	−393.4	−766.5	2.39
5,998	17	S35W	5,854	884.3	−415.3	−796.7	0.94
6,181	17.25	S34W	6,029	938.1	−445.8	−841.1	0.20
6,397	17.5	S38W	6,235	1,002.6	−483.7	−893.3	0.56
6,614	18.75	S36W	6,441	1,070.1	−524.4	−947.2	0.64
6,737	18.25	S39W	6,558	1,109.1	−548.1	−978.2	0.87
6,835	17.5	S38W	6,651	1,139.2	−566.8	−1,001.7	0.82
6,933	15.75	S37W	6,745	1,167.3	−583.9	−1,024.0	1.81
7,031	14.5	S37W	6,840	1,192.8	−599.3	−1,044.4	1.27
7,129	13.5	S36W	6,935	1,216.5	−613.4	−1,063.4	1.05
7,227	12	S35W	7,030	1,238.1	−626.0	−1,081.1	1.54
7,325	10.5	S37W	7,126	1,257.3	−637.2	−1,096.5	1.58
7,423	8.75	S36W	7,223	1,273.6	−646.9	−1,109.7	1.79
7,521	7	S37W	7,320	1,287.1	−654.9	−1,120.5	1.79
7,619	6.25	S37W	7,418	1,298.4	−661.7	−1,129.5	0.76
7,717	4.75	S38W	7,515	1,307.8	−667.4	−1,137.0	1.53
7,815	4	S42W	7,613	1,315.2	−672.3	−1,142.7	0.82
7,913	3.25	S43W	7,711	1,321.4	−676.4	−1,147.3	0.77
7,966	2.5	S47W	7,763	1,324.0	−678.3	−1,149.1	1.46
8,965	4	S80W	8,761	1,375.0	−729.0	−1,174.4	0.23
10,605	3	S65W	10,398	1,425.4	−828.3	−1,161.3	0.14

TABLE 8.29—BIT WALK DATA FOR PROBLEM 8.16

Depth (ft) (TVD)	Inclination (degrees)	Case 1 (°/100 ft)	Case 2 (°/100 ft)	Case 3 (°/100 ft)
0- 3,550	0-80	0.1° left	0°	0.1° right
3,550- 6,220	0-10	0.1° right	0.1° right	0.2° right
	10-30	0.2° right	0.2° right	0.4° right
	30-50	0.3° right	0.2° right	0.3° right
	50-80	0.0	0.0	0.0
6,220- 8,450	0-10	0.1° left	0.1° left	0.1° right
	10-30	0.2° left	0.1° left	0.2° right
	30-50	0.0	0.1° left	0.2° right
	50-80	0.0	0.0	0.0
8,450-10,130	0-10	0.1° right	0.2° right	0.2° right
	10-30	0.2° right	0.5° right	1.0° right
	30-50	0.3° right	1.0° right	0.7° right
	50-80	0.1° left	1.4° right	0.0

inclination is 30° and the direction is N60W. It is necessary to be 3° below the original wellbore. How much direction change can be obtained if the dogleg severity is to be kept below 4°/100 ft?

8.27. For a course length of 300 ft and a dogleg severity of 3.0°/100 ft, determine the tool-face setting for a jetting run whose purpose is to obtain maximum allowable drop while changing the original direction from S10E to S30E. The inclination angle before course change is 15°.

8.28. Derive Eq. 8.50 from Eq. 8.41 by making $\beta=1$.

8.29. Derive Eq. 8.51 from Eq. 8.45 by making $\beta=1.0$.

8.30. Derive Eq. 8.52 from Eq. 8.41 by making $\beta=1$.

8.31. Derive Eq. 8.53 from Eqs. 8.41, 8.45, and 8.48.

8.32. Derive Eq. 8.54 from Eq. 8.53.

8.33. An initial lead of 25° left of N33E has not turned right fast enough to hit the target. A motor must be run to correct the trajectory to N42E. At a measured depth of 4,835 ft, the inclination is 42° and the direction is N15E. How many feet must the motor drill to make the correction without losing inclination and maintaining the dogleg severity below 3.0°/100 ft? What should be the tool-face setting? (Assume no correction for reverse torque.)

and at a direction of S79W. You want to drop the inclination back to 7°. The new desired direction is N66W. What is the dogleg severity if this change is made over 250 ft of course length? What should be the tool-face setting to make this change? How much course length would be necessary if the maximum permissible dogleg severity were 3.5°/100 ft? (Assume no correction for reverse torque.) *Answer.* 266 ft of course length.

8.35. Repeat Problems 8.24, 8.25, and 8.26 using the Ouija Board nomograph.

8.36. Repeat Problems 8.33 and 8.34 using the Ragland diagram.

8.37. Derive Eq. 8.55 from Eq. 8.53. Hint: $\cos \beta = 1-2[\sin(\beta/2)^2]$ and $\cos \Delta\epsilon = 1-2[\sin(\Delta\epsilon/2)^2]$.

8.38. Calculate the total angle change of 11,050 ft of 5.0-in. Grade E 19.5-lbm/ft drillpipe and 1,200 ft of 8-in. drill collars ($2^{13}/_{16}$-in. ID) for a bit-generated torque of 650 ft-lbf. Assume the 8-in. motor has the same properties as the 8-in. collars. Use a shear modulus for steel. *Answer.* 181.97°.

8.39. How much reverse torque could be expected for Problem 8.33 if an 8-in. mud motor were used for the trajectory correction? (The expected maximum torque at the bit is 1,350 ft-lbf.) In what direction should the face be oriented to offset the reverse torque? The drillstring consists of 5-in., 19.5-lbm/ft Grade E drillpipe, 1,000 ft of 5-in. (3-in.-ID), 49.3-lbm/ft heavyweight drillpipe, and twelve 8-in. ($2^{13}/_{16}$-in.-ID) drill collars.

8.40. Generate figures similar to Figs. 8.37 and 8.38 for overall angle changes (β) of 2, 3, and 5° and for inclinations of 0 to 65°.

8.41. Determine the inclination, direction, and tool-face setting (if applicable) for the single-shot compass card pictures in Fig. 8.163. *Answer.* (1) Inclination=20°, direction=S45E, tool face=S34W; (4) Inclination=11°, direction=N22E, tool face=N55E.

8.42. Determine the inclination and direction from the single-shot compass card pictures in Fig. 8.164.

8.43. Determine the inclination and direction from the survey information found in Table 8.33 and from the multishot pictures in Fig. 8.165. The timer is set for 30 seconds between pictures.

8.44. Correct the single-shot survey data (Problem 8.41) for declination at the following locations: (1) McAllen, TX; (2) Marietta, OH; (3) Cut Bank, MT; (4) Long Beach, CA; (5) Evanston, WY. *Answer.* (1) S37E, (2) S28E, (3) none, (4) N30E, and (5) S47E.

8.45. Correct the multishot survey data (Problem 8.42) for declination at the following locations: Houma, LA; Tyler, TX; Rifle, CO; Casper, WY; Parkersburg, WV.

8.46. Determine how many nonmagnetic drill collars are necessary for drilling the directional wells described in Table 8.34. *Answer.* Zone I, one 30-ft drill collar; compass unit 3 to 4 ft below center.

8.47. Multishot gyroscope data were taken on a shallow directional well near Houston, TX. Table 8.35 supplies general information about the well and the start and end times for the gyroscope. Determine the inclinations and directions, and calculate the trajectory of the wellbore. Table 8.36 lists the tool-drift-check data, while Table 8.37 lists the multishot gyroscope data. *Answer.* measured depth=1,800 ft; TVD=1,590.9 ft; latitude=(N)645.4 ft; departure=191.7 ft (E).

8.48. Calculate the inclination and direction and the direction of the low sides of the hole from the readings taken with a rate gyroscope at a north latitude of 30°. The earth spin rate is 15.042°/hr, and north is positive. Earth's gravity is 1.0, and up is positive. Tool roll, or alpha, is measured clockwise positive from the top of the hole. Table 8.38 gives the measured data from the rate gyroscope.

8.49. It is noted that there is a compass error of 5.8° from a magnetic single-shot instrument in a well with an inclination of 66° and a direction of 165°. The measured magnetic strength at the location is 5.8 μT and the dip angle is 60°. What are the resultant magnetic strengths of the collars and drillstring at the location of the compass unit? *Answer.* 11.3 μT.

8.50. The measured compass error at an inclination of 32° and a direction of N75W is 10.5°; at 44° and N78W, the error is 13.7° (dip angle is 60°). Was the BHA changed between these surveys?

8.51. Determine the maximum survey error for the following directional well assuming a good magnetic survey.

0 to 2,000 ft	vertical
2,000 to 3,000 ft	10° constant build
3,000 to 5,000 ft	35° constant build
5,000 to 8,000 ft	40° average

**TABLE 8.30—SURVEY DATA FOR
PROBLEM 8.18**

Depth (md)	Inclination (degrees)	Direction	Depth (md)	Inclination (degrees)	Direction
100	1.20	N11.9W	5,300	46.47	N9.1E
200	1.12	N1.8E	5,400	46.62	N9.5E
300	0.56	N28E	5,500	46.95	N9.2E
400	0.52	N26.1E	5,600	47.49	N9.9E
500	0.65	N21.5E	5,700	47.93	N10.2E
600	0.67	N28.9E	5,800	48.55	N9.6E
700	0.58	N23E	5,900	49.18	N10E
800	0.60	N29.1E	6,000	49.69	N10.1E
900	0.50	N21.7E	6,100	49.79	N9.4E
1,000	0.67	N21.7E	6,200	48.95	N10.3E
1,100	0.74	N22.3E	6,300	50.12	N10.7E
1,200	0.58	N5.4E	6,400	49.95	N10.8E
1,300	0.60	N14.5W	6,500	50.35	N10.7E
1,400	0.51	N5.6W	6,600	50.93	N10.5E
1,500	0.45	N23.7W	6,700	50.19	N11.1E
1,600	0.62	N22.4W	6,800	51.87	N11.7E
1,700	1.04	N2.7W	6,900	53.01	N11.4E
1,800	2.39	N0.3E	7,000	53.72	N12.1E
1,900	4.10	N1.6E	7,100	53.47	N12.6E
2,000	5.85	N2.2W	7,200	53.72	N11.8E
2,100	7.44	N1W	7,300	52.62	N12.7E
2,200	9.60	N1.7E	7,400	53.75	N12E
2,300	11.9	N1E	7,500	52.41	N12.5E
2,400	13.86	N2.9E	7,600	52.90	N12.4E
2,500	16.47	N5.8E	7,700	52.44	N12.9E
2,600	19.8	N5E	7,800	51.59	N12.3E
2,700	22.56	N4.2E	7,900	51.74	N12.2E
2,800	24.88	N4E	8,000	50.56	N12.7E
2,900	26.19	N5.6E	8,100	50.65	N12.3E
3,000	29.28	N4.3E	8,200	49.93	N12.6E
3,100	32.52	N2.8E	8,300	50.06	N12.2E
3,200	35.7	N1.3E	8,400	50.42	N14E
3,300	38.64	N1.1E	8,500	50.75	N15.5E
3,400	41.16	N4E	8,600	51.15	N15.3E
3,500	42.75	N4.9E	8,700	51.79	N15.1E
3,600	42.45	N2.4E	8,800	51.72	N15E
3,700	42.43	N2.7E	8,900	52.13	N15E
3,800	42.47	N2.5E	9,000	52.60	N14.4E
3,900	43.10	N2.7E	9,100	52.90	N14.5E
4,000	43.66	N2.9E	9,200	53.02	N13.9E
4,100	44.23	N3.3E	9,300	53.18	N13.4E
4,200	44.73	N3.4E	9,400	53.67	N12.9E
4,300	45.03	N3.4E	9,500	54.09	N12.4E
4,400	45.55	N3.1E	9,600	54.14	N13.1E
4,500	45.92	N2.5E	9,700	53.05	N12.4E
4,600	46.16	N2E	9,800	52.18	N12.1E
4,700	46.55	N1.4E	9,900	51.44	N12.8E
4,800	46.9	N0.6E	10,000	50.84	N12.5E
4,900	47.23	N0.5E	10,100	50.79	N13.1E
5,000	47.41	N0.7E	10,200	50.93	N13.2E
5,100	46.05	N5.8E	10,300	50.94	N13.8E
5,200	46.08	N9.3E	10,400	51.10	N13.5E

TABLE 8.31—SURVEY DATA FOR WELL A—PROBLEM 8.23

Measured Depth (ft)	Inclination (degrees)	Direction	TVD (ft)	Vertical Section	East-West Coordinate	North-South Coordinate	Dogleg Severity (°/100 ft)
360	0.25	S25W	360	−0.3	−0.3	−0.7	0.0
416	0.25	S36W	416	−0.1	−0.6	−0.7	0.75
596	0.25	N3W	596	0.6	−0.8	0.0	0.06
903	1.5	N6E	903	3.9	−0.7	4.7	0.41
1,245	1.25	N52E	1,245	6.5	3.3	11.9	0.32
1,698	0.75	N34W	1,698	11.4	4.5	19.7	0.31
1,966	2.5	N32W	1,966	18.9	0.4	26.1	0.65
2,064	4.25	N30W	2,064	24.5	−2.6	31.0	1.79
2,162	5.25	N44W	2,161	32.6	−7.5	37.5	1.55
2,260	7	N54W	2,259	43.0	−15.4	44.3	2.08
2,358	8.25	N47W	2,356	55.9	−25.4	52.6	1.59
2,456	9.25	N42W	2,453	70.8	−35.9	63.2	1.28
2,554	10.5	N45W	2,549	87.6	−47.4	75.4	1.37
2,652	12	N47W	2,645	106.7	−61.2	88.7	1.58
2,750	14.75	N48W	2,741	129.3	−77.9	104.0	2.82
2,848	17	N47W	2,835	156.0	−97.7	122.1	2.31
2,946	19	N49W	2,928	186.1	−120.2	142.4	2.13
3,044	21	N52W	3,020	219.4	−146.0	163.7	2.29
3,142	22.5	N53W	3,111	255.1	−174.8	185.8	1.57
3,236	24	N52W	3,198	291.7	−204.3	208.4	1.65
3,330	23.75	N52W	3,283	330.0	−234.9	232.3	0.79
3,424	26.5	N50W	3,368	370.2	−266.4	257.9	2.08
3,517	26.75	N48W	3,451	411.7	−297.9	285.2	1.00
3,640	27	N46W	3,561	467.1	−338.6	323.2	0.76
3,734	26.75	N43W	3,645	509.6	−368.3	353.5	1.46
3,829	27.5	N41W	3,729	552.9	−397.3	385.6	1.24
3,927	26.5	N42W	3,817	597.4	−426.8	419.0	1.12
4,025	26.5	N44W	3,904	641.1	−456.6	450.9	0.91
4,123	26.5	N44W	3,992	684.8	−487.0	482.4	0.0
4,221	26.5	N46W	4,080	728.5	−517.9	513.3	0.91
4,319	25.75	N47W	4,168	771.6	−549.2	543.0	0.88
4,449	26.5	N47W	4,284	828.7	−591.1	582.1	0.57
4,663	26.25	N45W	4,476	923.6	−659.5	648.1	0.43
4,914	26.27	N42W	4,701	1,034.6	−735.9	728.7	0.53
5,012	27.25	N44W	4,789	1,078.7	−766.0	760.9	1.36
5,110	25.5	N49W	4,876	1,122.2	−797.6	790.9	2.88
5,208	25.5	N54W	4,965	1,163.9	−830.6	817.2	2.19
5,302	27.5	N57W	5,049	1,204.8	−865.2	840.9	2.56
5,549	27.75	N58W	5,268	1.315.5	−961.8	902.5	0.21
5,736	27.5	N56W	5,434	1.399.6	−1,034.5	949.7	0.51
5,985	27.5	N54W	5,654	1,511.9	−1,128.7	1,015.6	0.37
6,201	27.75	N53W	5,846	1,610.3	−1,209.2	1,075.2	0.24
6,455	26.5	N52W	6,072	1,724.5	−1,301.1	1,145.7	0.52
6,639	26.5	N50W	6,236	1,805.7	−1,364.9	1,197.4	0.48
6,734	26	N51W	6,322	1,847.4	−1,397.3	1,224.1	0.70
6,853	24.75	N51W	6,429	1,897.8	−1,436.9	1,256.2	1.05
6,944	24.25	N52W	6,512	1,935.1	−1,466.4	1,279.7	0.71
7,038	23	N50W	6,598	1,972.4	−1,495.7	1,303.4	1.58
7,130	21	N52W	6,683	2,006.5	−1,522.5	1,325.1	2.32
7,254	16.75	N51W	6,801	2,046.2	−1,553.9	1,350.1	3.44
7,379	13.25	N49W	6,922	2,078.3	−1,578.7	1,370.8	2.83
7,505	10	N49W	7,045	2,103.5	−1,597.8	1,387.5	2.58
7,630	7	N49W	7,169	2,121.9	−1,611.8	1,399.6	2.40
7,752	5.75	N48W	7,290	2,135.3	−1,621.9	1,408.6	1.03
7,967	4	N43W	7,504	2,153.6	−1,635.0	1,421.4	0.84
8,121	3	N39W	7,658	2,163.0	−1,641.1	1,428.5	0.67
8,439	3	N16W	7,975	2,179.0	−1,648.8	1,443.3	0.38
8,502	4.25	N19W	8,038	2,182.6	−1,650.0	1,447.1	2.00
8,596	5.5	N21W	8,132	2,190.0	−1,652.7	1,454.6	1.34
8,810	5.75	N15W	8,345	2,209.1	−1,659.2	1,474.5	0.30
8,967	4.75	N15W	8,501	2,221.8	−1,662.9	1,488.4	0.63
9,283	3.75	N13W	8,816	2,242.3	−1,668.6	1,511.1	0.32
10,835	3.75	N13W	10,365	2,330.4	−1,691.4	1,610.0	0.0

TABLE 8.32—SURVEY DATA FOR WELL B—PROBLEM 8.23

Measured Depth (ft)	Inclination (degrees)	Direction	TVD (ft)	Vertical Section	East-West Coordinate	North-South Coordinate	Dogleg Severity (°/100 ft)
454	0.25	N15W	454	0.9	−0.3	1.0	0.0
515	1.75	N32W	515	1.9	−0.7	1.9	2.47
606	2.75	N2W	606	5.1	−1.7	5.3	1.66
728	2.5	N34W	728	10.2	−3.5	10.7	1.20
910	2.75	N25W	910	18.3	−7.6	17.9	0.26
1,093	2.5	N30W	1,092	26.3	−11.4	25.4	0.18
1,310	1.5	N22W	1,309	33.6	−14.7	32.2	0.48
1,685	0.75	N30W	1,684	40.6	−18.0	38.8	0.20
1,787	0.5	N80W	1,786	41.7	−18.9	39.4	0.56
1,853	1.25	N55W	1,852	42.6	−19.8	39.8	1.25
1,951	2.5	N62W	1,950	45.7	−22.6	41.5	1.29
2,049	4.25	N69W	2,048	51.0	−27.8	43.9	1.83
2,147	6.25	N64W	2,146	59.3	−36.0	47.4	2.09
2,245	8.5	N58W	2,243	71.2	−47.0	53.5	2.42
2,343	10.5	N62W	2,339	86.7	−61.0	61.6	2.15
2,441	11.5	N57W	2,436	104.6	−77.1	71.1	1.41
2,539	13.25	N55W	2,531	125.1	−94.6	82.9	1.84
2,637	15.5	N55W	2,626	148.9	−114.5	96.8	2.29
2,735	17.5	N58W	2,720	176.0	−137.7	112.2	2.22
2,833	19.75	N59W	2,813	206.1	−164.4	128.5	2.32
2,931	21	N58W	2,905	239.0	−193.5	146.4	1.32
3,029	23.5	N59W	2,996	274.8	−225.1	165.8	2.58
3,089	24.75	N61W	3,050	298.2	−246.4	178.0	2.49
3,187	26.25	N61W	3,139	338.3	−283.3	198.5	1.53
3,286	26.5	N61W	3,228	380.2	−321.7	219.8	0.23
3,410	27.25	N60W	3,338	433.6	−370.5	247.4	0.70
3,533	28.25	N60W	3,447	488.4	−420.1	276.0	0.81
3,660	28.25	N55W	3,559	546.6	−470.8	308.3	1.86
3,782	29	N53W	3,666	604.0	−518.1	342.7	1.00
3,904	30	N52W	3,772	663.2	−565.7	379.3	0.91
4,028	30	N52W	3,880	724.4	−614.6	417.4	0.0
4,215	29.5	N50W	4,042	816.3	−686.7	475.8	0.59
4,401	29.5	N50W	4,204	907.2	−756.9	534.7	0.0
4,557	29.75	N50W	4,339	983.8	−815.9	584.3	0.14
4,746	29.5	N50W	4,504	1,076.5	−887.5	644.3	0.12
4,866	30	N50W	4,608	1,135.6	−933.1	682.6	0.41
4,960	29	N50W	4,690	1,181.6	−968.6	712.3	1.06
5,058	28.5	N46W	4,776	1,228.5	−1,003.6	743.9	2.03
5,156	28	N42W	4,862	1.274.9	−1,035.8	777.3	2.00
5,254	27.75	N39W	4,949	1.320.7	−1,065.6	812.1	1.45
5,352	27.25	N36W	5,035	1,365.7	−1,093.1	848.0	1.50
5,450	28.25	N33W	5,122	1,410.8	−1,119.0	885.6	1.75
5,548	27.75	N33W	5,209	1,456.1	−1,144.0	924.2	0.50
5,686	27.25	N33W	5,331	1,518.9	−1,178.8	977.6	0.35
5,904	27.75	N31W	5,525	1,617.7	−1,232.1	1,063.0	0.48
5,996	26.75	N31W	5,606	1,658.9	−1,253.8	1,099.1	1.08
6,141	26.5	N31W	5,736	1,722.5	−1,287.3	1,154.8	0.15
6,360	25.75	N28W	5,933	1,816.2	−1,334.7	1,238.7	0.69
6,577	25.25	N26W	6,128	1.906.0	−1,377.2	1,322.0	0.46
6,764	24.25	N24W	6,298	1,980.5	−1,410.2	1,392.9	0.70
6,862	23	N25W	6,388	2,017.7	−1,426.5	1,428.7	1.34
6,960	20.25	N23W	6,479	2,051.9	−1,441.2	1,461.6	2.90
7,058	18.75	N24W	6,571	2,082.7	−1,454.3	1,491.6	1.57
7,156	16.25	N24W	6,665	2,110.6	−1,466.2	1,518.6	2.55
7,254	15.25	N24W	6,759	2,135.7	−1,477.1	1,542.9	1.01
7,352	13.5	N21W	6,854	2,158.5	−1,486.4	1,565.4	1.94
7,450	11.75	N22W	6,950	2,178.5	−1,494.2	1,585.3	1.80
7,548	10.75	N19W	7,046	2,196.1	−1,500.9	1,603.2	1.18
7,646	9.25	N16W	7,142	2,211.5	−1,506.0	1,619.4	1.62
7,744	7.75	N15W	7,239	2,224.3	−1,509.9	1,633.4	1.54
7,842	6.5	N14W	7,337	2,235.0	−1,513.0	1,645.1	1.28
7,940	5.25	N13W	7,434	2,243.7	−1,515.3	1,654.9	1.28
8,322	3	N10W	7,815	2,267.2	−1,520.8	1,681.8	0.59
8,515	2.5	N5W	8,008	2,274.7	−1,522.0	1,691.0	0.28
9,288	1	N19E	8,781	2,289.9	−1,519.1	1,714.4	0.21
9,328	1	N19E	8,821	2,290.2	−1,518.9	1,715.1	0.0
10,325	2.5	N64E	9,817	2,293.1	−1,498.7	1,737.9	0.19
10,385	2.5	N64E	9,877	2,292.4	−1,496.3	1,739.0	0.0
11,105	2.5	N64E	10,596	2,283.2	−1,468.1	1,752.8	0.0

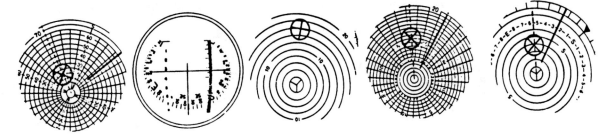

Fig. 8.163—Single-shot compass card pictures for Problem 8.41.

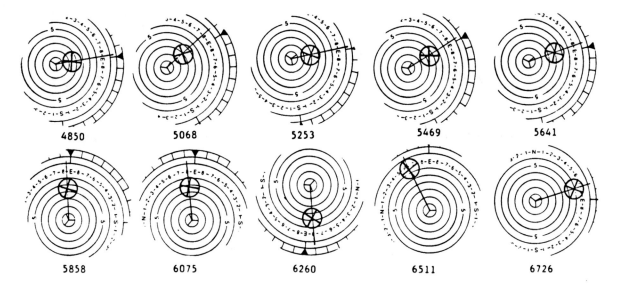

Fig. 8.164—Single-shot compass card pictures for Problem 8.42.

TABLE 8.33—SURVEY DATA FOR PROBLEM 8.43

Survey Time	Depth (ft)	Comments
12:45:00	0	Clock off
12:45:20	0	Clock on
12:45:25	0	Tool Dropped
12:45:55		
12:46:25		
12:53:25	8,914	On Bottom
12:53:55	8,914	Wait
12:54:25	8,884	Pull Stand ⅓ Out
12:54:55	8,854	Pull Stand ⅔ Out
12:55:25		Pull Stand Out
12:55:55	8,824	Wait
12:56:25	8,794	Pull Stand ⅓ Out
12:56:55	8,764	Pull Stand ⅔ Out
12:57:25		Pull Stand Out
12:57:55	8,734	Wait
12:58:25	8,704	Pull Stand ⅓ Out
12:58:55	8,674	Pull Stand ⅔ Out
12:59:25		Pull Stand Out
12:59:55	8,644	Wait
12:60:25	8,614	Pull Stand ⅓ Out
12:60:55	8,584	Pull Stand ⅔ Out
12:61:25		Pull Stand Out
12:61:55	8,554	Wait
12:62:25	8,554	Wait
12:62:55	8,524	Pull Stand ⅓ Out
12:53:25	8,914	On Bottom

TABLE 8.34—DIRECTIONAL WELL DATA FOR PROBLEM 8.46

Direction of Well	Maximum Inclination	Location
N60E	32°	Corpus Christi, TX
338°	62°	Long Beach, CA
25°	48°	North Sea
S25W	35°	North Slope, AK
N65W	70°	Boss Strait, Australia

8.52. Determine the probable survey error for the survey data of Problem 8.47. *Answer.* For a good gyroscope, 7.7 ft; for a poor gyroscope, 38.6 ft.

8.53. What is the probable survey error for Wells A and B (Problem 8.23), assuming Wells A and B were surveyed with a magnetic multishot?

8.54. You are drilling a directional well in a hot geothermal area where the heat makes turbines and positive displacement motors impractical to run. The kick-off depth is 1,650 ft and the hole size is 12¼ in. Design a kick-off program using an open-hole whipstock where the initial inclination is 2.5° and the direction is S45E. The acceptable dogleg severity is 4°/100 ft. In what direction should the toe of the whipstock be pointed to start the bit toward N75W? Can the orientation be achieved in one run? What is the final inclination? Write up a whipstock program for the drilling foreman.

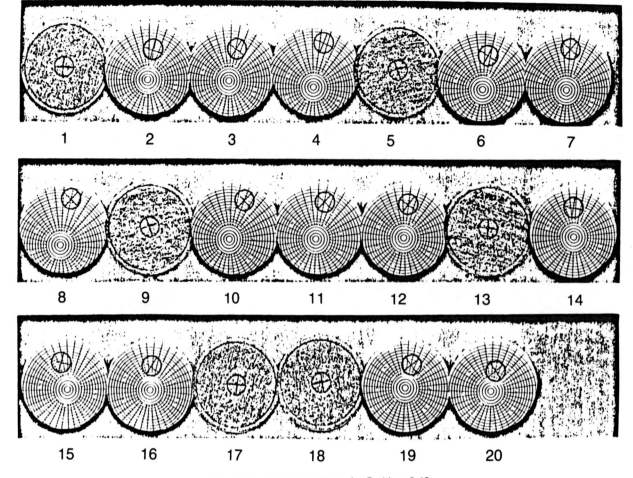

Fig. 8.165—Multishot picture for Problem 8.43.

TABLE 8.35—GENERAL DATA FOR THE WELL IN PROBLEM 8.47

True north azimuth is 197°   (SD)
The index north azimuth is 277°   (IS)
Gyro was started at 13-00
Gyro ended at 129-00

Gyro Start: 13-00	End: 129-00	Start: 277.00	End: 297.7	Corr: 20.7
Film Orient: 10-20	128-20	277.40	297.6	20.2
Case Orient: 0-0	127-00	277.80	297.5	19.7

8.55. You have been assigned to sidetrack out of Well B (Problem 8.23) at a measured depth of 5,000 ft and to head in a direction of S60W. The casing size is 10¾ in. The weight is 51 lbm/ft. The cement-bond log shows poor bonding from 4,800 to 3,200 ft. Above 3,200 ft, the bond log looks good. The formation from 3,600 to 6,100 ft is gumbo shale. Choose the best method for sidetracking the wellbore, and give the step-by-step procedure for the side-tracking operation.

8.56. Design a jetting program for kicking off in a 12¼-in. hole. The target direction is S72E; the horizontal departure is 8,500 ft; the TVD is 10,450 ft. Jetting information indicates that 250 hhp is necessary to jet a pocket for a 12¼-in. hole. The rig has a 4.5-in. FH, 16.60 lbm/ft drillpipe, and 7.75-in.-OD (2¹³/₁₆-in.-ID) drill collars. Both pumps are National 9-P-100 with 6- and 5½-in. liners and a 9¼-in. stroke length. Mud weight should be 9.8 lbm/gal. These are sands that can be jetted at 1,050 to 1,160 ft; 1,425 to 1,500 ft; 1,630 to 1,850 ft; and 2,010 to 2,090 ft. Plan the trajectory, hydraulics program, type of bit, jetting program, and BHA's.

TABLE 8.36—TOOL DRIFT CHECK FOR PROBLEM 8.47

Drift Check Number	Depth	Time	Gyro Azimuth	Inclination	Hole Direction
1	Case Surface	0-00	277.8		
		5-40	278.7	1°15'	171
2	25	17-40	60.5		
		21-20	60.8	25'	127
3	225	30-40	220.9		
		34-00	221.1	21'	102
4	475	44-20	60.0		
		49-40	60.9	8°05'	89
5	850	60-40	98.1		
		64-4	98.9	25°50'	102
6	1200	73-00	68.6		
		77-40	69.8	36°35'	100
7	1600	87-20	92.6		
		92-00	93.4	38°40'	118
8	1700	99-00	107.3		
		103-40	108.0	37°55'	131
9	500	113-40	91.2		
		117-40	91.8	9°30'	102
10	Case Surface	123-40	297.4		
		127-00	297.5	45'	58

TABLE 8.37—MULTISHOT GYRO DATA FOR PROBLEM 8.47

Corrected Time	Measured Depth	Observed Inclination		Observed Hole Azimuth	Gyro Azimuth
		Degrees	Minutes		
Centers					
3-20	(1)		10	97	350.0
4-20	(2)		05	88	56.9
5-00	(3)		05	99	119.9
6-00	(4)		05	106	178.7
6-40	(5)		10	98	238.0
7-40	(6)		10	97	292.5
Data					
17-40	25		25	127	60.5
22-00	50		25	132	24.8
23-40	75		25	140	344.1
25-20	100		20	136	332.5
26-40	125		20	135	326.0
27-40	150		25	127	312.2
29-00	175		15	128	271.5
30-00	200		20	124	247.8
30-40	225		20	102	220.9
35-00	250		25	101	187.4
36-00	275		05	92	147.3
37-20	300		07	76	114.8
38-20	325		05	158	108.0
39-20	350		30	138	100.8
40-40	375	1	40	108	96.5
42-00	400	3	20	106	78.7
43-00	425	5		96	63.4
43-40	450	6	30	92	58.0
44-20	475	8	05	89	60.0
51-20	500	9	20	90	74.5
52-40	550	12	25	94	93.1
53-40	600	14	25	93	89.8
55-20	650	17	30	92	84.7
56-40	700	20	15	94	92.5
58-00	750	22	35	94	91.2
59-40	800	24	20	99	97.0
60-40	850	25	50	102	98.1
65-40	900	27	30	100	87.6
67-00	950	29		96	98.5
68-00	1000	30	55	95	78.0
69-20	1050	32	30	98	89.0
70-20	1100	32	30	103	104.2
72-20	1150	34	25	102	92.6
73-00	1200	36	35	100	68.6
79-00	1250	38	15	98	65.9
80-40	1300	39	50	97	79.3
81-40	1350	42	05	98	77.9
83-20	1400	43		102	80.0
84-20	1450	43	10	106	75.1
85-40	1500	40	55	111	103.1
86-40	1550	36		117	97.2
87-20	1600	38	40	118	92.6
93-00	1650	38	15	124	93.3
94-20	1700	37	55	130	108.0
95-40	1750	39	10	132	99.3
97-40	1800	40	40	130	99.8
99-00	1700	37	55	131	107.3
106-00	1400	43		105	57.9
108-40	1100	32	25	110	79.9
111-40	800	34	20	109	118.4
113-40	500	9	30	102	91.2
120-00	200		20	142	69.6

8.57. If a 17½-in. bit requires 5,325 ft-lbf torque to drill at 25 ft/hr, what is the required pressure drop across a three-stage half-lobe with 62-in. rotor pitch, 1.125-in. eccentricity, and 4.555-in. diameter? *Answer.* 395 psi.

8.58. You are drilling at 3,250 ft with a 12-in. positive-displacement motor. The drillstring and the BHA consist of the following:

(1) 2,890 ft of 5½-in., 24.7-lbm/ft drillpipe,
(2) 360 ft of 9.5×3.125-in. drill collars, and
(3) 17½ Series 1-1-1 bit with two $^{14}/_{32}$-in. jets and one $^{12}/_{32}$-in. jet.

You have two pumps that are capable of pumping up to 1,250 gal/min. On bottom with 25,000-lbf WOB, the standpipe pressure is 2,240 psi and the pump rate is 948 gal/min. At 15,000-lbf WOB, the standpipe pressure is 2,040 psi, and the pump rate is 1,040 gal/min. With the bit off bottom and pumping 1,150 gal/min, the standpipe pressure is 1,840 psi.

What are the bit speeds at 25,000 lbf, at 15,000 lbf, and off bottom for mud weights of 9.5 and 13.5 lbm/gal? The plastic viscosity is 15 cp and the yield point is 10 lbm/100 sq ft.

8.59. What is the maximum torque of a multilobe motor with three teeth on the rotor and a stator diameter of 6.25 in.? The stator pitch is 11 in., and the pressure drop across the motor is 235 psi. *Answer.* 2,484 ft-lbf.

8.60. Compare the theoretical motor torque and bit speed of ¾- and ⅚-lobe PDM's.

8.61. Design a positive-displacement motor to meet the following specifications.

Length of motor: 16 to 18 ft.
Hole size: 12¼ to 17½ in.
Power range: 50 to 100 hp.
Maximum differential pressure: 500 psi.

Consider the following:

A. Is it a half-lobe or a multilobe motor? If multilobe, what kind?

B. What are the exact length, diameter, and pitch of the rotor?

C. What is the bit speed as a function of flow rate?

D. What is the maximum motor torque?

8.62. You have just completed drilling the Whale No. 1 to 15,840 ft. This is the first well in a total of 10 to be drilled from a man-made island in the Beaufort Sea. You are the drilling engineer in charge of optimizing the drilling program. While you are planning the second well, a vendor tries to persuade you to drill the second well— from surface to total depth—using a positive-displacement mud motor. Determine whether this would be economical by analyzing the data from this well and using the concepts previously developed. Analyze only the actual drilling data; ignore the intervals cored. The following information is available: round-trip time (t_R) is 1.1 times depth, and rig cost is $47,000/day.

Three PDM types are given along with the corresponding ratings. This information is provided as a possible solution, but it may be wise to analyze other types of PDM's and their corresponding performances. The choice of bits is not limited to those previously used. Table 8.39 lists the available data, and Table 8.40 provides information on the wellbore location.

8.63. What are the torque, power, pressure drop, and bit speed of a 7-in., 200-stage turbine operating at 300, 350, and 450 gal/min for a mud weight of 9.0 and 16 lbm/gal?

TABLE 8.38—SAMPLE DATA FOR PROBLEM 8.48

Station	Ay	Ax	Gy	Gx	Az	Inclination	Alpha
1	0.01543	0.01543	13.14122	0.11446	45	1.25	315
2	0.02072	0.01739	13.12538	−1.06336	45.26	1.55	319.99
3	0.02019	0.02406	13.12611	1.31285	45	1.80	310
4	0.02871	0.02678	13.22877	−0.28081	45.11	2.25	316.99
5	0.02745	0.03390	13.16444	1.53068	45.35	2.50	309
6	0.03675	0.03084	13.23515	−1.02337	45.5	2.75	320
7	0.03701	0.03701	13.29530	0.11037	45.7	3.00	315
8	0.04343	0.03644	13.28324	−0.97261	45.45	3.25	320
9	0.03924	0.04677	13.26710	1.40603	45.3	3.50	310
10	0.04783	0.04460	13.35932	−0.25742	45.55	3.75	317
11	0.04499	0.05556	13.29692	1.58048	45.8	4.10	309
12	0.05810	0.04875	13.37359	−1.00045	45.95	4.35	320
13	0.05671	0.05671	13.43029	0.15586	46.10	4.60	315
14	0.06610	0.05546	13.42072	−1.02263	46.25	4.95	320
15	0.05882	0.07009	13.42063	1.29363	46.5	5.25	310
16	0.07010	0.06537	13.49879	−0.34312	46.55	5.5	317
17	0.06305	0.07786	13.43055	1.59520	46.4	5.75	3.09
18	0.07874	0.06607	13.49760	−1.01953	46.55	5.90	320
19	0.07575	0.07575	13.55629	0.22624	46.35	6.15	315
20	0.08473	0.07109	13.53810	−0.97534	46.50	6.35	320

8.64. Design a turbine to drill with 8½-in. bits. The maximum power output is 1,500 rpm, and the pump rate is 400 gal/min (mud weight is 10.0 lbm/gal). You are designing for an overall mechanical efficiency of 0.68% and a hydraulic efficiency of 78%. This turbine should be able to drill with a new type of 8½-in. PDC bit that operates at torques of approximately 300 ft-lbf.

8.65. What are the overall power, torque, and bit speed of a combination 6½-in. PDM (Fig. 8.101A) and a 7-in. turbine (Fig. 8.118A) that has two connected shafts? The pump rate is 330 gal/min; the mud weight is 10.0 lbm/gal.

8.66. Your goal is to drill an 800-ft-thick shale section at a starting depth of 13,600 ft with an 8½-in. natural diamond bit on a 7-in., 180-stage turbodrill in 12-lbm/gal oil mud (PV 25, YV 35). The rig has a 10-P-130 pump, and you are using 5¾-in. liners, 5-in., 19.50-lbm/ft drillpipe, and a string of 6½×2-in. drill collars. The maximum allowable standpipe pressure is 3,825 psi. Laboratory tests at the research department indicate that the 0.45-sq-ft TFA bit will require about 25,000 lbf of true WOB to attain the torque that corresponds to peak turbine power. It is estimated that the bit will have an effective nozzle area of 0.37 sq in. and that the pump-off area will increase from about 12 to 16 sq in. over the 800-ft interval. The expected penetration rate is only 6 ft/hr and requires 133 hours of drilling time. Your supervisor thinks that a 133-hour diamond/turbine run will be too costly. He asks you to determine whether either of two options will reduce the drilling time to a more cost-effective time of 100 hours or less. Both options are aimed at reducing the parasitic power losses in the circulatory system, thus making available more hhp for the turbine.

Option 1. Dilute the 1,500-bbl mud system to what is believed to be a safe density of 10 lbm/gal (estimated PV 21, YV 32).

Option 2. Rent a string of 7-in. drill collars with a 3-in. bore.

Your preliminary investigation shows that it will cost no less than $52,400 to dilute and to treat the mud. The mud volume will increase excessively (to approximately 2,750 bbl), but the 1,450-psi reduction in bottomhole mud column pressure can effect a 20% improvement in penetration rate. Renting the collars will cost $14,700, including delivery, inspection, and rig time required for picking up and laying down. Other expenses for either option include rig and fuel costs at $500/hr, turbodrill cost at $500/hr, and a net bit cost of $18,000. The estimated round-trip time is 16 hours.

The turbine manufacturer has provided the information about the 180-stage tool found in Table 8.41. What do you recommend?

8.67. A drill-off test is performed with an 8½-in. natural diamond bit at a depth of 18,858 ft. The TFA of the bit is 0.40 sq in. The drilling mud is oil based, has a density of 13.5 lbm/gal, and is being pumped at a rate of 285 gal/min. The rotary table is turning at 100 rpm. At an indicated WOB of 14,000 lbf, the drawworks is locked and a drill-off test is run to measure the bit's pump-off force (see Table 8.42).

A. Did the bit drill off to the pump-off point or was the test stopped prematurely?

B. How much pump-off force is there at the pump-off point?

C. How much bit-pressure drop is there at the pump-off point?

D. What is the pump-off area of the bit?

E. What is the apparent nozzle area of the bit?

F. How much pump-off force is there at 14,000-lbf indicated WOB?

G. What is the true WOB at 14,000-lbf indicated WOB?

H. Estimate the bit-pressure drop at 14,000-lbf indicated WOB if the flow rate is increased to 300 gal/min.

I. Estimate the pump-off force at 14,000-lbf indicated WOB and 300 gal/min.

J. How much weight should be applied to get approximately 10,000-lbf true WOB at 300 gal/min?

8.68. A slack-off test is performed with an 8½-in. natural diamond bit at a depth of 10,680 ft. The TFA of the bit is 0.45 sq in. The drilling mud is water based, has a density of 9.0 lbm/gal, and is being pumped at a rate of 380 gal/min. The rotary table is turning at 75 rpm,

TABLE 8.39—AVAILABLE DATA FOR PROBLEM 8.62

PDM Type 1

Tool size O.D. (in.)	Recommended Hole Size (in.)	Pump Rate (gal/min) min.	Pump Rate (gal/min) max.	Bit Speed Range (rpm)	Maximum Differential Pressure (psi)	Maximum Torque (ft/lb)	Horsepower Range	Maximum Efficiency (percent)
6¾	7⅞- 9⅞	185	370	90-180	580	2,540	44- 87	70
8	9½-12¼	315	610	75-150	465	4,030	58-115	70
9½	12¼-17½	395	335	90-145	640	6,160	106-170	72
11¼	17½-26	525	1,055	70-140	520	8,850	118-236	73
9½	12¼-17½	1,500	2,400	90-145	44	8,350	79-127	72
11¼	17½-26	2,000	4,000	70-140	36	12,000	88-176	73

PDM Type 2

Tool size O.D. (in.)	Recommended Hole Size (in.)	Pump Rate (gal/min) min.	Pump Rate (gal/min) max.	Bit Speed Range (rpm)	Maximum Differential Pressure (psi)	Maximum Torque (ft/lb)	Horsepower Range	Maximum Efficiency (percent)	Length (ft)	Weight (lb)
4¾	6 - 7⅞	100	240	245-600	580	585	27- 67	83	20.0	840
6¾	8⅜- 9⅞	200	475	205-485	580	1,500	59-108	86	26.6	2,160
8	9½-12¼	245	635	145-380	465	2,090	59-152	88	26.9	2,800
9½	12¼-17½	395	740	195-365	695	3,890	145-271	90	32.8	5,200
11¼	17½-26	525	1055	120-250	465	5,380	123-256	90	32.2	7,300

PDM Type 3

Tool size O.D. (in.)	Recommended Hole Size (in.)	Pump Rate (gal/min) min.	Pump Rate (gal/min) max.	Bit Speed Range (rpm)	Maximum Differential Pressure (psi)	Maximum Torque (ft/lb)	Horsepower Range	Maximum Efficiency (percent)	Length (ft)	Weight (lb)
6¼	7⅞- 9⅞	170	345	200-510	580	1,015	39- 98	85	23.6	1,770
6¾	8⅜- 9⅞	160	395	140-480	465	995	27- 91	85	21.7	1,770
8	9½-12¼	200	475	160-400	465	1,475	46-113	87	23.6	2,430
9½	12¼-17½	240	610	130-340	465	2,280	56-148	90	24.6	3,970
9½	12¼-17½	395	900	140-325	290	2,210	59-137	90	24.6	3,970
11¼	17½-26	290	685	115-290	465	2,990	65-165	89	26.6	5,960

Data for Whale Number A-1:

Bit Program

Bit No.	Size (in.)	Type	Bit Cost ($)	Depth out (ft)	Footage (ft)	Drilling Time (Hours)	WOB (lb)	Bit Speed (rpm)	Flow Rate (gal/min)	Pump Pressure (psi)	Types of Nozzles
1	26.00	1-1-1	9,100	1,600	1,200	29	30,000	120	840	1,900	18-18-18
2	17.50	1-1-1	6,300	1,800	200	3	15,000	80	800	2,300	20-20-20 Drl Cement
3	17.50	1-1-6	6,300	2,280	480	5	15,000	350	700	1,800	20-20-20 PDM
4	17.50	1-1-4	6,300	4,300	2,020	19	45,000	200	1,000	3,400	18-18-18
5	17.50	1-1-4	6,300	6,300	2,000	21	40,000	125	950	3,450	16-18-18
6	12.25	1-1-1	5,700	6,500	200	8	15,000	80	480	2,700	12-12-12 Drl Cement
7	12.25	1-1-4	5,700	7,700	1,200	12	40,000	130	530	3,400	12-12-12
8	12.25	1-1-4	5,700	8,650	950	13	45,000	125	550	3,350	12-12-13
9	12.25	4-3-7	8,000	11,200	2,550	66	45,000	80	550	3,300	13-13-13
10	12.25	5-1-7	8,000	12,250	1,050	48	45,000	80	550	3,470	13-14-14
11	8.5	MC201	9,000	12,300	50	5	25,000	75	275	2,000	TFA = .5 Coring
12	8.5	MC201	9,000	12,350	50	5	25,000	75	275	2,000	TFA = .5 Coring/RR
13	8.5	MC201	9,000	12,400	50	5	25,000	75	275	2,000	TFA = .5 Coring/RR
14	12.25	2-1-5	4,000	12,400	0	15	25,000	80	550	3,470	14-14-14 Reaming
15	12.25	5-1-7	8,000	13,550	1,150	61	45,000	80	560	3,450	14-14-14
16	12.25	5-1-7	8,000	14,330	750	59	45,000	80	560	3,450	14-15-15
17	12.25	5-1-7	8,000	15,000	700	49	45,000	75	570	3,450	15-15-16
18	8.50	1-3-5	4,400	15,150	150	18	35,000	90		3,400	10-11-11 Drl Cement
19	8.50	MC201	9,000	15,200	50	5	35,000	75		2,200	TFA = .5 Coring/RR
20	8.50	MC201	9,000	15,250	50	6	25,000	75		2,200	TFA = .5 Coring/RR
21	8.50	MC201	9,000	15,300	50	10	25,000	75		2,200	TFA = .5 Coring/RR
22	8.50	MC201	9,000	15,350	50	10	25,000	75		2,200	TFA = .5 Coring/RR
23	8.50	MC201	9,000	15,400	50	10	25,000	75		2,200	TFA = .5 Coring/RR
24	8.50	MC201	9,000	15,450	50	10	25,000	75		2,200	TFA = .5 Coring/RR
25	8.50	MC201	9,000	15,500	50	10	25,000	75		2,200	TFA = .5 Coring/RR
26	8.50	MC201	9,000	15,500	50	20	25,000	75	275	2,200	TFA = .5 Coring/RR
27	8.50	5-1-7	4,400	15,840	340	56	35,000	90	310	3,400	10-11-11 11,800 TVD

TABLE 8.39—AVAILABLE DATA FOR PROBLEM 8.62 (cont.)

Bottomhole Assemblies

17½-in. Intermediate I

Building Interval 1 (1,800 to 2,280 ft)	Building Interval 2 (2,280 to 4,300 ft)	Building Interval 3 (4,300 to 6,100 ft)
17½ Bit	17½-in. Bit	17½-in. Bit
12 Dynadrill Delta 500 PDM	NW RWB Stab.	NB RWP Stab.
1½° Bent Sub	Shock Sub	Shock Sub
1 Int. 8-in. NMDC	1 Jnt. 8-in. NMDC	1 Jnt. 8-in. NMDC
MWD	MWD	MWD
1 Jnt. 8-in. NMDC	1 Jnt. 8-in. NMDC	1 Jnt. 8-in. NMDC
2 Jnt 8-in. DC	3 Jnt. 8-in. DC	3 Jn. 8-in. DC
3 Jnt 6.5-in. DC	3 Jnt. 6.5-in. DC	3 Jnt. 6.5-in. DC
35 Jnt. 5''-in. HWDP	35 Jnt. 5-in. HWDP	35 Jnt. 5-in. HWDP
Air Wt = 81,400 lb	1 Clamp-on NM Stab. 30 ft above first	1 Clamp-on NM Stab. 30 ft above first
	1 Clamp-on NM Stab. 65 ft above first	1 Clamp-on NM Stab. 40 ft above first
	Air Wt = 93,500 lbm	Air Wt. = 93,500 lbm

12¼-in. Intermediate II

Holding Interval 1 (6,100-19,950 ft)	8½-in. Production
12¼-in. Bit	Slight Dropping
NB RWP Stab.	8½ Bit
Shock Sub	8½ NB Stab.
1 Jnt. 8-in. NMDC	1 Jnt. 6½-in. short DC
MWD	1 Jnt. 8½-in. Stab.
1 Jnt. 8-in. NMDC	1 Jnt. 6½-in. DC
Full Gauge Stab.	1 Jnt. 8½-in. Stab
3 Jnt. 8-in. DC	5 Jnt. 6½-in. DC
36 Jnt. 5-in. HWDP	24 Jnt. 5-in. HWDP
3 Jnt. 6¾-in. DC	Jars
1 Clamp-on NM Stab 15 ft from first	12 Jnts 5-in. HWDP
1 Clamp-on NM Stab. 45 ft from first	Air Wt. = 73,000 lb
Air Wt. = 93,500 lb	

Casing Overview

Depth (ft)	Hole Size (in.)	Casing Program	Geologic Interval
0- 200	26.00	30-in. Conductor	
0- 1,600	17.50	20-in. Surface	Below Permafrost
0- 6,300	12.25	13⅜ Intermediate	Below T-5 Coals
0-15,000	8.50	9⅝ Intermediate	100 ft into Sag River
14,500-15,840		7-in. Prod. Liner	11,800 TVD

Geology Overview

Formation Maker	TVD Subsea (ft)	Estimated Measured Depth from kelly Bushing (ft)
Kelly Bushing	– 50	0
Bottom of Permafrost	1,600	1,650
Tertiary-Cretaceous Delta	1,600	1,650
T5 Coals	5,100	6,218
Cretaceous Pro Delta	6,400	8,395
Pebble Shale	8,925	11,991
Kuparck	8,050	12,303
Kingate	8,400	12,780
Sag River/Shublik	11,050	14,942
Ivishate	11,200	15,132
Lisburne	11,600	15,629

Mud Overview

Interval (ft)	Mud Wt (lb/gal)	PV (lb/100 sq ft)	YV (lb/100 sq ft)
0- 1,800	10	14	20
1,800- 2,280	9.9	14	20
2,280- 4,300	9.7	14	20
4,300- 6,300	9.5	14	20
6,300- 8,650	9.5	10	15
8,650-11,200	9.3	10	15
11,200-12,400	9.6	12	18
12,400-15,000	11.5	16	18
15,000-15,500	10.	12	18
15,500-15,800	9.5	12	18

TABLE 8.40—WELLBORE LOCATION FOR PROBLEM 8.62

Measured Depth (ft)	Inclination (degrees)	Direction	True Vertical Depth (ft)	Vertical Section	East-West	North-South	Dogleg Severity (°/100 ft)
100	0.00	N0.00E	100	0.0	0.0	0.0	0.00
1,000	0.00	N0.00E	1,000	0.0	0.0	0.0	0.00
1,000	0.00	N1.00E	1,800	0.0	0.0	0.0	0.00
1,000	3.00	N1.00E	1,899	2.5	0.0	2.6	0.00
2,000	6.00	N1.00E	1,999	10.1	0.2	10.5	3.00
2,100	9.00	N1.00E	2,098	22.6	0.4	23.5	3.00
2,200	12.00	N1.00E	2,197	40.1	0.7	41.7	3.00
2,250	13.50	N1.00E	2,245	50.7	8.9	52.8	2.99
2,350	15.50	N1.40E	2,342	74.8	1.4	77.8	2.00
2,450	17.50	N1.80E	2,430	102.1	2.2	106.2	2.00
2,550	19.50	N2.20E	2,533	132.8	3.3	137.9	2.00
2,650	21.50	N2.60E	2,627	166.6	4.8	172.9	2.00
2,750	23.50	N3.00E	2,709	203.7	6.7	211.1	2.00
2,850	25.50	N3.40E	2,800	244.0	9.0	252.5	2.00
2,950	27.50	N3.80E	2,899	287.4	11.8	297.0	2.00
3,050	29.50	N4.20E	2,987	333.8	15.1	344.6	2.01
3,150	31.00	N4.50E	3,074	383.0	16.9	394.9	1.51
3,250	23.50	N4.80E	3,159	434.4	23.2	447.3	1.51
3,350	34.00	N5.10E	3,242	487.9	27.9	501.9	1.51
3,450	35.50	N5.40E	3,324	543.7	33.2	558.7	1.51
3,550	37.00	N5.70E	3,405	601.7	38.9	617.5	1.51
3,650	38.50	N6.00E	3,484	661.7	45.1	678.4	1.51
3,750	40.00	N6.30E	3,562	723.8	51.9	741.3	1.51
3,850	41.50	N6.60E	3,637	788.0	59.2	806.2	1.51
3,950	43.00	N6.90E	3,711	854.1	67.1	873.0	1.51
4,050	44.50	N7.20E	3,784	922.2	75.6	941.6	1.51
4,150	46.00	N7.50E	3,854	992.2	84.7	1,012.0	1.51
4,250	46.50	N7.70E	3,923	1,063.2	94.3	1,083.6	0.51
4,350	47.00	N7.90E	3,992	1,135.3	104.1	1,155.8	0.51
4,450	47.50	N8.10E	4,060	1,207.8	114.4	1,228.5	0.51
4,550	48.00	N8.30E	4,127	1,280.9	124.9	1,301.8	0.51
4,650	48.50	N8.50E	4,193	1,354.7	135.8	1,375.6	0.51
4,750	49.00	N8.70E	4,259	1,429.1	147.1	1,449.9	0.51
4,850	49.50	N8.90E	4,325	1,504.0	158.7	1,524.8	0.51
4,950	50.00	N9.10E	4,389	1,579.6	170.6	1,600.2	0.51
5,050	50.50	N9.30E	4,453	1,655.7	182.9	1,676.1	0.51
5,150	51.00	N9.50E	4,516	1,732.5	195.5	1,752.5	0.51
5,250	51.50	N9.70E	4,579	1,809.8	208.5	1,829.4	0.51
5,350	52.00	N9.90E	4,641	1,887.7	221.9	1,906.7	0.51
5,450	52.50	N10.10E	4,702	1,966.1	235.6	1,984.6	0.51
5,550	53.00	N10.30E	4,763	2,045.2	429.7	2,063.0	0.51
5,650	53.50	N10.50E	4,822	2,174.7	264.2	2,141.8	0.52
5,750	54.00	N10.70E	4,882	2,204.9	279.0	2,221.0	0.52
5,850	54.50	N10.90E	4,940	2,285.5	294.2	2,300.7	0.52
5,950	55.00	N11.10E	4,998	2,366.7	309.8	2,380.9	0.52
6,050	55.50	N11.30E	5,055	2,448.5	235.8	2,461.5	0.52
6,100	55.75	N11.40E	5,083	2,489.5	333.9	2,502.0	0.50
6,200	55.55	N11.60E	5,139	2,571.7	350.4	2,582.9	0.25
6,300	55.35	N11.80E	5,196	2,653.7	367.1	2,663.5	0.25
6,400	55.15	N12.00E	5,253	2,735.5	384.0	2,743.9	0.25
6,500	54.95	N12.20E	5,310	2,817.2	401.2	2,824.1	0.25
6,600	54.75	N12.40E	5,368	2,898.6	418.6	4,904.0	0.25
6,700	54.55	N12.60E	5,426	2,979.9	436.3	2,983.6	0.25
6,800	54.35	N12.80E	5,484	3,061.1	454.1	3,063.0	0.25
6,900	54.15	N13.00E	5,542	3,142.0	472.3	3,142.1	0.25
7,000	53.95	N13.20E	5,601	3,222.0	490.6	3,220.9	0.25
7,100	53.75	N13.40E	5,660	3,303.3	509.2	3,299.5	0.25
7,200	53.55	N13.60E	5,719	3,383.7	528.0	3,377.0	0.25
7,300	53.35	N13.80E	5,779	3,463.9	547.0	3,455.9	0.25
7,400	53.15	N14.00E	5,839	3,543.9	566.3	3,533.6	0.25
7,500	52.95	N14.20E	5,899	3,623.7	585.7	3,611.1	0.25
7,600	52.75	N14.40E	5,959	3,703.3	605.4	3,688.4	0.25
7,700	52.55	N14.60E	6,020	3,782.7	625.3	3,765.3	0.25
7,800	52.35	N14.80E	6,081	3,861.9	645.4	3,842.0	0.25
7,900	52.15	N15.00E	6,142	3,941.0	665.8	3,918.4	0.25
8,000	51.95	N15.20E	6,204	6,019.8	686.3	3,994.6	0.25
8,100	51.75	N15.40E	6,265	4,098.4	707.1	4,070.4	0.25
8,200	51.55	N15.60E	6,327	4,176.7	728.0	4,146.0	0.25
8,300	51.35	N15.80E	6,390	4,254.9	749.2	4,221.3	0.25
8,400	51.15	N16.00E	6,452	4,332.9	770.6	4,296.3	0.25
8,500	50.95	N16.20E	6,515	4,410.7	792.1	4,371.0	0.25
8,600	50.75	N16.40E	6,578	4,488.2	813.9	4,445.4	0.25
8,700	50.55	N16.60E	6,642	4,565.5	835.8	4,519.6	0.25

TABLE 8.40—WELLBORE LOCATION FOR PROBLEM 8.62 (cont.)

8,800	50.35	N16.80E	6,705	4,642.6	858.0	4,593.4	0.25
8,900	50.15	N17.00E	6,769	4,719.5	880.4	4,667.0	0.25
9,000	49.95	N17.20E	6,833	4,796.2	902.9	4,740.2	0.25
9,100	49.75	N17.40E	6,898	4,872.6	925.6	4,813.2	0.25
9,200	49.55	N17.60E	6,963	4,948.8	948.5	4,885.9	0.25
9,300	49.35	N17.80E	7,028	5,024.8	971.6	4,958.3	0.25
9,400	49.15	N18.00E	7,093	5,100.5	994.9	5,030.4	0.25
9,500	48.95	N18.20E	7,159	5,176.0	1,018.4	5,102.2	0.25
9,600	48.75	N18.40E	7,224	5,251.3	1,042.0	5,173.7	0.25
9,700	48.55	N18.60E	7,290	5,326.4	1,065.9	5,244.0	0.22
9,800	48.35	N18.80E	7,357	5,401.2	1,089.9	5,315.7	0.22
9,900	48.15	N19.00E	7,423	5,475.7	1,114.0	5,386.3	0.22
10,000	47.95	N19.20E	7,490	5,550.1	1,138.4	5,456.6	0.22
10,100	47.75	N19.40E	7,557	5,624.1	1,162.9	5,526.6	0.22
10,200	47.55	N19.60E	7,625	5,698.0	1,187.5	5,596.2	0.22
10,300	47.35	N19.80E	7,692	5,771.6	1,212.4	5,665.6	0.22
10,400	47.15	N20.00E	7,760	5,844.9	1,237.4	5,734.6	0.22
10,500	46.95	N20.20E	7,828	5,918.0	1,262.5	5,803.4	0.22
10,600	46.75	N20.40E	7,897	5,990.8	1,287.8	5,871.8	0.22
10,700	46.55	N20.60E	7,965	6,063.4	1,313.3	5,939.9	0.22
10,800	46.35	N20.80E	8,034	6,135.8	1,338.9	6,007.7	0.22
10,900	46.15	N21.00E	8,103	6,207.8	1,364.7	6,075.2	0.22
11,000	45.95	N21.20E	8,173	6,279.7	1,390.6	6,142.3	0.22
11,100	45.75	N21.40E	8,242	6,351.2	1,416.7	6,209.2	0.22
11,200	45.55	N21.60E	8,312	6,422.5	1,442.9	6,275.7	0.22
11,300	45.35	N21.80E	8,382	6,493.5	1,469.2	6,341.9	0.22
11,400	45.15	N22.00E	8,453	6,464.3	1,495.7	6,407.8	0.22
11,500	44.95	N22.20E	8,524	6,634.8	1,522.3	6,473.4	0.22
11,600	44.75	N22.40E	8,594	6,705.0	1,549.1	6,438.6	0.22
11,700	44.55	N22.60E	8,666	6,775.0	1,576.0	6,603.6	0.22
11,800	44.35	N22.80E	8,737	6,844.7	1,603.0	6,668.2	0.22
11,900	44.15	N23.00E	8,809	6,914.1	1,630.2	6,732.4	0.22
12,000	43.95	N23.20E	8,880	6,983.3	1,657.4	6,796.4	0.22
12,100	43.75	N23.40E	8,953	7,052.1	1,684.8	6,860.0	0.22
12,200	43.55	N23.60E	9,025	7,120.7	1,712.4	6,923.3	0.22
12,300	43.35	N23.80E	9,097	7,189.1	1,740.0	6,986.3	0.22
12,400	43.15	N24.00E	9,170	7,257.1	1,767.0	7,048.9	0.22
12,500	42.95	N24.20E	9,243	7,324.8	1,795.6	7,111.2	0.22
12,600	42.75	N24.40E	9,317	7,392.3	1,823.6	7,173.2	0.22
12,700	42.55	N24.60E	9,390	7,459.5	1,051.7	7,234.9	0.22
12,800	42.35	N24.80E	9,464	7,526.4	1,879.9	7,296.2	0.22
12,900	42.15	N25.00E	9,538	7,593.0	1,908.2	7,357.2	0.22
13,000	41.95	N25.20E	9,612	7,659.3	1,936.6	7,417.8	0.22
13,100	41.75	N25.40E	7,687	7,725.4	1,965.2	7,478.1	0.22
13,200	41.55	N25.60E	9,762	7,791.1	1,993.8	7,538.1	0.22
13,300	41.35	N25.80E	9,836	7,856.6	2,022.5	7,597.8	0.22
13,400	41.15	N26.00E	9,912	7,921.7	2,051.3	7,657.1	0.22
13,500	40.95	N26.20E	9,987	7,986.6	2,080.2	7,716.0	0.22
13,600	40.75	N26.40E	10,063	8,051.1	2,109.2	7,774.7	0.22
13,700	40.55	N26.60E	10,139	8,115.4	2,138.2	7,833.0	0.22
13,800	40.35	N26.80E	10,215	8,179.4	2,167.4	7,890.9	0.22
13,900	40.15	N27.00E	10,291	8,243.0	2,196.6	2,948.6	0.22
14,000	39.95	N27.20E	10,368	8,306.4	2,225.9	8,005.8	0.22
14,100	39.75	N27.40E	10,444	8,369.5	2,255.3	8,067.8	0.22
14,200	39.55	N27.60E	10,521	8,432.2	2,784.8	8,119.4	0.22
14,300	39.35	N27.80E	10,599	8,494.7	2,314.3	8,175.6	0.22
14,400	39.15	N28.00E	10,676	8,556.8	2,343.9	8,231.5	0.22
14,500	39.95	N28.20E	10,754	8,618.7	2,373.6	8,287.1	0.22
14,600	38.75	N28.40E	10,831	8,680.2	2,403.3	8,342.3	0.22
14,700	38.55	N28.60E	10,910	8,741.4	2,433.1	8,397.2	0.22
14,800	38.55	N28.80E	10,998	8,802.3	2,463.0	8,451.6	0.22
14,900	38.15	N29.00E	11,066	8,862.9	2,492.9	8,506.0	0.22
15,000	37.95	N29.20E	11,145	8,923.2	2,522.9	8,559.8	0.22
15,100	37.75	N29.40E	11,224	8,983.2	2,552.9	8,613.3	0.22
15,132	37.68	N29.46E	11,249	9,002.3	2,562.5	8,630.4	0.00
15,232	37.18	N29.46E	11,329	9,061.7	2,592.4	8,683.3	0.49
15,332	36.68	N29.46E	11,409	9,120.4	2,622.0	8,735.6	0.49
15,432	36.18	N29.46E	11,489	9,178.4	2,651.2	8,787.3	0.49
15,532	35.68	N29.46E	11,570	9,235.7	2,680.0	8,838.4	0.49
15,632	35.18	N29.46E	11,652	9,292.3	2,708.5	8,888.9	0.49
15,732	34.68	N29.46E	11,734	9,348.3	2,736.7	8,938.7	0.49
15,832	34.18	N29.46E	11,816	9,403.5	2,764.5	8,987.9	0.49
15,840	34.14	N29.46E	11,823	9,407.9	2,766.7	8,991.8	0.00

while drilling ahead at 10 ft/hr with 22,000-lbf indicated WOB. The drillstring is raised about 1.0 ft off bottom, and then a slack-off test is run to measure the pump-off force.

WOB Indicator (lbf)	Standpipe Pressure (psi)
0.0	1,265
300	1,250
500	1,250
0.0	1,265
300	1,375
1,000	1,475
1,300	1,575
1,800	1,725
2,000	1,825
3,000	1,925
4,300	1,975
7,300	2,025
10,300	2,065
24,000	2,125

A. Locate the pump-off point.

B. How much bit-pressure drop is there at the pump-off point?

C. How much pump-off force is produced by the bit-pressure drop?

D. What is the pump-off area of the bit?

E. What is the apparent nozzle area of the bit?

F. How much pump-off force is there at 22,000-lbf indicated WOB?

G. What is the true WOB at 22,000-lbf indicated WOB?

H. An offset bit record shows that, in the same interval under comparable conditions, a radial-flow diamond bit with a pump-off area of 12 sq in. averaged 7 ft/hr, compared with 10 ft/hr with the current bit. Both bits have the same diameter and number of diamonds and typically are 50% salvageable after the entire interval is drilled. It appears that the bit that drills the faster is the better bit. Is this a valid conclusion? *Answer.* (B) 685 psi; (D) 4.0 sq in; (G) 18,650 lbf.

8.69. A slack-off test is performed with an 8½-in. matrix body polycrystalline diamond compact bit run on a 6¾-in. positive-displacement mud motor at a depth of 11,168 ft. The mud is oil based with a density of 7.8 lbm/gal and is being pumped at a rate of 380 gal/min. The indicated WOB is 24,000 lbf, and the standpipe pressure is 1,750 psi. The TFA of the bit was 1.0 sq in. when new and has drilled 493 ft. The mud motor is rated to produce maximum power at 580-psi differential pressure.

WOB Indicator (lbf)	Standpipe Pressure (psi)
0.0	1,175
1,000	1,190
2,500	1,190
4,000	1,175
5,000	1,175
6,500	1,200
8,000	1,325
9,000	1,410
11,000	1,450

TABLE 8.41—DATA ON 180-STAGE TOOL FOR PROBLEM 8.66

	Mud Weight			
	10 lbm/gal		12 lbm/gal	
Flow Rate (gal/min)	Pressure Drop (psi)	Axial Thrust (lbf)	Pressure Drop (psi)	Axial Thrust (lbf)
250	1,033	18,407	1,240	22,089
270	1,205	21,470	1,446	25,764
290	1,390	24,769	1,668	29,723
310	1,588	28,303	1,906	33,964
330	1,800	32,073	2,160	38,488
350	2,025	36,078	2,429	43,294

TABLE 8.42—DATA FOR PROBLEM 8.67

Elapsed Time (min)	WOB Indicator (lbf)	Standpipe Pressure (psi)
0	14,000	2,490
3	13,000	2,490
10	12,000	2,485
15	11,500	2,485
19	11,000	2,485
22	10,500	2,480
30	10,000	2,480
38	9,000	2,475
52	9,000	2,475
(off bottom)	9,000	1,650
13,000		1,510
15,500		1,550
18,000		1,610
20,000		1,650
22,000		1,690
24,000		1,740

A. We know that the on-bottom standpipe pressure is greater than the off-bottom because the mud motor Δp increased in proportion to bit torque. We do not expect much pressure drop (and, therefore, little pump-off force) from a bit rated at 1.0 sq in. TFA. Thus, it appears that the mud motor is being operated very close to its optimal condition of peak power—i.e., On-bottom SPP−Off-bottom SPP=1,740−1,175=565 psi; (565/580)×100 =97% of peak power. Is this a valid conclusion?

B. From the pump-off plot, determine whether the bit has a significant pressure drop and pump-off force.

C. Does the result from Part B mean that the bit has 9,000 lbf of pump-off force?

D. Explain how a bit rated at 1.0 sq in. TFA can have enough pressure drop to produce 4,000 lbf of pump-off force.

E. Initially (from the discussion in Problem 8.69A), it appeared that the mud motor was being operated 97% effectively. Now, with the benefit of the pump-off plot, determine how efficiently the motor was actually run.

F. The published literature for this particular mud motor tells us that the motor is rated at 2,540 ft-lbf maximum torque at 580-psi differential pressure. Determine the bit torque in this example at an applied WOB of 24,000 lbf.

8.70. You are drilling with a 12-in. PDM (Fig. 8.101B) with a circulating rate of 1,000 gal/min. The bit torque changes from 4,200 to 6,100 ft-lbf. What is the real WOB if you are drilling with 35,000 lbf? You have 4,000 ft

TABLE 8.43—DEFLECTIONS MEASURED IN TESTING A MUD MOTOR—PROBLEM 8.74

Bending Plane	Test 1				Test 2				Loading
	1	2	3	4	1	2	3	4	
XX	0.315	0.444	0.452	0.324	0.320	0.455	0.456	0.324	its own weight
	0.331	0.466	0.475	0.340	0.336	0.477	0.478	0.339	101 lbf at L
	0.347	0.448	0.498	0.357	0.351	0.498	0.499	0.355	202 lbf at L
	0.362	0.510	0.519	0.372	0.366	0.520	0.521	0.370	303 lbf at L
YY	0.340	0.478	0.485	0.348	0.340	0.478	0.485	0.348	its own weight
	0.354	0.497	0.505	0.363	0.355	0.498	0.506	0.364	101 lbf at L
	0.369	0.517	0.525	0.378	0.369	0.518	0.526	0.378	202 lbf at L
	0.383	0.538	0.547	0.393	0.383	0.539	0.547	0.393	303 lbf at L

*Turbine was rotated 90° and then releveled before the two tests were repeated.

of 6⅝-in. drillpipe (24.7 lbf/ft) and 45 joints of 10-in.-OD by 3-in.-ID steel drill collars. *Answer.* Assuming a bit with zero pumpoff force, the real WOB is 35,660 lbf.

8.71. Develop the strategy and economics for kicking off your wells for Problem 8.12. Will you jet, use a turbine, or use a PDM? Where will you kick off? Describe in detail the BHA you will use to kick off each well. Give a procedure for your kick-off, supplying details on whether you will use MWD as a steering tool or simply use single shots. Give estimated tool-fall headings and expected reverse torque. Note: You will have to research the current costs of the motors, MWD, steering, and single-shot tools to determine the best economics.

8.72. Derive the axial moment of inertia and the polar moment of inertia for a square drill collar. *Answer.* $J=(L^4/6)-(\pi d^4/32)$.

8.73. Derive the axial moment of inertia and polar moment of inertia for a hexagonal kelly.

8.74. Fig. 8.166 shows deflections measured in testing a mud motor. Calculate the equivalent stiffness, using the following information and Table 8.43.

Weight of turbine in air=2,980 lbm
Diameter=7.05 in.
Outside radius=3.525 in.
Circumference=πD=22.148228 in.
$\pi/4$ or 90° rotation is 5.5370570 in. of circumference for a solid circle $I=\pi r^4/4=\pi(3.53 \text{ in.})^4/4=121.26$ in⁴.
Structural steel, $E=30\times10^6$ psi.

Answer. 2.1707×10^9.

8.75. Determine the air weight and the weight for the following drillstring in a 16.2-lbm/gal mud.

Drillpipe:	2,520 ft of 5-in. Grade S135
Drillpipe:	6,520 ft of 5-in. Grade E
Heavy-weight drillpipe:	850 ft of 5-in. Range II
Drill collar:	620 ft of 6½-in. (H90 6⅝-in. tool joints)
Drill collar:	500 ft of 8 in. (H90 6⅝-in. tool joints)
Stabilizers:	Four 5.5 ft stainless steel 8.23-in. OD by 2¹³⁄₁₆-in. ID
Bit:	12¼-in. Series 5-2-4

8.76. Determine the static drag of the drillstring and BHA of Problem 8.75, assuming drag coefficients of 0.15, 0.18, and 0.22, where the wellbore is at a constant 30°

TABLE 8.44—FORCES FOR PROBLEM 8.85

Depth (ft)	Hole Diameter (in.)	Force (lbf)
0 to 4,000	26	−2,500 to −3,600
4,000 to 6,500	26	−5,000 to −6,500
6,500 to 8,500	17½	−3,500 to −5,000
8,500 to 14,500	17½	−2,000 to −25,000
14,500 to 18,000	17½	−4,000 to −7,000
18,000 to 22,000	8½	−1,000 to 2,000

TABLE 8.45—COORDINATES FOR PROBLEM 8.86

Well 12A	Sec. 12	1,320 FNL and 3,960 FEL
Well 13A	Sec. 11	200 FWL and 2,640 FSL
Well 14A	Sec. 11	500 FWL and 1,000 FEL
Well 15A	Sec. 1	440 FNL and 440 FEL
Well 16A	Sec. 2	2,640 FWL and 2,640 FSL
Well 17A	Sec. 12	400 FSL and 800 FWL
Well 18A	Sec. 12	3,960 FNL and 600 FEL
Well 19A	Sec. 1	1,320 FSL and 3,690 FEL
Well 20A	Sec. 11	1,320 FSL and 100 FEL
Well 21A	Sec. 2	2,640 FSL and 300 FSL

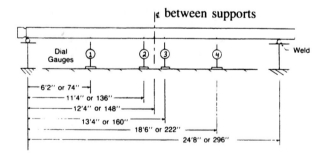

Fig. 8.166—Deflections measured in testing a mud motor—Problem 8.74.

inclination. The well was drilled with a slant-hole drilling rig. Mud weight is 12.8 lbm/gal. (Assume the load is equally distributed along each tool joint.)

8.77. You are drilling with a slick BHA at 1,650 ft where the inclination is at a constant 8° and the mud weight is 10.2 lbm/gal. The wellbore diameter is 12¼ in. and you have thirty-six 8¼-in.-OD by 2¹³/₁₆-in.-ID aluminum drill collars. If the formation force at this depth is 745 lbf, will the bit build or drop again with WOB's of 20,000, 30,000, and 45,000 lbf? At what WOB will it hold angle?

8.78 You are drilling at 7,450 ft with a 9⅞-in. bit. The inclination was built to 15°. If the formation force that is causing the bit to build angle is 385 lbf, where should the stabilizer be placed (assuming one stabilizer) to drill with 45,000 lbf and to maintain a 100-lbf bit side force? Assume 7¾-in.-OD by 2¹³/₁₆-in.-ID steel drill collars. The stabilizer is ¹/₃₂ in. under gauge. Mud weight is 9.8 lbm/gal.

8.79. What will be the side force at the bit in the solution of Problem 8.78 if a second stabilizer is used (if it is ¹/₃₂-in. under gauge and 344 ft above the first stabilizer)? All other parameters are the same.

8.80. It is desired to drill a high-angle directional well to obtain a horizontal departure of 8,500 ft at a TVD of 5,500 ft. You have a slant-hole rig that can start at 20° inclination. If you need 2,500 lbf of bit side force to achieve a build of 2°/100 ft or 4,500 lbf of bit side force to achieve 3.5°/100 ft, can you use a single-stabilizer BHA and drill with WOB's between 15,000 and 30,000 lbf to achieve your build? Would it be better to use a two-stabilizer building assembly? What would be an appropriate WOB? The hole size is 14¾ in. throughout the build portion; the mud weight is 10.5 lbm/gal, and you will use six 9½-in.-OD×2¹³/₁₆-in.-ID drill collars for the first two stands and twelve 8-in.-OD×2¹³/₁₅-in.-ID drill collars for the remainder of the BHA.

8.81. Using the same method as was used to derive the one- and two-stabilizer BHA side forces, derive the equations necessary to calculate the tangency point and side force for a three-stabilizer BHA.

8.82. For the following three BHA's, determine the side forces at the bit with WOB's of 10,000 and 20,000 lbf, inclinations of 10 and 30°, a 9⅞-in. bit, 7¾-in.-OD×2¹³/₁₆-in.-ID drill collar, and a 9.0-lbm/gal mud. All stabilizers are ¹/₃₂-in. under gauge.

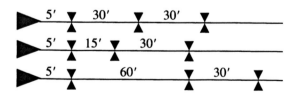

Answer. For a BHA with an inclination of 10°, WOB of 10,000 lbf, and length of 609.6 in., the side force is 237 lbf. For a BHA with an inclination of 10°, WOB of 20,000, and length of 603.7 in., the side force is 250 lbf. For a BHA with an inclination of 30°, a WOB of 10,000 lbf, and length of 466.7 in., the side force is 723 lbf. For a BHA with an inclination of 30°, WOB of 20,000, and length of 463.4, the side force is 741 lbf.

8.83. Design all the BHA's to drill your directional wells for Problem 8.12. State collar sizes, exact placement of stabilizers, number of drill collars, and your estimate of the operating conditions—i.e., WOB and rotary speed.

8.84. Determine the bit-walking tendency for the following situations, and give reasons for your decision.

A. Slick assembly, 10,000-lbf WOB, 60-rpm speed, 9⅞-in. Series 1-1-5 bit, 5° inclination, drilling gumbo shale. Hole is slightly overgauge (size is 12 to 13 in.).

B. Single-stabilizer building assembly (Fig. 8.136, No. 9), 35,000-lbf WOB, 110-rpm speed, 17½-in. Series 5-1-4 bit, 20° inclination, drilling soft sandstones and shales at 60 to 300 ft/hr. Hole is washing out (size is about 20 in.).

C. 70-ft building assembly (Fig. 8.136, No. 5), 30,000-lbf WOB, 90-rpm speed, 12¼-in. BHD bit, 15° inclination, drilling sandstones at 60 ft/hr and shales at 25 ft/hr. The hole is elliptical.

D. Holding assembly (Fig. 8.138, No. 4), 45,000-lbf WOB, 75-rpm speed, 12¼-in. Series 5-2-4 bit, 42° inclination, drilling siltstone and limestones at 12 ft/hr. The hole is gauge.

E. Dropping assembly (Fig. 8.142, No. 2), 3,500-lbf WOB, 60-rpm speed, 8½-in. Series 6-1-7 bit, 22° inclination, drilling limestone and dolomite at 6 ft/hr. The hole is gauge.

8.85. You are planning to drill a 22,000-ft exploration well, starting with a 26-in. hole to 4,000 ft, followed by 17½-in. hole to 14,500 ft, 12¼-in. hole to 18,500 ft, and 8½-in. hole to 22,000 ft. The well will be a deep test (Fig. 8.156A, 300 ft north of Well A). Design the drillstring, BHA's, bits, and operating conditions for the well to total depth, and give reasons for your selections. You must have the side forces listed in Table 8.44 to offset the formation forces to maintain less than 0.5°/1,000 ft to 4,000 ft and 1°/1,000 ft to 22,000 ft.

8.86. Calculate the dip at each surface location for the coordinates cited in Table 8.45 (see Fig. 8.155), and estimate where the bottomhole location would be if a well were drilled at each location.

8.87. You have been asked to develop a drilling plan to hit Target C in Fig. 8.156B. The top of the target is 14,000 ft TVD, and total depth is 15,100 ft TVD. Hole and casing sizes should be the same as those in Example 8.33. State where the surface location should be, determine the best trajectory plan to hit the target, and explain your strategy. Design the BHA's to accomplish your plan.

8.88. You are drilling ahead at 3,450 ft. A drilling break was experienced at 3,400 ft, after which the penetration rate increased to 75 ft/hr at a rotary speed of 70 rpm. Offset core information shows a formation bedding dip of 15.0° and an angle of internal friction of 29.0°. The current bit is a new milltooth with an individual wedge angle of 40°. The cohesive strength of the core was found to follow the variable cohesive strength equation,

$$\tau(p_i) = A - B[\cos 2(\alpha - p_i)]^n,$$

where A and B are empirical constants, n is a rock parameter constant, and α is the orientation angle between the bedding planes where the cohesive strength is a minimum. Tests show $A=6,035$, $B=2,035$, $\alpha=30°$, and $n=6$. Determine whether the tendency is to drill updip or downdip. *Answer.* A chip should form on the updip side of the hole, and the hole should deviate upward.

8.89. Refer to Fig. 8.160. Note that over the range of α_{dh} from 0° to 30°, the values of $(F_2 - F_1)/F_{min}$ for the 60° wedge are zero. This indicates that symmetrical chips should be formed and a bit with a 60° tooth should alleviate the natural hole deviation problem. What can be said about the 70° wedge curve?

8.90. Refer to Problem 8.88. Plot strength variance $(F_2 - F_1)/F_{min}$ for formation dips from 0° to 90°.

References

1. Craig, J.T. Jr. and Randall, B.V.: "Directional Survey Calculation," *Pet. Eng.* (March 1976).
2. Walstrom, J.E., Harvey, R.P., and Eddy, H.D.: "A Comparison of Various Directional Survey Models and an Approach To Model Error Analysis," *J. Pet. Tech.* (Aug. 1972) 935-43.
3. Wilson, G.J.: "An Improved Method for Computing Directional Survey," *J. Pet. Tech.* (Aug. 1968) 871-76; *Trans.*, AIME, **243**.
4. Taylor, H.L. and Mason, C.M.: "A Systematic Approach to Well Surveying Calculations," *Soc. Pet. Eng. J.* (Dec. 1972) 474-88.
5. Millheim, K.K., Gubler, F.H., and Zaremba, H.B.: "Evaluating and Planning Directional Wells Utilizing Post Analysis Techniques and Three Dimensional Bottomhole Assembly Program," paper SPE 8339 presented at the 1979 SPE Annual Technical Conference and Exhibition, Las Vegas, Sept. 23-26.
6. Lubinski, A.: "How Severe Is That Dogleg?" *World Oil* (Feb. 1, 1957) 95-104.
7. "Ragland Diagram," *Drilling Data Handbook*, Editions Technip (Paris) (1978) 334, 338.
8. Eickelberg, D. *et al.*: "Controlled Directional Drilling Instruments and Tools," paper presented at the 9th International Symposium of the firms Baker Oil Tools GmbH, Christensen Diamond Products GmbH, and Vetco Inspection GmbH (1982).
9. Gearhart, M., Zierner, K.H., and Knight, O.M.: "Mud Pulse MWD Systems Report," *J. Pet. Tech.* (Dec. 1981) 2301-06.
10. Uttecht, G.W. and DeWardt, J.P.: "Survey Accuracy is Improved by a New, Small OD Gyro," *World Oil* (March 1983) 61-66.
11. Wolff, C.J.M. and DeWardt, J.P.: "Borehole Position Uncertainty—Analysis of Measuring Methods and Derivation of Systematic Error Model," *J. Pet. Tech.* (Dec. 1981) 2339-50.
12. Cagle, *et al.*: "Casing Sidetracks Made Easy," *Pet. Eng.* (Sept. 1976).
13. Jürgens, R.: "Down-Hole Motors—Technological Status and Development Trends," paper presented at the 7th International Symposium of the firms Baker Oil Tools GmbH, Christensen Diamond Products GmbH, and Vetco Inspection GmbH, Celle, FDR (Sept. 8, 1978).
14. Warren, T.M.: "Factors Affecting Torque for a Roller Cone Bit," *J. Pet. Tech.* (Sept. 1984) 1500-08.
15. Winters, W. and Warren, T.M.: "Variations in Hydraulic Lift With Diamond Bits," paper SPE 10960 presented at the 1982 SPE Annual Technical Conference and Exhibition, New Orleans, Sept. 26-29.
16. Winters, W. and Warren, T.M.: "Determining the True Weight-on-Bit for Diamond Bits," paper SPE 11950 presented at the 1983 SPE Annual Technical Conference and Exhibition, San Francisco, Oct. 5-8.
17. Feenstra, R. and Kamp, W.A.: "A Technique for Continuously Controlled Directional Drilling," paper presented at the 1984 Drilling Technology Conference of the International Association of Oilwell Drilling Contractors, March 19-21.
18. Lubinski, A. and Woods, H.B.: "How to Determine Best Hole and Drill Collar Size," *Oil and Gas J.* (June 7, 1954) 104-05.
19. Millheim, K.K.: "Directional Drilling," *Oil and Gas J.* (eight-part series) beginning Nov. 6, 1978 and ending Feb. 12, 1979.
20. Jiazhi, B.: "Bottomhole Assembly Problems Solved by Beam-Column Theory," paper SPE 10561 presented at the 1982 SPE International Meeting on Petroleum Engineering, Beijing, March 19-22.
21. Timoshenko, S.: *Theory of Elastic Stability*, McGraw-Hill Book Co. Inc., New York City (1936).
22. Walker, H.B.: "Downhole Assembly Design Increases ROP," *World Oil* (June 1977) 59-65.
23. Millheim, K.K. and Apostal, M.C.: "The Effect of Bottomhole Assembly Dynamics on the Trajectory of a Bit," *J. Pet. Tech.* (Dec. 1981) 2323-38.
24. Lubinski, A. and Woods, H.B.: "Factors Affecting the Angle of Inclination and Dog-legging in Rotary Boreholes," *Drill. and Prod. Prac.*, API (1953) 222-42.
25. Rollins, H.M.: "Are 3° and 5° Straight Holes Worth Their Cost?" *Oil and Gas J.* (Nov. 1959) 163-64, 167-68, 171.
26. Murphey, C.E. and Cheatham, J.B. Jr.: "Hole Deviation and Drill String Behavior," *Soc. Pet. Eng. J.* (March 1966) 44-53; *Trans.*, AIME, **237**.
27. Knapp, S.R.: "New Bit Concept Helps Control Hole Deviation," *World Oil* (1965) 113-16.
28. Sultanov, R. and Shandalov, G.: "Effects of Geological Conditions on Well Deviation," (in Russian), *Izv. Vyssh. Ucheb. Zaved. Geol. Razved* (1961) **3**, 106.
29. Smith, M.B. and Cheatham, J.B.: "Deviation Forces Arising From Single Bit Tooth Indentation of an Anisotropic Porous Medium," *Trans.*, ASME, Series 399 (1977) 363-66.
30. McLamore, R.T.: "The Role of Rock Strength Anisotropy in Natural Hole Deviation," *J. Pet. Tech.* (Nov. 1971) 1313-21; *Trans.*, AIME, **251**.
31. Bradley, W.B.: "Deviation Forces From the Wedge of Penetration Failure of Anisotropic Rock," *J. Eng. Ind.* (Nov. 1973) 1093-99.

Nomenclature

A = area
A_{corr} = intercardinal correction
A_i = internal drillpipe area
A_s = cross-sectional area of steel in drillpipe
B_c = buoyancy correction factor
Δc = actual compass deflection
d_b = bit diameter
d_r = rotor diameter
D = depth
ΔD = depth increment
D_M = total measured depth
D_{MN} = new measured depth
D_s = stake direction
D_{tar} = total measured depth to target
D_x = depth of wedge penetration
D_2 = TVD at end of the buildup section
e = rotor eccentricity
E = Young's modulus
E_h = hydraulic efficiency
E_M = mechanical efficiency
E_R = Earth's spin vector
F = force
F_c = correction factor
F_d = ratio of drillpipe OD/ID
F_B = side force at bit
F_R = factor of the straight line section vs. section ratios
G = shear modulus
G = resultant bit curvature
h = height of vane
H_M = magnetic field
ΔH_M = strength of the magnetic error field
i = angle of index change
i_s = index setting
I = axial moment of inertia
J = moment of inertia
ℓ = clearance between bit and drill collar

L = north/south coordinate

ΔL = change in drillpipe length

L_c = course length between measured surveys

L_r = rotor lead

L_s = stator lead

L_{Dc} = length of arc section

L_T = tangency length

M = torque generated at the bit

M_b = bending moment

n = number of lobes or teeth on stator

n_r = number of lobes or teeth on rotor

n_s = number of stages

N_b = bit speed

Δp = pressure drop

p_b = axial or compressive load on bit

Δp_b = pressure drop across the bit

p_{do} = off-bottom standpipe pressures

Δp_i = change in internal drillpipe pressure

p_{ob} = drilled-off standpipe pressure

P_r = rotor pitch

P_s = stator pitch

q = circulation rate

Q = specific displacement

r_c = radius of curvature

s_c = east/west coordinate

s_{DC} = arc length

T_e = horizontal component of Earth's spin rate

W_b = weight on bit

ΔW_b = change in weight on bit

W_c = weight of the collars

W_m = mud weight

x = departure

α_{dh} = bedding dip angle relative to bottom of hole

α_q = inclination angle buildup

α_N = new inclination angle

β = exit blade angle

β = overall angle change

γ = tool-face angle

δ = dogleg severity

ϵ = azimuth

$\Delta\epsilon$ = change in azimuth

η = efficiency

θ = maximum inclination angle

μ = Poisson's ratio

τ_ρ = shear strength of the anisotropic rock along the plane of failure

ϕ = angle of internal friction of the anisotropic rock

Ω = included wedge half-angle

SI Metric Conversion Factors

ft $\times$ 3.048*	E$-$01	= m
ft-lbf $\times$ 1.355 818	E$-$03	= kJ
gal $\times$ 3.785 412	E$-$03	= m^3
hp $\times$ 7.460 43	E$-$01	= kW
in. $\times$ 2.54*	E$+$00	= cm
lbf $\times$ 4.448 222	E$+$00	= N
lbf/ft $\times$ 1.459 390	E$+$01	= N/m
lbf/sq ft $\times$ 4.788 026	E$-$02	= kPa
lbm/cu ft $\times$ 1.601 846	E$+$01	= kg/m^3
lbm/gal $\times$ 1.198 264	E$+$02	= kg/m^3
psi $\times$ 6.894 757	E$+$00	= kPa
sq in. $\times$ 6.451 6*	E$+$00	= cm^2

*Conversion factor is exact.

Appendix A

Development of Equations for Non-Newtonian Liquids in a Rotational Viscometer

Bingham Plastic Model

A fluid that follows the Bingham plastic model, unlike a Newtonian fluid, will not yield and begin to shear until a stress is applied that is large enough to break down the cohesive forces between the fluid particles. Initially, if a small torque is applied to the rotor such that the shear stress at the inner cylinder, τ_1, is less than the yield point of the fluid, τ_y, then no fluid movement will occur. If the torque is increased so that $\tau_1 > \tau_y > \tau_2$, movement will occur near the bob, but a portion of the fluid will remain solid and will move as a rigid body attached to the rotor. As can be seen from Eq. 4.45, the shear stress varies inversely with the square of the radius, so that the shear stress at the bob is always greater than the shear stress at the rotor. If the torque is increased further so that the shear stress at the rotor is also greater than the yield point of the fluid ($\tau_2 > \tau_y$), flow will occur throughout the entire region between the bob and the rotor. The equations developed in the following for characterizing fluids that follow the Bingham plastic model by means of a rotational viscometer apply only when the flow between the bob and the rotor is developed fully. This assumption is generally valid for drilling fluids and cement slurries at the 300- and 600-rpm rotor speeds.

The shear stress at any point in a fluid that follows the Bingham plastic model for which the shear stress is greater than the yield point is given by

$$\tau = \tau_y + \mu_p \dot{\gamma}.$$

Combining this equation with Eq. 4.45 yields

$$\frac{d\omega}{dr} = \frac{360.5\,\theta}{2\pi h r^3 \mu_p} - \frac{\tau_y}{\mu_p r}. \quad \ldots \ldots \ldots \ldots \ldots (A\text{-}1)$$

Assuming that no slip occurs at the walls of the viscometer, the angular velocity is zero at r_1 and ω_2 at r_2. Thus, separating variables in Eq. A-1 yields

$$\int_0^{\omega_2} d\Omega = \frac{360.5\,\theta}{2\pi h \mu_p} \int_{r_1}^{r_2} \frac{dr}{r^3} - \frac{\tau_y}{\mu_p} \int_{r_1}^{r_2} \frac{dr}{r}. \quad \ldots \ldots (A\text{-}2a)$$

Substituting the value $2\pi N/60$ for ω_2, the values of r_1, r_2, and h shown in Table 4.2, and changing the units of viscosity and yield point to field units of centipoise and pounds force per 100 sq ft yields

$$\mu_p = \frac{300\,\theta_N}{N} - \frac{300}{N}\tau_y. \quad \ldots \ldots \ldots \ldots \ldots \ldots (A\text{-}2b)$$

Two readings must be made with the rotational viscometer to solve for the two unknown fluid properties, μ_p and τ_y. Normally the 300- and 600-rpm readings are used. Substituting these values in Eq. A-2b gives

$$\mu_p = \frac{300\,\theta_{300}}{300} - \frac{300\,\tau_y}{300}$$

and

$$\mu_p = \frac{300\,\theta_{600}}{600} - \frac{300\tau_y}{600}.$$

Solving these two equations simultaneously for plastic viscosity and yield point leads to the following equations.

$$\mu_p = \theta_{600} - \theta_{300}. \quad \dots \dots \dots \dots \dots \dots \dots \text{(A-3)}$$

$$\tau_y = \theta_{300} - \mu_p. \quad \dots \dots \dots \dots \dots \dots \dots \dots \text{(A-4)}$$

Thus, if a standard torsion spring is used, the plastic viscosity is obtained simply by subtracting the 300-rpm dial reading from the 600-rpm dial reading, and the yield point is obtained by subtracting the plastic viscosity from the 300-rpm dial reading.

Example Problem A-1. A rotational viscometer containing a Bingham plastic fluid gives a dial reading of 12 at a rotor speed of 300 rpm and a dial reading of 20 at a rotor speed of 600 rpm. Compute the plastic viscosity and yield point of the fluid.

Solution. The plastic viscosity is given by Eq. A-3:

$$\mu_p = 20 - 12 = 8 \text{ cp.}$$

The yield point is given by Eq. A-4:

$$\tau_y = 12 - 8 = 4 \text{ lbf/100 sq ft.}$$

The shear rate present in a fluid that follows the Bingham plastic model at a given speed of rotation can be obtained by using Eq. A-1 and A-2a. Equation A-2a can be rearranged to give

$$\frac{360.5\,\theta}{\pi h \mu_p} = \frac{4\,\omega_2}{\left(\dfrac{1}{r_1{}^2} - \dfrac{1}{r_2{}^2}\right)} + \frac{4\,\tau_y \ln r_2/r_1}{\mu_p \left(\dfrac{1}{r_1{}^2} - \dfrac{1}{r_2{}^2}\right)}.$$

Combining this equation with Eq. A-1 yields

$$\frac{d\omega}{dr} = \frac{4\,\omega_2}{2r^3 \left(\dfrac{1}{r_1{}^2} - \dfrac{1}{r_1{}^2}\right)}$$

$$+ \frac{\tau_y}{r\mu_p}\left[\frac{4 \ln r_2/r_1}{2r^2 \left(\dfrac{1}{r_1{}^2} - \dfrac{1}{r_2{}^2}\right)} - 1\right].$$

Using the values for r_1, r_2, and h shown in Table 4.2, the value $2\pi N/60$ for ω_2, and changing the units of viscosity and yield point to field units of centipoise and pounds force per 100 sq ft yields

$$\frac{d\omega}{dr} = \frac{5.066N}{r^3} + \frac{479\,\tau_y}{r\mu_p}\left(\frac{3.174}{r^2} - 1\right).$$

Thus, the shear rate, $\dot\gamma$, is given by

$$\dot\gamma = r\frac{d\omega}{dr} = \frac{5.066N}{r^2} + \frac{479\,\tau_y}{\mu_p}\left(\frac{3.174}{r^2} - 1\right). \quad \dots \text{(A-5)}$$

The yield point computed from Eq. A-4 is an extrapolation of the 300- and 600-rpm shear stress values to a shear rate of zero. Usually, drilling fluids and cement slurries do not follow the Bingham plastic model closely at low rates of shear. Thus, the computed yield point generally does not correspond with the shear stress at which fluid movement begins. However, drilling fluids and cement slurries usually do tend to "gel" if left static for a period of time. It is common practice to measure the "gel strength" of a drilling fluid after a specified static waiting period. This is done by turning the rotor at a slow speed and noting the dial reading at which the gel structure is broken and fluid movement near the bob begins. The gel strength can be related to the dial reading using Eq. 4.45 evaluated at the bob radius of 1.7245:

$$\tau_g = \frac{360.5\,\theta}{2\pi(3.80)(1.7245)^2} = 5.08\,\theta. \quad \dots \dots \text{(A-6a)}$$

Converting the shear stress units to field units of pounds force per 100 sq ft yields

$$\tau_g = 1.06\,\theta. \quad \dots \dots \dots \dots \dots \dots \dots \dots \text{(A-6b)}$$

In practice, the factor 1.06 usually is truncated to 1.0 so that the dial reading of a viscometer containing a standard torsion spring is numerically equal to the gel strength in pounds force per 100 sq ft. The dial reading generally cannot be made very precisely, so the factor 1.06 is not considered significant.

Since drilling fluids quite often do not follow the Bingham plastic model at low shear rates, plastic viscosity and yield point values obtained from the 300- and 600-rpm dial readings may not characterize the fluid correctly at low rates of shear. Usually, the shear rate caused by viscous flow in the annulus will be much lower than the shear rate in a rotational viscometer at 300 or 600 rpm. If a six-speed rotational viscometer is available, the plastic viscosity and yield point also can be computed from the 3- and 6-rpm dial readings or the 100- and 200-rpm dial readings. Equations for this purpose, similar to Eqs. A-3 and A-4, can be obtained through use of Eq. A-2b. However, at low shear rates, the flow between the bob and the rotor may not be developed fully and thus Eq. A-2b may not apply. The rotor speed above which the assumption of fully developed flow is valid can be estimated if the yield point that applies at that rotor speed is known. The rigid body of fluid attached to the rotor will disappear when the shear stress in the fluid at r_2 is equal to the yield point of the fluid. Thus, we have

$$\tau_y = \frac{360.5\,\theta}{2\pi h r_2{}^2}.$$

Eq. A-2a can be rearranged to give

$$\frac{360.5\,\theta}{\pi h} = \frac{4(\omega_2 \mu_p + \tau_y \ln r_2/r_1)}{\left(\dfrac{1}{r_1{}^2} - \dfrac{1}{r_2{}^2}\right)}.$$

Combining these two equations and solving for the angular velocity, ω_2, yields

$$\omega_2 = \frac{\tau_y}{2\mu_p}\left[\left(\frac{r_2}{r_1}\right)^2 - 1 - 2\ln r_2/r_1\right]. \quad\ldots\ldots \text{(A-7a)}$$

Using the values for r_1 and r_2 shown in Table 4.2 and the value $2\pi N/60$ for ω_2, and changing the units of viscosity and yield point to field units of centipoise and pounds force per 100 sq ft yields

$$N = 20.6\frac{\tau_y}{\mu_p}. \quad\ldots\ldots\ldots\ldots\ldots\ldots\ldots\ldots\text{(A-7b)}$$

Since the drilling fluid may not follow the Bingham plastic model at low rates of shear, it is often difficult to obtain a meaningful estimate of the proper value of yield point to use in Eq. A-7b.

Power-Law Model

The shear stress in a power-law fluid between the bob and the rotor of a rotational viscometer is given by

$$\tau = K\left(r\frac{d\omega}{dr}\right)^n.$$

Combining this model of fluid behavior with Eq. 4.45 yields

$$\frac{360.5\,\theta}{2\pi hr^2} = Kr^n\left(\frac{d\omega}{dr}\right)^n. \quad\ldots\ldots\ldots\ldots\ldots\text{(A-8)}$$

Separating variables and noting that the angular velocity at r_1 is zero and the angular velocity at r_2 is the angular velocity of the rotor, ω_2, we obtain

$$\int_0^{\omega_2} d\omega = \left(\frac{360.5\,\theta}{2\pi hK}\right)^{1/n}\int_{r_1}^{r_2}\frac{dr}{r^{(2/n+1)}}. \quad\ldots\ldots\ldots\text{(A-9)}$$

Integration of this equation gives

$$\omega_2 = \left(\frac{360.5\,\theta}{2\pi hK}\right)^{1/n}\left(\frac{n}{2}\right)\left(\frac{1}{r_1^{2/n}} - \frac{1}{r_2^{2/n}}\right).$$

$$\ldots\ldots\ldots\ldots\ldots\ldots\text{(A-10)}$$

Eq. A-10 can be rearranged to give

$$\left(\frac{360.5\,\theta}{2\pi hK}\right)^{1/n} = \frac{2\omega_2}{n\left(\dfrac{1}{r_1^{2/n}} - \dfrac{1}{r_2^{2/n}}\right)}.$$

Combining this equation with Eq. A-8, substituting $2\pi N/60$ for ω_2, and solving for the shear rate, $\dot\gamma$, yields

$$\dot\gamma = r\frac{d\omega}{dr} = \frac{0.2094N\left(\dfrac{1}{r^{2/n}}\right)}{n\left(\dfrac{1}{r_1^{2/n}} - \dfrac{1}{r_2^{2/n}}\right)}. \quad\ldots\ldots\text{(A-11)}$$

For drilling fluids, n generally has a value between 0.5 and 1.0. The shear rate, $\dot\gamma$, is not very sensitive to the value of n for this range of n values. For example, at a

rotor speed of 300 rpm, the shear rate at the bob does not deviate greatly from the value of 511 seconds^{-1} that corresponds to a Newtonian fluid ($n = 1$).

Normally, the consistency index, K, and flow-behavior index, n, are obtained from dial readings of a rotational viscometer taken at 300 and 600 rpm. Substituting these two readings in Eq. A-10 yields

$$\frac{2\pi(300)}{60} = \left(\frac{360.5\,\theta_{300}}{2\pi hK}\right)^{1/n}$$

$$\cdot\left(\frac{n}{2}\right)\left[\left(\frac{1}{r_1^{2/n}}\right) - \left(\frac{1}{r_2^{2/n}}\right)\right]$$

and

$$\frac{2\pi(600)}{60} = \left(\frac{360.5\,\theta_{600}}{2\pi hK}\right)^{1/n}$$

$$\cdot\left(\frac{n}{2}\right)\left[\left(\frac{1}{r_1^{2/n}}\right) - \left(\frac{1}{r_2^{2/n}}\right)\right].$$

Dividing the second equation by the first equation and solving for the flow-behavior index, n, gives

$$n = 3.322\,\log(\theta_{600}/\theta_{300}). \quad\ldots\ldots\ldots\ldots\text{(A-12)}$$

Solving the first equation for the consistency index, K, gives

$$K = \frac{15.1\,\theta_{300}}{\left\{\dfrac{0.2094(300)}{n\left[\dfrac{1}{r_1^{2/n}} - \dfrac{1}{r_2^{2/n}}\right]}\right\}^n}$$

At first glance, it appears that the relation for computing K from the viscometer dial readings is quite complex. However, note that if the numerator and denominator of this equation are multiplied by $(1/r_1)^2$, then

$$K = \frac{15.1\left(\dfrac{1}{r_1}\right)^2\theta_{300}}{\dot\gamma_{300}^{\,n}}$$

It can be shown that the shear rate at 300 rpm for n values close to one is approximately equal to 511 seconds^{-1}. Using this value in the equation above and substituting the value for r_1 shown in Table 4.2 yields

$$K = \frac{5.1\,\theta_{300}}{511^n}, \quad\ldots\ldots\ldots\ldots\ldots\ldots\text{(A-13a)}$$

where K has units of dynes per second^n per square centimeter. If K is desired in dynes per second^n per 100 cm^2 or equivalent centipoise, then K is given by

$$K = \frac{510\,\theta_{300}}{511^n}. \quad\ldots\ldots\ldots\ldots\ldots\ldots\text{(A-13b)}$$

Similarly, if K is desired in units of pounds force per second^n per square feet, then K is given by

$$K = \frac{0.0106\,\theta_{300}}{511^n}. \quad\ldots\ldots\ldots\ldots\ldots\ldots\text{(A-13c)}$$

Appendix B

Development of Slot Flow Approximations for Annular Flow for Non-Newtonian Fluids

Bingham Plastic Model

Developing flow equations for a Bingham plastic fluid is complicated because portions of the fluid having a shear stress less than the yield point must move as a rigid plug down the conduit (Fig. B-1). At the inner layer of the plug, y_a, the shear stress, τ_a, must be equal to $-\tau_y$. Also, the shear stress in a slot must behave according to Eq. 4.57. Thus,

$$\tau_a = -\tau_y = \tau_o + y_a \frac{dp}{dL}. \quad \dots\dots\dots\dots\dots\dots (B\text{-}1)$$

Likewise, at the outer layer of the plug, the shear stress, τ_b, must be given by

$$\tau_b = +\tau_y = \tau_o + y_b \frac{dp}{dL}. \quad \dots\dots\dots\dots\dots\dots (B\text{-}2)$$

In the fluid region enclosed by the inner layer of the plug, the Bingham model is defined by Eq. 4.41c. Thus, the shear stress in this fluid region is given by

$$\tau = -\mu_p \frac{dv}{dy} - \tau_y = \tau_o + y \frac{dp}{dL}.$$

After separating variables and integrating, we obtain the following expression for fluid velocity.

$$v = -\frac{y^2}{2\mu_p} \frac{dp}{dL} - \frac{(\tau_y + \tau_o)y}{\mu_p} + v_o. \quad \dots\dots\dots\dots (B\text{-}3a)$$

The constant of integration, v_o, must be zero since the velocity is known to be zero at $y=0$. Also, Eq. B-1 can

be used to express $(\tau_y + \tau_o)$ in terms of y_a and dp_f/dL. Thus, Eq. B-3a also can be expressed by

$$v = \frac{1}{2\mu_p} \frac{dp_f}{dL}(-y^2 + 2y_a y), \ (y \leq y_a). \quad \dots\dots\dots (B\text{-}3b)$$

The velocity of the plug region can be obtained by evaluating Eq. B-3b at $y=y_a$. The velocity of the plug is given by

$$v_p = \frac{y_a{}^2}{2\mu_p} \frac{dp_f}{dL}, \ (y_b \geq y \geq y_a). \quad \dots\dots\dots\dots (B\text{-}3c)$$

In the fluid region enclosing the outer layer of the plug, the Bingham model is defined by Eq. 4.41a. Thus,

$$\tau = -\mu_p \frac{dv}{dy} + \tau_y = \tau_o + y \frac{dp_f}{dL}.$$

After separating variables and integrating, we obtain the following expression for fluid velocity.

$$v = -\frac{y^2}{2\mu_p} \frac{dp_f}{dL} - \frac{(\tau_o - \tau_y)y}{\mu_p} + v_c. \quad \dots\dots\dots\dots (B\text{-}4a)$$

The constant of integration can be evaluated using the boundary condition of zero velocity at $y=h$. Applying this condition to Eq. B-4a and using Eq. B-2 to define $(\tau_o - \tau_y)$ yields

$$0 = -\frac{h^2}{2\mu_p} \frac{dp_f}{dL} + \frac{y_b h}{\mu_p} \frac{dp_f}{dL} + v_c.$$

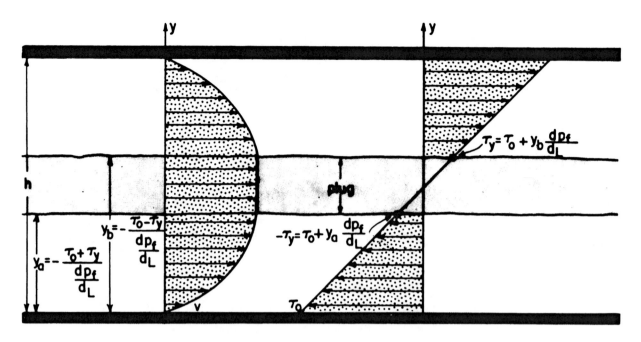

Fig. B-1—Laminar flow of Bingham plastic in a slot.

Solving this expression for v_c gives

$$v_c = \frac{1}{2\mu_p} \frac{dp_f}{dL}(h^2 - 2y_b h).$$

Substituting this expression for v_c into Eq. B-4a and simplifying yields

$$v = \frac{1}{2\mu_p} \frac{dp_f}{dL}(h^2 - y^2 + 2y_b y - 2y_b h), \quad (y \geq y_b).$$

$$\dotfill \text{(B-4b)}$$

The velocity of the plug region also must be given by Eq. B-4b evaluated at $y = y_b$. Thus, the velocity of the plug is given by

$$v_p = \frac{(y_b^2 - 2y_b h + h^2)}{2\mu_p} \frac{dp_f}{dL}, \quad (y_b \geq y \geq y_a).$$

$$\dotfill \text{(B-4c)}$$

Equating Eqs. B-3c and B-4c gives

$$y_a^2 = y_b^2 - 2y_b h + h^2.$$

In terms of τ_o and τ_y, this equation becomes

$$\frac{(\tau_o + \tau_y)^2}{\left(\frac{dp_f}{dL}\right)^2} = \frac{(\tau_o - \tau_y)^2}{\left(\frac{dp_f}{dL}\right)^2} + \frac{2(\tau_o - \tau_y)h}{\left(\frac{dp_f}{dL}\right)} + h^2.$$

Solving this expression for τ_o, the shear stress at $y = 0$ yields

$$\tau_o = \frac{h}{2} \frac{dp_f}{dL}. \dotfill \text{(B-5)}$$

The total flow rate through the slot is defined by

$$q = W \int_0^h v\,dy = W \int_0^{y_a} v\,dy + W v_p \int_{y_a}^{y_b} dy + W \int_{y_b}^h v\,dy.$$

Each of the three integrals on the right can be evaluated and summed to obtain the flow rate. However, less algebra is required if the improper integral on the left is first integrated by parts. Integrating by parts gives

$$q = W \int_0^h v\,dy = [vy]_0^h - \int_0^h y \frac{dv}{dy}dy = 0 - \int_0^h y \frac{dv}{dy}dy$$

$$= \frac{W}{2\mu_p} \frac{dp_f}{dL}\left[\int_0^{y_a}(2y^2 - 2y_a y)dy + \int_{y_b}^h(2y^2 - 2y_b y)dy\right].$$

Integration of this equation gives

$$q = \frac{W}{2\mu_p} \frac{dp_f}{dL}\left(\frac{y_b^3}{3} - y_b h^2 - \frac{y_a^3}{3} + \frac{2h^3}{3}\right).$$

Expressing this equation in terms of τ_o and τ_y using Eqs. B-1, B-2, and B-5 yields

$$q = \frac{Wh^3}{12\mu_p} \frac{dp_f}{dL}\left[1 + \frac{3}{2}\frac{\tau_y}{\tau_o} - \frac{1}{2}\left(\frac{\tau_y}{\tau_o}\right)^3\right]. \quad \dots \text{(B-6a)}$$

For the conditions usually encountered in rotary drilling applications, the shear stress at the wall, τ_o, is more than

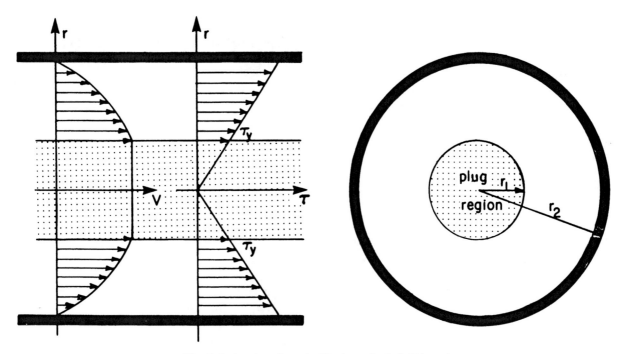

Fig. B-2—Laminar flow of a Bingham plastic fluid in a pipe.

twice the yield point, τ_y, and the last term in Eq. B-6 can be neglected without introducing a significant error. Dropping this term and substituting the expression for τ_o given by Eq. B-5 yields

$$q = \frac{Wh^3}{12\mu_p}\frac{dp_f}{dL} - \frac{Wh^2}{4\mu_p}\tau_y. \quad \ldots\ldots\ldots\ldots\ldots (B\text{-}6b)$$

Combining the expressions for Wh and for h given by Eqs. 4.55a and 4.55b with Eq. B-6b gives

$$q = \frac{\pi}{12\mu_p}\frac{dp_f}{dL}(r_2{}^2 - r_1{}^2)(r_2 - r_1)^2$$

$$- \frac{\pi}{4\mu_p}\tau_y(r_2{}^2 - r_1{}^2)(r_2 - r_1). \quad \ldots\ldots\ldots (B\text{-}6c)$$

Expressing the flow rate in terms of the mean flow velocity, $\bar{v}$, and solving for the frictional pressure gradient, dp_f/dL, gives

$$\frac{dp_f}{dL} = \frac{12\mu_p\bar{v}}{(r_2 - r_1)^2} + \frac{3\tau_y}{(r_2 - r_1)}. \quad \ldots\ldots\ldots\ldots (B\text{-}6d)$$

Converting from consistent units to more convenient field units of pounds per square inch per foot, centipoise, feet per second, inches, and pounds force per 100 sq ft, we obtain

$$\frac{dp_f}{dL} = \frac{\mu_p\bar{v}}{1{,}000(d_2 - d_1)^2} + \frac{\tau_y}{200(d_2 - d_1)}. \quad \ldots (B\text{-}6e)$$

The derivation of the equations for the laminar flow of Bingham plastic fluids through a pipe is quite similar to

the derivation of the slot flow equation. As in the case of a slot, the portions of the fluid flowing near the center of the conduit that have a shear stress less than the yield point must move as a rigid plug down the conduit (Fig. B-2). At the radius of the plug, r_p, the shear stress, τ_p, must be equal to the yield point, τ_y. Also, the shear stress in a circular tube must behave according to Eq. 4.51. Thus,

$$\tau_p = \tau_y = \frac{r_p}{2}\frac{dp_f}{dL}. \quad \ldots\ldots\ldots\ldots\ldots\ldots\ldots (B\text{-}7)$$

In the fluid region enclosing the circular plug, the Bingham model is defined by Eq. 4.41a. Thus, the shear stress in this fluid region is given by

$$\tau = -\mu_p\frac{dv}{dr} + \tau_y = \frac{r}{2}\frac{dp_f}{dL}.$$

After separating variables and integrating, we obtain the following expressions for fluid velocity.

$$v = \frac{r^2}{4\mu_p}\frac{dp_f}{dL} + \frac{\tau_y}{\mu_p}r + v_w. \quad \ldots\ldots\ldots\ldots (B\text{-}8a)$$

The constant of integration, v_w, is obtained using the boundary condition of zero velocity at the pipe wall. Substituting this boundary condition in Eq. B-8a gives

$$0 = \frac{r_w{}^2}{4\mu_p}\frac{dp_f}{dL} + \frac{\tau_y}{\mu_p}r_w + v_w.$$

Solving this equation for v_w gives

$$v_w = \frac{r_w^2}{4\mu_p}\frac{dp_f}{dL} - \frac{\tau_y}{\mu_p}r_w.$$

Substituting this expression for v_w in Eq. B-8a yields

$$v = \frac{1}{4\mu_p}\frac{dp_f}{dL}(r_w^2 - r^2) - \frac{\tau_y}{\mu_p}(r_w - r), \quad (r \geq r_p).$$

$$.......................... (B-8b)$$

The velocity of the plug region can be obtained by evaluating Eq. B-8b at $r=r_p$. Substituting the expression for τ_y given by Eq. B-7 into Eq. B-8b and evaluating at $r=r_p$ yields

$$v_p = \frac{1}{4\mu_p}\frac{dp_f}{dL}(r_w - r_p)^2, \quad (r \leq r_p). \quad ....(B-8c)$$

The total flow rate through the pipe is defined by

$$q = \int v\,dA = 2\pi\int_0^{r_w} vr\,dr = \pi[r^2 v]\Big|_0^{r_w} - \pi\int_0^{r_w} r^2\frac{dv}{dr}dr.$$

The term $[r^2 v]$ is zero at $r=0$ and at $r=r_w$. Also, the integral

$$\int_0^{r_p} r^2\frac{dv}{dr}dr$$

is zero because dv/dr is zero for $r \leq r_p$. Thus, the flow rate is given by

$$q = \frac{\pi}{2\mu_p}\frac{dp_f}{dL}\int_{r_p}^{r_w} r^3\,dr - \frac{\pi\tau_y}{\mu_p}\int_{r_p}^{r_w} r^2\,dr.$$

Integration of this equation gives

$$q = \frac{\pi}{8\mu_p}\frac{dp_f}{dL}(r_w^4 - r_p^4) - \frac{\pi\tau_y}{3\mu_p}(r_w^3 - r_p^3).$$

Expressing this equation in terms of τ_w and τ_y using Eqs. 4.61 and B-7 yields

$$q = \frac{\pi r_w^4}{8\mu_p}\frac{dp_f}{dL}\left[1 - \frac{4}{3}\left(\frac{\tau_y}{\tau_w}\right) + \frac{1}{3}\left(\frac{\tau_y}{\tau_w}\right)^4\right],$$

$$(\tau_y \leq \tau_o). \quad .... (B-9a)$$

Note that this expression reduces to the familiar Hagen-Poiseuille law for τ_y equal to zero. Note also that as τ_y approaches τ_w, the term in brackets approaches zero. This means that the shear stress at the wall must exceed the yield point to cause flow.

For the conditions usually encountered in rotary drilling applications, the shear stress at the wall is more than

twice the yield point, and the last term in Eq. B-9a can be neglected without introducing a significant error. Dropping this term and substituting the expression for τ_w given by Eq. 4.61 evaluated at r_w yields

$$q = \frac{\pi r_w^4}{8\mu_p}\frac{dp_f}{dL} - \frac{\pi r_w^3}{3\mu_p}\tau_y. \quad .... (B-9b)$$

Expressing the flow rate in terms of the mean flow velocity, $\bar{v}$, and solving for the frictional pressure gradient, dp_f/dL, gives

$$\frac{dp_f}{dL} = \frac{8\mu_p\bar{v}}{r_w^2} + \frac{8\tau_y}{3r_w}. \quad .... (B-9c)$$

Converting from consistent units to more convenient field units, of pounds per square inch, centipoise, feet per second, inches, and pounds force per 100 sq ft, we obtain

$$\frac{dp_f}{dL} = \frac{\mu_p\bar{v}}{1,500d^2} + \frac{\tau_y}{225d}. \quad .... (B-9d)$$

The shear rate in a Bingham plastic fluid at the pipe wall can be obtained from the shear stress at the pipe wall and the appropriate frictional pressure gradient equation. For a circular pipe,

$$\dot{\gamma}_w = \frac{\tau_w - \tau_y}{\mu_p} = \frac{r_w}{2\mu_p}\frac{dp_f}{dL} - \frac{\tau_y}{\mu_p}$$

$$= \frac{r_w}{2\mu_p}\left(\frac{8\mu_p\bar{v}}{r_w^2} + \frac{8}{3}\frac{\tau_y}{r_w}\right) - \frac{\tau_y}{\mu_p}.$$

After simplifying and changing from consistent units to more convenient field units of seconds^{-1}, feet per second, inches, and pounds force per 100 sq ft, we obtain

$$\dot{\gamma}_w = \frac{96\bar{v}}{d} + 159.7\frac{\tau_y}{\mu_p} \quad .... (B-10)$$

for circular pipe. Similarly, for an annulus, the shear rate at the wall is given by

$$\dot{\gamma}_w = \frac{h}{2\mu_p}\frac{dp_f}{dL} - \frac{\tau_y}{\mu_p} = \frac{(r_2 - r_1)}{2\mu_p}\left[\frac{12\mu_p\bar{v}}{(r_2 - r_1)^2}\right.$$

$$\left. + \frac{3\tau_y}{(r_2 - r_1)}\right] - \frac{\tau_y}{\mu_p}.$$

After simplifying and changing to field units, we obtain

$$\dot{\gamma}_w = \frac{144\bar{v}}{(d_2 - d_1)} + 239.5\frac{\tau_y}{\mu_p} \quad .... (B-11)$$

for the annulus.

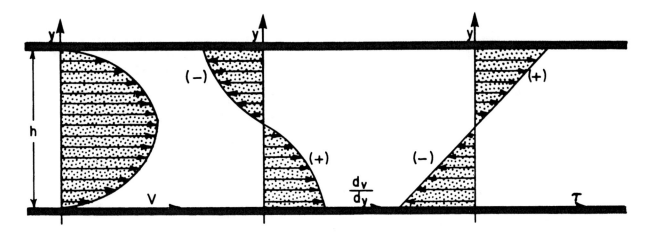

Fig. B-3—Laminar flow of a power-law fluid in a slot.

Power-Law Model

The annular flow of power-law fluids also can be approximated closely using the less-complex flow equations for a narrow slot. The shear stress for a power-law fluid in a narrow slot is given by Eqs. 4.42, 4.57, and 4.58.

$$\tau = K\left(-\frac{dv}{dy}\right)\left|-\frac{dv}{dy}\right|^{n-1}$$

$$= y\frac{dp_f}{dL} + \tau_o.$$

As discussed previously for Newtonian and Bingham plastic models, shear stress, τ, changes linearly with distance, y, in a slot. This is a consequence of the flow geometry and is independent of the rheological models used to describe the flow. Also, by inspection of Fig. B-3, we see that for a slot, τ is zero at $y=h/2$. Thus,

$$\tau_o = \frac{h}{2}\frac{dp_f}{dL}.$$

Also, as shown in Fig. B-3, the velocity gradient $(dv)/(dy)$ is positive for $0 \le y \le h/2$ and is negative for $h/2 < y \le h$. This means that the absolute value,

$$\left|-\frac{dv}{dy}\right|^{n-1},$$

does not have a continuous first derivative at $y=h/2$, and the shear stress equation must be integrated piecewise over each half of the slot. For the $0 \le y \le h/2$ region, the shear stress equation can be simplified to

$$\tau = -K\left(\frac{dv}{dy}\right)^n$$

$$= y\frac{dp_f}{dL} - \frac{h}{2}\frac{dp_f}{dL}.$$

Separating variables and integrating gives

$$v = \frac{-K}{(1+1/n)\dfrac{dp_f}{dL}}\left[\frac{1}{K}\frac{dp_f}{dL}\left(\frac{h}{2}-y\right)\right]^{1+1/n} + v_o.$$

$$\dots\dots\dots\dots\dots\dots\dots \text{(B-12a)}$$

The constant of integration, v_o, can be evaluated using the boundary condition that $v=0$ at $y=0$:

$$0 = \frac{-K}{(1+1/n)\dfrac{dp_f}{dL}}\left[\frac{1}{K}\frac{dp_f}{dL}\left(\frac{h}{2}-0\right)\right]^{1+1/n} + v_o.$$

Solving for v_o and substituting the result in Eq. B-12a yields

$$v = \frac{-K}{(1+1/n)\dfrac{dp_f}{dL}}\left\{\left[\frac{1}{K}\frac{dp_f}{dL}\left(\frac{h}{2}-y\right)\right]^{1+1/n}\right.$$

$$\left.-\left(\frac{1}{K}\frac{dp_f}{dL}\frac{h}{2}\right)^{1+1/n}\right\}.$$

This expression can be simplified to

$$v = \frac{\left(\dfrac{1}{K}\dfrac{dp_f}{dL}\right)^{1/n}}{(1+1/n)}\left[\left(\frac{h}{2}\right)^{1+1/n}-\left(\frac{h}{2}-y\right)^{1+1/n}\right],$$

$$0 \le y \le h/2. \dots\dots\dots\dots\dots\dots\text{(B-12b)}$$

The flow rate in the $0 \le y \le h/2$ region, which is half of the total flow rate, is given by

$$q/2 = vdA = W\int_0^{h/2} vdy$$

$$\frac{q}{2} = \frac{W\left(\dfrac{1}{K}\dfrac{dp_f}{dL}\right)^{1/n}}{(1+1/n)} \int_0^{h/2} \left[\left(\frac{h}{2}\right)^{1+1/n}\right.$$

$$\left. - \left(\frac{h}{2}-y\right)^{1+1/n}\right] dy.$$

Integrating and simplifying this equation yields

$$q = \frac{\left(\dfrac{1}{K}\dfrac{dp_f}{dL}\right)^{1/n} Wh^{(2+1/n)}}{2^{1/n}(4+2/n)}. \qquad \ldots\ldots\ldots\ldots (B\text{-}13a)$$

Substituting the expressions for Wh and for h given by Eqs. 4.55a and 4.55b into Eq. B-13a gives

$$q = \frac{\pi\left(\dfrac{1}{K}\dfrac{dp_f}{dL}\right)^{1/n}}{2^{1/n}(4+2/n)} (r_2{}^2 - r_1{}^2)(r_2 - r_1)^{1+1/n}.$$

$$\ldots\ldots\ldots\ldots\ldots\ldots\ldots (B\text{-}13b)$$

Expressing the flow rate in terms of the mean flow velocity, $\bar{v}$, and solving for the frictional pressure gradient, dp_f/dL, gives

$$\frac{dp_f}{dL} = \frac{2K(4+2/n)^n \bar{v}^n}{(r_2 - r_1)^{n+1}}. \qquad \ldots\ldots\ldots\ldots (B\text{-}14a)$$

Converting from consistent units to more convenient field units, we obtain

$$\frac{dp_f}{dL} = \frac{K\bar{v}^n}{144,000(d_2 - d_1)^{1+n}}\left(\frac{2+1/n}{0.0208}\right)^n.$$

$$\ldots\ldots\ldots\ldots\ldots\ldots\ldots (B\text{-}14b)$$

The derivation of the pressure loss equation for the laminar flow of a power-law fluid in a pipe is quite similar to the derivation of the slot-flow pressure-loss equation. The shear stress for a power-law fluid in a circular pipe is given by Eqs. 4.42, 4.51, and 4.52.

$$\tau = K\left(-\frac{dv}{dr}\right)\left|-\frac{dv}{dr}\right|^{n-1} = \frac{r}{2}\frac{dp_f}{dL}.$$

Separating variables and integrating gives

$$v = \frac{\left(\dfrac{1}{2K}\dfrac{dp_f}{dL}\right)^{1/n}}{1+1/n}(r_w{}^{1+1/n} - r^{1+1/n}). \quad \ldots\ldots (B\text{-}15b)$$

where v_o is the constant of integration corresponding to the fluid velocity at $r=0$. Since the fluid wets the pipe walls, the velocity, v, is zero for $r=r_w$. Applying this boundary condition to Eq. B-15a yields

$$0 = \frac{-\left(\dfrac{1}{2K}\dfrac{dp_f}{dL}\right)^{1/n}}{1+1/n} r_w{}^{1+1/n} + v_o.$$

Thus, v_o is given by

$$v_o = \frac{\left(\dfrac{1}{2K}\dfrac{dp_f}{dL}\right)^{1/n}}{1+1/n} r_w{}^{1+1/n}.$$

Substituting this value for v_o in Eq. B-15a yields

$$v_o = \frac{\left(\dfrac{1}{2K}\dfrac{dp_f}{dL}\right)^{1/n}}{1+1/n}(r_w{}^{1+1/n} - r^{1+1/n}). \quad \ldots (B\text{-}15b)$$

The flow rate, q, is given by

$$q = v\,dA = v(2\pi r)dr$$

$$= \frac{2\pi\left(\dfrac{1}{2K}\dfrac{dp_f}{dL}\right)^{1/n}}{1+1/n} \int_0^{r_w} (r_w{}^{1+1/n}r - r^{2+1/n})dr.$$

Integrating this equation yields

$$q = \frac{\pi\left(\dfrac{1}{2K}\dfrac{dp_f}{dL}\right)^{1/n}}{(3+1/n)} r_w{}^2 r_w{}^{1+1/n}. \qquad \ldots\ldots\ldots (B\text{-}16a)$$

Expressing the flow rate in terms of the mean flow velocity, $\bar{v}$, and solving for the frictional pressure gradient, dp_f/dL, gives

$$\frac{dp_f}{dL} = \frac{2K\bar{v}^n(3+1/n)^n}{r_w{}^{n+1}}. \qquad \ldots\ldots\ldots\ldots (B\text{-}16b)$$

Converting from consistent units to more convenient field units, we obtain

$$\frac{dp_f}{dL} = \frac{K\bar{v}^n}{144,000 d^{n+1}}\left(\frac{3+1/n}{0.0416}\right)^n. \qquad \ldots\ldots\ldots (B\text{-}16c)$$

The shear rate in a power law fluid at the pipe wall can be obtained from the shear stress at the pipe wall and the appropriate frictional pressure gradient equation. For a circular pipe,

$$\dot{\gamma}_w = \left(\frac{\tau_w}{K} \right)^{1/n} = \left(\frac{r_w}{2K} \frac{dp_f}{dL} \right)^{1/n}$$

$$= \left[\frac{r_w}{2K} \frac{2K \bar{v}^n (3 + 1/n)^n}{r_w^{n+1}} \right]^{1/n} .$$

After simplifying and changing from consistent units to more convenient field units, we obtain

$$\dot{\gamma}_w = \frac{24 \bar{v} (3 + 1/n)}{d} . \quad \ldots\ldots\ldots\ldots\ldots\ldots (B\text{-}17)$$

Similarly, for an annulus, the shear rate at the wall is given by

$$\dot{\gamma}_w = \left(\frac{h}{2K} \frac{dp_f}{dL} \right)^{1/n}$$

$$= \left[\frac{r_2 - r_1}{2K} \frac{2K(4 + 2/n)^n \bar{v}^n}{(r_2 - r_1)^{n+1}} \right]^{1/n}$$

After simplifying and changing from consistent units to more convenient field units, we obtain

$$\dot{\gamma}_w = \frac{48 \bar{v} (2 + 1/n)}{d_2 - d_1} . \quad \ldots\ldots\ldots\ldots\ldots\ldots (B\text{-}18)$$

Author Index

Subject Index

A

Abbreviations in describing bit condition, 213
Abnormal bit wear, 213, 214
Abnormal formation pressure, 246, 247, 250-252, 300
Abnormal pressure gradients, 260, 300
Abrasiveness:
  Average for formation, 218, 219
  Constant, 218-220, 238
  Of formation, definition, 210
Absolute roughness of pipe, 146-148
Acceleration method for calculating well trajectory, 365
Accelerator for cement, 98, 99
Accelerometers, 377, 382, 386, 398-400
Accumulators, 23, 24, 31
Acidic, definition of, 45
Acoustic actuators, 31
Acoustic energy travel in cased wells, 109
Acoustic log, effect of shale hydration on ITT, 277
Acoustic logging tools, 110
Acoustic transmission method, 383
Acoustic-type position indicator, 28
Active solids in muds, definition, 41
Activity, definition of, 77
Activity of water, 250
Additives for cements:
  Barite, 97
  Bentonite, 94-96
  Calcium chloride, 97, 98
  Density control, 92-94
  Diatomaceous earth, 93-96, 99
  Expanded perlite, 93-96, 101
  Filtration control, 101, 102
  Gilsonite, 93-95, 101
  Gypsum, 98, 99, 101
  Hematite, 93, 94, 96-98
  Hydrazine, 102
  Ilmenite, 94, 97
  Lost circulation control, 101
  Nylon, 102
  Paraformaldehyde, 102
  Pozzolan, 93, 96
  Radioactive tracers, 102
  Retarders, 98-100, 102
  Sand, 97, 98
  Settling time control, 98
  Silica flour, 93, 94, 102
  Sodium chloride, 98, 99, 102
  Sodium chromate, 102
  Sodium silicate, 99
  Solid hydrocarbons, 93, 94
  Viscosity control, 101, 102
Additives for drilling fluids:
  Chemical, 60, 61
  Density control, 66, 67
  Filtration control, 64, 65, 69, 80
  Lost circulation control, 65
Adiabatic processes, 128
Adjustable choke, 25-27, 162
Adsorption of water by shale, 76
Adsorptive pressure, 77, 250
Air drilling, 16, 17
Alaska, 332, 392, 400
Algorithms, 237, 431, 434, 438, 441
Alignment indicators, 28
Alkaline, definition of, 45
Alkaline lignite, definition of, 61
Alkalinity, definition of, 49

Alkalinity control of oil muds, 81
Alloys, typical properties of, 428
Alternate turbulence criteria, 148, 149
Alumina octahedral sheet, 55, 56
American Petroleum Institute (API):
  Barite, 57, 58, 66-71, 81
  BOP stack, arrangement of, 24, 25
  Bull. 5C1, 330, 350
  Bull. 5C2, 39, 349
  Bull. 5C3, 349
  Bull. D7, 350
  Bull. D10, 8, 39
  Burst pressure rating, 308
  Care and use of casing, 330
  Casing cementing, 95, 99, 100
  Casing grade, 302
  Casing landing methods, 340, 344-348
  Casing performance properties, 305-330
  Casing strings, 325, 331
  Cement classes, 89-92, 98
  Cement types, physical requirements of, 91
  Chock manifold arrangements, 25-27
  Collapse pressure equations, 310, 325
  Connectors, 303-306
  Derrick specifications and ratings, 8, 39
  Diagnostic tests, 42-53, 70
  Drilling cost data, 33
  Drilling mud report form, 42, 43, 48
  Drillpipe specifications, 19
  Equations, cement, 86
  Filter press, 45
  Filtration loss, 65, 101
  Formulas for computing performance properties, 302, 305, 310, 327
  Joint strength formulas, 327, 328
  Joint strength, round thread couplings, 328
  Prod. Bull., Sec. 4, 111
  Recommended practice for care and use of casing, 330
  RP. 9B, 11, 40
  RP. 10B, 89, 111
  RP. 13B, 84, 274
  RP. 53, 25, 39
  Slip and cut program, drilling line, 11
  Specifications for casing connectors, 303
  Spec. 5A, 39
  Spec. 5AC, 349
  Spec. 5AX, 39
  Spec. 9A, 39
  Spec. 10B, 111
  Spec. 13A, 84
  Squeeze cementing, 95, 99, 100
  Standards for casing and other tubular goods, 302
  Std. 4A, 8, 39
  Std. 4D, 8, 39
  Std. 4E, 40
  Std. 7, 8, 39
  Study committee, 344
  Testing liquid drilling fluids, 42
  Tests for well cements, 86-99
  Water content of cement, 89, 92
  Water loss, 46, 47
  Yield strength, 302
API casing performance properties:

Bending, effect of, 325-328
Burst pressure, 307, 308
Collapse pressure, 308-310
Field handling, effect of, 329, 330
Hydrogen sulfide, effect of, 328, 329
Stress, combined, effect of, 310, 311, 325
Summary of, 310-324
Tension, 305-307
American Society for Testing and Materials (ASTM), 89, 91, 111
Amine-treated bentonite, 80
Amine-treated lignite, 81
Analysis of mud logging data, 268-273
Analysis of multistabilizer BHA, 437
Anchor piles, 27
Angle averaging method, 359, 363, 364, 366
Aniline point of oil, 76
Anisotropy index, 446
Annular:
  Cementing through tubing, 106
  Flow for Non-Newtonian fluids, 477-483
  Fluid element, free-body diagram, 139
  Fluid velocity, 154, 170, 180, 181
  Friction loss, 162
  Frictional pressure gradient, 101
  Frictional pressure loss in, 163
  Geometry, extension of pipe flow equations, 149, 150
  Flow for Non-Newtonian fluids, 477-483
  Flow of power-law fluid in a slot, 481
  Pressure during well control operations, 119-122
  Pressure prediction, 121, 122
  Pressure profiles, 119, 162
  Preventer, 23, 24
  Velocity, effective mean, 169, 170
  Velocity, equivalent, 177
  Velocity, mean, 140, 155, 166, 167
Annulus:
  Equivalent diameter of, 149
  Frictional pressure loss in, 141
  Laminar flow in, 141-145, 151, 152, 155, 165-169
  Represented as slot, 141-143
  Shear rate at wall, 143, 144
  Shear stress at wall, 143
  Velocity profile of, 139, 165
API (see American Petroleum Institute)
Apparent:
  Dispersion rigs, 80
  Matrix transit time, 255
  Newtonian viscosity, 42, 44, 133, 150, 152, 153, 176, 181, 182
Archimedes' relation, 122, 126
Arctic wells, 332, 344
Asbestos as a clay substitute, 65, 66
ASTM cement types, 90, 92
Atlas Bradford connectors, 305
Attapulgite, 55, 56, 65
Austenitic steels, 389
Autoclave, 86
Average:
  Bulk density data, 249
  Sediment porosity, 248
  Shale porosity, 273
  Weight for casing threads and couplings, 303

CPSIA information can be obtained
at www.ICGtesting.com
Printed in the USA
FSOW01n1201220116
16024FS